Medical Mycology: Current Trends and Future Prospects

Medical Mycology: Current Trends and Future Prospects

Editor: Perry Jacobs

AMERICAN
MEDICAL PUBLISHERS
www.americanmedicalpublishers.com

AMERICAN
MEDICAL PUBLISHERS
www.americanmedicalpublishers.com

Cataloging-in-Publication Data

Medical mycology : current trends and future prospects / edited by Perry Jacobs.
p. cm.
Includes bibliographical references and index.
ISBN 979-8-88740-254-3
1. Medical mycology. 2. Medical microbiology. 3. Pathogenic fungi. 4. Mycoses.
5. Diagnostic mycology. I. Jacobs, Perry.
QR245 .M43 2023
616.015--dc23

American Medical Publishers,
41 Flatbush Avenue,
1st Floor, New York,
NY 11217, USA

ISBN 979-8-88740-254-3 (Hardback)

Contents

Preface

Mycology refers to the branch of biology, which is related with the study of fungi along with their taxonomy, human uses, and biochemical and genetic properties. It seeks to find the uses of fungi for various purposes including food, tinder, entheogens and traditional medicine, along with the risks associated with them like infection or toxicity. Fungi play crucial roles for life on Earth as symbionts, in the form of lichens, mycorrhizae and insect symbionts. Various fungi can break down complicated organic biomolecules like polycyclic aromatic hydrocarbons, lignin, petroleum and pollutants like xenobiotics. Fungi play an important role in the global carbon cycle through the decomposition of these molecules. The field of medical mycology has become significant in the current time, where the incidence of invasive fungal infections is on rise, which poses therapeutic and diagnostic challenges in the modern medical practice. The book studies, analyses and upholds the pillars of medical mycology and its utmost significance in modern times. It will unravel the recent studies in this field. Researchers and students engaged in medical mycology will be assisted by this book.

This book is a result of research of several months to collate the most relevant data in the field.

When I was approached with the idea of this book and the proposal to edit it, I was overwhelmed. It gave me an opportunity to reach out to all those who share a common interest with me in this field. I had 3 main parameters for editing this text:

1. Accuracy - The data and information provided in this book should be up-to-date and valuable to the readers.

2. Structure - The data must be presented in a structured format for easy understanding and better grasping of the readers.

3. Universal Approach - This book not only targets students but also experts and innovators in the field, thus my aim was to present topics which are of use to all.

Thus, it took me a couple of months to finish the editing of this book.

I would like to make a special mention of my publisher who considered me worthy of this opportunity and also supported me throughout the editing process. I would also like to thank the editing team at the back-end who extended their help whenever required.

Editor

1

Contribution of IL-1RI Signaling to Protection against *Cryptococcus neoformans* 52D in a Mouse Model of Infection

Mitra Shourian[1,2], *Ben Ralph*[3,4], *Isabelle Angers*[2,5], *Donald C. Sheppard*[3,4,6] *and Salman T. Qureshi*[1,2,5,6]*

[1] *Division of Experimental Medicine, McGill University, Montreal, QC, Canada,* [2] *Meakins-Christie Laboratories, McGill University, Montreal, QC, Canada,* [3] *Program in Infectious Diseases and Immunology in Global Health, Centre for Translational Biology, The Research Institute of the McGill University Health Center (RI-MUHC), Montreal, QC, Canada,* [4] *Department of Microbiology and Immunology, McGill University, Montreal, QC, Canada,* [5] *Program in Translational Research in Respiratory Diseases, Department of Critical Care, The Research Institute of the McGill University Health Center (RI-MUHC), Montreal, QC, Canada,* [6] *Department of Medicine, McGill University, Montreal, QC, Canada*

Correspondence:
Salman T. Qureshi
salman.qureshi@mcgill.ca

Interleukin-1 alpha (IL-1α) and interleukin-1 beta (IL-1β) are pro-inflammatory cytokines that are induced after *Cryptococcus neoformans* infection and activate the interleukin-1 receptor type I (IL-1RI). To establish the role of IL-1RI signaling in protection against cryptococcal infection, we analyzed wild-type (WT) and IL-1RI-deficient (IL-1RI$^{-/-}$) mice on the BALB/c background. IL-1RI$^{-/-}$ mice had significantly reduced survival compared to WT mice after intratracheal challenge with *C. neoformans* 52D. Microbiological analysis showed a significant increase in the lung and brain fungal burden of IL-1RI$^{-/-}$ compared to WT mice beginning at weeks 1 and 4 postinfection, respectively. Histopathology showed that IL-1RI$^{-/-}$ mice exhibit greater airway epithelial mucus secretion and prominent eosinophilic crystals that were absent in WT mice. Susceptibility of IL-1RI$^{-/-}$ mice was associated with significant induction of a Th2-biased immune response characterized by pulmonary eosinophilia, M2 macrophage polarization, and recruitment of CD4^{+} IL-13^{+} T cells. Expression of pro-inflammatory [IL-1α, IL-1β, TNFα, and monocyte chemoattractant protein 1 (MCP-1)], Th1-associated (IFNγ), and Th17-associated (IL-17A) cytokines was significantly reduced in IL-1RI$^{-/-}$ lungs compared to WT. WT mice also had higher expression of KC/CXCL1 and sustained neutrophil recruitment to the lung; however, antibody-mediated depletion of these cells showed that they were dispensable for lung fungal clearance. In conclusion, our data indicate that IL-1RI signaling is required to activate a complex series of innate and adaptive immune responses that collectively enhance host defense and survival after *C. neoformans* 52D infection in BALB/c mice.

Keywords: *Cryptococcus neoformans*, fungal pneumonia, interleukin-1, interleukin-1 receptor, lung inflammation, cytokines, macrophage polarization, lymphocyte differentiation

INTRODUCTION

Cryptococcus neoformans is an encapsulated yeast that is estimated to cause approximately 223,000 cases of meningitis each year and is responsible for 15% of AIDS-related deaths (1). In healthy individuals, inhalation of infectious propagules is usually contained in the lung, but among those with a defective immune response, uncontrolled replication may result in dissemination to other parts of the body with a tropism for the brain (2, 3). Severe cryptococcal disease occurs primarily in patients with uncontrolled HIV/AIDS and is also found in solid organ transplant recipients, those receiving exogenous immunosuppression, patients with primary or acquired immunodeficiency, and increasingly among immunologically normal hosts (4–7).

The pattern of cytokine expression is a crucial determinant of the pathogenesis of cryptococcal infection (3, 8–11). Th1-type cytokines [interleukin (IL)-12 and IFNγ] promote phagocytosis by dendritic cells (DCs) and polarize macrophages toward a classically activated phenotype (M1), thereby increasing fungal clearance (12–15). On the other hand, Th2-type cytokines (IL-4, IL-5, and IL-13) are associated with a significant eosinophil chemotaxis to the lungs and induction of alternatively activated (M2) macrophages that facilitate cryptococcal proliferation and dissemination (16–18). There is some evidence that Th17-type cytokines (IL-17A and IL-23) contribute to protection against infection with wild-type (WT) *C. neoformans*; however, they appear to be less effective compared to Th1-type cytokines (19–23). Inhibition of IL-17A expression or signaling had no significant effect on M1 macrophage polarization, resolution of infection, or survival in mice infected with *C. neoformans* H99 that has been engineered to express IFNγ (24, 25). Finally, a prospective analysis of HIV-infected humans suggested a potential role for IL-17 in the immunopathogenesis of cryptococcal meningitis; however, further studies are required to confirm this hypothesis (26).

The mechanisms that initiate and regulate the innate immune response against *C. neoformans* infection are not fully understood. The interaction of *C. neoformans* with host cells triggers production of several pro-inflammatory cytokines including TNFα, IL-6, and IL-1 (27–30). Both interleukin-1 alpha (IL-1α) and interleukin-1 beta (IL-1β) are induced during cryptococcal infection *in vitro* (27, 28, 31–34) and *in vivo* (35–40) in a NLRP3-dependent manner, and internalization of opsonized encapsulated cryptococci has been shown to activate the canonical NLRP3–ASC–caspase-1 and non-canonical NLRP3–ASC–caspase-8 inflammasome (34, 41). The magnitude of IL-1 expression between inbred mice with different genetic backgrounds has also been associated with natural resistance or susceptibility to progressive cryptococcal infection (35). After intratracheal infection with *C. neoformans* 52D, the level of IL-1β expression was 11-fold higher in the lungs of resistant SJL/J inbred mice compared to the susceptible C57BL/6 inbred strain. A subsequent analysis of WT and interleukin-1 receptor (IL-1R)-deficient mice on the C57BL/6 genetic background did not identify significant differences in survival or fungal dissemination after intranasal infection with *C. neoformans* H99; however, at day 12 postinfection, the IL-1R$^{-/-}$ mice had a modest elevation of lung fungal burden (37).

Given the essential role for cytokine-mediated inflammation and the evidence for IL-1α and IL-1β induction in response to *C. neoformans*, we hypothesized that the contribution of IL-1R-dependent signaling to host defense may have been underestimated by infection of WT and IL-1R$^{-/-}$ mice on the susceptible C57BL/6 genetic background with a highly virulent *C. neoformans* strain. To test this hypothesis, we performed intratracheal inoculation of inbred BALB/c mice and IL-1R$^{-/-}$ mice on the same genetic background with *C. neoformans* 52D and analyzed fungal burden and immune responses at serial time points. This approach was chosen to model the process of natural infection in a relatively resistant host with a moderately virulent cryptococcal strain. Our findings demonstrate that IL-1RI$^{-/-}$ mice had a significantly higher fungal burden in the lungs and brains as well as a significantly higher mortality compared to BALB/c mice. In IL-1RI$^{-/-}$ mice, *C. neoformans* 52D infection was associated with heightened lung eosinophilia, elevated airway mucus secretion, and a greater percentage of M2 macrophages and CD4$^+$ Th2 cells along with significantly fewer lung neutrophils, DCs, Th1, and Th17 cells. Taken together, this study shows that IL-1R-dependent signaling contributes to protection against *C. neoformans* 52D infection in BALB/c mice by triggering a complex innate and adaptive immune response and raises the possibility that modulation of this signaling axis could be a potential therapeutic strategy.

MATERIALS AND METHODS

Mice

Inbred BALB/c mice were purchased from Charles River and maintained in our facility. IL-1RI$^{-/-}$ mice were purchased from Jackson Labs and backcrossed to BALB/c for 10 generations. Mice were provided with sterile food and water and cared for according to the Canadian Council on Animal Care guidelines. All experiments were performed using 7- to 9-week-old male and female mice. Mice were humanely euthanized with CO_2 upon completion of experiments, and every effort was made to minimize suffering. All experimental protocols were reviewed and approved by the McGill University Animal Care Committee.

Cryptococcus neoformans

Cryptococcus neoformans 52D (ATCC 24067) was grown and maintained on Sabouraud dextrose agar (SDA; BD, Becton Dickinson and Company). To prepare an infectious dose, a single colony was suspended in Sabouraud dextrose broth (BD) and grown to early stationary phase (48 h) at room temperature on a rotator. The stationary phase culture was then washed with sterile phosphate-buffered saline (PBS), counted on a hemocytometer, and diluted to 2×10^5 CFU/ml in sterile PBS. The fungal concentration of the experimental dose was confirmed by plating a dilution of the inoculum on SDA and counting the CFU after 72 h of incubation at room temperature.

Intratracheal Infection with *C. neoformans*

For intratracheal administration of *C. neoformans*, mice were anesthetized with 150 mg/kg of ketamine (Ayerst Veterinary Laboratories) and 10 mg/kg of xylazine (Bayer) intraperitoneally. A small skin incision was made below the jaw along the trachea, and the underlying glands and muscle were separated. Infection was performed by intratracheal injection of 10^4 *C. neoformans* in 50 μl PBS through a 22-gauge catheter *via* a 1-ml tuberculin syringe. The incision was closed using the 9-mm EZ clip wound closing kit (Stoelting CO), and mice were monitored daily after surgery.

Tissue Isolation and CFU Assay

After mice were euthanized with CO_2, their lungs, spleen, and brain were excised and placed in sterile, ice-cold PBS. Tissues were then homogenized using a glass tube and pestle attached to a mechanical tissue homogenizer (Glas-Col) and plated at various dilutions on SDA. Plates were incubated at 37°C for 72 h, and CFU were counted. For survival analyses, mice were inoculated as stated above and monitored twice daily for up to 110 days postinfection.

Histopathological Analysis

After euthanasia, lungs were perfused with ice-cold PBS *via* the right ventricle of the heart. Using 10% buffered formalin acetate (Fisher Scientific), the lungs were inflated to a pressure of 25 cm H_2O and fixed overnight. Subsequently, lungs were embedded in paraffin, sectioned at 5 μm, and stained with hematoxylin–eosin (H&E), periodic acid–Schiff (PAS), or mucicarmine reagents at the Histology Facility of the Goodman Cancer Research Centre (McGill University). Representative photographs of lung sections were taken using a BX51 microscope (Olympus), QICAM Fast 1394 digital charge-coupled device camera (QImaging), and Image-Pro Plus software version 7.0.1.658 (Media Cybernetics).

Flow Cytometry

Lungs were excised using sterile technique and placed in RPMI (Gibco, Invitrogen) supplemented with 10% fetal bovine serum (Wisent). Subsequently lungs were minced using surgical blades and incubated with 1 mg/ml collagenase (Sigma) at 37°C for 1 h. After incubation, lung pieces were passed through a 16-gauge needle and filtered through a 70-μm cell strainer (BD). Red blood cells were removed using ACK lysis buffer, cells were counted with a hemacytometer using trypan blue dye, and 5×10^6 cells in 100 μl FACS buffer/well were dispensed in 96-well plates. Fc receptors were blocked with the addition of unlabeled anti-CD16/32 antibodies [93; eBioscience (eBio)], and single-cell suspensions were stained with the following fluorescence-conjugated anti-mouse monoclonal antibodies purchased from eBio, BD, and BioLegend: CD45 (30-F11), B220 (RA3-6B2), CD3e (145-2C11), CD4 (GK1.5), CD8 (53-6.7), CD49b (DX5), γδ TCR (GL3), CD11b (M1/70), CD11c (N418), MHCII (M5/114.15.2), Ly6G (1A8), CD86 (GL1), CD80 (16-10A1), CD64 (X54-5/7.1), CD24 (M1/69), SiglecF (E50-2440), CD103 (2E7), Ly6C (HK1.4), and CD206 (C068C2). Non-viable cells were excluded using a fixable viability dye reagent (eBio). Lineage negative cells (Lin^-) were defined as $CD45^+$ cells that did not express any other surface markers in this panel. Data were acquired using a LSRFortessa flow cytometer (BD) and analyzed using Flow Jo software. The absolute number of leukocytes was determined by multiplying the percentage of $CD45^+$ cells by the total number of counted cells.

Intracellular Staining

Lungs were processed as described above, and 5×10^6 cells/well were dispensed in 96-well plates. For cytokine analysis, cells were stimulated for 4 h with phorbol 12-myristate 13-acetate (PMA) and calcium ionophore (ionomycin) in the presence of brefeldin A (GolgiPlug) for the final 3 h. Cells were then washed, blocked with anti-CD16/32 antibodies, and stained with the surface antibodies. Cells were then fixed, permeabilized, and stained with IL-13 (eBio13A), IFNγ (XMG1.2), and IL-17A (17B7). Intracellular staining for Nos2 (CXNFT) was done as described for cytokines without PMA and ionomycin stimulation. Data were acquired using a LSRFortessa flow cytometer with gating determined by fluorescence-minus-one controls and analyzed using FlowJo software.

Total Lung Cytokine and Chemokine Production

Mice were euthanized and lungs flushed with 10 ml of ice-cold PBS. Whole lungs were homogenized in 2 ml PBS with Halt protease and phosphatase inhibitor cocktail (Fisher Scientific) using a sterilized glass tube and pestle attached to a mechanical tissue homogenizer (Glas-Col) and spun at 12,000 rpm for 20 min. Supernatants were collected, and aliquots were stored at −80°C for further analysis. The following cytokines and chemokines in whole-lung protein samples were analyzed using DuoSet enzyme-linked immunosorbent assay kits (R&D Systems): TNFα (DY410), IL-6 (DY406), IL-1β (DY401), IL-1α (DY400), monocyte chemoattractant protein 1 (MCP-1; MJE00), IL-12/IL-23P40 (DY2398), IFNγ (DY485), CXCL1/KC (DY453), IL-17A (DY421), and IL-13 (DY413).

Neutrophil Depletion

BALB/c mice received an intratracheal inoculum of 1×10^4 CFU of *C. neoformans* strain 52D. Mice were treated with 100 μl of PBS or 200 μg of anti-1A8 antibody (Bio X Cell) in a volume of 100 μl, 1 day before infection and daily during the study. At day 12 postinfection, lungs were excised, and fungal burden was analyzed.

Statistical Analysis

To test the significance of single comparisons, an unpaired Student's *t*-test was applied with a threshold $P \leq 0.05$. For all experiments, the mean and SEM is shown. Survival curves were analyzed by the log-rank test. All statistical analysis was performed with GraphPad Prism software version 6 (GraphPad Software Inc.).

RESULTS

IL-1RI$^{-/-}$ Mice Have Impaired Survival and an Increased Fungal Burden in the Lung, Brain, and Spleen following *C. neoformans* 52D Infection

To investigate the role of IL-1RI-mediated signaling after *C. neoformans* 52D infection, we constructed IL-1RI$^{-/-}$ mice on the BALB/c background by repeated backcrossing. We challenged mice with *C. neoformans* 52D and measured the survival rate and tissue fungal burden. No deaths were observed in WT mice; however, IL-1RI$^{-/-}$ mice started to die at 40 days postinfection and had a 73% mortality rate at 100 days postinfection (**Figure 1A**). Microbiological analysis also showed a significant increase of fungal burden in IL-1RI$^{-/-}$ mice compared to the WT strain at all time points tested (**Figure 1B**). Importantly, a significant difference in lung fungal burden was observed at 7 days postinfection, suggesting that the IL-1RI signaling affects the initial host response to *C. neoformans* infection. At 35 days postinfection, there was almost a 20-fold increase of lung CFU in the IL-1RI$^{-/-}$ compared to the WT strain. Analysis of the spleen showed a trend toward higher CFU in the IL-1RI$^{-/-}$ mice compared to the WT strain that reached statistical significance at day 14 postinfection (**Figure 1C**). Analysis of the brain showed comparable CFU in both strains at 14 days postinfection; however, at 35 days postinfection, all of the WT mice had cleared the infection, while 10 of 16 (62%) of IL-1RI$^{-/-}$ mice still had detectable fungal growth (**Figure 1D**). Taken together, these data establish a role for IL-1R-mediated signaling in controlling fungal growth in the lungs and brain, limiting organ dissemination, and increasing survival after *C. neoformans* 52D infection.

An Altered Pattern of Pulmonary Inflammation Is Present in IL-1RI$^{-/-}$ Lungs following *C. neoformans* 52D Infection

The significant differences in survival and fungal burden between WT and IL-1RI$^{-/-}$ mice prompted us to investigate the effect of IL-1RI signaling on lung pathology after infection with *C. neoformans* 52D. Histopathological analysis was conducted at 35 days postinfection to correspond with the greatest difference in fungal burden prior to the onset of mortality (**Figures 2A–C**). H&E staining revealed that WT mice displayed abundant lung leukocyte infiltration that was almost absent in the IL-1RI$^{-/-}$ strain. Notably, eosinophilic crystals that have been associated with alternatively activated macrophages in *C. neoformans* 52D infection were clearly observed in IL-1RI$^{-/-}$ lung sections but were absent in the WT. Mucicarmine staining of the cryptococcal cell wall showed that most fungi were located within WT phagocytes with only a few visible extracellular organisms in the parenchyma or airways. In contrast, IL-1RI$^{-/-}$ sections showed lung parenchyma that was filled with heavily encapsulated extracellular cryptococci. PAS staining clearly revealed mucus

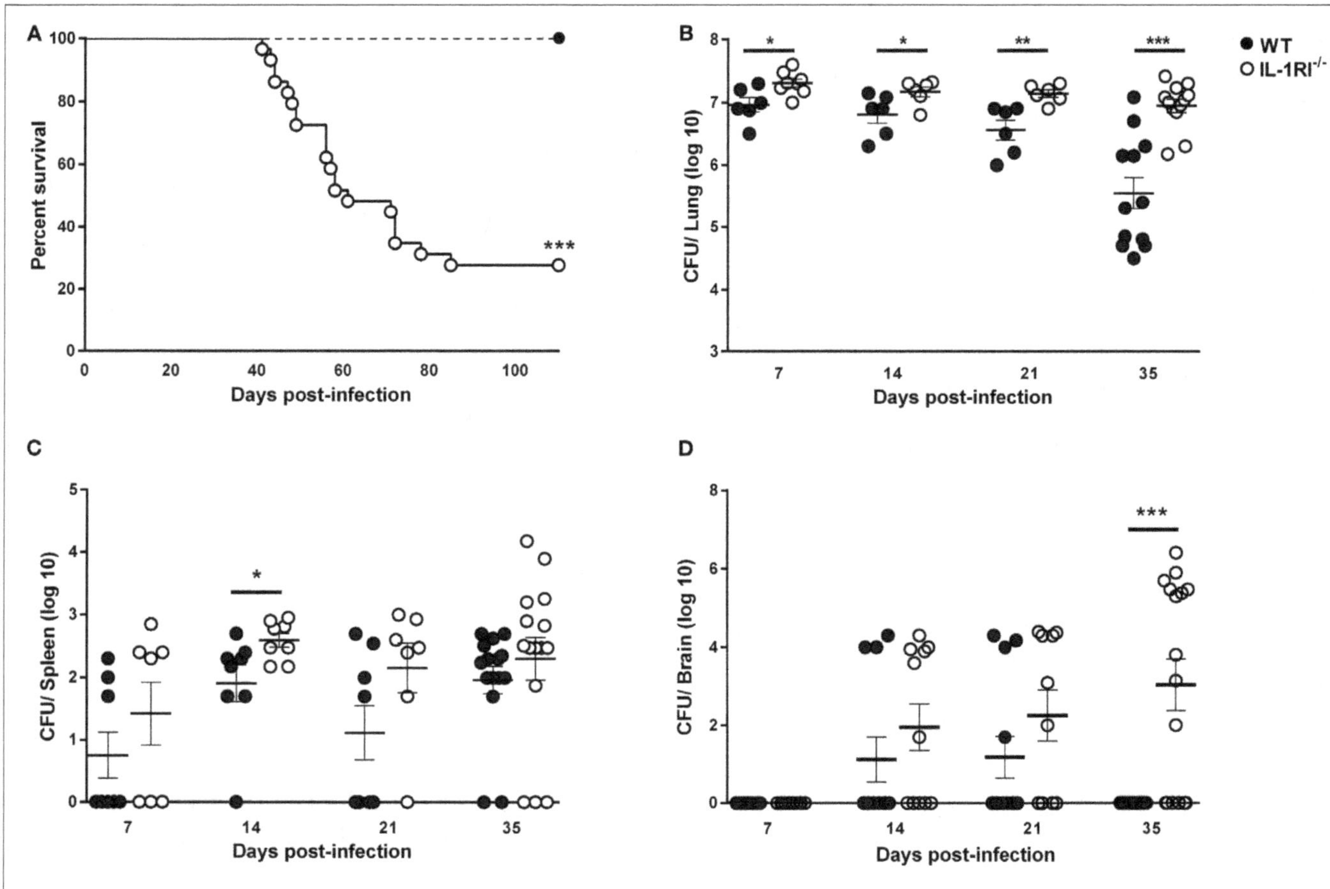

FIGURE 1 | Interleukin-1 receptor type I (IL-1RI) signaling is required for survival and control of fungal burden after infection with *Cryptococcus neoformans* 52D. Wild-type (WT) and IL-1RI-deficient (IL-1RI$^{-/-}$) mice were infected intratracheally with 10^4 CFU of *C. neoformans* strain 52D. **(A)** Mice were observed for up to 110 days for survival analysis (n = 12 mice/strain, using a log-rank test). **(B–D)** Fungal burden in the lung, brain, and spleen at serial time intervals was determined by plating tissue homogenates on Sabouraud dextrose agar. CFU data are shown as mean ± SEM and representative of two independent experiments (n = 6–15 mice/strain/time point). $^*P \leq 0.05$, $^{**}P \leq 0.01$, and $^{***}P \leq 0.001$.

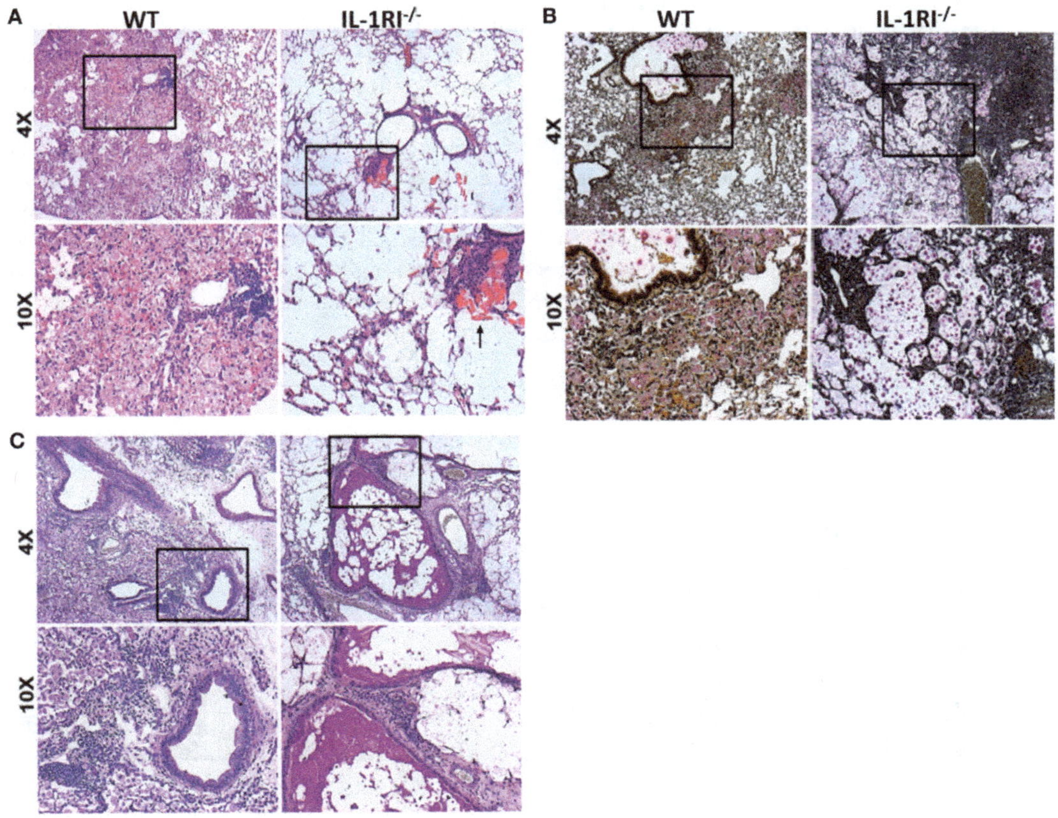

FIGURE 2 | Decreased inflammation in the lungs of interleukin-1 receptor type I-deficient (IL-1RI$^{-/-}$) mice following infection with *Cryptococcus neoformans* 52D. Wild-type (WT) and IL-1RI$^{-/-}$ mice were infected intratracheally with 10^4 CFU of *C. neoformans* 52D. Lungs were harvested at day 35 postinfection; perfused with phosphate-buffered saline; embedded in paraffin; and stained with hematoxylin–eosin (H&E), mucicarmine, or periodic acid–Schiff (PAS). Representative H&E images **(A)** show a significant reduction of inflammation in IL-1RI$^{-/-}$ compared to WT mice; the black arrow points to eosinophilic crystals in IL-1R$^{-/-}$ lung. Mucicarmine staining **(B)** shows numerous heavily encapsulated extracellular *C. neoformans* in the airspaces of IL-1RI$^{-/-}$ mice compared to WT mice. Representative images of lungs stained with PAS **(C)** show goblet cell hyperplasia and mucus in the airways of infected IL-1RI$^{-/-}$ mice compared to WT mice. Each image is representative of 100 fields examined (n = 4 mice/strain from two independent experiments).

secretion by airway epithelial cells in IL-1RI$^{-/-}$ mice that was not observed in the airways of WT mice. Taken together, this histopathological analysis confirmed the results of the lung fungal burden studies and demonstrated reduced inflammation with signs of Th2 polarization in IL-1RI$^{-/-}$ mice compared to the WT strain.

Inflammatory Cytokine and Chemokine Production Is Decreased in the Lungs of IL-1RI$^{-/-}$ Mice following *C. neoformans* 52D Infection

To determine the effect of IL-1RI signaling on the production of soluble inflammatory mediators, WT and IL-1RI$^{-/-}$ mice were infected with *C. neoformans* 52D, and the concentration of pro-inflammatory cytokines (IL-1α, IL-1β, TNFα, and IL-6), chemokines (MCP-1 and KC), Th1-associated cytokines (IFNγ and IL-12), and representative Th2-associated (IL-13) and Th17-associated (IL-17A) cytokines was measured in whole-lung homogenates at serial time points (**Figure 3**). No significant differences in the level of these mediators were observed between two strains prior to infection. In WT mice, both IL-1α and IL-1β were produced in the lungs at day 7 postinfection and continued to increase until day 14 postinfection. Compared to WT, IL-1RI$^{-/-}$ mice had significantly lower production of these two cytokines at day 14 postinfection. The production of TNFα, MCP-1, and KC was significantly higher in WT compared to IL-1RI$^{-/-}$ mice at day 14 postinfection. Significantly greater production of IFNγ and IL-17A was also observed in the lungs of WT mice compared to IL-1RI$^{-/-}$ at day 14 postinfection. IL-13 production did not differ between strains at day 7 and day 14 postinfection, although a modest increase was observed in IL-1R$^{-/-}$ mice compared to WT at day 21 postinfection. In summary, BALB/c mice exhibited significantly greater production of pro-inflammatory, Th1, and Th17 cytokines, as well as chemokines, compared to IL-1RI$^{-/-}$ mice; these findings demonstrate a broad effect of IL-1RI signaling on the lung inflammatory response after *C. neoformans* 52D infection.

IL-1RI$^{-/-}$ Mice Exhibit Reduced Neutrophil and Increased Eosinophil Recruitment to the Lungs following *C. neoformans* 52D Infection

To characterize the effect of IL-1RI signaling on the cellular immune response after *C. neoformans* infection, flow cytometry analysis of whole-lung digests was performed on WT and IL-1RI$^{-/-}$ mice at serial time points postinfection. A comprehensive gating strategy was used for the identification of resident and recruited myeloid cell subsets (**Figure 4**) (42–44). Prior to infection, no significant difference was observed in the total number of lung leukocytes between the two strains. The total number of CD45^{+} cells peaked at day 14 in both strains; however, it was significantly higher in WT compared to IL-1RI$^{-/-}$ mice at 14 and 21 days postinfection (**Figure 5A**). At 7 days postinfection neutrophils (CD11c^{-}, CD11b^{+}, and Ly6G^{high}) were the most frequent leukocyte subset in both strains; however, their percentage and total number was significantly higher in the WT compared to the IL-1RI$^{-/-}$ at 14 and 21 days postinfection (**Figures 5B,C**). Conversely, the percentage and number of lung eosinophils (CD11c^{-}, CD11b^{+}, Siglec F^{+}, and CD24^{+})

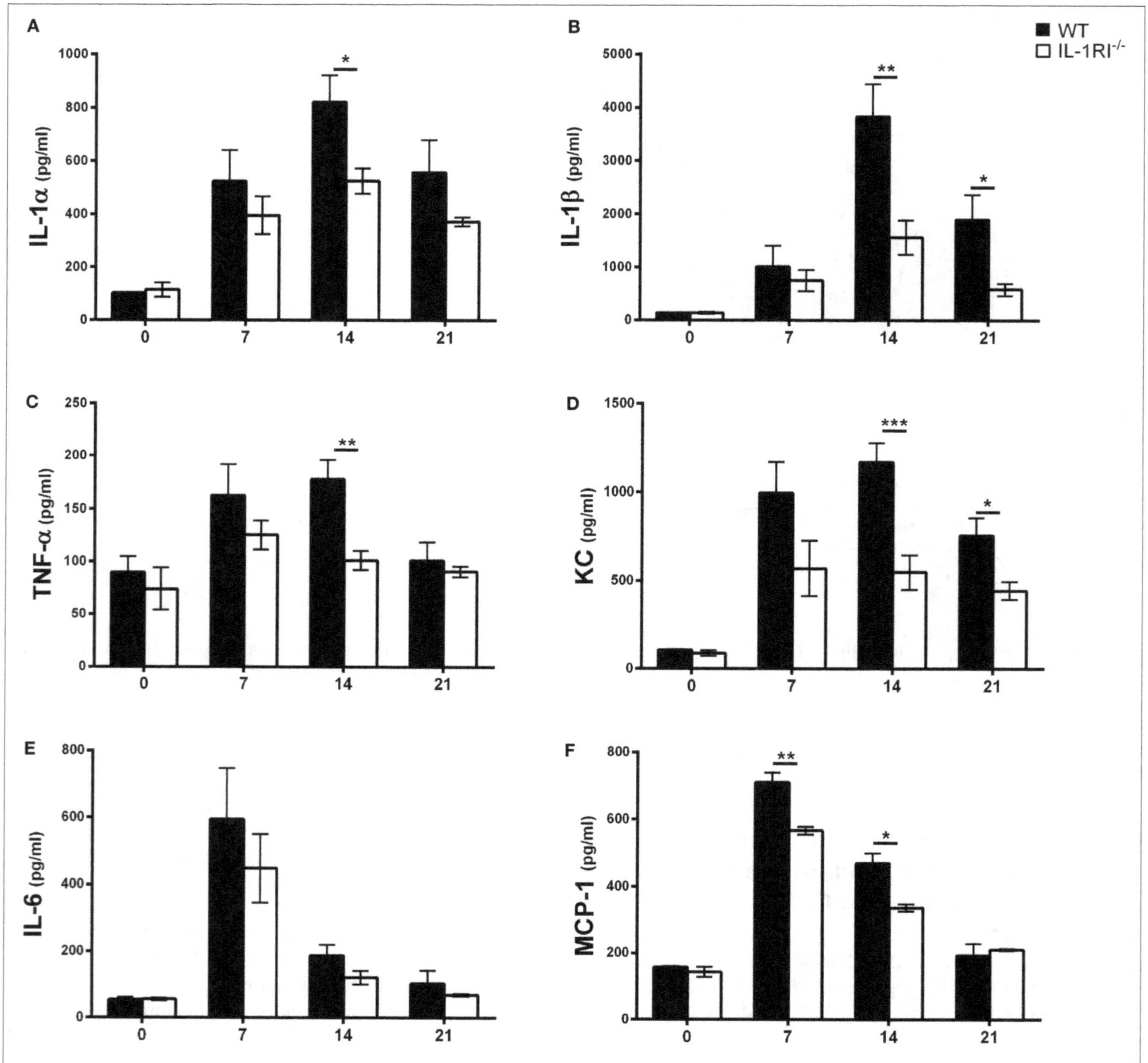

FIGURE 3 | Continued

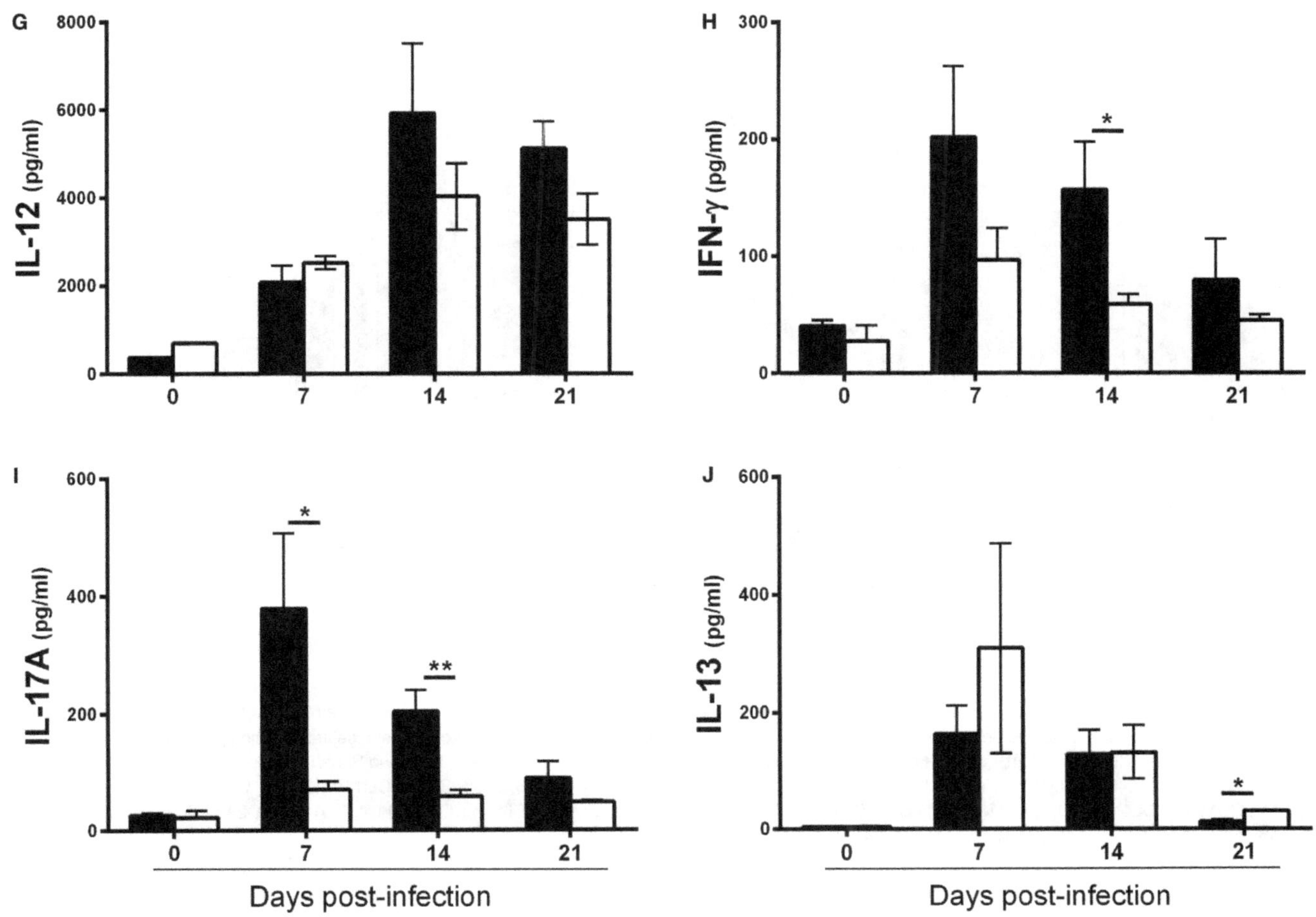

FIGURE 3 | Interleukin-1 receptor type I-deficient (IL-1RI$^{-/-}$) lungs have decreased inflammatory cytokine and chemokine production after *Cryptococcus neoformans* 52D infection. Whole-lung protein was collected at 0, 7, 14, and 21 days postinfection with 10^4 CFU of *C. neoformans* 52D. Enzyme-linked immunosorbent assay was performed to determine the level of pro-inflammatory cytokines **(A–C, E)**, chemokines **(D, F)**, and Th1/Th2/Th17-type cytokines **(G–J)**. Data are shown as mean ± SEM and representative of two independent experiments (n = 4 mice/strain/time point). $^*P \le 0.05$, $^{**}P \le 0.01$, and $^{***}P \le 0.001$. MCP-1, monocyte chemoattractant protein 1.

was significantly higher in IL-1RI$^{-/-}$ mice compared to the WT strain at 14 and 21 days postinfection (**Figures 5D,E**). These data suggest that IL-1R signaling plays an important role in recruitment of neutrophils during the host response to *C. neoformans* 52D infection. In the absence of IL-1R, mice develop significant and sustained lung eosinophilia that is associated with a higher fungal burden.

To evaluate the functional significance of early and sustained neutrophil recruitment to the lungs of BALB/c mice after infection with *C. neoformans* 52D, the effect of antibody-mediated depletion on tissue fungal burden and lung cell infiltration was characterized. Briefly, WT mice received 200 µg of anti-Ly6G antibody (clone 1A8) in a volume of 100 µl *via* intraperitoneal injection 24 h before infection and daily thereafter. To capture the overall effect of neutrophil depletion during the innate and adaptive phases of immunity, lung fungal burden was determined at 12 days postinfection. Interestingly, this analysis showed that neutrophil-depleted mice had a significantly lower cryptococcal burden in the lungs compared to control mice (**Figures 5F–G**).

IL-1RI$^{-/-}$ Mice Recruit Fewer Monocyte-Derived DC and Macrophages to the Lung following *C. neoformans* 52D Infection

Inflammatory monocyte-derived macrophages (ExMs) and DCs are important for protection against *C. neoformans* infection (45, 46). We investigated the effect of IL-1RI signaling on the number of resident and monocyte-derived myeloid cells by harvesting lungs at different times postinfection and analyzing cells by flow cytometry. No significant difference in the percentage of pDCs, CD103$^+$ DCs, and CD11b$^+$ cDCs was observed in the lungs of BALB/c and IL-1R$^{-/-}$ mice after *C. neoformans* 52D infection; however, at day 21 postinfection, there was a significantly higher percentage and number of AMs in BALB/c compared to IL-1R$^{-/-}$ mice (**Figures 6 A,B,D**). As both monocyte-derived ExMs and

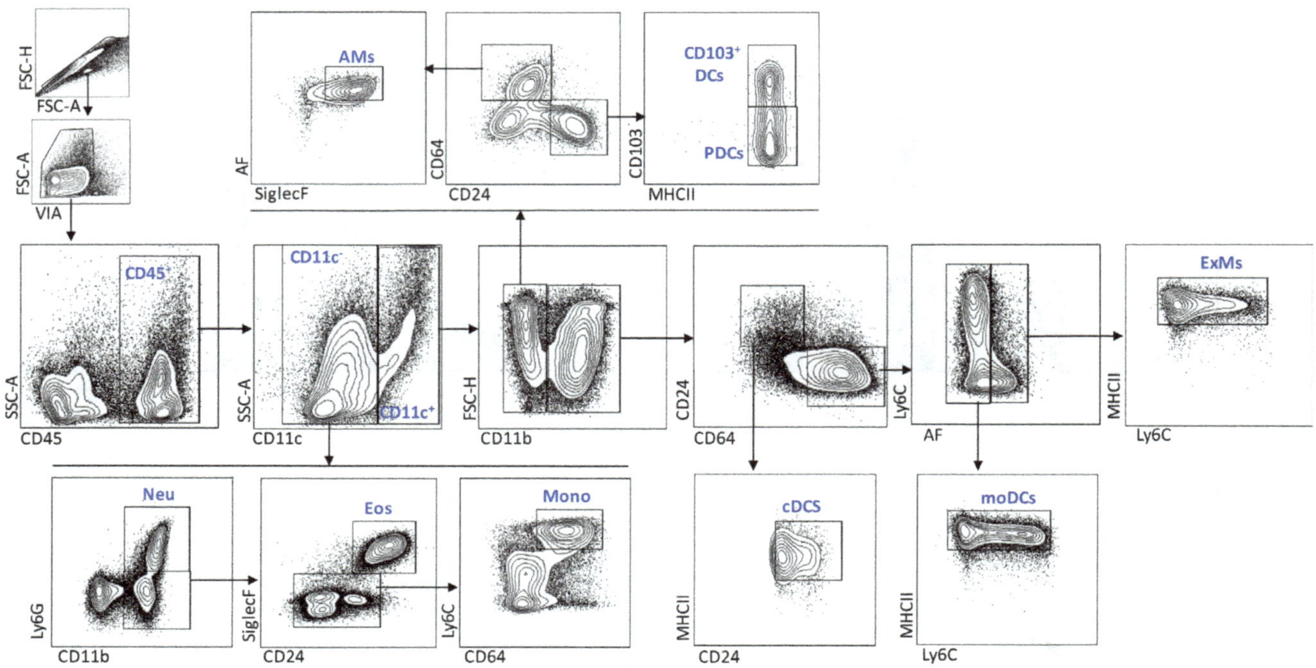

FIGURE 4 | Flow cytometry gating strategy used to identify myeloid cell subsets in the lungs after *Cryptococcus neoformans* 52D infection. Lung cell suspensions were stained with fluorochrome-labeled antibodies and analyzed as described in Section "Materials and Methods." Representative gating plots for $CD45^+$ subsets are shown; neutrophils ($CD11c^-$, $CD11b^+$, and $Ly6G^{high}$), eosinophils ($CD11c^-$, $CD11b^+$, Ly6G low/negative, $CD24^+$, and $SiglecF^+$), monocytes ($CD11c^-$, $CD11b^+$, Ly6G low/negative, $CD64^+$, and $Ly6C^{high}$), AMs ($CD11c^+$, $CD11b^-$, $CD64^+$, $SiglecF^+$, and AF^+), $CD103^+$DCs ($CD11c^+$, $CD11b^-$, $CD24^+$, $CD103^+$, and $MHCII^+$), pDCs ($CD11c^+$, $CD11b^-$, $CD24^+$, $CD103^-$, and $MHCII^+$), cDCs ($CD11c^+$, $CD11b^+$, $CD64^-$, $CD24^+$, and $MHCII^+$), mDCs ($CD11c^+$, $CD11b^+$, $CD24^-$, $CD64^+$, AF^-, $MHCII^{high}$, and $Ly6C^{+/-}$), and ExMs ($CD11c^+$, $CD11b^+$, $CD24^-$, $CD64^+$, AF^+, $MHCII^+$, and $Ly6C^{+/-}$).

DCs are $CD11b^+$, $CD11c^+$, $CD24^-$, $MHCII^+$, and $CD64^+$, we used autofluorescence to distinguish macrophages from DCs (42, 45, 47) (**Figure 4**). This analysis showed comparable recruitment of both cell types between the two strains at day 7 postinfection; however, WT mice had a significantly higher number of inflammatory DCs (days 14 and 21) and ExMs (day 21) compared to IL-1RI$^{-/-}$ mice (**Figures 6C,E**).

The macrophage polarization pattern is also important for protection against cryptococcal infection (8, 48). Classically activated macrophages (M1) that express high levels of pro-inflammatory cytokines and costimulatory molecules, produce high levels of reactive nitrogen and oxygen intermediates, and promote strong IL-12-mediated Th1 responses are efficient killers of *C. neoformans*. In contrast, alternatively activated macrophages (M2) that express chitinase-like 3 (Ym1), found in inflammatory zone (FIZZ1), mannose receptor (CD206), and arginase-1 (Argl), have reduced pro-inflammatory cytokine secretion and are less microbicidal (3, 9, 42, 43, 47, 49–53). As the number of recruited macrophages peaked at day 14 postinfection in both strains, we characterized polarization at this time point using iNOS and CD206 as representative markers for M1 and M2 macrophages, respectively. At 14 days postinfection, the percentage of M1 macrophages was significantly greater in WT mice compared to IL-1R$^{-/-}$ mice, while the percentage of M2 macrophages was greater in IL-1R$^{-/-}$ compared to WT mice (**Figures 6F,G**). Notably, IL-1RI$^{-/-}$ macrophages showed greater upregulation of the M2-associated marker CD206 at 14 days postinfection (**Figures 6H,I**), while WT macrophages displayed higher expression of the M1-associated marker CD80 (43) at 14 and 21 days postinfection (**Figures 6J,K**). Taken together, these results indicate that IL-1RI signaling has an important role in recruitment of inflammatory DCs and macrophages and increases the ratio of M1/M2-polarized macrophages after *C. neoformans* 52D infection.

T Cells Are the Predominant Sources of IL-17A and IFNγ in WT Lungs Infected with *C. neoformans* 52D

To characterize the mechanism of differential IL-17A and IFNγ expression between WT and IL-1RI$^{-/-}$ lungs, we identified the main sources of these cytokines after *C. neoformans* 52D infection. Compared to IL-1RI$^{-/-}$ mice, WT mice showed significantly more IL-17A-producing cells at 7, 14, and 21 days postinfection and a trend toward a higher number of IFNγ-producing cells at day 21 postinfection (**Figures 7A–C**). Several immune cell types including $CD4^+$ (Th17), $CD8^+$ T (Tc17) cells, NK cells, iNKT cells, γδT cells, B cells, ILCs, DCs, and neutrophils have been shown to produce IL-17 during fungal infection (25, 54–56). In our study, at day 7 postinfection, intracellular cytokine staining of WT lymphocytes showed that $CD4^+$ and γδT cells were the most common IL-17A$^+$ subsets (**Figure 7D**). A similar pattern was observed at

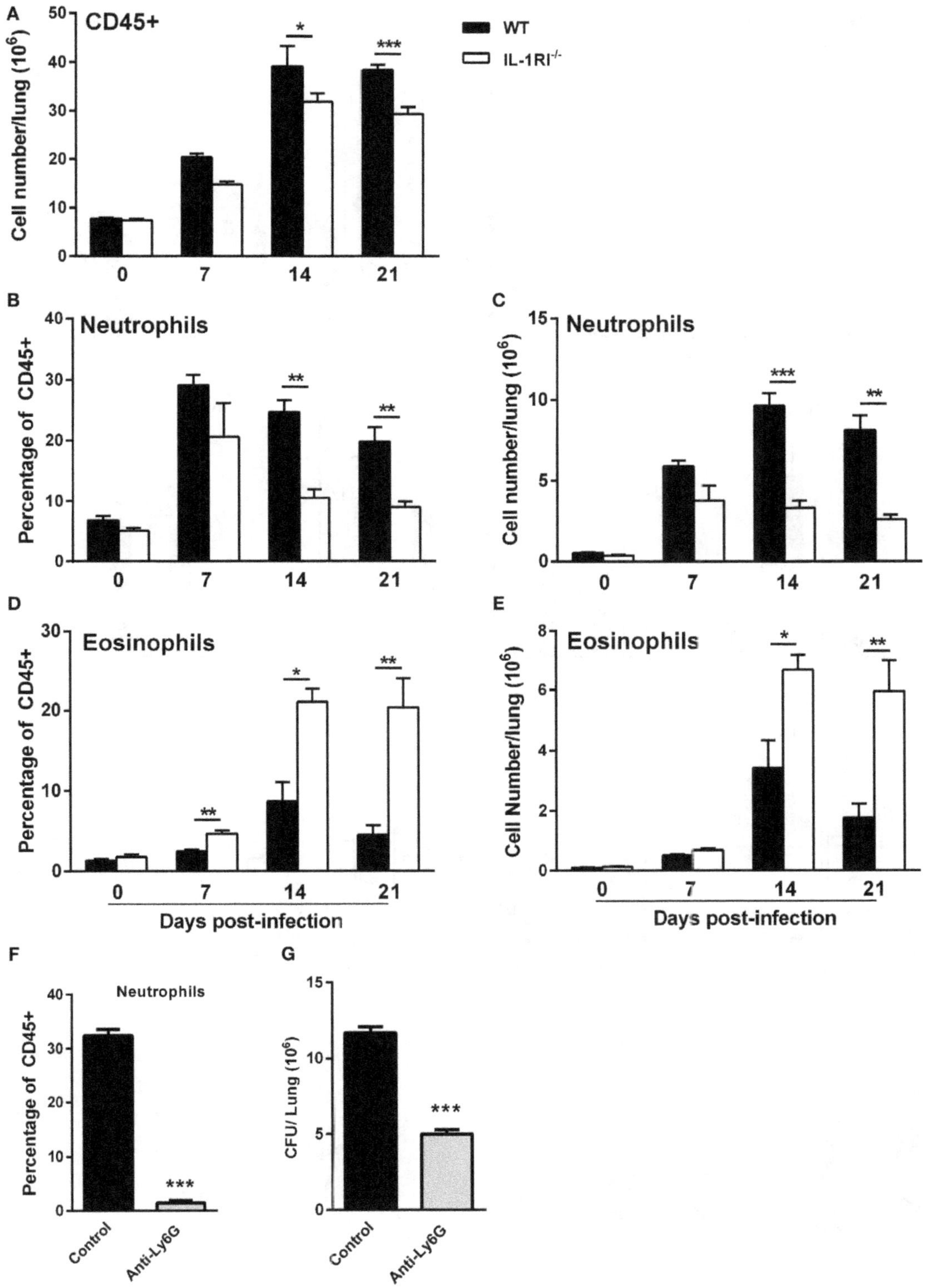

FIGURE 5 | Interleukin-1 receptor type I-deficient (IL-1RI$^{-/-}$) mice have decreased neutrophil and increased eosinophil recruitment to the lungs after *Cryptococcus neoformans* 52D infection. Lung cell suspensions from uninfected and infected wild-type (WT) and IL-1RI$^{-/-}$ mice were stained with fluorochrome-labeled antibodies and analyzed by flow cytometry as described in Section "Materials and Methods." **(A)** Absolute numbers of total CD45$^+$ cells in the lungs at 0, 7, 14, and 21 days postinfection. **(B–E)** Percentage and total number of neutrophils and eosinophils at 0, 7, 14, and 21 days postinfection. Data are shown as mean ± SEM and representative of two independent experiments (*n* = 4 mice/strain/time point). $^*P \leq 0.05$, $^{**}P \leq 0.01$, and $^{***}P \leq 0.001$. **(F,G)** BALB/c mice underwent intratracheal infection with 1 × 10^4 CFU of *C. neoformans* strain 52D. Mice were treated with phosphate-buffered saline or anti-Ly6G antibody 1 day prior to infection and daily during the study. At 12 days postinfection, lungs were excised for analysis of neutrophil recruitment and CFU. **(F)** The number of neutrophils and **(G)** fungal burden is shown. Data are pooled from two independent experiments and shown as mean ± SEM (*n* = 8 mice/group). $^{***}P \leq 0.001$.

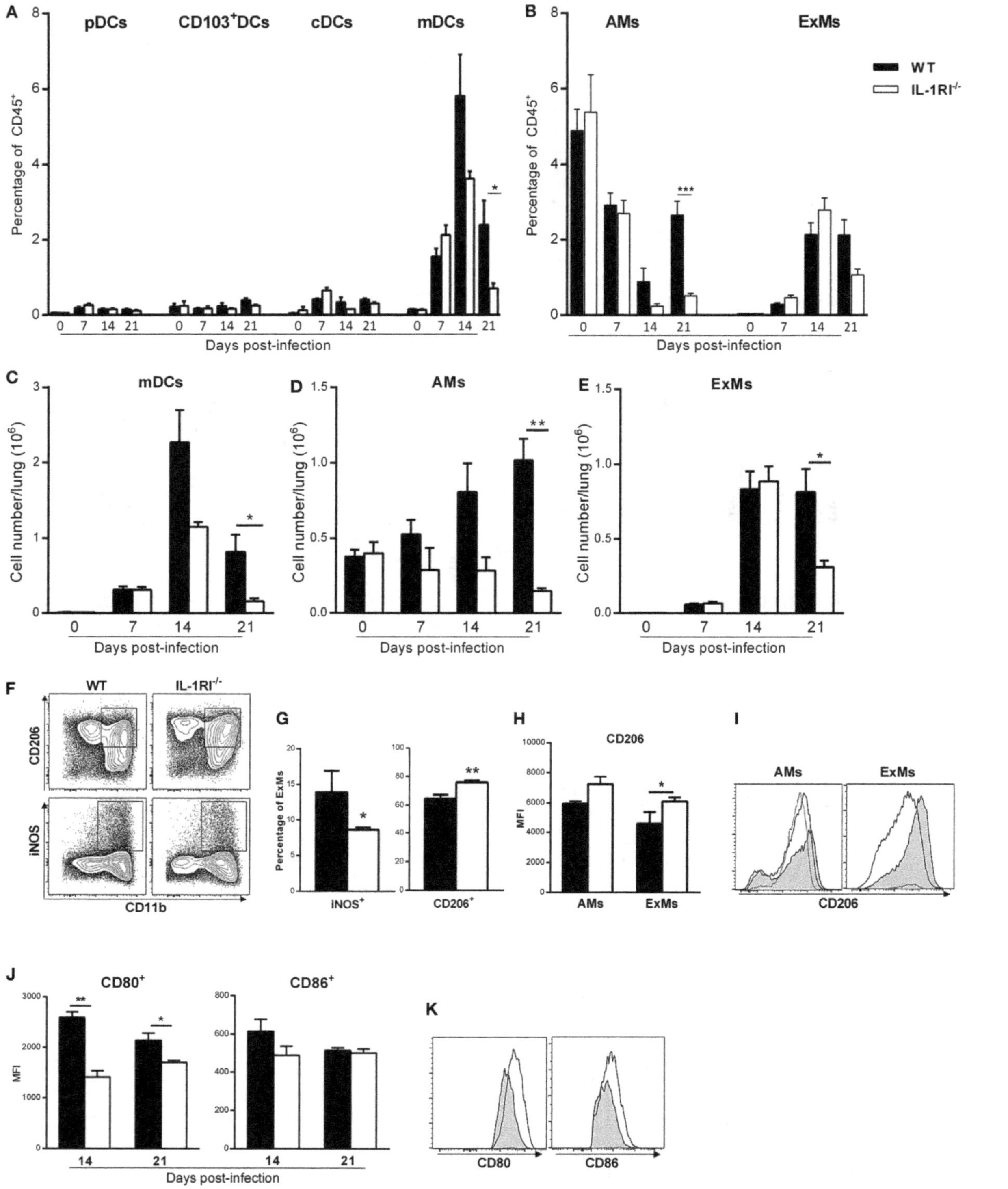

FIGURE 6 | Interleukin-1 receptor type I-deficient (IL-1RI$^{-/-}$) mice have fewer monocyte-derived dendritic cells (DCs) and macrophages in the lungs after *Cryptococcus neoformans* 52D infection. Lung cell suspensions from uninfected and infected mice were stained with fluorochrome-labeled antibodies and analyzed by flow cytometry as described in Section "Materials and Methods." **(A,B)** Percentage of DC and macrophage subsets at 0, 7, 14, and 21 days postinfection. **(C–E)** Total number of mDCs, AMs, and ExMs at 0, 7, 14, and 21 days postinfection. **(F)** Representative plots of M1 (CD11b$^+$, iNOS$^+$) and M2 (CD11b$^+$, CD206$^+$) polarized macrophages in wild-type (WT) and IL-1RI$^{-/-}$ mice at 14 days postinfection. **(G)** Percentage of M1- and M2-polarized macrophages in WT and IL-1RI$^{-/-}$ mice at 14 days postinfection. **(H)** Mean fluorescence intensity (MFI) of CD206 expression on macrophages in WT compared to IL-1RI$^{-/-}$ mice and **(I)** Upregulation of CD206 in AMs and ExMs in IL-1RI$^{-/-}$ compared to WT mice at 14 days postinfection. **(J)** MFI and **(K)** upregulation of CD80- and CD86-positive cells derived from ExMs at 14 days postinfection is shown; **(I,K)** IL-1RI$^{-/-}$, gray filled lines; WT, white filled solid lines; uninfected mice, dashed lines. Data are shown as mean ± SEM and representative of two independent experiments (n = 4 mice/strain/time point). $^*P \leq 0.05$, $^{**}P \leq 0.01$, and $^{***}P \leq 0.001$.

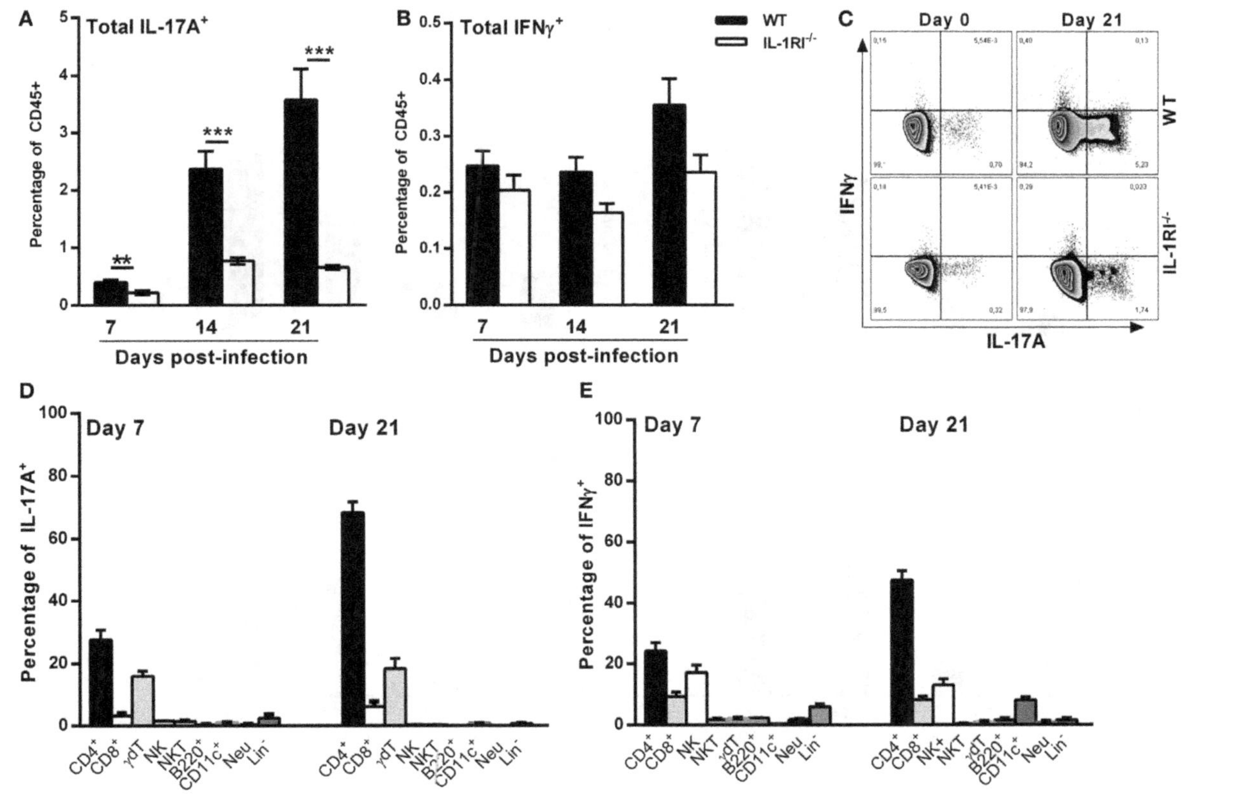

FIGURE 7 | T cells are the predominant sources of interleukin (IL)-17A and IFNγ in BALB/c lungs infected with *Cryptococcus neoformans* 52D. **(A)** Lung cell suspensions from uninfected and infected mice were harvested and restimulated with phorbol 12-myristate 13-acetate (PMA)–ionomycin followed by intracellular staining for IL-17A and IFNγ. **(A,B)** Percentage of total $CD45^+IFN\gamma^+$ and $CD45^+IL\text{-}17A^+$ cells at 7, 14, and 21 days postinfection. **(C)** Representative flow cytometry plots of lung $CD45^+$ cells from individual mice harvested at 21 days postinfection. **(D,E)** Percentage of IL-17A- and IFNγ-producing cell types in WT mice at 7 and 21 days postinfection is shown. Data are shown as mean ± SEM and representative of two independent experiments ($n = 4$ mice/strain/time point). $^{**}P \leq 0.01$ and $^{***}P \leq 0.001$.

day 21 postinfection with $CD4^+$ T cells and γδT cells accounting for 60 and 20%, respectively, of $IL\text{-}17A^+$ cells. $CD4^+$ and $CD8^+$ T-cells, NK cells, γδT cells, and neutrophils have been shown to produce IFNγ during fungal infection (57–59). In our study, $CD4^+$ T and NK cells were the predominant IFNγ-producing subsets at day 7 and day 21 postinfection (**Figure 7E**).

Effect of IL-1RI Signaling on the Lung Lymphocyte Infiltration following *C. neoformans* 52D Infection

As lymphocytes are necessary for effective clearance of *C. neoformans,* we compared the recruitment of $CD4^+$ or $CD8^+$ T cells, γδT cells, and B cells to the lungs of WT and IL-1RI$^{-/-}$ mice at different time points after infection. Flow cytometry analysis showed that WT mice recruit a significantly higher number of $CD4^+$ cells compared to the IL-1RI$^{-/-}$ strain at 14 and 21 days postinfection (**Figure 8A**). Recruitment of $CD8^+$ T cells was comparable between the two strains at all time points, although WT mice showed a trend toward a higher number of $CD8^+$ T cells at day 21 compared to IL-1RI$^{-/-}$ mice (**Figure 8B**). WT mice demonstrated increased recruitment of γδT cells at day 14 and day 21 postinfection compared to uninfected mice; in contrast, there was no significant increase of this cell type in IL-1RI$^{-/-}$ mice during infection (**Figure 8C**). No differences in the number of B cells recruited to the lungs during infection were observed between the two strains (**Figure 8D**). Taken together, this analysis demonstrates that IL-1RI signaling selectively regulates T lymphocyte recruitment to the lungs during the adaptive phase of immunity against *C. neoformans* 52D infection.

Pulmonary $CD4^+$ T Cells from IL-1RI$^{-/-}$ Mice Display Diminished Th17 and Increased Th2 Cytokine Production following *C. neoformans* Infection

It has been clearly shown that a Th1/Th17 response is protective and a Th2 response is detrimental, respectively, against *C. neoformans* infection (60). To analyze the effect of IL-1R signaling on T cell differentiation during infection, we harvested lungs at serial time points, restimulated the cells with PMA/ionomycin, and stained for intracellular IFNγ, IL-13, and IL-17A as representative cytokines for Th1, Th2, and Th17 polarization states, respectively (**Figure 9**). The results

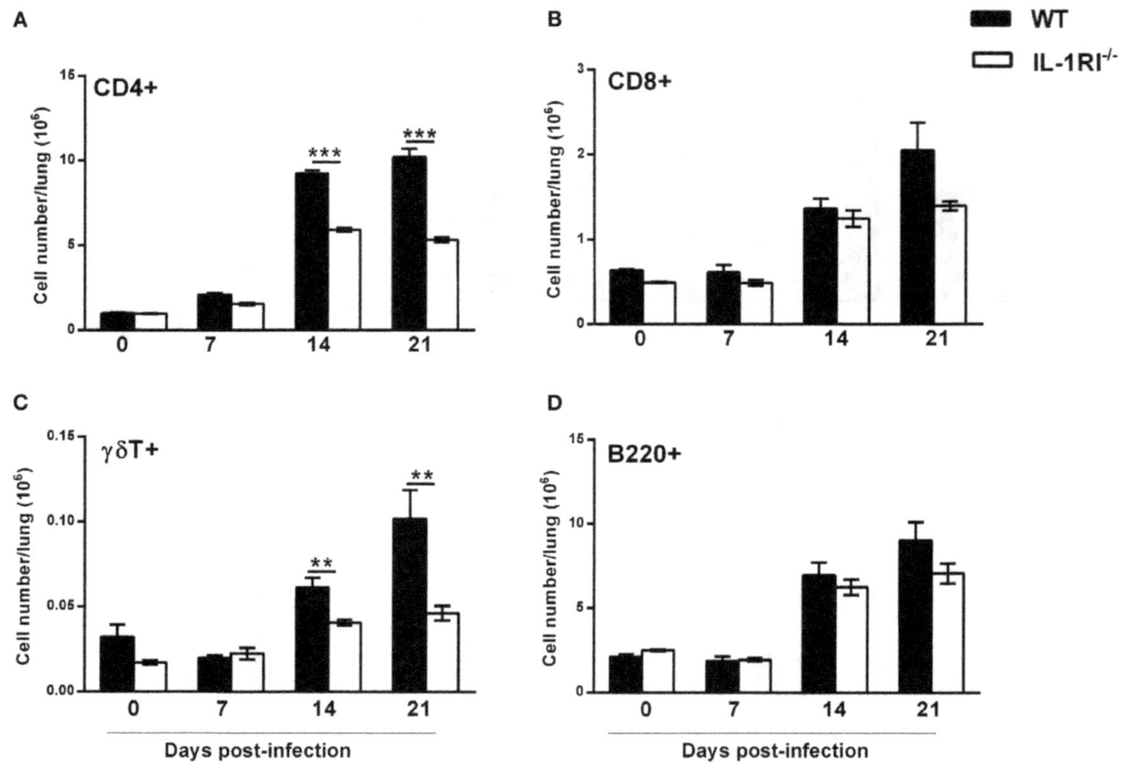

FIGURE 8 | Lungs of interleukin-1 receptor type I-deficient (IL-1RI$^{-/-}$) mice display fewer CD4$^+$ and γδT$^+$ lymphocytes during the adaptive phase of immunity after *Cryptococcus neoformans* 52D infection. Lung cell suspensions from uninfected and infected mice were stained with fluorochrome-labeled antibodies and analyzed by flow cytometry as described in Section "Materials and Methods." **(A–D)** Total number of CD3$^+$CD4$^+$, CD3$^+$CD8$^+$, CD3$^+$γδ$^+$, and B220$^+$ cells in the lungs at 0, 7, 14, and 21 days postinfection. Data are shown as mean ± SEM and representative of two independent experiments (*n* = 4 mice/strain/time point). $^{**}P \leq 0.01$ and $^{***}P \leq 0.001$.

demonstrated a significantly higher number of CD4$^+$ IFNγ$^+$ cells in the lungs of WT compared to IL-1RI$^{-/-}$ mice at 7 days postinfection with a trend toward more CD4$^+$ IFNγ$^+$ cells at days 14 and 21. Compared to the IL-1RI$^{-/-}$ strain, WT mice showed a trend toward more CD4$^+$ IL-17A$^+$ cells at day 7 with a significant increase of this cell type at days 14 and 21. In contrast, IL-1RI$^{-/-}$ lungs contained a significantly higher percentage of CD4$^+$ IL13$^+$ cells compared to WT lungs at 14 and 21 days postinfection. In summary, these findings demonstrate that after *C. neoformans* infection, IL-1RI signaling significantly increased Th1 differentiation during the early phase of infection and strongly promoted Th17 differentiation during the late phase of infection.

DISCUSSION

Induction of IL-1α/β during mouse cryptococcal infection has been reported, but a clear role for IL-1R-dependent signaling in the host immune response has not been demonstrated (27, 32, 34, 35, 37). Here, we provide evidence that IL-1RI deficiency on the BALB/c background has deleterious effects on the outcome of pulmonary *C. neoformans* 52D infection. The most significant findings of this study are as follows: (1) IL-1RI$^{-/-}$ mice cannot clear moderately virulent *C. neoformans* 52D and develop progressive infection of the lungs and brain resulting in death starting at day 40 postinfection; (2) susceptibility of IL-1RI$^{-/-}$ mice is associated with reduced levels of pro-inflammatory, Th1, and Th17 cytokines; (3) IL-1RI signaling in response to *C. neoformans* 52D infection regulates the recruitment of inflammatory DCs to the lung, contributes to recruitment and M1 polarization of macrophages, and promotes Th1/Th17 differentiation of CD4$^+$ T cells; and (4) lung neutrophil recruitment associated with IL-1R signaling is dispensable for protection against *C. neoformans* 52D infection. Taken together, these data clearly demonstrate that IL-1R-dependent signaling plays a complex and essential role in the control of progressive *C. neoformans* 52D infection.

Previously, intranasal infection of C57BL/6 and IL-1RI$^{-/-}$ mice with 2 × 10^4 CFU of the virulent *C. neoformans* H99 strain was shown to cause >90% mortality in both groups (37). In the same report, mice lacking MyD88, an intracellular adaptor for IL-1RI, IL-18R, and several Toll-like receptors, had a trend toward reduced survival but no significant difference in fungal burden compared to WT mice after *C. neoformans* challenge (37). Notably, two earlier studies showed that MyD88$^{-/-}$ mice have a significantly shorter survival time and a higher lung fungal burden compared to WT, TLR2$^{-/-}$, and TLR4$^{-/-}$ mice after *C. neoformans* infection (61, 62). These differences may be attributable, at least in part, to variation in the experimental methods that were used including the dose,

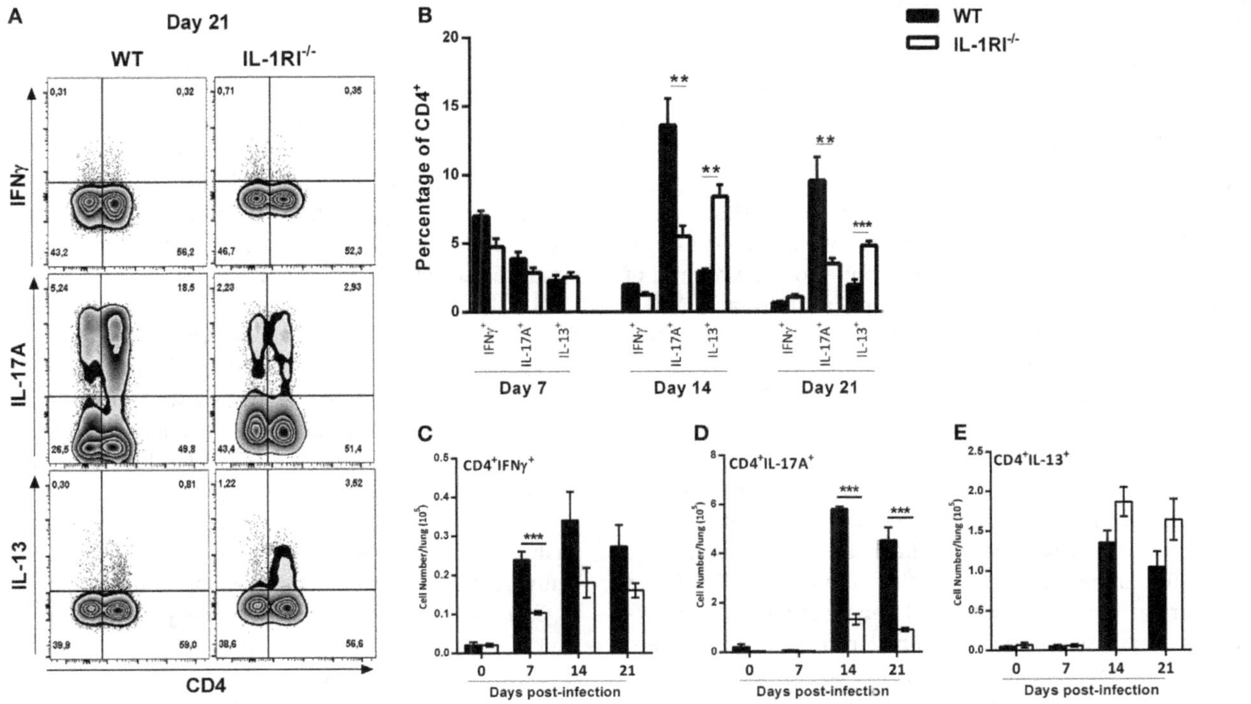

FIGURE 9 | Decreased Th1/Th17 type cytokine expression by CD4+ T cells from interleukin-1 receptor type I-deficient (IL-1RI$^{-/-}$) lungs infected with *Cryptococcus neoformans* 52D. **(A)** Representative flow cytometry plots of lung lymphocytes from individual mice harvested at 14 days postinfection and restimulated with phorbol 12-myristate 13-acetate (PMA)–ionomycin, followed by intracellular staining for IFNγ, IL-17A, and IL-13. **(B)** Percentage and **(C–E)** total numbers of CD4+IFN+, CD4+IL-17A+ and CD4+IL13+ cells are shown. Data are shown as mean ± SEM and representative of two independent experiments (n = 4 mice/strain/time point). $^{*}P \leq 0.05$, $^{**}P \leq 0.01$, and $^{***}P \leq 0.001$. WT, wild type.

route, and strain of *C. neoformans* (60, 63–65). Furthermore, inbred mouse strains also display marked differences in resistance or susceptibility to a standardized cryptococcal infection, highlighting the importance of the host genetic background in disease pathogenesis (66–68). Our data are consistent with other studies showing that BALB/c mice have a naturally resistant phenotype after respiratory infection with the moderately virulent *C. neoformans* 52D strain. Specifically, BALB/c mice progressively clear pulmonary *C. neoformans* 52D infection in association with numerous hallmarks of a protective Th1 response including tight mononuclear cell infiltrates and classically activated macrophages and do not develop central nervous system dissemination (18, 66, 69, 70). Our observation that both IL-1α and IL-1β were induced in the lungs of BALB/c mice after intratracheal infection with *C. neoformans* 52D is also consistent with earlier reports that associated the induction of IL-1β in lung and brain with resistance to cryptococcal infection (35, 71, 72).

Interleukin-1 is a central mediator of inflammation and links innate and adaptive immune response mechanisms (73). Binding of IL-1α or IL-1β to IL-1RI is followed by the recruitment of the IL-1 receptor accessory protein (IL-1RAcP) and activation of signal transduction pathways that induce the expression of IL-1 responsive genes including IL-6, MCP-1, and TNFα (74–77). Induction of pro-inflammatory cytokines followed by generation of a Th1 adaptive immune response is critical for control of cryptococcosis (8, 11, 78). Compared to the BALB/c strain, IL-1RI$^{-/-}$ mice had significantly reduced expression of KC, TNFα, and MCP-1 that was associated with increased lung fungal burden at day 7 after infection. TNFα is one of the main target genes of the IL-1 signaling cascade (76, 77), and both mediators share downstream pathways that induce pro-inflammatory gene expression (79, 80). TNFα signaling in the afferent phase of cryptococcal infection is associated with optimal DC activation and induction of Th1/Th17 polarization and protective immunity (78, 81–84). MCP1/CCR2 signaling is also responsible for the recruitment of inflammatory DCs and macrophages after cryptococcal infection (45, 46, 85). Thus, the reduced expression of pro-inflammatory cytokines and chemokines is one mechanism that could explain the susceptibility of IL-1RI$^{-/-}$ mice to progressive cryptococcosis.

After *C. neoformans* infection, DCs phagocytose and kill cryptococci by oxidative and non-oxidative mechanisms, play an important role in antigen presentation, and drive protective immune responses by secreting cytokines and chemokines (86–89). Compared to other innate cell types, lung DCs express a high level of IL-1RI and signaling *via* this receptor has been shown to promote the maturation and survival of pulmonary DCs and their CCR7-dependent migration to lymph nodes after Influenza A infection (90). At 21 days postinfection with *C. neoformans*, the total number of moDCs in the lung was significantly lower in IL-1RI$^{-/-}$ compared to WT mice, suggesting that recruitment and activation of DCs in the LALNs may be regulated by IL-1R signaling in this model. In addition to DCs, inflammatory macrophages that strongly express microbicidal enzymes such as iNOS play a significant role in fungal clearance (45, 46, 69, 91). After *C. neoformans* 52D infection, we observed that lung macrophages of IL-1R$^{-/-}$ mice had reduced expression of the classical activation markers CD80 and iNOS and increased expression of the alternative activation marker CD206 compared to WT, a pattern that is associated with reduced fungal killing capacity. Our findings are similar to a recent study in BALB/c mice infected with *C. neoformans* 52D that correlated an elevated ratio of Arg1/iNOS expression with an increase in fungal burden and showed a reversal of this ratio during the subsequent period of fungal clearance (48).

In addition to monocyte-derived macrophages and DCs, significantly greater neutrophil recruitment was observed in WT compared to IL-1R$^{-/-}$ lungs. Both IL-1α and IL-1β can promote neutrophil migration (92–95), and diminished neutrophil recruitment to the site of infection due to IL-1R deficiency has been associated with increased susceptibility to several bacterial and fungal infections including *Legionella pneumophila*, Group B *Streptococcus*, *Citrobacter rodentium*, and *Candida albicans* (55, 96–100). Inbred mouse strains including SJL/J, CBA/J, and BALB/c are naturally resistant to pulmonary cryptococcal infection and exhibit substantial neutrophil recruitment the lungs; however, the importance of these cells in host protection is not clear (35, 67, 68). For example, an early study of BALB/c mice given a single injection of anti-Gr-1 (anti-Ly6C/6G) antibody showed less inflammatory damage and significantly longer survival compared to controls after *C. neoformans* 52D infection (101). A subsequent study of BALB/c mice that had undergone prior immunization with *C. neoformans* strain H99γ showed that neutrophil depletion with a specific anti-Ly6G antibody did not affect pulmonary fungal burden (102). Finally, a recent report showed that profound neutrophilia in type 2-deficient STAT6$^{-/-}$ mice on a C57BL/6 background was associated with immunopathology and exacerbation of cryptococcal disease (103). To specifically analyze the contribution of neutrophils to resistance against *C. neoformans* 52D, we used anti-Ly6G to deplete these cells in WT BALB/c mice throughout the course of infection (104, 105). In the absence of neutrophil recruitment, we observed a significantly lower lung fungal burden at 12 days postinfection compared to controls. This finding suggests that, despite their abundance in the lung, neutrophils may have a detrimental effect on host defense against moderately virulent *C. neoformans* 52D (101). Several mechanisms may explain this observation, including competition for cryptococcal antigen between neutrophils and antigen-presenting cells, neutrophil secretion of the immunosuppressive cytokine TGFβ1, or production of IL-1 receptor antagonist, a molecule that inhibits IL-1R signaling (100, 106–110). Further research is necessary to precisely establish the physiological mechanisms that control neutrophil recruitment during cryptococcal infection and to determine whether these cells make a positive contribution to host resistance.

Along with reduced pro-inflammatory cytokines, IL-1R$^{-/-}$ mice showed diminished levels of lung IFNγ compared to WT mice at the early (day 7) and late (days 14 and 21) phases of infection. Intracellular cytokine staining identified CD4^{+} lymphocytes as the most prominent IFNγ-producing cell type. As very few studies have identified IL-1R expression on Th1 cells (111), induction of IFNγ expression by CD4^{+} T cells appears to be an indirect effect of IL-1RI signaling on DCs and possibly other cell types (90). IFNγ plays a central role in host defense against cryptococci by enhancing the fungal internalization and killing by phagocytes (78, 83). An important role for early IFNγ secretion and the development of a Th1 response against *C. neoformans* 52D infection was previously shown in resistant C.B-17 mice (a BALB/c strain congenic for C57BL/6 immunoglobulin heavy chain gene segment), whereas the absence of this response in the C57BL/6 strain correlated with susceptibility (11).

IL-1 is known to regulate the expression of the transcription factors IRF4 and RORγt, both of which play a major role in the induction of CD4^{+}IL-17^{+} (Th17) cells in mice and humans (112–114). IL-1 signaling has been shown to be essential for the development of Th17 immunity to infection with *Coccidioides sp* (115), and mice with deletions of IL-17 or IL-17R are susceptible to candidiasis, pulmonary aspergillosis, and histoplasmosis (55). The role of IL-17 during cryptococcal infection has been analyzed using mice with a C57BL/6 genetic background. In one study, IL-17RA deficiency did not impair pulmonary clearance of *C. neoformans* 52D at 1 or 6 weeks postinfection nor did it alter survival compared to WT mice (116). Another study using IL-17A-deficient mice showed that this cytokine does contribute to fungal clearance from the lung but was not essential for 8-week survival (19). In contrast, administration of IL-23, which is essential for the differentiation of Th17 lymphocytes, led to prolonged survival and reduced fungal burden in C57BL/6 mice (22). A Th17-polarized immune response appears to facilitate the resolution of *C. neoformans* 52D infection through several mechanisms including lung recruitment of activated DCs and inflammatory macrophages, induction of IFNγ-producing CD4^{+} and CD8^{+} T cells, and enhanced fungal containment within macrophages (19–22). Compared to BALB/c, IL-1R$^{-/-}$ mice display several phenotypes that may be attributable to a diminished Th17 response including reduced recruitment of DCs and inflammatory macrophages and increased recruitment of eosinophils and CD4^{+}IL-13^{+} cells to the lungs. On the basis of marked difference between WT and IL-1R$^{-/-}$ mice, we speculate that IL-17 plays a non-redundant role in survival after *C. neoformans* 52D infection; however, studies of BALB/c mice that are deficient for IL-17 or IL-17RA would be required to formally test this hypothesis.

In mouse models, IL-1 signaling is protective against infection with a wide spectrum of intracellular pathogens including *Leishmania amazonensis*, *Mycobacterium avium*, *Toxoplasma gondii*, and *Listeria monocytogenes* (117–121). IL-1RI-deficient mice are also highly susceptible to pulmonary challenge with *Aspergillus fumigatus*; in this model, IL-1α has been shown to be crucial for optimal leukocyte recruitment and IL-1β has been shown to be essential for optimal activation of macrophage antifungal activity (122). It has been suggested that polymorphisms in the IL-1 gene cluster might be important in susceptibility or resistance to invasive pulmonary aspergillosis in humans (123, 124). Both IL-1α and IL-1β have also been shown to play an important role in disseminated candidiasis (125–128), and IL-1 signaling has shown to contribute to host resistance against pulmonary histoplasmosis and *Coccidioides sp.* infection (115, 129). This study expands the role of IL-1 in host defense by demonstrating that IL-1R$^{-/-}$ mice on the Balb/c background are highly susceptible to progressive *C. neoformans* 52D infection of the lungs and brain. IL-1R deficiency in this model results in impaired Th1/Th17 responses and the development of a Th2-biased adaptive immune response. As IL-1α and IL-1β are equally potent activators of IL-1RI signaling yet have different tissue distribution and activation kinetics, future studies that characterize mice that are deficient in either IL-1α or IL-1β or the study of specific cytokine-deficient animals would provide valuable insights into the specific contributions of each cytokine to the development of protective immunity against *C. neoformans* infection.

AUTHOR CONTRIBUTIONS

MS conceived and performed experiments and wrote/edited the manuscript for important intellectual content. BR and IA performed experiments and edited the manuscript for important intellectual content. DS provided reagents and edited the manuscript for important intellectual content. SQ conceived, designed, and supervised the study and wrote/edited the manuscript for important intellectual content.

ACKNOWLEDGMENTS

We thank the histology core facility of the Bellini Life Sciences Complex for assistance with lung tissue preparation and staining.

FUNDING

This work was supported by grants to the Research Institute of the McGill University Health Centre and a Research Chair (DS) from the Fonds de Recherche Santé Quebec, the Canadian Institutes of Health Research to SQ (MOP-102494) and DS (MOP-81361 and MOP-123306), the Costello Memorial Research Fund (SQ), and the Research Institute and Department of Critical Care of the McGill University Health Centre (SQ).

REFERENCES

1. Rajasingham R, Smith RM, Park BJ, Jarvis JN, Govender NP, Chiller TM, et al. Global burden of disease of HIV-associated cryptococcal meningitis: an updated analysis. *Lancet Infect Dis* (2017) 17(8):873–81. doi:10.1016/S1473-3099(17)30243-8
2. Adams P. Cryptococcal meningitis: a blind spot in curbing AIDS. *Lancet* (2016) 387(10028):1605–6. doi:10.1016/S0140-6736(16)30250-1
3. Hole C, Wormley FL Jr. Innate host defenses against *Cryptococcus neoformans*. *J Microbiol* (2016) 54(3):202–11. doi:10.1007/s12275-016-5625-7
4. Neofytos D, Fishman JA, Horn D, Anaissie E, Chang CH, Olyaei A, et al. Epidemiology and outcome of invasive fungal infections in solid organ transplant recipients. *Transpl Infect Dis* (2010) 12(3):220–9. doi:10.1111/j.1399-3062.2010.00492.x
5. Pappas PG. Cryptococcal infections in non-HIV-infected patients. *Trans Am Clin Climatol Assoc* (2013) 124:61–79.
6. Brizendine KD, Baddley JW, Pappas PG. Predictors of mortality and differences in clinical features among patients with Cryptococcosis according to immune status. *PLoS One* (2013) 8(3):e60431. doi:10.1371/journal.pone.0060431
7. Panackal AA, Wuest SC, Lin YC, Wu T, Zhang N, Kosa P, et al. Paradoxical immune responses in non-HIV cryptococcal meningitis. *PLoS Pathog* (2015) 11(5):e1004884. doi:10.1371/journal.ppat.1004884
8. Arora S, Olszewski MA, Tsang TM, McDonald RA, Toews GB, Huffnagle GB. Effect of cytokine interplay on macrophage polarization during chronic pulmonary infection with *Cryptococcus neoformans*. *Infect Immun* (2011) 79(5):1915–26. doi:10.1128/IAI.01270-10
9. Leopold Wager CM, Wormley FL Jr. Classical versus alternative macrophage activation: the Ying and the Yang in host defense against pulmonary fungal infections. *Mucosal Immunol* (2014) 7(5):1023–35. doi:10.1038/mi.2014.65
10. Koguchi Y, Kawakami K. Cryptococcal infection and Th1-Th2 cytokine balance. *Int Rev Immunol* (2002) 21(4–5):423–38. doi:10.1080/08830180213274
11. Hoag KA, Street NE, Huffnagle GB, Lipscomb MF. Early cytokine production in pulmonary *Cryptococcus neoformans* infections distinguishes susceptible and resistant mice. *Am J Respir Cell Mol Biol* (1995) 13(4):487–95. doi:10.1165/ajrcmb.13.4.7546779
12. Chen GH, McDonald RA, Wells JC, Huffnagle GB, Lukacs NW, Toews GB. The gamma interferon receptor is required for the protective pulmonary inflammatory response to *Cryptococcus neoformans*. *Infect Immun* (2005) 73(3):1788–96. doi:10.1128/IAI.73.3.1788-1796.2005
13. Hoag KA, Lipscomb MF, Izzo AA, Street NE. IL-12 and IFN-gamma are required for initiating the protective Th1 response to pulmonary cryptococcosis in resistant C.B-17 mice. *Am J Respir Cell Mol Biol* (1997) 17(6):733–9. doi:10.1165/ajrcmb.17.6.2879
14. Kawakami K, Qureshi MH, Zhang T, Koguchi Y, Shibuya K, Naoe S, et al. Interferon-gamma (IFN-gamma)-dependent protection and synthesis of chemoattractants for mononuclear leucocytes caused by IL-12 in the lungs of mice infected with *Cryptococcus neoformans*. *Clin Exp Immunol* (1999) 117(1):113–22. doi:10.1046/j.1365-2249.1999.00955.x
15. Pietrella D, Lupo P, Bistoni F, Vecchiarelli A. An early imbalance of interleukin 12 influences the adjuvant effect of mannoproteins of *Cryptococcus neoformans*. *Cell Microbiol* (2004) 6(9):883–91. doi:10.1111/j.1462-5822.2004.00411.x
16. Blackstock R, Murphy JW. Role of interleukin-4 in resistance to *Cryptococcus neoformans* infection. *Am J Respir Cell Mol Biol* (2004) 30(1):109–17. doi:10.1165/rcmb.2003-0156OC
17. Muller U, Stenzel W, Kohler G, Werner C, Polte T, Hansen G, et al. IL-13 induces disease-promoting type 2 cytokines, alternatively activated macrophages and allergic inflammation during pulmonary infection of mice with *Cryptococcus neoformans*. *J Immunol* (2007) 179(8):5367–77. doi:10.4049/jimmunol.179.8.5367
18. Jain AV, Zhang Y, Fields WB, McNamara DA, Choe MY, Chen GH, et al. Th2 but not Th1 immune bias results in altered lung functions in a murine model of pulmonary *Cryptococcus neoformans* infection. *Infect Immun* (2009) 77(12):5389–99. doi:10.1128/IAI.00809-09
19. Murdock BJ, Huffnagle GB, Olszewski MA, Osterholzer JJ. Interleukin-17A enhances host defense against cryptococcal lung infection through effects mediated by leukocyte recruitment, activation, and gamma interferon production. *Infect Immun* (2014) 82(3):937–48. doi:10.1128/IAI.01477-13
20. Valdez PA, Vithayathil PJ, Janelsins BM, Shaffer AL, Williamson PR, Datta SK. Prostaglandin E2 suppresses antifungal immunity by inhibiting interferon regulatory factor 4 function and interleukin-17 expression in T cells. *Immunity* (2012) 36(4):668–79. doi:10.1016/j.immuni.2012.02.013
21. Zhang Y, Wang F, Tompkins KC, McNamara A, Jain AV, Moore BB, et al. Robust Th1 and Th17 immunity supports pulmonary clearance but cannot prevent systemic dissemination of highly virulent *Cryptococcus*

neoformans H99. *Am J Pathol* (2009) 175(6):2489–500. doi:10.2353/ajpath.2009.090530

22. Kleinschek MA, Muller U, Schutze N, Sabat R, Straubinger RK, Blumenschein WM, et al. Administration of IL-23 engages innate and adaptive immune mechanisms during fungal infection. *Int Immunol* (2010) 22(2):81–90. doi:10.1093/intimm/dxp117
23. Kleinschek MA, Muller U, Brodie SJ, Stenzel W, Kohler G, Blumenschein WM, et al. IL-23 enhances the inflammatory cell response in *Cryptococcus neoformans* infection and induces a cytokine pattern distinct from IL-12. *J Immunol* (2006) 176(2):1098–106. doi:10.4049/jimmunol.176.2.1098
24. Hardison SE, Wozniak KL, Kolls JK, Wormley FL Jr. Interleukin-17 is not required for classical macrophage activation in a pulmonary mouse model of *Cryptococcus neoformans* infection. *Infect Immun* (2010) 78(12):5341–51. doi:10.1128/IAI.00845-10
25. Wozniak KL, Hardison SE, Kolls JK, Wormley FL. Role of IL-17A on resolution of pulmonary *C. neoformans* infection. *PLoS One* (2011) 6(2):e17204. doi:10.1371/journal.pone.0017204
26. Marais S, Meintjes G, Lesosky M, Wilkinson KA, Wilkinson RJ. Interleukin-17 mediated differences in the pathogenesis of HIV-1-associated tuberculous and cryptococcal meningitis. *AIDS* (2016) 30(3):395–404. doi:10.1097/QAD.0000000000000904
27. Vecchiarelli A, Retini C, Pietrella D, Monari C, Tascini C, Beccari T, et al. Downregulation by cryptococcal polysaccharide of tumor necrosis factor alpha and interleukin-1 beta secretion from human monocytes. *Infect Immun* (1995) 63(8):2919–23.
28. Retini C, Vecchiarelli A, Monari C, Tascini C, Bistoni F, Kozel TR. Capsular polysaccharide of *Cryptococcus neoformans* induces proinflammatory cytokine release by human neutrophils. *Infect Immun* (1996) 64(8):2897–903.
29. Delfino D, Cianci L, Lupis E, Celeste A, Petrelli ML, Curro F, et al. Interleukin-6 production by human monocytes stimulated with *Cryptococcus neoformans* components. *Infect Immun* (1997) 65(6):2454–6.
30. Levitz SM, Tabuni A, Kornfeld H, Reardon CC, Golenbock DT. Production of tumor necrosis factor alpha in human leukocytes stimulated by *Cryptococcus neoformans*. *Infect Immun* (1994) 62(5):1975–81.
31. Vecchiarelli A, Retini C, Monari C, Tascini C, Bistoni F, Kozel TR. Purified capsular polysaccharide of *Cryptococcus neoformans* induces interleukin-10 secretion by human monocytes. *Infect Immun* (1996) 64(7):2846–9.
32. Schoffelen T, Illnait-Zaragozi MT, Joosten LA, Netea MG, Boekhout T, Meis JF, et al. *Cryptococcus gattii* induces a cytokine pattern that is distinct from other cryptococcal species. *PLoS One* (2013) 8(1):e55579. doi:10.1371/journal.pone.0055579
33. Marroni M, Pericolini E, Cenci E, Bistoni F, Vecchiarelli A. Functional defect of natural immune system in an apparent immunocompetent patient with pulmonary cryptococcosis. *J Infect* (2007) 54(1):e5–8. doi:10.1016/j.jinf.2006.03.018
34. Chen M, Xing Y, Lu A, Fang W, Sun B, Chen C, et al. Internalized *Cryptococcus neoformans* activates the canonical caspase-1 and the noncanonical caspase-8 inflammasomes. *J Immunol* (2015) 195(10):4962–72. doi:10.4049/jimmunol.1500865
35. Guillot L, Carroll SF, Homer R, Qureshi ST. Enhanced innate immune responsiveness to pulmonary *Cryptococcus neoformans* infection is associated with resistance to progressive infection. *Infect Immun* (2008) 76(10):4745–56. doi:10.1128/IAI.00341-08
36. Subramaniam KS, Datta K, Quintero E, Manix C, Marks MS, Pirofski LA. The absence of serum IgM enhances the susceptibility of mice to pulmonary challenge with *Cryptococcus neoformans*. *J Immunol* (2010) 184(10):5755–67. doi:10.4049/jimmunol.0901638
37. Wang JP, Lee CK, Akalin A, Finberg RW, Levitz SM. Contributions of the MyD88-dependent receptors IL-18R, IL-1R, and TLR9 to host defenses following pulmonary challenge with *Cryptococcus neoformans*. *PLoS One* (2011) 6(10):e26232. doi:10.1371/journal.pone.0026232
38. Nakamura K, Miyagi K, Koguchi Y, Kinjo Y, Uezu K, Kinjo T, et al. Limited contribution of Toll-like receptor 2 and 4 to the host response to a fungal infectious pathogen, *Cryptococcus neoformans*. *FEMS Immunol Med Microbiol* (2006) 47(1):148–54. doi:10.1111/j.1574-695X.2006.00078.x
39. Upadhya R, Lam WC, Maybruck B, Specht CA, Levitz SM, Lodge JK. Induction of protective immunity to cryptococcal infection in mice by a heat-killed, chitosan-deficient strain of *Cryptococcus neoformans*. *MBio* (2016) 7(3):e547–516. doi:10.1128/mBio.00547-16
40. Uicker WC, Doyle HA, McCracken JP, Langlois M, Buchanan KL. Cytokine and chemokine expression in the central nervous system associated with protective cell-mediated immunity against *Cryptococcus neoformans*. *Med Mycol* (2005) 43(1):27–38. doi:10.1080/13693780410001731510
41. Lei G, Chen M, Li H, Niu JL, Wu S, Mao L, et al. Biofilm from a clinical strain of *Cryptococcus neoformans* activates the NLRP3 inflammasome. *Cell Res* (2013) 23(7):965–8. doi:10.1038/cr.2013.49
42. Roussey JA, Viglianti SP, Teitz-Tennenbaum S, Olszewski MA, Osterholzer JJ. Anti-PD-1 antibody treatment promotes clearance of persistent cryptococcal lung infection in mice. *J Immunol* (2017) 199(10):3535–46. doi:10.4049/jimmunol.1700840
43. Misharin AV, Morales-Nebreda L, Mutlu GM, Budinger GR, Perlman H. Flow cytometric analysis of macrophages and dendritic cell subsets in the mouse lung. *Am J Respir Cell Mol Biol* (2013) 49(4):503–10. doi:10.1165/rcmb.2013-0086MA
44. Guilliams M, Lambrecht BN, Hammad H. Division of labor between lung dendritic cells and macrophages in the defense against pulmonary infections. *Mucosal Immunol* (2013) 6(3):464–73. doi:10.1038/mi.2013.14
45. Osterholzer JJ, Chen GH, Olszewski MA, Zhang YM, Curtis JL, Huffnagle GB, et al. Chemokine receptor 2-mediated accumulation of fungicidal exudate macrophages in mice that clear cryptococcal lung infection. *Am J Pathol* (2011) 178(1):198–211. doi:10.1016/j.ajpath.2010.11.006
46. Osterholzer JJ, Curtis JL, Polak T, Ames T, Chen GH, McDonald R, et al. CCR2 mediates conventional dendritic cell recruitment and the formation of bronchovascular mononuclear cell infiltrates in the lungs of mice infected with *Cryptococcus neoformans*. *J Immunol* (2008) 181(1):610–20. doi:10.4049/jimmunol.181.1.610
47. Xu J, Flaczyk A, Neal LM, Fa Z, Cheng D, Ivey M, et al. Exploitation of scavenger receptor, macrophage receptor with collagenous structure, by *Cryptococcus neoformans* promotes alternative activation of pulmonary lymph node CD11b(+) conventional dendritic cells and non-protective Th2 bias. *Front Immunol* (2017) 8:1231. doi:10.3389/fimmu.2017.01231
48. Davis MJ, Tsang TM, Qiu Y, Dayrit JK, Freij JB, Huffnagle GB, et al. Macrophage M1/M2 polarization dynamically adapts to changes in cytokine microenvironments in *Cryptococcus neoformans* infection. *MBio* (2013) 4(3):e264–213. doi:10.1128/mBio.00264-13
49. Sica A, Mantovani A. Macrophage plasticity and polarization: in vivo veritas. *J Clin Invest* (2012) 122(3):787–95. doi:10.1172/JCI59643
50. Murray PJ, Allen JE, Biswas SK, Fisher EA, Gilroy DW, Goerdt S, et al. Macrophage activation and polarization: nomenclature and experimental guidelines. *Immunity* (2014) 41(1):14–20. doi:10.1016/j.immuni.2014.06.008
51. May RC, Stone NR, Wiesner DL, Bicanic T, Nielsen K. Cryptococcus: from environmental saprophyte to global pathogen. *Nat Rev Microbiol* (2016) 14(2):106–17. doi:10.1038/nrmicro.2015.6
52. Mantovani A, Sica A, Sozzani S, Allavena P, Vecchi A, Locati M. The chemokine system in diverse forms of macrophage activation and polarization. *Trends Immunol* (2004) 25(12):677–86. doi:10.1016/j.it.2004.09.015
53. Mantovani A, Sozzani S, Locati M, Allavena P, Sica A. Macrophage polarization: tumor-associated macrophages as a paradigm for polarized M2 mononuclear phagocytes. *Trends Immunol* (2002) 23(11):549–55. doi:10.1016/S1471-4906(02)02302-5
54. Veldhoen M. Interleukin 17 is a chief orchestrator of immunity. *Nat Immunol* (2017) 18(6):612–21. doi:10.1038/ni.3742
55. Underhill DM, Pearlman E. Immune interactions with pathogenic and commensal fungi: a two-way street. *Immunity* (2015) 43(5):845–58. doi:10.1016/j.immuni.2015.10.023
56. Gladiator A, Wangler N, Trautwein-Weidner K, LeibundGut-Landmann S. Cutting edge: IL-17-secreting innate lymphoid cells are essential for host defense against fungal infection. *J Immunol* (2013) 190(2):521–5. doi:10.4049/jimmunol.1202924
57. Gozalbo D, Gil ML. IFN-gamma in *Candida albicans* infections. *Front Biosci (Landmark Ed)* (2009) 14:1970–8. doi:10.2741/3356
58. Romani L. Immunity to fungal infections. *Nat Rev Immunol* (2011) 11(4):275–88. doi:10.1038/nri2939
59. Rodrigues DR, Fernandes RK, Balderramas Hde A, Penitenti M, Bachiega TF, Calvi SA, et al. Interferon-gamma production by human neutrophils upon stimulation by IL-12, IL-15 and IL-18 and challenge with *Paracoccidioides brasiliensis*. *Cytokine* (2014) 69(1):102–9. doi:10.1016/j.cyto.2014.05.009

60. Voelz K, May RC. Cryptococcal interactions with the host immune system. *Eukaryot Cell* (2010) 9(6):835–46. doi:10.1128/EC.00039-10
61. Biondo C, Midiri A, Messina L, Tomasello F, Garufi G, Catania MR, et al. MyD88 and TLR2, but not TLR4, are required for host defense against *Cryptococcus neoformans*. *Eur J Immunol* (2005) 35(3):870–8. doi:10.1002/eji.200425799
62. Yauch LE, Mansour MK, Shoham S, Rottman JB, Levitz SM. Involvement of CD14, toll-like receptors 2 and 4, and MyD88 in the host response to the fungal pathogen *Cryptococcus neoformans* in vivo. *Infect Immun* (2004) 72(9):5373–82. doi:10.1128/IAI.72.9.5373-5382.2004
63. Olszewski MA, Zhang Y, Huffnagle GB. Mechanisms of cryptococcal virulence and persistence. *Future Microbiol* (2010) 5(8):1269–88. doi:10.2217/fmb.10.93
64. Urai M, Kaneko Y, Ueno K, Okubo Y, Aizawa T, Fukazawa H, et al. Evasion of innate immune responses by the highly virulent *Cryptococcus gattii* by altering capsule glucuronoxylomannan structure. *Front Cell Infect Microbiol* (2015) 5:101. doi:10.3389/fcimb.2015.00101
65. Montone KT. Regulating the T-cell immune response toward the H99 strain of *Cryptococcus neoformans*. *Am J Pathol* (2009) 175(6):2255–6. doi:10.2353/ajpath.2009.090891
66. Chen GH, McNamara DA, Hernandez Y, Huffnagle GB, Toews GB, Olszewski MA. Inheritance of immune polarization patterns is linked to resistance versus susceptibility to *Cryptococcus neoformans* in a mouse model. *Infect Immun* (2008) 76(6):2379–91. doi:10.1128/IAI.01143-07
67. Carroll SF, Lafferty EI, Flaczyk A, Fujiwara TM, Homer R, Morgan K, et al. Susceptibility to progressive *Cryptococcus neoformans* pulmonary infection is regulated by loci on mouse chromosomes 1 and 9. *Infect Immun* (2012) 80(12):4167–76. doi:10.1128/IAI.00417-12
68. Shourian M, Flaczyk A, Angers I, Mindt BC, Fritz JH, Qureshi ST. The Cnes2 locus on mouse chromosome 17 regulates host defense against cryptococcal infection through pleiotropic effects on host immunity. *Infect Immun* (2015) 83(12):4541–54. doi:10.1128/IAI.00697-15
69. Zhang Y, Wang F, Bhan U, Huffnagle GB, Toews GB, Standiford TJ, et al. TLR9 signaling is required for generation of the adaptive immune protection in *Cryptococcus neoformans*-infected lungs. *Am J Pathol* (2010) 177(2):754–65. doi:10.2353/ajpath.2010.091104
70. Zaragoza O, Alvarez M, Telzak A, Rivera J, Casadevall A. The relative susceptibility of mouse strains to pulmonary *Cryptococcus neoformans* infection is associated with pleiotropic differences in the immune response. *Infect Immun* (2007) 75(6):2729–39. doi:10.1128/IAI.00094-07
71. Blasi E, Barluzzi R, Mazzolla R, Pitzurra L, Puliti M, Saleppico S, et al. Biomolecular events involved in anticryptococcal resistance in the brain. *Infect Immun* (1995) 63(4):1218–22.
72. Maffei CM, Mirels LF, Sobel RA, Clemons KV, Stevens DA. Cytokine and inducible nitric oxide synthase mRNA expression during experimental murine cryptococcal meningoencephalitis. *Infect Immun* (2004) 72(4):2338–49. doi:10.1128/IAI.72.4.2338-2349.2004
73. Garlanda C, Dinarello CA, Mantovani A. The interleukin-1 family: back to the future. *Immunity* (2013) 39(6):1003–18. doi:10.1016/j.immuni.2013.11.010
74. Weber A, Wasiliew P, Kracht M. Interleukin-1 (IL-1) pathway. *Sci Signal* (2010) 3(105):cm1. doi:10.1126/scisignal.3105cm1
75. Casanova JL, Abel L, Quintana-Murci L. Human TLRs and IL-1Rs in host defense: natural insights from evolutionary, epidemiological, and clinical genetics. *Annu Rev Immunol* (2011) 29:447–91. doi:10.1146/annurev-immunol-030409-101335
76. O'Neill LA, Dinarello CA. The IL-1 receptor/toll-like receptor superfamily: crucial receptors for inflammation and host defense. *Immunol Today* (2000) 21(5):206–9. doi:10.1016/S0167-5699(00)01611-X
77. Bethea JR, Gillespie GY, Benveniste EN. Interleukin-1 beta induction of TNF-alpha gene expression: involvement of protein kinase C. *J Cell Physiol* (1992) 152(2):264–73. doi:10.1002/jcp.1041520207
78. Herring AC, Lee J, McDonald RA, Toews GB, Huffnagle GB. Induction of interleukin-12 and gamma interferon requires tumor necrosis factor alpha for protective T1-cell-mediated immunity to pulmonary *Cryptococcus neoformans* infection. *Infect Immun* (2002) 70(6):2959–64. doi:10.1128/IAI.70.6.2959-2964.2002
79. Dinarello CA. Anti-cytokine therapies in response to systemic infection. *J Investig Dermatol Symp Proc* (2001) 6(3):244–50. doi:10.1046/j.0022-202x.2001.00046.x
80. Saperstein S, Chen L, Oakes D, Pryhuber G, Finkelstein J. IL-1beta augments TNF-alpha-mediated inflammatory responses from lung epithelial cells. *J Interferon Cytokine Res* (2009) 29(5):273–84. doi:10.1089/jir.2008.0076
81. Kawakami K, Qifeng X, Tohyama M, Qureshi MH, Saito A. Contribution of tumour necrosis factor-alpha (TNF-alpha) in host defence mechanism against *Cryptococcus neoformans*. *Clin Exp Immunol* (1996) 106(3):468–74. doi:10.1046/j.1365-2249.1996.d01-870.x
82. Milam JE, Herring-Palmer AC, Pandrangi R, McDonald RA, Huffnagle GB, Toews GB. Modulation of the pulmonary type 2 T-cell response to *Cryptococcus neoformans* by intratracheal delivery of a tumor necrosis factor alpha-expressing adenoviral vector. *Infect Immun* (2007) 75(10):4951–8. doi:10.1128/IAI.00176-07
83. Huffnagle GB, Toews GB, Burdick MD, Boyd MB, McAllister KS, McDonald RA, et al. Afferent phase production of TNF-alpha is required for the development of protective T cell immunity to *Cryptococcus neoformans*. *J Immunol* (1996) 157(10):4529–36.
84. Xu J, Eastman AJ, Flaczyk A, Neal LM, Zhao G, Carolan J, et al. Disruption of early tumor necrosis factor alpha signaling prevents classical activation of dendritic cells in lung-associated lymph nodes and development of protective immunity against cryptococcal infection. *MBio* (2016) 7(4):e510–6. doi:10.1128/mBio.00510-16
85. Traynor TR, Kuziel WA, Toews GB, Huffnagle GB. CCR2 expression determines T1 versus T2 polarization during pulmonary *Cryptococcus neoformans* infection. *J Immunol* (2000) 164(4):2021–7. doi:10.4049/jimmunol.164.4.2021
86. Leopold Wager CM, Wormley FL Jr. Is development of a vaccine against *Cryptococcus neoformans* feasible? *PLoS Pathog* (2015) 11(6):e1004843. doi:10.1371/journal.ppat.1004843
87. Wozniak KL, Vyas JM, Levitz SM. In vivo role of dendritic cells in a murine model of pulmonary cryptococcosis. *Infect Immun* (2006) 74(7):3817–24. doi:10.1128/IAI.00317-06
88. Siegemund S, Alber G. *Cryptococcus neoformans* activates bone marrow-derived conventional dendritic cells rather than plasmacytoid dendritic cells and down-regulates macrophages. *FEMS Immunol Med Microbiol* (2008) 52(3):417–27. doi:10.1111/j.1574-695X.2008.00391.x
89. Syme RM, Spurrell JC, Amankwah EK, Green FH, Mody CH. Primary dendritic cells phagocytose *Cryptococcus neoformans* via mannose receptors and Fcgamma receptor II for presentation to T lymphocytes. *Infect Immun* (2002) 70(11):5972–81. doi:10.1128/IAI.70.11.5972-5981.2002
90. Pang IK, Ichinohe T, Iwasaki A. IL-1R signaling in dendritic cells replaces pattern-recognition receptors in promoting CD8(+) T cell responses to influenza A virus. *Nat Immunol* (2013) 14(3):246–53. doi:10.1038/ni.2514
91. Qiu Y, Zeltzer S, Zhang Y, Wang F, Chen GH, Dayrit J, et al. Early induction of CCL7 downstream of TLR9 signaling promotes the development of robust immunity to cryptococcal infection. *J Immunol* (2012) 188(8):3940–8. doi:10.4049/jimmunol.1103053
92. Dinarello CA. Immunological and inflammatory functions of the interleukin-1 family. *Annu Rev Immunol* (2009) 27:519–50. doi:10.1146/annurev.immunol.021908.132612
93. Lee PY, Kumagai Y, Xu Y, Li Y, Barker T, Liu C, et al. IL-1alpha modulates neutrophil recruitment in chronic inflammation induced by hydrocarbon oil. *J Immunol* (2011) 186(3):1747–54. doi:10.4049/jimmunol.1001328
94. Schmitz N, Kurrer M, Bachmann MF, Kopf M. Interleukin-1 is responsible for acute lung immunopathology but increases survival of respiratory influenza virus infection. *J Virol* (2005) 79(10):6441–8. doi:10.1128/JVI.79.10.6441-6448.2005
95. Mascarenhas DP, Pereira MS, Manin GZ, Hori JI, Zamboni DS. Interleukin 1 receptor-driven neutrophil recruitment accounts to MyD88-dependent pulmonary clearance of *Legionella pneumophila* infection in vivo. *J Infect Dis* (2015) 211(2):322–30. doi:10.1093/infdis/jiu430
96. Barry KC, Fontana MF, Portman JL, Dugan AS, Vance RE. IL-1alpha signaling initiates the inflammatory response to virulent *Legionella pneumophila* in vivo. *J Immunol* (2013) 190(12):6329–39. doi:10.4049/jimmunol.1300100
97. Biondo C, Mancuso G, Midiri A, Signorino G, Domina M, Lanza Cariccio V, et al. Essential role of interleukin-1 signaling in host defenses against group B streptococcus. *MBio* (2014) 5(5):e1428–1414. doi:10.1128/mBio.01428-14
98. Lee YS, Yang H, Yang JY, Kim Y, Lee SH, Kim JH, et al. Interleukin-1 (IL-1) signaling in intestinal stromal cells controls KC/CXCL1 secretion, which correlates with recruitment of IL-22-secreting neutrophils at early stages

of *Citrobacter rodentium* infection. *Infect Immun* (2015) 83(8):3257–67. doi:10.1128/IAI.00670-15
99. Altmeier S, Toska A, Sparber F, Teijeira A, Halin C, LeibundGut-Landmann S. IL-1 coordinates the neutrophil response to *C. albicans* in the oral mucosa. *PLoS Pathog* (2016) 12(9):e1005882. doi:10.1371/journal.ppat.1005882
100. Mantovani A, Cassatella MA, Costantini C, Jaillon S. Neutrophils in the activation and regulation of innate and adaptive immunity. *Nat Rev Immunol* (2011) 11(8):519–31. doi:10.1038/nri3024
101. Mednick AJ, Feldmesser M, Rivera J, Casadevall A. Neutropenia alters lung cytokine production in mice and reduces their susceptibility to pulmonary cryptococcosis. *Eur J Immunol* (2003) 33(6):1744–53. doi:10.1002/eji.200323626
102. Wozniak KL, Kolls JK, Wormley FL Jr. Depletion of neutrophils in a protective model of pulmonary cryptococcosis results in increased IL-17A production by gammadelta T cells. *BMC Immunol* (2012) 13:65. doi:10.1186/1471-2172-13-65
103. Wiesner DL, Smith KD, Kashem SW, Bohjanen PR, Nielsen K. Different lymphocyte populations direct dichotomous eosinophil or neutrophil responses to pulmonary cryptococcus infection. *J Immunol* (2017) 198(4):1627–37. doi:10.4049/jimmunol.1600821
104. Daley JM, Thomay AA, Connolly MD, Reichner JS, Albina JE. Use of Ly6G-specific monoclonal antibody to deplete neutrophils in mice. *J Leukoc Biol* (2008) 83(1):64–70. doi:10.1189/jlb.0407247
105. Hasenberg A, Hasenberg M, Mann L, Neumann F, Borkenstein L, Stecher M, et al. Catchup: a mouse model for imaging-based tracking and modulation of neutrophil granulocytes. *Nat Methods* (2015) 12(5):445–52. doi:10.1038/nmeth.3322
106. Barquero-Calvo E, Martirosyan A, Ordonez-Rueda D, Arce-Gorvel V, Alfaro-Alarcon A, Lepidi H, et al. Neutrophils exert a suppressive effect on Th1 responses to intracellular pathogen *Brucella abortus*. *PLoS Pathog* (2013) 9(2):e1003167. doi:10.1371/journal.ppat.1003167
107. Yang CW, Strong BS, Miller MJ, Unanue ER. Neutrophils influence the level of antigen presentation during the immune response to protein antigens in adjuvants. *J Immunol* (2010) 185(5):2927–34. doi:10.4049/jimmunol.1001289
108. Eken C, Gasser O, Zenhaeusern G, Oehri I, Hess C, Schifferli JA. Polymorphonuclear neutrophil-derived ectosomes interfere with the maturation of monocyte-derived dendritic cells. *J Immunol* (2008) 180(2):817–24. doi:10.4049/jimmunol.180.2.817
109. Gasser O, Schifferli JA. Activated polymorphonuclear neutrophils disseminate anti-inflammatory microparticles by ectocytosis. *Blood* (2004) 104(8):2543–8. doi:10.1182/blood-2004-01-0361
110. Bourke E, Cassetti A, Villa A, Fadlon E, Colotta F, Mantovani A. IL-1 beta scavenging by the type II IL-1 decoy receptor in human neutrophils. *J Immunol* (2003) 170(12):5999–6005. doi:10.4049/jimmunol.170.12.5999
111. Santarlasci V, Cosmi L, Maggi L, Liotta F, Annunziato F. IL-1 and T helper immune responses. *Front Immunol* (2013) 4:182. doi:10.3389/fimmu.2013.00182
112. Chung Y, Chang SH, Martinez GJ, Yang XO, Nurieva R, Kang HS, et al. Critical regulation of early Th17 cell differentiation by interleukin-1 signaling. *Immunity* (2009) 30(4):576–87. doi:10.1016/j.immuni.2009.02.007
113. Sutton C, Brereton C, Keogh B, Mills KH, Lavelle EC. A crucial role for interleukin (IL)-1 in the induction of IL-17-producing T cells that mediate autoimmune encephalomyelitis. *J Exp Med* (2006) 203(7):1685–91. doi:10.1084/jem.20060285
114. Sims JE, Smith DE. The IL-1 family: regulators of immunity. *Nat Rev Immunol* (2010) 10(2):89–102. doi:10.1038/nri2691
115. Hung CY, Jimenez-Alzate Mdel P, Gonzalez A, Wuthrich M, Klein BS, Cole GT. Interleukin-1 receptor but not toll-like receptor 2 is essential for MyD88-dependent Th17 immunity to Coccidioides infection. *Infect Immun* (2014) 82(5):2106–14. doi:10.1128/IAI.01579-13
116. Szymczak WA, Sellers RS, Pirofski LA. IL-23 dampens the allergic response to *Cryptococcus neoformans* through IL-17-independent and -dependent mechanisms. *Am J Pathol* (2012) 180(4):1547–59. doi:10.1016/j.ajpath.2011.12.038
117. Xin L, Li Y, Soong L. Role of interleukin-1beta in activating the CD11c(high) CD45RB-dendritic cell subset and priming *Leishmania amazonensis*-specific CD4+ T cells in vitro and in vivo. *Infect Immun* (2007) 75(10):5018–26. doi:10.1128/IAI.00499-07
118. Hunter CA, Chizzonite R, Remington JS. IL-1 beta is required for IL-12 to induce production of IFN-gamma by NK cells. A role for IL-1 beta in the T cell-independent mechanism of resistance against intracellular pathogens. *J Immunol* (1995) 155(9):4347–54.
119. Denis M, Ghadirian E. Interleukin-1 is involved in mouse resistance to *Mycobacterium avium*. *Infect Immun* (1994) 62(2):457–61.
120. Chen W, Havell EA, Moldawer LL, McIntyre KW, Chizzonite RA, Harmsen AG. Interleukin 1: an important mediator of host resistance against *Pneumocystis carinii*. *J Exp Med* (1992) 176(3):713–8. doi:10.1084/jem.176.3.713
121. Juffermans NP, Florquin S, Camoglio L, Verbon A, Kolk AH, Speelman P, et al. Interleukin-1 signaling is essential for host defense during murine pulmonary tuberculosis. *J Infect Dis* (2000) 182(3):902–8. doi:10.1086/315771
122. Caffrey AK, Lehmann MM, Zickovich JM, Espinosa V, Shepardson KM, Watschke CP, et al. IL-1alpha signaling is critical for leukocyte recruitment after pulmonary *Aspergillus fumigatus* challenge. *PLoS Pathog* (2015) 11(1):e1004625. doi:10.1371/journal.ppat.1004625
123. Sainz J, Perez E, Gomez-Lopera S, Jurado M. IL1 gene cluster polymorphisms and its haplotypes may predict the risk to develop invasive pulmonary aspergillosis and modulate C-reactive protein level. *J Clin Immunol* (2008) 28(5):473–85. doi:10.1007/s10875-008-9197-0
124. Wojtowicz A, Gresnigt MS, Lecompte T, Bibert S, Manuel O, Joosten LA, et al. IL1B and DEFB1 polymorphisms increase susceptibility to invasive mold infection after solid-organ transplantation. *J Infect Dis* (2015) 211(10):1646–57. doi:10.1093/infdis/jiu636
125. Romani L. Immunity to fungal infections. *Nat Rev Immunol* (2004) 4(1):1–23. doi:10.1038/nri1255
126. Mayer-Barber KD, Yan B. Clash of the Cytokine Titans: counter-regulation of interleukin-1 and type I interferon-mediated inflammatory responses. *Cell Mol Immunol* (2017) 14(1):22–35. doi:10.1038/cmi.2016.25
127. Van't Wout JW, Van der Meer JW, Barza M, Dinarello CA. Protection of neutropenic mice from lethal *Candida albicans* infection by recombinant interleukin 1. *Eur J Immunol* (1988) 18(7):1143–6. doi:10.1002/eji.1830180728
128. Vonk AG, Netea MG, van Krieken JH, Iwakura Y, van der Meer JW, Kullberg BJ. Endogenous interleukin (IL)-1 alpha and IL-1 beta are crucial for host defense against disseminated candidiasis. *J Infect Dis* (2006) 193(10):1419–26. doi:10.1086/503363
129. Deepe GS Jr, McGuinness M. Interleukin-1 and host control of pulmonary histoplasmosis. *J Infect Dis* (2006) 194(6):855–64. doi:10.1086/506946

2

Autoimmune Regulator Deficiency Results in a Decrease in STAT1 Levels in Human Monocytes

*Ofer Zimmerman[1], Lindsey B. Rosen[1], Muthulekha Swamydas[1], Elise M. N. Ferre[1], Mukil Natarajan[1], Frank van de Veerdonk[2], Steven M. Holland[1] and Michail S. Lionakis[1]**

[1] Laboratory of Clinical Infectious Diseases, National Institute of Allergy and Immunology, National Institutes of Health, Bethesda, MD, United States, [2] Department of Internal Medicine, Radboud University Medical Center, Radboud Institute for Molecular Life Sciences (RILMS), Nijmegen, Netherlands

***Correspondence:**
Michail S. Lionakis
lionakism@niaid.nih.gov

Autoimmune polyendocrinopathy-candidiasis-ectodermal dystrophy (APECED) is a rare primary immunodeficiency disorder typically caused by biallelic autoimmune regulator (*AIRE*) mutations that manifests with chronic mucocutaneous candidiasis (CMC) and autoimmunity. Patients with *STAT1* gain-of-function (GOF) mutations also develop CMC and autoimmunity; they exhibit increased STAT1 protein levels at baseline and STAT1 phosphorylation (pSTAT1) upon interferon (IFN)-γ stimulation relative to healthy controls. AIRE interacts functionally with a protein that directly regulates STAT1, namely protein inhibitor of activated STAT1, which inhibits STAT1 activation. Given the common clinical features between patients with *AIRE* and *STAT1* GOF mutations, we sought to determine whether APECED patients also exhibit increased levels of STAT1 protein and phosphorylation in $CD14^+$ monocytes. We obtained peripheral blood mononuclear cells from 8 APECED patients and 13 healthy controls and assessed the levels of STAT1 protein and STAT1 tyrosine phosphorylation at rest and following IFN-γ stimulation, as well as the levels of *STAT1* mRNA. The mean STAT1 protein levels in $CD14^+$ monocytes exhibited a ~20% significant decrease in APECED patients both at rest and after IFN-γ stimulation relative to that of healthy donors. Similarly, the mean peak value of IFN-γ-induced pSTAT1 level was ~20% significantly lower in APECED patients compared to that in healthy controls. The decrease in STAT1 and peak pSTAT1 in APECED patients was not accompanied by decreased *STAT1* mRNA or anti-IFN-γ autoantibodies; instead, it correlated with the presence of autoantibodies to type I IFN and decreased $AIRE^{-/-}$ monocyte surface expression of IFN-γ receptor 2. Our data show that, in contrast to patients with *STAT1* GOF mutations, APECED patients show a moderate but consistent and significant decrease in total STAT1 protein levels, associated with lower peak levels of pSTAT1 molecules after IFN-γ stimulation.

Keywords: STAT1, phosphorylation, chronic mucocutaneous candidiasis, *AIRE*, APECED, $CD14^+$ monocytes, IFN-γ

Abbreviations: APECED, autoimmune polyendocrinopathy-candidiasis-ectodermal dystrophy; APS-1, autoimmune polyglandular syndrome type-1; CC, coiled-coil; CMC, chronic mucocutaneous candidiasis; DB, DNA binding; GOF, gain-of-function; LOF, loss-of-function; PBMC, peripheral blood mononuclear cells; STAT, signal transducer and activator of transcription.

INTRODUCTION

Autoimmune polyendocrinopathy-candidiasis-ectodermal dystrophy (APECED) or autoimmune polyglandular syndrome type-1 (APS-1; OMIM 240300) is a monogenic disorder caused by biallelic mutations in the autoimmune regulator (*AIRE*) gene, a thymus-enriched transcription regulator that promotes central immune tolerance *via* the expression of peripheral tissue self-antigens in medullary thymic epithelial cells (1, 2). Additional AIRE functions have recently been proposed to also contribute to immunological tolerance (3–6). In addition, heterozygous dominant-negative *AIRE* mutations in the plant homeodomain 1 domain have also been described, associated with organ-specific APECED-associated autoimmune manifestations and/or chronic mucocutaneous candidiasis (CMC) (7–9). APECED patients manifest with a characteristic constellation of CMC and autoimmunity that involves both endocrine and non-endocrine tissues (10–12). In fact, APECED is the only CMC-associated primary immunodeficiency disorder in which CMC is the sole consistent infectious disease phenotype. In addition to the breakdown in mechanisms of T-cell tolerance, AIRE-deficient patients also have high titers of neutralizing autoantibodies against Th17 cytokines and tissue-specific autoantigens, which have been shown to correlate with the development of CMC and organ-specific autoimmune manifestations, respectively (13–15). In addition, APECED patients exhibit a decreased frequency of peripheral blood IL-17-producing CD4$^+$ T cells following PMA/ionomycin stimulation [(3, 15), Lionakis, unpublished observations].

Heterozygous *STAT1* gain-of-function (GOF) mutations were initially implicated in causing autosomal-dominant CMC (16, 17) but have thereafter also been associated with the development of autoimmunity that can involve endocrine and non-endocrine tissues (16–19); beyond these common clinical features between APECED and *STAT1* GOF mutations, patients with *STAT1* GOF mutations also develop a broad-spectrum of infectious, inflammatory, and vascular manifestations not seen in APECED (19). These *STAT1* mutations are considered GOF because of enhanced phosphorylated STAT1 molecules upon interferon (IFN)-γ stimulation (17, 20). Impaired production of Th17 cytokines by T-cells has been implicated in the pathogenesis of CMC in these patients (17, 19, 20).

Due to the overlap in CMC and other clinical features between patients with APECED and *STAT1* GOF mutations, and because AIRE interacts functionally with a protein inhibitor of activated STAT1 (PIAS1), which inhibits STAT1 activation (21, 22), we aimed to study STAT1 protein level and phosphorylation upon IFN-γ stimulation in patients with *AIRE* mutations and determine whether human AIRE deficiency phenocopies the cell-intrinsic enhanced STAT1 levels seen in patients with *STAT1* GOF mutations.

MATERIALS AND METHODS

Study Participants

Eight APECED patients were enrolled (2015–2017) on a NIAID IRB-approved protocol and provided written informed consent. All eight had a clinical diagnosis of APECED based on the development of any two manifestations within the classic triad of CMC, hypoparathyroidism, and adrenal insufficiency. The most common clinical manifestations among the eight APECED patients included CMC (88%), hypoparathyroidism (100%), adrenal insufficiency (88%), and enamel hypoplasia (100%). The full list and frequencies of clinical manifestations of the eight APECED patients are outlined in **Table 1**. The most common *AIRE* mutant allele in these eight patients was c.967_979del13 (60%), followed by c.769C>T (13%). Two patients were compound heterozygous for c.967_979del13 and c.769C>T, while the remaining six patients had six different *AIRE* genotypes. Samples from a patient carrying the c.1057G>A E353K *STAT1* GOF mutation and a patient with the autosomal dominant form of IFN-γ receptor 1 deficiency carrying the 818del4 mutation were also collected under a NIAID IRB-approved protocol and provided written informed consent. Healthy volunteer blood samples from 13 individuals were obtained for STAT1 protein evaluation and from 10 individuals for *STAT1* mRNA level determination under IRB-approved protocols through the Department of Transfusion Medicine, at the NIH Clinical Center. The study was conducted in accordance with the Helsinki Declaration.

TABLE 1 | Clinical manifestations of the eight autoimmune polyendocrinopathy-candidiasis-ectodermal dystrophy patients included in this study.

Manifestation[a]	Number of affected patients	% of affected patients
Adrenal insufficiency	7	88
Asplenia	1	13
Autoimmune hepatitis	2	25
Alopecia	1	13
B12 deficiency	3	38
Chronic mucocutaneous candidiasis	7	88
Enamel hypoplasia	8	100
Gastritis	3	38
Gonadal failure	3	38
Early-onset hypertension	4	50
Hypoparathyroidism	8	100
Hypothyroidism	3	38
Intestinal dysfunction	6	75
Keratoconjunctivitis	1	13
Nail dystrophy	1	13
Tubulointerstitial nephritis	1	13
Pneumonitis	2	25
Sjogren's-like syndrome	3	38
Urticarial eruption	3	38
Vitiligo	1	13

[a] *The manifestations are presented in alphabetical order.*

Peripheral Blood Mononuclear Cells (PBMC) Isolation and Intracellular Staining for STAT1 and pSTAT1

STAT1 protein and pSTAT1 levels were examined using flow cytometry in paired APECED patients and healthy control individuals in seven independent experiments at rest and up to 60 min after IFN-γ stimulation. Each patient and healthy donor was evaluated only once. In six of the seven independent experiments, a single APECED patient was evaluated along with two accompanying healthy donors, and in one experiment two APECED patients were evaluated together with one accompanying healthy donor.

PBMC were isolated by density-gradient centrifugation using lymphocyte separation media (Lonza) and resuspended in RPMI culture media (Gibco), supplemented with pyruvate (100 mM, Sigma Aldrich), glutamate (200 mM, Life Technologies), penicillin/streptomycin (100 U/100 μg/ml, Life Technologies), 10% fetal bovine serum (Serum Source International) and HEPES (20 mM, General Electric).

Intracellular phosphorylated STAT1 (pSTAT1) and total STAT1 were determined by FACS analysis, as previously described (18, 23). Freshly isolated PBMC were resuspended at 10^6 cells per 100 μl in RPMI and were serum starved for 30 min, in polystyrene round-bottom tubes (Becton Dickinson Falcon). Cells were then incubated with FITC-conjugated anti-human CD14 (Becton Dickinson cat# 555397). Cells were then stimulated with IFN-γ (800 U/ml) for 15, 30, or 60 min at 37°C, fixed with 2% Paraformaldehyde (Electron Microscopy Sciences) at 37°C for 10 min, permeabilized with 100% methanol in dark on ice for 30 min, washed with PBS/2%FBS, and incubated for 1 h in the dark at 4°C with combinations of PerCP–Cy5.5-conjugated anti-human pSTAT1 (Y701) (Becton Dickinson cat# 560113) and Alexa Fluor 647-conjugated anti-human N-terminus STAT1 (Becton Dickinson cat# 558560) with Fix and Perm Medium B (Life technologies). Alexa Fluor 647-conjugated IgG1 isotype control (Becton Dickinson cat# 557783) was used. Baseline pSTAT1 levels were used as a control for the specificity of the PerCP–Cy5.5-conjugated anti-human pSTAT1 antibody, by comparing between pSTAT1 levels as expressed in geometric mean of fluorescence-at rest and after stimulation. Samples were washed once with PBS/2%FBS and resuspended in 1% Paraformaldehyde. All data were collected with LSR Fortessa or LSRII (Becton Dickinson) and analyzed with FlowJo software (Treestar, Ashland, OR, USA).

IFN-α, IFN-ω, and IFN-γ Autoantibody Detection

A particle-based multiplex assay was used to detect IFN-α, IFN-ω, and IFN-γ autoantibodies in the serum or plasma samples from the eight APECED patients and compared with healthy control subjects ($n = 100$) enrolled through the NIH Blood Bank, as previously described (24).

IFN-γ Receptors 1 and 2 Expression on Monocytes

Frozen PBMC from eight APECED patients and eight healthy donors were used for measuring IFN-γ receptors 1 and 2 levels on CD14⁺ monocytes. Cells were resuspended at 10^6 cells per 100 μl in PBS and incubated with LIVE/DEAD® Fixable Aqua Dead Cell Stain Kit (ThermoFisher) in 4°C followed by staining with FITC-conjugated anti-human CD14 (Becton Dickinson cat# 555397), PE-conjugated anti CD119 (IFN-γ receptor 1; IFN-γR1; Becton Dickinson cat# 558937), or APC-conjugated anti-IFN-γ receptor 2 (IFN-γR2; R&D cat# FAB773A) for 30 min. PE-conjugated IgG2b κ isotype control (Cat# 555058) and APC-conjugated IgG isotype control (R&D cat# IC108A) were used for control staining. Cells were then washed with PBS/2%FBS and fixed with 1% PFA. Data were collected with LSRII (Becton Dickinson) and analyzed with FlowJo software (Treestar, Ashland, OR, USA). The geometric mean fluorescence intensity on monocytes for each receptor was calculated after subtracting the geometric mean fluorescence intensity of the corresponding isotype control staining.

STAT1 mRNA Expression Determination

Frozen PBMC from the 8 APECED patients and 10 healthy donors were used for measuring *STAT1* mRNA expression by quantitative PCR (qPCR). To determine the percent of live $CD14^+$ monocytes within PBMC, an aliquot of the PBMC was incubated with LIVE/DEAD® Fixable Violet Dead Cell Stain Kit (ThermoFisher) in 4°C followed by staining with FITC-conjugated anti-human CD14 (Becton Dickinson cat# 555397). Data were collected with LSRII (Becton Dickinson) and analyzed with FlowJo software (Treestar, Ashland, OR, USA). Among the live PBMC, the mean percentage of $CD14^+$ monocytes was similar in the patient and healthy donor groups: 7.2 ± 0.8 vs. 7.7 ± 1.7, respectively ($p = 0.75$). For mRNA extraction, the RNeasy kit (Qiagen) was used, according to the manufacturer's instructions. To convert mRNA to cDNA, the high-capacity cDNA reverse transcription kit (Applied Biosystems) was used. qPCR was then performed with TaqMan detection (TaqMan® Universal Master Mix II, with UNG; ThermoFisher), using the 7,500 real-time PCR system (Applied Biosystems) and predesigned primer and probe mixes for glyceraldehyde-3-phosphate dehydrogenase (GAPDH; ThermoFisher) or STAT1 (ThermoFisher). All qPCR assays were performed in duplicate and results were normalized to GAPDH transcript levels using the threshold cycle (Ct) method.

AIRE Sequencing

Genomic DNA was extracted from whole blood, amplified and sequenced for *AIRE* exons and flanking splice sites as previously described (25).

Statistical Analysis

The geometric mean of fluorescence for pSTAT1 or STAT1 protein levels was calculated using the FlowJo software and the values obtained from APECED patients were normalized to the values obtained from the same-day healthy control samples. All results were expressed as mean ± SEM unless otherwise indicated. Statistical analyses were performed by Student's *t*-test or Mann–Whitney U test, where appropriate, using the GraphPad Prism 7 software (La Jolla, CA, USA). A p value of less than 0.05 was considered significant.

RESULTS

We enrolled eight APECED patients from eight non-consanguineous families with clinical and genetic diagnosis of APECED. Two (25%) were male and six (75%) were female. The mean patient age was 22 years (range, 9–56 years); 3 (38%) were children, with a mean age of 11 years. Thirteen healthy donors were enrolled for same-day harvesting and comparative analyses of STAT1 and pSTAT1 protein levels. A *STAT1* GOF patient was tested as control for enhanced STAT1 protein and pSTAT1 levels [patient 1 in Ref. (20)], and a patient with the autosomal dominant form of IFN-γR1 deficiency [patient described in the case report in Ref. (26)] was tested as control for absent IFN-γ-induced pSTAT1 stimulation (**Figures 1A,B**).

We examined STAT1 protein and pSTAT1 levels in paired APECED patients and healthy control individuals, at rest and up to 60 min after IFN-γ stimulation. We focused on IFN-γ because it induces STAT1 phosphorylation and homo-dimerization without the involvement of other STAT molecules as in the case of STAT1–STAT2 heterodimer formation induced by IFN-α stimulation (27, 28). We also focused on $CD14^+$ monocytes because of their relatively high levels of IFN-γ receptors 1 and 2, which allows for detection of rapid activation of STAT1 (29).

In all the eight tested APECED patients, STAT1 protein levels were lower than the same-day healthy donors mean levels (**Figure 1C**). The eight patients' mean $CD14^+$ monocyte STAT1 protein level at rest was ~80% of that observed in healthy donors ($p = 0.003$). After IFN-γ stimulation, the significant decrease in $CD14^+$ monocyte mean STAT1 protein level of APECED patients persisted at all examined time-points throughout the 60 min of the experiment (**Figure 1C**).

Consistent with previous reports (23), we found that pSTAT1 induction peaked 15 min after IFN-γ stimulation and started to decline toward baseline at 30 min after stimulation in all 13 healthy donor monocytes (**Figure 1A**). Similar kinetics of pSTAT1 induction peak and decline were observed in all eight APECED patients (**Figure 1A**). APECED patient monocytes' mean pSTAT1 levels at rest were not significantly different compared to healthy

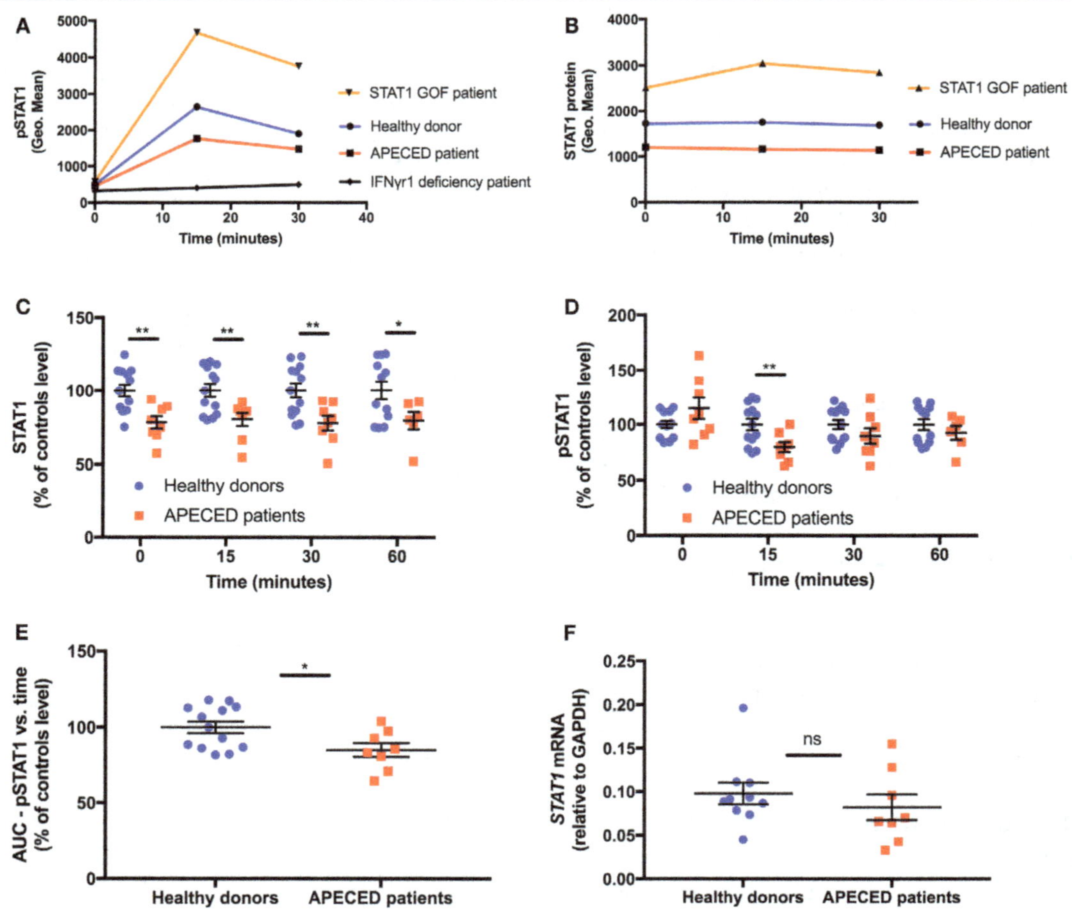

FIGURE 1 | Autoimmune regulator deficiency results in a decrease in STAT1 protein levels in human monocytes. **(A)** Representative depiction of pSTAT1 level at rest and up to 30 min after interferon (IFN)-γ stimulation in $CD14^+$ cells of a *STAT1* gain-of-function (GOF) patient (orange), an autoimmune polyendocrinopathy-candidiasis-ectodermal dystrophy (APECED) patient (red), a patient with the autosomal dominant form of IFN-γR1 deficiency (black) and a healthy donor (blue) **(B)** Representative depiction of STAT1 protein level at rest and up to 30 min after IFN-γ stimulation in $CD14^+$ cells of a *STAT1* GOF patient (orange), an APECED patient (red), and a healthy donor (blue). Protein and phosphorylation levels are expressed in geometric mean of fluorescence (Geo. Mean), as measured by flow cytometry. STAT1 total protein **(C)** and pSTAT1 **(D)** levels in $CD14^+$ cells of APECED patients ($n = 8$; red dots) and healthy donors ($n = 13$; blue dots) at rest (time 0) and up to 60 min after IFN-γ stimulation. Total protein and phosphorylation levels are expressed in % of the same-day control average values, for each time point—0, 15, 30, and 60 min, separately. **(E)** Area under the curve of $CD14^+$ cells STAT1 phosphorylation vs. time in APECED patients ($n = 8$; red dots) and healthy donors ($n = 13$; blue dots). **(F)** *STAT1* mRNA level, relative to glyceraldehyde-3-phosphate dehydrogenase, in peripheral blood mononuclear cells of healthy donors ($n = 10$) and APECED patients ($n = 8$) at rest. ns, not significant. $^*p < 0.05$; $^{**}p < 0.01$, by *t*-test. Quantitative data represent mean ± SEM.

donor levels at rest (108.7 ± 9.2% vs. 100 ± 2.83%, respectively, $p = 0.29$; **Figure 1D**). However, the mean peak pSTAT1 level of the APECED patients was 20% decreased compared to that seen in healthy donors at 15 min (80 ± 4.5% vs. 100 ± 4.5%, respectively. $p = 0.008$). Thereafter, at 30 min after IFN-γ stimulation, the patients' mean pSTAT1 level was lower than healthy donors (87 ± 5.1% vs. 100 ± 3.9%, respectively), and the difference was close to statistical significance ($p = 0.06$), whereas at 60 min after IFN-γ stimulation the differences in pSTAT1 levels between the patients and healthy donors groups were not significant (91 ± 5.5% vs. 100 ± 4.3%, respectively, $p = 0.25$; **Figure 1D**). We also calculated the area under the curve (AUC) of STAT1 phosphorylation (expressed as geometric mean of fluorescence) vs. time (expressed as minutes) for up to 60 min after IFN-γ stimulation in both groups and compared patient AUC level to the same-day healthy donor average. Consistent with the mean peak pSTAT1 values, APECED patients mean AUC levels were ~15% decreased relative to those seen in healthy controls ($p = 0.025$, **Figure 1E**). Because Th17 cells are decreased in peripheral blood of APECED [(3, 15), Lionakis, unpublished observations] and STAT1 GOF patients (17, 19, 20), our findings collectively indicate that the Th17 frequency decrease in AIRE deficiency is not caused by a STAT1 GOF state in peripheral blood monocytes.

Because among the eight APECED patients, seven different *AIRE* genotypes were observed, reflecting the greater genetic diversity of APECED in North America (25), we were not able to examine whether there is a correlation between specific *AIRE* mutations and the degree of pSTAT1 and total STAT1 levels in AIRE-deficient monocytes.

Given the decreased STAT1 protein levels seen in APECED patients, we asked whether *STAT1* is also decreased at the mRNA level in APECED patients. Thus, we compared *STAT1* mRNA levels by qPCR in PBMC of 8 APECED patients and 10 healthy donors and found them to be similar ($p = 0.32$) (**Figure 1F**).

Because APECED patients are known to have autoantibodies against cytokines, we examined whether autoantibodies to type I IFNs or IFN-γ were present in the serum of the eight APECED patients, which could potentially affect the levels of pSTAT1 in the patients group, after IFN-γ stimulation. All eight tested patients had autoantibodies to IFN-α and IFN-ω, while no patient had autoantibodies to IFN-γ (**Figure 2A**).

Given the decreased peak pSTAT1 level after IFN-γ stimulation in AIRE-deficient monocytes, we also asked whether the surface expression of IFN-γ receptors 1 and 2 is decreased in the monocytes of the eight APECED patients using flow cytometry. We found no difference in the expression of IFN-γR1 between healthy donor and APECED monocytes but a significant ~40% decrease in surface expression of IFN-γR2 in patient monocytes (mean fluorescence intensity 186.4 ± 14.61 vs. 117.3 ± 10.68, $p = 0.002$, respectively, **Figure 2B**).

DISCUSSION

Because of the similarities in clinical presentation of CMC and autoimmunity and decreased peripheral blood Th17 frequencies between patients with APECED and *STAT1* GOF mutations, our study examined STAT1 protein and phosphorylation levels in patients with APECED and shows that APECED patient monocytes do not share the same cell-intrinsic increase in STAT1 protein and phosphorylation levels as cells from patients with *STAT1* GOF mutations. In fact, our findings reveal a moderate but consistent and significant decrease in STAT1 protein and phosphorylation levels in APECED CD14+ monocytes at rest and after IFN-γ stimulation.

Until recently, it had been postulated that impaired nuclear dephosphorylation is the underlying cause of increase in pSTAT1 levels in patients with *STAT1* GOF mutations (17, 19, 20). Nonetheless, recent reports have described normal dephosphorylation rates in some patients with a *STAT1* GOF mutations; for example, Sobh et al. (30), Meesilpavikkai et al. (31), and Weinacht et al. (32) described three new *STAT1* GOF mutations in the SH2 (p.H629Y and p.V653I) and linker (p.E545K) domains with enhanced pSTAT1 levels but normal dephosphorylation. In addition, Tabellini and colleagues recently reported high levels of STAT1 protein in NK cells from seven patients carrying five different CC or DNA binding (DB) domain *STAT1* GOF mutations (23). Similarly, high STAT1 protein levels were found in CD14+ monocytes and CD3+ T cells of 12 patients with 10 different CC, DB, or SH2 domain *STAT1* GOF mutations (Zimmerman and Holland, submitted manuscript). After controlling the *STAT1* GOF patients' pSTAT1 levels for the corresponding total STAT1 protein levels, STAT1 phosphorylation levels were normal (Zimmerman and Holland, submitted manuscript). These findings collectively support the hypothesis that high resting and IFN-γ-stimulated total STAT1 protein levels may serve as a background against which high pSTAT1 levels occur in some patients with *STAT1* GOF mutations. Alternatively, the pSTAT1 levels may not directly correlate with total STAT1 levels, as indicated in other reports (33). Future studies using more STAT1 GOF and STAT1 loss-of-function (LOF) patients will be needed to determine the temporal dynamics and correlation of total and pSTAT1 molecules.

In our study, based on the presence of CMC and other shared clinical features between APECED and *STAT1* GOF mutation patients, we tested the hypothesis that AIRE deficiency enhances STAT1 cellular responses as a result of enhanced STAT1 protein and phosphorylation levels. Instead, we found a ~20% decrease in STAT1 protein and peak pSTAT1 levels in APECED patients; the decrease in total STAT1 levels could indicate that the decreased STAT1 protein level may be a determining factor for the decreased peak STAT1 phosphorylation levels observed in APECED CD14+ monocytes, although it could be accounted for by alternative factors that remain to be elucidated.

The molecular mechanism behind our finding is currently unknown. One possible explanation could have been that AIRE is involved in the regulation of STAT1 transcription; however, we found no significant difference in *STAT1* mRNA levels between healthy donor and APECED PMBCs. Alternatively, AIRE may regulate STAT1 post-transcriptionally at the level of degradation or SUMOylation or other process directly or indirectly *via* known (e.g., PIAS1) or yet-unknown protein partners (21, 22). Alternatively, APECED patient autoantibodies against type I IFNs, which have been shown to exhibit the ability to neutralize and block IFN-α activity, STAT1 phosphorylation, and the expression of interferon-stimulated genes (34, 35), may adversely affect tonic

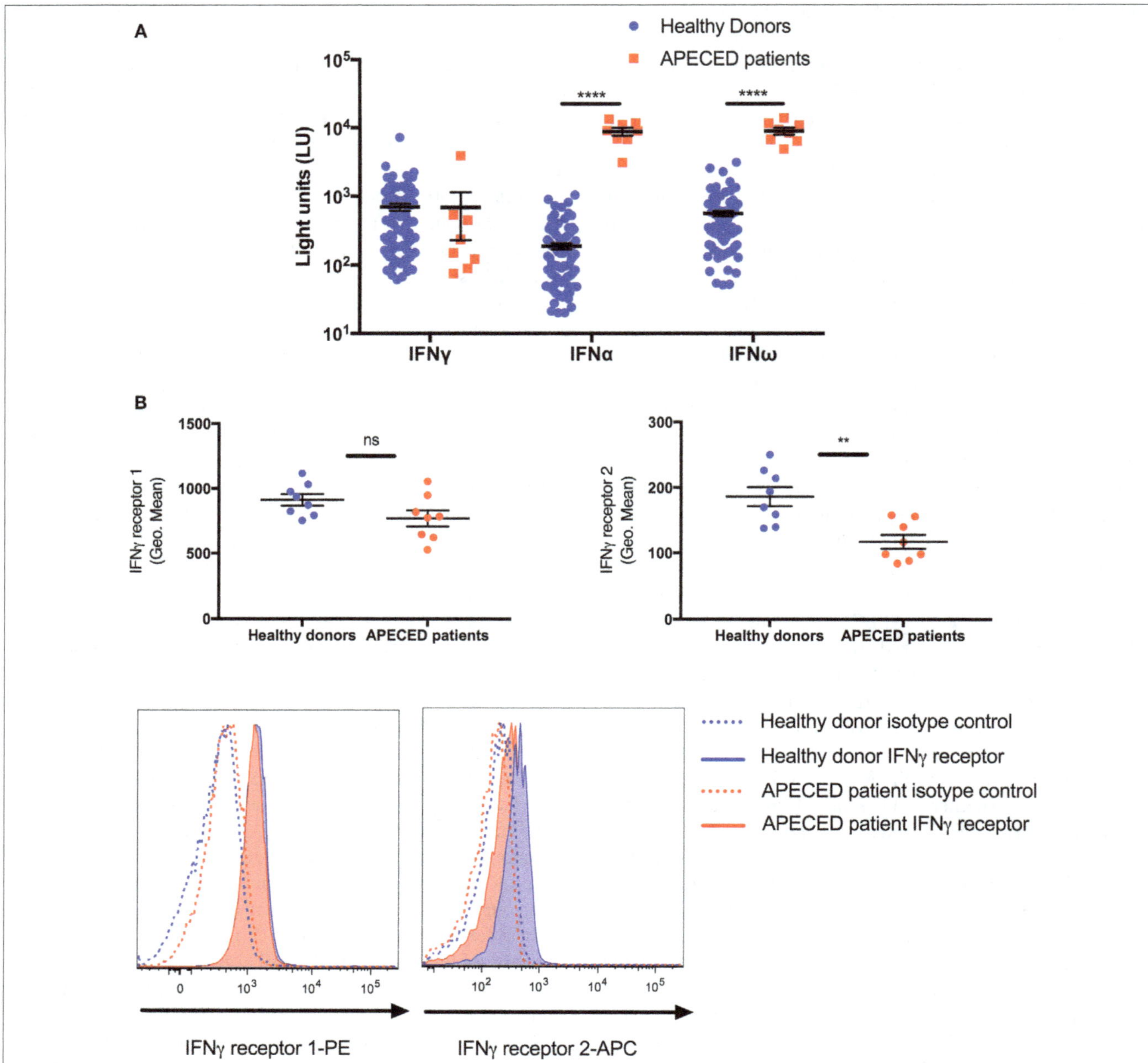

FIGURE 2 | The decrease in STAT1 protein and pSTAT1 levels seen in autoimmune polyendocrinopathy-candidiasis-ectodermal dystrophy (APECED) patients correlates with lower monocyte interferon (IFN)-γR2 receptor expression and the presence of type I IFN autoantibodies. **(A)** Sera from 8 APECED patients and 100 healthy donors were evaluated for the presence of autoantibodies against IFN-γ, IFN-α, and IFN-ω. Shown is autoantibody immunoreactivity against the indicated cytokines expressed as fluorescence intensity using a particle-based multiplex assay. **(B)** IFN-γ receptor 1 and 2 levels were measured in CD14+ monocytes of healthy donors ($n = 8$) and APECED patients ($n = 8$). Summary data on mean fluorescence intensity and representative histogram FACS plots are shown. $^{**}p < 0.01$; $^{****}p < 0.0001$, by Mann–Whitney U test **(A)** or *t*-test **(B)**. Quantitative data represent median ± SEM.

STAT1 expression in blood monocytes. The decrease in IFN-γR2 expression on CD14+ monocytes may be a contributor to the decreased peak pSTAT1 levels in APECED patients' cells. Future studies will be required to further examine this hypothesis as well as molecules and functional responses downstream of STAT1.

Impaired STAT1-dependent responses in patients with LOF *STAT1* mutations or in patients with defects in the IFN-γ receptor signaling axis, underlie Mendelian susceptibility to mycobacterial disease (36). The defect in STAT1 levels post-IFN-γ stimulation that we observed in our APECED patients is modest relative to the complete absence of STAT1 phosphorylation following IFN-γ stimulation in patients with IFN-γ receptor deficiency [(23); **Figure 1A**]. Therefore, significant residual STAT1 signaling is functional in APECED patients to prevent the development of mycobacterial infections in these patients. Whether the decrease in STAT1 protein and phosphorylation that we identified is a contributing factor to the development of viral infections in some APECED patients [(37); Lionakis, unpublished observations] remains to be elucidated. However, the decrease in STAT1 levels is unlikely to contribute to the pathogenesis of CMC as patients with defects in the IFN-γ receptor signaling axis do not have Th17 defects and do not develop CMC (36). Hence, at this point, the biological and clinical implications of the decreased STAT1 levels in patients' monocytes are unclear.

In summary, we describe an association between AIRE deficiency and a decrease in STAT1 protein level in primary human monocytes that is not accompanied by decreased *STAT1* mRNA levels but correlates with the presence of type I IFN autoantibodies and decreased monocyte surface expression of IFN-γR2. The mechanism behind this finding and its clinical and biological implications in APECED patients require further investigation.

ETHICS STATEMENT

Eight APECED patients were enrolled (2015–2017) on a NIAID IRB-approved protocol and provided written informed consent. Samples from a patient carrying the c.1057G>A E353K STAT1 GOF mutation and a patient with the autosomal dominant form of IFN-γ receptor 1 deficiency carrying the 818del4 mutation were also collected under a NIAID IRB-approved protocol and provided written informed consent. Healthy volunteer blood samples from 13 individuals were obtained for STAT1 protein evaluation and from 10 individuals for STAT1 mRNA level determination under IRB-approved protocols through the Department of Transfusion Medicine, at the NIH Clinical Center. The study was conducted in accordance with the Helsinki Declaration.

AUTHOR CONTRIBUTIONS

ML conceived the project and contributed to the design and supervision of the experiments. OZ and ML wrote the manuscript. OZ, LR, and MS conducted experiments and generated manuscript figures. OZ, LR, MN, FV, and ML contributed to the interpretation of the data. EF contributed to healthy donor enrollment. SH provided patient care contributed patients' samples critical for the study execution.

FUNDING

This research was supported by the Division of Intramural Research, NIAID, NIH. The content of this article are those of the authors and do not reflect the official policy of the Department of Health and Human Services, nor does mention of trade names, commercial products, or organizations imply endorsement by the US government.

REFERENCES

1. Anderson MS, Venanzi ES, Klein L, Chen Z, Berzins SP, Turley SJ, et al. Projection of an immunological self shadow within the thymus by the aire protein. *Science* (2002) 298(5597):1395–401. doi:10.1126/science.1075958
2. Mathis D, Benoist C. Aire. *Annu Rev Immunol* (2009) 27:287–312. doi:10.1146/annurev.immunol.25.022106.141532
3. Fujikado N, Mann AO, Bansal K, Romito KR, Ferre EM, Rosenzweig SD, et al. Aire inhibits the generation of a perinatal population of interleukin-17A-producing γδ T cells to promote immunologic tolerance. *Immunity* (2016) 45(5):999–1012. doi:10.1016/j.immuni.2016.10.023
4. Yang S, Fujikado N, Kolodin D, Benoist C, Mathis D. Immune tolerance. Regulatory T cells generated early in life play a distinct role in maintaining self-tolerance. *Science* (2015) 348(6234):589–94. doi:10.1126/science.aaa7017
5. Gardner JM, Devoss JJ, Friedman RS, Wong DJ, Tan YX, Zhou X, et al. Deletional tolerance mediated by extrathymic Aire-expressing cells. *Science* (2008) 321(5890):843–7. doi:10.1126/science.1159407
6. Malchow S, Leventhal DS, Lee V, Nishi S, Socci ND, Savage PA. Aire enforces immune tolerance by directing autoreactive T cells into the regulatory T cell lineage. *Immunity* (2016) 44(5):1102–13. doi:10.1016/j.immuni.2016.02.009
7. Oftedal BE, Hellesen A, Erichsen MM, Bratland E, Vardi A, Perheentupa J, et al. Dominant mutations in the autoimmune regulator AIRE are associated with common organ-specific autoimmune diseases. *Immunity* (2015) 42(6):1185–96. doi:10.1016/j.immuni.2015.04.021
8. Cetani F, Barbesino G, Borsari S, Pardi E, Cianferotti L, Pinchera A, et al. A novel mutation of the autoimmune regulator gene in an Italian kindred with autoimmune polyendocrinopathy-candidiasis-ectodermal dystrophy, acting in a dominant fashion and strongly cosegregating with hypothyroid autoimmune thyroiditis. *J Clin Endocrinol Metab* (2001) 86(10):4747–52. doi:10.1210/jcem.86.10.7884
9. Su MA, Giang K, Zumer K, Jiang H, Oven I, Rinn JL, et al. Mechanisms of an autoimmunity syndrome in mice caused by a dominant mutation in Aire. *J Clin Invest* (2008) 118(5):1712–26. doi:10.1172/JCI34523
10. Ahonen P, Myllärniemi S, Sipilä I, Perheentupa J. Clinical variation of autoimmune polyendocrinopathy-candidiasis-ectodermal dystrophy (APECED) in a series of 68 patients. *N Engl J Med* (1990) 322(26):1829–36. doi:10.1056/NEJM199006283222601
11. Perheentupa J. Autoimmune polyendocrinopathy-candidiasis-ectodermal dystrophy. *J Clin Endocrinol Metab* (2006) 91(8):2843–50. doi:10.1210/jc.2005-2611
12. Wolff AS, Erichsen MM, Meager A, Magitta NF, Myhre AG, Bollerslev J, et al. Autoimmune polyendocrine syndrome type 1 in Norway: phenotypic variation, autoantibodies, and novel mutations in the autoimmune regulator gene. *J Clin Endocrinol Metab* (2007) 92(2):595–603. doi:10.1210/jc.2006-1873
13. Consortium TF-GA. An autoimmune disease, APECED, caused by mutations in a novel gene featuring two PHD-type zinc-finger domains. *Nat Genet* (1997) 17(4):399–403. doi:10.1038/ng1297-399
14. Puel A, Doffinger R, Natividad A, Chrabieh M, Barcenas-Morales G, Picard C, et al. Autoantibodies against IL-17A, IL-17F, and IL-22 in patients with chronic mucocutaneous candidiasis and autoimmune polyendocrine syndrome type I. *J Exp Med* (2010) 207(2):291–7. doi:10.1084/jem.20091983
15. Kisand K, Boe Wolff AS, Podkrajsek KT, Tserel L, Link M, Kisand KV, et al. Chronic mucocutaneous candidiasis in APECED or thymoma patients correlates with autoimmunity to Th17-associated cytokines. *J Exp Med* (2010) 207(2):299–308. doi:10.1084/jem.20091669
16. van de Veerdonk FL, Plantinga TS, Hoischen A, Smeekens SP, Joosten LA, Gilissen C, et al. STAT1 mutations in autosomal dominant chronic mucocutaneous candidiasis. *N Engl J Med* (2011) 365(1):54–61. doi:10.1056/NEJMoa1100102
17. Liu L, Okada S, Kong XF, Kreins AY, Cypowyj S, Abhyankar A, et al. Gain-of-function human STAT1 mutations impair IL-17 immunity and underlie chronic mucocutaneous candidiasis. *J Exp Med* (2011) 208(8):1635–48. doi:10.1084/jem.20110958
18. Uzel G, Sampaio EP, Lawrence MG, Hsu AP, Hackett M, Dorsey MJ, et al. Dominant gain-of-function STAT1 mutations in FOXP3 wild-type immune dysregulation-polyendocrinopathy-enteropathy-X-linked-like syndrome. *J Allergy Clin Immunol* (2013) 131(6):1611–23. doi:10.1016/j.jaci.2012.11.054
19. Toubiana J, Okada S, Hiller J, Oleastro M, Lagos Gomez M, Aldave Becerra JC, et al. Heterozygous STAT1 gain-of-function mutations underlie an unexpectedly broad clinical phenotype. *Blood* (2016) 127(25):3154–64. doi:10.1182/blood-2015-11-679902
20. Sampaio EP, Hsu AP, Pechacek J, Bax HI, Dias DL, Paulson ML, et al. Signal transducer and activator of transcription 1 (STAT1) gain-of-function

mutations and disseminated coccidioidomycosis and histoplasmosis. *J Allergy Clin Immunol* (2013) 131(6):1624–34. doi:10.1016/j.jaci.2013.01.052

21. Liu B, Liao J, Rao X, Kushner SA, Chung CD, Chang DD, et al. Inhibition of Stat1-mediated gene activation by PIAS1. *Proc Natl Acad Sci U S A* (1998) 95(18):10626–31. doi:10.1073/pnas.95.18.10626
22. Ilmarinen T, Kangas H, Kytömaa T, Eskelin P, Saharinen J, Seeler JS, et al. Functional interaction of AIRE with PIAS1 in transcriptional regulation. *Mol Immunol* (2008) 45:1847–62. doi:10.1016/j.molimm.2007.10.045
23. Tabellini G, Vairo D, Scomodon O, Tamassia N, Ferraro RM, Patrizi O, et al. Impaired natural killer cell functions in patients with signal transducer and activator of transcription 1 (STAT1) gain-of-function mutations. *J Allergy Clin Immunol* (2017). doi:10.1016/j.jaci.2016.10.051
24. Ding L, Mo A, Jutivorakool K, Pancholi M, Holland SM, Browne SK. Determination of human anticytokine autoantibody profiles using a particle-based approach. *J Clin Immunol* (2012) 32(2):238–45. doi:10.1007/s10875-011-9621-8
25. Ferre EMN, Rose SR, Rosenzweig SD, Burbelo PD, Romito KR, Niemela JE, et al. Redefined clinical features and diagnostic criteria in autoimmune polyendocrinopathy-candidiasis-ectodermal dystrophy. *JCI Insight* (2016) 1(13):e88782. doi:10.1172/jci.insight.88782
26. Vinh DC, Masannat F, Dzioba RB, Galgiani JN, Holland SM. Refractory disseminated coccidioidomycosis & mycobacteriosis in interferon-γ receptor 1 deficiency. *Clin Infect Dis* (2009) 49(6):e62–5. doi:10.1086/605532
27. Shuai K, Horvath CM, Huang LH, Qureshi SA, Cowburn D, Darnell JE Jr. Interferon activation of the transcription factor Stat1 involves dimerization through SH2-phosphotyrosyl peptide interactions. *Cell* (1994) 76(5):821–8. doi:10.1016/0092-8674(94)90357-3
28. Au-Yeung N, Mandhana R, Horvath CM. Transcriptional regulation by STAT1 and STAT2 in the interferon JAK-STAT pathway. *JAKSTAT* (2013) 2(3):e23931. doi:10.4161/jkst.23931
29. Bernabei P, Coccia EM, Rigamonti L, Bosticardo M, Forni G, Pestka S, et al. Interferon-γ receptor 2 expression as the deciding factor in human T, B, and myeloid cell proliferation or death. *J Leukoc Biol* (2001) 70(6):950–60.
30. Sobh A, Chou J, Schneider L, Geha RS, Massaad MJ. Chronic mucocutaneous candidiasis associated with an SH2 domain gain-of-function mutation that enhances STAT1 phosphorylation. *J Allergy Clin Immunol* (2016) 138(1):297–9. doi:10.1016/j.jaci.2015.12.1320
31. Meesilpavikkai K, Dik WA, Schrijver B, Nagtzaam NM, van Rijswijk A, Driessen GJ, et al. A novel heterozygous mutation in the STAT1 SH2 domain causes chronic mucocutaneous candidiasis, atypically diverse infections, autoimmunity, and impaired cytokine regulation. *Front Immunol* (2017) 8:274. doi:10.3389/fimmu.2017.00274
32. Weinacht KG, Charbonnier LM, Alroqi F, Plant A, Qiao Q, Wu H, et al. Ruxolitinib reverses dysregulated T helper cell responses and controls autoimmunity caused by a novel signal transducer and activator of transcription 1 (STAT1) gain-of-function mutation. *J Allergy Clin Immunol* (2017) 139:1629–40.e2. doi:10.1016/j.jaci.2016.11.022
33. Kong XF, Ciancanelli M, Al-Hajjar S, Alsina L, Zumwalt T, Bustamante J, et al. A novel form of human STAT1 deficiency impairing early but not late responses to interferons. *Blood* (2010) 116(26):5895–906. doi:10.1182/blood-2010-04-280586
34. Kisand K, Link M, Wolff AS, Meager A, Tserel L, Org T, et al. Interferon autoantibodies associated with AIRE deficiency decrease the expression of IFN-stimulated genes. *Blood* (2008) 112(7):2657–66. doi:10.1182/blood-2008-03-144634
35. Meyer S, Woodward M, Hertel C, Vlaicu P, Haque Y, Kärner J, et al. AIRE-deficient patients harbor unique high-affinity disease-ameliorating autoantibodies. *Cell* (2016) 166(3):582–95. doi:10.1016/j.cell.2016.06.024
36. Boisson-Dupuis S, Kong XF, Okada S, Cypowyj S, Puel A, Abel L, et al. Inborn errors of human STAT1: allelic heterogeneity governs the diversity of immunological and infectious phenotypes. *Curr Opin Immunol* (2012) 24(4):364–78. doi:10.1016/j.coi.2012.04.011
37. Nagafuchi S, Umene K, Yamanaka F, Oohashi S, Shindo M, Kurisaki H, et al. Recurrent herpes simplex virus infection in a patient with autoimmune polyendocrinopathy-candidiasis-ectodermal dystrophy associated with L29P and IVS9-1G>C compound heterozygous autoimmune regulator gene mutations. *J Intern Med* (2007) 261(6):605–10. doi:10.1111/j.1365-2796.2007.01786.x

3

Induction of Broad-Spectrum Protective Immunity against Disparate *Cryptococcus* Serotypes

Marley C. Caballero Van Dyke[1,2], Ashok K. Chaturvedi[1,2], Sarah E. Hardison[1,2], Chrissy M. Leopold Wager[1,2], Natalia Castro-Lopez[1,2], Camaron R. Hole[1,2], Karen L. Wozniak[1,2] and Floyd L. Wormley Jr.[1,2]*

[1] Department of Biology, The University of Texas at San Antonio, San Antonio, TX, United States, [2] The South Texas Center for Emerging Infectious Diseases, The University of Texas at San Antonio, San Antonio, TX, United States

***Correspondence:**
Floyd L. Wormley Jr.
floyd.wormley@utsa.edu

Cryptococcosis is a fungal disease caused by multiple *Cryptococcus* serotypes; particularly *C. neoformans* (serotypes A and D) and *C. gattii* (serotypes B and C). To date, there is no clinically available vaccine to prevent cryptococcosis. Mice given an experimental pulmonary vaccination with a *C. neoformans* serotype A strain engineered to produce interferon-γ, denoted H99γ, are protected against a subsequent otherwise lethal experimental infection with *C. neoformans* serotype A. Thus, we determined the efficacy of immunization with *C. neoformans* strain H99γ to elicit broad-spectrum protection in BALB/c mice against multiple disparate *Cryptococcus* serotypes. We observed significantly increased survival rates and significantly decreased pulmonary fungal burden in H99γ immunized mice challenged with *Cryptococcus* serotypes A, B, or D compared to heat-killed H99γ (HKH99γ) immunized mice. Results indicated that prolonged protection against *Cryptococcus* serotypes B or D in H99γ immunized mice was CD4$^+$ T cell dependent and associated with the induction of predominantly Th1-type cytokine responses. Interestingly, immunization with H99γ did not elicit greater protection against challenge with the *Cryptococcus* serotype C tested either due to low overall virulence of this strain or enhanced capacity of this strain to evade host immunity. Altogether, these studies provide "proof-of-concept" for the development of a cryptococcal vaccine that provides cross-protection against multiple disparate serotypes of *Cryptococcus*.

Keywords: *Cryptococcus neoformans*, *Cryptococcus gattii*, cryptococcosis, host–fungal interaction, fungal vaccines, fungal immunology

INTRODUCTION

Cryptococcosis is a worldwide fungal disease caused by species in the *Cryptococcus neoformans/Cryptococcus gattii* species complex (1). *C. neoformans* and *C. gattii* cause pneumonia in immunocompromized and immunocompetent individuals and can disseminate to the central nervous system resulting in life-threatening meningoencephalitis. Cryptococcal meningoencephalitis is the most common disseminated fungal disease in AIDS patients (2) and is responsible for 15% of AIDS-related deaths (3). The former species *C. neoformans* is divided into two species, *C. neoformans* and *C. deneoformans*. *C. neoformans* (serotype A) is distributed worldwide with the highest amount of disease cases occurring in sub-Saharan Africa where approximately 21.7 million people currently live with AIDS, and *C. deneoformans* (serotype D) is predominantly geographically restricted to Europe and Latin America (3–5). Conversely, *C. gattii* (serotype B *C. deuterogattii* and serotype C *C. bacillisporus*) predominantly occurs in tropical and subtropical climates (6–8). However, cryptococcosis due to *C. gattii* is also observed in more temperate climates including British Columbia, Canada, and southwest, southeast, northwest, and northeast regions of the USA and in Mediterranean Europe (9–16). Species in the *C. neoformans*/*C. gattii* complex can cause disease in both apparently healthy individuals and immunocompromized hosts such as AIDS patients, individuals on prolonged treatment with corticosteroids, and in patients on immunosuppressive drugs to prevent rejection of solid organ transplants [reviewed by Kwon-Chung et al. (17)] (7, 8, 18–20). There is a higher occurrence of *C. gattii* disease in immunocompetent individuals compared to *C. neoformans* (20–22). Currently, there is no vaccine clinically available to prevent cryptococcosis and current drug therapies are often rendered ineffective due to the development of drug resistance by *Cryptococcus* or drug toxicity (23). Thus, the overall incidence of disease and mortality associated with cryptococcosis and the potential expanded geographic distribution of the pathogen indicates an urgent need for immunotherapies and/or vaccines to combat cryptococcosis.

Results from multiple studies in both humans and animal models suggest that cell-mediated immunity (CMI) by Th1-type $CD4^+$ T cells is the primary host defense against cryptococcosis (24, 25). However, studies suggest that the host immune response against *C. gattii* and *C. neoformans* differ in that *C. gattii* infection appears to be more immune suppressive, which may explain the disparate clinical presentation displayed by these different species (26–31). Differences in the host response to various *Cryptococcus* serotypes complicate efforts to devise a vaccine that can provide broad-spectrum protection against cryptococcosis caused by disparate *Cryptococcus* serotypes.

Historically, several approaches have been employed to develop an anti-*Cryptococcus* vaccine that generates protective antibody mediated and/or cell-mediated immune responses against *Cryptococcus* [reviewed in Ref. (32).]. Recently, studies have demonstrated that mice given an immunization with glucan particles (GPs) packaged with alkaline extracts from mutant *C. neoformans* or *C. gattii* strains develop increased survival against cryptococcosis (33). Other, studies showed cross-protection of mice immunized with a heat-killed chitosan deficient *C. neoformans* strain against challenge with *C. neoformans* or *C. gattii* (34). Rella and colleagues showed that mice vaccinated with a *C. neoformans* mutant strain that lacks the enzyme sterylglucosidase (Δ*sgl*1) are protected against challenge with both *C. neoformans* H99 and *C. gattii* R265 strains (35).

Studies have shown that mice immunized with a *C. neoformans* strain H99, serotype A, engineered to produce murine interferon (IFN)-γ, denoted H99γ, develop protective Th1-type immune responses against a subsequent otherwise lethal pulmonary challenge with wild-type *C. neoformans* strain H99 (25, 36–40). The presence of $CD4^+$ and/or $CD8^+$ T cells are required for protection during the immunization phase with H99γ but not during subsequent challenge with *C. neoformans* strain H99 (40). These results and those by Rella et al. highlight that protection against pulmonary cryptococcosis in immunocompetent hosts can be maintained in immunosuppressed hosts and that the development of a prophylactic vaccine against cryptococcosis is feasible. Nonetheless, no studies have shown that protection elicited following immunization with one *Cryptococcus* serotype provides similar levels of cross-protection against all four *Cryptococcus* serotypes.

Consequently, the objective of these studies was to determine the potential for an anti-cryptococcal vaccine to provide broad-spectrum protection against multiple disparate *Cryptococcus* serotypes. We therefore determined the efficacy of immunization with *C. neoformans* strain H99γ to elicit protection against experimental pulmonary challenge with disparate cryptococcal serotypes in mice. We demonstrate that immunization with *C. neoformans* strain H99γ, serotype A, elicits a predominantly protective Th1-type immune response against challenge with different serotypes of *Cryptococcus*. These data support the premise that development of a broad-spectrum prophylactic vaccine against cryptococcosis is achievable.

MATERIALS AND METHODS

Mice

Female BALB/c (National Cancer Institute/Charles River Laboratories) 4–6 weeks of age were used throughout these studies. Mice were housed at the University of Texas at San Antonio Small Animal Laboratory Vivarium and all animal experiments were conducted following NIH guidelines for housing and care of laboratory animals and in accordance with protocols approved by the Institutional Animal Care and Use Committee (protocol number MU021) of the University of Texas at San Antonio.

Strains and Media

Cryptococcus neoformans strains H99 (serotype A, mating type α), *C. deuterogattii* strain R265 (serotype B; a kind gift from Dr. Joseph Heitman of Duke University Medical Center in Durham, NC, USA), *C. bacillisporus* strain WSA87 (serotype C), and *C. deneoformans* strain R4247 (serotype D) (each kind gifts from Dr. Brian Wickes of the UT Health San Antonio, San Antonio, TX, USA), and H99γ [an IFN-gamma producing *C. neoformans* strain derived from H99 (38)] were recovered from 15% glycerol stocks stored at −80°C prior to use in the experiments described herein. The strains were maintained on yeast-extract-peptone-dextrose (YPD) media (1% yeast extract, 2% peptone, 2% dextrose, and 2% Bacto agar). Yeast cells were grown for 16 h at 30°C with shaking in YPD broth, collected by centrifugation, washed three times with sterile PBS, and viable yeasts were quantified using trypan blue dye exclusion on a hemacytometer.

Pulmonary Cryptococcal Infections

Cryptococcal infections were initiated as previously described (25, 41). Briefly, mice were anesthetized with 2% isoflurane using a rodent anesthesia device (Eagle Eye Anesthesia, Jacksonville, FL, USA) and given an intranasal inoculation with 1×10^4 CFU of *Cryptococcus* strains H99, R265, WSA87, or R4247 in 50 μl of sterile PBS. Alternatively, mice were given an intranasal immunization with 1×10^4 CFU of *C. neoformans* strain H99γ or heat-killed H99γ (HKH99γ) yeasts in 50 μl of sterile PBS, allowed 70 days to resolve the infection and subsequently given an intranasal challenge with 1×10^4 CFU of H99, R265, WSA87, or R4247 in 50 μl of sterile PBS. The inocula used for immunizations and challenges were verified by quantitative culture on YPD agar. The mice were monitored by inspection twice daily. Mice were euthanized on pre-determined days post challenge and lung tissues excised using aseptic technique. Tissues were homogenized in 1 ml of sterile PBS followed by culture of 10-fold dilutions of each tissue on YPD agar supplemented with chloramphenicol (Mediatech Inc., Herndon, VA, USA). CFU were enumerated following incubation at 30°C for 48 h. Alternatively, mice intended for survival analysis were monitored by inspection twice daily and euthanized if they appeared to be in pain or moribund (weight loss, ataxia, listlessness, or failure to groom). Mice were euthanized using CO_2 inhalation followed by cervical dislocation.

T Cell Depletion

Mice were immunized with 1×10^4 CFU of *C. neoformans* strain H99γ as described above. After 70 days of rest, mice were depleted of $CD4^+$ T cells, $CD8^+$ T cells, or both $CD4^+/CD8^+$ T cell subsets *via* intraperitoneal administration of anti-CD4 (clone GK1.5) and anti-CD8 (clone 2.43) antibodies (**Table 1**) (each from Cell Culture Company/NCCC, Minneapolis, MN, USA) or given isotype control antibody (rat IgG_{2b}) (BioXCell, Lebanon, NH, USA). Each mouse received 200 μg of GK1.5 and/or 2.43 or control rat IgG_{2b} antibodies in a volume of 200 μl PBS 48 h prior to challenge with H99, R265, WSA87, or R4247 and weekly thereafter during the observation period. The efficiency of T cell depletion in lungs and spleens was assessed by flow cytometric analysis using anti-CD4 and anti-CD8 antibodies that bind epitopes of the CD4 and CD8 protein at locations distinct from GK1.5 and 2.43. Efficiency was determined to be >98% at each anatomic location for each depletion via comparison of T cell subsets in treated mice with those in control animals.

TABLE 1 | Antibodies used in flow cytometry.

mAb	Clone	Manufacturer
CD45R (B220)-PE	RA3-6B2	BD Biosciences
CD11c-Pe-Cy7	N418	eBiosciences
CD11c-PE	N418	eBiosciences
CD11b-Pe-Cy7	M1170	eBiosciences
CD45-APC	30-F11	eBiosciences
CD45-Pe-Cy7	30-F11	eBiosciences
F4/80-APC	BM8	Life Technologies
Ly6G-PE	1A8	BD Biosciences
T cell mix-CD8a APC, CD3e, Pe-Cy7, CD4 PE	CD8a 53-6.7, CD3e-145-2C11, CD4 RM4-5	BD Biosciences
CD19-PE	1D3	BD Biosciences
Siglec-F-PE	E50-2440	BD Biosciences
CD3e-Pe-Cy7	145-2C11	eBiosciences
CD4-PE	RM4-5	BD Biosciences
Intracellular flow		
Interferon-γ-APC	XMG1.2	eBiosciences
IL-4-APC	11B11	eBiosciences
IL-17A-APC	eBio17B7	eBiosciences
Depletion antibodies		
Anti-CD4	GK1.5	Cell Culture Company
Anti-CD8	2.43	Cell Culture Company

Cytokine Analysis

As previously described (25, 41), cytokine levels in lung tissues were analyzed using the Bio-Plex protein array system (Luminex-based technology; Bio-Rad Laboratories, Hercules, CA, USA). Briefly, the left lobe of the lung was excised and homogenized in ice-cold sterile PBS (1 ml). An aliquot (50 μl) was taken to quantify the pulmonary fungal burden and an anti-protease buffer solution containing PBS, protease inhibitors (inhibiting cysteine, serine, and other metalloproteinases), and 0.05% Triton X-100 was added to the homogenate and then clarified by centrifugation ($800 \times g$) for 10 min. Supernatants from pulmonary homogenates were assayed for the presence of IL-1α, IL-1β, IL-1, IL-3, IL-4, IL-5, IL-6, IL-9, IL-10, IL-12(p40), IL-12(p70), IL-13, IL-17A, CCL5/RANTES, CCL11/Eotaxin, CXCL1/KC, CCL3/MIP-1α, CCL4/MIP-1β, CCL2/MCP-1, G-CSF, GM-CSF, TNF-α, and IFN-γ.

Pulmonary Leukocyte Isolation

Lungs were excised on days 3, 7, and 14 post challenge and digested enzymatically at 37°C for 30 min in 10 ml of digestion buffer (RPMI 1640 and 1 mg/ml of collagenase type IV) (Sigma Chemical Co., St. Louis, MO, USA) with intermittent (every 10 min) stomacher homogenizations as previously described (25, 41). The digested tissues were then filtered through sterile nylon filters of various pore sizes (70 and 40 μm; BD Biosciences) and washed with sterile HBSS to enrich for leukocytes. Erythrocytes were lysed by incubation in NH_4Cl buffer (0.859% NH_4Cl, 0.1% $KHCO_3$, 0.0372% Na_2EDTA pH 7.4) for 3 min on ice followed by the addition of a twofold excess of PBS. The leukocyte population was then washed twice with sterile PBS, suspended in sterile PBS + 2% heat-inactivated fetal bovine serum (FACS buffer), and enumerated in a hemacytometer using trypan blue dye exclusion. Flow cytometry analysis was used to determine the percentage of each leukocyte population as well as the absolute number of total leukocytes ($CD45^+$) within the lung cell suspension for standardization of hemacytometer counts.

Flow Cytometry

Standard methodology was employed for the direct immunofluorescence of pulmonary leukocytes (25, 41). Briefly, in 96-well U-bottom plates, 100 μl containing 1×10^6 cells in PBS plus 2% FBS (FACS buffer) were incubated with Fc block (BD Biosciences) diluted in FACS buffer for 5 min to block nonspecific binding of antibodies to cellular Fc receptors. An optimal concentration of fluorochrome-conjugated antibodies (0.125—1 μg/1×10^6 cells) as listed in **Table 1** was added in various combinations to allow for dual or triple staining, and cells were then incubated for 30 min at 4°C. Cells were washed three times with FACS buffer and fixed in 200 μl of 2% ultrapure formaldehyde (Polysciences) diluted in FACS buffer (fixation buffer). Cells analyzed for intracellular cytokine staining were fixed in fixation buffer for 10 min in the dark at room temp. The cells were then permeabilized with 0.1% saponin (diluted in FACS buffer) and incubated for 10 min in the dark at room temperature. Antibodies listed in

Table 1 were subsequently added at optimal concentrations, and cells incubated at 4°C for 30 min. Following incubation, cells were washed three times with 0.1% saponin buffer and resuspended in fixation buffer. Cells incubated with either FACS buffer alone or single fluorochrome-conjugated Abs were used to determine positive staining and spillover/compensation calculations, and background fluorescence determined with FlowJo Software (FlowJo, LLC, Ashland, OR, USA). Raw data were collected with the BD FACSArray flow cytometer (BD Biosciences) and then analyzed using FlowJo Software. Dead cells were excluded on the basis of forward angle and 90° light scatter. For data analyses, 30,000 events (cells) were evaluated from a predominantly leukocyte population identified by back gating from $CD45^+$ stained cells. The absolute number of total leukocytes was quantified by multiplying the total number of cells observed by hemacytometer counting by the percentage of $CD45^+$ cells determined by flow cytometry. The absolute number of leukocytes ($CD45^+$ cells), T cells ($CD4^+/CD3^+$ and $CD8^+/CD3^+$), $CD19^+/CD45^+$, $Ly6G^+/CD45^+$, $F4/80^+/CD45^+$, $CD11b^{int}/CD11c^+/CD45^+$, $CD11b^{hi}CD11c^+/CD45^+$, and $Siglec\text{-}F^+/CD11b^+$ was determined by multiplying the percentage of each gated population by the total number of $CD45^+$ cells. Gating strategies are provided as supplemental materials. Intracellular cytokine staining is represented in histograms with unstained cells as a control.

Statistical Analysis

The unpaired Student's *t* test (two-tailed) was used to analyze the immunization studies fungal burden, pulmonary cell populations, and cytokine/chemokine data where appropriate using GraphPad Prism version 5.00 for Windows (GraphPad Prism Software, San Diego, CA, USA). Survival data were analyzed using the log-rank test (GraphPad Software). Significant differences were defined as $^*p < 0.05$, $^{**}p < 0.005$, or $^{***}p < 0.0001$.

RESULTS

Immunization with *C. neoformans* Strain H99γ Elicits Protection against Disparate *Cryptococcus* Serotypes

Previous studies in our lab demonstrated that experimental pulmonary infection with *C. neoformans* strain H99γ in mice results in clearance of the acute infection and induction of protective immunity against a subsequent otherwise lethal challenge with a WT *C. neoformans* strain H99 that does not produce IFN-γ (25, 36–40). Nevertheless, studies to date have not demonstrated that protective immunity generated against one serotype of *Cryptococcus* affords protection against multiple disparate *Cryptococcus* serotypes. Consequently, we sought to determine the level of cross-protection that is generated in mice immunized with *C. neoformans* strain H99γ, a serotype A strain, against pulmonary infection with disparate serotypes of *Cryptococcus*. To establish a baseline prior to evaluating if immunization with *Cryptococcus* strain H99γ, serotype A, enhanced protection against challenge with disparate *Cryptococcus* serotypes, we first determined the relative virulence of each serotype in non-immunized mice. To do this, BALB/c mice were given an experimental pulmonary infection with either *Cryptococcus* strain H99 (serotype A), R265 (serotype B), WSA87 (serotype C), or R4247 (serotype D) and evaluated for mortality (**Figure 1A**). As shown in **Figure 1A**, experimental pulmonary infection with *Cryptococcus* strains H99, R265, and R4247 resulted in 100% mortality with median survival times of 28, 32, and 33 days, respectively. In contrast, we observed 87.5% survival in mice given an experimental pulmonary infection with WSA87. *Cryptococcus* strain WSA87, serotype C, appeared to be significantly less virulent compared to all the other strains tested during our analysis.

Next, we tested the potential of mice immunized with *C. neoformans* strain H99γ (serotype A) to be protected against challenge with *Cryptococcus* serotypes B, C, or D. For this, mice were immunized with *C. neoformans* strain H99γ or heat-killed H99γ (HKH99γ), rested for 70 days, and separate groups subsequently given an experimental pulmonary challenge with each *Cryptococcus* serotype. We observed a 100% survival rate in mice immunized with H99γ and challenged with H99, while the HKH99γ immunized mice challenged with H99 demonstrated a median survival time of 28 days post challenge (**Figure 1B**), confirming our previous observations (38, 40). Mice immunized with H99γ and then challenged with either R265 (*C. gattii* serotype B) or R4247 (*C. deneoformans* serotype D) exhibited median survival times of 59 and 72 days post challenge, respectively, and significantly increased survival rates compared to HKH99γ immunized mice challenged with the same strain (median survival of 33 and 34.5 days in R265 and R4247 challenged mice, respectively; **Figure 1B**). Additionally, we observed a 20% survival rate in protectively immunized mice challenged with R4247 at 100 days post challenge. **Figure 1B** shows that no significant difference in mortality was observed in H99γ immunized mice challenged with WSA87 (median survival time of 90.5 days) compared to HKH99γ immunized mice challenged with the same strain (median survival time of 87.5 days). The mortality of mice immunized with HKH99γ and challenged with each *Cryptococcus* serotype mirrored the mortality rates of non-immunized mice given an experimental infection with each serotype and provided no additional protection (**Figures 1A,B**).

Pulmonary fungal burden was also quantified at days 3, 7, and 14 post challenge in H99γ or HKH99γ immunized mice challenged with each individual serotype (**Figure 1C**). Mice immunized with HKH99γ and challenged with each species showed progressive growth of each serotype in the lungs. In contrast, we observed significantly less pulmonary fungal burden as early as day 3 in H99γ immunized mice challenged with R265 and WSA87 compared to HKH99γ immunized mice (**Figure 1C**). Additionally, H99γ immunized mice showed significantly less pulmonary fungal burden following challenge with all serotypes on days 7 and 14 post challenge compared to HKH99γ immunized mice challenged with the same strain. Brain fungal burden was also quantified at day 14 post challenge and a trend (although not statistically significant) toward a reduction of brain fungal burden together with the number of mice negative for *Cryptococcus* in the brain was observed in

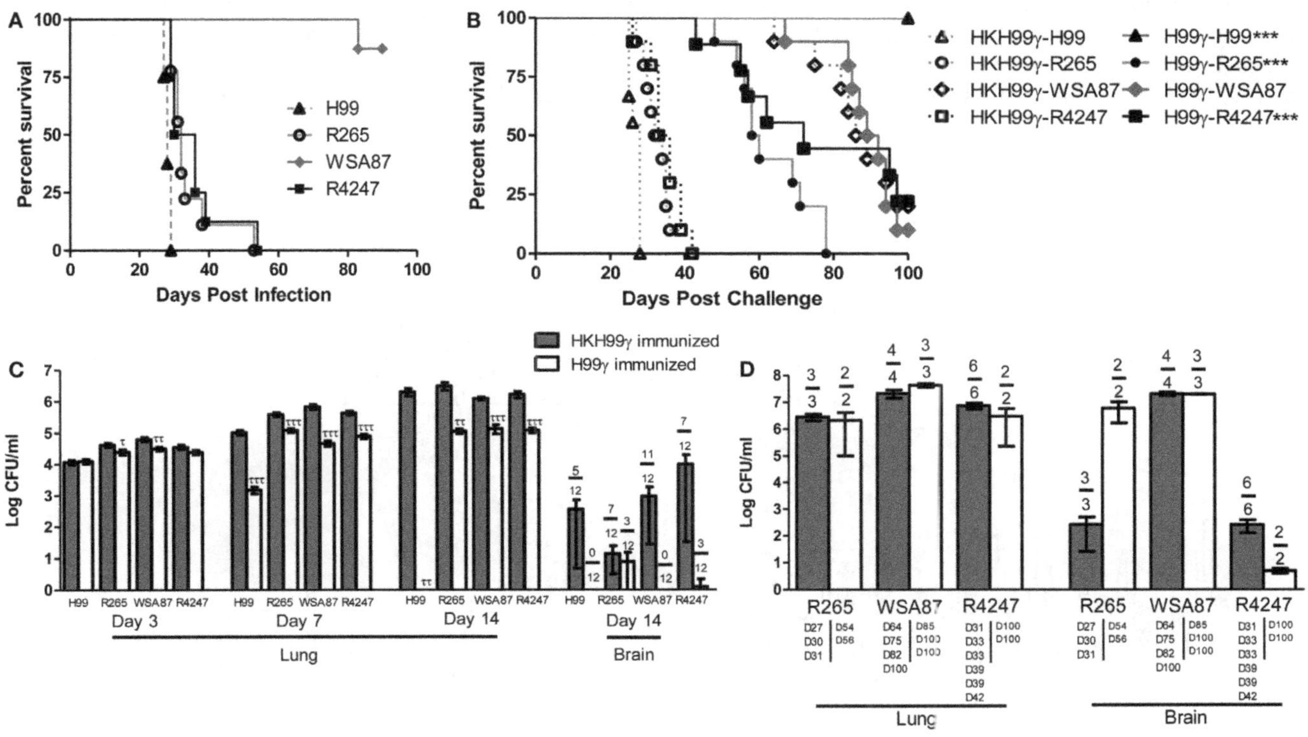

FIGURE 1 | Virulence of disparate *Cryptococcus* serotypes in non-immunized or serotype A immunized mice. BALB/c mice received an intranasal inoculation **(A)** with 1 × 10⁴ CFU of *Cryptococcus* strain H99 (serotype A), R265 (serotype B), WSA87 (serotype C), or R4247 (serotype D). Mice were observed for up to 90 days for mortality analysis. Alternatively, BALB/c mice received an intranasal immunization **(B)** with 1 × 10⁴ CFU of either H99γ or HKH99γ, allowed 70 days to resolve the infection, and subsequently challenged with individual serotypes of *Cryptococcus*. Mice were observed for up to 100 days post challenge for survival analysis **(B)** or, alternatively pulmonary and brain fungal burden determined at days 3, 7, and 14 post challenge **(C)**. Pulmonary and brain fungal burden were analyzed from mice that appeared moribund or sacrificed at 100 days post challenge with the day the mice were sacrificed indicated below the bars **(D)**. Survival data shown are from one experiment, using 8–10 mice per group. Fungal burden data are cumulative of three experiments using 4 mice per group per time point. Numbers above the bars represent the number of mice positive for *Cryptococcus*. *$p < 0.05$, **$p < 0.005$, ***$p < 0.0001$. *Significant increase compared to HKH99γ; ${}^{\tau}p < 0.05$, ${}^{\tau\tau}p < 0.005$, ${}^{\tau\tau\tau}p < 0.0001$. ${}^{\tau}$Significant decrease compared to HKH99γ.

H99γ immunized mice challenged with each serotype compared to their HKH99γ immunized counterparts (**Figure 1C**). Overall, our data show that mice immunized with H99γ, serotype A, display significant protection against challenge with serotypes A, B, or D of *Cryptococcus*.

Figure 1D demonstrates the fungal burden of moribund mice or mice sacrificed at the conclusion of the study. The HKH99γ immunized mice challenged with R265 appeared to have similar pulmonary fungal burden and less brain fungal burden at time of death compared to H99γ immunized mice challenged with R265 (**Figure 1D**). No significant differences in pulmonary or brain fungal burden were observed between H99γ and HKH99γ immunized mice challenged with WSA87 (**Figure 1D**). Both groups of mice immunized with HKH99γ or H99γ and challenged with R4247 appear to have similar pulmonary fungal burden and lower brain fungal burden at sacrifice with the H99γ immunized mice surviving longer. However, we cannot definitively state if the mice succumbed to pulmonary and/or brain fungal burden. Overall, immunization with H99γ appeared to prolong the survival of mice challenged with each serotype.

Pulmonary Leukocyte Recruitment during Pulmonary Cryptococcosis Is Increased in H99γ Immunized Mice Challenged with Each *Cryptococcus* Serotype

Our next goal was to examine the pulmonary inflammatory response in immunized mice during infection with the various representative *Cryptococcus* serotypes. We observed a significant increase in the total number of leukocytes recruited to the lungs of H99γ immunized mice challenged with H99 at days 3 and 7 post challenge compared to HKH99γ immunized mice (**Figure 2A**). Additionally, we observed significant increases in leukocyte recruitment to the lungs of H99γ immunized mice challenged with R265 on days 7 and 14 post challenge, H99γ immunized mice challenged with WSA87 on days 3 and 7 post challenge, and H99γ immunized mice challenged with R4247 at each time point post challenge compared to their HKH99γ immunized counterparts (**Figure 2A**). We observed an increased total cell number of F4/80⁺ cells in the lungs of H99γ immunized mice challenged with H99 at day 3, R4247 at days 3 and 7, and WSA87 at day 14 post challenge compared to HKH99γ immunized mice (**Figure 2B**). We observed a significant increase in the total number of CD11b^{hi}/CD11c^{+} cells recruited to the lungs of H99γ immunized mice challenged with H99, WSA87, or R4247 at day 3, each serotype at day 7 post challenge, and in H99γ immunized mice challenged with WSA87 at day 14 post challenge compared to HKH99γ immunized mice challenged with the same serotype (**Figure 2C**). **Figure 2C** also shows a significant increase in the percentage

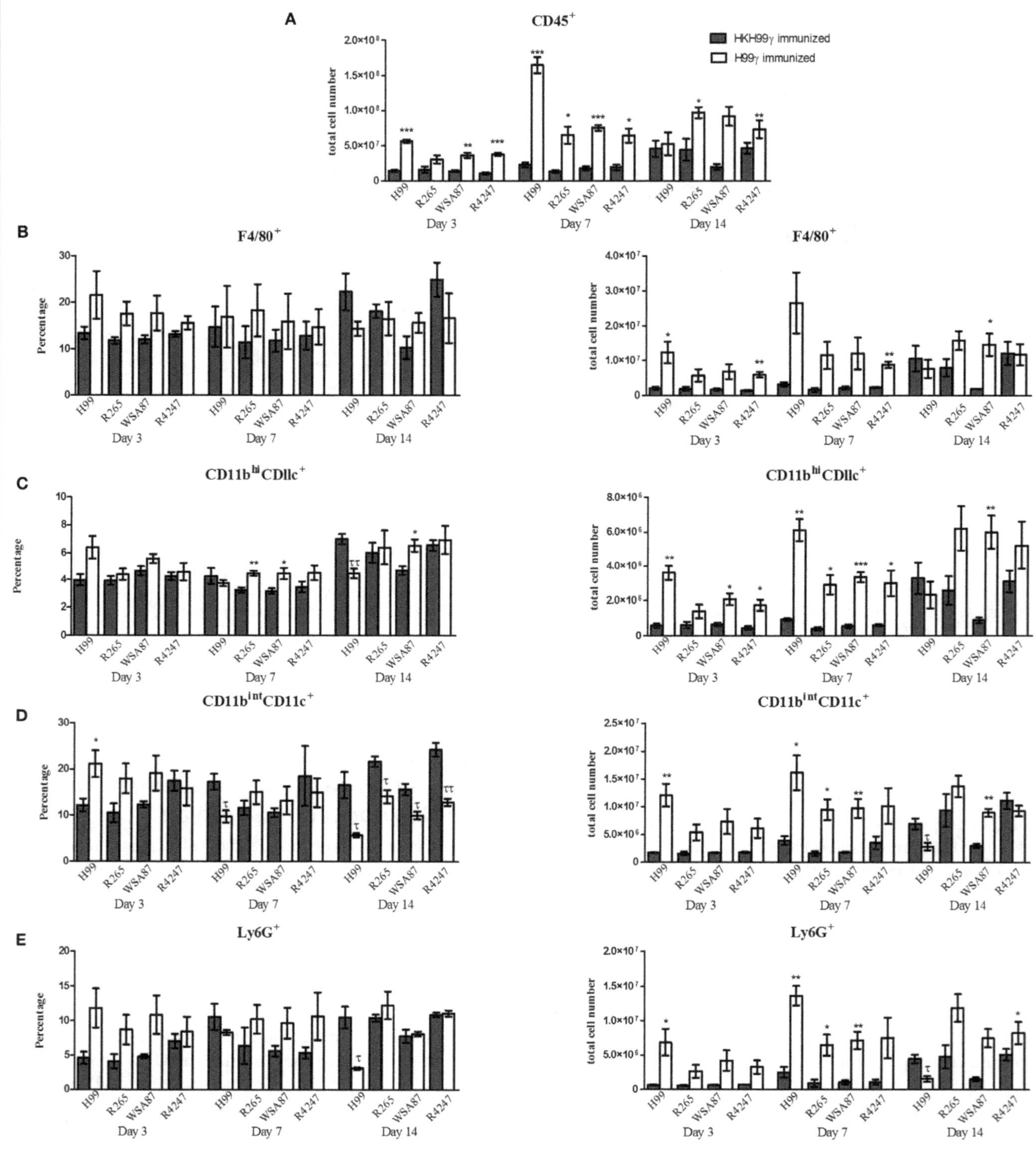

FIGURE 2 | Continued

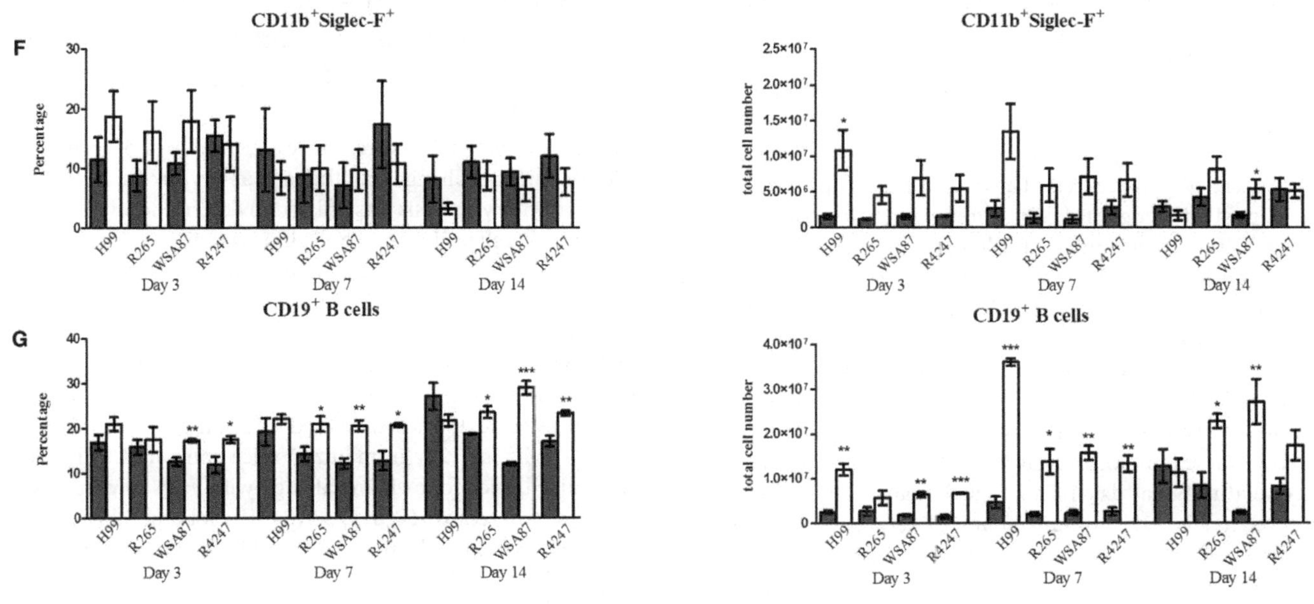

FIGURE 2 | Pulmonary leukocyte populations in immunized mice challenged with different serotypes of *Cryptococcus*. BALB/c mice were immunized intranasally with 1×10^4 CFU of either H99γ or HKH99γ, allowed 70 days to resolve the infection, and subsequently challenged with individual serotypes of *Cryptococcus*. At days 3, 7, and 14 post challenge, lungs were excised, tissues digested, and pulmonary infiltrates were analyzed by flow cytometry. Leukocytes were labeled with anti-CD45 **(A)** or dual labeled with anti-CD45 and antibodies for specific cell types **(B–G)** and analyzed by flow cytometry. Data shown are the mean ± SEM of absolute cell numbers or percentage from three independent experiments performed using 4 mice per group per experiment. *$p < 0.05$, **$p < 0.005$, ***$p < 0.0001$. *Significant increase compared to HKH99γ; $^{\tau}p < 0.05$, $^{\tau\tau}p < 0.005$, $^{\tau\tau\tau}p < 0.0001$. $^{\tau}$Significant decrease compared to HKH99γ.

of CD11b^{hi}/CD11c$^+$ cells in H99γ immunized mice challenged with R265 or WSA87 at day 7, and WSA87 at day 14 post challenge. There was also a significant increase in the total number of CD11b^{int}CD11c$^+$ cells in H99γ immunized mice challenged with H99 at days 3 and 7, R265 at day 7 post challenge, and WSA87 at days 7 and 14 post challenge (**Figure 2D**). **Figure 2D** also shows a significant increase in CD11b^{int}CD11c$^+$ cells in H99γ immunized mice challenged with H99 at day 3, while we observe a significant decrease in these cells at day 7 in H99 challenged mice, and across each serotype at day 14 post challenge. The recruitment of Ly6G$^+$ cells to the lungs was significantly higher in H99γ immunized mice challenged with H99 at days 3 and 7 post challenge, R265 or WSA87 at day 7 post challenge, and in H99γ immunized mice challenged with R4247 on day 14 post challenge compared to their HKH99γ immunized counterparts (**Figure 2E**). While mice immunized with H99γ and challenged with H99 are clearing the infection, these mice show a significant decrease in the total number and percentage of Ly6G$^+$ cells at day 14 post infection compared to HKH99γ immunized mice (**Figure 2E**). We observed a significant increase in the total number of Siglec-F$^+$ cells in mice immunized with H99γ and challenged with H99 at day 3 post challenge and WSA87 challenged mice at day 14 post challenge (**Figure 2F**). We observed a significant increase in the total number of B cells in mice immunized with H99γ and challenged with H99, WSA87, or R4247 at day 3 post challenge, across each serotype at day 7 post challenge, and R265 or WSA87 at day 14 post challenge (**Figure 2G**). For the percentage of B cells, we observed a significant increase in H99γ immunized mice challenged with WSA87 or R4247 across each time point and in mice challenged with R265 at days 7 and 14 post challenge (**Figure 2G**). Overall, our data show significantly increased pulmonary recruitment of F4/80$^+$ cells, CD11b$^+$/CD11c$^+$ cells, B cells, and Ly6G$^+$ cells in mice immunized with the serotype A *Cryptococcus* strain H99γ following challenge with various other *Cryptococcus* serotypes which, altogether, may contribute to the increased protective responses observed in these mice.

Immunization with H99γ Elicits Elevated Proinflammatory and a Predominant Th1-Type Cytokine Response following Challenge with Disparate *Cryptococcus* Serotypes

The pulmonary fungal burden and mortality analysis suggested that mice immunized with H99γ were able to mount protective responses against challenge with disparate *Cryptococcus* serotypes. Therefore, we evaluated cytokine levels in the lungs of H99γ or HKH99γ immunized mice following challenge with each *Cryptococcus* serotype. Lung homogenates were prepared from pulmonary tissues of mice on days 3, 7, and 14 post challenge with *Cryptococcus* serotypes A, B, C, or D, and the levels of Th1 associated (IFN-γ, IL-2, IL-12p70), Th2 associated (IL-4 and IL-5), immunoregulatory (IL-10 and G-CSF), and proinflammatory (IL-1α, IL-1β, IL-17A, and TNF-α) cytokines and chemokines (CCL3, CCL4, CCL2, CXCL1, and CCL5) were determined. We observed that H99γ immunized mice challenged with *C. neoformans* strain H99 exhibited a significant increase in Th1 associated cytokines (IFN-γ, IL-2, and IL-12p70) as early as day 3 post challenge that remained significantly elevated, except IL-2, through day 7 post challenge compared to HKH99γ immunized mice challenged with H99 (**Tables 2** and **3**). However, we observed significantly less Th1 associated cytokine (IL-12p70 and IFN-γ) production at day 14 post challenge in H99γ immunized mice compared to HKH99γ immunized mice (**Table 4**). Also, proinflammatory (IL-1α and IL-1β) and immunoregulatory (G-CSF) cytokines and chemokine levels (CCL3, CCL4, CCL2, and CXCL1) in H99γ immunized mice challenged with H99 were significantly less compared to levels observed in HKH99γ immunized mice challenged with H99 at day 14 post challenge (**Table 4**). The significant decrease in Th1 associated and proinflammatory cytokine and chemokine levels in H99γ immunized mice challenged with H99 at day 14 post challenge compared to levels observed in HKH99γ immunized mice likely coincides with the significant reduction in pulmonary fungal burden observed in the H99γ immunized mice; whereas, the HKH99γ immunized mice are experiencing progressive disease (**Figure 1C**). Th1 associated cytokine levels in the lungs of H99γ immunized mice challenged with each other serotype were generally elevated at day 3 post challenge and remained significantly elevated at days 7 and 14 post challenge, particularly cytokines IFN-γ and IL-12p70, compared to their HKH99γ immunized counterparts (**Tables 2–4**). For Th2-associated cytokines, we observed a significant increase at day 3 post challenge in mice that were immunized with H99γ and challenged with each serotype [H99 (IL-4 and IL-5), R265, WSA87, or R4247 (IL-5)] (**Table 2**). At day 7 post challenge in mice immunized with H99γ, we observed an increase in IL-4 in mice challenged with R265 or R4247 (**Table 3**). By day 14 post challenge, there was a statistically significant decrease in Th2 associated cytokines (IL-4 and IL-5) across each serotype except for WSA87 challenged mice that showed significantly increased IL-5 levels across days 7 and 14 post challenge (**Tables 3** and **4**). We also observed an overall trend toward significantly increased proinflammatory cytokine (IL-1α, IL-1β, and IL-17A), immunoregulatory cytokine (G-CSF), and chemokine (CCL3, CCL4, CCL2, CXCL1, and CCL5) production as early as day 3 post challenge that remained elevated through day 7 post challenge in H99γ immunized mice challenged with each serotype compared to similarly challenged mice immunized using HKH99γ (**Tables 2** and **3**). We observed an overall trend toward significantly higher levels of proinflammatory cytokine (IL-1α, IL-1β, IL-17A, and TNF-α), immunoregulatory cytokine (G-CSF), and chemokine levels (CCL3, CCL4, CCL2, CXCL1, and RANTES) was observed in H99γ immunized mice challenged with R265, WSA87, or R4247 compared to similarly challenged HKH99γ immunized at day 14 post challenge except for CCL3 and CXCL1 for R265 challenged mice, and CCL3 and CCL2 for R4247 challenged mice (**Table 4**). Altogether, the cytokine results suggest that immunization with *C. neoformans* strain H99γ elicits a heightened cytokine anamnestic response against multiple disparate serotypes of *Cryptococcus* compared to HKH99γ immunized mice.

Vaccine-Mediated Protection against Pulmonary Cryptococcosis Is Associated with Enhanced Th1-Type Immune Responses

We next evaluated T cell responses in the lungs of immunized mice following challenge with the various *Cryptococcus* serotypes. Flow cytometry analysis was utilized to determine levels of CD4$^+$ and CD8$^+$ T cell infiltration to the lungs on days 3, 7, and 14 post challenge. Also, intracellular cytokine staining of CD4$^+$ T cells for representative T helper (Th) 1, Th2, and Th17-type cytokines IFN-γ, IL-4, and IL-17A, respectively, and subsequent flow cytometry analysis was used to access the CD4$^+$ Th cell phenotype during the anamnestic response to pulmonary cryptococcosis. We observed an overall trend toward increased total numbers of CD4$^+$ and CD8$^+$ T cells in H99γ immunized mice compared to HKH99γ immunized mice following challenge with each serotype as the infection progressed (**Figures 3A,B**). The total number of CD4$^+$ T cells was significantly increased in H99γ immunized mice challenged with H99, R265, or R4247 at day 3 post challenge, and in mice challenged with H99 at day 7 post challenge compared to their HKH99γ immunized counterparts (**Figure 3A**). However, there was no significant difference in percentage of CD4$^+$T cells between groups. We also observed an increase in the total number of CD8$^+$ T cells in H99γ immunized mice challenged with H99, WSA87, or R4247 at day 3 post challenge, in mice challenged with H99, R265, or WSA87 at day 7 post challenge, and in R265 or WSA87 challenged mice at day 14 post challenge compared to HKH99γ immunized mice challenged with their equivalent serotype (**Figure 3B**). A significant increase in the percentage of CD8$^+$ T cells was observed in mice immunized with H99γ and challenged with R265 or R4247 at day 14 post challenge compared to their HKH99γ immunized counterparts.

Overall, we observed significant increases in the total number of CD4$^+$/IFN-γ$^+$ T cells followed by CD4$^+$/IL-17A$^+$ T cells and then CD4$^+$/IL-4$^+$ T cells at day 7 post challenge in H99γ immunized mice challenged with each serotype compared to HKH99γ immunized mice challenged with the corresponding serotype (**Figures 3C–E**). We observed greater total CD4$^+$/IFN-γ$^+$ T cell numbers in H99γ immunized mice challenged with H99 compared to H99γ immunized mice challenged with R265, WSA87, or R4247 at day 7 post challenge (**Figure 3C**). These results suggest that mice immunized with H99γ are capable of eliciting a predominantly Th1-type anamnestic response upon challenge with each disparate *Cryptococcus* serotype. These data show that mice immunized with the serotype A strain exhibit an increased putatively protective Th1-type immune response following challenge with other disparate serotypes.

Protection Afforded by Immunization with H99γ against Serotypes B and D of *Cryptococcus* Requires CD4$^+$ T Cells

Previous studies in our lab demonstrated that H99γ immunized mice depleted of CD4$^+$ or CD8$^+$ T cells were completely protected against an otherwise lethal challenge with the non-IFN-γ producing *C. neoformans* strain H99 (40). Also, H99γ immunized

2 | Cytokine levels within lung homogenates of H99γ and HKH99γ immunized mice infected with each serotype.

	Day 3 post challenge							
	HKH99γ	**H99γ**	**HKH99γ**	**H99γ**	**HKH99γ**	**H99γ**	**HKH99γ**	**H99γ**
	H99		**R265**		**WSA87**		**R4247**	
sociated								
n (IFN)-γ	5.88 ± 1.45	**12.57 ± 1.84***	4.26 ± 1.07	6.461 ± 0.9474	5.59 ± 1.55	5.94 ± 0.68	3.22 ± 0.71	**6.40 ± 0.89**
	26.03 ± 6.49	**109.30 ± 18.96****	17.98 ± 3.66	21.3 ± 2.35	22.15 ± 6.41	35.64 ± 4.94	15.09 ± 3.28	**27.58 ± 4.31**
0	13.51 ± 0.72	**69.90 ± 8.56*****	14.95 ± 0.58	**27.9 ± 1.754*****	16.04 ± 4.42	**34.21 ± 3.23****	9.91 ± 1.08	**29.20 ± 3.97**
sociated								
	4.15 ± 1.40	**34.58 ± 6.29*****	4.36 ± 1.33	5.873 ± 1.514	4.58 ± 1.31	9.25 ± 2.39	4.22 ± 1.27	6.02 ± 1.42
	2.49 ± 0.24	**10.96 ± 0.94*****	2.26 ± 0.30	**4.729 ± 0.4244*****	2.91 ± 0.63	**6.32 ± 0.52*****	2.00 ± 0.19	**5.43 ± 0.27**
ammatory								
	142.40 ± 25.02	**20.86 ± 3.38**†††	16.60 ± 1.85	18.53 ± 0.9751	18.45 ± 3.08	**37.21 ± 4.70****	15.58 ± 1.69	**23.42 ± 2.61**
	45.79 ± 4.72	**815.70 ± 124.40*****	46.04 ± 6.80	**105.9 ± 8.724*****	45.37 ± 2.90	**248.90 ± 40.52*****	37.11 ± 1.94	**120.40 ± 18.9**
	2.56 ± 0.41	**501.70 ± 83.71*****	2.72 ± 0.25	**44.56 ± 8.306*****	3.09 ± 0.44	**195.50 ± 43.31*****	2.41 ± 0.30	**52.92 ± 12.5**
	45.64 ± 5.51	64.35 ± 7.17	40.39 ± 4.26	48.37 ± 3.266	38.08 ± 4.85	**53.17 ± 4.03***	30.81 ± 2.78	**48.19 ± 4.99**
oregulatory								
	5.15 ± 0.61	**23.05 ± 1.96*****	4.25 ± 0.76	**11.18 ± 1.105*****	4.30 ± 0.51	**12.78 ± 0.81*****	3.93 ± 0.55	**8.90 ± 1.09**
	5.22 ± 1.25	**510.90 ± 64.82*****	5.27 ± 1.19	**45.88 ± 5.85*****	5.62 ± 1.30	**135.90 ± 24.62*****	4.76 ± 1.21	**64.73 ± 12.3**
kine								
MIP-1α	43.41 ± 5.17	**452.80 ± 49.90*****	45.01 ± 3.49	**111.60 ± 9.96*****	68.20 ± 24.17	**194.70 ± 18.11*****	37.94 ± 1.68	**122.20 ± 12.1**
MIP-1β	6.64 ± 0.81	**79.97 ± 13.09*****	6.27 ± 0.53	**16.58 ± 1.97*****	6.35 ± 0.70	**35.92 ± 6.91*****	4.72 ± 0.21	**17.78 ± 1.94**
MCP-1	57.51 ± 8.79	**934.30 ± 88.77*****	64.73 ± 4.93	**189.30 ± 21.11*****	64.70 ± 6.91	**411.70 ± 95.02****	56.65 ± 6.29	**183.00 ± 17.5**
KC	26.76 ± 1.74	**971.50 ± 94.97*****	34.39 ± 4.76	**189.70 ± 31.80*****	32.66 ± 2.49	**460.40 ± 79.29*****	26.03 ± 1.98	**211.60 ± 31.0**
ANTES	152.90 ± 19.47	**1241.00 ± 127.50*****	166.60 ± 11.56	**321.00 ± 42.19****	273.70 ± 118.80	**560.70 ± 49.99***	156.00 ± 10.44	**375.60 ± 38.2**

own are expressed as mean ± SEM and are cumulative of three experiments using four mice per group per time point.
values indicate significance either increased or decreased compared to HKH99γ.
5, $^{**}p < 0.005$, $^{***}p < 0.0001$ significant increase compared to HKH99γ.
0001 significant decrease compared to HKH99γ.

3 | Cytokine levels within lung homogenates of H99γ and HKH99γ immunized mice infected with each serotype.

	Day 7 post challenge							
	HKH99γ	H99γ	HKH99γ	H99γ	HKH99γ	H99γ	HKH99γ	H99γ
	H99		R265		WSA87		R4247	
ociated								
	5.23 ± 0.79	**13.28 ± 0.88*****	4.83 ± 0.54	**10.36 ± 1.08*****	5.29 ± 0.95	**10.40 ± 0.42*****	5.70 ± 1.46	**9.61 ± 0.62***
	19.69 ± 5.12	31.41 ± 3.14	24.52 ± 8.36	35.83 ± 4.22	20.02 ± 4.53	**37.75 ± 5.55***	24.48 ± 6.46	38.61 ± 4.57
0	27.22 ± 5.78	**139.60 ± 30.20****	14.19 ± 3.31	**77.97 ± 15.14****	13.12 ± 2.53	**75.36 ± 5.15*****	16.58 ± 4.11	**61.84 ± 6.50****
ociated								
	30.90 ± 14.04	12.87 ± 3.31	1.98 ± 1.30	**8.46 ± 1.58****	0.33 ± 0.12	5.19 ± 2.53	0.45 ± 0.16	**4.86 ± 1.29****
	12.51 ± 2.74	8.41 ± 0.72	8.08 ± 2.36	6.89 ± 0.39	0.96 ± 0.37	**5.73 ± 1.47****	7.05 ± 0.83	6.97 ± 1.16
mmatory								
	31.91 ± 5.40	**353.40 ± 95.66****	20.30 ± 5.94	**109.40 ± 17.98*****	15.39 ± 1.78	**152.30 ± 35.52****	18.42 ± 2.24	**95.36 ± 17.63**
	123.10 ± 23.21	**1816.00 ± 409.20****	46.24 ± 4.47	**782.00 ± 137.60*****	29.71 ± 3.71	**942.80 ± 206.40*****	49.95 ± 7.70	**647.60 ± 121.6**
	6.84 ± 2.27	**663.20 ± 90.46*****	0.62 ± 0.24	**454.60 ± 83.45*****	0.47 ± 0.18	**669.40 ± 210.30****	0.77 ± 0.23	**481.50 ± 112.0**
	39.17 ± 3.51	**84.68 ± 5.97*****	34.34 ± 1.70	**66.83 ± 3.35*****	37.01 ± 3.59	**66.90 ± 3.61*****	40.08 ± 4.15	**59.52 ± 3.14****
oregulatory								
	7.31 ± 0.84	**40.08 ± 5.68*****	5.30 ± 0.92	**25.05 ± 3.14*****	4.02 ± 0.51	**27.08 ± 2.55*****	3.87 ± 0.78	**21.09 ± 2.47****
	64.74 ± 13.03	**408.90 ± 99.61****	5.92 ± 0.85	**211.10 ± 44.00*****	3.22 ± 0.34	**298.50 ± 89.68****	13.12 ± 2.22	**189.40 ± 53.74**
kine								
IP-1α	138.00 ± 31.56	**354.40 ± 36.93*****	44.37 ± 3.16	**226.90 ± 15.24*****	29.94 ± 2.05	**255.40 ± 53.29*****	34.75 ± 5.91	**207.70 ± 37.53**
IP-1β	29.95 ± 6.78	**66.32 ± 14.73***	8.82 ± 1.72	**40.40 ± 4.68*****	6.86 ± 1.17	**43.55 ± 6.54*****	6.77 ± 1.42	**31.96 ± 3.67****
CP-1	372.60 ± 95.97	**962.50 ± 152.70****	89.06 ± 14.96	**713.00 ± 127.20*****	38.46 ± 5.58	**636.00 ± 64.17*****	116.90 ± 28.72	**474.90 ± 58.73**
KC	268.20 ± 47.52	**1055.00 ± 169.70*****	58.81 ± 7.29	**774.70 ± 118.20*****	45.73 ± 14.61	**782.60 ± 140.60*****	86.95 ± 15.98	**643.10 ± 111.4**
ANTES	246.70 ± 31.91	**1407.00 ± 156.60*****	135.20 ± 6.51	**621.90 ± 59.09*****	105.80 ± 8.11	**880.10 ± 183.90*****	86.45 ± 11.71	**607.50 ± 110.2**

wn are expressed as mean ± SEM and are cumulative of three experiments using four mice per group per time point.
alues indicate significance either increased or decreased compared to HKH99γ.
5, $^{**}p < 0.005$, $^{***}p < 0.0001$ significant increase compared to HKH99γ

4 | Cytokine levels within lung homogenates of H99γ and HKH99γ immunized mice infected with each serotype.

	Day 14 post challenge							
	HKH99γ	**H99γ**	**HKH99γ**	**H99γ**	**HKH99γ**	**H99γ**	**HKH99γ**	**H99γ**
	H99		**R265**		**WSA87**		**R4247**	
ociated								
	8.37 ± 0.60	**6.33 ± 0.54**$^{\tau}$	7.09 ± 0.74	**11.71 ± 0.88****	4.61 ± 0.57	**10.52 ± 0.66*****	4.80 ± 0.23	**10.82 ± 0.94****
	15.67 ± 1.19	15.87 ± 3.17	14.14 ± 0.43	**28.83 ± 2.17*****	13.21 ± 4.47	**28.20 ± 1.48****	9.67 ± 1.21	**28.49 ± 1.09****
0	78.16 ± 10.72	**30.11 ± 2.46**$^{\tau\tau\tau}$	62.74 ± 9.46	**130.30 ± 23.33***	21.28 ± 2.23	**127.10 ± 20.38*****	41.41 ± 3.31	**135.10 ± 21.07**
ociated								
	190.30 ± 33.23	**1.37 ± 0.37**$^{\tau\tau\tau}$	102.40 ± 17.90	**14.11 ± 4.43**$^{\tau\tau\tau}$	6.38 ± 2.57	4.96 ± 0.96	68.98 ± 7.51	**12.23 ± 2.89**$^{\tau\tau}$
	23.74 ± 5.57	**1.83 ± 0.70**$^{\tau\tau}$	73.75 ± 21.66	**8.24 ± 0.98**$^{\tau\tau}$	4.95 ± 1.27	**8.88 ± 1.09***	74.35 ± 13.94	**7.35 ± 0.86**$^{\tau\tau}$
mmatory								
	151.60 ± 18.96	**20.36 ± 2.50**$^{\tau\tau\tau}$	122.20 ± 12.17	**380.80 ± 84.73****	29.50 ± 6.28	**337.90 ± 46.85*****	105.90 ± 14.39	**471.00 ± 39.58**
	862.80 ± 218.50	**87.04 ± 8.71**$^{\tau\tau}$	398.20 ± 71.61	**2434.00 ± 519.10****	82.79 ± 21.65	**2346.00 ± 383.20*****	315.20 ± 53.03	**2645.00 ± 322.5**
	55.84 ± 10.26	33.87 ± 6.83	17.10 ± 4.86	**959.40 ± 178.60*****	3.24 ± 1.27	**897.10 ± 105.60*****	6.65 ± 2.03	**968.90 ± 83.29**
	52.10 ± 2.22	46.05 ± 3.00	45.81 ± 2.05	**80.03 ± 6.04*****	32.51 ± 2.89	**76.46 ± 4.20*****	38.77 ± 1.57	**71.21 ± 4.29****
regulatory								
	26.14 ± 1.78	**11.61 ± 0.93**$^{\tau\tau\tau}$	27.59 ± 3.00	**40.29 ± 4.54***	7.96 ± 0.98	**39.11 ± 3.23*****	17.97 ± 1.63	**40.08 ± 3.30****
	62.52 ± 15.47	**5.02 ± 0.42**$^{\tau\tau}$	42.48 ± 4.68	**81.41 ± 13.82***	13.80 ± 2.37	**110.50 ± 20.14*****	35.52 ± 4.65	**88.34 ± 10.39**
kine								
P-1α	1023.00 ± 252.10	**99.74 ± 21.77**$^{\tau\tau}$	515.30 ± 77.06	362.80 ± 47.70	73.85 ± 15.04	**346.10 ± 41.46*****	286.90 ± 44.40	366.10 ± 23.44
P-1β	40.83 ± 8.61	**8.71 ± 1.24**$^{\tau\tau}$	21.83 ± 4.15	**40.93 ± 4.74****	8.96 ± 2.82	**36.65 ± 4.08*****	18.59 ± 1.98	**40.43 ± 2.62****
CP-1	542.50 ± 117.10	**93.71 ± 11.92**$^{\tau\tau}$	1244.00 ± 213.40	**516.20 ± 71.17**$^{\tau\tau}$	183.80 ± 51.99	**585.40 ± 61.80*****	827.00 ± 188.20	486.40 ± 35.50
KC	533.10 ± 57.85	**80.39 ± 10.84**$^{\tau\tau\tau}$	496.70 ± 65.65	655.90 ± 70.33	297.00 ± 67.81	**751.40 ± 77.80*****	345.20 + 55.87	**625.50 ± 31.55**
ANTES	582.60 ± 114.80	305.40 ± 71.84	285.20 ± 27.58	**1402.00 ± 240.60*****	201.50 ± 39.46	**1623.00 ± 219.30*****	214.70 ± 45.55	**1377.00 ± 162.9**

wn are expressed as mean ± SEM and are cumulative of three experiments using four mice per group per time point.
alues indicate significance either increased or decreased compared to HKH99γ.
, $^{**}p < 0.005$, $^{***}p < 0.0001$ significant increase compared to HKH99γ.
$^{\tau\tau}p < 0.005$, $^{\tau\tau\tau}p < 0.0001$ significant decrease compared to HKH99γ.

mice that were subsequently depleted of both CD4⁺ and CD8⁺ T cells were shown to have an 80% survival rate following challenge with *C. neoformans* strain H99. Therefore, we sought to determine the necessity of T cells in protection against cryptococcosis caused by the different serotypes of *Cryptococcus* in H99γ immunized mice. For this, BALB/c mice were immunized with *C. neoformans* strain H99γ and allowed 70 days to resolve the infection. Mice were then depleted of CD4⁺ and/or CD8⁺ T cells or received isotype control antibody two days prior to challenge and weekly thereafter during the observation period. H99γ immunized mice depleted of both CD4⁺ and CD8⁺ T cells prior to and during challenge with H99 demonstrated an 87.5% survival rate upon the conclusion of the study (**Figure 4A**) as previously demonstrated (40). **Figure 4** shows that H99γ immunized mice treated with isotype control antibodies and challenged with R265 (**Figure 4B**) or R4247 (**Figure 4D**) had similar prolonged survival (median survival times of 65 and 64.5 days for R265 and R4247 challenged mice, respectively); similar to observations demonstrated in **Figure 1B**. In contrast, we observed 100% mortality in H99γ immunized mice depleted of CD4⁺ T cells alone during challenge with R265 (median survival of 35.5 days) or depleted of CD4⁺ and CD8⁺ T cells (median survival of 34 days; **Figure 4B**). Interestingly, H99γ immunized mice depleted of CD8⁺ T cells alone and challenged with R265 showed increased, although not statistically significant, survival (median survival of 80 days) compared to that observed in isotype control-treated mice (median survival of 65 days). We also observed that H99γ immunized mice challenged with R4247 while depleted of CD4⁺ T cells alone experienced 87.5% mortality (median survival time of 30.5 days) and 100% mortality upon depletion of both CD4⁺ and CD8⁺ T cells (median survival time of 31.5 days; **Figure 4D**). Additionally, H99γ immunized mice depleted of CD8⁺ T cells alone and challenged with R4247 had increased survival although not statistically significant (median survival of 75.5 days) compared to the isotype control-treated mice (median survival of 64.5 days). Mice immunized with H99γ and subsequently depleted of CD4⁺ T and/or CD8⁺ T cells prior to and during challenge with WSA87 did not show any difference in survival compared to the isotype control-treated mice (**Figure 4C**). Overall, we observed that depletion of CD4⁺ T cells resulted in a significant loss of protection against challenge with R265 (serotype B) or R4247 (serotype D) in H99γ (serotype A) immunized mice, indicating that CD4⁺ T cells are required for broad-spectrum protection against these serotypes.

DISCUSSION

While the overwhelming majority of *Cryptococcus* exposures do not progress to life-threatening illness, the ubiquitous presence of *Cryptococcus* in the environment world-wide indicates that exposure to persons predicted to be at an exceptionally high risk for developing cryptococcosis (i.e., patients scheduled to receive organ transplants, otherwise healthy HIV⁺ persons, and immune competent persons in areas observed to contain *C. gattii*) is inevitable and supports the need for development of a prophylactic and/or therapeutic vaccine that provides broad-spectrum protection against multiple *Cryptococcus* serotypes (9–16, 42). We have previously shown that mice immunized with *C. neoformans* strain H99γ, serotype A, are protected against a subsequent otherwise lethal challenge with wild-type *C. neoformans* strain H99 (25, 36–38, 40, 41, 43–45). Herein, we demonstrate the potential for achieving broad-spectrum protection against multiple clinically relevant strains of *Cryptococcus* serotypes following vaccination with a *Cryptococcus* serotype A strain previously shown to elicit protective immunity against *C. neoformans*. Specifically, we demonstrated that mice immunized with *Cryptococcus* strain H99γ, serotype A, develop significant protection against challenge with serotypes A, B, or D of *Cryptococcus*. Our data are in line with a similar study showing that mice vaccinated with an avirulent chitosan-deficient *C. neoformans* strain also created from a serotype A background strain exhibited significantly delayed mortality when challenged with serotype B *C. gattii* strains R265 or WM276 (34). Altogether, these studies clearly demonstrate that significant cross-protection can be developed following vaccination with a *Cryptococcus* strain of a disparate serotype and provide further proof-of-principle for the development of a prophylactic vaccine that can protect against multiple disparate *Cryptococcus* serotypes.

Both clinical and experimental studies have shown that Th1-type CD4⁺ T CMI is critical for protection against cryptococcosis (24, 25, 46). We observed a predominantly Th1-type and proinflammatory cytokine profile and increased survival in H99γ immunized mice following challenge with representative *Cryptococcus* serotypes B or D. Additionally, intracellular cytokine staining of CD4⁺ T cells showed that H99γ immunized mice were capable of eliciting a predominantly Th1-type anamnestic immune response upon subsequent challenge with the other *Cryptococcus* serotypes. These studies indicated that a Th1-type cell-mediated immune response was responsible for the increased survival observed in H99γ immunized mice challenged with *Cryptococcus* strains from serotypes B or D. We also observed increases in some Th2-type cytokines in H99γ immunized mice; however, the overall low levels observed are unlikely to be biologically relevant and were not as significant as the levels observed in HKH99γ immunized mice at day 14 post challenge. These findings do suggest that a Th1-Th2 balance may be important to achieve an optimal host immune response against *Cryptococcus* (47). Additionally, we observed an overall increase in the total number of infiltrating leukocytes in H99γ immunized mice compared to HKH99γ immunized mice challenged with their equivalent serotypes. Dendritic cells are known to play a vital role in the initial response against *C. neoformans* with the ability to phagocytose and kill the fungal organisms (48, 49). The significant increase in the infiltration of CD11b⁺/CD11c⁺ cells in mice given the H99γ immunization prior to challenge with the different serotypes compared to HKH99γ immunized mice correlates with reduced fungal burden in these mice. We did observe an increase in F4/80⁺ cells in H99γ immunized mice. However, previous studies in our laboratory has demonstrated that quantity of macrophages does not correlate with clearance, rather the ability of F4/80⁺ cells to polarize to a classically activated, M1 phenotype promotes clearance of *C. neoformans* (36, 41, 43, 50). The increase in infiltrating Ly6G⁺ cells correlates with significant increases in known neutrophil chemoattractants

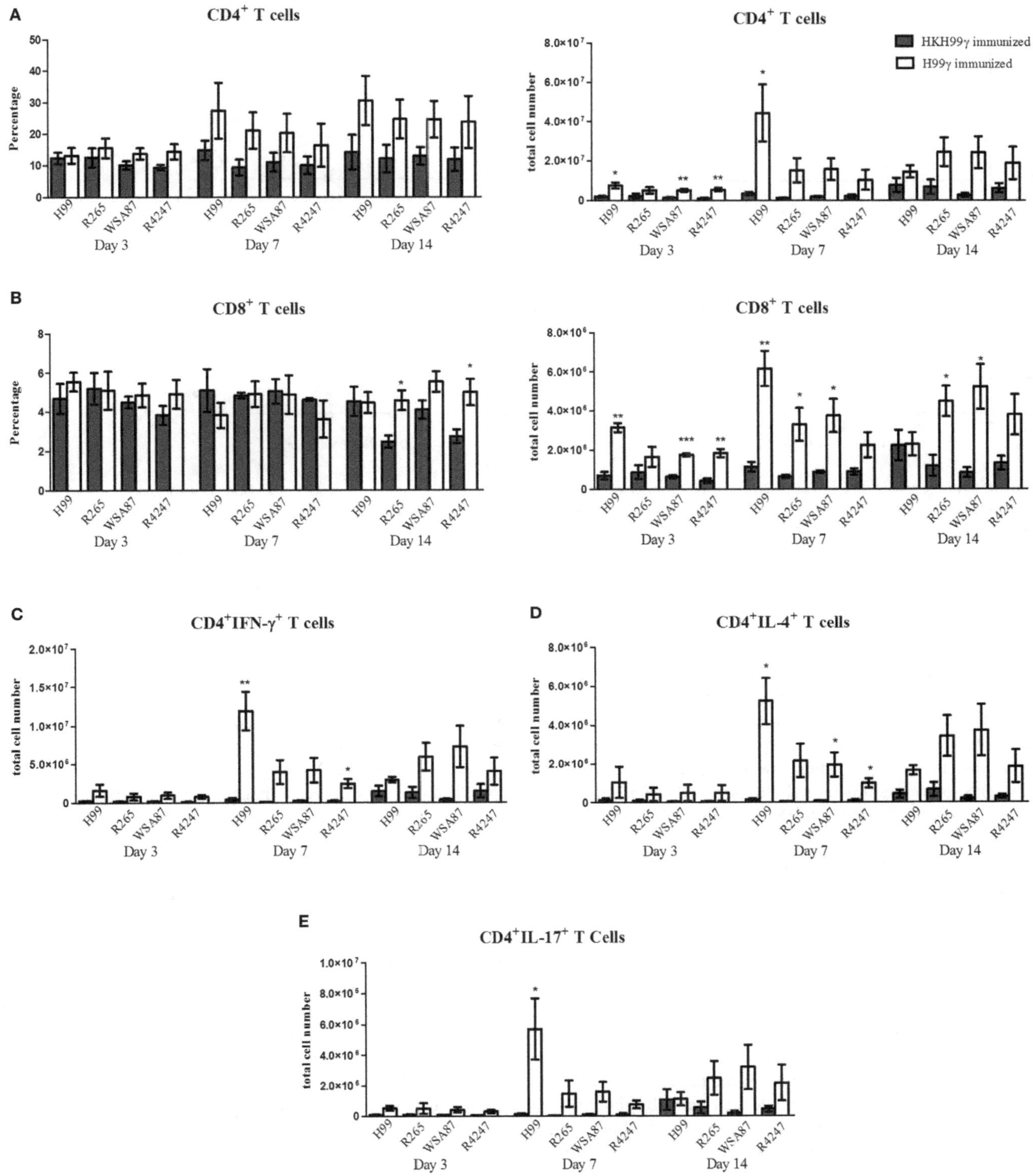

FIGURE 3 | Pulmonary T cell recruitment in immunized mice challenged with disparate serotypes of *Cryptococcus*. BALB/c mice were immunized intranasally with 1×10^4 CFU of either H99γ or HKH99γ, allowed 70 days to resolve the infection, and subsequently challenged with one serotype of *Cryptococcus*. At days 3, 7, and 14 post challenge, lungs were excised, tissues digested, and pulmonary immune cell infiltrates were analyzed by flow cytometry **(A,B)** as described previously in **Figure 2** or further permeabilized to stain for intracellular cytokines **(C–E)**. Data shown are the mean ± SEM of absolute cell numbers or percentage from three independent experiments performed using 4 mice per group per experiment. *$p < 0.05$, **$p < 0.005$, ***$p < 0.0001$. *Significant increase compared to HKH99γ.

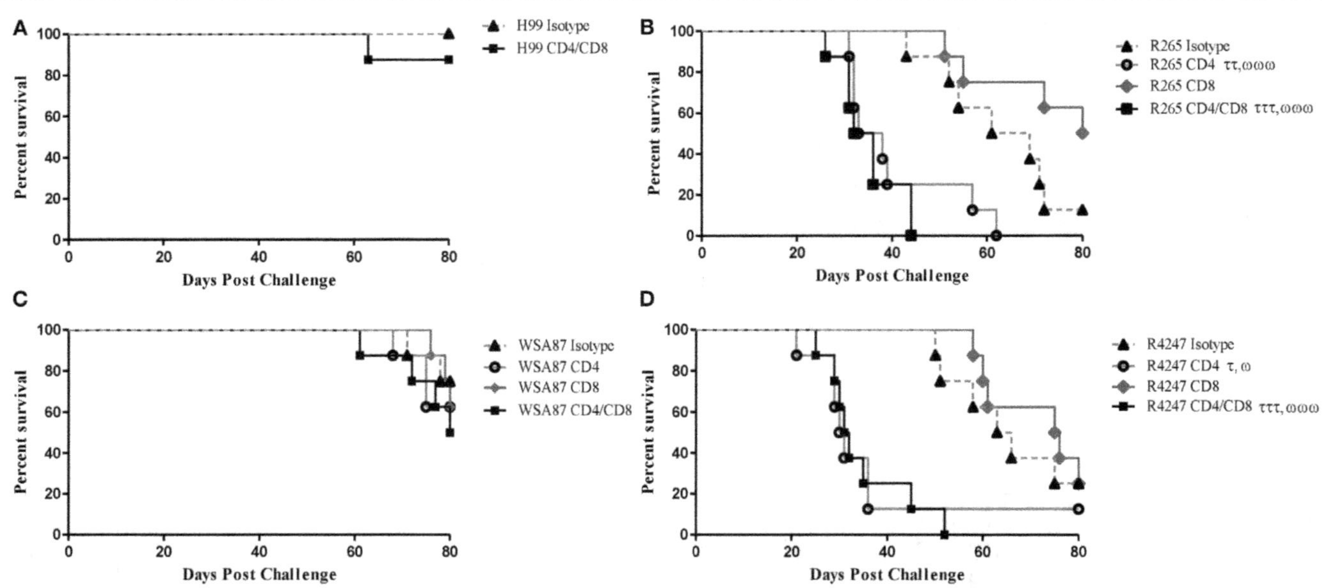

FIGURE 4 | Protection mediated by immunization with *C. neoformans* strain H99γ, serotype A, appears to be predominantly CD4+ T cell-mediated. BALB/c mice were immunized intranasally with 1 × 10^4 CFU of H99γ and allowed 70 days to rest. Mice were depleted of CD4+ T cells, CD8+ T cells, or both CD4+/CD8+ T cell subsets or received isotype control antibody two days prior to challenge with one serotype of *Cryptococcus* and weekly thereafter during the observation period. Mice were observed for up to 80 d post challenge for survival analysis. Survival data shown are from one experiment each, using 8 mice per group. However, the results are shown for individual groups and mice challenged with H99 **(A)**, R265 **(B)**, WSA87 **(C)** or R4247 **(D)** for simplicity. $^{\tau}p < 0.05$, $^{\tau\tau}p < 0.005$, $^{\tau\tau\tau}p < 0.0001$. τ = significant decrease compared to isotype; $^{\omega}p < 0.05$, $^{\omega\omega}p < 0.005$, $^{\omega\omega\omega}p < 0.0001$. ω = significant decrease compared to CD8+ T cell depleted.

[IL-17A, G-CSF (indirect chemotaxis), and CXCL1/KC] in the H99γ immunized mice (51, 52). However, previous studies from our lab demonstrate that neutrophils are not critical for clearance of *C. neoformans* strain H99 in H99γ immunized mice (44).

The vast majority of patients who acquire cryptococcosis due to *C. neoformans* are severely immunocompromized while those who acquire *C. gattii* infections appear to have little to no known immunodeficiency (53–55). Nonetheless, the protection observed in H99γ immunized mice challenged with serotypes B or D appeared to be CD4+ T cell-dependent. Interestingly, depletion of CD8+ T cells appeared to partially enhance survival in H99γ immunized mice challenged with serotypes B or D. We hypothesize that this additional protection may be attributed to a reduction in putative deleterious CD8+ T cell-mediated inflammatory responses or regulatory function in CD8+ T cell depleted mice challenged with serotypes B or D; however, this hypothesis will need to be confirmed in follow-up studies. In contrast, significant protection was evident in H99γ immunized mice rendered CD4+ and CD8+ T cell deficient and challenged with the serotype A *Cryptococcus* strain, H99. These studies demonstrate that immunization with H99γ produces memory T cells that allow for a rapid response against the invading cryptococcal strain as previously shown in our studies (40). Subtle species-specific differences in an antigen's peptide sequence and/or structure in the different serotypes tested may account for the lack of complete protection observed in our studies. Without the memory T cells present, there is no delay in onset of disease and the CD4+ T cell deficient mice challenged with serotypes B or D succumb to infection at a rate similar to the mice that were immunized with HKH99γ. Recent studies have shown increased protection against *C. neoformans* and *C. gattii* using a vaccine formulation comprised of antigens extracted from capsule or chitosan deficient *Cryptococcus* strains by treatment with an alkaline solution and packaged into GPs (33). The source of antigens for each vaccine formulation was specifically tailored to each challenge (i.e., *C. neoformans* antigens for *C. neoformans* challenge), and thus, the efficacy for the formulations to provide cross-protection was not determined. However, these studies also suggested that the protection induced following immunization with the *Cryptococcus* antigen/GP formulation against *C. neoformans* challenge was dependent on T CMI (33). The protection observed in T cell-deficient, H99γ immunized mice against a subsequent serotype A challenge may be antibody-mediated or may be suggestive of a trained innate cell population that compensates for the lack of T cells. Although we did observe an increase in the total number of B cells in H99γ immunized mice compared to HKH99γ immunized mice on days 7 and 14 post challenge, previous studies from our lab show that B cells are not required for the generation of protective immunity against cryptococcosis in H99γ immunized mice suggesting that antibody-mediated protection in the absence of T cells is unlikely (25). Overall, immunization with H99γ resulted in a non-T cell dependent protective immune response to challenge with a serotype A *Cryptococcus* strain showing that long-term protection in immunocompromized hosts can be generated. However, our results showing that CD4+ T cells are required for the protection observed in H99γ immunized mice challenged with serotypes B and D suggests that a vaccine formulation designed to induce complete protection against multiple *Cryptococcus* serotypes should include antigens specific to each serotype or contain antigens with significant homology across the serotypes.

Our studies also showed that the serotype C *Cryptococcus* strain used herein, WSA87, is less virulent or remains latent in mice compared to all the other strains tested. When H99γ immunized mice were subsequently challenged with WSA87, we observed a statistically significant increase in immune cell infiltrates at day 7 post challenge compared to HKH99γ immunized mice which correlated with the significant decrease of fungal burden in H99γ immunized mice. However, survival during challenge with serotype C was similar regardless of the immunization strategy most likely due to the initial low virulence of this strain or the strain's ability to remain latent with mice not succumbing to infection until almost day 90 post challenge. Also, depletion of $CD4^+$ and/or $CD8^+$ T cells was observed to have no impact on the survival of H99γ immunized mice challenged with WSA87. The results suggest that WSA87 suppresses T CMI or that protection against WSA87 is not dependent on T CMI. The serotype C strain chosen for this study was a clinical isolate that caused human disease; however, virulence due to this strain appeared to be relatively low in mice. *Cryptococcus* strains of the same serotype are known to exhibit significant variability in virulence within murine models. Thus, it is inappropriate to suggest that the serotypes utilized in our study are completely representative of all *Cryptococcus* strains within their serotype. However, these specific results suggest an unfortunate and perhaps unavoidable reality that developing vaccine formulations that target the more pathogenic *Cryptococcus* strains may still allow for less virulent strains of *Cryptococcus* to slip under the host's immunological radar resulting in disease and mortality.

While no fungal vaccines are clinically available, an ideal vaccine would mirror the results of the currently available vaccines against other microbial pathogens that can protect against multiple serotypes. The quadrivalent meningococcal conjugate vaccine (MenACWY-D; Menactra®, Sanofi Pasteur, Swiftwater, PA, USA), for example, contains polysaccharides from serotypes A, C, Y, and W-135 meningococci conjugated to a diphtheria protein carrier (56) and provides broad-spectrum protection against those serotypes that cause the majority of meningococcal disease. While the studies presented herein determined the efficacy of vaccination with one serotype to provide cross-protection against other serotypes, our results suggest that the identification and inclusion of multiple cross-protective $CD4^+$ T cell-dependent antigens will be critical for the development of a vaccine formulation that elicits broad-spectrum protection against *Cryptococcus* strains that cause the majority of clinical disease. Although much work remains toward the development of a cross-protective cryptococcal vaccine, these studies provide a foundation for understanding the protective host immune response to multiple clinically relevant strains of *Cryptococcus* and the need to identify protective $CD4^+$ T cell epitopes that can be incorporated into a vaccine formulation that provides broad-spectrum protection against cryptococcosis.

AUTHOR CONTRIBUTIONS

FW designed study, performed statistical analysis, interpreted study results, and participated in drafting and editing of manuscript. MD assisted in study design, performed experiments and statistical analysis, participated in interpretation of results, and drafted manuscript. AC, SH, CW, NC-L, CH, and KW performed experiments and statistical analysis, participated in interpretation of results and manuscript revision.

ACKNOWLEDGMENTS

The funders had no role in study design, data collection and analysis, decision to publish, or preparation of the manuscript.

FUNDING

Supported by research grants 2RO1AI071752 (FW) and 2RO1AI071752-08S1 (MD) from the National Institute of Allergy and Infectious Diseases (NIAID) of the National Institutes of Health (NIH) and grant GM060655 (MD) from the National Institute of General Medical Sciences of the NIH.

REFERENCES

1. Hagen F, Khayhan K, Theelen B, Kolecka A, Polacheck I, Sionov E, et al. Recognition of seven species in the *Cryptococcus gattii/Cryptococcus neoformans* species complex. *Fungal Genet Biol* (2015) 78:16–48. doi:10.1016/j.fgb.2015.02.009
2. Vibhagool A, Sungkanuparph S, Mootsikapun P, Chetchotisakd P, Tansuphaswaswadikul S, Bowonwatanuwong C, et al. Discontinuation of secondary prophylaxis for cryptococcal meningitis in human immunodeficiency virus-infected patients treated with highly active antiretroviral therapy: a prospective, multicenter, randomized study. *Clin Infect Dis* (2003) 36(10):1329–31. doi:10.1086/374849
3. Rajasingham R, Smith RM, Park BJ, Jarvis JN, Govender NP, Chiller TM, et al. Global burden of disease of HIV-associated cryptococcal meningitis: an updated analysis. *Lancet Infect Dis* (2017) 17(8):873–81. doi:10.1016/S1473-3099(17)30243-8
4. Maziarz EK, Perfect JR. Cryptococcosis. *Infect Dis Clin North Am* (2016) 30(1):179–206. doi:10.1016/j.idc.2015.10.006
5. Perez C, Dolande M, Moya M, Rosello A, de Capriles CR, Landaeta ME, et al. *Cryptococcus neoformans, Cryptococcus gattii*: serotypes in Venezuela. *Mycopathologia* (2008) 166(3):149–53. doi:10.1007/s11046-008-9132-1
6. Kwon-Chung KJ, Bennett JE. Epidemiologic differences between the two varieties of *Cryptococcus neoformans*. *Am J Epidemiol* (1984) 120(1):123–30. doi:10.1093/oxfordjournals.aje.a113861
7. Lizarazo J, Escandon P, Agudelo CI, Firacative C, Meyer W, Castaneda E. Retrospective study of the epidemiology and clinical manifestations of *Cryptococcus gattii* infections in Colombia from 1997-2011. *PLoS Negl Trop Dis* (2014) 8(11):e3272. doi:10.1371/journal.pntd.0003272

8. MacDougall L, Fyfe M, Romney M, Starr M, Galanis E. Risk factors for *Cryptococcus gattii* infection, British Columbia, Canada. *Emerg Infect Dis* (2011) 17(2):193–9. doi:10.3201/eid1702.101020
9. Byrnes EJ, Heitman J. *Cryptococcus gattii* outbreak expands into the Northwestern United States with fatal consequences. *F1000 Biol Rep* (2009) 1:62. doi:10.3410/B1-62
10. Byrnes EJ III, Li W, Lewit Y, Perfect JR, Carter DA, Cox GM, et al. First reported case of *Cryptococcus gattii* in the Southeastern USA: implications for travel-associated acquisition of an emerging pathogen. *PLoS One* (2009) 4(6):e5851. doi:10.1371/journal.pone.0005851
11. Datta K, Bartlett KH, Baer R, Byrnes E, Galanis E, Heitman J, et al. Spread of *Cryptococcus gattii* into pacific northwest region of the United States. *Emerg Infect Dis* (2009) 15(8):1185–91. doi:10.3201/eid1508.081384
12. Datta K, Bartlett KH, Marr KA. *Cryptococcus gattii*: emergence in Western North America: exploitation of a novel ecological niche. *Interdiscip Perspect Infect Dis* (2009) 2009:176532. doi:10.1155/2009/176532
13. Galanis E, Macdougall L, Kidd S, Morshed M; British Columbia Cryptococcus gattii Working Group. Epidemiology of *Cryptococcus gattii*, British Columbia, Canada, 1999-2007. *Emerg Infect Dis* (2010) 16(2):251–7. doi:10.3201/eid1602.090900
14. Hagen F, Colom MF, Swinne D, Tintelnot K, Iatta R, Montagna MT, et al. Autochthonous and dormant *Cryptococcus gattii* infections in Europe. *Emerg Infect Dis* (2012) 18(10):1618–24. doi:10.3201/eid1810.120068
15. McCulloh RJ, Phillips R, Perfect JR, Byrnes EJ III, Heitman J, Dufort E. *Cryptococcus gattii* genotype VGI infection in New England. *Pediatr Infect Dis J* (2011) 30(12):1111–4. doi:10.1097/INF.0b013e31822d14fd
16. Walraven CJ, Gerstein W, Hardison SE, Wormley F, Lockhart SR, Harris JR, et al. Fatal disseminated *Cryptococcus gattii* infection in New Mexico. *PLoS One* (2011) 6(12):e28625. doi:10.1371/journal.pone.0028625
17. Kwon-Chung KJ, Fraser JA, Doering TL, Wang Z, Janbon G, Idnurm A, et al. *Cryptococcus neoformans* and *Cryptococcus gattii*, the etiologic agents of cryptococcosis. *Cold Spring Harb Perspect Med* (2014) 4(7):a019760. doi:10.1101/cshperspect.a019760
18. Chen SC, Slavin MA, Heath CH, Playford EG, Byth K, Marriott D, et al. Clinical manifestations of *Cryptococcus gattii* infection: determinants of neurological sequelae and death. *Clin Infect Dis* (2012) 55(6):789–98. doi:10.1093/cid/cis529
19. Ngamskulrungroj P, Chang Y, Sionov E, Kwon-Chung KJ. The primary target organ of *Cryptococcus gattii* is different from that of *Cryptococcus neoformans* in a murine model. *mBio* (2012) 3(3):e00103-12. doi:10.1128/mBio.00103-12
20. Speed B, Dunt D. Clinical and host differences between infections with the two varieties of *Cryptococcus neoformans*. *Clin Infect Dis* (1995) 21(1):28–34; discussion 5–6. doi:10.1093/clinids/21.1.28
21. Lomes NR, Melhem MS, Szeszs MW, Martins Mdos A, Buccheri R. Cryptococcosis in non-HIV/non-transplant patients: a Brazilian case series. *Med Mycol* (2016) 54(7):669–76. doi:10.1093/mmy/myw021
22. Mitchell DH, Sorrell TC, Allworth AM, Heath CH, McGregor AR, Papanaoum K, et al. Cryptococcal disease of the CNS in immunocompetent hosts: influence of cryptococcal variety on clinical manifestations and outcome. *Clin Infect Dis* (1995) 20(3):611–6. doi:10.1093/clinids/20.3.611
23. Perfect JR, Cox GM. Drug resistance in *Cryptococcus neoformans*. *Drug Resist Updat* (1999) 2(4):259–69. doi:10.1054/drup.1999.0090
24. Olszewski MA, Zhang Y, Huffnagle GB. Mechanisms of cryptococcal virulence and persistence. *Future Microbiol* (2010) 5(8):1269–88. doi:10.2217/fmb.10.93
25. Wozniak KL, Ravi S, Macias S, Young ML, Olszewski MA, Steele C, et al. Insights into the mechanisms of protective immunity against *Cryptococcus neoformans* infection using a mouse model of pulmonary cryptococcosis. *PLoS One* (2009) 4(9):e6854. doi:10.1371/journal.pone.0006854
26. Angkasekwinai P, Sringkarin N, Supasorn O, Fungkrajai M, Wang YH, Chayakulkeeree M, et al. *Cryptococcus gattii* infection dampens Th1 and Th17 responses by attenuating dendritic cell function and pulmonary chemokine expression in the immunocompetent hosts. *Infect Immun* (2014) 82(9):3880–90. doi:10.1128/IAI.01773-14
27. Cheng PY, Sham A, Kronstad JW. Cryptococcus gattii isolates from the British Columbia cryptococcosis outbreak induce less protective inflammation in a murine model of infection than *Cryptococcus neoformans*. *Infect Immun* (2009) 77(10):4284–94. doi:10.1128/IAI.00628-09
28. Dong ZM, Murphy JW. Effects of the two varieties of *Cryptococcus neoformans* cells and culture filtrate antigens on neutrophil locomotion. *Infect Immun* (1995) 63(7):2632–44.
29. Huston SM, Li SS, Stack D, Timm-McCann M, Jones GJ, Islam A, et al. *Cryptococcus gattii* is killed by dendritic cells, but evades adaptive immunity by failing to induce dendritic cell maturation. *J Immunol* (2013) 191(1):249–61. doi:10.4049/jimmunol.1202707
30. Wright L, Bubb W, Davidson J, Santangelo R, Krockenberger M, Himmelreich U, et al. Metabolites released by *Cryptococcus neoformans* var. *neoformans* and var. *gattii* differentially affect human neutrophil function. *Microbes Infect* (2002) 4(14):1427–38. doi:10.1016/S1286-4579(02)00024-2
31. Urai M, Kaneko Y, Ueno K, Okubo Y, Aizawa T, Fukazawa H, et al. Evasion of innate immune responses by the highly virulent *Cryptococcus gattii* by altering capsule glucuronoxylomannan structure. *Front Cell Infect Microbiol* (2015) 5:101. doi:10.3389/fcimb.2015.00101
32. Hole CR, Wormley FL Jr. Vaccine and immunotherapeutic approaches for the prevention of cryptococcosis: lessons learned from animal models. *Frontiers Microbiol* (2012) 3:291. doi:10.3389/fmicb.2012.00291
33. Specht CA, Lee CK, Huang H, Tipper DJ, Shen ZT, Lodge JK, et al. Protection against experimental cryptococcosis following vaccination with glucan particles containing *Cryptococcus* alkaline extracts. *mBio* (2015) 6(6):e1905–15. doi:10.1128/mBio.01905-15
34. Upadhya R, Lam WC, Maybruck B, Specht CA, Levitz SM, Lodge JK. Induction of protective immunity to cryptococcal infection in mice by a heat-killed, chitosan-deficient strain of *Cryptococcus neoformans*. *mBio* (2016) 7(3):e00547-16. doi:10.1128/mBio.00547-16
35. Rella A, Mor V, Farnoud AM, Singh A, Shamseddine AA, Ivanova E, et al. Role of sterylglucosidase 1 (Sgl1) on the pathogenicity of *Cryptococcus neoformans*: potential applications for vaccine development. *Frontiers Microbiol* (2015) 6:836. doi:10.3389/fmicb.2015.00836
36. Hardison SE, Herrera G, Young ML, Hole CR, Wozniak KL, Wormley FL Jr. Protective immunity against pulmonary cryptococcosis is associated with STAT1-mediated classical macrophage activation. *J Immunol* (2012) 189(8):4060–8. doi:10.4049/jimmunol.1103455
37. Hardison SE, Ravi S, Wozniak KL, Young ML, Olszewski MA, Wormley FL Jr. Pulmonary infection with an interferon-gamma-producing *Cryptococcus neoformans* strain results in classical macrophage activation and protection. *Am J Pathol* (2010) 176(2):774–85. doi:10.2353/ajpath.2010.090634
38. Wormley FL Jr, Perfect JR, Steele C, Cox GM. Protection against cryptococcosis by using a murine gamma interferon-producing *Cryptococcus neoformans* strain. *Infect Immun* (2007) 75(3):1453–62. doi:10.1128/IAI.00274-06
39. Wozniak KL, Hardison S, Olszewski M, Wormley FL Jr. Induction of protective immunity against cryptococcosis. *Mycopathologia* (2012) 173(5–6):387–94. doi:10.1007/s11046-011-9505-8
40. Wozniak KL, Young ML, Wormley FL Jr. Protective immunity against experimental pulmonary cryptococcosis in T cell-depleted mice. *Clin Vaccine Immunol* (2011) 18(5):717–23. doi:10.1128/CVI.00036-11
41. Leopold Wager CM, Hole CR, Wozniak KL, Olszewski MA, Mueller M, Wormley FL Jr. STAT1 signaling within macrophages is required for antifungal activity against *Cryptococcus neoformans*. *Infect Immun* (2015) 83(12):4513–27. doi:10.1128/IAI.00935-15
42. Park BJ, Wannemuehler KA, Marston BJ, Govender N, Pappas PG, Chiller TM. Estimation of the current global burden of cryptococcal meningitis among persons living with HIV/AIDS. *AIDS* (2009) 23(4):525–30. doi:10.1097/QAD.0b013e328322ffac
43. Leopold Wager CM, Hole CR, Wozniak KL, Olszewski MA, Wormley FL Jr. STAT1 signaling is essential for protection against *Cryptococcus neoformans* infection in mice. *J Immunol* (2014) 193(8):4060–71. doi:10.4049/jimmunol.1400318
44. Wozniak KL, Kolls JK, Wormley FL Jr. Depletion of neutrophils in a protective model of pulmonary cryptococcosis results in increased IL-17A production by gammadelta T cells. *BMC Immunol* (2012) 13:65. doi:10.1186/1471-2172-13-65
45. Wozniak KL, Hardison SE, Kolls JK, Wormley FL. Role of IL-17A on resolution of pulmonary *C. neoformans* infection. *PLoS One* (2011) 6(2):e17204. doi:10.1371/journal.pone.0017204
46. Chaturvedi AK, Wormley FL Jr. *Cryptococcus* antigens and immune responses: implications for a vaccine. *Expert Rev Vaccines* (2013) 12(11):1261–72. doi:10.1586/14760584.2013.840094

47. Koguchi Y, Kawakami K. Cryptococcal infection and Th1-Th2 cytokine balance. *Int Rev Immunol* (2002) 21(4–5):423–38. doi:10.1080/08830180213274
48. Osterholzer JJ, Milam JE, Chen GH, Toews GB, Huffnagle GB, Olszewski MA. Role of dendritic cells and alveolar macrophages in regulating early host defense against pulmonary infection with *Cryptococcus neoformans*. *Infect Immun* (2009) 77(9):3749–58. doi:10.1128/IAI.00454-09
49. Wozniak KL, Levitz SM. *Cryptococcus neoformans* enters the endolysosomal pathway of dendritic cells and is killed by lysosomal components. *Infect Immun* (2008) 76(10):4764–71. doi:10.1128/IAI.00660-08
50. Arora S, Olszewski MA, Tsang TM, McDonald RA, Toews GB, Huffnagle GB. Effect of cytokine interplay on macrophage polarization during chronic pulmonary infection with *Cryptococcus neoformans*. *Infect Immun* (2011) 79(5):1915–26. doi:10.1128/IAI.01270-10
51. Mellett M, Atzei P, Horgan A, Hams E, Floss T, Wurst W, et al. Orphan receptor IL-17RD tunes IL-17A signalling and is required for neutrophilia. *Nat Commun* (2012) 3:1119. doi:10.1038/ncomms2127
52. Yong KL. Granulocyte colony-stimulating factor (G-CSF) increases neutrophil migration across vascular endothelium independent of an effect on adhesion: comparison with granulocyte-macrophage colony-stimulating factor (GM-CSF). *Br J Haematol* (1996) 94(1):40–7. doi:10.1046/j.1365-2141.1996.d01-1752.x
53. Chen SC-A, Meyer W, Sorrell TC. *Cryptococcus gattii* infections. *Clin Microbiol Rev* (2014) 27(4):980–1024. doi:10.1128/cmr.00126-13
54. Rosen LB, Freeman AF, Yang LM, Jutivorakool K, Olivier KN, Angkasekwinai N, et al. Anti-GM-CSF autoantibodies in patients with cryptococcal meningitis. *J Immunol* (2013) 190(8):3959–66. doi:10.4049/jimmunol.1202526
55. Saijo T, Chen J, Chen SC, Rosen LB, Yi J, Sorrell TC, et al. Anti-granulocyte-macrophage colony-stimulating factor autoantibodies are a risk factor for central nervous system infection by *Cryptococcus gattii* in otherwise immunocompetent patients. *mBio* (2014) 5(2):e912–4. doi:10.1128/mBio.00912-14
56. Pace D, Pollard AJ, Messonier NE. Quadrivalent meningococcal conjugate vaccines. *Vaccine* (2009) 27(Suppl 2):B30–41. doi:10.1016/j.vaccine.2009.05.003

Th17-Inducing Cytokines IL-6 and IL-23 are Crucial for Granuloma Formation During Experimental Paracoccidioidomycosis

*Fabrine Sales Massafera Tristão†, Fernanda Agostini Rocha†, Daniela Carlos, Natália Ketelut-Carneiro, Camila Oliveira Silva Souza, Cristiane Maria Milanezi and João Santana Silva**

Department of Biochemistry and Immunology, School of Medicine of Ribeirão Preto, University of São Paulo, Ribeirão Preto, Brazil

Correspondence:
João Santana Silva
jsdsilva@fmrp.usp.br

†These authors have contributed equally to this work.

Paracoccidioidomycosis (PCM), a chronic granulomatous disease caused by the thermally dimorphic fungus *Paracoccidioides brasiliensis* and *Paracoccidioides lutzii*, has the highest mortality rate among systemic mycosis. The T helper 1-mediated immunity is primarily responsible for acquired resistance during *P. brasiliensis* infection, while susceptibility is associated with Th2 occurrence. Th17 is a population of T $CD4^+$ cells that, among several chemokines and cytokines, produces IL-17A and requires the presence of IL-1, IL-6, and TGF-β for differentiation in mice and IL-23 for its maintenance. Th17 has been described as an arm of the immune system that enhances host protection against several bacterial and fungal infections, as *Pneumocystis carinii* and *Candida albicans*. In this study, we aimed to evaluate the Th17 immune response and the role of Th17-associated cytokines (IL-6, IL-23, and IL-17A) during experimental PCM. First, we observed that *P. brasiliensis* infection [virulent yeast strain 18 of *P. brasiliensis* (Pb18)] increased the IL-17A production *in vitro* and all the evaluated Th17-associated cytokines in the lung tissue from C57BL/6 wild-type mice. In addition, the deficiency of IL-6, IL-23, or IL-17 receptor A (IL-17RA) impaired the compact granuloma formation and conferred susceptibility during infection, associated with reduced tumor necrosis factor-α, IFN-γ, and inducible nitric oxide synthase enzyme expression. Our data suggest that IL-6 production by bone marrow-derived macrophages (BMDMs) is important to promote the Th17 differentiation during Pb18 infection. In accordance, the adoptive transfer of BMDMs from C57BL/6 to infected $IL\text{-}6^{-/-}$ or $IL\text{-}17RA^{-/-}$ mice reduced the fungal burden in the lungs compared to nontransferred mice and reestablished the pulmonary granuloma formation. Taken together, these results suggest that Th17-associated cytokines are involved in the modulation of immune response and granuloma formation during experimental PCM.

Keywords: paracoccidioidomycosis, Th17, IL-17A, IL-6, IL-23

INTRODUCTION

Paracoccidioidomycosis (PCM), a systemic mycosis characterized by chronic and granulomatous inflammation, is caused by the thermally dimorphic fungus *Paracoccidioides brasiliensis* and *Paracoccidioides lutzii* (1). PCM has high incidence in several countries of Latin America and represents the eighth cause of death among infectious and parasitic diseases. Moreover, it induces the highest mortality rate (1.45 per million inhabitants) among systemic mycosis, being considered an occupational disease and a serious social problem (2). It affects mainly farm workers and usually leads to the formation of pulmonary dysfunction as a consequence of primary infection progression or reactivation of a latent focus (3).

The infection is acquired after inhalation of fungal propagules that convert into the invasive yeast form once in the lungs (4). The resistance to infection is associated with a potent T helper 1 (Th1) response, which induces activation of macrophages that actively control the fungal growth (5, 6). Otherwise, resistance is associated with a predominantly Th2 and Th9 response, inducing eosinophilia and humoral immune responses (7).

Besides the mentioned Th subtypes, classified by defined phenotypic characteristics and specialized functions in immunity, T cells can also differentiate into Th17, a distinct cell lineage that produces various chemokines and cytokines, such as tumor necrosis factor (TNF)-α, IL-6, IL-21, IL-22, IL-17F, and IL-17A (8, 9). IL-1 signaling in T cells is required for the early Th17 differentiation *in vitro* and *in vivo*, and after polarization, IL-1 also allowed Th17 cells to maintain their cytokine secretion profile (10–12). The murine differentiation of Th17 requires the presence of IL-6, IL-21, and TGF-β, which activate Stat3 and induce the expression of the transcription factor retinoic acid-related orphan receptor (RORγt) (10, 13–15). Although IL-23 is not necessary for Th17 differentiation, it is essential for the maintenance of the differentiated Th17 cells (16).

Notably, Th17 cells are described as an arm of the immune system that enhances host protection against several intracellular and extracellular bacterial infections (17). In addition, fungal infections also have been associated with induction of Th17 immune response, triggering effector mechanisms, as production of antimicrobial peptides (8, 18, 19) and factors important for neutrophil function and recruitment (20–22), promoting the resistance to infection (23–25). In fact, the IL-17A and IL-23 production was important in the protective response during experimental *Pneumocystis carinii* infection (26). In addition, *Candida albicans* recognition by antigen-presenting cells promoted the IL-6 and IL-23 secretion, leading to Th17 differentiation and enhancement of host resistance (27–29). Similarly, *P. brasiliensis* was able to induce Th17 and IL-17-producing CD8$^+$ T cell (Tc17) differentiation (30, 31).

The cytokine IL-17A coordinates tissue inflammation through higher expression of pro-inflammatory cytokines and chemokines, which collectively determine the magnitude of the inflammatory response. The IL-17A receptors (IL-17RAs) are found in the surface of leukocytes, keratinocytes, fibroblasts, epithelial, mesothelial, and vascular endothelial cells, and its action includes granulopoiesis, neutrophil recruitment, and inflammatory responses (21, 22, 32–34). Moreover, IL-17A plays a critical role in the induction of mature granuloma formation during mycobacterial infection (35). Granulomas are immunological structures important in the host defense against fungi. In the course of granuloma maturation, the recruitment of phagocytes and lymphocytes is triggered by various cytokines and chemokines that are initially produced by infected macrophages, as TNF-α (36). In corroboration, our group showed that TNF-α-deficient mice are highly susceptible to *P. brasiliensis* infection, are not able to mount organized granulomas, and have a great amount of fungus in the lesions and high mortality rates (37). Moreover, IL-17-expressing cells have been detected within and around the granulomas in the skin and oral mucosal lesions from PCM patients (38).

Because little is known about the mechanisms involved in the immune response during PCM, we sought here to investigate the role of IL-6, IL-23, and IL-17A during the experimental *P. brasiliensis* infection. Our data suggest that IL-6-production by bone marrow-derived macrophages (BMDMs) is important to promote the IL-17A production and consequent induction of a protective immune response and granuloma formation during *P. brasiliensis* experimental infection.

MATERIALS AND METHODS

Mice

Male 6- to 7-week-old C57BL/6 mice, and genetically deficient ($^{-/-}$) for IL-6, IL-23p19, and IL-17RA, were obtained from our Isogenic Breeding Unit and maintained under specific pathogen-free conditions in microisolator cages in the animal housing facility of the Department of Biochemistry and Immunology, School of Medicine of Ribeirão Preto, São Paulo University, Ribeirão Preto, Brazil. Mice were supplied with sterilized food and water *ad libitum*. Experiments were conducted according to the ethical principles of animal research adopted by the Brazilian College of Animal Experimentation and approved by the Ethical Commission in Animal Research (protocol 095/2010).

Fungus, *In Vitro* and *In Vivo* Infections

The virulent yeast strain 18 of *P. brasiliensis* (Pb18) was cultured (7 days at ~36°C) in Brain Heart Infusion (BHI) agar medium (Oxoid Basingstoke, Hampshire, England) supplemented with gentamicin (100 μg/mL) and 5% fetal bovine serum (FBS; Gibco BRL, Life Technologies, Inc.). After, Pb18 colonies were grown for 24 h (37°C, 150 rpm) in F12 Coon's modified medium (Sigma-Aldrich, St. Louis, MO, USA) containing gentamicin (100 μg/mL). The Pb18 yeast cells were then harvested and washed in sterile phosphate-buffered saline (PBS; pH 7.2), and the fungal viability was determined by fluorescein diacetate-ethidium bromide staining (39). Only high viability suspensions (≥90%) were used in this study. For *in vivo* infection, the concentration of fungal cells was adjusted to 1×10^7 yeasts/mL in PBS and 100 μL of this solution (1×10^6 cells) was intravenously (i.v.) inoculated in each mice. *In vitro* fungal stimulation was performed with a multiplicity of infection (MOI) of 1.

Splenocyte and Pulmonary Cells Isolation and Activation

The spleen from uninfected male C57BL/6 mice were removed, and single-cell suspensions were obtained after red blood cells lysis (1 min) with buffer containing NH_4Cl (0.16 M, 9 parts) and Tris-HCl (0.17 M, 1 part). Splenocytes (1×10^6 cells/well) were cultured for 5 days with 1×10^6 Pb18 yeasts cells in RPMI 1640 supplemented with 5% FBS (Gibco), 0.1 mM non-essential amino acids, 1 mM sodium pyruvate, L-glutamine (2 mM), 100 U of penicillin/mL, and 100 μg of streptomycin/mL (all from Sigma-Aldrich). Lung lobules were excised from uninfected and Pb18-infected mice, washed in PBS, minced with scissors, and enzyme digested at 37°C for 30–35 min in 1 mL of digestion buffer [RPMI 1640, 2 mg/mL collagenase IV (Sigma) and 1 mg/mL DNase (Sigma)]. Tissue fragments were further homogenized using a 1-mL pipette, crushed through a 50-μm pore size nylon filter (BD Biosciences, San Jose, CA, USA) and then centrifuged (1,300 rpm, 10 min, 4°C). Next, red blood cells were eliminated using lysis buffer, and remaining cells were washed in PBS, centrifuged, and resuspended in RPMI 1640 containing 5% FBS. For splenocytes and pulmonary cells activation, 1×10^6 cells/well were cultured for 4 h with PMA (50 ng/mL), ionomycin (500 ng/mL), and brefeldin A (5 mg/mL), followed by staining for extracellular and intracellular markers and analysis by flow cytometry.

Flow Cytometry

The expression of CD4, CD8, CD19, CD49b, γδ, F4/80, IL-6, and IL-17A was assessed in splenocytes or pulmonary cells by flow cytometry as previously reported (40, 41). The antibodies were conjugated to different fluorochromes (BD Biosciences, eBiosciences, San Diego, CA, USA, and Santa Cruz Biotechnologies, Santa Cruz, CA, USA). For intracellular cytokine staining, cells were previously permeabilized using PBS containing 1% FBS, 0.1% sodium azide, and 0.2% saponin. Data acquisition was performed using a FACSCanto II flow cytometer and FACSDiva software (BD Biosciences). Data were plotted and analyzed using FlowJo software (Tree Star, Ashland, OR, USA).

Measurement of Cytokines in Lung Supernatant

The left lung (42) from uninfected and Pb18-infected mice were removed, weighed, homogenized in sterile PBS-containing protease inhibitor (Complete, Roche), and centrifuged (1,300 rpm, 10 min, 4°C). Supernatants were collected and stored at −20°C. The levels of IL-17A, IL-6, IL-23, TNF-α, and IFN-γ were measured by enzyme-linked immunosorbent assay (ELISA) according to manufacturers' recommendations (BD Pharmingen, San Jose, CA, USA). The reading was held in eMax ELISA reader (Molecular Devices, Sunnyvale, CA, USA) at 450 nm.

Recovery of Colony-Forming Units (CFUs)

The amount of viable yeast cells in the lungs from Pb18-infected mice (3–7 animals per group) was determined after 15 and 30 days postinfection (dpi). Briefly, lung lobules were aseptically collected, weighed, and homogenized using a sterile tissue grinder (IKA®-Werke, Deutschland, Germany). Each homogenate was diluted 1:10 in sterile PBS and plated (100 μL) on BHI agar supplemented with 5% FBS and gentamicin (100 μg/mL). Plates were incubated at 35–37°C for 7 days, and the amount of CFU per gram of tissue was calculated.

Histopathology

Comparative histopathology was conducted with excised lung lobules from C57BL/6, IL-6$^{-/-}$, IL-23$^{-/-}$, and IL-17RA$^{-/-}$ mice at 30 dpi. Tissue was fixed with 10% formalin for 24 h and embedded in paraffin. Next, lung sections (5 μm) were stained by standard procedures with hematoxylin and eosin for lesion analysis, Grocott methenamine silver stain for fungal cell labeling, or silver impregnation (Gomori method) to demonstrate reticulum fibers. Sections were examined with HLIMAGE++ 97 Application Western Vision Software using an optical microscope coupled to a digital camera. Granulomas were visually scored in mature or immature structures according to cellular and reticulin organization and distribution, as described before (43, 44). The percentage of mature granulomas and the granulomatous diameter were calculated. Reticulin measurement was performed in the halo surrounding the granulomas as previously described (45).

Immunohistochemistry

Lungs from C57BL/6, IL-6$^{-/-}$, IL-23$^{-/-}$, and IL-17RA$^{-/-}$ mice were immersed in OCT medium (Sakura Finetek), snap-frozen in liquid nitrogen, and stored at −80°C until analysis. Tissue sections (5 μm) were submitted to an immunohistochemical reaction. Briefly, slides were incubated with anti-mouse IgG as control or rabbit IgG anti-mouse inducible nitric oxide synthase (iNOS) enzyme (Santa Cruz Biotechnology, Santa Cruz, CA, USA) diluted 100 times in PBS 0.01% saponin. At the end, the reaction was followed by Mayer's hematoxylin counterstaining. We measured immunostained areas using Image J software as previously described (41). Briefly, the range of positivity was defined using the IHC Tool box. Next, the images were converted to 8-bit, and the grayscale was converted to binary (black and white). The threshold was adjusted, and the labeled areas became the black portions. Finally, the percentage of stained area was analyzed.

BMDMs Differentiation

Bone marrow cells isolated from femurs and tibias from 7-week-old C57BL/6, IL-6$^{-/-}$, IL-23$^{-/-}$, and IL-17RA$^{-/-}$ naive mice were cultured (7 days, 37°C, 5% CO_2) in RPMI 1640 medium supplemented with 20% FBS and 30% L-929 cell conditioned media, as previously described (46). After differentiation, cells were harvested and infected with *P. brasiliensis* (MOI of 0.04).

Isolation of Naive T CD4$^+$ Cells and Co-Culture with BMDMs

Naive T CD4$^+$ cells were purified from C57BL/6 lymph nodes by using the CD4$^+$ MicroBeads Isolation Kit (Miltenyi Biotec) in conjunction with an AutoMacs separator. Purified naive T cells were cultured for 5 days with *P. brasiliensis*-infected BMDMs from C57BL/6 or IL-6$^{-/-}$ mice (MOI of 0.04). Next, the amount of T CD4$^+$IL-17A$^+$ was analyzed by flow cytometry.

Adoptive Transfer

After adoptive transference of C57BL/6 BMDMs (1×10^6, i.v.) into IL-6$^{-/-}$, IL-23$^{-/-}$, and IL-17RA$^{-/-}$ animals, mice were infected with 1×10^6 *P. brasiliensis* yeast cells, and the CFU was determined at day 30 as described above.

Statistical Analysis

The differences observed between uninfected or Pb18-infected wild-type (WT) group (C57BL/6) and IL-6-, IL-17RA-, or

IL-23-deficient mice were analyzed by applying the ANOVA test followed by multiple comparison using the Bonferroni method (GraphPad Prism, GraphPad Software version 5.0, San Diego, CA, USA). Values are expressed as mean ± SEM. All values were considered significant when $p < 0.05$.

RESULTS

P. brasiliensis Infection Increases IL-17A Production

To evaluate whether fungal cells modulate the IL-17A production, splenocytes from C57BL/6 mice were isolated, cultivated in the presence of or not of *P. brasiliensis*, and the frequency of IL-17A-producing cells was determined 5 days later. Our results showed that the stimulation of spleen cells with viable Pb18 increased the IL-17A expression (**Figure 1A**), and the frequency of IL-17A-producing CD4 T cells (5.5-fold, $p < 0.05$; **Figure 1B**) compared to medium alone. Next, to evaluate whether *P. brasiliensis* modulates the Th17 profile *in vivo*, C57BL/6 mice were infected with Pb18 yeast cells, and the IL-17A production was determined in the lungs at day 30 postinfection. As expected, we observed that Pb18-infected mice exhibited a significant increase (~9.5-fold) in IL-17A production compared to uninfected mice (**Figure 1C**). In addition, at 30 dpi, there was an increase in the number of IL-17A-producing cells (~4.5-fold; **Figure 1D**).

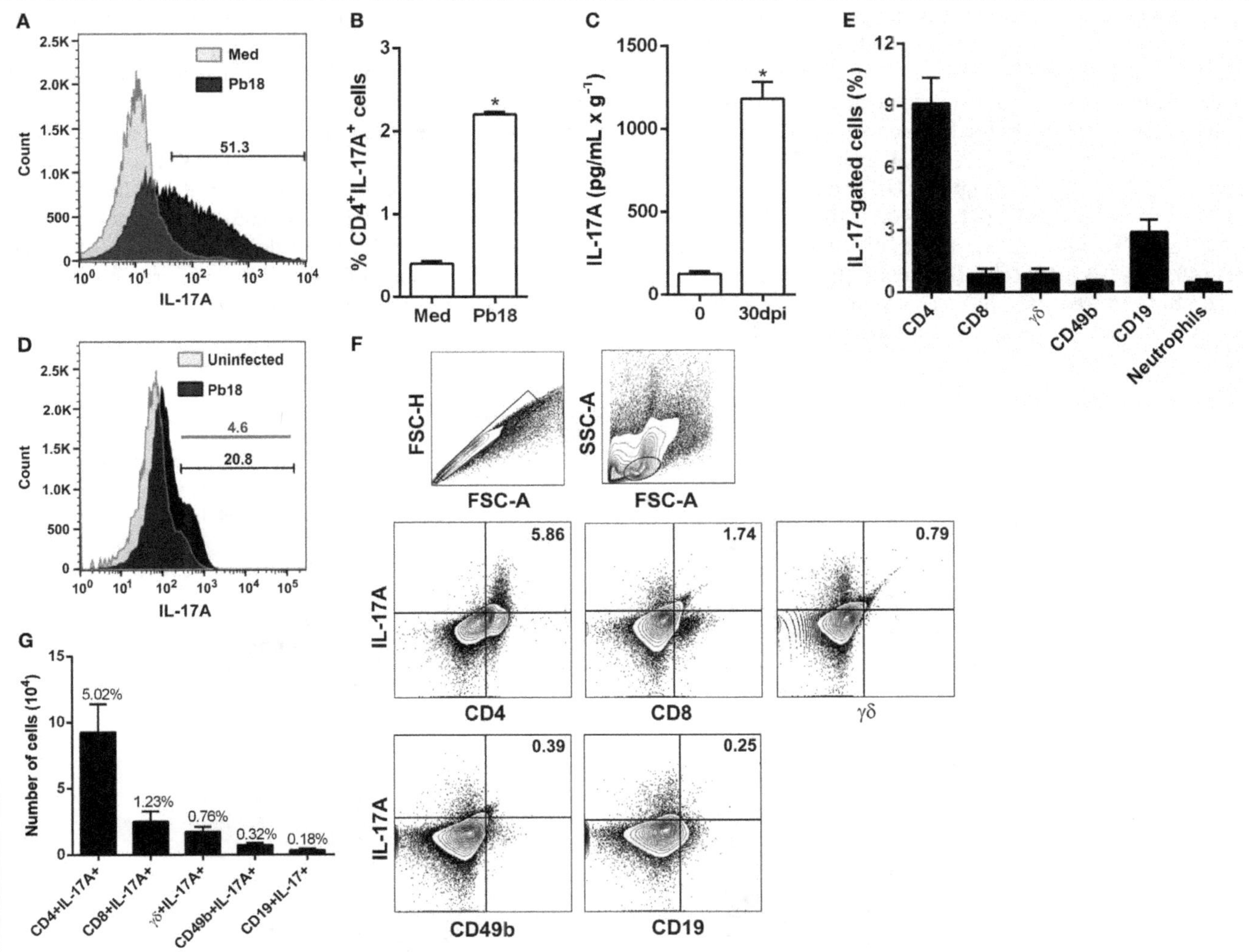

FIGURE 1 | *Paracoccidioides brasiliensis* infection increases the *in vitro* and *in vivo* IL-17A production. **(A,B)** Splenocytes from C57BL/6 mice were isolated and cultivated (1×10^6) in the presence of or not of viable virulent yeast strain 18 of *P. brasiliensis* (Pb18) during 5 days (multiplicity of infection of 1). **(A)** The frequency of IL-17A⁺-producing cells and **(B)** Th17 (CD4⁺IL-17A⁺) cells was determined by flow cytometry. **(C–E)** C57BL/6 mice were infected with 1×10^6 Pb18 yeast cells (intravenously). **(C)** The level of IL-17A was measured by enzyme-linked immunosorbent assay in the whole lung homogenate after 30 days postinfection (dpi). **(D)** The frequency of IL-17A-producing cells was evaluated in the lungs from uninfected and infected (30 dpi) mice. **(E–G)** The frequency of natural killer (CD49b⁺IL-17A⁺), Tγδ (γδ⁺IL-17A⁺), B (CD19⁺IL-17A⁺), IL-17-producing CD8⁺ T (CD8⁺IL-17A⁺), and Th17 cells was assessed in the lungs after 30 dpi. **(E)** Cells were analyzed inside the IL-17-positive gate. **(F)** Lymphocytes were gated inside the singlet population. One representative mice is shown. **(G)** Absolute number (bars) and mean percentage (above bars) of each IL-17-producing cell population gated inside lymphocytes. Bars represent the mean ± SEM of 5 mice. *$p < 0.05$ compared to non-infected control. Similar results were obtained in three independent experiments.

Despite the fact that most of the evidence describes IL-17A as a cytokine secreted by T cells, part of the IL-17A is produced by different cell populations, including innate immune cells (47). To determine the major source of IL-17A in the lungs during the experimental PCM, we evaluated the frequency and absolute number of Th17, Tc17, $\gamma\delta^+$IL-17A$^+$, CD49b$^+$IL-17A$^+$ [natural killer (NK)], and CD19$^+$IL-17A$^+$ (B cells). Initially, we excluded the doublets; delineated a gate comprising lymphocytes, monocytes, and neutrophils; and then a subgate in IL-17-positive cells, which allowed us to evaluate the percentage of each cell subtype (**Figure 1E**). The Th17 population was significantly increased in comparison with other IL-17-producing cells ($p < 0.05$; **Figure 1E**). For better and clear results, we next used a second gate strategy, which consisted of singlets selection in forward scatter height (FSC-H) against forward scatter area (FSC-A) (**Figure 1F**, upper left panel). Next, lymphocytes were gated inside the previous population in side scatter area (SSC-A) against FSC-A (**Figure 1F**, upper right panel), and all analyzed cell populations were gated inside the lymphocyte population. Our data clearly showed the predominance of IL-17A-producing CD4$^+$ cells, both in percentage (~5%, **Figures 1F,G**) or absolute number (~9.2×10^4 cells; **Figure 1G**), compared with other IL-17A$^+$ cells subtypes: CD8 (~1.2%, ~2.5×10^4 cells), $\gamma\delta$ (~0.7%, ~1.7×10^4 cells), NK (~0.3%, ~0.7×10^4 cells), and B lymphocytes (~0.1%, ~0.3×10^4 cells) (**Figures 1F,G**). After evaluating the number and frequency of IL-17-producing neutrophils, we could not see an increase of this cell subtype in the lungs from Pb18-infected mice (data not shown). These findings suggest that Th17 cells are the main source of IL-17A during *P. brasiliensis* infection.

The Absence of IL-6 and IL-23 Impairs the Th17 Immune Response during Experimental *P. brasiliensis* Infection

The Th17 subset differentiation requires the presence of IL-6, and its maintenance is sustained by IL-23 (13, 14, 16). Thus, we evaluated the production of these cytokines in the lungs from naive and Pb18-infected mice at 15 and 30 dpi. Our data showed that at both time points, the IL-6 and IL-23 levels were significantly increased when compared to uninfected mice (**Figure 2A**). To confirm these findings, IL-6$^{-/-}$ and IL-23$^{-/-}$ mice were infected with *P. brasiliensis* yeast cells, and the IL-17A levels were evaluated in the pulmonary tissue after 15 and 30 days. We found that the deficiency of IL-6 or IL-23 impaired the IL-17A-production, which was ~2.5 times smaller than C57BL/6 mice at the same period of infection ($p < 0.05$; **Figure 2B**). Moreover, when compared to WT mice, we found that the frequency and absolute number of IL-17A-producing CD4$^+$ T cells were decreased ($p < 0.05$) in the lung tissue from IL-6$^{-/-}$ and IL-23$^{-/-}$ mice at 30 dpi (**Figure 2C**). These data establish the importance of IL-6 and IL-23 in promoting the late Th17-immune response during the *P. brasiliensis* experimental infection.

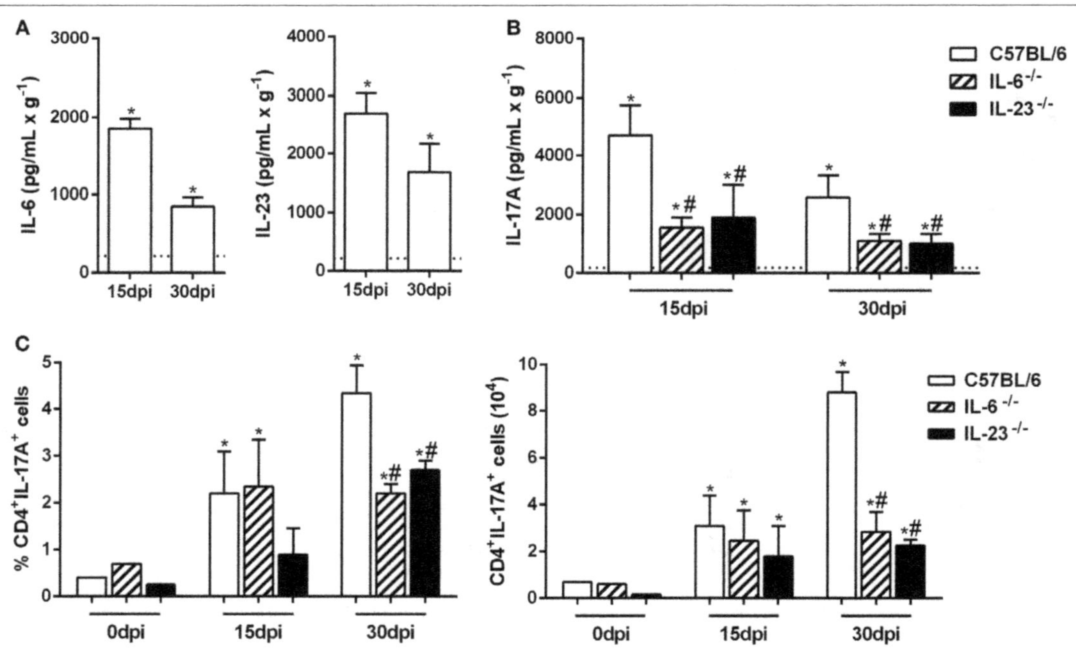

FIGURE 2 | Virulent yeast strain 18 of *Paracoccidioides brasiliensis* (Pb18)-induced IL-6 and IL-23 production improves the establishment of the Th17 profile. **(A–C)** Male C57BL/6, IL-6$^{-/-}$ and IL-23$^{-/-}$ mice were infected with 1×10^6 Pb18 yeast cells (intravenously). **(A)** IL-6 and IL-23 productions were quantified in the lung homogenate from uninfected and Pb18-infected C57BL/6 mice, at 15 and 30 days postinfection (dpi). **(B)** The IL-17A production was measured in the lung tissue from C57BL/6-, IL-6-, and IL-23-deficient mice. **(C)** The frequency and absolute number of Th17 (CD4$^+$IL-17A$^+$) cells were evaluated in the pulmonary tissue at 0, 15, and 30 dpi. The data represent the mean ± SEM of five mice and are representative of three independent experiments. Uninfected mice are represented by the dashed line. $^*p < 0.05$ compared to uninfected mice. $^\#p < 0.05$ compared to Pb18-infected C57BL/6 mice at the same time point.

Th17-Associated Cytokines Contribute to Compact Granuloma Formation during the Experimental *P. brasiliensis* Infection

To verify whether *P. brasiliensis*-induced Th17-associated cytokines are important to control the fungal growth and pulmonary infection, we analyzed the amount of CFU and the histopathology of lung tissue from Pb18-infected IL-6-, IL-23-, and IL-17RA-deficient mice. Our data revealed that all deficient mice showed impaired fungal control at both 15 and 30 days after *P. brasiliensis* inoculation when compared to C57BL/6 mice ($p < 0.05$; **Figure 3A**). Likewise, the Grocott stain at 30 dpi corroborated a higher fungal load in the lungs of all deficient mice when compared to infected C57BL/6 mice (**Figure 3B**), suggesting that Th17-associated cytokines, as IL-6 and IL-23, are important to control the fungal growth during the experimental PCM in mice. To better assess the inflammatory process developed during *P. brasiliensis* infection, we analyzed

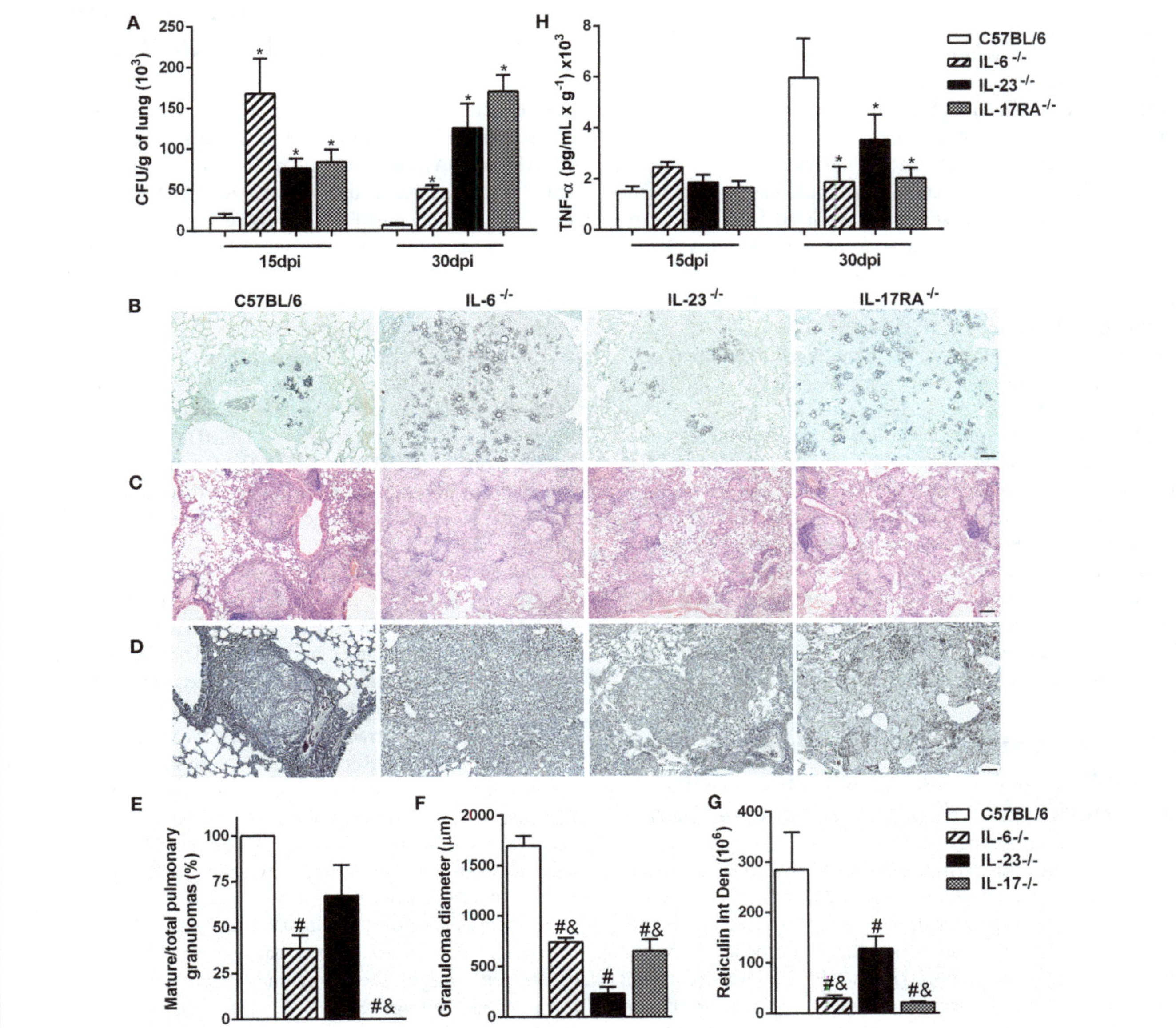

FIGURE 3 | The Th17-associated cytokines contribute to development and organization of granuloma during experimental *Paracoccidioides brasiliensis* infection. **(A–E)** C57BL/6, IL-6$^{-/-}$, IL-23$^{-/-}$, and IL-17 receptor A (IL-17RA)$^{-/-}$ mice were infected with virulent yeast strain 18 of *P. brasiliensis* (Pb18) (1×10^6 cells, intravenously). **(A)** The number of colony-forming units (CFUs) was determined in the pulmonary tissue after 15 and 30 days postinfection (dpi). **(B–D)** Lung sections of infected mice were stained with **(B)** Grocott methenamine silver, **(C)** hematoxylin and eosin, or **(D)** silver impregnation (Gomori method) at 30 dpi and analyzed by light microscopy. Scale bar: 50 µm **(B,D)** and 100 µm **(C)**. **(E)** The percentage of mature granulomas and **(F)** diameter of mature and immature granulomas were calculated. **(G)** Integrated density of reticulin present in the halo surrounding the mature granulomas. **(H)** The tumor necrosis factor (TNF)-α production was measured in the lung homogenate from Pb18-infected mice at 15 and 30 dpi. Similar results were obtained in three independent experiments. Data represent the mean ± SEM of five mice. Uninfected mice are represented by the dashed line. $^*p < 0.05$ compared to uninfected mice. $^{\#}p < 0.05$ compared to Pb18-infected C57BL/6 mice at the same time point. $^{\&}p < 0.05$ compared to Pb18-infected IL-23$^{-/-}$ mice at the same time point.

the histology of pulmonary tissue at 30 dpi. The C57BL/6 mice showed bigger and mature granulomas characterized by epithelioid and giant cells surrounded by a peripheral mononuclear layer (**Figures 3C,E,F**). In contrast, most of the granulomas from IL-$6^{-/-}$ and IL-17RA$^{-/-}$ mice exhibited an intermediate size and immature profile with few lymphocytes forming the peripheral ring, associated with a diffuse inflammatory process (**Figures 3C,E,F**). Specifically, the IL-23 deficiency induced the formation of numerous well-formed granulomas, but with a smaller diameter compared to all Pb18-infected groups (**Figures 3C,E,F**).

Reticulin fibers are composed of type III collagen; besides very delicate and fine, they form a firm layer linking the connective to the surrounding tissue, being considered a main element of alveolar organization (48). We observed a well-organized and dense reticulin ring contouring the granulomas in the lungs from Pb18-infected C57BL/6 mice (**Figures 3D,G**). In addition, the diminished number of compact granulomas in IL-$6^{-/-}$ and IL-17RA$^{-/-}$ pulmonary tissue was associated with a defective reticulin halo accumulation and formation (**Figures 3D,G**). Moreover, a tiny reticulin layer surrounded the small mature and immature granulomas in the lungs from IL-23-deficient mice at 30 days post-Pb18 infection (**Figures 3D,G**).

Since TNF-α is important to the granuloma formation (36, 37), we also evaluated the expression of this cytokine in the lung tissue from Pb18-infected C57BL/6, IL-$6^{-/-}$, IL-$23^{-/-}$, and IL-17RA$^{-/-}$ mice. Our data showed that *P. brasiliensis* infection increased the TNF-α production at both evaluated time points (**Figure 3H**). There were no differences in TNF-α levels in the various non-infected knock-out mice (data not shown). At 15 dpi, all groups showed similar TNF-α production. However, when compared to C57BL/6 mice, the levels of TNF-α were reduced in all deficient mice at 30 dpi ($p < 0.05$; **Figure 3H**). These results suggest that IL-6, IL-17A, and in less extent, IL-23 modulate the formation of organized and compact granulomas during the experimental *P. brasiliensis* infection.

Th17-Associated Cytokines Promote IFN-γ Production and iNOS Expression during the Experimental *P. brasiliensis* Infection

Several studies have demonstrated that the Th1-type cytokine IFN-γ is implicated in resistance during *P. brasiliensis* infection, inducing cell migration and activation (37, 49). Then, we next evaluated the importance of IL-6, IL-17A, and IL-23 in modulating the IFN-γ-production during the experimental PCM. All naive knock-out mice showed similar IFN- γ levels (data not shown), and these levels were increased after the *P. brasiliensis* infection (**Figure 4A**). At both evaluated time points, 15 and 30 dpi, all deficient mice showed a reduction of IFN-γ levels in the pulmonary tissue in comparison to C57BL/6 group (**Figure 4A**). These findings suggest that Th17-associated cytokines are involved in the IFN-γ production during the experimental PCM in mice.

It was previously demonstrated that IFN-γ is able to induce fungicidal activity in macrophages through nitric oxide (NO) production (50, 51), which is indirectly measured by iNOS expression. At 15 dpi, there was increased iNOS expression inside the granulomatous structure from C57BL/6 mice, in comparison with all analyzed deficient mice at the same time point (**Figures 4B,C**).

Together, these results suggest that the deficiency of Th17-associated cytokines (IL-6 and IL-23) modulates the IFN-γ production and iNOS expression and dampens the fungal death during an experimental PCM model.

Adoptive Transfer of F4/80$^+$ Cells Restores Resistance to *P. brasiliensis* Infection in IL-6- and IL-17RA-Deficient Mice

We first showed that ~38% of F4/80$^+$ cells (macrophages) from splenocytes cultured with Pb18 yeast cells are able to produce IL-6 (**Figure 5A**). Gate strategy consisted of doublets exclusion in FSC-H against FSC-A (data not shown). Next, monocytes were gated in SSC-A against FSC-A (**Figure 5A**, left panel), and then the IL-6-positive cells were selected (**Figure 5A**, middle panel). Finally, the IL-6$^+$F4/80$^+$ population was delimited (**Figure 5A**, right panel). We next evaluated the role of IL-6 in Th17 differentiation during Pb18 infection. The BMDMs from C57BL/6 and IL-$6^{-/-}$ mice were co-cultured with naive CD4$^+$ lymphocytes isolated from C57BL/6 WT mice, in the presence of or not of Pb18 yeast cells. We observed that the frequency and absolute number of Th17 cells induced by WT BMDMs were significantly higher than that of IL-$6^{-/-}$ BMDMs ($p < 0.05$; **Figures 5B,C**, respectively). Overall, these data suggest that the IL-6 produced by F4/80$^+$ macrophages induces Th17 cells during the *P. brasiliensis* infection.

To verify the relevance of Th17-associated cytokines over the macrophage fungicidal activity during *P. brasiliensis* infection, BMDMs from C57BL/6 mice were transferred to IL-6-, IL-23-, and IL-17RA-deficient mice, and the fungal burden was evaluated after 30 dpi. To confirm that transferred cells arrive in the lungs, we labeled BMDMs from GFP$^+$ mice with F4/80, and at day 7 after transference, we tracked these cells. We verified a frequency of ~7.3% of F4/80$^+$GFP$^+$ cells in the pulmonary tissue (data not shown). The IL-$6^{-/-}$ and IL-17RA$^{-/-}$ transferred mice showed diminished number of Pb18 yeast cells in comparison with same genotypes without transfer ($p < 0.05$; **Figure 6A**). However, the adoptive transfer of BMDMs did not change the fungal burden in IL-$23^{-/-}$ mice (**Figure 6A**).

Since resistance to *P. brasiliensis* infection is characterized by the presence of mature granulomas, we evaluated the influence of macrophages in the granuloma formation. Corroborating our previous findings (**Figure 3C**), the Pb18-infected IL-6- and IL-17RA-deficient mice showed diminished capacity to mount organized granulomas at 30 dpi when compared to C57BL/6 group (**Figures 6B,C**). Surprisingly, the adoptive transfer of WT BMDMs restored the ability of IL-$6^{-/-}$ and IL-17RA$^{-/-}$ mice to form mature granulomas in the lung tissue (**Figures 6B,C**). We could not observe histological differences in Pb18-infected IL-23-deficient mice transferred or not with C57BL/6 BMDMs, which continued showing small well-formed granulomas in the lungs after transference (**Figures 6B,D**).

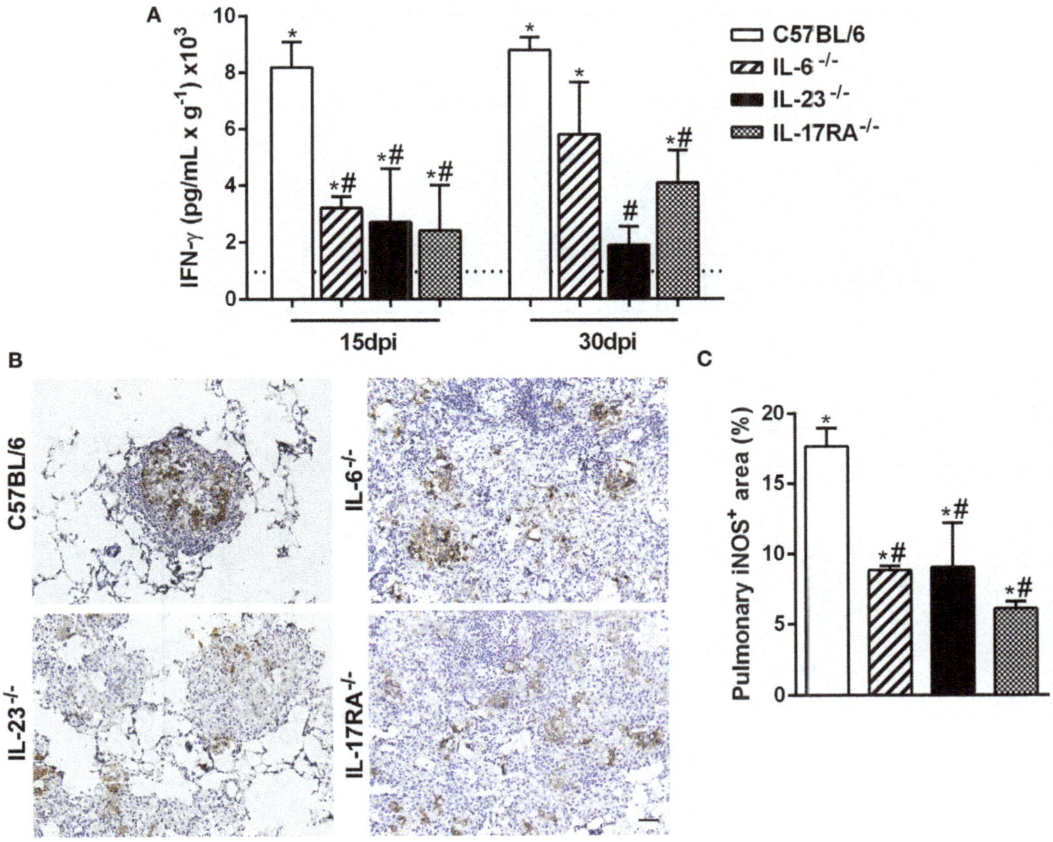

FIGURE 4 | Th17-associated cytokines regulate the IFN-γ production and inducible nitric oxide synthase enzyme (iNOS) expression during the experiment of virulent yeast strain 18 of *Paracoccidioides brasiliensis* (Pb18) infection. **(A,B)** C57BL/6, IL-6$^{-/-}$, IL-23$^{-/-}$, and IL-17 receptor A (IL-17RA)$^{-/-}$ mice were infected (intravenously) with 1×10^6 Pb18 yeast cells. **(A)** IFN-γ levels were measured by enzyme-linked immunosorbent assay in the lung homogenate at 15 and 30 days postinfection (dpi). **(B)** Immunohistochemistry was performed in lung tissue to analyze the iNOS expression at 30 dpi. **(C)** The percentage of the pulmonary area containing iNOS$^+$ cells is indicated. Lung sections were stained for iNOS at 30 dpi using immunohistochemistry. Scale bar: 50 μm. Similar results were obtained in three independent experiments. Data represent the mean ± SEM of five mice. Uninfected mice are represented by the dashed line. $^*p < 0.05$ compared to uninfected mice. $^\#p < 0.05$ compared to Pb18-infected C57BL/6 mice at the same time point.

Taken together, our data suggest that Th17-associated cytokines are involved in the modulation of immune response and granuloma formation during experimental PCM.

DISCUSSION

The generation of a protective immune response is essential for human resistance against infectious diseases, including the systemic mycosis caused by *P. brasiliensis*. If the host innate immune response is unable to suppress the infection foci, it is followed by cell-mediated immunity (52). Several studies have demonstrated that the severe PCM occurs due to the inability to develop an effective Th1 response and, therefore, unsuitable for the formation of dense granulomas, while Th2 response is inefficient to contain the spread of infection (53). Since Th17 discovery, these cells have been associated with immune response to fungal diseases and considered one of the main mechanisms that cause resistance to mycosis (23).

In fact, the production of IL-17A is instrumental for protective response in an experimental model of *P. carinii* infection (26). In addition, Th17-associated cytokines improve the host control against *C. albicans* and *Aspergillus fumigatus*, leading to the migration of potentially fungicide neutrophils (54). Moreover, the presence of IL-17A-producing cells in lesions of PCM patients was demonstrated, in an attempt to increase the host immune defenses against fungal cells (38, 55). In agreement, here, we demonstrated the increased expression of Th17-associated cytokines (IL-6, IL-23, and IL-17A) in the lung tissue after the *P. brasiliensis* infection in mice.

Besides Th17, IL-17A is synthesized by NK cells, γδ T cells, neutrophils, and innate lymphoid cells, in a more rapid way than T cells due to the constitutive expression of the Th17-associated transcription factor RORγt in all these cells subtypes (47). During experimental PCM, our results showed that Th17 lymphocytes are the main source of IL-17A in the lungs, followed by CD8$^+$ cells. The presence of Tc17s in the lungs from *P. brasiliensis*-infected mice or in cutaneous and mucosal lesions from PCM patients was showed before (30, 38). Moreover, Tc17 differentiation of naive T lymphocytes was demonstrated *in vitro* after Pb18 infection (31). During *A. fumigatus* infection, neutrophils were positive for

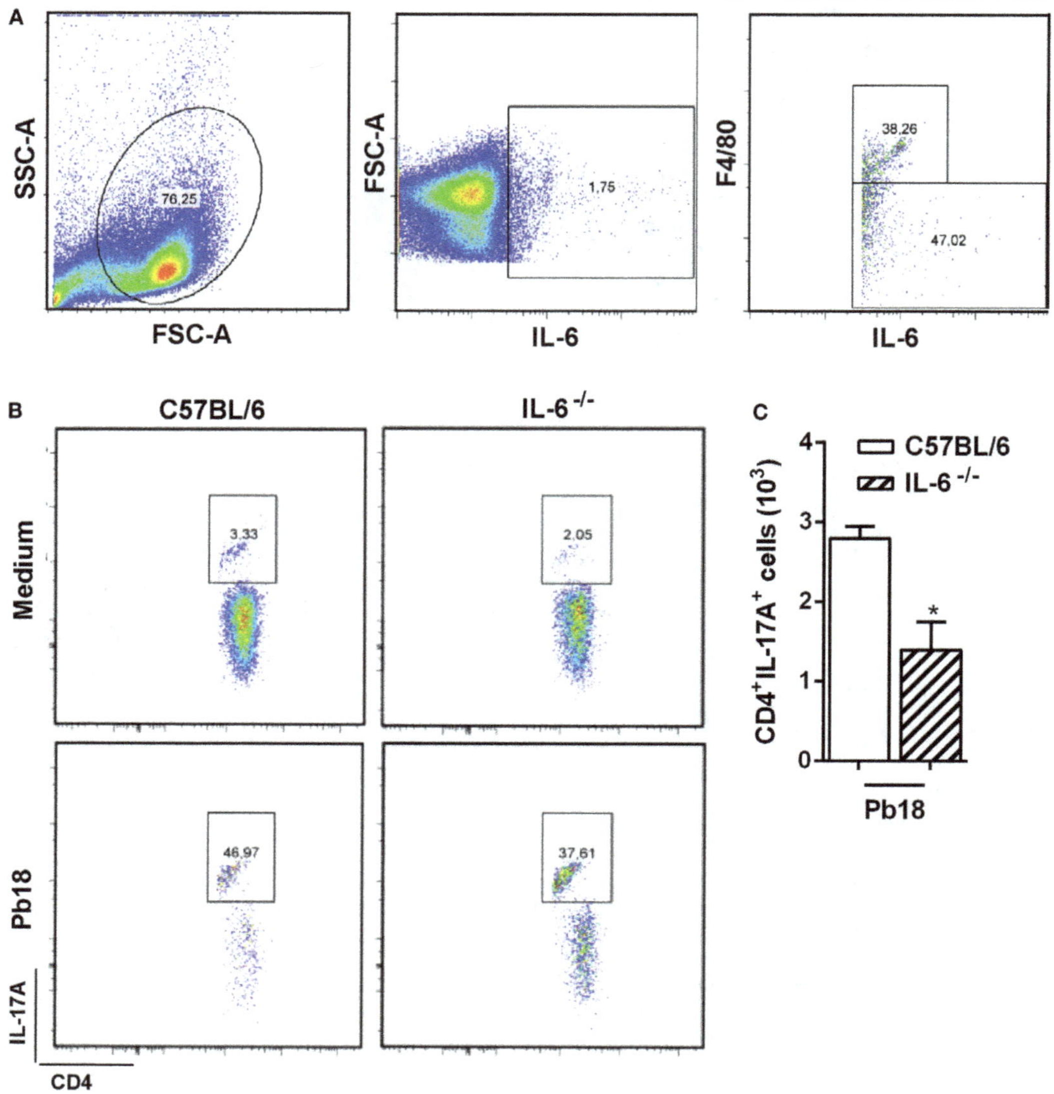

FIGURE 5 | Macrophage-derived IL-6 improves Th17 differentiation during virulent yeast strain 18 of *Paracoccidioides brasiliensis* (Pb18) experimental infection. **(A)** Splenocytes from C57BL/6 mice were isolated and cultivated in the presence of or not of viable Pb18 yeasts cells during 5 days, and the frequency of IL-6-producing F4/80+ macrophages was analyzed by flow cytometry. **(B,C)** Bone marrow-derived macrophages were differentiated from C57BL/6 and IL-6$^{-/-}$ mice and co-cultured with naive T CD4+ cells from C57BL/6 wild-type mice in the presence of or not of Pb18 yeast cells (multiplicity of infection of 0.04) during 5 days. Next, the frequency **(B)** and absolute number **(C)** of Th17 (CD4+IL-17A+) cells were analyzed by flow cytometry. Similar results were obtained in three independent experiments. Data represent the mean ± SEM of five mice. *$p < 0.05$ compared with C57BL/6 mice.

intracellular IL-17A expression and also produced this cytokine in a Dectin-1- and IL-23-dependent manner (56). However, the *P. brasiliensis* infection did not modulate the IL-17-producing neutrophils in the lungs. Moreover, we found that IL-6 is important to the protective response during the infection. Similarly, previous work demonstrated that IL-6 increases murine resistance to *P. brasiliensis* yeast cells (57). Previous work highlighted the critical roles of IL-1 during early Th17 differentiation and expansion/maintenance (11). Our group partially elucidated the mechanisms modulated by IL-1 during PCM (58). We have found that *P. brasiliensis* promoted a NLRP3 inflammasome-dependent caspase-1 activation to release the bioactive IL-1β, and IL-1R1$^{-/-}$ mice displayed a slightly increased survival rate to infection compared with the C57BL/6 mice. Moreover, significantly higher amounts of the fungus were recovered at 30 dpi in the lungs of the IL-1R1-deficient mice, which were contained in organized and compact granulomas (58). These data corroborate the idea that the absence of cytokines associated with Th17 differentiation may increase the susceptibility to experimental PCM.

Granulomas are formed as a consequence of chronic antigen persistence, and their formation involves the interaction between the antigenic organism and host immune cells (48). During primary tuberculosis, IL-17-producing cells are induced, leading to IL-17A synthesis, which is a potent inflammatory cytokine capable of increase chemokines expression that promotes cell recruitment and granuloma organization (59). Similarly, it is

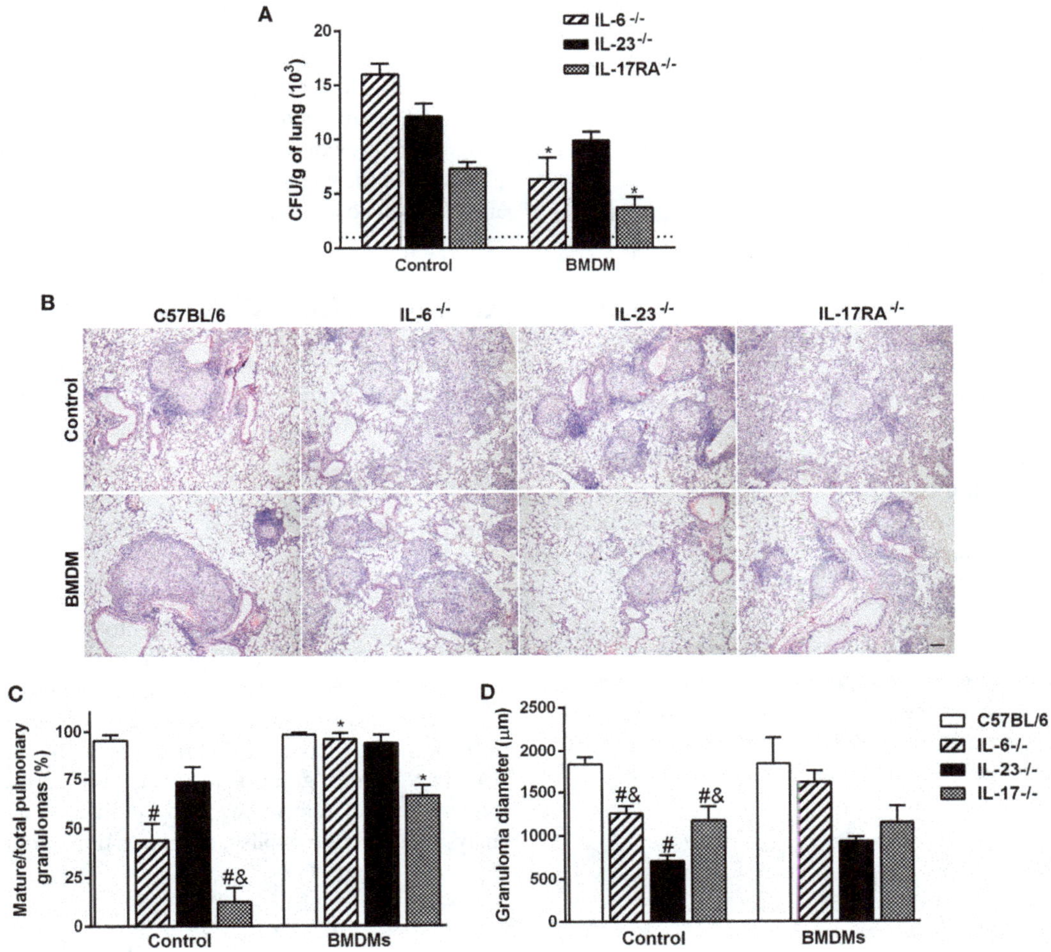

FIGURE 6 | Adoptive transfer of wild-type F4/80$^+$ macrophages restores the granuloma formation and resistance during *Paracoccidioides brasiliensis* infection in susceptible IL-6- and IL-17 receptor A (IL-17RA)-deficient mice. **(A,B)** Macrophages (1 × 10^6), previously differentiated *in vitro* from C57BL/6 bone marrow-derived macrophages (BMDMs), were adoptively transferred to C57BL/6, IL-6$^{-/-}$, IL-23$^{-/-}$, and IL-17RA$^{-/-}$ mice, followed by the intravenous infection with 1 × 10^6 virulent yeast strain 18 of *P. brasiliensis* (Pb18). **(A)** The number of colony-forming unit (CFU) was determined in the pulmonary tissue at day 30 postinfection. Dashed line represents the amount of CFU in C57BL/6 mice. **(B)** Lung sections of Pb18-infected mice, transferred or not with wild-type BMDMs, were stained with hematoxylin and eosin. **(C)** The percentage of mature granulomas and **(D)** diameter of mature and immature granulomas was calculated. Scale bar: 100 µm. Similar results were obtained in three independent experiments. Data represent the mean ± SEM of five mice. $^{\#}p < 0.05$ compared to Pb18-infected C57BL/6 mice without BMDMs transfer (control). $^{\&}p < 0.05$ compared to control IL-23$^{-/-}$. $^{*}p < 0.05$ compared with same genotypes without transfer.

possible that IL-17A is leading to cell migration during experimental PCM, contributing to granuloma formation. Moreover, we showed that the Th17-associated cytokine IL-6 constitutes an essential molecule involved in granuloma development during the *P. brasiliensis* infection. Here, we established for the first time the importance of IL-6 and IL-23 for the Th17 induction/expansion during experimental PCM. According to our results, these cytokines play a critical role in the protection against *P. brasiliensis* yeast cells through the induction of mature granuloma formation and control of fungal growth.

The resistance observed in some PCM patients is dependent on cellular activities mediated by IFN-γ and TNF-α. An efficient host immune response and a potent fungicidal activity against *P. brasiliensis* are determined by the synergistic effect between these two cytokines (37, 49). Several studies have demonstrated that Th17 cells can turn into IFN-γ-expressing T cells (60–62). IFN-γ stimulates *P. brasiliensis*-infected macrophages to secrete TNF-α, required for the development and persistence of well-formed granulomas (35) and is important for host resistance against fungal cells (63). Our data showed a diminished IFN-γ and TNF-α production in the pulmonary tissue from IL-6$^{-/-}$, IL-23$^{-/-}$, and IL-17RA$^{-/-}$ mice when compared to C57BL/6 group. The lower production of IFN-γ and TNF-α correlated with impaired granuloma formation in the lung tissue after *P. brasiliensis* infection. In addition, IFN-γ induces inflammatory cells to produce NO, which plays a well-documented role in fungal clearance (64). Besides IFN-γ, the NO production in chronic inflammation is supported by IL-17A (65, 66). This statement justify the decreased iNOS expression inside the granulomatous structures in the lung tissue from *P. brasiliensis*-infected IL-6-, IL-23-, and IL-17RA-deficient mice, associated with increased fungal load.

It is known that IL-17A increases the macrophage activity and survival (67). However, the importance of macrophages in Th17 cell responses is still poorly understood. An inflammatory stimulus is able to increase the expression of IL-17A receptors in macrophages both *in vitro* and *in vivo* (68). Interestingly, mycobacterial infection induces a new type of macrophage population, different from M1/M2 macrophages, that downregulates T cell production of both Th1- and Th2-type cytokines but markedly increases the production of IL-17A and IL-22 through upregulation of Th17 cell expansion, *via* IL-6 and TGF-β, but not IL-21 and IL-23 (69). Similarly, we found that *P. brasiliensis* experimental infection induced IL-6 synthesis even in the early infection, and this cytokine is essential to induce a Th17 immune profile. In corroboration, we showed that IL-6 deficiency in macrophages is associated with abrogated Th17 cell generation, diminished fungal load in the lungs, and absence of organized granulomas. The adoptive transfer of IL-6-competent macrophages restored the resistance in *P. brasiliensis*-infected IL-6- and IL-17RA-deficient mice, but did not change the granuloma formation and fungal clearance in IL-23$^{-/-}$ group.

Taken together, our data suggest that macrophage-produced IL-6 and IL-23 are essential to induce a Th17 immune profile, which positively regulates the TNF-α, IFN-γ, and iNOS expression, contributing to mature granuloma formation and consequent better control of *P. brasiliensis* experimental infection.

AUTHOR CONTRIBUTIONS

FT, FR, DC, NK-C, CS, and CM performed the *in vitro* and *in vivo* experiments. FT, FR, DC, NK-C, and JS designed protocols. FT, FR, and DC conducted data analysis. FT, FR, DC, NK-C, CS, CM, and JS provided scientific input. JS supervised the project. FT reviewed the literature and wrote the manuscript. DC, NK-C, and JS provided comments and corrections of the manuscript.

ACKNOWLEDGMENTS

We are grateful to doctor Marcos Antônio Rossi (*in memoriam*) for precious contribution and to Wander C. R. da Silva and Maria Elena Riul for invaluable technical assistance.

FUNDING

The research leading to these results received funding from Conselho Nacional de Desenvolvimento Científico e Tecnológico (CNPq), Coordenação de Aperfeiçoamento Pessoal de Nível Superior (CAPES), São Paulo Research Foundation (FAPESP) under grant agreement no. 2013/08216-2 (Center for Research in Inflammatory Disease), and from the University of São Paulo NAP-DIN under grant agreement no. 11.1.21625.01.0. The funders had no role in the study design, data collection and analysis, decision to publish, or preparation of this manuscript.

REFERENCES

1. de Oliveira HC, Michaloski JS, da Silva JF, Scorzoni L, de Paula ESAC, Marcos CM, et al. Peptides derived from a phage display library inhibit adhesion and protect the host against infection by *Paracoccidioides brasiliensis* and *Paracoccidioides lutzii*. *Front Pharmacol* (2016) 7:509. doi:10.3389/fphar.2016.00509
2. Coutinho ZF, Silva D, Lazera M, Petri V, Oliveira RM, Sabroza PC, et al. Paracoccidioidomycosis mortality in Brazil (1980–1995). *Cad Saude Publica* (2002) 18(5):1441–54. doi:10.1590/S0102-311X2002000500037
3. Marques SA. Paracoccidioidomycosis. *Clin Dermatol* (2012) 30(6):610–5. doi:10.1016/j.clindermatol.2012.01.006
4. McEwen JG, Bedoya V, Patino MM, Salazar ME, Restrepo A. Experimental murine paracoccidiodomycosis induced by the inhalation of conidia. *J Med Vet Mycol* (1987) 25(3):165–75. doi:10.1080/02681218780000231
5. Calich VL, Vaz CA, Burger E. Immunity to *Paracoccidioides brasiliensis* infection. *Res Immunol* (1998) 149(4–5):407–417; discussion 499–500. doi:10.1016/S0923-2494(98)80764-5
6. Calich VL, da Costa TA, Felonato M, Arruda C, Bernardino S, Loures FV, et al. Innate immunity to *Paracoccidioides brasiliensis* infection. *Mycopathologia* (2008) 165(4–5):223–36. doi:10.1007/s11046-007-9048-1
7. Mamoni RL, Nouer SA, Oliveira SJ, Musatti CC, Rossi CL, Camargo ZP, et al. Enhanced production of specific IgG4, IgE, IgA and TGF-beta in sera from patients with the juvenile form of paracoccidioidomycosis. *Med Mycol* (2002) 40(2):153–9. doi:10.1080/mmy.40.2.153.159
8. Liang SC, Tan XY, Luxenberg DP, Karim R, Dunussi-Joannopoulos K, Collins M, et al. Interleukin (IL)-22 and IL-17 are coexpressed by Th17 cells and cooperatively enhance expression of antimicrobial peptides. *J Exp Med* (2006) 203(10):2271–9. doi:10.1084/jem.20061308
9. Zambrano-Zaragoza JF, Romo-Martinez EJ, Duran-Avelar Mde J, Garcia-Magallanes N, Vibanco-Perez N. Th17 cells in autoimmune and infectious diseases. *Int J Inflam* (2014) 2014:651503. doi:10.1155/2014/651503
10. Chen Z, O'Shea JJ. Th17 cells: a new fate for differentiating helper T cells. *Immunol Res* (2008) 41(2):87–102. doi:10.1007/s12026-007-8014-9
11. Chung Y, Chang SH, Martinez GJ, Yang XO, Nurieva R, Kang HS, et al. Critical regulation of early Th17 cell differentiation by interleukin-1 signaling. *Immunity* (2009) 30(4):576–87. doi:10.1016/j.immuni.2009.02.007
12. Guo L, Wei G, Zhu J, Liao W, Leonard WJ, Zhao K, et al. IL-1 family members and STAT activators induce cytokine production by Th2, Th17, and Th1 cells. *Proc Natl Acad Sci U S A* (2009) 106(32):13463–8. doi:10.1073/pnas.0906988106
13. Bettelli E, Korn T, Kuchroo VK. Th17: the third member of the effector T cell trilogy. *Curr Opin Immunol* (2007) 19(6):652–7. doi:10.1016/j.coi.2007.07.020
14. Stockinger B, Veldhoen M. Differentiation and function of Th17 T cells. *Curr Opin Immunol* (2007) 19(3):281–6. doi:10.1016/j.coi.2007.04.005
15. Wilson NJ, Boniface K, Chan JR, McKenzie BS, Blumenschein WM, Mattson JD, et al. Development, cytokine profile and function of human interleukin 17-producing helper T cells. *Nat Immunol* (2007) 8(9):950–7. doi:10.1038/ni1497
16. Langrish CL, Chen Y, Blumenschein WM, Mattson J, Basham B, Sedgwick JD, et al. IL-23 drives a pathogenic T cell population that induces autoimmune inflammation. *J Exp Med* (2005) 201(2):233–40. doi:10.1084/jem.20041257
17. Graeber KE, Olsen NJ. Th17 cell cytokine secretion profile in host defense and autoimmunity. *Inflamm Res* (2012) 61(2):87–96. doi:10.1007/s00011-011-0419-1
18. Chan YR, Liu JS, Pociask DA, Zheng M, Mietzner TA, Berger T, et al. Lipocalin 2 is required for pulmonary host defense against *Klebsiella* infection. *J Immunol* (2009) 182(8):4947–56. doi:10.4049/jimmunol.0803282

19. Ferreira MC, Whibley N, Mamo AJ, Siebenlist U, Chan YR, Gaffen SL. Interleukin-17-induced protein lipocalin 2 is dispensable for immunity to oral candidiasis. *Infect Immun* (2014) 82(3):1030–5. doi:10.1128/IAI.01389-13
20. Laan M, Cui ZH, Hoshino H, Lotvall J, Sjostrand M, Gruenert DC, et al. Neutrophil recruitment by human IL-17 via C-X-C chemokine release in the airways. *J Immunol* (1999) 162(4):2347–52.
21. Hoshino H, Laan M, Sjostrand M, Lotvall J, Skoogh BE, Linden A. Increased elastase and myeloperoxidase activity associated with neutrophil recruitment by IL-17 in airways in vivo. *J Allergy Clin Immunol* (2000) 105(1 Pt 1):143–9. doi:10.1016/S0091-6749(00)90189-1
22. Ye P, Rodriguez FH, Kanaly S, Stocking KL, Schurr J, Schwarzenberger P, et al. Requirement of interleukin 17 receptor signaling for lung CXC chemokine and granulocyte colony-stimulating factor expression, neutrophil recruitment, and host defense. *J Exp Med* (2001) 194(4):519–27. doi:10.1084/jem.194.4.519
23. Romani L. Immunity to fungal infections. *Nat Rev Immunol* (2011) 11(4): 275–88. doi:10.1038/nri2939
24. Hernandez-Santos N, Gaffen SL. Th17 cells in immunity to *Candida albicans*. *Cell Host Microbe* (2012) 11(5):425–35. doi:10.1016/j.chom.2012.04.008
25. Perez-Nazario N, Rangel-Moreno J, O'Reilly MA, Pasparakis M, Gigliotti F, Wright TW. Selective ablation of lung epithelial IKK2 impairs pulmonary Th17 responses and delays the clearance of *Pneumocystis*. *J Immunol* (2013) 191(9):4720–30. doi:10.4049/jimmunol.1301679
26. Rudner XL, Happel KI, Young EA, Shellito JE. Interleukin-23 (IL-23)-IL-17 cytokine axis in murine *Pneumocystis carinii* infection. *Infect Immun* (2007) 75(6):3055–61. doi:10.1128/IAI.01329-06
27. Netea MG, Brown GD, Kullberg BJ, Gow NA. An integrated model of the recognition of *Candida albicans* by the innate immune system. *Nat Rev Microbiol* (2008) 6(1):67–78. doi:10.1038/nrmicro1815
28. Conti HR, Gaffen SL. Host responses to *Candida albicans*: Th17 cells and mucosal candidiasis. *Microbes Infect* (2010) 12(7):518–27. doi:10.1016/j.micinf.2010.03.013
29. Pandiyan P, Conti HR, Zheng L, Peterson AC, Mathern DR, Hernandez-Santos N, et al. CD4(+)CD25(+)Foxp3(+) regulatory T cells promote Th17 cells in vitro and enhance host resistance in mouse *Candida albicans* Th17 cell infection model. *Immunity* (2011) 34(3):422–34. doi:10.1016/j.immuni.2011.03.002
30. Loures FV, Araujo EF, Feriotti C, Bazan SB, Costa TA, Brown GD, et al. Dectin-1 induces M1 macrophages and prominent expansion of CD8+IL-17+ cells in pulmonary paracoccidioidomycosis. *J Infect Dis* (2014) 210(5):762–73. doi:10.1093/infdis/jiu136
31. Loures FV, Araujo EF, Feriotti C, Bazan SB, Calich VL. TLR-4 cooperates with Dectin-1 and mannose receptor to expand Th17 and Tc17 cells induced by *Paracoccidioides brasiliensis* stimulated dendritic cells. *Front Microbiol* (2015) 6:261. doi:10.3389/fmicb.2015.00261
32. Nakae S, Saijo S, Horai R, Sudo K, Mori S, Iwakura Y. IL-17 production from activated T cells is required for the spontaneous development of destructive arthritis in mice deficient in IL-1 receptor antagonist. *Proc Natl Acad Sci U S A* (2003) 100(10):5986–90. doi:10.1073/pnas.1035999100
33. Koenders MI, Lubberts E, Oppers-Walgreen B, van den Bersselaar L, Helsen MM, Kolls JK, et al. Induction of cartilage damage by overexpression of T cell interleukin-17A in experimental arthritis in mice deficient in interleukin-1. *Arthritis Rheum* (2005) 52(3):975–83. doi:10.1002/art.20885
34. Lu YJ, Gross J, Bogaert D, Finn A, Bagrade L, Zhang Q, et al. Interleukin-17A mediates acquired immunity to pneumococcal colonization. *PLoS Pathog* (2008) 4(9):e1000159. doi:10.1371/journal.ppat.1000159
35. Okamoto Yoshida Y, Umemura M, Yahagi A, O'Brien RL, Ikuta K, Kishihara K, et al. Essential role of IL-17A in the formation of a mycobacterial infection-induced granuloma in the lung. *J Immunol* (2010) 184(8):4414–22. doi:10.4049/jimmunol.0903332
36. Egen JG, Rothfuchs AG, Feng CG, Winter N, Sher A, Germain RN. Macrophage and T cell dynamics during the development and disintegration of mycobacterial granulomas. *Immunity* (2008) 28(2):271–84. doi:10.1016/j.immuni.2007.12.010
37. Souto JT, Figueiredo F, Furlanetto A, Pfeffer K, Rossi MA, Silva JS. Interferon-gamma and tumor necrosis factor-alpha determine resistance to *Paracoccidioides brasiliensis* infection in mice. *Am J Pathol* (2000) 156(5):1811–20. doi:10.1016/S0002-9440(10)65053-5
38. Pagliari C, Fernandes ER, Stegun FW, da Silva WL, Seixas Duarte MI, Sotto MN. Paracoccidioidomycosis: cells expressing IL17 and Foxp3 in cutaneous and mucosal lesions. *Microb Pathog* (2011) 50(5):263–7. doi:10.1016/j.micpath.2010.12.008
39. Calich VL, Purchio A, Paula CR. A new fluorescent viability test for fungi cells. *Mycopathologia* (1979) 66(3):175–7. doi:10.1007/BF00683967
40. Moreira AP, Cavassani KA, Massafera Tristao FS, Campanelli AP, Martinez R, Rossi MA, et al. CCR5-dependent regulatory T cell migration mediates fungal survival and severe immunosuppression. *J Immunol* (2008) 180(5):3049–56. doi:10.4049/jimmunol.180.5.3049
41. Tristao FS, Rocha FA, Moreira AP, Cunha FQ, Rossi MA, Santana Silva J. 5-lipoxygenase activity increases susceptibility to *Paracoccidioides brasiliensis* experimental infection. *Infect Immun* (2013) 81(4):1256–66. doi:10.1128/IAI.01209-12
42. Tristao FS, Rocha FA, Dias FC, Rossi MA, Silva JS. The left lung is preferentially targeted during experimental paracoccidioidomycosis in C57BL/6 mice. *Braz J Med Biol Res* (2013) 46(10):839–43. doi:10.1590/1414-431X20133140
43. Montenegro MR, Franco M. Pathology. In: Franco ML, da Silva Lacaz C, Restrepo-Moreno A, Del Negro G, editors. *Paracoccidioidomycosis*. Boca Raton: CRC Press (1994). p. 131–50.
44. Mariano M. The experimental granuloma. A hypothesis to explain the persistence of the lesion. *Rev Inst Med Trop Sao Paulo* (1995) 37(2):161–76. doi:10.1590/S0036-46651995000200012
45. Lucero HA, Patterson S, Matsuura S, Ravid K. Quantitative histological image analyses of reticulin fibers in a myelofibrotic mouse. *J Biol Methods* (2016) 3(4):e60. doi:10.14440/jbm.2016.152
46. Marim FM, Silveira TN, Lima DS Jr, Zamboni DS. A method for generation of bone marrow-derived macrophages from cryopreserved mouse bone marrow cells. *PLoS One* (2010) 5(12):e15263. doi:10.1371/journal.pone.0015263
47. Cua DJ, Tato CM. Innate IL-17-producing cells: the sentinels of the immune system. *Nat Rev Immunol* (2010) 10(7):479–89. doi:10.1038/nri2800
48. Regev D, Surolia R, Karki S, Zolak J, Montes-Worboys A, Oliva O, et al. Heme oxygenase-1 promotes granuloma development and protects against dissemination of mycobacteria. *Lab Invest* (2012) 92(11):1541–52. doi:10.1038/labinvest.2012.125
49. Alegre AC, Oliveira AF, Dos Reis Almeida FB, Roque-Barreira MC, Hanna ES. Recombinant paracoccin reproduces the biological properties of the native protein and induces protective Th1 immunity against *Paracoccidioides brasiliensis* infection. *PLoS Negl Trop Dis* (2014) 8(4):e2788. doi:10.1371/journal.pntd.0002788
50. Gonzalez A, de Gregori W, Velez D, Restrepo A, Cano LE. Nitric oxide participation in the fungicidal mechanism of gamma interferon-activated murine macrophages against *Paracoccidioides brasiliensis* conidia. *Infect Immun* (2000) 68(5):2546–52. doi:10.1128/IAI.68.5.2546-2552.2000
51. Nishikaku AS, Molina RF, Ribeiro LC, Scavone R, Albe BP, Cunha CS, et al. Nitric oxide participation in granulomatous response induced by *Paracoccidioides brasiliensis* infection in mice. *Med Microbiol Immunol* (2009) 198(2):123–35. doi:10.1007/s00430-009-0113-x
52. Fortes MR, Miot HA, Kurokawa CS, Marques ME, Marques SA. Immunology of paracoccidioidomycosis. *An Bras Dermatol* (2011) 86(3):516–24. doi:10.1590/S0365-05962011000300014
53. Pagliari C, Sotto MN. Dendritic cells and pattern of cytokines in paracoccidioidomycosis skin lesions. *Am J Dermatopathol* (2003) 25(2):107–12. doi:10.1097/00000372-200304000-00003
54. Zelante T, De Luca A, Bonifazi P, Montagnoli C, Bozza S, Moretti S, et al. IL-23 and the Th17 pathway promote inflammation and impair antifungal immune resistance. *Eur J Immunol* (2007) 37(10):2695–706. doi:10.1002/eji.200737409
55. de Castro LF, Ferreira MC, da Silva RM, Blotta MH, Longhi LN, Mamoni RL. Characterization of the immune response in human paracoccidioidomycosis. *J Infect* (2013) 67(5):470–85. doi:10.1016/j.jinf.2013.07.019
56. Werner JL, Gessner MA, Lilly LM, Nelson MP, Metz AE, Horn D, et al. Neutrophils produce interleukin 17A (IL-17A) in a dectin-1- and IL-23-dependent manner during invasive fungal infection. *Infect Immun* (2011) 79(10):3966–77. doi:10.1128/IAI.05493-11
57. Pina A, de Araujo EF, Felonato M, Loures FV, Feriotti C, Bernardino S, et al. Myeloid dendritic cells (DCs) of mice susceptible to paracoccidioidomycosis

suppress T cell responses whereas myeloid and plasmacytoid DCs from resistant mice induce effector and regulatory T cells. *Infect Immun* (2013) 81(4):1064–77. doi:10.1128/IAI.00736-12

58. Ketelut-Carneiro N, Silva GK, Rocha FA, Milanezi CM, Cavalcanti-Neto FF, Zamboni DS, et al. IL-18 triggered by the Nlrp3 inflammasome induces host innate resistance in a pulmonary model of fungal infection. *J Immunol* (2015) 194(9):4507–17. doi:10.4049/jimmunol.1402321
59. Torrado E, Cooper AM. IL-17 and Th17 cells in tuberculosis. *Cytokine Growth Factor Rev* (2010) 21(6):455–62. doi:10.1016/j.cytogfr.2010.10.004
60. Bending D, De la Pena H, Veldhoen M, Phillips JM, Uyttenhove C, Stockinger B, et al. Highly purified Th17 cells from BDC2.5NOD mice convert into Th1-like cells in NOD/SCID recipient mice. *J Clin Invest* (2009) 119(3):565–72. doi:10.1172/JCI37865
61. Lee YK, Turner H, Maynard CL, Oliver JR, Chen D, Elson CO, et al. Late developmental plasticity in the T helper 17 lineage. *Immunity* (2009) 30(1):92–107. doi:10.1016/j.immuni.2008.11.005
62. Shimizu J, Takai K, Fujiwara N, Arimitsu N, Ueda Y, Wakisaka S, et al. Excessive CD4+ T cells co-expressing interleukin-17 and interferon-gamma in patients with Behcet's disease. *Clin Exp Immunol* (2012) 168(1):68–74. doi:10.1111/j.1365-2249.2011.04543.x
63. Figueiredo F, Alves LM, Silva CL. Tumour necrosis factor production in vivo and in vitro in response to *Paracoccidioides brasiliensis* and the cell wall fractions thereof. *Clin Exp Immunol* (1993) 93(2):189–94. doi:10.1111/j.1365-2249.1993.tb07964.x
64. Nascimento FR, Calich VL, Rodriguez D, Russo M. Dual role for nitric oxide in paracoccidioidomycosis: essential for resistance, but overproduction associated with susceptibility. *J Immunol* (2002) 168(9):4593–600. doi:10.4049/jimmunol.168.9.4593
65. Mazzoni A, Bronte V, Visintin A, Spitzer JH, Apolloni E, Serafini P, et al. Myeloid suppressor lines inhibit T cell responses by an NO-dependent mechanism. *J Immunol* (2002) 168(2):689–95. doi:10.4049/jimmunol.168.2.689
66. Miljkovic D, Trajkovic V. Inducible nitric oxide synthase activation by interleukin-17. *Cytokine Growth Factor Rev* (2004) 15(1):21–32. doi:10.1016/j.cytogfr.2003.10.003
67. Sergejeva S, Ivanov S, Lotvall J, Linden A. Interleukin-17 as a recruitment and survival factor for airway macrophages in allergic airway inflammation. *Am J Respir Cell Mol Biol* (2005) 33(3):248–53. doi:10.1165/rcmb.2004-0213OC
68. Barin JG, Baldeviano GC, Talor MV, Wu L, Ong S, Quader F, et al. Macrophages participate in IL-17-mediated inflammation. *Eur J Immunol* (2012) 42(3):726–36. doi:10.1002/eji.201141737
69. Tatano Y, Shimizu T, Tomioka H. Unique macrophages different from M1/M2 macrophages inhibit T cell mitogenesis while upregulating Th17 polarization. *Sci Rep* (2014) 4:4146. doi:10.1038/srep04146

5

The Interface between Fungal Biofilms and Innate Immunity

John F. Kernien[1], Brendan D. Snarr[2], Donald C. Sheppard[2,3] and Jeniel E. Nett[1,4]*

[1] Department of Medicine, University of Wisconsin, Madison, WI, United States, [2] Department of Microbiology and Immunology, McGill University, Montreal, QC, Canada, [3] Department of Medicine, McGill University, Montreal, QC, Canada, [4] Department of Medical Microbiology and Immunology, University of Wisconsin, Madison, WI, United States

***Correspondence:**
Jeniel E. Nett
jenett@medicince.wisc.edu

Fungal biofilms are communities of adherent cells surrounded by an extracellular matrix. These biofilms are commonly found during infection caused by a variety of fungal pathogens. Clinically, biofilm infections can be extremely difficult to eradicate due to their resistance to antifungals and host defenses. Biofilm formation can protect fungal pathogens from many aspects of the innate immune system, including killing by neutrophils and monocytes. Altered immune recognition during this phase of growth is also evident by changes in the cytokine profiles of monocytes and macrophages exposed to biofilm. In this manuscript, we review the host response to fungal biofilms, focusing on how these structures are recognized by the innate immune system. Biofilms formed by *Candida*, *Aspergillus*, and *Cryptococcus* have received the most attention and are highlighted. We describe common themes involved in the resilience of fungal biofilms to host immunity and give examples of biofilm defenses that are pathogen-specific.

Keywords: biofilm, matrix, fungi, neutrophil extracellular trap, innate immunity, neutrophil, *Aspergillus*, *Candida*

INTRODUCTION

Fungi frequently flourish as biofilms, which are aggregated communities encased in a protective extracellular matrix (1, 2). Many clinically relevant fungi have been shown to form biofilms, such as *Candida* spp., *Aspergillus* spp., *Cryptococcus neoformans*, *Fusarium* spp., *Blastoschizomyces capitatus*, *Malassezia pachydermatis*, *Pneumocystis* spp., *Trichosporon asahii*, *Rhizopus* spp., and *Rhizomucor* spp (3–13). In the clinical setting, fungal biofilms can propagate on artificial medical devices, such as catheters, as well as on epithelial and endothelial surfaces (3, 14–19) (**Figure 1A**). In addition, during invasive infection, fungal pathogens can proliferate as non-surface associated microcolonies embedded in extracellular matrix (18) (**Figure 1B**). Biofilms resist antifungal therapies and host defenses, making them notoriously difficult to eradicate (4, 20–36). One defining trait of biofilm formation is the production of a microbial-produced extracellular matrix, or the "glue" necessary for adhesion, which also serves as a shield that creates protected reservoirs of infection (4, 5, 20, 33, 37). As investigations reveal the complex composition of the extracellular matrix for several fungal pathogens, it has become increasing clear that this material is distinct from the cell wall (4, 5, 18, 38–41). Therefore, biofilm growth provides a means to present unique moieties and conceal cell wall ligands. Studies are just beginning to shed light on how biofilm formation and matrix production influence host recognition (5, 15, 27, 29–32, 42–50). In this review, we examine the innate immune response to fungal biofilms, highlighting how production of the extracellular matrix alters immunity. While a variety of fungal pathogens have been shown to produce biofilms, investigations examining host interactions have primarily utilized model pathogens *Candida albicans*, *Aspergillus fumigatus*, and *C. neoformans*, which will be the focus of our discussion.

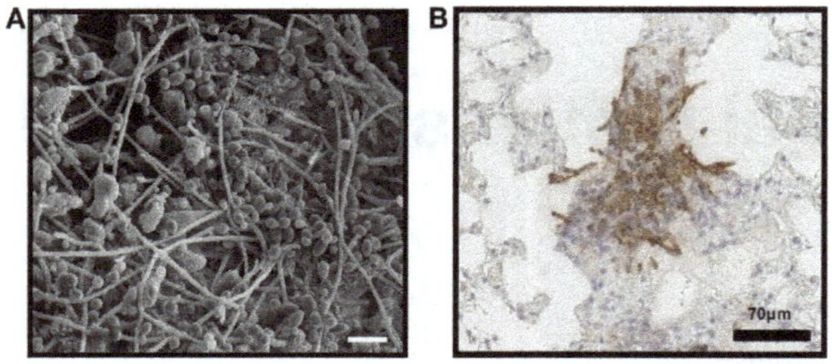

FIGURE 1 | *In vivo* fungal biofilms. **(A)** *Candida albicans* biofilm growing on the luminal surface of a rat venous catheter for 24 h. Scanning electron microscopy reveals adherent organisms growing within in an extracellular matrix. **(B)** Immunohistochemistry of pulmonary tissue from an immunocompromised mouse infected with *Aspergillus fumigatus* and stained with an anti-galactosaminogalactan antibody. Brown indicates accumulation of galactosaminogalactan-containing biofilm matrix surrounding hyphae growing within pulmonary tissues.

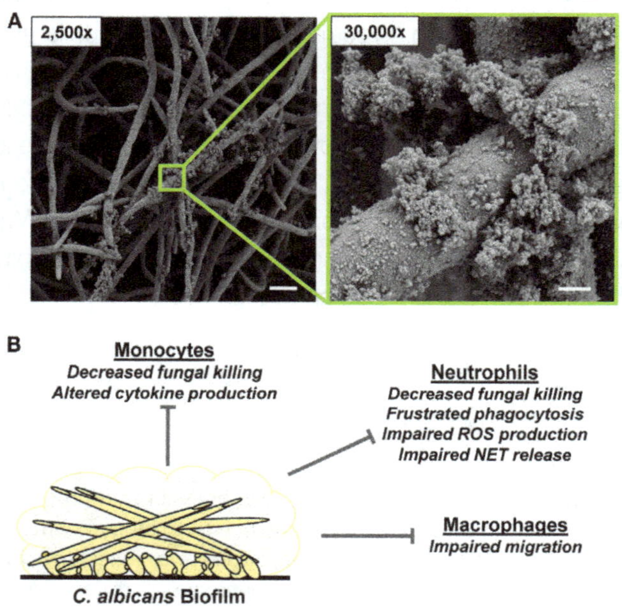

FIGURE 2 | *Candida albicans* biofilm formation and innate immune response. **(A)** Scanning electron microscopy images reveal *C. albicans* biofilms grown on coverslips. Biofilms were grown for 48 h. Measurement bars represent 10 and 1 μm for 2,500x and 30,000x, respectively. **(B)** Summary of innate immune responses impaired by *C. albicans* biofilms.

Candida BIOFILMS

Biofilm Formation

Candida spp., commensal fungi of the gastrointestinal tract, can cause severe, disseminated disease with high mortality, particularly in patients with implanted medical devices or compromised immune systems (17, 51–57). The vast majority of these infections involve biofilm formation on either an artificial or a biotic surface (13, 58, 59) (**Figure 2**). Clinical *Candida* biofilms grow on diverse medical devices, including central venous catheters, urinary catheters, prosthetic valves, left ventricular assist devices, and oral devices, such as dentures (3, 60). Both vaginal and oral mucosal surfaces promote biofilm formation as well (16, 61). The majority of the *in vitro* and *in vivo* biofilm studies have utilized *C. albicans*, the most widespread species. However, non-*albicans* species, including *Candida tropicalis*, *Candida parapsilosis*, and *Candida glabrata*, similarly produce clinically relevant biofilms (3, 62–68). In addition, this virulence trait has been described for the emerging pathogen *Candida auris* (69). *Candida* biofilm formation involves the adherence of yeast to a substrate, the proliferation of cells to form a fungal community, and the production of an extracellular matrix (37, 70–72). *C. albicans* biofilm development often involves the production of hyphae, although the degree of filamentation varies among strains and niches. Non-*albicans* strains lacking the ability to filament produce biofilms composed of layers of yeast embedded in an extracellular matrix (73).

Matrix Composition

Upon encounter with biofilm, immune cells are first confronted with the extracellular matrix covering the fungal cells. A combination of both *in vitro* and *in vivo* models has been integral for the dissection of *Candida* biofilm matrix assembly and composition (20, 22, 39, 40, 58, 67, 70, 74–79). For *C. albicans*, *in vitro* studies have revealed that the mature biofilm matrix consists of a variety of macromolecules, including protein (55%), carbohydrate (25%), lipid (15%), and DNA (5%) (23, 38, 40, 80). Interestingly, many of the matrix components vastly differ from the cell wall components that would be initially recognized by immune cells in the absence of biofilm (40). For example, the main polysaccharide of *C. albicans* biofilm matrix is α-1,2-branched α-1,6 mannan, which is found in high molecular weight structures of approximately 12,000 residues. In contrast, mannans of the outer cell wall layer are 5- to 10-fold smaller. Furthermore, the matrix mannans associate with β-1,6 glucans, forming a mannan–glucan complex that assembles extracellularly, and this structure has not been identified in the cell wall of *C. albicans* (40, 75). In addition, matrix β-1,6 glucan exists in a linear conformation, while the β-1,6 glucan of the cell wall is highly branched (81). Proteomic analysis of *in vitro* biofilms shows some similarities between matrix-associated proteins and those released into the media during planktonic growth (79). However, the extracellular matrix lacks many of the proteins associated with the cell wall (40, 82). *In vivo*, host proteins contribute to the construction of biofilm, with an astonishing majority (>95%) of the matrix-associated proteins of host origin (76). This finding demonstrates variation in the content of fungal biofilm matrix *in vivo* and highlights the importance of including animal models for investigation of *Candida* biofilms.

Neutrophil-*Candida* Biofilm Interactions

Neutrophils are primary responders to *C. albicans* infection, with neutropenic patients prone to severe, lethal candidiasis (83–86). Neutrophils respond to chemokines and other signals during recruitment to the site of infection. Upon pathogen encounter, neutrophils elicit various responses important for control of infection, including phagocytosis, degranulation, reactive oxygen species (ROS) production, and neutrophil extracellular trap (NET) release (87). In the context of *C. albicans* biofilms, neutrophils are the primary leukocyte recruited to the site of infection (15, 76, 88–90). In fact, neutrophil recruitment has been observed in animal models mimicking diverse clinical biofilms, including catheter-related infections (vascular and urinary), denture stomatitis, oral candidiasis, and vaginal candidiasis (15, 76, 91). Despite their presence at the site of infection, neutrophils fail to eradicate *Candida* biofilms. Pioneering studies with human neutrophils revealed that *C. albicans* biofilms resist neutrophil attack, in comparison to their planktonic counterparts (29, 30). Intact biofilm structure is required for this resistance, as resuspension of the biofilm cells reverses the phenotype (29). Furthermore, biofilm impairment of neutrophils is robust, persisting despite neutrophil priming by IFN-γ or G-CSF (92).

The ability of *Candida* biofilms to withstand immune attack appears to vary by strain and species. For *C. albicans*, biofilms are approximately twofold to fivefold more resistant to killing when compared to planktonic cells (29–31, 92). Investigation of *C. parapsilosis* found a similar trend, but did not identify a significant difference in susceptibility to neutrophils between biofilms and planktonic organisms (43, 93). The lack of statistical significance was attributed to the heterogeneity of biofilm formation, resulting in high assay variability. A recent investigation found *C. glabrata* biofilms also resist neutrophil killing, exhibiting a threefold higher resistance for biofilm over planktonic organisms (45).

For killing and containment of *C. albicans*, neutrophils release NETs, which are structures of DNA, histones, and antimicrobial proteins (94, 95). These structures are particularly well-suited to combat large organisms, such as hyphae, which are unable to be ingested by phagocytosis (95). As *C. albicans* biofilms consist of aggregated cells and hyphal elements, NETosis would seemingly be an efficient method of attack. However, a recent study has revealed that neutrophils fail to release NETs in the presence of *C. albicans* biofilms (31). This phenomenon is conserved across a variety of *C. albicans* strains exhibiting differing degrees of filamentation and biofilm architecture (44). This inhibitory pathway appears to be closely linked to the production of an extracellular matrix, as physical or genetic disruption of this process restores NET release (31). Remarkably, when neutrophils are induced to generate NETs prior to biofilms exposure, biofilm inhibition is observed (31). This suggests that inhibition of NETosis is an adaptation by *C. albicans* biofilms to prevent killing by neutrophils. Other species appear to employ this mechanism as well. For example, *C. glabrata* biofilms also impair NET release, although the inhibition is not as pronounced (45).

Recent studies have begun to shed light on the planktonic *C. albicans* cell surface components that induce NET release. The process appears to be multifactorial, as β-glucan, mannan, and secreted aspartic proteases all variably trigger NETosis (93, 96, 97). NET inhibition by biofilm likely involves concealment of cell surface ligands by the extracellular matrix, as disruption of this process permits NET release (31). In particular, disruption of the matrix mannan–glucan complex in a *pmr1*Δ/Δ mutant strain reverses the NET inhibition phenotype, suggesting a role for this unique polysaccharide complex (31, 75). In addition, studies by Zawrotniak et al. show that NET induction by cell wall mannan is concentration-dependent, with higher concentrations failing to trigger NETosis *in vitro* (97). Further investigation would be of interest to explore a similar pattern for polysaccharides of the biofilm matrix in NET inhibition.

Upon encounter with *C. albicans* biofilms, the generation of ROS by neutrophils is dampened compared to the response observed for planktonic organisms (30, 31, 44). Multiple pathways govern NET release and a subset of these depend on ROS production (96, 98–103). In response to planktonic *C. albicans*, both ROS-dependent and ROS-independent pathways trigger NETosis, which may not be surprising given the numerous cell surface ligands expected to be involved (96, 97, 103). While further investigation is needed to dissect pathways impairing neutrophil function by biofilms, inhibition of ROS production is likely involved. A study of *C. glabrata*–neutrophil interactions demonstrates a similar neutrophil response, with reduced ROS production upon encounter with biofilm (45). Taken together, these studies show that *C. albicans* biofilms inhibit the release of NETs and resist killing by neutrophils. The pathway appears to involve the production of an extracellular matrix and dampening of neutrophil ROS production. Based on studies with *C. glabrata*, it may be conserved, in part, among *Candida* spp.

Monocytes and Macrophages Interactions with *Candida* Biofilms

Chandra et al. first demonstrated that *Candida* biofilms resist attack by monocytes and can alter their cytokine profile (42). Peripheral blood mononuclear cells (PBMCs) fail to phagocytose biofilm-associated *C. albicans*, in contrast to planktonic organisms (42). However, these cells remain viable, migrating within the biofilm, even providing a stimulus for biofilm proliferation through an unknown mechanism (29, 42). Compared to planktonic organisms, *C. albicans* biofilms are twofold to threefold more resistant to killing by monocytes (29).

Encounter with *Candida* biofilms influences cytokine release by mononuclear cells. One of the more intriguing alterations is the downregulation of TNF-α, a cytokine which facilitates phagocyte activation. Compared to planktonic organisms, exposure to *C. albicans* biofilms significantly diminishes the production of TNF-α by monocytic cell line THP-1 (29). Not only is this predicted to impact phagocyte function in the host but the alteration in production of TNF-α may also have a direct impact on the biofilm. Application of exogenous TNF-α has been shown to prevent *C. albicans* biofilm formation, through a TNF receptor-independent pathway (104). Furthermore, this activity is blocked by preincubation of TNF-α with *N,N'*-diacetylchitobiose, a major carbohydrate component of *C. albicans* cell wall (104). Therefore, inhibition of TNF-α by biofilms may represent an evolutionary adaption and mechanism of immune evasion. However, much remains a mystery about how the cytokine response influences the host response to biofilm infection. For example, when compared to planktonic cells, PBMCs exposed to *C. albicans* biofilms produce elevated levels of IL-1β, IL-10, and MCP-1 and reduced levels of IL-6 and MIP1β (42). How these combinations of both pro- and anti-inflammatory cytokines are triggered and their influence on host response to biofilm infection is unknown.

Recent work by Alonso et al. revealed that formation of *C. albicans* biofilm impairs the migratory capacity of macrophages (105). Upon exposure to biofilm, the migration of murine macrophages (J774.1 cell line) is reduced approximately twofold when compared to encounter with planktonic organisms. A *pmr1*Δ/Δ mutant similarly impaired macrophage migration during biofilm growth. As this biofilm is deficient in matrix mannan production, the macrophage inhibition is likely related to another factor, such as physical structure (75, 105). Therefore, *Candida* biofilms may elicit distinct inhibitory pathways for neutrophils and macrophages (31, 105).

Aspergillus BIOFILMS

Biofilm Formation

Aspergillus spp. grow ubiquitously in the environment and individuals are constantly exposed to their spores, which are released into the air (106). Immunocompetent individuals clear these spores after inhalation, but those with impaired immunity are at risk for development of severe disease. *Aspergillus* spp. can cause a variety of clinical diseases, including invasive, chronic, and allergic forms (106). The chronic form of disease typically involves formation of an aspergilloma, or fungal ball, in the sinus or lung cavity. These dense structures consist of agglutinated hyphae with occasional conidial heads growing as a biofilm encased in an adhesive extracellular matrix (18) (**Figure 3**). As the community matures, the inner cells loose viability, likely due to starvation. *A. fumigatus* also produces extracellular matrix material during invasive pulmonary aspergillosis (18). However, during this mode of growth, the hyphae remain separated without an inner core of decaying fungal mass. *In vitro*, *A. fumigatus* forms biofilms on agar medium in aerial, static conditions that mimic the host niches for aspergilloma formation (4).

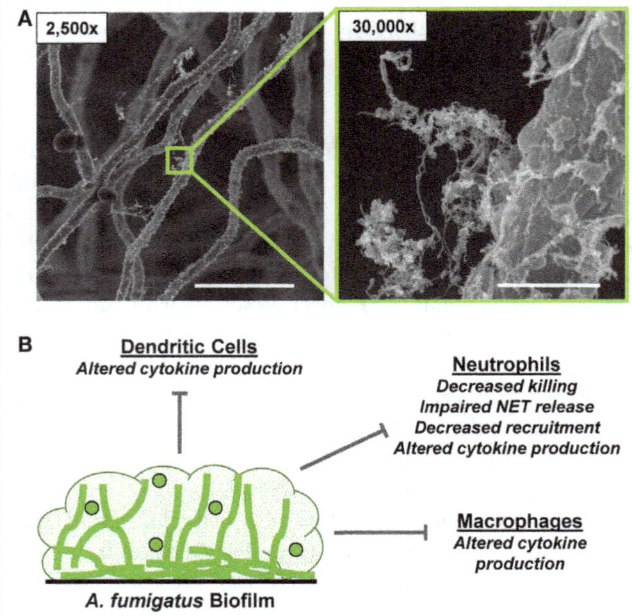

FIGURE 3 | *Aspergillus fumigatus* biofilm formation and innate immune response. **(A)** Scanning electron microscopy images reveal *A. fumigatus* biofilms grown on coverslips. Biofilms were grown for 24 h. Measurement bars represent 10 and 1 μm for 2,500x and 30,000x, respectively. **(B)** Summary of innate immune responses impaired by *A. fumigatus* biofilms.

Matrix Production

Aspergillus fumigatus produces a unique extracellular matrix during biofilm growth *in vitro* and *in vivo* (4, 18, 107). By biochemical analysis, this material consists of 40% protein, 43% carbohydrates, 14% lipids, and 3% aromatic-containing compounds, as well as DNA (108–110). The polysaccharides of the extracellular matrix exhibit cohesion properties and provide immune protection. The main matrix polysaccharides include galactomannan and galactosaminogalactan (GAG), of which, GAG has received the most attention (18). *A. fumigatus* strains deficient in GAG production lack the capacity to form biofilms or produce extracellular matrix (47). GAG is an α-1,4-linked linear heteroglycan composed of variable combinations of galactose and *N*-acetyl-galactosamine (GalNAc) (111, 112). GalNAc residues within the GAG polymer are partially deacetyated by the secreted enzyme Agd3, rendering mature GAG polycationic (113). Deacetylation is required for GAG to mediate adhesion between hyphae and other anionic surfaces such as host cells, plastic, and glass (113). GAG production has also been reported for other *Aspergillus* spp., including *Aspergillus parasiticus*, *Aspergillus niger*, and *Aspergillus nidulans*, although the relative proportion of galactose and GalNAc varies between strains and likely influences matrix function as detailed below (114–117). While galactomannan and GAG are universally present in *A. fumigatus* matrix, key differences exist between the clinical niche biofilms. For example, aspergilloma biofilms produce a thicker extracellular matrix, which contains α-glucan, a polysaccharide absent in biofilms formed during invasive aspergillosis (18). Aspergilloma biofilms also produce melanin, an immune modulator (4, 18, 33, 118). However, its specific role during biofilm formation is unknown.

Innate Immunity to *Aspergillus* Biofilms

Neutrophils are key players in the innate immune defense against *Aspergillus*. Neutropenia, often in the face of chemotherapy or hematologic malignancy, places patients at high risk factor for invasive pulmonary aspergillosis, which can progress to disseminated lethal disease (119, 120). Neutrophils are recruited to *Aspergillus* spores *in vivo* and are critical for their engulfment through phagocytosis (121–123). Neutrophils also release NETs in response to hyphal elements (46, 124–126). While NETs lack significant activity against conidia, they exhibit modest inhibitory activity against the larger hyphal forms of *A. fumigatus* (46, 126). Production of an extracellular matrix shields *A. fumigatus* from neutrophil attack and much of this protection is attributed to GAG (32). *A. fumigatus* mutants deficient in GAG synthesis or deacetylation exhibit attenuated virulence (47, 113), and treatment with recombinant glycoside hydrolases that degrade GAG reduces fungal growth in a murine model of aspergillosis (127).

Studies of differences in GAG composition between *A. fumigatus* and *A. nidulans* have suggested that protection against neutrophil attack is mediated by hyphal-associated GalNAc-rich GAG (32). Unlike *A. fumigatus*, *A. nidulans* produces GalNAc-poor GAG, which contains over fivefold higher levels of galactose and produces poorly adherent biofilms containing minimal extracellular matrix. *A. nidulans* is also less virulent in a murine model of pulmonary aspergillosis and more than twofold more susceptible to killing by human neutrophils. Heterologous expression of the *A. fumigatus uge3* gene encoding a GalNAc-epimerase in *A. nidulans* results in the production of *A. fumigatus*-like GalNAc-rich GAG (32). Unlike wild-type *A. nidulans*, the GalNAc-rich GAG-producing strain of *A. nidulans* forms biofilm, produces extracellular matrix, and resists killing by neutrophils. The protective effects of GAG are dependent on NADPH oxidase and likely involve defense against NETs released through activation of this pathway. It has been hypothesized that GAG-mediated protection against NETs is mediated by electrostatic repulsion between this partially deacetylated cationic polysaccharide and cationic antimicrobial peptides or histones contained within NETs (32).

The immunomodulatory effects of GAG during biofilm formation are likely multifactorial. Hyphal-associated GAG masks β-glucans on the cell wall of hyphae (47, 128) and alters recognition by murine bone marrow-derived dendritic cells *in vitro* (47). Genetic disruption of GAG synthesis leads to increased pro-inflammatory cytokine release through Dectin-1 signaling (47). In a non-neutropenic murine model of pulmonary aspergillosis, genetic disruption of GAG synthesis results in production of a non-protective, hyper-inflammatory response marked by increased neutrophil recruitment (47). This observation suggests that GAG impairs neutrophil recruitment during biofilm growth and is consistent with the reports that soluble GAG can modulate immunity through induction of apoptosis in neutrophils and stimulation of anti-inflammatory IL-1Ra production by macrophages *in vitro* (49, 129). Further, *in vivo* studies are needed to evaluate the relative role of these functions of GAG in invasive and chronic *Aspergillus* infections.

Recent studies have begun to shed light on the mechanisms involved in NET release in response to fungi (32, 46, 93, 97, 124–126, 130). Specific ligands triggering this response to *Aspergillus* remain largely unknown, and how the extracellular matrix may influence these pathways is of great interest. For example, NET production is reduced in response to resting conidia when compared to hyphae (46). This inhibition is linked to RodA, a hydrophobin on the surface of conidia that masks pathogen-associated molecular patterns, including β-glucan (46, 48). As transcriptional analysis shows abundance of *RodA* during *A. fumigatus* biofilm growth when compared to planktonic conditions, it is interesting to postulate a role for RodA production in immune evasion during biofilm growth (131). Also, as melanin production has been described for some *Aspergillus* biofilms, investigation of a role for this immune modulator is also intriguing (4, 18, 33, 118).

Cryptococcus BIOFILMS

Biofilm Formation

Cryptococcus spp. are opportunistic environmental fungal pathogens that cause life-threatening meningoencephalitis, particularly in patients with suppressed immunity in the setting of HIV or organ transplantation (132). Following inhalation of spores from the environment, *C. neoformans* disseminates from the lungs, with a propensity for the central nervous system. *C. neoformans* also exhibits a predilection for artificial surfaces and forms biofilms on medical devices, such as cerebrospinal fluid shunts, vascular catheters, and prosthetic dialysis fistulae (133–136). These adherent communities are composed of yeast encased in an extracellular matrix (5, 41, 137). *In vitro*, *C. neoformans* biofilms mature in 24–48 h and display a multiple-drug-resistance phenotype (5).

Matrix Production

Cryptococcus neoformans produces a protective polysaccharide capsule composed of glucuronoxylomannan (GXM), galactoxylomannan, and mannoprotein (137, 138). During biofilm growth, these capsular polysaccharides are shed into the surrounding milieu, ultimately providing extracellular matrix material for surface adhesion and cell–cell cohesion (5). Acapsular *C. neoformans* mutants are unable to form biofilms (5). Martinez and Casadevall identified GXM as the principle polysaccharide of the *C. neoformans* biofilm matrix (41). This polysaccharide has received the most attention due to its immunomodulatory properties and high abundance in the biofilm matrix (5, 138–141). However, biochemical analysis also shows the presence of sugars not found in GXM, suggesting that the biofilm matrix contains additional polysaccharides (41). Little is known about the structure of these polysaccharides and how they may influence immunity to *Cryptococcus* biofilms.

Innate Immunity to *Cryptococcus* Biofilms

While studies have begun to dissect the impact of biofilm formation on immunity to *Cryptococcus*, much of this host–fungal interaction remains a mystery. Production of a GXM-rich extracellular matrix appears to be the key defense against host immunity. Genetic or antibody-mediated disruption of GXM impairs biofilm formation and diminishes virulence (5, 141). As a capsule polysaccharide, GXM is responsible for a multifaceted inhibition of neutrophil function, impeding chemotaxis, phagocytosis, NET production, and antifungal activity (138, 139, 142, 143). Similar mechanisms of diminished neutrophil function are anticipated in response to *C. neoformans* biofilms and may even be augmented given the high GXM content of biofilm matrix (41). Furthermore, capsular GXM can impair phagocytosis by monocytes and macrophages (138). However, it is unknown if phagocytosis would be an effective response against *Cryptococcus* biofilms, given the large structure of cohesive, aggregated yeast (41).

In addition to the immunomodulatory activity of the extracellular matrix, *C. neoformans* biofilms also resist antimicrobial host defenses. Compared to planktonic *C. neoformans*, biofilms tolerate higher concentrations of defensins, including PG-1, β-defensin-1, and β-defensin-3 (27). This resistance is even further augmented when biofilms are induced to produce melanin through L-Dopa supplementation. Biofilm formation also protects *C. neoformans* from oxidative stress induced by a variety of stimuli (27). Taken together, these studies show that *Cryptococcus* biofilms withstand innate immunity through both immune modulation and resistance to immune attack.

CONCLUSION

Adoption of a biofilm lifestyle during fungal infection is increasingly recognized as a mechanism to avoid host immune attack and provide a protective niche. In this environment, the extracellular matrix can shield the fungal cell wall from host cellular recognition, modulating the immune response. In addition, the extracellular matrix can provide protection from antimicrobial defenses, such as defensins, oxidative stress, and NETs. Furthermore, biofilm formation produces an aggregated community that may resist engulfment by phagocytosis.

While it is clear that biofilm formation significantly influences immunity, studies are just beginning to shed light on the many mechanisms underlying this modulation of host response. As biofilms are heterogeneous structures with variations in architecture and composition based on their environmental niche, mechanisms impairing immunity likely vary among clinical biofilms. Therefore, inclusion of conditions closely mimicking the host and animal models of biofilm infection remains critical for future studies. While recent studies have revealed the influence of biofilm formation on the innate immune response, still little is known about how these structures may modulate adaptive immunity.

Fungal biofilms are among the most difficult infections to treat due to their high tolerance of antifungals and immune evasion strategies. The incidence of fungal biofilm infections is likely to rise given the growing number of patients with artificial medical devices and immunocompromising conditions. Anti-biofilm therapies are urgently needed. Understanding the dynamics of biofilm formation, matrix production, and how these processes induce resistance to multiple facets of the innate immune system may lead to biofilm-specific antifungal strategies.

AUTHOR CONTRIBUTIONS

JK, DS, and JN wrote the manuscript. JK and BS constructed the figures.

FUNDING

JN is supported by the National Institutes of Health (K08 AI108727), the Burroughs Wellcome Fund (1012299), and the Doris Duke Charitable Foundation (112580130). BS has been supported by graduate scholarships from CFC and CIHR. DS is supported by a Research Chair from the Fonds de Recherche Quebec Santé.

REFERENCES

1. Donlan RM. Biofilm formation: a clinically relevant microbiological process. *Clin Infect Dis* (2001) 33(8):1387–92. doi:10.1086/322972
2. Hoyle BD, Jass J, Costerton JW. The biofilm glycocalyx as a resistance factor. *J Antimicrob Chemother* (1990) 26(1):1–5. doi:10.1093/jac/26.1.1
3. Kojic EM, Darouiche RO. *Candida* infections of medical devices. *Clin Microbiol Rev* (2004) 17(2):255–67. doi:10.1128/CMR.17.2.255-267.2004
4. Beauvais A, Schmidt C, Guadagnini S, Roux P, Perret E, Henry C, et al. An extracellular matrix glues together the aerial-grown hyphae of *Aspergillus fumigatus*. *Cell Microbiol* (2007) 9(6):1588–600. doi:10.1111/j.1462-5822.2007.00895.x
5. Martinez LR, Casadevall A. Specific antibody can prevent fungal biofilm formation and this effect correlates with protective efficacy. *Infect Immun* (2005) 73(10):6350–62. doi:10.1128/iai.73.10.6350-6362.2005
6. Imamura Y, Chandra J, Mukherjee PK, Lattif AA, Szczotka-Flynn LB, Pearlman E, et al. *Fusarium* and *Candida albicans* biofilms on soft contact lenses: model development, influence of lens type, and susceptibility to lens care solutions. *Antimicrob Agents Chemother* (2008) 52(1):171–82. doi:10.1128/aac.00387-07
7. D'Antonio D, Parruti G, Pontieri E, Di Bonaventura G, Manzoli L, Sferra R, et al. Slime production by clinical isolates of *Blastoschizomyces capitatus* from patients with hematological malignancies and catheter-related fungemia. *Eur J Clin Microbiol Infect Dis* (2004) 23(10):787–9. doi:10.1007/s10096-004-1207-4
8. Cannizzo FT, Eraso E, Ezkurra PA, Villar-Vidal M, Bollo E, Castella G, et al. Biofilm development by clinical isolates of *Malassezia pachydermatis*. *Med Mycol* (2007) 45(4):357–61. doi:10.1080/13693780701225767
9. Cushion MT, Collins MS, Linke MJ. Biofilm formation by *Pneumocystis* spp. *Eukaryot Cell* (2009) 8(2):197–206. doi:10.1128/ec.00202-08
10. Di Bonaventura G, Pompilio A, Picciani C, Iezzi M, D'Antonio D, Piccolomini R. Biofilm formation by the emerging fungal pathogen *Trichosporon asahii*: development, architecture, and antifungal resistance. *Antimicrob Agents Chemother* (2006) 50(10):3269–76. doi:10.1128/aac.00556-06
11. Singh R, Shivaprakash MR, Chakrabarti A. Biofilm formation by zygomycetes: quantification, structure and matrix composition. *Microbiology* (2011) 157(Pt 9):2611–8. doi:10.1099/mic.0.048504-0
12. Sun Y, Chandra J, Mukherjee P, Szczotka-Flynn L, Ghannoum MA, Pearlman E. A murine model of contact lens-associated *Fusarium* keratitis. *Invest Ophthalmol Vis Sci* (2010) 51(3):1511–6. doi:10.1167/iovs.09-4237
13. Nobile CJ, Johnson AD. *Candida albicans* biofilms and human disease. *Annu Rev Microbiol* (2015) 69:71–92. doi:10.1146/annurev-micro-091014-104330
14. Escande W, Fayad G, Modine T, Verbrugge E, Koussa M, Senneville E, et al. Culture of a prosthetic valve excised for streptococcal endocarditis positive for *Aspergillus fumigatus* 20 years after previous *A. fumigatus* endocarditis. *Ann Thorac Surg* (2011) 91(6):e92–3. doi:10.1016/j.athoracsur.2011.01.102
15. Dongari-Bagtzoglou A, Kashleva H, Dwivedi P, Diaz P, Vasilakos J. Characterization of mucosal *Candida albicans* biofilms. *PLoS One* (2009) 4(11):e7967. doi:10.1371/journal.pone.0007967
16. Harriott MM, Lilly EA, Rodriguez TE, Fidel PL Jr, Noverr MC. *Candida albicans* forms biofilms on the vaginal mucosa. *Microbiology* (2010) 156(Pt 12):3635–44. doi:10.1099/mic.0.039354-0
17. Ganguly S, Mitchell AP. Mucosal biofilms of *Candida albicans*. *Curr Opin Microbiol* (2011) 14(4):380–5. doi:10.1016/j.mib.2011.06.001
18. Loussert C, Schmitt C, Prevost MC, Balloy V, Fadel E, Philippe B, et al. In vivo biofilm composition of *Aspergillus fumigatus*. *Cell Microbiol* (2010) 12(3):405–10. doi:10.1111/j.1462-5822.2009.01409.x
19. Seidler MJ, Salvenmoser S, Muller FM. *Aspergillus fumigatus* forms biofilms with reduced antifungal drug susceptibility on bronchial epithelial cells. *Antimicrob Agents Chemother* (2008) 52(11):4130–6. doi:10.1128/aac.00234-08
20. Al-Fattani MA, Douglas LJ. Biofilm matrix of *Candida albicans* and *Candida tropicalis*: chemical composition and role in drug resistance. *J Med Microbiol* (2006) 55(Pt 8):999–1008. doi:10.1099/jmm.0.46569-0
21. Mukherjee PK, Chandra J, Kuhn DM, Ghannoum MA. Mechanism of fluconazole resistance in *Candida albicans* biofilms: phase-specific role of efflux pumps and membrane sterols. *Infect Immun* (2003) 71(8):4333–40. doi:10.1128/IAI.71.8.4333-4340.2003

22. Nett J, Lincoln L, Marchillo K, Massey R, Holoyda K, Hoff B, et al. Putative role of beta-1,3 glucans in *Candida albicans* biofilm resistance. *Antimicrob Agents Chemother* (2007) 51(2):510–20. doi:10.1128/aac.01056-06
23. Ramage G, Vandewalle K, Wickes BL, Lopez-Ribot JL. Characteristics of biofilm formation by *Candida albicans*. *Rev Iberoam Micol* (2001) 18(4):163–70.
24. Mowat E, Butcher J, Lang S, Williams C, Ramage G. Development of a simple model for studying the effects of antifungal agents on multicellular communities of *Aspergillus fumigatus*. *J Med Microbiol* (2007) 56(Pt 9):1205–12. doi:10.1099/jmm.0.47247-0
25. Seidler MJ, Salvenmoser S, Müller FMC. *Aspergillus fumigatus* forms biofilms with reduced antifungal drug susceptibility on bronchial epithelial cells. *Antimicrob Agents Chemother* (2008) 52(11):4130–6. doi:10.1128/aac.00234-08
26. Mowat E, Lang S, Williams C, McCulloch E, Jones B, Ramage G. Phase-dependent antifungal activity against *Aspergillus fumigatus* developing multicellular filamentous biofilms. *J Antimicrob Chemother* (2008) 62(6):1281–4. doi:10.1093/jac/dkn402
27. Martinez LR, Casadevall A. *Cryptococcus neoformans* cells in biofilms are less susceptible than planktonic cells to antimicrobial molecules produced by the innate immune system. *Infect Immun* (2006) 74(11):6118–23. doi:10.1128/iai.00995-06
28. Ramage G, Rajendran R, Sherry L, Williams C. Fungal biofilm resistance. *Int J Microbiol* (2012) 2012:528521. doi:10.1155/2012/528521
29. Katragkou A, Kruhlak MJ, Simitsopoulou M, Chatzimoschou A, Taparkou A, Cotten CJ, et al. Interactions between human phagocytes and *Candida albicans* biofilms alone and in combination with antifungal agents. *J Infect Dis* (2010) 201(12):1941–9. doi:10.1086/652783
30. Xie Z, Thompson A, Sobue T, Kashleva H, Xu H, Vasilakos J, et al. *Candida albicans* biofilms do not trigger reactive oxygen species and evade neutrophil killing. *J Infect Dis* (2012) 206(12):1936–45. doi:10.1093/infdis/jis607
31. Johnson CJ, Cabezas-Olcoz J, Kernien JF, Wang SX, Beebe DJ, Huttenlocher A, et al. The extracellular matrix of *Candida albicans* biofilms impairs formation of neutrophil extracellular traps. *PLoS Pathog* (2016) 12(9):1–23. doi:10.1371/journal.ppat.1005884
32. Lee MJ, Liu H, Barker BM, Snarr BD, Gravelat FN, Al Abdallah Q, et al. The fungal exopolysaccharide galactosaminogalactan mediates virulence by enhancing resistance to neutrophil extracellular traps. *PLoS Pathog* (2015) 11(10):e1005187. doi:10.1371/journal.ppat.1005187
33. Beauvais A, Latge JP. *Aspergillus* biofilm in vitro and in vivo. *Microbiol Spectr* (2015) 3(4):1–10. doi:10.1128/microbiolspec.MB-0017-2015
34. Chandra J, Mukherjee PK. *Candida* biofilms: development, architecture, and resistance. *Microbiol Spectr* (2015) 3(4):1–14. doi:10.1128/microbiolspec.MB-0020-2015
35. Desai JV, Mitchell AP, Andes DR. Fungal biofilms, drug resistance, and recurrent infection. *Cold Spring Harb Perspect Med* (2014) 4(10):1–18. doi:10.1101/cshperspect.a019729
36. Nett JE. Future directions for anti-biofilm therapeutics targeting *Candida*. *Expert Rev Anti Infect Ther* (2014) 12(3):375–82. doi:10.1586/14787210.2014.885838
37. Fox EP, Nobile CJ. A sticky situation: untangling the transcriptional network controlling biofilm development in *Candida albicans*. *Transcription* (2012) 3(6):315–22. doi:10.4161/trns.22281
38. Baillie GS, Douglas LJ. Matrix polymers of *Candida* biofilms and their possible role in biofilm resistance to antifungal agents. *J Antimicrob Chemother* (2000) 46(3):397–403. doi:10.1093/jac/46.3.397
39. Hawser SP, Baillie GS, Douglas LJ. Production of extracellular matrix by *Candida albicans* biofilms. *J Med Microbiol* (1998) 47(3):253–6. doi:10.1099/00222615-47-3-253
40. Zarnowski R, Westler WM, Lacmbouh GA, Marita JM, Bothe JR, Bernhardt J, et al. Novel entries in a fungal biofilm matrix encyclopedia. *MBio* (2014) 5(4):e1333-1314. doi:10.1128/mBio.01333-14
41. Martinez LR, Casadevall A. *Cryptococcus neoformans* biofilm formation depends on surface support and carbon source and reduces fungal cell susceptibility to heat, cold, and UV light. *Appl Environ Microbiol* (2007) 73(14):4592–601. doi:10.1128/aem.02506-06
42. Chandra J, McCormick TS, Imamura Y, Mukherjee PK, Ghannoum MA. Interaction of *Candida albicans* with adherent human peripheral blood mononuclear cells increases *C. albicans* biofilm formation and results in differential expression of pro- and anti-inflammatory cytokines. *Infect Immun* (2007) 75(5):2612–20. doi:10.1128/iai.01841-06
43. Katragkou A, Chatzimoschou A, Simitsopoulou M, Georgiadou E, Roilides E. Additive antifungal activity of anidulafungin and human neutrophils against *Candida parapsilosis* biofilms. *J Antimicrob Chemother* (2011) 66(3):588–91. doi:10.1093/jac/dkq466
44. Kernien J, Johnson C, Nett J. Conserved inhibition of neutrophil extracellular trap release by clinical *Candida albicans* biofilms. *J Fungi (Basel)* (2017) 3(3):49. doi:10.3390/jof3030049
45. Johnson CJ, Kernien JF, Hoyer AR, Nett JE. Mechanisms involved in the triggering of neutrophil extracellular traps (NETs) by *Candida glabrata* during planktonic and biofilm growth. *Sci Rep* (2017) 7(1):13065. doi:10.1038/s41598-017-13588-6
46. Bruns S, Kniemeyer O, Hasenberg M, Aimanianda V, Nietzsche S, Thywissen A, et al. Production of extracellular traps against *Aspergillus fumigatus* in vitro and in infected lung tissue is dependent on invading neutrophils and influenced by hydrophobin RodA. *PLoS Pathog* (2010) 6(4):e1000873. doi:10.1371/journal.ppat.1000873
47. Gravelat FN, Beauvais A, Liu H, Lee MJ, Snarr BD, Chen D, et al. *Aspergillus* galactosaminogalactan mediates adherence to host constituents and conceals hyphal beta-glucan from the immune system. *PLoS Pathog* (2013) 9(8):e1003575. doi:10.1371/journal.ppat.1003575
48. de Jesus Carrion S, Leal SM, Ghannoum MA, Pearlman E. The RodA hydrophobin on *Aspergillus fumigatus* spores masks Dectin-1 and Dectin-2 dependent responses and enhances fungal survival in vivo. *J Immunol* (2013) 191(5):2581–8. doi:10.4049/jimmunol.1300748
49. Gresnigt MS, Bozza S, Becker KL, Joosten LA, Abdollahi-Roodsaz S, van der Berg WB, et al. A polysaccharide virulence factor from *Aspergillus fumigatus* elicits anti-inflammatory effects through induction of interleukin-1 receptor antagonist. *PLoS Pathog* (2014) 10(3):e1003936. doi:10.1371/journal.ppat.1003936
50. Chai LY, Netea MG, Sugui J, Vonk AG, Van De Sande WW, Warris A, et al. *Aspergillus fumigatus* conidial melanin modulates host cytokine response. *Immunobiology* (2010) 215(11):915–20. doi:10.1016/j.imbio.2009.10.002
51. Achkar JM, Fries BC. *Candida* infections of the genitourinary tract. *Clin Microbiol Rev* (2010) 23(2):253–73. doi:10.1128/cmr.00076-09
52. Kennedy MJ, Volz PA. Ecology of *Candida albicans* gut colonization: inhibition of *Candida* adhesion, colonization, and dissemination from the gastrointestinal tract by bacterial antagonism. *Infect Immun* (1985) 49(3):654–63.
53. Kumamoto CA. *Candida* biofilms. *Curr Opin Microbiol* (2002) 5(6):608–11. doi:10.1016/S1369-5274(02)00371-5
54. Kumamoto CA. Inflammation and gastrointestinal *Candida* colonization. *Curr Opin Microbiol* (2011) 14(4):386–91. doi:10.1016/j.mib.2011.07.015
55. Webb BC, Thomas CJ, Willcox MD, Harty DW, Knox KW. *Candida*-associated denture stomatitis. Aetiology and management: a review. Part 2. Oral diseases caused by *Candida* species. *Aust Dent J* (1998) 43(3):160–6. doi:10.1111/j.1834-7819.1998.tb00172.x
56. Vazquez JA. Options for the management of mucosal candidiasis in patients with AIDS and HIV infection. *Pharmacother J Hum Pharmacol Drug Ther* (1999) 19(1):76–87. doi:10.1592/phco.19.1.76.30509
57. Launay O, Lortholary O, Bouges-Michel C, Jarrousse B, Bentata M, Guillevin L. Candidemia: a nosocomial complication in adults with late-stage AIDS. *Clin Infect Dis* (1998) 26(5):1134–41. doi:10.1086/520291
58. Douglas LJ. *Candida* biofilms and their role in infection. *Trends Microbiol* (2003) 11(1):30–6. doi:10.1016/S0966-842X(02)00002-1
59. Kumamoto CA, Vinces MD. Alternative *Candida albicans* lifestyles: growth on surfaces. *Annu Rev Microbiol* (2005) 59:113–33. doi:10.1146/annurev.micro.59.030804.121034
60. Uppuluri P, Pierce CG, Lopez-Ribot JL. *Candida albicans* biofilm formation and its clinical consequences. *Future Microbiol* (2009) 4(10):1235–7. doi:10.2217/fmb.09.85
61. Dongari-Bagtzoglou A. Mucosal biofilms: challenges and future directions. *Expert Rev Anti Infect Ther* (2008) 6(2):141–4. doi:10.1586/14787210.6.2.141
62. Shin JH, Kee SJ, Shin MG, Kim SH, Shin DH, Lee SK, et al. Biofilm production by isolates of *Candida* species recovered from nonneutropenic patients: comparison of bloodstream isolates with isolates from other sources. *J Clin Microbiol* (2002) 40(4):1244–8. doi:10.1128/jcm.40.4.1244-1248.2002

63. Bizerra FC, Nakamura CV, de Poersch C, Estivalet Svidzinski TI, Borsato Quesada RM, Goldenberg S, et al. Characteristics of biofilm formation by *Candida tropicalis* and antifungal resistance. *FEMS Yeast Res* (2008) 8(3):442–50. doi:10.1111/j.1567-1364.2007.00347.x
64. Kuhn DM, Chandra J, Mukherjee PK, Ghannoum MA. Comparison of biofilms formed by *Candida albicans* and *Candida parapsilosis* on bioprosthetic surfaces. *Infect Immun* (2002) 70(2):878–88. doi:10.1128/IAI.70.2.878-888.2002
65. Jain N, Kohli R, Cook E, Gialanella P, Chang T, Fries BC. Biofilm formation by and antifungal susceptibility of *Candida* isolates from urine. *Appl Environ Microbiol* (2007) 73(6):1697–703. doi:10.1128/aem.02439-06
66. Lewis RE, Kontoyiannis DP, Darouiche RO, Raad II, Prince RA. Antifungal activity of amphotericin B, fluconazole, and voriconazole in an in vitro model of *Candida* catheter-related bloodstream infection. *Antimicrob Agents Chemother* (2002) 46(11):3499–505. doi:10.1128/AAC.46.11.3499-3505.2002
67. Thein ZM, Samaranayake YH, Samaranayake LP. In vitro biofilm formation of *Candida albicans* and non-*albicans Candida* species under dynamic and anaerobic conditions. *Arch Oral Biol* (2007) 52(8):761–7. doi:10.1016/j.archoralbio.2007.01.009
68. Tumbarello M, Posteraro B, Trecarichi EM, Fiori B, Rossi M, Porta R, et al. Biofilm production by *Candida* species and inadequate antifungal therapy as predictors of mortality for patients with candidemia. *J Clin Microbiol* (2007) 45(6):1843–50. doi:10.1128/jcm.00131-07
69. Sherry L, Ramage G, Kean R, Borman A, Johnson EM, Richardson MD, et al. Biofilm-forming capability of highly virulent, multidrug-resistant *Candida auris*. *Emerg Infect Dis* (2017) 23(2):328–31. doi:10.3201/eid2302.161320
70. Chandra J, Kuhn DM, Mukherjee PK, Hoyer LL, McCormick T, Ghannoum MA. Biofilm formation by the fungal pathogen *Candida albicans*: development, architecture, and drug resistance. *J Bacteriol* (2001) 183(18):5385–94. doi:10.1128/JB.183.18.5385-5394.2001
71. Ramage G, Mowat E, Jones B, Williams C, Lopez-Ribot J. Our current understanding of fungal biofilms. *Crit Rev Microbiol* (2009) 35(4):340–55. doi:10.3109/10408410903241436
72. Ramage G, Saville SP, Thomas DP, López-Ribot JL. *Candida* biofilms: an update. *Eukaryot Cell* (2005) 4(4):633–8. doi:10.1128/ec.4.4.633-638.2005
73. Silva S, Henriques M, Martins A, Oliveira R, Williams D, Azeredo J. Biofilms of non-*Candida albicans Candida* species: quantification, structure and matrix composition. *Med Mycol* (2009) 47(7):681–9. doi:10.3109/13693780802549594
74. Mitchell KF, Taff HT, Cuevas MA, Reinicke EL, Sanchez H, Andes DR. Role of matrix β-1,3 glucan in antifungal resistance of non-*albicans Candida* biofilms. *Antimicrob Agents Chemother* (2013) 57(4):1918–20. doi:10.1128/aac.02378-12
75. Mitchell KF, Zarnowski R, Sanchez H, Edward JA, Reinicke EL, Nett JE, et al. Community participation in biofilm matrix assembly and function. *Proc Natl Acad Sci U S A* (2015) 112(13):4092–7. doi:10.1073/pnas.1421437112
76. Nett JE, Zarnowski R, Cabezas-Olcoz J, Brooks EG, Bernhardt J, Marchillo K, et al. Host contributions to construction of three device-associated *Candida albicans* biofilms. *Infect Immun* (2015) 83(12):4630–8. doi:10.1128/iai.00931-15
77. Martins M, Henriques M, Lopez-Ribot JL, Oliveira R. Addition of DNase improves the in vitro activity of antifungal drugs against *Candida albicans* biofilms. *Mycoses* (2012) 55(1):80–5. doi:10.1111/j.1439-0507.2011.02047.x
78. Nett JE, Sanchez H, Cain MT, Ross KM, Andes DR. Interface of *Candida albicans* biofilm matrix-associated drug resistance and cell wall integrity regulation. *Eukaryot Cell* (2011) 10(12):1660–9. doi:10.1128/EC.05126-11
79. Thomas DP, Bachmann SP, Lopez-Ribot JL. Proteomics for the analysis of the *Candida albicans* biofilm lifestyle. *Proteomics* (2006) 6(21):5795–804. doi:10.1002/pmic.200600332
80. Martins M, Uppuluri P, Thomas DP, Cleary IA, Henriques M, Lopez-Ribot JL, et al. Presence of extracellular DNA in the *Candida albicans* biofilm matrix and its contribution to biofilms. *Mycopathologia* (2010) 169(5):323–31. doi:10.1007/s11046-009-9264-y
81. Klis FM, de Groot P, Hellingwerf K. Molecular organization of the cell wall of *Candida albicans*. *Med Mycol* (2001) 39(Suppl 1):1–8. doi:10.1080/mmy.39.1.1.8-0
82. Chaffin WL, López-Ribot JL, Casanova M, Gozalbo D, Martínez JP. Cell wall and secreted proteins of *Candida albicans*: identification, function, and expression. *Microbiol Mol Biol Rev* (1998) 62(1):130–80.
83. Edwards JE Jr, Lehrer RI, Stiehm ER, Fischer TJ, Young LS. Severe candidal infections: clinical perspective, immune defense mechanisms, and current concepts of therapy. *Ann Intern Med* (1978) 89(1):91–106. doi:10.7326/0003-4819-89-1-91
84. Fidel PL Jr. Immunity to *Candida*. *Oral Dis* (2002) 8(Suppl 2):69–75. doi:10.1034/j.1601-0825.2002.00015.x
85. Erwig LP, Gow NA. Interactions of fungal pathogens with phagocytes. *Nat Rev Microbiol* (2016) 14(3):163–76. doi:10.1038/nrmicro.2015.21
86. Pappas PG, Kauffman CA, Andes DR, Clancy CJ, Marr KA, Ostrosky-Zeichner L, et al. Clinical practice guideline for the management of candidiasis: 2016 update by the Infectious Diseases Society of America. *Clin Infect Dis* (2016) 62(4):e1–50. doi:10.1093/cid/civ933
87. Kolaczkowska E, Kubes P. Neutrophil recruitment and function in health and inflammation. *Nat Rev Immunol* (2013) 13(3):159–75. doi:10.1038/nri3399
88. Nett JE. The host's reply to *Candida* biofilm. *Pathogens* (2016) 5(1):1–10. doi:10.3390/pathogens5010033
89. Yano J, Noverr MC, Fidel PL Jr. Cytokines in the host response to *Candida* vaginitis: identifying a role for non-classical immune mediators, S100 alarmins. *Cytokine* (2012) 58(1):118–28. doi:10.1016/j.cyto.2011.11.021
90. Nieminen MT, Hernandez M, Novak-Frazer L, Kuula H, Ramage G, Bowyer P, et al. DL-2-hydroxyisocaproic acid attenuates inflammatory responses in a murine *Candida albicans* biofilm model. *Clin Vaccine Immunol* (2014) 21(9):1240–5. doi:10.1128/CVI.00339-14
91. Fidel PL, Barousse M, Espinosa T, Ficarra M, Sturtevant J, Martin DH, et al. An intravaginal live *Candida* challenge in humans leads to new hypotheses for the immunopathogenesis of vulvovaginal candidiasis. *Infect Immun* (2004) 72(5):2939–46. doi:10.1128/iai.72.5.2939-2946.2004
92. Katragkou A, Simitsopoulou M, Chatzimoschou A, Georgiadou E, Walsh TJ, Roilides E. Effects of interferon-gamma and granulocyte colony-stimulating factor on antifungal activity of human polymorphonuclear neutrophils against *Candida albicans* grown as biofilms or planktonic cells. *Cytokine* (2011) 55(3):330–4. doi:10.1016/j.cyto.2011.05.007
93. Branzk N, Lubojemska A, Hardison SE, Wang Q, Gutierrez MG, Brown GD, et al. Neutrophils sense microbe size and selectively release neutrophil extracellular traps in response to large pathogens. *Nat Immunol* (2014) 15(11):1017–25. doi:10.1038/ni.2987
94. Brinkmann V, Reichard U, Goosmann C, Fauler B, Uhlemann Y, Weiss DS, et al. Neutrophil extracellular traps kill bacteria. *Science* (2004) 303(5663):1532–5. doi:10.1126/science.1092385
95. Urban CF, Reichard U, Brinkmann V, Zychlinsky A. Neutrophil extracellular traps capture and kill *Candida albicans* yeast and hyphal forms. *Cell Microbiol* (2006) 8(4):668–76. doi:10.1111/j.1462-5822.2005.00659.x
96. Byrd AS, O'Brien XM, Johnson CM, Lavigne LM, Reichner JS. An extracellular matrix-based mechanism of rapid neutrophil extracellular trap formation in response to *Candida albicans*. *J Immunol* (2013) 190(8):4136–48. doi:10.4049/jimmunol.1202671
97. Zawrotniak M, Bochenska O, Karkowska-Kuleta J, Seweryn-Ozog K, Aoki W, Ueda M, et al. Aspartic proteases and major cell wall components in *Candida albicans* trigger the release of neutrophil extracellular traps. *Front Cell Infect Microbiol* (2017) 7:414. doi:10.3389/fcimb.2017.00414
98. Kenny EF, Herzig A, Kruger R, Muth A, Mondal S, Thompson PR, et al. Diverse stimuli engage different neutrophil extracellular trap pathways. *Elife* (2017) 6:e24437. doi:10.7554/eLife.24437
99. Bjornsdottir H, Welin A, Michaelsson E, Osla V, Berg S, Christenson K, et al. Neutrophil NET formation is regulated from the inside by myeloperoxidase-processed reactive oxygen species. *Free Radic Biol Med* (2015) 89:1024–35. doi:10.1016/j.freeradbiomed.2015.10.398
100. Stoiber W, Obermayer A, Steinbacher P, Krautgartner WD. The role of reactive oxygen species (ROS) in the formation of extracellular traps (ETs) in humans. *Biomolecules* (2015) 5(2):702–23. doi:10.3390/biom5020702
101. Gray RD, Lucas CD, Mackellar A, Li F, Hiersemenzel K, Haslett C, et al. Activation of conventional protein kinase C (PKC) is critical in the generation of human neutrophil extracellular traps. *J Inflamm (Lond)* (2013) 10(1):12. doi:10.1186/1476-9255-10-12

102. Hogan D, Wheeler RT. The complex roles of NADPH oxidases in fungal infection. *Cell Microbiol* (2014) 16(8):1156–67. doi:10.1111/cmi.12320
103. Metzler KD, Fuchs TA, Nauseef WM, Reumaux D, Roesler J, Schulze I, et al. Myeloperoxidase is required for neutrophil extracellular trap formation: implications for innate immunity. *Blood* (2011) 117(3):953–9. doi:10.1182/blood-2010-06-290171
104. Rocha FAC, Alves A, Rocha MFG, Cordeiro RA, Brilhante RSN, Pinto A, et al. Tumor necrosis factor prevents *Candida albicans* biofilm formation. *Sci Rep* (2017) 7(1):1206. doi:10.1038/s41598-017-01400-4
105. Alonso M, Gow N, Erwig L, Bain J. Macrophage migration is impaired within *Candida albicans* biofilms. *J Fungi* (2017) 3(3):31. doi:10.3390/jof3030031
106. Patterson TF, Thompson GR III, Denning DW, Fishman JA, Hadley S, Herbrecht R, et al. Practice guidelines for the diagnosis and management of aspergillosis: 2016 update by the infectious diseases society of America. *Clin Infect Dis* (2016) 63(4):e1–60. doi:10.1093/cid/ciw326
107. Sheppard DC, Howell PL. Biofilm exopolysaccharides of pathogenic fungi: lessons from bacteria. *J Biol Chem* (2016) 291(24):12529–37. doi:10.1074/jbc.R116.720995
108. Reichhardt C, Ferreira JA, Joubert LM, Clemons KV, Stevens DA, Cegelski L. Analysis of the *Aspergillus fumigatus* biofilm extracellular matrix by solid-state nuclear magnetic resonance spectroscopy. *Eukaryot Cell* (2015) 14(11):1064–72. doi:10.1128/ec.00050-15
109. Rajendran R, Williams C, Lappin DF, Millington O, Martins M, Ramage G. Extracellular DNA release acts as an antifungal resistance mechanism in mature *Aspergillus fumigatus* biofilms. *Eukaryot Cell* (2013) 12(3):420–9. doi:10.1128/EC.00287-12
110. Shopova I, Bruns S, Thywissen A, Kniemeyer O, Brakhage AA, Hillmann F. Extrinsic extracellular DNA leads to biofilm formation and colocalizes with matrix polysaccharides in the human pathogenic fungus *Aspergillus fumigatus*. *Front Microbiol* (2013) 4:141. doi:10.3389/fmicb.2013.00141
111. Fontaine T, Delangle A, Simenel C, Coddeville B, van Vliet SJ, van Kooyk Y, et al. Galactosaminogalactan, a new immunosuppressive polysaccharide of *Aspergillus fumigatus*. *PLoS Pathog* (2011) 7(11):e1002372. doi:10.1371/journal.ppat.1002372
112. Lee MJ, Gravelat FN, Cerone RP, Baptista SD, Campoli PV, Choe SI, et al. Overlapping and distinct roles of *Aspergillus fumigatus* UDP-glucose 4-epimerases in galactose metabolism and the synthesis of galactose-containing cell wall polysaccharides. *J Biol Chem* (2014) 289(3):1243–56. doi:10.1074/jbc.M113.522516
113. Lee MJ, Geller AM, Bamford NC, Liu H, Gravelat FN, Snarr BD, et al. Deacetylation of fungal exopolysaccharide mediates adhesion and biofilm formation. *MBio* (2016) 7(2):1–14. doi:10.1128/mBio.00252-16
114. Ruperez P, Leal JA. Extracellular galactosaminogalactan from *Aspergillus parasiticus*. *Trans Br Mycol Soc* (1981) 77(3):621–5. doi:10.1016/S0007-1536(81)80111-8
115. Bardalaye PC, Nordin JH. Galactosaminogalactan from cell walls of *Aspergillus niger*. *J Bacteriol* (1976) 125(2):655–69.
116. Gorin PA, Eveleigh DE. Extracellular 2-acetamido-2-deoxy-D-galacto-D-galactan from *Aspergillus nidulans*. *Biochemistry* (1970) 9(25):5023–7. doi:10.1021/bi00827a029
117. Leal JA, Rupérez P. Extracellular polysaccharide production by *Aspergillus nidulans*. *Trans Br Mycol Soc* (1978) 70(1):115–20. doi:10.1016/S0007-1536(78)80180-6
118. Akoumianaki T, Kyrmizi I, Valsecchi I, Gresnigt MS, Samonis G, Drakos E, et al. *Aspergillus* cell wall melanin blocks LC3-associated phagocytosis to promote pathogenicity. *Cell Host Microbe* (2016) 19(1):79–90. doi:10.1016/j.chom.2015.12.002
119. Feldmesser M. Role of neutrophils in invasive aspergillosis. *Infect Immun* (2006) 74(12):6514–6. doi:10.1128/iai.01551-06
120. Gerson SL, Talbot GH, Hurwitz S, Strom BL, Lusk EJ, Cassileth PA. Prolonged granulocytopenia: the major risk factor for invasive pulmonary aspergillosis in patients with acute leukemia. *Ann Intern Med* (1984) 100(3):345–51. doi:10.7326/0003-4819-100-3-345
121. Bonnett CR, Cornish EJ, Harmsen AG, Burritt JB. Early neutrophil recruitment and aggregation in the murine lung inhibit germination of *Aspergillus fumigatus* conidia. *Infect Immun* (2006) 74(12):6528–39. doi:10.1128/iai.00909-06
122. Sturtevant J, Latge JP. Participation of complement in the phagocytosis of the conidia of *Aspergillus fumigatus* by human polymorphonuclear cells. *J Infect Dis* (1992) 166(3):580–6. doi:10.1093/infdis/166.3.580
123. Behnsen J, Narang P, Hasenberg M, Gunzer F, Bilitewski U, Klippel N, et al. Environmental dimensionality controls the interaction of phagocytes with the pathogenic fungi *Aspergillus fumigatus* and *Candida albicans*. *PLoS Pathog* (2007) 3(2):e13. doi:10.1371/journal.ppat.0030013
124. Jaillon S, Peri G, Delneste Y, Fremaux I, Doni A, Moalli F, et al. The humoral pattern recognition receptor PTX3 is stored in neutrophil granules and localizes in extracellular traps. *J Exp Med* (2007) 204(4):793–804. doi:10.1084/jem.20061301
125. Bianchi M, Hakkim A, Brinkmann V, Siler U, Seger RA, Zychlinsky A, et al. Restoration of NET formation by gene therapy in CGD controls aspergillosis. *Blood* (2009) 114(13):2619–22. doi:10.1182/blood-2009-05-221606
126. McCormick A, Heesemann L, Wagener J, Marcos V, Hartl D, Loeffler J, et al. NETs formed by human neutrophils inhibit growth of the pathogenic mold *Aspergillus fumigatus*. *Microbes Infect* (2010) 12(12–13):928–36. doi:10.1016/j.micinf.2010.06.009
127. Snarr BD, Baker P, Bamford NC, Sato Y, Liu H, Lehoux M, et al. Microbial glycoside hydrolases as antibiofilm agents with cross-kingdom activity. *Proc Natl Acad Sci U S A* (2017) 114(27):7124–9. doi:10.1073/pnas.1702798114
128. Gravelat FN, Ejzykowicz DE, Chiang LY, Chabot JC, Urb M, Macdonald KD, et al. *Aspergillus fumigatus* MedA governs adherence, host cell interactions and virulence. *Cell Microbiol* (2010) 12(4):473–88. doi:10.1111/j.1462-5822.2009.01408.x
129. Robinet P, Baychelier F, Fontaine T, Picard C, Debre P, Vieillard V, et al. A polysaccharide virulence factor of a human fungal pathogen induces neutrophil apoptosis via NK cells. *J Immunol* (2014) 192(11):5332–42. doi:10.4049/jimmunol.1303180
130. Rohm M, Grimm MJ, D'Auria AC, Almyroudis NG, Segal BH, Urban CF. NADPH oxidase promotes neutrophil extracellular trap formation in pulmonary aspergillosis. *Infect Immun* (2014) 82(5):1766–77. doi:10.1128/IAI.00096-14
131. Gibbons JG, Beauvais A, Beau R, McGary KL, Latge JP, Rokas A. Global transcriptome changes underlying colony growth in the opportunistic human pathogen *Aspergillus fumigatus*. *Eukaryot Cell* (2012) 11(1):68–78. doi:10.1128/ec.05102-11
132. Perfect JR, Dismukes WE, Dromer F, Goldman DL, Graybill JR, Hamill RJ, et al. Clinical practice guidelines for the management of cryptococcal disease: 2010 update by the infectious diseases society of America. *Clin Infect Dis* (2010) 50(3):291–322. doi:10.1086/649858
133. Walsh TJ, Schlegel R, Moody MM, Costerton JW, Salcman M. Ventriculoatrial shunt infection due to *Cryptococcus neoformans*: an ultrastructural and quantitative microbiological study. *Neurosurgery* (1986) 18(3):373–5. doi:10.1227/00006123-198603000-00025
134. Bach MC, Tally PW, Godofsky EW. Use of cerebrospinal fluid shunts in patients having acquired immunodeficiency syndrome with cryptococcal meningitis and uncontrollable intracranial hypertension. *Neurosurgery* (1997) 41(6):1280–2; discussion 2–3. doi:10.1097/00006123-199712000-00008
135. Braun DK, Janssen DA, Marcus JR, Kauffman CA. Cryptococcal infection of a prosthetic dialysis fistula. *Am J Kidney Dis* (1994) 24(5):864–7. doi:10.1016/S0272-6386(12)80683-4
136. Banerjee U, Gupta K, Venugopal P. A case of prosthetic valve endocarditis caused by *Cryptococcus neoformans* var. *neoformans*. *J Med Vet Mycol* (1997) 35(2):139–41. doi:10.1080/02681219780001031
137. Martinez LR, Casadevall A. Biofilm formation by *Cryptococcus neoformans*. *Microbiol Spectr* (2015) 3(3):1–11. doi:10.1128/microbiolspec.MB-0006-2014
138. Vecchiarelli A. Immunoregulation by capsular components of *Cryptococcus neoformans*. *Med Mycol* (2000) 38(6):407–17. doi:10.1080/714030973
139. Dong ZM, Murphy JW. Intravascular cryptococcal culture filtrate (CneF) and its major component, glucuronoxylomannan, are potent inhibitors of leukocyte accumulation. *Infect Immun* (1995) 63(3):770–8.
140. Baena-Monroy T, Moreno-Maldonado V, Franco-Martinez F, Aldape-Barrios B, Quindos G, Sanchez-Vargas LO. *Candida albicans*, *Staphylococcus aureus* and *Streptococcus mutans* colonization in patients wearing dental prosthesis. *Med Oral Patol Oral Cir Bucal* (2005) 10(Suppl 1):E27–39.
141. Martinez LR, Christaki E, Casadevall A. Specific antibody to *Cryptococcus neoformans* glucuronoxylomannan antagonizes antifungal drug action

against cryptococcal biofilms in vitro. *J Infect Dis* (2006) 194(2):261–6. doi:10.1086/504722

142. Rocha JD, Nascimento MT, Decote-Ricardo D, Corte-Real S, Morrot A, Heise N, et al. Capsular polysaccharides from *Cryptococcus neoformans* modulate production of neutrophil extracellular traps (NETs) by human neutrophils. *Sci Rep* (2015) 5:8008. doi:10.1038/srep08008
143. Monari C, Casadevall A, Retini C, Baldelli F, Bistoni F, Vecchiarelli A. Antibody to capsular polysaccharide enhances the function of neutrophils from patients with AIDS against *Cryptococcus neoformans*. *AIDS* (1999) 13(6):653–60. doi:10.1097/00002030-199904160-00005

6

The Elusive Anti-*Candida* Vaccine: Lessons from the Past and Opportunities for the Future

*Gloria Hoi Wan Tso†, Jose Antonio Reales-Calderon† and Norman Pavelka**

*Singapore Immunology Network (SIgN), Agency of Science, Technology and Research (A*STAR), Singapore*

***Correspondence:** Norman Pavelka norman_pavelka@immunol.a-star.edu.sg*

†These authors have contributed equally as first authors.

Candidemia is a bloodstream fungal infection caused by *Candida* species and is most commonly observed in hospitalized patients. Even with proper antifungal drug treatment, mortality rates remain high at 40–50%. Therefore, prophylactic or preemptive antifungal medications are currently recommended in order to prevent infections in high-risk patients. Moreover, the majority of women experience at least one episode of vulvovaginal candidiasis (VVC) throughout their lifetime and many of them suffer from recurrent VVC (RVVC) with frequent relapses for the rest of their lives. While there currently exists no definitive cure, the only available treatment for RVVC is again represented by antifungal drug therapy. However, due to the limited number of existing antifungal drugs, their associated side effects and the increasing occurrence of drug resistance, other approaches are greatly needed. An obvious prevention measure for candidemia or RVVC relapse would be to immunize at-risk patients with a vaccine effective against *Candida* infections. In spite of the advanced and proven techniques successfully applied to the development of antibacterial or antiviral vaccines, however, no antifungal vaccine is still available on the market. In this review, we first summarize various efforts to date in the development of anti-*Candida* vaccines, highlighting advantages and disadvantages of each strategy. We next unfold and discuss general hurdles encountered along these efforts, such as the existence of large genomic variation and phenotypic plasticity across *Candida* strains and species, and the difficulty in mounting protective immune responses in immunocompromised or immunosuppressed patients. Lastly, we review the concept of "trained immunity" and discuss how induction of this rapid and nonspecific immune response may potentially open new and alternative preventive strategies against opportunistic infections by *Candida* species and potentially other pathogens.

Keywords: *Candida*, candidemia, candidiasis, vaccine, trained immunity, opportunistic infections, immunocompromised patients

INTRODUCTION

Due to advances in medicine and surgery, over the past century there has been a rising population of immunocompromised patients, which are at an elevated risk to suffer from opportunistic infections (1, 2). While hospital-acquired fungal infections are less frequent than bacterial ones, they disproportionately account for higher mortality rates, longer hospitalization times and increased healthcare costs (3). Risk factors for fungal infections are very broad and nonspecific, and they include chronic respiratory disease, cancer, HIV infection, organ transplantation, neutropenia, presence of central venous catheters, prolonged hospital stay, administration of total parenteral nutrition, exposure to invasive procedures, chemotherapy, hemodialysis, gastric acid suppression, and use of broad-spectrum antibiotics (4, 5). Candidiasis (caused by *Candida* species) is the most common opportunistic fungal infection and the fourth most common nosocomial bloodstream infection (6, 7). Despite the best available standards of care, the incidence of candidemia, a sign of invasive or systemic candidiasis, is on the rise in the US (8) and mortality rates often exceed 50% despite use of antifungal drugs (9). This is especially true in intensive care units and in immunocompromised patients, where *Candida* bloodstream infections are estimated to strike ~400,000 patients a year, with an associated mortality of 46–75% (10). For all these reasons, and because the hospitalization and treatment cost of these patients is very high (3, 11), CDC guidelines recommend prophylactic or preemptive antifungal treatment in patients considered at high risk of candidemia (12).

Mucosal fungal infections, though almost never life-threatening, are much more common than invasive ones and can be associated with high morbidity, socioeconomic impact, and low quality of life. The most common infection sites are the oral cavity and the genitourinary tract (13). Vulvovaginal candidiasis (VVC) is the most common mucosal fungal infection, with different studies estimating that 50–70% of women suffer an episode of VVC at least once in their lifetime and 5–8% of women suffer from recurrent VVC (RVVC) (14). To this date, there still exists no definitive cure for RVVC, which is commonly treated with over-the-counter antifungal medications but continues to relapse ≥4 times a year for the rest of a woman's life (15).

Antifungal agents available for the treatment of systemic or mucosal candidiasis are restricted to only a few classes of compounds, such as azoles, polyenes, allylamines, and echinocandins (16). However, adverse side effects, toxicity, and emergence of drug resistance limit the use of these drugs (10). In particular, various *Candida* species can acquire resistance to different antifungals or, even worse, to more than one drug (17–20). In fact, the emergence of drug resistance in *Candida* spp. has been on a growing trend over the past decade (21), with multidrug-resistant *Candida* spp. now being reported all over the world (22, 23). Finally, the recent discovery of a novel multidrug resistant species of *Candida* (*C. auris*) poses a further threat to our ability to use antifungal drugs to treat candidiasis (24).

To reduce the incidence and mortality of opportunistic fungal infections, experts agree that we need (i) improved diagnostics to allow more rapid implementation of appropriate therapies, (ii) more effective antifungal agents associated with less severe side effects, and (iii) to develop immunotherapies based on our mounting knowledge of antifungal immunity (10, 25). These same strategies could be used also to reduce the recurrence and severity of RVVC episodes. We hereby argue that an anti-*Candida* vaccine, as long as it was effective in the populations of patients it is intended to protect, could help reduce the global burden of both systemic and vaginal candidiasis bringing about substantial socioeconomic benefits.

In this review, we first provide an up-to-date overview of various efforts that have been attempted so far in the quest of an anti-*Candida* vaccine, which has remained elusive. We next critically assess potential common problems that might have hindered such efforts and finally propose a few tentative solutions to overcoming those problems.

CURRENT ANTI-*CANDIDA* VACCINE LANDSCAPE

In the last decades, several different anti-*Candida* vaccines have been proposed but only few of them have been tested in clinical trials (**Table 1**). Here, we will summarize the current anti-*Candida* vaccine landscape and discuss advantages and disadvantages of each vaccination strategy.

TABLE 1 | Summary of past and current anti-*Candida* vaccine candidates.

Vaccine category	Description	Clinical trial stage	Protection (*C. albicans* infection site)	Cross-protection (non-*C. albicans* organisms)	Reference
LA	Hyphal-defective *C. albicans* strain PCA-2		SYS	*Staphylococcus aureus*	(26)
LA	MAP kinase-defective *hog1Δ C. albicans* strain CNC13		SYS		(27)
LA	Cell wall-defective *ecm33Δ C. albicans* strain RML2U		SYS		(28)
LA	Filamentation-repressible *C. albicans* strain tet-NRG1		SYS		(29)
LA	Yeast-locked *C. albicans* strain *cph1Δ/efg1Δ*		SYS		(30)
LA	Heat-killed or live *Saccharomyces cerevisiae* clinical strains		SYS	*Aspergillus fumigatus, Cryptococcus grubii, and Coccidioides posadasii*	(31–34)
REC	Recombinant N-terminus of *C. albicans* Als1p		SYS, VAG, OR		(35, 36)
REC	Recombinant N-terminus of *C. albicans* Als3p		SYS, VAG	*S. aureus*	(37, 38)
REC	Recombinant N-terminus of *C. albicans* Als3p plus alum adjuvant (NDV-3)	Phase II	SYS, VAG	*S. aureus*	(39)
REC	Recombinant *C. albicans* Sap2p		VAG		(40)
REC	Recombinant *C. albicans* Sap2p (PEV-7)	Phase I	VAG		(41)
REC	Recombinant *C. albicans* Hsp90p (Mycograb)		SYS		(42)
REC	Recombinant *C. albicans* Hyr1p		SYS	*C. glabrata, C. krusei, C. parapsilosis and C. tropicalis*	(43)
EX	Crude cell wall extract		SYS		(44)
GC	Mannans and peptide conjugates		SYS		(45, 46)
GC	β-glucan conjugated with MF59 adjuvant		VAG		(47)
GC	Laminarin conjugated with diphtheria toxoid (Lam-CRM197)		SYS, VAG	*A. fumigatus*	(48)

LA, live attenuated; REC, recombinant; EX, extract; GC, glycoconjugate; SYS, systemic; VAG, vaginal; OR, oral.

Historically, vaccines are developed to boost antigen-specific immune responses and elicit protective immunological memory against specific pathogenic strains. In the case of *Candida* species, most of the efforts have been focused around the most common species, *C. albicans*. Different strategies have been used to immunize hosts against candidiasis: (i) live-attenuated strains, (ii) recombinant proteins, and (iii) glycoconjugates.

Live-Attenuated Vaccines

The first ever reported vaccine, Edward Jenner's famous vaccinia virus, was a live virus causing cowpox in cattle but protecting humans against smallpox (49). By all definitions, it was a live-attenuated vaccine, which caused only a very mild form of the disease in humans while at the same time raising long-lasting protective immune memory. Up to this date, most currently available antiviral vaccines (e.g., yellow fever, measles, varicella) and some antibacterial ones (e.g., BCG) are essentially live-attenuated versions of their pathogenic counterparts. Because of this long history of success, it is not surprising that several studies have ascribed various attenuated strains of *C. albicans* the ability to confer protection against candidiasis.

The morphogenetic transition between the conidial (yeast) and hyphal (filamentous) form is one of the most important virulence factors in *C. albicans* (50); for this reason, the absence of filamentation is a common characteristic of attenuated strains of this species. Several *C. albicans* mutants associated with no or low virulence, including PCA-2 (incapable of yeast-hyphae conversion), CNC13 (deleted in the MAP kinase *HOG1*), RML2U [deleted in the cell wall protein (CWP) gene *ECM33*], and tet-NRG1 (in which the filamentation repressor *NRG1* can be overexpressed by doxycycline), have been used to immunize mice and shown to protect them from a subsequent lethal systemic infection with a virulent strain (26–29). Also, the *C. albicans* double-mutant *cph1/efg1*, which is severely impaired in hyphal morphogenesis, partially protects mice from systemic infection by a wild-type *C. albicans* strain (30).

Not only *C. albicans* attenuated strains have been used to protect against systemic candidiasis, but also the generally regarded-as-safe baker's yeast *Saccharomyces cerevisiae* has been successfully used to this end. Stevens and colleagues have described the use of heat-killed or live *S. cerevisiae* as a protective vaccine against *C. albicans* infection, but also against *A. fumigatus*, *Cryptococcus grubii*, and *Coccidioides* infection in a dose-dependent manner (31–34).

Despite these success stories at the basic research level, none of these vaccine candidates has progressed to clinical trials. Possible reasons include the fact that characterization of these strains is complex, that the stability of the virulence-attenuating mutations is not guaranteed, and that the use of live-attenuated vaccines is currently not recommended in immunocompromised patients due to their weaker immune defense and the higher probability to develop a disease.

Recombinant Proteins

In contrast to live *Candida* vaccines (how attenuated they may be), recombinant vaccines are generally perceived as potentially safer to the human host, because they do not contain any infectious agent. This characteristic makes recombinant protein vaccines more suitable especially for the immunocompromised individuals that would most benefit from such a vaccine. In general, this strategy has focused on proteins expressed on the surface of the fungus, such as CWPs or adhesion proteins, to ensure that epitopes would be easily accessible and "visible" by the immune system. Efforts also often concentrated on proteins critical to virulence, such as hyphal-specific effectors, in order to mount immune responses specifically against the pathogenic forms of the fungus. Besides, current work has shown that the strongest adaptive immune memory responses are often directed toward hyphal-associated proteins.

Agglutinin-Like Sequence Proteins

Agglutinin-like sequence (Als) proteins are located at the surface of *C. albicans* and play important roles in the adhesion to human endothelial cells and in the development of invasive candidiasis (51). Because these proteins are both exposed on the fungal surface and important for invasion of host cells, several vaccines have been proposed against invasive candidiasis using recombinant versions of various Als proteins, including Als1p and Als3p, formulated with or without adjuvants.

The recombinant N-terminus of Als1 (rAls1p-N) was produced and purified from *S. cerevisiae* and used in combination with Complete Freund's Adjuvant (CFA) subcutaneously and boosted with different doses of rAls1p-N with Incomplete Freund's Adjuvant (IFA) at day 21 post-immunization; lethal challenge with live *C. albicans* showed a survival rate of 50–57% and a decrease in fungal burden (36). This vaccine was deemed effective and improved survival in both immunocompetent and neutropenic mice and in murine models of oropharyngeal candidiasis and *Candida* vaginitis (35).

Vaccination with the recombinant N-terminus of Als3 (rAls3p-N) induced a stronger antibody response and survival rate compared with the rAls1p-N vaccine, and was more effective than rAls1p-N in the murine model of oropharyngeal and vaginal candidiasis (37). Interestingly, it showed also protection against *S. aureus* infection (38), suggesting the existence of shared immune epitopes between these distantly related species; consistently, *Candida* Als3p is structurally similar to a clumping factor of *S. aureus* (52).

Bar et al. identified an antigenic peptide by immunoproteomic approaches called pAls, widely conserved in many non-*Candida* species; this epitope mediates T-cell-dependent protection from invasive candidiasis. This pAls epitope is found in the Als3 protein and the authors suggest an implication of this epitope in the protective effect of the Als3 vaccine (53).

NDV-3A, a rAls3p-N vaccine formulated with Alhydrogel adjuvant, has been tested in a phase-I clinical trial. The trial was performed on 40 healthy volunteers and showed an increase in antibody titers at two different doses, as well as increased cytokine responses and IgG and IgA_1 titers in the revaccinated subjects (39). A few adverse events have been described, but overall the vaccine was well tolerated by the subjects and hence the results were deemed as promising. Like the version of rAls3p-N vaccine without adjuvant, NDV-3 also showed activity against *S. aureus* infection, further corroborating the idea that the vaccination with *Candida* antigens containing epitope homologs found in other organisms can be harnessed to generate "convergent immunity." NDV-3A is now in a phase-II clinical trial to test the immunotherapeutic effect of the vaccine in women with RVVC.[1]

Secreted Aspartil Proteases

Secreted Aspartil Proteases (SAP) constitute a family of 10 proteins secreted by *C. albicans*, which are required for adhesion, epithelial, and endothelial invasion and fungal cell metabolism (54, 55). Sap2p is the most abundantly expressed SAP in *C. albicans* and its recombinant form has been used to immunize rats intravaginally or intranasally, either using cholera toxin as an adjuvant or without an adjuvant; the vaccination resulted in the clearance of the *Candida* vaginal infection (40). The same research group also developed PEV7, a truncated version of Sap2p (amino acids 77–400) incorporated into the lipid bilayer of influenza virosomes. Because preclinical data demonstrated that intramuscularly vaccinated rats showed protection against vaginal candidiasis (41), PEV7 has progressed to clinical trials as a therapeutic vaccine for the treatment of RVVC. In a randomized phase-I trial, the subjects, whether immunized *via* intramuscular injections or by intravaginal capsules, showed a strong B-cell-mediated immune response in vaginal and cervical samples. All volunteers showed a mucosal immune response with consistently high titers across the groups, the response was dose-dependent and no serious adverse events were reported.[2]

Heat Shock Protein 90 (Hsp90p)

Heat shock protein 90 is a highly conserved stress-induced chaperone, with key functions in setting cellular responses to stressful stimuli, which is indispensable for yeast viability and located in the cell wall of *C. albicans*. The 47-kDa carboxyl fragment of *C. albicans* Hsp90 had been identified as a *Candida* antigen in different studies (56). This antigen is very abundant and immunogenic, and the presence of antibodies against Hsp90 correlates with good prognosis, whereas low levels are associated with mortality (57). Various Hsp90 epitopes (LSREM, LKVIRK, and DEPAGE) have been identified using a phage-display library and shown to induce Hsp90-specific serum antibodies and to prolong survival in a mouse model of systemic candidiasis (42). A recombinant protein called Mycograb, consisting of cross-linked Hsp90 NKILKVIRKNIVKK peptide-binding variable domains of human antibody heavy and light chains, was constructed and expressed in *Escherichia coli*. When combined with amphotericin B, Mycograb produced significant improvement in patients with invasive candidiasis (58, 59). However, the drug was refused by the European Medicines Agency on the grounds of product safety and quality, and its ability to potentiate the effects of amphotericin B were later found to be non Hsp90-specific (60).

Hyphally Regulated Proteins

Hyphally regulated 1 (Hyr1) is a glycosyl phosphatidylinositol (GPI)-anchor mannoprotein that is expressed during hyphal formation on the cell wall of *C. albicans*. A recombinant version of the N-terminus of Hyr1 (rHyr1p-N) was produced in *E. coli* and used to immunize mice *via* subcutaneous injection with either CFA or aluminum hydroxide and boosted on day 21 with IFA. After 2 weeks, vaccinated mice were challenged with a lethal dose of *C. albicans* and non-albicans *Candida* (NAC) species. The vaccine was effective against infections by *C. albicans*, *C. glabrata*, *C. krusei*, *C. parapsilosis*, and *C. tropicalis* in both immunocompetent and neutropenic mice (43).

CWP Extracts

Not only purified recombinant proteins, but also crude *C. albicans* cell wall extracts, seem to be effective to protect against invasive candidiasis. In fact, subcutaneous immunization with β-mercaptoethanol-extracted *C. albicans* CWPs in combination with Ribi Adjuvant System (RAS) R-700 followed by a booster injected 21 days after the first immunization, conferred protection in 75% of immunized mice after a lethal challenge with *C. albicans* (44). However, such crude preparations are unlikely to progress to clinical trials due to the difficulty in characterizing and standardizing such complex vaccine formulations.

Glycoconjugates

An alternative to protein-based vaccines is to use glycans commonly found in fungal cell walls but absent in host cells. In fact, the *Candida* cell wall represents a hub of pathogen-associated molecular pattern (PAMP)–pathogen recognition receptor (PRR) interactions that dictate downstream immune responses (61). For these reasons, various cell wall polysaccharides have been tested as vaccine targets against systemic candidiasis.

Mannans and Derivative Peptide Conjugates

Mannans are polymers of *O*-linked or *N*-linked mannosides attached to CWPs and constitute the outermost, and hence most accessible, layer of the *C. albicans* cell wall. While *O*-mannans are predominately recognized by TLR4 (62), *N*-mannans are recognized by a multitude of receptors, including the Mannose Receptor (MR), DC-SIGN, Dectin-2, Galectin-3, and Mincle (63). Classically, *Candida* mannans have been attributed immunesuppressive properties (64); but the discovery that *N*-linked mannans are critically required for recognition by DC-SIGN and MR on human dendritic cells (DCs) has established a rational for conjugating them to antigenic peptides. Consistent with this idea, mannosylated antigens were shown to be presented more effectively than non-mannosylated forms (65).

In one example, researchers have used computational epitope searches to select 6 peptides in *C. albicans* CWPs (fructose-bisphosphate aldolase; methyltetrahydropteroyltriglutamate; hyphal wall protein-1; enolase; glyceraldehyde-3-phosphate dehydrogenase; and phosphoglycerate kinase) and conjugated these peptides with β-mannans to generate the first series of fully synthetic glycopeptide vaccines against systemic candidiasis. Depending on the specific peptide, the immunized mice showed 80–100% survival and reduced kidney fungal burden after the challenge with *C. albicans* (45).

More recent studies showed than the vaccination with BSA-based conjugates bearing synthetic α-1,6-branched oligomannosides induced humoral responses in mice and induced production of potentially protective antibodies (46). Immunization with

[1] https://www.clinicaltrials.gov/ct2/show/NCT01926028 (Accessed: April 17, 2018).
[2] https://www.clinicaltrials.gov/ct2/show/NCT01067131 (Accessed: April 17, 2018).

Candida mannan-derived branched 3,6-di-*O*-substituted α-oligomannosides conjugated to BSA (M5-BSA and M6-BSA conjugates) induced similar levels of CD19$^+$ B lymphocytes, but the candidacidal activity of polymorphonuclear leukocytes induced by opsonization with M6-BSA antisera was higher than that with M5-BSA, thus revealing some important differences in their ability to induce an effective and protective immune response against *Candida* infection (66). A better mechanistic understanding of the type of immune responses elicited by various *Candida* mannans is therefore required before these compounds can progress to clinical studies.

β-Glucan and Derivative Conjugate Vaccines

Directly underneath mannans lies a hidden layer of β-glucans, which is an essential component of the *Candida* cell wall (67) and elicits Dectin-1-dependent innate immune responses that are critical for host protection against fungi (68). Moreover, β-glucans from the *C. albicans* cell wall are highly immunogenic, leading to epigenetic and metabolic reprogramming of monocytes/macrophages, elevated cytokine responses and to innate host protection against systemic candidiasis (69–71). This phenomenon was termed "trained immunity" (72) and is discussed further in Section "Trained Immunity—Toward the Next Generation of Vaccines?"

When used in combination with the MF59 adjuvant (oil-in-water emulsion of squalene oil), fungal-derived β-glucan showed protection against murine vaginal candidiasis but is currently not approved for use in humans (47). The conjugation of the β-glucan preparation Laminarin (a soluble Dectin-1 ligand derived from seaweed) with diphtheria toxoid (Lam-CRM197) resulted, in murine models, in significant protection against systemic candidiasis, as well as cross-protection against aspergillosis (48). In further studies, researchers conjugated the diphtheria toxoid with various other β-glucan preparations, such as curdlan (a high-molecular-weight β-1,3-linked polymer produced by bacteria) or synthetic oligosaccharides that contained either only linear β-1,3 linkages or also branched β-1,6 linkages, and combined it with the MF59 adjuvant. Interestingly, conjugates raising antibodies against β-1,3-glucans were protective in a mouse model of systemic candidiasis, while anti-β-1,6-antibodies appeared to reverse this effect (73). Overall these studies reveal the high immunogenicity of β-glucans but also a surprisingly complex relationship between antifungal immune responses and host protection against fungal infections. Moreover, different *Candida* species expose β-glucan on their surface to different degrees (74), questioning how broadly protective such vaccines would be against various candidiasis-causing agents. These complexities will need to be better understood, before these polysaccharides can be translated into safe and effective antifungal vaccines for human use.

COMMON CHALLENGES FACED SO FAR

The fact that *C. albicans* is thought to have coevolved with humans for at least the past 2,000 years (75), that it is a lifelong inhabitant of humans that colonizes the gastrointestinal (GI) tract since birth (76) and that is then transmitted between family members (77), poses a few conceptual and technical challenges in the development of an anti-*Candida* vaccine. On one hand, this long interaction history with the human GI tract suggests that these fungi might have evolved a series of mechanisms to escape various host immune defenses, including a high genetic, phenotypic and morphological plasticity (78). On the other hand, humans might have learned to recognize *Candida* species as commensals and might have therefore developed immune tolerance toward them—breaking this tolerance is expected to be both difficult and in some cases even counterproductive (79). Lastly, the highest risk group for *Candida*-related infections is represented by immunocompromised and immunosuppressed patients, i.e., a class of individuals that is intrinsically less responsive to immunization. Here, we will outline each of these issues in greater detail, and later we will attempt to offer some possible ways forward based on recent learning points.

Zeroing in on a Moving Target

Morphological and Phenotypic Plasticity of *C. albicans*

A first difficulty encountered when designing an anti-*Candida* vaccine lies in the morphological and phenotypic plasticity of these fungi (**Figure 1**), which makes them almost "moving targets." In fact, *C. albicans* is a polymorphic fungus that can reversibly transition between yeast, pseudohyphal and hyphal forms—a property that is closely linked to its evolutionary adaptation to life inside the human host. When *C. albicans* grows in its unicellular yeast form, it is commonly regarded as a harmless colonizer, while its switching to the hyphal form is related to pathogenesis, in that hyphae adhere to and invade epithelial cells resulting in extensive damage to host cells (80).

Interestingly, *C. albicans* isolated from the blood of patients with candidiasis often matches those recovered from the GI tract of the same patients (81), suggesting that *C. albicans* invades the bloodstream directly from the GI tract. How exactly the GI barrier is crossed is currently not completely understood, but it is thought to be mediated by or facilitated by epithelial tissue damage; and that *C. albicans* hyphal formation might be critical for this ability (82). Therefore, hyphal-specific or hyphal-associated CWPs such as Hyr1, Hwp2, Plb5, and Sod5 have all been proposed as vaccine target candidates (83).

However, vaccines against hyphal CWPs alone may not be sufficient to provide immune protection. While direct penetration into the mucosal surfaces by the hyphal form is important for invasion, widespread dissemination through the bloodstream and to the organs is thought to be facilitated by the yeast form (84). As both yeast and hyphae are detected in patients with candidiasis (85), and some mutants defective in filamentation displayed similar virulence as the wild type (86), it is the plasticity of this morphological switch, rather than the hyphal form alone, that appears to critically underlie the virulence of *C. albicans*. The variety of different forms of *C. albicans* that exist in various organs may represent a further challenge in vaccine development. For example, in mouse infection models, while *C. albicans* is mainly observed in the hyphal form in the kidney, it actually persists as a yeast in the spleen and the liver (87). To achieve sterile protection, the ideal vaccine should be able to stimulate clearance of the fungus not only in the former but also in the latter organs.

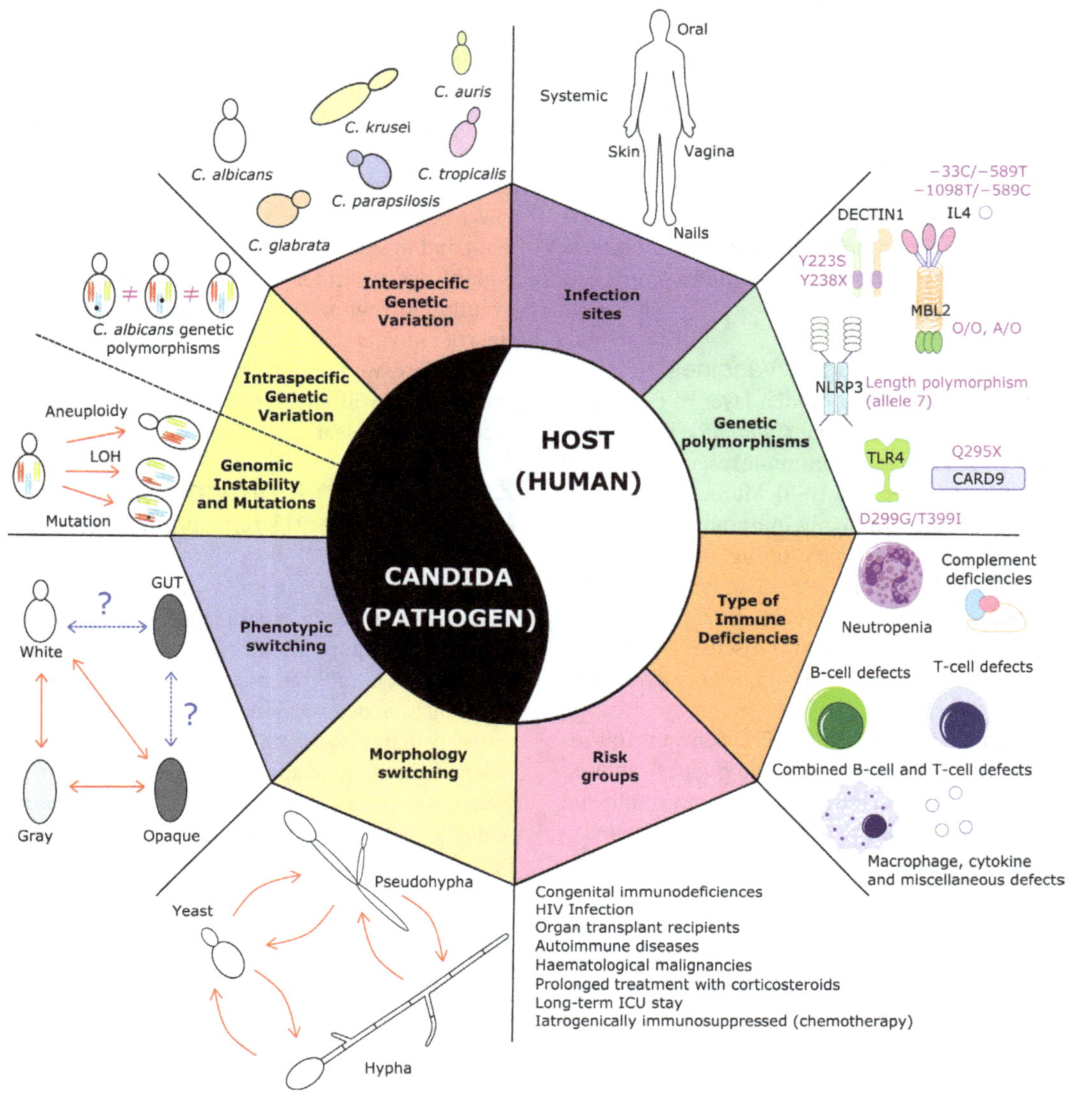

FIGURE 1 | Sources of variability in *Candida*-related infections. The figure illustrates the plethora of variations at the level of the pathogen and the host, which pose serious challenges in the development of an antifungal vaccine. On the pathogen side, *Candida* species display large phenotypic, morphological, and genetic variation and dynamics; on the host side, candidiasis can occur in different types of patients, with different types of immune system dysfunction or carrying different genetic polymorphisms, and in different anatomical sites. Taken together, this broad and diverse clinical spectrum of the disease makes it difficult to design a "one-size-fits-all" vaccine that would protect all these different patients from all these different types of fungal infection.

A second form of plasticity in *C. albicans* is a heritable white-to-opaque phenotypic switch. Smooth and white colonies are observed during the white phase, while elongated and rod-like cells giving rise to flattened gray colonies appear in the opaque phase (88). Opaque cells are less virulent than white cells in mouse bloodstream infection model (89, 90) and undergo filamentation in response to stimuli that are distinct form white cells (91). The switching between white and opaque phenotypes may facilitate *C. albicans* to evade from host immune response, as opaque cells are less susceptible to be phagocytosed by macrophages (92) and also capable to evade killing by neutrophils (93).

Another phenotypic switch, known as gastrointestinally induced transition (GUT) and found as a result of *WOR1* overexpression, was recently described to confer increased fitness of *C. albicans* in the GI tract of antibiotic-treated mice (94). While GUT cells share some resemblance with opaque cells, such as a reduced virulent in mouse bloodstream infections, they appear to be transcriptionally distinct from both white and opaque cells, therefore likely representing an independent epigenetic state. Whether GUT cells also arise in mucosal or systemic infections in humans and whether these cells express specific factors that are important for disease remains to be determined, but their discovery nonetheless sheds light on additional layers of complexity in the morphological and phenotypic plasticity of this fungal pathogen.

As a whole, morphological and phenotypic plasticity renders *C. albicans* a "skilled transformer" (78). Therefore, vaccines directed against single targets expressed only in a particular fungal form may not be sufficient to raise protective immune responses against the entire arsenal of antigens and virulence factors that are dynamically presented to the host in different organs at different times.

Genomic Plasticity of *Candida albicans*

Candida albicans, apart from morphological and phenotypic transitions, also displays a high degree of genomic plasticity including gross chromosome rearrangement, aneuploidy and loss of heterozygosity when exposed to different stresses. Such genomic changes equip *C. albicans* to rapidly adapt to an adverse environment by changing the copy number of specific genes on a given chromosome (95). One of the well-known examples is the identification of an isochromosome composed of two copies of the left arm of chromosome 5, which is associated with azole resistance (17) due to amplification of two resistance genes, *ERG11* and *TAC1* (96). Genomic variations are commonly observed in clinical *C. albicans* isolates, which includes copy number variations, chromosomal inversions, subtelomeric hypervariation, loss of heterozygosity and aneuploidy (97). Moreover, *C. albicans* with increased DNA content (aneuploidy) occurs at a higher frequency in clinical blood isolates when compared with mucosal isolates (98). This suggests that acquisition of aneuploidy, and the consequent copy number alteration of specific genes, may provide a rapid mechanism to modify the expression pattern of certain antigens that were originally targeted by a certain vaccine.

Non-albicans *Candida* Species

The genus *Candida* represents a highly heterogeneous group of >50 known species. Nevertheless, >90% of the invasive *Candida* infections are caused by *C. albicans*, *C. glabrata*, *C. parapsilosis*, *C. tropicalis* or *C. krusei* (99). While this figure has remained largely unchanged, the relative ranking between these species has seen significant shifts in different regions over the past decades (100). While *C. albicans* is still considered the most common species causing candidemia, increasing rates of NAC species in candidemia have been reported worldwide (18, 101, 102). Moreover, significant variations of *Candida* species are detected in different patient groups and geographical regions (103). For instance, *C. glabrata* has emerged as an important pathogen in northern Europe, the United States and Canada with a higher rate of incidence in adults than in children, and lower in neonates (104). *C. parapsilosis*, instead, is more prominent in southern Europe, Asia, and South America and is mostly associated with low-birth-weight neonates and transplant recipients (105). Finally, *C. tropicalis* constitutes 20–45% of *Candida* isolates in the Asia-Pacific region (18) and invasive candidiasis due to this species is commonly associated with patients with neutropenia and malignancy (106). Recently, the emergence of *C. auris*, first reported in 2009 in Japan, highlights a new challenge of antifungal treatment, as *C. auris* is often multi-drug resistant and also difficult to be diagnosed with standard laboratory methods, which is contributing to its rapid spreading over multiple countries (107). In addition to NAC, multispecies candidemia is also emerging as a threat (108). Taken together, these observations suggest that vaccines specifically against *C. albicans* alone may not be sufficient to provide protection against candidiasis caused by other emerging NAC species or mixed *Candida* species infections.

Breaking Immune Tolerance Toward Fungi

A microbial ecosystem in the human intestine, known as the gut microbiome, harbors more than 100 trillion microorganisms and thus immunological tolerogenic responses are required in order to maintain gut homeostasis and prevent chronic inflammation (109). Immune tolerance toward human gut commensals, such as *Bacteroides fragilis* and certain *Clostridia* species, is maintained by regulatory T (Treg) cell responses (110–112). Apart from bacteria, *Candida* species are the most common fungal species found in the GI tract; it is therefore reasonable to assume that similar tolerance mechanisms might have evolved to regulate the commensal relationship between humans and fungi. One such mechanism, which likely resulted from the coevolution of bacterial microbiota, commensal fungi and host immune system, relies on the metabolic tryptophan-AhR pathway and 2,3-indoleamine dioxygenase (IDO) (113, 114). *C. albicans*, being a lifelong inhabitant of humans that colonizes the GI tract since birth (76), is able to induce IDO expression in DCs. IDO-expressing DCs then promote tolerogenic Treg responses, probably facilitating the switch from pathogenicity to commensalism in *C. albicans* (115).

The existence of immunological tolerance toward *C. albicans* and probably other *Candida* species poses two serious challenges for anti-*Candida* vaccine development. First, unlike obligate pathogens with no commensal relationship with humans, this tolerance represents an additional hurdle toward establishment of effective and protective immunological memory. Second, much of the clinical manifestations of *Candida*-related infections are more due to host-derived immunopathology than to pathogen-derived host damage (116, 117). For instance, mice lacking the chemokine receptor Ccr1, which is critical for neutrophil recruitment, display improved renal function during invasive candidiasis (118) and administration of Ccr1 antagonists reduces renal tissue injury and improves survival in mice challenged with systemic candidiasis (119).

It has thus been argued that a careful balance between immunity and tolerance must be established to maintain commensalism (79). Breaking host tolerance against fungi and the self-regulated equilibrium between Th17 and Treg responses might lead to undesired consequences, such as exacerbating fungal disease progression or other underlying inflammatory or autoimmune conditions.

Vaccinating Patients With No Immunity

Fungal infections, especially invasive ones, most frequently occur in individuals with compromised or suppressed immunity. Hence, it appears that the patients that would mostly benefit from a future antifungal vaccine are also those least likely to respond to it. Is such an effort then even justified? Or is it doomed to fail from the start? Answers to these questions are more complex than one might expect.

Efficacy and safety are always a primary concern in any vaccine development effort, and especially so in vulnerable subjects such as patients with lowered immune defenses. As mentioned, the risk of *Candida* infection is especially high in patients with neutropenia, hematological malignancies, solid-organ transplants, prolonged treatment with corticosteroids, long-term ICU stay, chemotherapy and HIV infection (120). The ideal anti-*Candida* vaccine should possess all of the following properties: (i) zero risk of causing a *Candida*-related infection or to exacerbate the immunopathology associated with an ongoing or future fungal infection, (ii) be immunogenic enough to elicit protection in patients with little or no immunity, (iii) do not increase the risk of aggravating any underlying disease. While this may sound like a "catch-22," vaccinating immunocompromised patients has been attempted with various degrees of success in the past for other kinds of infection.

For instance, a single-dose of the pneumococcal vaccine was recently deemed safe and immunogenic in children under active immunosuppressive therapy (121). However, low immunogenicity of the meningococcal vaccine was reported in solid-organ transplant recipients (122). Lower immunogenicity against the influenza vaccine is generally reported in immunocompromised patients when compared with healthy individuals (123). However, using a high-dose influenza vaccine was reported to be more immunogenic than the standard dose in children and young adults with leukemia or solid tumors, although not in those with HIV (124).

What is progressively becoming clearer is that a "one-size-fits-all" anti-*Candida* vaccine may never see the light. For example, a vaccine effective in preventing candidemia in neutropenic cancer patients might not be effective in T-cell-deficient HIV or transplant patients and *vice versa*. As a whole, one of the greatest challenges of any antifungal vaccine will be to deal with the large diversity of underlying disease states and of the associated types of immunosuppression that characterizes this highly heterogeneous risk group (**Figure 1**).

BACK TO THE FUTURE: NEW VACCINE STRATEGIES ON THE HORIZON?

Multivalent Vaccines

Some monovalent vaccines, i.e., those directed against a single strain or a single antigen, are very effective. A good example is represented by the measles vaccine, which has almost completely eradicated the disease in countries where it has been employed consistently throughout the population (125). However, the most recent vaccines tend to be multivalent, i.e., they carry multiple antigens of two or more strains/serotypes of the same pathogen. An example is the quadrivalent meningococcal vaccine that protects against 4 serogroups (A, C, W-135, and Y) of meningococci. The quadrivalent vaccine was shown to be more effective in reducing invasive meningococcal disease incidence when compared with the monovalent C vaccine (126). Another example is the 13-valent pneumococcal conjugate vaccine (PCV13) containing antigens from 13 serotypes of pneumococci. A replacement of the seven-valent pneumococcal conjugate vaccine (PCV7) with PCV13 covering additional six of the most prevalent serotypes that are not included in PCV7 successfully reduced the burden of pneumococcal disease in pediatric populations (127).

As *C. albicans* itself expresses a range of virulence factors and several NAC species also carry their own species-specific virulence factors, the recent development of univalent subunit vaccines may face practical obstacles. Considering above success stories of multivalent vaccines applied to bacterial infections, it has been argued that a better approach toward the development of an anti-*Candida* vaccine would be to simultaneously target multiple unrelated virulence-associated antigens (78). The predicted advantages include the lower probability of selecting "escape mutants" and a higher selectively against the pathogenic form of *C. albicans*, thus sparing the commensal—and perhaps beneficial—form of this gut inhabitant. In particular, it was proposed to combine a few of the univalent vaccines that have so far progressed furthest in clinical trials, such as the Als3 and the Sap2 vaccines (78).

More recently, a study reported that a newly identified cytolytic peptide toxin, Candidalysin, is secreted from *C. albicans* to induce epithelial cell damage and inflammation and to facilitate tissue invasion (128). Intriguingly, the mechanism of action of this novel virulence factor is partially uncoupled from the hyphal morphogenesis program, suggesting that a vaccine targeting both the former and the latter would be more effective than one targeting only one or the other. In general, as more knowledge is gained on the pathogenesis of *C. albicans*, further targets could be added to the list of antigens to be included in a multivalent formulation. At the same time, we can also predict that as the complexity of the vaccine increases, so will regulatory hurdles and manufacturing challenges. Nevertheless, testing the efficacy and safety of a combination of more and more antigens is likely going to be crucial for better vaccines against *Candida* infections.

Trained Immunity—Toward the Next Generation of Vaccines?

The classic vaccination paradigm is based on adaptive immunity and the raising of long-lasting, protective B- and/or T-cell memory responses (129). As mentioned above, mounting protective antibodies or T-cell responses across different *Candida* strains and species may be challenging due to their large genetic, phenotypic and morphological variation and plasticity. In addition, the capacity to mount B- and T-memory responses may be impaired in immunocompromised or immunosuppressed patients such as HIV/AIDS or transplant recipients who are among the highest risk groups for candidiasis. Though innate immunity lacks most of the properties of classical immunological memory, various studies have demonstrated that certain vaccines induce a type of innate immune memory known as "trained immunity" (**Figure 2**), which is mediated by monocytes, macrophages or NK cells (72). For instance, BCG vaccination reduces mortality in low-birth-weight children (130, 131) and vaccination against measles reduces all-cause mortality in childhood (132), suggesting non-specific cross-protection against other infections (133).

Testing whether such innate memory responses could be stimulated also in adults, PBMCs obtained 2 weeks or 3 months after BCG vaccination were found to produce higher levels of pro-inflammatory cytokines such as TNF-α and IL-1β when restimulated *in vitro* with *C. albicans* (134). Such increased cytokine production capacity involved epigenetic changes mediated by increased H3K4 trimethylation in monocytes and was dependent on NOD2 and Rip2. Similarly, SCID mice vaccinated by BCG are protected against lethal *C. albicans* infection through a T and B lymphocyte-independent mechanism (134). Apart from the increased production of pro-inflammatory cytokines in monocytes, another study reported that BCG vaccination also

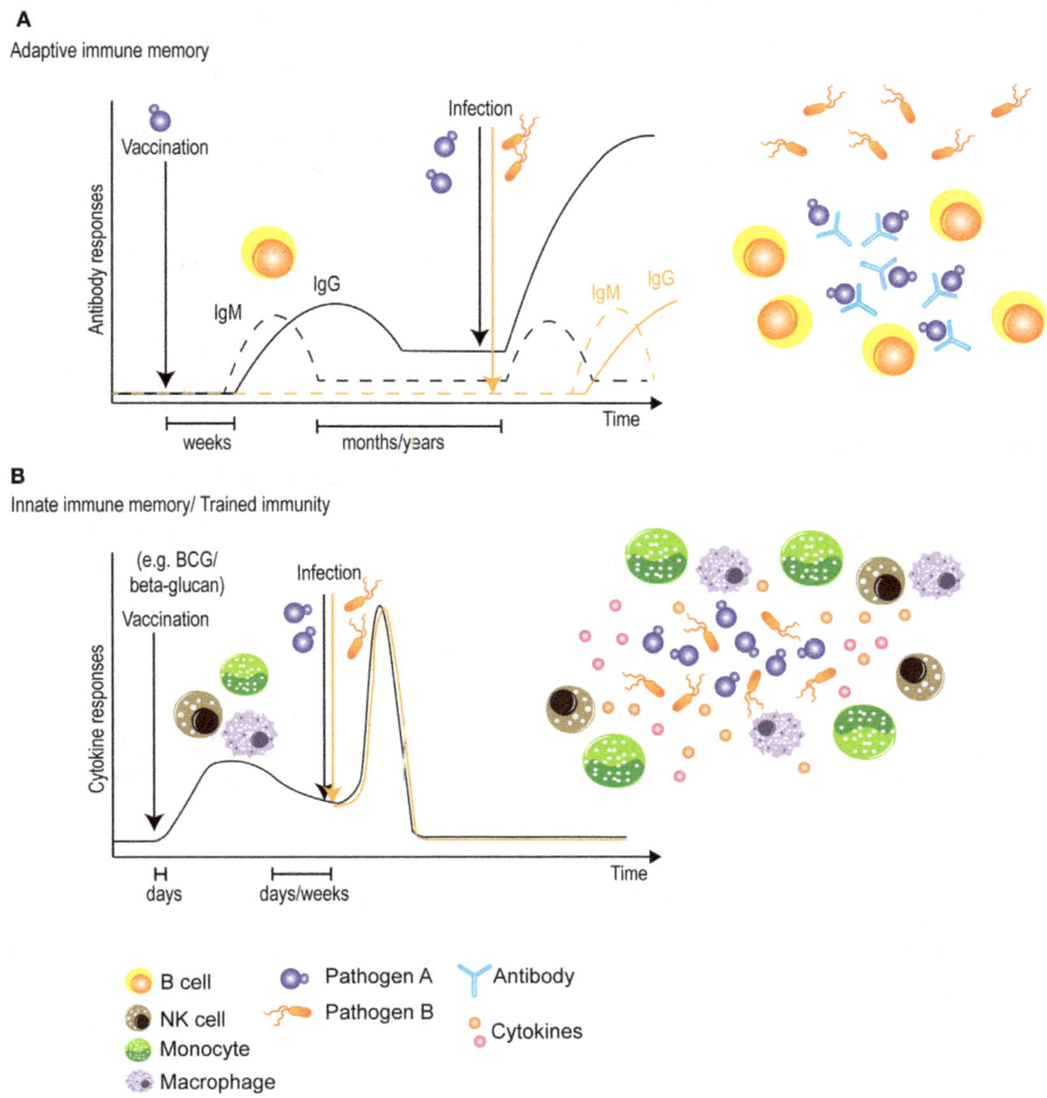

FIGURE 2 | Vaccination strategies targeting adaptive or innate immune memory. **(A)** Classical vaccines initially induce a slow adaptive immune response. During this primary response, however, adaptive immune memory is raised, allowing for a more rapid and stronger secondary response upon encounter of the targeted pathogen. This protection is very long-lived, but is limited to the antigens present in the vaccine. **(B)** Most vaccines also induce a very rapid innate immune response. During this primary response, some vaccines also "train" innate immune cells, allowing for a stronger secondary innate response against pathogens. One advantage of this innate immune memory is the non-specificity of the secondary response, which could potentially lead to broadly cross-protective vaccines. A potential disadvantage would be the relatively shorter time window of protection.

enhances IL-1β production in human NK cells when stimulated with *C. albicans*. BCG-induced protection against disseminated *C. albicans* infection was shown to be partially dependent on NK cells in NOD-SCID-IL2Rγ$^{-/-}$ mice (135). These studies warrant clinical trials to test if BCG vaccination could reduce the risk of candidiasis in high-risk patients.

Consistent with what reviewed in Section "Live-Attenuated Vaccines," systemic infection of mice with an avirulent *C. albicans* strain was shown by Antonio Cassone's group to confer protection against a subsequent challenge with a pathogenic *C. albicans* strain (26). The protection was shown to be non-specific, as cross-protection against *C. tropicalis* and *Staphylococcus aureus* was equally observed, and to be mediated by macrophage-like cells, as adoptive transfer of "plastic-adherent" cells was sufficient to confer protection against subsequent challenge with a virulent strain of *C. albicans*. Several years later, Mihai Netea's group demonstrated that vaccination of both wild-type and RAG1$^{-/-}$ mice, but not CCR2$^{-/-}$ mice that lack circulating monocytes, with a sublethal dose of a virulent *C. albicans* strain similarly affords protection against reinfection with a lethal dose of the same strain, and dubbed the phenomenon "trained immunity" (69). Increased TNF-α and IL-6 levels from the trained monocytes were induced by β-glucan found in the *C. albicans* cell wall and sensed by Dectin-1 on host cells. In another study, mice trained with *Candida*-derived β-glucan showed increased serum levels of TNF-α and IL-6 when challenged by LPS four days later, but the response was transient and diminished after 20 days (136). To test the possibility of using β-glucan as a vaccine or "immune trainer" in humans, a pilot study was conducted by orally administrating β-glucan to healthy volunteers and subsequently testing innate immune responses in PBMCs restimulated *in vitro* with *C. albicans* (137). Enhanced innate immune responses as in the mouse model, however, were unfortunately not observed in this human study. This may probably be due to the solubility of β-glucan and the absorbability of β-glucan in the human GI tract. Further studies using different administration routes are still worth exploring, in order to establish whether β-glucan could one day be used as a vaccine to induce trained immunity against subsequent *Candida* infections (**Figure 2**).

CONCLUDING REMARKS

With the increasing reports of multidrug resistance in several *Candida* species, prophylactic vaccination of at-risk patients likely represents a more effective long-term measure to reduce the growing incidence of *Candida*-related infections. The main challenges faced by *Candida* vaccine developers are the large variation and plasticity of these fungi, the existence of preestablished immunological tolerance and the difficulty in raising protective memory responses in patients with impaired adaptive immunity. In addition to multivalent vaccines, we propose that future vaccine development efforts should harness the growing mechanistic understanding of trained innate immunity, which might provide not only protection against candidiasis but also potentially cross-protection against a wide range of opportunistic infections. Further research on the mechanism, efficacy, and safety of raising trained immunity especially in immunocompromised patients would pave the way toward the development of a new generation of vaccines against *Candida*-related and other nosocomial infections.

AUTHOR CONTRIBUTIONS

GT and JR-C contributed equally as first authors. GT and JR-C conducted literature review and drafted the manuscript. NP conceptualized and oversaw the study and revised the manuscript. All authors read and approved the submitted version.

ACKNOWLEDGMENTS

The authors are thankful to all authors of the cited literature for their invaluable contributions and apologize for any omissions that were due to lack of space.

FUNDING

This work was supported by an A*STAR Investigatorship to NP (JCO/1437a00117) and by core funding by the Singapore Immunology Network (SIgN), A*STAR.

REFERENCES

1. Fishman JA. Infection in solid-organ transplant recipients. *N Engl J Med* (2007) 357:2601–14. doi:10.1056/NEJMra064928
2. Teoh F, Pavelka N. How chemotherapy increases the risk of systemic candidiasis in cancer patients: current paradigm and future directions. *Pathogens* (2016) 5:6. doi:10.3390/pathogens5010006
3. Menzin J, Meyers JL, Friedman M, Perfect JR, Langston AA, Danna RP, et al. Mortality, length of hospitalization, and costs associated with invasive fungal infections in high-risk patients. *Am J Health Syst Pharm* (2009) 66:1711–7. doi:10.2146/ajhp080325
4. Pfaller MA, Diekema DJ. Epidemiology of invasive candidiasis: a persistent public health problem. *Clin Microbiol Rev* (2007) 20:133–63. doi:10.1128/CMR.00029-06
5. Richardson M, Lass-Florl C. Changing epidemiology of systemic fungal infections. *Clin Microbiol Infect* (2008) 14(Suppl 4):5–24. doi:10.1111/j.1469-0691.2008.01978.x
6. Wisplinghoff H, Bischoff T, Tallent SM, Seifert H, Wenzel RP, Edmond MB. Nosocomial bloodstream infections in US hospitals: analysis of 24,179 cases from a prospective nationwide surveillance study. *Clin Infect Dis* (2004) 39:309–17. doi:10.1086/421946
7. Peman J, Canton E, Quindos G, Eraso E, Alcoba J, Guinea J, et al. Epidemiology, species distribution and in vitro antifungal susceptibility of fungaemia in a Spanish multicentre prospective survey. *J Antimicrob Chemother* (2012) 67:1181–7. doi:10.1093/jac/dks019
8. Cleveland AA, Farley MM, Harrison LH, Stein B, Hollick R, Lockhart SR, et al. Changes in incidence and antifungal drug resistance in candidemia: results from population-based laboratory surveillance in Atlanta and Baltimore, 2008-2011. *Clin Infect Dis* (2012) 55:1352–61. doi:10.1093/cid/cis697
9. Eliakim-Raz N, Babaoff R, Yahav D, Yanai S, Shaked H, Bishara J. Epidemiology, microbiology, clinical characteristics, and outcomes of candidemia in internal medicine wards-a retrospective study. *Int J Infect Dis* (2016) 52:49–54. doi:10.1016/j.ijid.2016.09.018
10. Brown GD, Denning DW, Gow NA, Levitz SM, Netea MG, White TC. Hidden killers: human fungal infections. *Sci Transl Med* (2012) 4:165rv113. doi:10.1126/scitranslmed.3004404
11. Smith PB, Morgan J, Benjamin JD, Fridkin SK, Sanza LT, Harrison LH, et al. Excess costs of hospital care associated with neonatal candidemia. *Pediatr Infect Dis J* (2007) 26:197–200. doi:10.1097/01.inf.0000253973.89097.c0
12. Pappas PG, Kauffman CA, Andes DR, Clancy CJ, Marr KA, Ostrosky-Zeichner L, et al. Clinical practice guideline for the management of candidiasis: 2016 update by the Infectious Diseases Society of America. *Clin Infect Dis* (2016) 62:e1–50. doi:10.1093/cid/civ1194
13. Havlickova B, Czaika VA, Friedrich M. Epidemiological trends in skin mycoses worldwide. *Mycoses* (2008) 51(Suppl 4):2–15. doi:10.1111/j.1439-0507.2008.01606.x
14. Sobel JD. Vulvovaginal candidosis. *Lancet* (2007) 369:1961–71. doi:10.1016/S0140-6736(07)60917-9
15. Sobel JD, Faro S, Force RW, Foxman B, Ledger WJ, Nyirjesy PR, et al. Vulvovaginal candidiasis: epidemiologic, diagnostic, and therapeutic considerations. *Am J Obstet Gynecol* (1998) 178:203–11. doi:10.1016/S0002-9378(98)80001-X
16. Cannon RD, Lamping E, Holmes AR, Niimi K, Baret PV, Keniya MV, et al. Efflux-mediated antifungal drug resistance. *Clin Microbiol Rev* (2009) 22: 291–321. doi:10.1128/CMR.00051-08
17. Selmecki A, Forche A, Berman J. Aneuploidy and isochromosome formation in drug-resistant *Candida albicans*. *Science* (2006) 313:367–70. doi:10.1126/science.1128242
18. Pfaller MA, Diekema DJ, Gibbs DL, Newell VA, Ellis D, Tullio V, et al. Results from the ARTEMIS DISK Global Antifungal Surveillance Study, 1997 to 2007: a 10.5-year analysis of susceptibilities of *Candida* species to fluconazole and voriconazole as determined by CLSI standardized disk diffusion. *J Clin Microbiol* (2010) 48:1366–77. doi:10.1128/JCM.02117-09
19. Alexander BD, Johnson MD, Pfeiffer CD, Jimenez-Ortigosa C, Catania J, Booker R, et al. Increasing echinocandin resistance in *Candida glabrata*: clinical failure correlates with presence of FKS mutations and elevated minimum inhibitory concentrations. *Clin Infect Dis* (2013) 56:1724–32. doi:10.1093/cid/cit136
20. Reales-Calderon JA, Molero G, Gil C, Martinez JL. The fungal resistome: a risk and an opportunity for the development of novel antifungal therapies. *Future Med Chem* (2016) 8:1503–20. doi:10.4155/fmc-2016-0051
21. Fadda ME, Podda GS, Pisano MB, Deplano M, Cosentino S. Prevalence of *Candida* species in different hospital wards and their susceptibility to

antifungal agents: results of a three year survey. *J Prev Med Hyg* (2008) 49:69–74. doi:10.15167/2421-4248/jpmh2008.49.2.119

22. Vermitsky JP, Edlind TD. Azole resistance in *Candida glabrata*: coordinate upregulation of multidrug transporters and evidence for a Pdr1-like transcription factor. *Antimicrob Agents Chemother* (2004) 48:3773–81. doi:10.1128/AAC.48.10.3773-3781.2004
23. Vermitsky JP, Earhart KD, Smith WL, Homayouni R, Edlind TD, Rogers PD. Pdr1 regulates multidrug resistance in *Candida glabrata*: gene disruption and genome-wide expression studies. *Mol Microbiol* (2006) 61:704–22. doi:10.1111/j.1365-2958.2006.05235.x
24. Navalkele BD, Revankar S, Chandrasekar P. *Candida auris*: a worrisome, globally emerging pathogen. *Expert Rev Anti Infect Ther* (2017) 15(9):819–27. doi:10.1080/14787210.2017.1364992
25. Denning DW, Perlin DS, Muldoon EG, Colombo AL, Chakrabarti A, Richardson MD, et al. Delivering on antimicrobial resistance agenda not possible without improving fungal diagnostic capabilities. *Emerg Infect Dis* (2017) 23:177–83. doi:10.3201/eid2302.152042
26. Bistoni F, Vecchiarelli A, Cenci E, Puccetti P, Marconi P, Cassone A. Evidence for macrophage-mediated protection against lethal *Candida albicans* infection. *Infect Immun* (1986) 51:668–74.
27. Fernandez-Arenas E, Molero G, Nombela C, Diez-Orejas R, Gil C. Low virulent strains of *Candida albicans*: unravelling the antigens for a future vaccine. *Proteomics* (2004) 4:3007–20. doi:10.1002/pmic.200400929
28. Martinez-Lopez R, Nombela C, Diez-Orejas R, Monteoliva L, Gil C. Immunoproteomic analysis of the protective response obtained from vaccination with *Candida albicans* ecm33 cell wall mutant in mice. *Proteomics* (2008) 8:2651–64. doi:10.1002/pmic.200701056
29. Saville SP, Lazzell AL, Chaturvedi AK, Monteagudo C, Lopez-Ribot JL. Efficacy of a genetically engineered *Candida albicans* tet-NRG1 strain as an experimental live attenuated vaccine against hematogenously disseminated candidiasis. *Clin Vaccine Immunol* (2009) 16:430–2. doi:10.1128/CVI.00480-08
30. Yang YL, Wang CW, Chen CT, Wang MH, Hsiao CF, Lo HJ. Non-lethal *Candida albicans* cph1/cph1 efg1/efg1 mutant partially protects mice from systemic infections by lethal wild-type cells. *Mycol Res* (2009) 113:388–90. doi:10.1016/j.mycres.2008.11.016
31. Capilla J, Clemons KV, Liu M, Levine HB, Stevens DA. *Saccharomyces cerevisiae* as a vaccine against coccidioidomycosis. *Vaccine* (2009) 27:3662–8. doi:10.1016/j.vaccine.2009.03.030
32. Liu M, Capilla J, Johansen ME, Alvarado D, Martinez M, Chen V, et al. *Saccharomyces* as a vaccine against systemic aspergillosis: 'the friend of man' a friend again? *J Med Microbiol* (2011) 60:1423–32. doi:10.1099/jmm.0.033290-0
33. Liu M, Clemons KV, Johansen ME, Martinez M, Chen V, Stevens DA. *Saccharomyces* as a vaccine against systemic candidiasis. *Immunol Invest* (2012) 41:847–55. doi:10.3109/08820139.2012.692418
34. Majumder T, Liu M, Chen V, Martinez M, Alvarado D, Clemons KV, et al. Killed *Saccharomyces cerevisiae* protects against lethal challenge of *Cryptococcus grubii*. *Mycopathologia* (2014) 178:189–95. doi:10.1007/s11046-014-9798-5
35. Spellberg BJ, Ibrahim AS, Avenissian V, Filler SG, Myers CL, Fu Y, et al. The anti-*Candida albicans* vaccine composed of the recombinant N terminus of Als1p reduces fungal burden and improves survival in both immunocompetent and immunocompromised mice. *Infect Immun* (2005) 73:6191–3. doi:10.1128/IAI.73.9.6191-6193.2005
36. Ibrahim AS, Spellberg BJ, Avanesian V, Fu Y, Edwards JE Jr. The anti-*Candida* vaccine based on the recombinant N-terminal domain of Als1p is broadly active against disseminated candidiasis. *Infect Immun* (2006) 74:3039–41. doi:10.1128/IAI.74.5.3039-3041.2006
37. Spellberg BJ, Ibrahim AS, Avanesian V, Fu Y, Myers C, Phan QT, et al. Efficacy of the anti-*Candida* rAls3p-N or rAls1p-N vaccines against disseminated and mucosal candidiasis. *J Infect Dis* (2006) 194:256–60. doi:10.1086/504691
38. Lin L, Ibrahim AS, Xu X, Farber JM, Avanesian V, Baquir B, et al. Th1-Th17 cells mediate protective adaptive immunity against *Staphylococcus aureus* and *Candida albicans* infection in mice. *PLoS Pathog* (2009) 5:e1000703. doi:10.1371/journal.ppat.1000703
39. Schmidt CS, White CJ, Ibrahim AS, Filler SG, Fu Y, Yeaman MR, et al. NDV-3, a recombinant alum-adjuvanted vaccine for *Candida* and *Staphylococcus aureus*, is safe and immunogenic in healthy adults. *Vaccine* (2012) 30:7594–600. doi:10.1016/j.vaccine.2012.10.038
40. De Bernardis F, Boccanera M, Adriani D, Girolamo A, Cassone A. Intravaginal and intranasal immunizations are equally effective in inducing vaginal antibodies and conferring protection against vaginal candidiasis. *Infect Immun* (2002) 70:2725–9. doi:10.1128/IAI.70.5.2725-2729.2002
41. De Bernardis F, Amacker M, Arancia S, Sandini S, Gremion C, Zurbriggen R, et al. A virosomal vaccine against candidal vaginitis: immunogenicity, efficacy and safety profile in animal models. *Vaccine* (2012) 30:4490–8. doi:10.1016/j.vaccine.2012.04.069
42. Yang Q, Wang L, Lu DN, Gao RJ, Song JN, Hua PY, et al. Prophylactic vaccination with phage-displayed epitope of *C. albicans* elicits protective immune responses against systemic candidiasis in C57BL/6 mice. *Vaccine* (2005) 23:4088–96. doi:10.1016/j.vaccine.2004.07.005
43. Luo G, Ibrahim AS, French SW, Edwards JE Jr, Fu Y. Active and passive immunization with rHyr1p-N protects mice against hematogenously disseminated candidiasis. *PLoS One* (2011) 6:e25909. doi:10.1371/journal.pone.0025909
44. Thomas DP, Viudes A, Monteagudo C, Lazzell AL, Saville SP, Lopez-Ribot JL. A proteomic-based approach for the identification of *Candida albicans* protein components present in a subunit vaccine that protects against disseminated candidiasis. *Proteomics* (2006) 6:6033–41. doi:10.1002/pmic.200600321
45. Xin H, Dziadek S, Bundle DR, Cutler JE. Synthetic glycopeptide vaccines combining beta-mannan and peptide epitopes induce protection against candidiasis. *Proc Natl Acad Sci U S A* (2008) 105:13526–31. doi:10.1073/pnas.0803195105
46. Paulovicova L, Paulovicova E, Karelin AA, Tsvetkov YE, Nifantiev NE, Bystricky S. Effect of branched alpha-oligomannoside structures on induction of anti-*Candida* humoral immune response. *Scand J Immunol* (2013) 77:431–41. doi:10.1111/sji.12044
47. Pietrella D, Rachini A, Torosantucci A, Chiani P, Brown AJ, Bistoni F, et al. A beta-glucan-conjugate vaccine and anti-beta-glucan antibodies are effective against murine vaginal candidiasis as assessed by a novel in vivo imaging technique. *Vaccine* (2010) 28:1717–25. doi:10.1016/j.vaccine.2009.12.021
48. Torosantucci A, Bromuro C, Chiani P, De Bernardis F, Berti F, Galli C, et al. A novel glyco-conjugate vaccine against fungal pathogens. *J Exp Med* (2005) 202:597–606. doi:10.1084/jem.20050749
49. Smith KA. Edward Jenner and the small pox vaccine. *Front Immunol* (2011) 2:21. doi:10.3389/fimmu.2011.00021
50. Lo HJ, Kohler JR, Didomenico B, Loebenberg D, Cacciapuoti A, Fink GR. Nonfilamentous *C. albicans* mutants are avirulent. *Cell* (1997) 90:939–49. doi:10.1016/S0092-8674(00)80358-X
51. Hoyer LL, Cota E. *Candida albicans* agglutinin-like sequence (Als) family vignettes: a review of Als protein structure and function. *Front Microbiol* (2016) 7:280. doi:10.3389/fmicb.2016.00280
52. Spellberg B, Ibrahim AS, Yeaman MR, Lin L, Fu Y, Avanesian V, et al. The antifungal vaccine derived from the recombinant N terminus of Als3p protects mice against the bacterium *Staphylococcus aureus*. *Infect Immun* (2008) 76:4574–80. doi:10.1128/IAI.00700-08
53. Bar E, Gladiator A, Bastidas S, Roschitzki B, Acha-Orbea H, Oxenius A, et al. A novel Th cell epitope of *Candida albicans* mediates protection from fungal infection. *J Immunol* (2012) 188:5636–43. doi:10.4049/jimmunol.1200594
54. Hube B, Naglik J. *Candida albicans* proteinases: resolving the mystery of a gene family. *Microbiology* (2001) 147:1997–2005. doi:10.1099/00221287-147-8-1997
55. Naglik JR, Challacombe SJ, Hube B. *Candida albicans* secreted aspartyl proteinases in virulence and pathogenesis. *Microbiol Mol Biol Rev* (2003) 67:400–28, table of contents. doi:10.1128/MMBR.67.3.400-428.2003
56. Matthews R, Burnie J. Cloning of a DNA sequence encoding a major fragment of the 47 kilodalton stress protein homologue of *Candida albicans*. *FEMS Microbiol Lett* (1989) 51:25–30. doi:10.1016/0378-1097(89)90071-2
57. Matthews RC, Burnie JP, Tabaqchali S. Isolation of immunodominant antigens from sera of patients with systemic candidiasis and characterization of serological response to *Candida albicans*. *J Clin Microbiol* (1987) 25:230–7.
58. Matthews RC, Rigg G, Hodgetts S, Carter T, Chapman C, Gregory C, et al. Preclinical assessment of the efficacy of mycograb, a human recombinant antibody against fungal HSP90. *Antimicrob Agents Chemother* (2003) 47:2208–16. doi:10.1128/AAC.47.7.2208-2216.2003

59. Pachl J, Svoboda P, Jacobs F, Vandewoude K, van der Hoven B, Spronk P, et al. A randomized, blinded, multicenter trial of lipid-associated amphotericin B alone versus in combination with an antibody-based inhibitor of heat shock protein 90 in patients with invasive candidiasis. *Clin Infect Dis* (2006) 42:1404–13. doi:10.1086/503428
60. Richie DL, Ghannoum MA, Isham N, Thompson KV, Ryder NS. Nonspecific effect of Mycograb on amphotericin B MIC. *Antimicrob Agents Chemother* (2012) 56:3963–4. doi:10.1128/AAC.00435-12
61. Netea MG, Brown GD, Kullberg BJ, Gow NA. An integrated model of the recognition of *Candida albicans* by the innate immune system. *Nat Rev Microbiol* (2008) 6:67–78. doi:10.1038/nrmicro1815
62. Netea MG, Gow NA, Munro CA, Bates S, Collins C, Ferwerda G, et al. Immune sensing of *Candida albicans* requires cooperative recognition of mannans and glucans by lectin and toll-like receptors. *J Clin Invest* (2006) 116:1642–50. doi:10.1172/JCI27114
63. Hall RA, Gow NA. Mannosylation in *Candida albicans*: role in cell wall function and immune recognition. *Mol Microbiol* (2013) 90:1147–61. doi:10.1111/mmi.12426
64. Nelson RD, Shibata N, Podzorski RP, Herron MJ. *Candida* mannan: chemistry, suppression of cell-mediated immunity, and possible mechanisms of action. *Clin Microbiol Rev* (1991) 4:1–19. doi:10.1128/CMR.4.1.1
65. Tan MC, Mommaas AM, Drijfhout JW, Jordens R, Onderwater JJ, Verwoerd D, et al. Mannose receptor-mediated uptake of antigens strongly enhances HLA class II-restricted antigen presentation by cultured dendritic cells. *Eur J Immunol* (1997) 27:2426–35. doi:10.1002/eji.1830270942
66. Paulovicova L, Paulovicova E, Karelin AA, Tsvetkov YE, Nifantiev NE, Bystricky S. Immune cell response to *Candida* cell wall mannan derived branched alpha-oligomannoside conjugates in mice. *J Microbiol Immunol Infect* (2015) 48:9–19. doi:10.1016/j.jmii.2013.08.020
67. Douglas CM, D'ippolito JA, Shei GJ, Meinz M, Onishi J, Marrinan JA, et al. Identification of the FKS1 gene of *Candida albicans* as the essential target of 1,3-beta-D-glucan synthase inhibitors. *Antimicrob Agents Chemother* (1997) 41:2471–9.
68. Taylor PR, Tsoni SV, Willment JA, Dennehy KM, Rosas M, Findon H, et al. Dectin-1 is required for beta-glucan recognition and control of fungal infection. *Nat Immunol* (2007) 8:31–8. doi:10.1038/ni1408
69. Quintin J, Saeed S, Martens JHA, Giamarellos-Bourboulis EJ, Ifrim DC, Logie C, et al. *Candida albicans* infection affords protection against reinfection via functional reprogramming of monocytes. *Cell Host Microbe* (2012) 12:223–32. doi:10.1016/j.chom.2012.06.006
70. Cheng SC, Quintin J, Cramer RA, Shepardson KM, Saeed S, Kumar V, et al. mTOR- and HIF-1alpha-mediated aerobic glycolysis as metabolic basis for trained immunity. *Science* (2014) 345:1250684. doi:10.1126/science.1250684
71. Saeed S, Quintin J, Kerstens HH, Rao NA, Aghajanirefah A, Matarese F, et al. Epigenetic programming of monocyte-to-macrophage differentiation and trained innate immunity. *Science* (2014) 345:1251086. doi:10.1126/science.1251086
72. Netea MG, Joosten LA, Latz E, Mills KH, Natoli G, Stunnenberg HG, et al. Trained immunity: a program of innate immune memory in health and disease. *Science* (2016) 352:aaf1098. doi:10.1126/science.aaf1098
73. Bromuro C, Romano M, Chiani P, Berti F, Tontini M, Proietti D, et al. Beta-glucan-CRM197 conjugates as candidates antifungal vaccines. *Vaccine* (2010) 28:2615–23. doi:10.1016/j.vaccine.2010.01.012
74. Sem X, Le GT, Tan AS, Tso G, Yurieva M, Liao WW, et al. Beta-glucan exposure on the fungal cell wall tightly correlates with competitive fitness of *Candida* species in the mouse gastrointestinal tract. *Front Cell Infect Microbiol* (2016) 6:186. doi:10.3389/fcimb.2016.00186
75. Barnett JA. A history of research on yeasts 12: medical yeasts part 1, *Candida albicans*. *Yeast* (2008) 25:385–417. doi:10.1002/yea.1595
76. Fujimura KE, Sitarik AR, Havstad S, Lin DL, Levan S, Fadrosh D, et al. Neonatal gut microbiota associates with childhood multisensitized atopy and T cell differentiation. *Nat Med* (2016) 22:1187–91. doi:10.1038/nm.4176
77. Bougnoux ME, Diogo D, Francois N, Sendid B, Veirmeire S, Colombel JF, et al. Multilocus sequence typing reveals intrafamilial transmission and microevolutions of *Candida albicans* isolates from the human digestive tract. *J Clin Microbiol* (2006) 44:1810–20. doi:10.1128/JCM.44.5.1810-1820.2006
78. Cassone A. Development of vaccines for *Candida albicans*: fighting a skilled transformer. *Nat Rev Microbiol* (2013) 11:884–91. doi:10.1038/nrmicro3156
79. Carvalho A, Cunha C, Bozza S, Moretti S, Massi-Benedetti C, Bistoni F, et al. Immunity and tolerance to fungi in hematopoietic transplantation: principles and perspectives. *Front Immunol* (2012) 3:156. doi:10.3389/fimmu.2012.00156
80. Kumamoto CA, Vinces MD. Contributions of hyphae and hypha-co-regulated genes to *Candida albicans* virulence. *Cell Microbiol* (2005) 7:1546–54. doi:10.1111/j.1462-5822.2005.00616.x
81. Miranda LN, van der Heijden IM, Costa SF, Sousa AP, Sienra RA, Gobara S, et al. *Candida* colonisation as a source for candidaemia. *J Hosp Infect* (2009) 72:9–16. doi:10.1016/j.jhin.2009.02.009
82. Koh AY, Kohler JR, Coggshall KT, Van Rooijen N, Pier GB. Mucosal damage and neutropenia are required for *Candida albicans* dissemination. *PLoS Pathog* (2008) 4:e35. doi:10.1371/journal.ppat.0040035
83. Heilmann CJ, Sorgo AG, Siliakus AR, Dekker HL, Brul S, De Koster CG, et al. Hyphal induction in the human fungal pathogen *Candida albicans* reveals a characteristic wall protein profile. *Microbiology* (2011) 157:2297–307. doi:10.1099/mic.0.049395-0
84. Saville SP, Lazzell AL, Monteagudo C, Lopez-Ribot JL. Engineered control of cell morphology in vivo reveals distinct roles for yeast and filamentous forms of *Candida albicans* during infection. *Eukaryot Cell* (2003) 2:1053–60. doi:10.1128/EC.2.5.1053-1060.2003
85. Di Carlo P, Di Vita G, Guadagnino G, Cocorullo G, D'arpa F, Salamone G, et al. Surgical pathology and the diagnosis of invasive visceral yeast infection: two case reports and literature review. *World J Emerg Surg* (2013) 8:38. doi:10.1186/1749-7922-8-38
86. Noble SM, French S, Kohn LA, Chen V, Johnson AD. Systematic screens of a *Candida albicans* homozygous deletion library decouple morphogenetic switching and pathogenicity. *Nat Genet* (2010) 42:590–8. doi:10.1038/ng.605
87. Lionakis MS, Lim JK, Lee CC, Murphy PM. Organ-specific innate immune responses in a mouse model of invasive candidiasis. *J Innate Immun* (2011) 3:180–99. doi:10.1159/000321157
88. Lohse MB, Johnson AD. White-opaque switching in *Candida albicans*. *Curr Opin Microbiol* (2009) 12:650–4. doi:10.1016/j.mib.2009.09.010
89. Kvaal CA, Srikantha T, Soll DR. Misexpression of the white-phase-specific gene WH11 in the opaque phase of *Candida albicans* affects switching and virulence. *Infect Immun* (1997) 65:4468–75.
90. Kvaal C, Lachke SA, Srikantha T, Daniels K, Mccoy J, Soll DR. Misexpression of the opaque-phase-specific gene PEP1 (SAP1) in the white phase of *Candida albicans* confers increased virulence in a mouse model of cutaneous infection. *Infect Immun* (1999) 67:6652–62.
91. Si H, Hernday AD, Hirakawa MP, Johnson AD, Bennett RJ. *Candida albicans* white and opaque cells undergo distinct programs of filamentous growth. *PLoS Pathog* (2013) 9:e1003210. doi:10.1371/journal.ppat.1003210
92. Lohse MB, Johnson AD. Differential phagocytosis of white versus opaque *Candida albicans* by *Drosophila* and mouse phagocytes. *PLoS One* (2008) 3:e1473. doi:10.1371/journal.pone.0001473
93. Sasse C, Hasenberg M, Weyler M, Gunzer M, Morschhauser J. White-opaque switching of *Candida albicans* allows immune evasion in an environment-dependent fashion. *Eukaryot Cell* (2013) 12:50–8. doi:10.1128/EC.00266-12
94. Pande K, Chen C, Noble SM. Passage through the mammalian gut triggers a phenotypic switch that promotes *Candida albicans* commensalism. *Nat Genet* (2013) 45:1088–91. doi:10.1038/ng.2710
95. Selmecki A, Forche A, Berman J. Genomic plasticity of the human fungal pathogen *Candida albicans*. *Eukaryot Cell* (2010) 9:991–1008. doi:10.1128/EC.00060-10
96. Selmecki A, Gerami-Nejad M, Paulson C, Forche A, Berman J. An isochromosome confers drug resistance in vivo by amplification of two genes, ERG11 and TAC1. *Mol Microbiol* (2008) 68:624–41. doi:10.1111/j.1365-2958.2008.06176.x
97. Hirakawa MP, Martinez DA, Sakthikumar S, Anderson MZ, Berlin A, Gujja S, et al. Genetic and phenotypic intra-species variation in *Candida albicans*. *Genome Res* (2015) 25:413–25. doi:10.1101/gr.174623.114
98. Mandelblat M, Frenkel M, Abbey D, Ben Ami R, Berman J, Segal E. Phenotypic and genotypic characteristics of *Candida albicans* isolates from bloodstream and mucosal infections. *Mycoses* (2017) 60:534–45. doi:10.1111/myc.12623
99. Pfaller MA, Diekema DJ. Epidemiology of invasive mycoses in North America. *Crit Rev Microbiol* (2010) 36:1–53. doi:10.3109/10408410903241444
100. Falagas ME, Apostolou KE, Pappas VD. Attributable mortality of candidemia: a systematic review of matched cohort and case-control studies. *Eur J Clin Microbiol Infect Dis* (2006) 25:419–25. doi:10.1007/s10096-006-0159-2

101. Eggimann P, Garbino J, Pittet D. Epidemiology of *Candida* species infections in critically ill non-immunosuppressed patients. *Lancet Infect Dis* (2003) 3:685–702. doi:10.1016/S1473-3099(03)00801-6
102. Bassetti M, Merelli M, Righi E, Diaz-Martin A, Rosello EM, Luzzati R, et al. Epidemiology, species distribution, antifungal susceptibility, and outcome of candidemia across five sites in Italy and Spain. *J Clin Microbiol* (2013) 51:4167–72. doi:10.1128/JCM.01998-13
103. Pfaller MA, Castanheira M, Messer SA, Moet GJ, Jones RN. Variation in *Candida* spp. distribution and antifungal resistance rates among bloodstream infection isolates by patient age: report from the SENTRY Antimicrobial Surveillance Program (2008-2009). *Diagn Microbiol Infect Dis* (2010) 68:278–83. doi:10.1016/j.diagmicrobio.2010.06.015
104. Guinea J. Global trends in the distribution of *Candida* species causing candidemia. *Clin Microbiol Infect* (2014) 20(Suppl 6):5–10. doi:10.1111/1469-0691.12539
105. Trofa D, Gacser A, Nosanchuk JD. *Candida* parapsilosis, an emerging fungal pathogen. *Clin Microbiol Rev* (2008) 21:606–25. doi:10.1128/CMR.00013-08
106. Nucci M, Colombo AL. Candidemia due to *Candida tropicalis*: clinical, epidemiologic, and microbiologic characteristics of 188 episodes occurring in tertiary care hospitals. *Diagn Microbiol Infect Dis* (2007) 58:77–82. doi:10.1016/j.diagmicrobio.2006.11.009
107. Spivak ES, Hanson KE. *Candida auris*: an emerging fungal pathogen. *J Clin Microbiol* (2018) 56:e01588-17. doi:10.1128/JCM.01588-17
108. Nace HL, Horn D, Neofytos D. Epidemiology and outcome of multiple-species candidemia at a tertiary care center between 2004 and 2007. *Diagn Microbiol Infect Dis* (2009) 64:289–94. doi:10.1016/j.diagmicrobio.2009.03.010
109. Nutsch KM, Hsieh CS. T cell tolerance and immunity to commensal bacteria. *Curr Opin Immunol* (2012) 24:385–91. doi:10.1016/j.coi.2012.04.009
110. Round JL, Mazmanian SK. Inducible Foxp3+ regulatory T-cell development by a commensal bacterium of the intestinal microbiota. *Proc Natl Acad Sci U S A* (2010) 107:12204–9. doi:10.1073/pnas.0909122107
111. Atarashi K, Tanoue T, Oshima K, Suda W, Nagano Y, Nishikawa H, et al. Treg induction by a rationally selected mixture of clostridia strains from the human microbiota. *Nature* (2013) 500:232–6. doi:10.1038/nature12331
112. Furusawa Y, Obata Y, Fukuda S, Endo TA, Nakato G, Takahashi D, et al. Commensal microbe-derived butyrate induces the differentiation of colonic regulatory T cells. *Nature* (2013) 504:446–50. doi:10.1038/nature12721
113. Zelante T, Iannitti RG, Cunha C, De Luca A, Giovannini G, Pieraccini G, et al. Tryptophan catabolites from microbiota engage aryl hydrocarbon receptor and balance mucosal reactivity via interleukin-22. *Immunity* (2013) 39:372–85. doi:10.1016/j.immuni.2013.08.003
114. Romani L, Zelante T, De Luca A, Iannitti RG, Moretti S, Bartoli A, et al. Microbiota control of a tryptophan-AhR pathway in disease tolerance to fungi. *Eur J Immunol* (2014) 44:3192–200. doi:10.1002/eji.201344406
115. Bonifazi P, Zelante T, D'angelo C, De Luca A, Moretti S, Bozza S, et al. Balancing inflammation and tolerance in vivo through dendritic cells by the commensal *Candida albicans*. *Mucosal Immunol* (2009) 2:362–74. doi:10.1038/mi.2009.17
116. Duggan S, Leonhardt I, Hunniger K, Kurzai O. Host response to *Candida albicans* bloodstream infection and sepsis. *Virulence* (2015) 6:316–26. doi:10.4161/21505594.2014.988096
117. Carpino N, Naseem S, Frank DM, Konopka JB. Modulating host signaling pathways to promote resistance to infection by *Candida albicans*. *Front Cell Infect Microbiol* (2017) 7:481. doi:10.3389/fcimb.2017.00481
118. Lionakis MS, Fischer BG, Lim JK, Swamydas M, Wan W, Richard Lee CC, et al. Chemokine receptor Ccr1 drives neutrophil-mediated kidney immunopathology and mortality in invasive candidiasis. *PLoS Pathog* (2012) 8:e1002865. doi:10.1371/journal.ppat.1002865
119. Lionakis MS, Albert ND, Swamydas M, Lee CR, Loetscher P, Kontoyiannis DP. Pharmacological blockade of the chemokine receptor CCR1 protects mice from systemic candidiasis of hematogenous origin. *Antimicrob Agents Chemother* (2017) 61:e02365-16. doi:10.1128/AAC.02365-16
120. Kullberg BJ, Arendrup MC. Invasive candidiasis. *N Engl J Med* (2015) 373:1445–56. doi:10.1056/NEJMra1315399
121. Hung TY, Kotecha RS, Blyth CC, Steed SK, Thornton RB, Ryan AL, et al. Immunogenicity and safety of single-dose, 13-valent pneumococcal conjugate vaccine in pediatric and adolescent oncology patients. *Cancer* (2017) 123:4215–23. doi:10.1002/cncr.30764
122. Wyplosz B, Derradji O, Hong E, Francois H, Durrbach A, Duclos-Vallee JC, et al. Low immunogenicity of quadrivalent meningococcal vaccines in solid organ transplant recipients. *Transpl Infect Dis* (2015) 17:322–7. doi:10.1111/tid.12359
123. Zbinden D, Manuel O. Influenza vaccination in immunocompromised patients: efficacy and safety. *Immunotherapy* (2014) 6:131–9. doi:10.2217/imt.13.171
124. Hakim H, Allison KJ, Van de Velde LA, Tang L, Sun Y, Flynn PM, et al. Immunogenicity and safety of high-dose trivalent inactivated influenza vaccine compared to standard-dose vaccine in children and young adults with cancer or HIV infection. *Vaccine* (2016) 34:3141–8. doi:10.1016/j.vaccine.2016.04.053
125. Holzmann H, Hengel H, Tenbusch M, Doerr HW. Eradication of measles: remaining challenges. *Med Microbiol Immunol* (2016) 205:201–8. doi:10.1007/s00430-016-0451-4
126. Vickers DM, Anonychuk AM, De Wals P, Demarteau N, Bauch CT. Evaluation of serogroup C and ACWY meningococcal vaccine programs: projected impact on disease burden according to a stochastic two-strain dynamic model. *Vaccine* (2015) 33:268–75. doi:10.1016/j.vaccine.2013.09.034
127. Daniels CC, Rogers PD, Shelton CM. A review of pneumococcal vaccines: current polysaccharide vaccine recommendations and future protein antigens. *J Pediatr Pharmacol Ther* (2016) 21:27–35. doi:10.5863/1551-6776-21.1.27
128. Moyes DL, Wilson D, Richardson JP, Mogavero S, Tang SX, Wernecke J, et al. Candidalysin is a fungal peptide toxin critical for mucosal infection. *Nature* (2016) 532:64–8. doi:10.1038/nature17625
129. Zepp F. Principles of vaccine design-lessons from nature. *Vaccine* (2010) 28(Suppl 3):C14–24. doi:10.1016/j.vaccine.2010.07.020
130. Aaby P, Roth A, Ravn H, Napirna BM, Rodrigues A, Lisse IM, et al. Randomized trial of BCG vaccination at birth to low-birth-weight children: beneficial nonspecific effects in the neonatal period? *J Infect Dis* (2011) 204:245–52. doi:10.1093/infdis/jir240
131. Biering-Sorensen S, Aaby P, Napirna BM, Roth A, Ravn H, Rodrigues A, et al. Small randomized trial among low-birth-weight children receiving bacillus Calmette-Guerin vaccination at first health center contact. *Pediatr Infect Dis J* (2012) 31:306–8. doi:10.1097/INF.0b013e3182458289
132. Aaby P, Martins CL, Garly ML, Bale C, Andersen A, Rodrigues A, et al. Non-specific effects of standard measles vaccine at 4.5 and 9 months of age on childhood mortality: randomised controlled trial. *BMJ* (2010) 341: c6495. doi:10.1136/bmj.c6495
133. Benn CS, Netea MG, Selin LK, Aaby P. A small jab--a big effect: non-specific immunomodulation by vaccines. *Trends Immunol* (2013) 34:431–9. doi:10.1016/j.it.2013.04.004
134. Kleinnijenhuis J, Quintin J, Preijers F, Joosten LA, Ifrim DC, Saeed S, et al. Bacille Calmette-Guerin induces NOD2-dependent nonspecific protection from reinfection via epigenetic reprogramming of monocytes. *Proc Natl Acad Sci U S A* (2012) 109:17537–42. doi:10.1073/pnas.1202870109
135. Kleinnijenhuis J, Quintin J, Preijers F, Joosten LA, Jacobs C, Xavier RJ, et al. BCG-induced trained immunity in NK cells: role for non-specific protection to infection. *Clin Immunol* (2014) 155:213–9. doi:10.1016/j.clim.2014.10.005
136. Garcia-Valtanen P, Guzman-Genuino RM, Williams DL, Hayball JD, Diener KR. Evaluation of trained immunity by beta-1, 3 (d)-glucan on murine monocytes in vitro and duration of response in vivo. *Immunol Cell Biol* (2017) 95:601–10. doi:10.1038/icb.2017.13
137. Leentjens J, Quintin J, Gerretsen J, Kox M, Pickkers P, Netea MG. The effects of orally administered beta-glucan on innate immune responses in humans, a randomized open-label intervention pilot-study. *PLoS One* (2014) 9: e108794. doi:10.1371/journal.pone.0108794

7

A Multifaceted Role of Tryptophan Metabolism and Indoleamine 2,3-Dioxygenase Activity in *Aspergillus fumigatus*–Host Interactions

*Tsokyi Choera[1], Teresa Zelante[2], Luigina Romani[2] and Nancy P. Keller[1,3]**

[1] Department of Medical Microbiology and Immunology, University of Wisconsin-Madison, Madison, WI, United States, [2] Department of Experimental Medicine, University of Perugia, Perugia, Italy, [3] Department of Bacteriology, University of Wisconsin-Madison, Madison, WI, United States

***Correspondence:**
Nancy P. Keller
npkeller@wisc.edu

Aspergillus fumigatus is the most prevalent filamentous fungal pathogen of humans, causing either severe allergic bronchopulmonary aspergillosis or often fatal invasive pulmonary aspergillosis (IPA) in individuals with hyper- or hypo-immune deficiencies, respectively. Disease is primarily initiated upon the inhalation of the ubiquitous airborne conidia—the initial inoculum produced by *A. fumigatus*—which are complete developmental units with an ability to exploit diverse environments, ranging from agricultural composts to animal lungs. Upon infection, conidia initially rely on their own metabolic processes for survival in the host's lungs, a nutritionally limiting environment. One such nutritional limitation is the availability of aromatic amino acids (AAAs) as animals lack the enzymes to synthesize tryptophan (Trp) and phenylalanine and only produce tyrosine from dietary phenylalanine. However, *A. fumigatus* produces all three AAAs through the shikimate–chorismate pathway, where they play a critical role in fungal growth and development and in yielding many downstream metabolites. The downstream metabolites of Trp in *A. fumigatus* include the immunomodulatory kynurenine derived from indoleamine 2,3-dioxygenase (IDO) and toxins such as fumiquinazolines, gliotoxin, and fumitremorgins. Host IDO activity and/or host/microbe-derived kynurenines are increasingly correlated with many *Aspergillus* diseases including IPA and infections of chronic granulomatous disease patients. In this review, we will describe the potential metabolic cross talk between the host and the pathogen, specifically focusing on Trp metabolism, the implications for therapeutics, and the recent studies on the coevolution of host and microbe IDO activation in regulating inflammation, while controlling infection.

Keywords: *Aspergillus fumigatus*, tryptophan metabolism, IDO, kynurenines, toxins, non-ribosomal peptides, peripheral tolerance, Th17 cells

INTRODUCTION

Aspergillus fumigatus is a saprophytic fungus that has a worldwide distribution. The asexual spores (called conidia) are ubiquitous and individuals inhale hundreds of spores daily. While most inhaled conidia are cleared by individuals with a healthy immune system, *A. fumigatus* can act as an opportunistic human pathogen in individuals with altered immune functions. Disease presentation can vary on the status of the host's immune system; *A. fumigatus* can cause allergic bronchopulmonary aspergillosis, a severe allergenic response, in the hyper-immune, or the fatal invasive growth invasive pulmonary aspergillosis (IPA) in the hypo-immune, or in individuals with other susceptibilities such as patients unable to mount the necessary oxidative defenses such as in individuals with chronic granulomatous disease (CGD) (1).

The manifestation of disease is dependent not only on the host's immune status but also fungal factors including strain heterogeneity (2). *A. fumigatus* growth in the mammalian lung, following survival of resident pulmonary defenses, requires the fungus to adapt to a hypoxic and nutritionally scarce environment. *Aspergillus* mutants unable to synthesize primary metabolites necessary for growth are generally impaired in virulence. For example, deletion of *cpcA*, a transcription factor that globally modulates amino acid biosynthesis in the fungus led to a less virulent phenotype in a murine model of IPA (3). Additional studies have shown that mutants in sulfur utilization (4), uracil/uridine synthesis (5), zinc uptake, iron acquisition, and many more (6, 7) are also decreased in virulence.

To complicate disease progression further, there is an alarming rise in antifungal resistance strains of *A. fumigatus* (8, 9). Therefore, an understanding of *A. fumigatus* and host metabolic pathways is important in identifying nutrient limitations. One critical metabolic pathway is the biosynthesis of aromatic amino acids [AAAs, tryptophan (Trp), phenylalanine, and tyrosine], which are required not only for growth of *A. fumigatus* but are also precursors for several toxins (**Table 1**). The host relies on dietary sources for all AAAs while *A. fumigatus* synthesizes all three. However, the host and *A. fumigatus* both possess AAA catabolic enzymes. In particular, one key enzyme important in immune homeostasis is indoleamine 2,3-dioxygenase (IDO), which converts Trp to kynurenine and related metabolites in both organisms. Historically, host IDOs activity has been described as an effective antimicrobial control for pathogens that are natural Trp auxotrophs such as *Staphylococcus aureus*, *Chlamydia* spp., and *Toxoplasma gondii*, presumably by Trp starvation (10). However, *A. fumigatus* can synthesize its own Trp and thus the Trp starvation may not be an effective pathogen control for those microbes able to synthesize their own Trp pools. Although, IDOs also play additional roles in host defenses through modifying kynurenine levels and subsequent cytokine responses as described below. In this review, we will summarize the recent studies describing the anabolic and the catabolic pathways of Trp metabolism, the implications for therapeutics, and the host–pathogen interaction.

TABLE 1 | *Aspergillus fumigatus* non-ribosomal peptides containing aromatic amino acids (AAAs) in their peptide structure.

Toxin	AAA[a]	Interaction with host[b-i]
Fumigaclavine C[b]	Tryptophan (Trp)	Downregulation of Th1 cytokines including TNF-α, IL-1β, and IL-17A. Induction of host cell apoptosis Decrease activation of caspase-1
Fumiquinazoline C[c]	Anthranilate Trp	Cytotoxic
Fumisoquin[d]	Tyrosine	Nothing reported
Fumitremorgin[e] Tryprostatin A[f] Verruculogen[g]	Trp	Neurotoxic and produces tremors in mice Tryprostatin causes the inhibition of microtubule assembly Verruculogen causes modification of the electrophysiological properties of HNEC
Gliotoxin[h]	Phenylalanine	Virulent in a steroid murine model of IPA Induction of host cell apoptosis and causes epithelial cell damage Inhibition of phagocytosis and oxidative bursts
Hexadehydroastechrome[i]	Trp	Overexpression resulted in a significantly higher virulence in a neutropenic murine model of IPA[b]

IPA, invasive pulmonary aspergillosis; HNEC, human nasal epithelial cells.
[a]*Only the AAA is designated. Other amino acids are also in the structure of these metabolites.*
[b–i]*Biosynthesis and host interactions based from the following sources:* [b]*(11, 12);* [c]*(13);* [d]*(14);* [e]*(15, 16);* [f]*(17);* [g]*(18, 19);* [h]*(7, 20);* [i]*(21, 22).*

Trp SYNTHESIS AND POTENTIAL THERAPEUTIC TARGETING

Chorismic acid derived from the shikimic acid pathway is a key intermediate in producing Trp, phenylalanine (Phe), and tyrosine (Tyr) in microorganisms including *A. fumigatus* (**Figure 1**). Trp and Phe are classified as essential amino acids, whereas mammals acquire them from diet, whereas Tyr is synthesized *via* the hydroxylation of Phe (23, 24). The absence of the AAA biosynthetic enzymes and the low bioavailability of Trp in humans makes the Trp biosynthetic enzymes attractive targets for antifungals (25).

Fungal Trp Anabolic Pathway

Aromatic amino acid synthesis has been extensively studied in *Saccharomyces cerevisiae* and provides the basis for the functional characterization of orthologous enzymes in filamentous fungi (23, 24, 26–28). The shikimic acid pathway is a 7-enzymatic step reaction that initiates with two substrates, phosphoenolpyruvate (PEP) and erythrose-4-phosphate (E4P), which are intermediates of glycolysis and pentose phosphate pathways, respectively (29) (**Figure 1**). The first step of the shikimic acid pathway is catalyzed by 3-deoxy-D-arabinoheptulosonate 7-phosphate (DAHP) synthase to convert PEP and E4P to DAHP. In *S. cerevisiae* and *A. nidulans*, there are two DAHP synthases, Aro3 and Aro4, which are allosterically inhibited by phenylalanine and tyrosine, respectively (24). Steps 2–6 in filamentous fungi such as *A. nidulans* and *A. fumigatus* are completed by the pentafunctional enzyme AroM, or Aro1 in the model organism *S. cerevisiae* (30). The shikimate pathway culminates in the production of chorismic acid synthesized by the enzyme chorismate synthase (AroB) from 5-enolpyruvylshikimate-3-phosphate (EPSP) (31).

The synthesis of Trp from chorismate is initiated by an anthranilate synthase (AS), which converts chorismate to anthranilate, followed by three enzymatic steps as presented in **Figure 1** with the respective functions outlined in **Table 2**. AS(s) in *S. cerevisiae* have been characterized, and it consists of two subunits: anthranilate synthase subunit I (AAS-I), which binds chorismate and is subject to feedback inhibition by Trp and anthranilate synthase

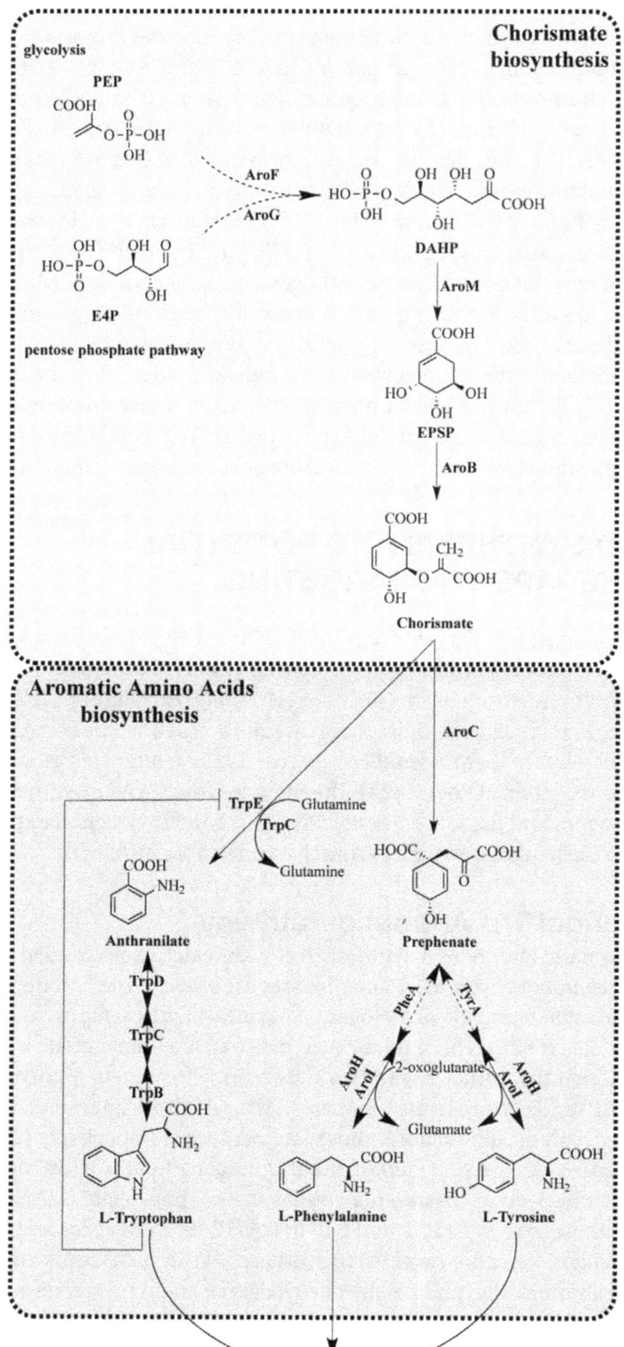

FIGURE 1 | Tryptophan anabolism of *Aspergillus fumigatus*. [Modified from that of Wang et al. (28)]. Solid arrows indicate characterized reaction as being present in *A. fumigatus* with product detected. Dashed arrows indicate uncharacterized reactions; however, putative orthologs are present in *A. fumigatus*.

subunit II (AAS-II) which is a glutamine amidotransferase (32). The *A. nidulans trpC*, an AAS-II encoding gene, was characterized in 1977 (33) and found exchangeable with an *A. fumigatus trpC* in 1994 (34). Wang et al. (28) recently characterized *trpE*, the AAS-I encoding gene in *A. fumigatus*. Wang et al. (28) explored the functions of two putative AAS-Is termed *trpE* and *icsA* by creating null mutants. The deletion of *trpE* led to a Trp auxotrophic strain, whereas the deletion of *icsA* did not. To ensure that *icsA* did not serve a redundant role for Trp synthesis, the group overexpressed *icsA* in a *trpE* deletion and showed that the overexpression of *icsA* does not reverse the Trp auxotrophy concluding that TrpE is the only AS in *A. fumigatus*. Interestingly, the group showed that IcsA is an active enzyme in *A. fumigatus* as the precursor-chorismate pool is altered in the absence or overproduction of IcsA; however, the product is not known (28). Sasse et al. also confirmed that deletion of *trpE* (they termed *trpA*) results in an *A. fumigatus* Trp auxotrophy (31).

The Shikimate Pathway As Potential Antifungal Targets

Currently, there are four major classes of antifungals: azoles and amphotericin B targeting ergosterol, 5-fluorocytosine targeting DNA synthesis, and echinocandins targeting cell wall synthesis. These antifungals either exhibit high toxicity to the mammalian cell (particularly amphotericin B and 5-fluorocytosine) or lose efficacy due to the emergence of drug resistant strains (azoles and echinocandins) (9). With *A. fumigatus* being a eukaryotic pathogen and sharing many proteins with mammalian hosts, there are limitations to developing effective and safe antifungals and therefore a great need for treatments that are fungal specific. Since Trp is a human-essential amino acid and the enzymes in the biosynthesis are fungal specific, several studies have suggested utilizing and finding drugs to target the enzymes of this pathway (35–38).

Targeting essential amino acid pathways have already shown potential for new classes of antifungals. Several groups have explored inhibitors of genes or enzymes involved in methionine biosynthesis. Azoxybacillin, a compound isolated from *B. cereus* targets methionine biosynthesis by interfering with expression of homoserine transacetylase and sulfite reductase encoding genes (39–41). Whereas azoxybacillin displayed a broad spectrum antifungal activity *in vitro*, *in vivo* activity was low possibly due to bioavailability in the host (41). R1-331, a natural product from *Streptomyces akiyoshiensis*, is an effective inhibitor of homoserine dehydrogenase involved in both methionine and threonine biosynthesis (42, 43). Yamaguchi et al. show that R1-331 was active against medically important fungi such as *Candida albicans* and *Cryptoccocus neoformans* and proved to be effective in the treatment of systemic murine candidiasis (42, 44).

Compounds targeting AAA pathways are limited with the most famous being the herbicide Roundup, where the active ingredient glyphosate inhibits EPSP synthase, one of the first enzymes initiating the shikimate pathway (45) (**Figure 1**). Glyphosate has shown to inhibit growth of several fungi including *Candida maltose* (46), *Pneumocystis* (47), and *Cryptococcus neoformans* where glyphosate delayed fungal melanization *in vitro* and *in vivo* and prolonged mice survival during infection (48). Another inhibitor of AAA pathway is a fluorinated anthranilate moiety, 6-FABA, which targets the TrpE enzyme and showed bactericidal activity when used on *Mycobacterium tuberculosis* (49). The studies of

TABLE 2 | *Aspergillus fumigatus* tryptophan (Trp) metabolism genes and putative protein function.

Protein	Ortholog in *Saccharomyces cerevisiae*	Gene name in *A. fumigatus*	Protein function[a]	Ortholog in mammals
Chorismate biosynthesis				
AroF	Aro3	Afu1g02110	DAHP synthase	–
AroG	Aro4	Afu7g04070		
AroM	Aro1	Afu1g13740	EPSP synthase	–
AroB	Aro2	Afu1g06940	Chorismate synthase	–
Aromatic amino acid (AAA) biosynthesis				
TrpE	Trp2	Afu6g12580	Anthranilate synthase	–
TrpC	Trp3	Afu1g13090		
TrpD	Trp4	Afu4g11980	Anthranilate phosphoribosyltransferase	–
TrpC	Trp1	Afu1g13090	Phosphoribosylanthranilate isomerase	–
TrpB	Trp5	Afu2g13250	Trp synthase	–
IcsA	–	Afu6g12110	Isochorismate synthase	–
AroC	Aro7	Afu5g13130	Chorismate mutase	–
PheA	Pha2	Afu5g05690	Prephenate dehydratase	–
TyrA	Tyr1	Afu2g10450	Prephenate dehydrogenase	–
AroH	Aro8	Afu2g13630	AAA transaminase	–
AroI	Aro9	Afu5g02290		
Trp degradation				
IdoA	Bna2	Afu3g14250	Indoleamine 2,3-dioxygenases	IDO1
IdoB	Bna2	Afu4g09830		IDO2
IdoC	Bna2	Afu7g02010		TDO
FmdS	Bna7	Afu1g09960	Kynurenine formamidase	AFMID
Bna4	Bna4	Afu6g07340	Kynureninase	
Bna5	Bna5	Afu4g09840		KYNU
AroH	Aro8	Afu2g13630	AAA transaminase	LAAO
AroI	Aro9	Afu5g02290		
AadA	–	Afu3g02240	Trp carboxylase	AADC
MaoN	–	Afu3g00100	Monoamine oxidase	MAOA
AldA	Ald4	Afu2g00720	Aldehyde dehydrogenase	ALDH
	Ald5	Afu7g01000		
		Afu6g11430		

[a]Prediction of protein function based on AspGD (http://www.aspgd.org/) and KEGG (http://www.genome.jp/kegg/kegg2.html).
DAHP synthase, 3-deoxy-D-arabinoheptulosonate 7-phosphate synthase; EPSP synthase, enolpyruvylshikimate-3-phosphate synthase; AFMID, arylformamidase; KYNU, kynureninase; LAAO, L-amino-acid oxidase; AADC, aromatic-L-amino-acid decarboxylase, MAOA, monoamine oxidase; ALDH2, aldehyde dehydrogenase family.

these inhibitors suggest that the Trp biosynthetic pathway could be fruitful in future antifungal drug design.

The value of AAA pathways as drug targets is supported by the findings that AAA auxotrophic mutants are less virulent in animal infection models. Sasse et al. explored the possibility of these pathways as potential drug target by testing the virulence of several AAA auxotrophic mutants in a murine IPA model (31). This study demonstrated that AroM (**Figure 1**) was required for *A. fumigatus* viability. The group also constructed a conditional AroB repression strain that was attenuated virulence. Both a Trp auxotroph (TrpE mutant) and Tyr/Phe auxotroph (AroC mutant) were severely attenuated in virulence for pulmonary infection. Interestingly, the group also unveiled a putative difference in AAA distribution within the host by conducting a systemic infection showing that in a bloodstream infection the TrpE and the AroC mutants although less virulent, can establish some infection (31). Taken together, these results suggest that inhibitors of AAA biosynthetic pathways can potentially be used against *A. fumigatus* as a standalone treatment in a localized pulmonary infection or as an additive treatment in a systemic infection. The result of the bloodstream infection observed by Sasse et al. also suggests that there are mechanisms for the fungus to sense Trp in its environment and utilize it. In *S. cerevisiae*, the Trp specific permease, Tat2, is required for Trp uptake in yeast (50) and its closest homolog in *A. fumigatus* (Afu7g04290) is upregulated during fungal encounters with neutrophils (51) and dendritic cells (52). In *A. nidulans*, the G-protein coupled receptor (Gpr) H may be responsible for sensing Trp and glucose and GprH is conserved in *A. fumigatus* (53, 54). Perhaps, for a systemic infection, inhibition of specific permeases or development of a GprH antagonist would be useful in reducing infections by *Aspergillus*.

CATABOLIC Trp METABOLITES IN FUNGI AND HOST

Although the anabolic Trp pathway is absent in mammals, the common catabolic pathways exist in the mammalian host with possession of the same enzymes as *A. fumigatus* (**Figure 2**). Through the expression of Trp degradation enzymes, immune cells are both controlling inflammation and combating microbial infection. In addition to Trp degradation pathways conserved with animals, *A. fumigatus* can also direct Trp pools to secondary metabolites that may impact host health and response (**Table 1**).

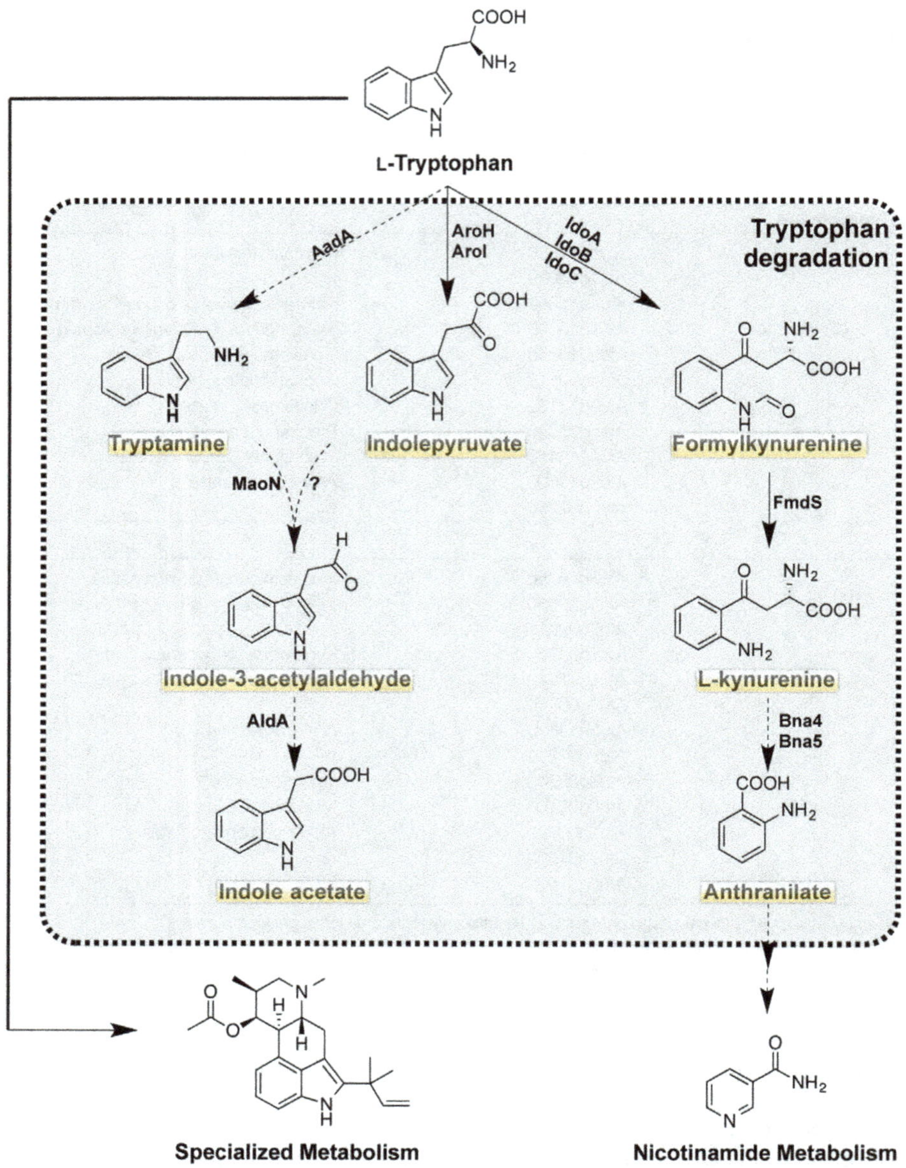

FIGURE 2 | Tryptophan catabolism of *Aspergillus fumigatus*. Highlighted products are putatively produced in the mammalian host as the orthologous enzymes are present in the host (http://www.genome.jp/kegg). Solid arrows indicate characterized reaction as being present in *A. fumigatus* with product detected. Dashed arrows indicate uncharacterized reactions, however putative orthologs are present in *A. fumigatus*.

Fungal Trp Catabolism

There are three putative pathways (**Figure 2**; **Table 2**) for the degradation of Trp in *A. fumigatus*. The kynurenine branch is catalyzed by IDOs that convert Trp into formylkynurenine. In *A. fumigatus*, there are 3 putative *ido* genes: *idoA*, *idoB*, and *idoC*, the orthologs of *A. oryzae idoα*, *idoβ*, and *idoγ*, respectively (55). Enzymatic studies of *Aspergillus oryzae* IDO enzymes suggest two of the three enzymes, IDOα, and IDOβ, may participate in Trp degradation as they have a higher affinity of its substrate. However the recent study by Wang et al., suggested IDOb might be the more dominant enzyme than IDOa as determined by gene expression of *A. fumigatus* grown on Trp amended media (28). Additionally, *idoC* gene expression was slightly induced by the addition of Trp, but the authors note that IDOc had a closer relationship to bacterial IDOs than to fungal IDOs (28, 55). In *S. cerevisiae*, formylkynurenine is further oxidized to the immunomodulatory product kynurenine by a kynurenine formamidase denoted as Bna7, which has been described in *A. nidulans* and is predicted to be involved in NAD (+), biosynthesis (56). The kynurenine branch in fungi is involved in the *de novo* biosynthesis of NAD (+), a coenzyme that is required for oxidation-reduction reactions (57).

Tryptophan can also be metabolized via the indole pyruvate pathway, initiated through the transamination of Trp by aromatic aminotransferases (termed Aro8 and Aro9 in *S. cerevisiae*). These aromatic aminotransferases are also involved in the synthesis of Phe and Tyr in *S. cerevisiae*, and their orthologs are present in *A. fumigatus* (28). In *S. cerevisiae*, the deletion of both Aro8 and Aro9 results in Phe and Tyr auxotrophies (24, 58, 59). In *Candida* spp., where filamentation and pigment production play a role in virulence, the products of these enzymes have been described to influence both phenotypes. The deletion of *aro8* in *C. glabrata* results in a reduced pigment production and leads to an increased sensitivity to hydrogen peroxide (27). The *aro8* and *aro9* mutants of *C. albicans* results in a decreased conversion of Trp to indole acetaldehyde, which is formed via decarboxylation of indole pyruvate. Filamentation of *C. albicans* increased with the exposure to indole acetaldehyde (60).

Indole acetaldehyde can also be produced via the third putative product of Trp degradation: tryptamine. Tryptamine is most famously known as the active compound in psilocybin and for its similarity to serotonins (61). Although the production of tryptamine has yet to be described in *A. fumigatus*, the downstream product of the tryptamine and indole pyruvate pathway, indole acetic acid has been described in several *Aspergillus* spp. including *A. fumigatus* (62, 63). Downstream metabolism of kynurenine, indole pyruvate, and tryptamine has not been explored further, but *A. fumigatus* does possess putative enzymes for the re-synthesis of anthranilate, the precursor to Trp (as denoted in **Figures 1** and **2**).

AAA Incorporation into *Aspergillus* Toxins

Many filamentous fungi, including *A. fumigatus*, produce bioactive small molecules that can have detrimental impacts on human health. Subsets of these toxins are small peptides synthesized by non-ribosomal peptide synthetases (NRPS). Several pathogenic *Aspergillus* species synthesize AAA derived peptides including gliotoxin (*Phe* and serine) (64), fumiquinazoline (*Trp*, *Anthranilate*, and Alanine) (13), fumigaclavine (*Trp*) (65), fumitremorgin (*Trp* and Proline) (66), hexadehydroastechrome (*Trp* and Alanine) (21), fumisoquin (*Tyr*, Serine, and Methionine) (14), DPP-IV inhibitor WYK-1 (*Trp*, *Tyr*, and Leucine) (67), cyclopiazonic acid (*Trp*) (68) and benzomalvin (*Phe* and *Anthranilate*) (69). **Table 1** summarizes the known AAA derived secondary metabolites of *A. fumigatus* and their effect on the host.

Although, fumiquinazolines have yet to be assessed for virulence in an animal model, they are known to have cytotoxic properties (13). Fumigaclavines have been described to have immunosuppressive properties in several studies including the suppression of antifungal cytokines such as TNFα, IL-17, and IFN-γ (12). Fumitremorgin, verruculogen, and tryprostatin—all related products of the fumitremorgin pathway—induce tremorgenic activity in mice and act on the central nervous system (15, 66, 70, 71). Mutants in the hexadehydroastechrome pathway (21) and gliotoxin (20, 21) pathways have altered virulence in murine IPA models. Decreased virulence of the *gliP* mutant (GliP is the NRPS required for gliotoxin synthesis) is dependent on host immune status [reviewed in Ref. (7)]. Overexpression of *hasA* encoding the hexadehydroastechrome transcription factor and thus leading to increased hexadehydroastechrome production was more virulent than wild type *A. fumigatus* in a neutropenic model of IPA (21). Although the exact mechanism underlying the increased virulence of the OE::*hasA* strain is unknown, iron homeostasis and cross talk between metabolic pathways may contribute to the increased virulence of OE::*hasA* (22). These studies highlight the potential contribution of AAA derived toxins in virulence of *A. fumigatus*.

Host Trp Catabolism *via* IDO

The function of host IDO during mammalian infection was originally thought to center on the anti-proliferative effects of pathogenic microorganism via deprivation of Trp exerted by the host. IDO is up-regulated by interferon gamma (IFNγ) and depletes Trp (the least abundant essential amino acid) to inhibit pathogen expansion (72, 73), as demonstrated in the constraint of chlamydial growth (74). Numerous studies have since implicated IDO activity as important in fungal infections and have reported the relative outcomes of IDO expression on disease progression (**Table 3**). Accumulating data continues to support that IDO participates in the host–pathogen interaction in human epithelial cells; therefore, the co-evolution of host and microbe Trp metabolism has been investigated (75). The current consensus is that IDO activation is pivotal in regulating inflammatory processes directly

TABLE 3 | Summary of studies where IDO enzyme activity was found to be implicated in fungal infections.

Fungus	Mouse model	IDO	Outcome	Reference
Candida albicans	Gastrointestinal inf.	Upregulation	Protection of the host against fungus	(76)
C. albicans	Gastrointestinal inf.	Upregulation	Protection of the host against fungus	(77)
C. albicans	Gastrointestinal inf.	Upregulation	Protection of the host against fungus	(78)
C. albicans	*In vitro*	Downregulation	Not done	(79)
Aspergillus fumigatus	Keratitis	Upregulation	Protection of the host against fungus	(80)
A. fumigatus	Allergy	Overexpression	Protection of the host against fungus	(81)
A. fumigatus	IPA	Upregulation	Protection of the host against fungus	(82)
A. fumigatus	IPA in chronic granulomatous disease mice	Upregulation	Protection of the host against fungus	(83)
A. fumigatus	IPA in CF mice	Upregulation	Protection of the host against fungus	(84)
Paracoccidioides brasiliensis	Pulmonary infection	Upregulation	Protection of the host against fungus	(85)
P. brasiliensis	Pulmonary infection	Upregulation	Protection of the host against fungus	(86)
P. brasiliensis	Pulmonary infection	Upregulation	Protection of the host against fungus	(87)
Histoplasma capsulatum	Pulmonary infection	Downregulation	Protection of the host against fungus	(88)
H. capsulatum	Pulmonary infection	Upregulation	Protection of the host against fungus	(89)

via Trp depletion and indirectly via the IDO-mediated release of Trp catabolic secondary metabolites (namely, kynurenines).

Dietary Trp is catabolized by two different IDO protein isoforms, IDO1 and IDO2 that are expressed by immune cells, and TDO (Trp 2,3-dioxygenase) that is mainly expressed in the liver. Cells involved in the innate processes of the anti-microbial defense, such as dendritic cells (DCs), neutrophils, and macrophages express IDO1 upon microbial encounter mainly *via* toll-like receptor stimulation. How fungi specifically induce IDO expression is not known; however, induction by other pathogens is associated with pathogen associated molecular patterns, including lipopolysaccharides and CpG oligodeoxynucleotides (78, 90–92), underlining a role for kynurenine metabolism in microbial-induced inflammatory processes.

IDO-Mediated Tolerance: Impacts on Antimicrobial Responses

Evolutionary studies have shown that the host immune defense against microbes is characterized by three different mechanisms: avoidance, resistance, and tolerance (93). Modules of immunity provide resistance to limit pathogen burden and tolerance and host damage caused by the immune reaction *per se*. However, the inflammatory reaction, although largely considered beneficial for its antimicrobial functions, may also contribute to pathogenicity. Thus, rescue from infection pathology may not only depend on microbial colonization (and inactivation of resistance mechanisms) but also on the resolution of tissue inflammatory pathology through tolerogenic responses to pathogens (94).

Studies using a mouse model of mucosal or invasive *C. albicans* infection found that systemic inhibition of IDO *in vivo* reduced gastrointestinal inflammation and unexpectedly, elevated the levels of fungal colonization compared to control mice (77). Notably, tolerogenic responses toward *C. albicans* were abrogated when IDO was antagonized *in vivo*, as shown in various models of inflammatory disorders (95, 96). As with *C. albicans*, IDO and kynurenine production during *A. fumigatus* infection contributes to fungal pathogen eradication and the regulation of an unacceptable level of tissue damage (97). Indeed, IDO can increase kynurenine host levels to induce adaptive Treg expansion while limiting Th17 polarization (83, 96). In this context, the Th17 pathway, which downregulates Trp catabolism, may instead favor pathology and better explain the paradoxical correlation between fungal infection and chronic inflammation (98).

Another example of this paradox was demonstrated in the context of CGD, in which an NADPH oxidase defect results in reduced host production of antimicrobial ROS and extreme susceptibility to *Aspergillus* infections (1, 83). Although human studies have excluded a role for IDO in CGD (99), further investigations into the IDO pathway are warranted as such studies have failed to demonstrate functional IDO activity at sites of chronic inflammation. Measures of IDO functional activity during IPA have, however, been made in mouse models, and implicate defective IDO activity as a key mediator of chronic inflammation in CGD (83). An exaggerated Th17 pulmonary response was associated with reduced fungal clearance in mouse models of CGD that develop IPA. Here, reduced IDO function was directly related to NADPH/ROS deficiency, as ROS is essential for IDO catalytic activity in mammals (100). ROS deficiency as a result of reduced NADPH function, significantly enhanced IL-17 inflammation and fungal germination in the lung, thus further reducing neutrophil-mediated antimicrobial activities (83).

Since regulation of homeostasis and peripheral tolerance are extremely important in prevention of invasive Aspergillosis or allergy to *Aspergillus* antigens (97, 101), the role of IDO has been extensively studied in this model of fungal infection (81, 82, 102). These studies highlight the induction of the IDO metabolic pathway at different site of fungal colonization as keratinocytes or lung as well as the important anti-inflammatory activity of IDO in the tissue microenvironment (80–82, 84).

Aryl Hydrocarbon Receptor (AhR) Activation by IDO Metabolites in Mammals: Biological Consequences

The AhR is a ligand-activated transcription factor first identified for its role during embryonic development and induction of xenobiotic metabolizing enzymes as a response to environmental toxins, such as dioxin (103). More recently, AhR has been shown to play a critical role in immunity by acting as an immune modulator during fungal infection (85). The connection between the AhR and the immune response lies in part in the endogenous Ahr ligands, which comprise many Trp metabolites, including kynurenine (104). Microbial Trp-derived metabolites can activate the AhR, leading to adjustments in the immune response that may hinder disease development (105). The AhR–IDO axis has been recently demonstrated in fungal infection, highlighting a role for IDO-derived metabolites to trigger AhR target genes (85, 106). For example, one AhR target gene, *Il22*, has been widely studied in the context of fungal/microbial infections (105, 107–109). AhR activation by IDO metabolites can also mediate the expansion of peripheral Treg with anti-inflammatory properties. Using IDO-deficient mice, increased pulmonary disease caused by *Paracoccidioides brasiliensis* was associated with decreased Treg expansion and reduced AhR protein expression (85). In murine models of IPA, distinct Treg populations capable of mediating anti-inflammatory effects expand following exposure to *Aspergillus* conidia (97). Late in infection, tolerogenic adaptive Treg (with shared phenotypic identity with the Treg controlling autoimmune diseases or diabetes) produce IL-10 and TGFβ, inhibit Th2 cells, and prevent an allergic reaction to *Aspergillus* (97).

CONCLUSION

The interplay of Trp metabolic pathways and fungal/host interactions is intriguing with many unanswered questions of the exact nature of crosstalk of shared metabolites and consequences of activation of Trp degradative pathways. In *Aspergillus* infections in particular, not only does the pathogen synthesize and degrade Trp but it also can utilize this amino acid (and its precursor anthranilate or the other two AAAs Tyr and Phe) to yield several potentially damaging toxins (**Table 1**). Also, as both host and *Aspergillus* share catabolic IDO pathways, it is unclear which organism may generate immunomodulatory Trp degradation products and if they respond to each other's products (e.g., kynurenine). Development of *A. fumigatus* IDO mutants for investigation of disease development could yield valuable information on this front. The research on the expression host IDOs exhibit the importance of an extremely coordinated immune response to mount the right inflammatory response for clearance of spores. However, while an increased IDO expression in the host can control inflammation, the suppression of the IDO-regulated antifungal Th17 responses can favor fungal growth. In this context, it will be critical to explore the entire IDO-mediated innate response, including the specific T cell regulatory subsets affected by IDO activity.

Although the Trp catabolic pathway is shared between host and pathogen, the anabolic pathway is unique to *A. fumigatus*. The antifungals currently used in treatment are becoming increasingly ineffective with emerging drug resistant strains; therefore, drugs targeting essential fungal specific pathways are needed. A proposal for fungal treatment has been highlighted through the studies of essential amino acids. As AAA mutants are auxotrophs and decreased in virulence (28, 31), investigations of drugs targeting these pathway enzymes could lead to novel antifungal compounds. Indeed, a few compounds have exhibited some efficacy in targeting Trp metabolic pathways in *M. tuberculosis* and several fungi and efforts to identify additional inhibitors are warranted.

AUTHOR CONTRIBUTIONS

TC, TZ, and NK have made a substantial, direct, and intellectual contribution to the work and LR provided intellectual insights to the revision. All authors have approved it for publication.

ACKNOWLEDGMENTS

The authors would like to thank support from Dalai Lama Trust MSN178745 to TC, The Italian Grant "Programma per Giovani Ricercatori - Rita Levi Montalcini 2013" to TZ, and NIH 5R01AI065728-10 to NK.

REFERENCES

1. Segal BH, Romani LR. Invasive aspergillosis in chronic granulomatous disease. *Med Mycol* (2009) 47:S282–90. doi:10.1080/13693780902736620
2. Keller NP. Heterogeneity confounds establishment of "a" model microbial strain. *MBio* (2017) 8:e00135-17. doi:10.1128/mBio.00135-17
3. Krappmann S, Bignell EM, Reichard U, Rogers T, Haynes K, Braus GH. The *Aspergillus fumigatus* transcriptional activator CpcA contributes significantly to the virulence of this fungal pathogen. *Mol Microbiol* (2004) 52:785–99. doi:10.1111/j.1365-2958.2004.04015.x
4. Amich J, Dumig M, O'keeffe G, Binder J, Doyle S, Beilhack A, et al. Exploration of sulfur assimilation of *Aspergillus fumigatus* reveals biosynthesis of sulfur-containing amino acids as a virulence determinant. *Infect Immun* (2016) 84:917–29. doi:10.1128/IAI.01124-15
5. D'Enfert C, Diaquin M, Delit A, Wuscher N, Debeaupuis JP, Huerre M, et al. Attenuated virulence of uridine-uracil auxotrophs of *Aspergillus fumigatus*. *Infect Immun* (1996) 64:4401–5.
6. Rhodes JC. *Aspergillus fumigatus*: growth and virulence. *Med Mycol* (2006) 44(Suppl 1):S77–81. doi:10.1080/13693780600779419
7. Dagenais TR, Keller NP. Pathogenesis of *Aspergillus fumigatus* in invasive aspergillosis. *Clin Microbiol Rev* (2009) 22:447–65. doi:10.1128/CMR.00055-08
8. Sanglard D. Emerging threats in antifungal-resistant fungal pathogens. *Front Med* (2016) 3:11. doi:10.3389/fmed.2016.00011
9. Perlin DS, Rautemaa-Richardson R, Alastruey-Izquierdo A. The global problem of antifungal resistance: prevalence, mechanisms, and management. *Lancet Infect Dis* (2017) 17(12):e383–92. doi:10.1016/S1473-3099(17)30316-X
10. Mackenzie CR, Heseler K, Muller A, Daubener W. Role of indoleamine 2,3-dioxygenase in antimicrobial defence and immuno-regulation: tryptophan depletion versus production of toxic kynurenines. *Curr Drug Metab* (2007) 8:237–44. doi:10.2174/138920007780362518
11. Zhu YX, Yao LY, Jiao RH, Lu YH, Tan RX. Enhanced production of fumigaclavine C in liquid culture of *Aspergillus fumigatus* under a two-stage process. *Bioresour Technol* (2014) 152:162–8. doi:10.1016/j.biortech.2013.10.089
12. Guo W, Hu S, Elgehama A, Shao F, Ren R, Liu W, et al. Fumigaclavine C ameliorates dextran sulfate sodium-induced murine experimental colitis via NLRP3 inflammasome inhibition. *J Pharmacol Sci* (2015) 129:101–6. doi:10.1016/j.jphs.2015.05.003
13. Lim FY, Ames B, Walsh CT, Keller NP. Co-ordination between BrlA regulation and secretion of the oxidoreductase FmqD directs selective accumulation of fumiquinazoline C to conidial tissues in *Aspergillus fumigatus*. *Cell Microbiol* (2014) 16:1267–83. doi:10.1111/cmi.12284
14. Baccile JA, Spraker JE, Le HH, Brandenburger E, Gomez C, Bok JW, et al. Plant-like biosynthesis of isoquinoline alkaloids in *Aspergillus fumigatus*. *Nat Chem Biol* (2016) 12:419–24. doi:10.1038/nchembio.2061
15. Suzuki S, Kikkawa K, Yamazaki M. Abnormal behavioral effects elicited by a neurotropic mycotoxin, fumitremorgin A in mice. *J Pharmacobiodyn* (1984) 7:935–42. doi:10.1248/bpb1978.7.935
16. Frisvad JC, Larsen TO. Extrolites of *Aspergillus fumigatus* and other pathogenic species in *Aspergillus* section fumigati. *Front Microbiol* (2015) 6:1485. doi:10.3389/fmicb.2015.01485
17. Usui T, Kondoh M, Cui CB, Mayumi T, Osada H. Tryprostatin A, a specific and novel inhibitor of microtubule assembly. *Biochem J* (1998) 333(Pt 3):543–8. doi:10.1042/bj3330543
18. Khoufache K, Puel O, Loiseau N, Delaforge M, Rivollet D, Coste A, et al. Verruculogen associated with *Aspergillus fumigatus* hyphae and conidia modifies the electrophysiological properties of human nasal epithelial cells. *BMC Microbiol* (2007) 7:5. doi:10.1186/1471-2180-7-5
19. Feng Y, Holte D, Zoller J, Umemiya S, Simke LR, Baran PS. Total synthesis of verruculogen and fumitremorgin A enabled by ligand-controlled C-H borylation. *J Am Chem Soc* (2015) 137:10160–3. doi:10.1021/jacs.5b07154
20. Sugui JA, Pardo J, Chang YC, Zarember KA, Nardone G, Galvez EM, et al. Gliotoxin is a virulence factor of *Aspergillus fumigatus*: *gliP* deletion attenuates virulence in mice immunosuppressed with hydrocortisone. *Eukaryot Cell* (2007) 6:1562–9. doi:10.1128/EC.00141-07
21. Yin WB, Baccile JA, Bok JW, Chen Y, Keller NP, Schroeder FC. A nonribosomal peptide synthetase-derived iron(III) complex from the pathogenic fungus *Aspergillus fumigatus*. *J Am Chem Soc* (2013) 135:2064–7. doi:10.1021/ja311145n
22. Wiemann P, Lechner BE, Baccile JA, Velk TA, Yin WB, Bok JW, et al. Perturbations in small molecule synthesis uncovers an iron-responsive secondary metabolite network in *Aspergillus fumigatus*. *Front Microbiol* (2014) 5:530. doi:10.3389/fmicb.2014.00530
23. Lingens F. Regulation of aromatic amino acid biosynthesis in microorganisms. *Acta Microbiol Acad Sci Hung* (1976) 23:161–6.
24. Braus GH. Aromatic amino acid biosynthesis in the yeast *Saccharomyces cerevisiae*: a model system for the regulation of a eukaryotic biosynthetic pathway. *Microbiol Rev* (1991) 55:349–70.
25. Tagliamonte A, Gessa R, Biggio G, Vargiu L, Gessa GL. Daily changes of free serum tryptophan in humans. *Life Sci* (1974) 14:349–54. doi:10.1016/0024-3205(74)90065-4
26. Pereira SA, Livi GP. Aromatic amino-acid biosynthesis in *Candida albicans*: identification of the ARO4 gene encoding a second DAHP synthase. *Curr Genet* (1996) 29:441–5. doi:10.1007/BF02221512
27. Brunke S, Seider K, Almeida RS, Heyken A, Fleck CB, Brock M, et al. *Candida glabrata* tryptophan-based pigment production via the Ehrlich pathway. *Mol Microbiol* (2010) 76:25–47. doi:10.1111/j.1365-2958.2010.07052.x

28. Wang PM, Choera T, Wiemann P, Pisithkul T, Amador-Noguez D, Keller NP. TrpE feedback mutants reveal roadblocks and conduits toward increasing secondary metabolism in *Aspergillus fumigatus*. *Fungal Genet Biol* (2016) 89:102–13. doi:10.1016/j.fgb.2015.12.002
29. Hawkins AR, Lamb HK, Moore JD, Charles IG, Roberts CF. The pre-chorismate (shikimate) and quinate pathways in filamentous fungi: theoretical and practical aspects. *J Gen Microbiol* (1993) 139:2891–9. doi:10.1099/00221287-139-12-2891
30. Duncan K, Edwards RM, Coggins JR. The pentafunctional AroM enzyme of *Saccharomyces cerevisiae* is a mosaic of monofunctional domains. *Biochem J* (1987) 246:375–86. doi:10.1042/bj2460375
31. Sasse A, Hamer SN, Amich J, Binder J, Krappmann S. Mutant characterization and *in vivo* conditional repression identify aromatic amino acid biosynthesis to be essential for *Aspergillus fumigatus* virulence. *Virulence* (2016) 7:56–62. doi:10.1080/21505594.2015.1109766
32. Graf R, Mehmann B, Braus GH. Analysis of feedback-resistant anthranilate synthases from *Saccharomyces cerevisiae*. *J Bacteriol* (1993) 175:1061–8. doi:10.1128/jb.175.4.1061-1068.1993
33. Kafer E. The anthranilate synthetase enzyme complex and the trifunctional *trpC* gene of *Aspergillus*. *Can J Genet Cytol* (1977) 19:723–38. doi:10.1139/g77-079
34. Borgia PT, Dodge CL, Eagleton LE, Adams TH. Bidirectional gene transfer between *Aspergillus fumigatus* and *Aspergillus nidulans*. *FEMS Microbiol Lett* (1994) 122:227–31. doi:10.1111/j.1574-6968.1994.tb07172.x
35. Hu W, Sillaots S, Lemieux S, Davison J, Kauffman S, Breton A, et al. Essential gene identification and drug target prioritization in *Aspergillus fumigatus*. *PLoS Pathog* (2007) 3:e24. doi:10.1371/journal.ppat.0030024
36. Blanco B, Prado V, Lence E, Otero JM, Garcia-Doval C, Van Raaij MJ, et al. *Mycobacterium tuberculosis* shikimate kinase inhibitors: design and simulation studies of the catalytic turnover. *J Am Chem Soc* (2013) 135:12366–76. doi:10.1021/ja405853p
37. Ianiri G, Idnurm A. Essential gene discovery in the basidiomycete *Cryptococcus neoformans* for antifungal drug target prioritization. *MBio* (2015) 6:e02334-14. doi:10.1128/mBio.02334-14
38. Kaltdorf M, Srivastava M, Gupta SK, Liang C, Binder J, Dietl AM, et al. Systematic identification of anti-fungal drug targets by a metabolic network approach. *Front Mol Biosci* (2016) 3:22. doi:10.3389/fmolb.2016.00022
39. Aoki Y, Kondoh M, Nakamura M, Fujii T, Yamazaki T, Shimada H, et al. A new methionine antagonist that has antifungal activity: mode of action. *J Antibiot (Tokyo)* (1994) 47:909–16. doi:10.7164/antibiotics.47.909
40. Aoki Y, Kamiyama T, Fujii T, Yamamoto M, Ohwada J, Arisawa M. Design of an antifungal methionine inhibitor not antagonized by methionine. *Biol Pharm Bull* (1995) 18:1267–71. doi:10.1248/bpb.18.1267
41. Aoki Y, Yamamoto M, Hosseini-Mazinani SM, Koshikawa N, Sugimoto K, Arisawa M. Antifungal azoxybacilin exhibits activity by inhibiting gene expression of sulfite reductase. *Antimicrob Agents Chemother* (1996) 40:127–32.
42. Yamaguchi M, Yamaki H, Shinoda T, Tago Y, Suzuki H, Nishimura T, et al. The mode of antifungal action of (S)2-amino-4-oxo-5-hydroxypentanoic acid, RI-331. *J Antibiot (Tokyo)* (1990) 43:411–6. doi:10.7164/antibiotics.43.411
43. Jacques SL, Ejim LJ, Wright GD. Homoserine dehydrogenase from *Saccharomyces cerevisiae*: kinetic mechanism and stereochemistry of hydride transfer. *Biochim Biophys Acta* (2001) 1544:42–54. doi:10.1016/S0167-4838(00)00202-8
44. Yamaki H, Yamaguchi M, Nishimura T, Shinoda T, Yamaguchi H. Unique mechanism of action of an antifungal antibiotic RI-331. *Drugs Exp Clin Res* (1988) 14:467–72.
45. Tan S, Evans R, Singh B. Herbicidal inhibitors of amino acid biosynthesis and herbicide-tolerant crops. *Amino Acids* (2006) 30:195–204. doi:10.1007/s00726-005-0254-1
46. Bode R, Melo C, Birnbaum D. Mode of action of glyphosate in *Candida maltosa*. *Arch Microbiol* (1984) 140:83–5. doi:10.1007/BF00409776
47. Chin K, Wyder MA, Kaneshiro ES. Glyphosate reduces organism viability and inhibits growth *in vitro* of *Pneumocystis*. *J Eukaryot Microbiol* (1999) 46:139S–41S.
48. Fernandes JD, Martho K, Tofik V, Vallim MA, Pascon RC. The role of amino acid permease and tryptophan biosynthesis in *Cryptococcus neoformans* survival. *PLoS One* (2015) 10:e0132369. doi:10.1371/journal.pone.0132369
49. Abdel-Rahman HM, El-Koussi NA, Hassan HY. Fluorinated 1,2,4-triazolo[1,5-a]pyrimidine-6-carboxylic acid derivatives as antimycobacterial agents. *Arch Pharm (Weinheim)* (2009) 342:94–9. doi:10.1002/ardp.200800113
50. Forsberg H, Gilstring CF, Zargari A, Martinez P, Ljungdahl PO. The role of the yeast plasma membrane SPS nutrient sensor in the metabolic response to extracellular amino acids. *Mol Microbiol* (2001) 42:215–28. doi:10.1046/j.1365-2958.2001.02627.x
51. Sugui JA, Kim HS, Zarember KA, Chang YC, Gallin JI, Nierman WC, et al. Genes differentially expressed in conidia and hyphae of *Aspergillus fumigatus* upon exposure to human neutrophils. *PLoS One* (2008) 3:e2655. doi:10.1371/journal.pone.0002655
52. Morton CO, Varga JJ, Hornbach A, Mezger M, Sennefelder H, Kneitz S, et al. The temporal dynamics of differential gene expression in *Aspergillus fumigatus* interacting with human immature dendritic cells *in vitro*. *PLoS One* (2011) 6:e16016. doi:10.1371/journal.pone.0016016
53. Grice CM, Bertuzzi M, Bignell EM. Receptor-mediated signaling in *Aspergillus fumigatus*. *Front Microbiol* (2013) 4:26. doi:10.3389/fmicb.2013.00026
54. Brown NA, Dos Reis TF, Ries LN, Caldana C, Mah JH, Yu JH, et al. G-protein coupled receptor-mediated nutrient sensing and developmental control in *Aspergillus nidulans*. *Mol Microbiol* (2015) 98:420–39. doi:10.1111/mmi.13135
55. Yuasa HJ, Ball HJ. The evolution of three types of indoleamine 2,3 dioxygenases in fungi with distinct molecular and biochemical characteristics. *Gene* (2012) 504:64–74. doi:10.1016/j.gene.2012.04.082
56. Fraser JA, Davis MA, Hynes MJ. The formamidase gene of *Aspergillus nidulans*: regulation by nitrogen metabolite repression and transcriptional interference by an overlapping upstream gene. *Genetics* (2001) 157:119–31.
57. Ohashi K, Kawai S, Murata K. Secretion of quinolinic acid, an intermediate in the kynurenine pathway, for utilization in NAD+ biosynthesis in the yeast *Saccharomyces cerevisiae*. *Eukaryot Cell* (2013) 12:648–53. doi:10.1128/EC.00339-12
58. Urrestarazu A, Vissers S, Iraqui I, Grenson M. Phenylalanine- and tyrosine-auxotrophic mutants of *Saccharomyces cerevisiae* impaired in transamination. *Mol Gen Genet* (1998) 257:230–7. doi:10.1007/s004380050643
59. Bulfer SL, Brunzelle JS, Trievel RC. Crystal structure of *Saccharomyces cerevisiae* Aro8, a putative alpha-aminoadipate aminotransferase. *Protein Sci* (2013) 22:1417–24. doi:10.1002/pro.2315
60. Rao RP, Hunter A, Kashpur O, Normanly J. Aberrant synthesis of indole-3-acetic acid in *Saccharomyces cerevisiae* triggers morphogenic transition, a virulence trait of pathogenic fungi. *Genetics* (2010) 185:211–20. doi:10.1534/genetics.109.112854
61. Tylš F, Páleníček T, Horáček J. Psilocybin – summary of knowledge and new perspectives. *Eur Neuropsychopharmacol* (2014) 24:342–56. doi:10.1016/j.euroneuro.2013.12.006
62. Bilkay IS, Karakoç S, Aksoz N. Indole-3-acetic acid and gibberellic acid production in *Aspergillus niger*. *Turk J Biol* (2010) 34:313–8. doi:10.3906/biy-0812-15
63. Pattaeva M, Bakhtiyor R. Growth and phytohormones production by thermophilic *Aspergillus fumigatus* 2 and thermotolerant *Aspergillus terreus* 8 strains in salt stress. *Br J Appl Sci Technol* (2015) 8:305–12. doi:10.9734/BJAST/2015/12292
64. Macdonald JC, Slater GP. Biosynthesis of gliotoxin and mycelianamide. *Can J Biochem* (1975) 53:475–8. doi:10.1139/o75-066
65. Mulinti P, Allen NA, Coyle CM, Gravelat FN, Sheppard DC, Panaccione DG. Accumulation of ergot alkaloids during conidiophore development in *Aspergillus fumigatus*. *Curr Microbiol* (2014) 68:1–5. doi:10.1007/s00284-013-0434-2
66. Yamazaki M, Suzuki S. Toxicology of tremorgenic mycotoxins, fumitremorgin A and B. *Dev Toxicol Environ Sci* (1986) 12:273–82.
67. Imamura K, Tsuyama Y, Hirata T, Shiraishi S, Sakamoto K, Yamada O, et al. Identification of a gene involved in the synthesis of a dipeptidyl peptidase IV inhibitor in *Aspergillus oryzae*. *Appl Environ Microbiol* (2012) 78:6996–7002. doi:10.1128/AEM.01770-12
68. Liu X, Walsh CT. Cyclopiazonic acid biosynthesis in *Aspergillus* sp.: characterization of a reductase-like R* domain in cyclopiazonate synthetase that forms and releases cyclo-acetoacetyl-L-tryptophan. *Biochemistry* (2009) 48:8746–57. doi:10.1021/bi901123r

69. Clevenger KD, Bok JW, Ye R, Miley GP, Verdan MH, Velk T, et al. A scalable platform to identify fungal secondary metabolites and their gene clusters. *Nat Chem Biol* (2017) 13:895–901. doi:10.1038/nchembio.2408
70. Yamazaki M, Fujimoto H, Kawasaki T. Chemistry of tremorogenic metabolites. Fumitremorgin A from *Aspergillus fumigatus*. *Chem Pharm Bull (Tokyo)* (1980) 28:245–54. doi:10.1248/cpb.28.245
71. Cole RJ. Fungal tremorogens. *Prikl Biokhim Mikrobiol* (1993) 29:44–50.
72. Colangelo EJ. Cervicocranium and the aviator's protective helmet. *Aviat Space Environ Med* (1975) 46:1263–4.
73. Zelante T, Fallarino F, Bistoni F, Puccetti P, Romani L. Indoleamine 2,3-dioxygenase in infection: the paradox of an evasive strategy that benefits the host. *Microbes Infect* (2009) 11:133–41. doi:10.1016/j.micinf.2008.10.007
74. Thomas SM, Garrity LF, Brandt CR, Schobert CS, Feng GS, Taylor MW, et al. IFN-gamma-mediated antimicrobial response. Indoleamine 2,3-dioxygenase-deficient mutant host cells no longer inhibit intracellular *Chlamydia* spp. or *Toxoplasma* growth. *J Immunol* (1993) 150:5529–34.
75. Aldajani WA, Salazar F, Sewell HF, Knox A, Ghaemmaghami AM. Expression and regulation of immune-modulatory enzyme indoleamine 2,3-dioxygenase (IDO) by human airway epithelial cells and its effect on T cell activation. *Oncotarget* (2016) 7:57606–17. doi:10.18632/oncotarget.11586
76. De Luca A, Montagnoli C, Zelante T, Bonifazi P, Bozza S, Moretti S, et al. Functional yet balanced reactivity to *Candida albicans* requires TRIF, MyD88, and IDO-dependent inhibition of Rorc. *J Immunol* (2007) 179:5999–6008. doi:10.4049/jimmunol.179.9.5999
77. Bozza S, Fallarino F, Pitzurra L, Zelante T, Montagnoli C, Bellocchio S, et al. A crucial role for tryptophan catabolism at the host/*Candida albicans* interface. *J Immunol* (2005) 174:2910–8. doi:10.4049/jimmunol.174.5.2910
78. Bonifazi P, Zelante T, D'angelo C, De Luca A, Moretti S, Bozza S, et al. Balancing inflammation and tolerance *in vivo* through dendritic cells by the commensal *Candida albicans*. *Mucosal Immunol* (2009) 2:362–74. doi:10.1038/mi.2009.17
79. Cheng SC, Van De Veerdonk F, Smeekens S, Joosten LA, Van Der Meer JW, Kullberg BJ, et al. *Candida albicans* dampens host defense by downregulating IL-17 production. *J Immunol* (2010) 185:2450–7. doi:10.4049/jimmunol.1000756
80. Jiang N, Zhao G, Lin J, Hu L, Che C, Li C, et al. Indoleamine 2,3-dioxygenase is involved in the inflammation response of corneal epithelial cells to *Aspergillus fumigatus* infections. *PLoS One* (2015) 10:e0137423. doi:10.1371/journal.pone.0137423
81. Paveglio SA, Allard J, Foster Hodgkins SR, Ather JL, Bevelander M, Campbell JM, et al. Airway epithelial indoleamine 2,3-dioxygenase inhibits CD4+ T cells during *Aspergillus fumigatus* antigen exposure. *Am J Respir Cell Mol Biol* (2011) 44:11–23. doi:10.1165/rcmb.2009-0167OC
82. De Luca A, Bozza S, Zelante T, Zagarella S, D'angelo C, Perruccio K, et al. Non-hematopoietic cells contribute to protective tolerance to *Aspergillus fumigatus* via a TRIF pathway converging on IDO. *Cell Mol Immunol* (2010) 7:459–70. doi:10.1038/cmi.2010.43
83. Romani L, Fallarino F, De Luca A, Montagnoli C, D'angelo C, Zelante T, et al. Defective tryptophan catabolism underlies inflammation in mouse chronic granulomatous disease. *Nature* (2008) 451:211–5. doi:10.1038/nature06471
84. Iannitti RG, Carvalho A, Cunha C, De Luca A, Giovannini G, Casagrande A, et al. Th17/Treg imbalance in murine cystic fibrosis is linked to indoleamine 2,3-dioxygenase deficiency but corrected by kynurenines. *Am J Respir Crit Care Med* (2013) 187:609–20. doi:10.1164/rccm.201207-1346OC
85. Araujo EF, Feriotti C, Galdino NAL, Preite NW, Calich VLG, Loures FV. The IDO-AhR axis controls Th17/Treg immunity in a pulmonary model of fungal infection. *Front Immunol* (2017) 8:880. doi:10.3389/fimmu.2017.00880
86. Araujo EF, Medeiros DH, Galdino NA, Condino-Neto A, Calich VL, Loures FV. Tolerogenic plasmacytoid dendritic cells control *Paracoccidioides brasiliensis* infection by inducting regulatory T cells in an IDO-dependent manner. *PLoS Pathog* (2016) 12:e1006115. doi:10.1371/journal.ppat.1006115
87. Araujo EF, Loures FV, Bazan SB, Feriotti C, Pina A, Schanoski AS, et al. Indoleamine 2,3-dioxygenase controls fungal loads and immunity in paracoccidioidomicosis but is more important to susceptible than resistant hosts. *PLoS Negl Trop Dis* (2014) 8:e3330. doi:10.1371/journal.pntd.0003330
88. Hage CA, Horan DJ, Durkin M, Connolly P, Desta Z, Skaar TC, et al. *Histoplasma capsulatum* preferentially induces IDO in the lung. *Med Mycol* (2013) 51:270–9. doi:10.3109/13693786.2012.710857
89. George MM, Subramanian Vignesh K, Landero Figueroa JA, Caruso JA, Deepe GS Jr. Zinc induces dendritic cell tolerogenic phenotype and skews regulatory T cell-Th17 balance. *J Immunol* (2016) 197:1864–76. doi:10.4049/jimmunol.1600410
90. Gorski JP, Howard JB. Effect of methylamine on the structure and function of the fourth component of human complement, C4. *J Biol Chem* (1980) 255:10025–8.
91. Jung ID, Lee C-M, Jeong Y-I, Lee JS, Park WS, Han J, et al. Differential regulation of indoleamine 2,3-dioxygenase by lipopolysaccharide and interferon gamma in murine bone marrow derived dendritic cells. *FEBS Lett* (2007) 581:1449–56. doi:10.1016/j.febslet.2007.02.073
92. Fallarino F, Volpi C, Zelante T, Vacca C, Calvitti M, Fioretti MC, et al. IDO mediates TLR9-driven protection from experimental autoimmune diabetes. *J Immunol* (2009) 183:6303–12. doi:10.4049/jimmunol.0901577
93. Rivas FV, Chervonsky AV, Medzhitov R. ART and immunology. *Trends Immunol* (2014) 35:451. doi:10.1016/j.it.2014.09.002
94. Soares MP, Gozzelino R, Weis S. Tissue damage control in disease tolerance. *Trends Immunol* (2014) 35:483–94. doi:10.1016/j.it.2014.08.001
95. Romani L. Cell mediated immunity to fungi: a reassessment. *Med Mycol* (2008) 46:515–29. doi:10.1080/13693780801971450
96. Zelante T, Bozza S, De Luca A, D'angelo C, Bonifazi P, Moretti S, et al. Th17 cells in the setting of *Aspergillus* infection and pathology. *Med Mycol* (2009) 47(Suppl 1):S162–9. doi:10.1080/13693780802140766
97. Montagnoli C, Fallarino F, Gaziano R, Bozza S, Bellocchio S, Zelante T, et al. Immunity and tolerance to *Aspergillus* involve functionally distinct regulatory T cells and tryptophan catabolism. *J Immunol* (2006) 176:1712–23. doi:10.4049/jimmunol.176.3.1712
98. De Luca A, Pariano M, Cellini B, Costantini C, Villella VR, Jose SS, et al. The IL-17F/IL-17RC axis promotes respiratory allergy in the proximal airways. *Cell Rep* (2017) 20:1667–80. doi:10.1016/j.celrep.2017.07.063
99. De Ravin SS, Zarember KA, Long-Priel D, Chan KC, Fox SD, Gallin JI, et al. Tryptophan/kynurenine metabolism in human leukocytes is independent of superoxide and is fully maintained in chronic granulomatous disease. *Blood* (2010) 116:1755–60. doi:10.1182/blood-2009-07-233734
100. Macchiarulo A, Camaioni E, Nuti R, Pellicciari R. Highlights at the gate of tryptophan catabolism: a review on the mechanisms of activation and regulation of indoleamine 2,3-dioxygenase (IDO), a novel target in cancer disease. *Amino Acids* (2009) 37:219–29. doi:10.1007/s00726-008-0137-3
101. Zelante T, De Luca A, Bonifazi P, Montagnoli C, Bozza S, Moretti S, et al. IL-23 and the Th17 pathway promote inflammation and impair antifungal immune resistance. *Eur J Immunol* (2007) 37:2695–706. doi:10.1002/eji.200737409
102. Jiang N, Zhao GQ, Lin J, Hu LT, Che CY, Li C, et al. Expression of indoleamine 2,3-dioxygenase in a murine model of *Aspergillus fumigatus* keratitis. *Int J Ophthalmol* (2016) 9:491–6. doi:10.18240/ijo.2016.04.03
103. Mackowiak B, Wang H. Mechanisms of xenobiotic receptor activation: direct vs. indirect. *Biochim Biophys Acta* (2016) 1859:1130–40. doi:10.1016/j.bbagrm.2016.02.006
104. Hubbard TD, Murray IA, Perdew GH. Indole and tryptophan metabolism: endogenous and dietary routes to Ah receptor activation. *Drug Metab Dispos* (2015) 43:1522–35. doi:10.1124/dmd.115.064246
105. Zelante T, Iannitti RG, Cunha C, De Luca A, Giovannini G, Pieraccini G, et al. Tryptophan catabolites from microbiota engage aryl hydrocarbon receptor and balance mucosal reactivity via interleukin-22. *Immunity* (2013) 39:372–85. doi:10.1016/j.immuni.2013.08.003
106. Bessede A, Gargaro M, Pallotta MT, Matino D, Servillo G, Brunacci C, et al. Aryl hydrocarbon receptor control of a disease tolerance defence pathway. *Nature* (2014) 511:184–90. doi:10.1038/nature13323
107. Goupil M, Cousineau-Cote V, Aumont F, Senechal S, Gaboury L, Hanna Z, et al. Defective IL-17- and IL-22-dependent mucosal host response to *Candida albicans* determines susceptibility to oral candidiasis in mice expressing the HIV-1 transgene. *BMC Immunol* (2014) 15:49. doi:10.1186/s12865-014-0049-9
108. Guo X, Qiu J, Tu T, Yang X, Deng L, Anders RA, et al. Induction of innate lymphoid cell-derived interleukin-22 by the transcription factor STAT3 mediates protection against intestinal infection. *Immunity* (2014) 40:25–39. doi:10.1016/j.immuni.2013.10.021
109. Xu X, Weiss ID, Zhang HH, Singh SP, Wynn TA, Wilson MS, et al. Conventional NK cells can produce IL-22 and promote host defense in *Klebsiella pneumoniae* pneumonia. *J Immunol* (2014) 192:1778–86. doi:10.4049/jimmunol.1300039

8

Fungal Microbiota in Chronic Airway Inflammatory Disease and Emerging Relationships with the Host Immune Response

Irene Zhang[1]*, Steven D. Pletcher*[2]*, Andrew N. Goldberg*[2]*, Bridget M. Barker*[1] *and Emily K. Cope*[1*]

[1] Pathogen and Microbiome Institute, Northern Arizona University, Flagstaff, AZ, United States, [2] Department of Otolaryngology Head and Neck Surgery, University of California, San Francisco, San Francisco, CA, United States

***Correspondence:**
Emily K. Cope
emily.cope@nau.edu*

The respiratory tract is a complex system that is inhabited by niche-specific communities of microbes including bacteria, fungi, and viruses. These complex microbial assemblages are in constant contact with the mucosal immune system and play a critical role in airway health and immune homeostasis. Changes in the composition and diversity of airway microbiota are frequently observed in patients with chronic inflammatory diseases including chronic rhinosinusitis (CRS), cystic fibrosis, allergy, and asthma. While the bacterial microbiome of the upper and lower airways has been the focus of many recent studies, the contribution of fungal microbiota to inflammation is an emerging research interest. Within the context of allergic airway disease, fungal products are important allergens and fungi are potent inducers of inflammation. In addition, murine models have provided experimental evidence that fungal microbiota in peripheral organs, notably the gastrointestinal (GI) tract, influence pulmonary health. In this review, we explore the role of the respiratory and GI microbial communities in chronic airway inflammatory disease development with a specific focus on fungal microbiome interactions with the airway immune system and fungal-bacterial interactions that likely contribute to inflammatory disease. These findings are discussed in the context of clinical and immunological features of fungal-mediated disease in CRS, allergy, and asthmatic patients. While this field is still nascent, emerging evidence suggests that dysbiotic fungal and bacterial microbiota interact to drive or exacerbate chronic airway inflammatory disease.

Keywords: airway microbiome, airway fungal microbiome, mycobiome, host-microbiome interactions, bacterial-fungal interactions, biofilm

INTRODUCTION

Microbial communities associated with host mucosal surfaces play essential and diverse roles in biological processes from metabolism to immune regulation and homeostasis. In return, host inflammatory responses can shape the microbial community by supporting the growth of some microbes while inhibiting others (Kumamoto, 2016). Most studies of host-associated microbiota in health and disease have focused on the bacterial microbiota, despite fungal diseases incurring a substantial infectious disease burden (Huffnagle and Noverr, 2013). Because of this, relatively little is known about the importance and function of the fungal microbiota (mycobiota). Recent advances in culture-independent sequencing approaches have helped elucidate the fungal diversity, burden, and functions associated with mucosal surfaces such as the human intestinal tract, oral cavity, skin, and respiratory tract.

In virtually every body habitat studied to date, niche-specific microbial communities have been identified, including the gastrointestinal and respiratory tracts (Human Microbiome Project Consortium, 2012). Fungi live in host niches as commensals and interact with bacteria and the host as members of the healthy microbiome. While fungal microbiota are numerically small, they are diverse, have large genomes, and potentially act as important keystone species in the microbiome (Huffnagle and Noverr, 2013). Studies have shown that disrupting commensal fungi can affect both local and peripheral immune responses and enhance disease states. In addition, members of the mycobiome can switch from commensalism to pathogenicity, often dependent upon co-colonizing microbial taxa (Cohen et al., 1969; Peleg et al., 2010). Recent studies suggest a role for airway microbial communities in maintaining airway health (Abreu et al., 2012). In this perspective, we focus on the potential role of fungal interactions with the host and bacterial co-colonizers chronic rhinosinusitis (CRS) and lower airway comorbidities.

RESPIRATORY TRACT PHYSIOLOGY

Like the GI tract (Jeraldo et al., 2012), the respiratory tract is composed of distinct environments that vary according to mucosal architecture and immune responses. Reflective of this, the microbial communities that inhabit the respiratory tract exhibit niche specificity; compositionally distinct communities exist between the nares, sinuses, oral cavity, oropharynx and lower airway in healthy individuals (Lemon et al., 2010; Bassis et al., 2015). The sinonasal cavity, nasopharynx, and trachea are lined with a pseudostratified columnar ciliated epithelium and mucus-producing goblet cells. Lung bronchi and bronchioles are covered by stratified columnar ciliated epithelia and few goblet cells. Mucus secretion and immune effector molecules protect the host from insults by trapping and removing inhaled pathogens and irritants, a process orchestrated by ciliary beating (Cohen, 2006). Airway epithelial cells express pattern recognition receptors (PRRs), including toll-like receptors (TLR) and non-TLR PRRs such as Dectin-1, that can recognize and respond to components of microorganisms (Ryu et al., 2013; Liu et al., 2015).

CHRONIC INFLAMMATORY DISEASES OF THE UPPER AND LOWER AIRWAYS

Chronic airway inflammatory diseases including asthma, rhinosinusitis, and rhinitis are substantial health problems resulting in significant healthcare expenditures. Chronic rhinosinusitis (CRS) alone affects approximately 15% of the adult population annually; recent estimates suggest that costs associated with CRS exceed $65 billion each year accounting for 5% of the US healthcare budget (Caulley et al., 2015). Several epidemiological studies suggest that concomitant upper and lower airway disease result from shared pathophysiologic mechanisms (Passalacqua et al., 2001; Ciprandi et al., 2012; Licari et al., 2014).

CRS is clinically defined by duration of symptoms and endoscopic or radiographic evidence of sinus inflammation. Multiple subtypes of CRS have been described based on clinical (Akdis et al., 2013; Orlandi et al., 2016), immunological (Tomassen et al., 2016) and microbiological (Cope et al., 2017) heterogeneity. The role of fungi in the pathophysiology of CRS has long been controversial. While some authors argued that fungi drove inflammation in nearly all CRS patients (Ponikau et al., 2000), this finding has been recently disputed (Orlandi and Marple, 2010; Fokkens et al., 2012). Fungi clearly play a central role in several discrete subtypes of CRS, such as allergic fungal sinusitis and development of fungus balls (deShazo et al., 1997). Fungi found in culture-based studies of CRS patients include members of *Aspergillus*, *Alternaria*, *Candida*, *Cladosporium*, and *Penicillium* (Ponikau et al., 1999). Molecular studies have expanded this list to include fastidious fungi *Malassezia*, *Curvularia*, *Schizophyllum*, and *Neocosmospora* (Cleland et al., 2014; Gelber et al., 2016; Zhao et al., 2017). While there are few studies on host-fungal interactions in the upper airways, recent research suggests that defects in epithelial genes might compromise immune barrier function and lead to dysfunctional host immune responses to bacterial or fungal colonization (Tieu et al., 2009). Additionally, fungal glycans (e.g., β-glucan) interact with airway epithelial cells via surface receptors and can act as determinants of allergenicity or induce inflammation (Roy and Klein, 2013). In allergic bronchopulmonary aspergillosis (ABPA), structural abnormalities in the airway epithelium allow fungal spores to breach immune defenses and germinate into hyphae (Chaudhary and Marr, 2011). Upon germination, fungal spores secrete proteases which can disrupt epithelial barrier integrity (Chen et al., 2011). Epithelial barrier dysfunction is frequently observed in CRS and asthma (Tieu et al., 2009; Lambrecht and Hammad, 2012). The presence of bacterial and fungal biofilms increased mucosal IgE in CRS patients, suggesting that these biofilms interact with the host immune system and perpetuate chronic inflammation (Foreman et al., 2012).

CRS is phenotypically classified into two groups based on the presence or absence of nasal polyps. This clinical classification of CRS patients was initially reflected immunologically with a predominance of T_H2 cells and eosinophils in CRS patients with polyps (CRSwNP) and a predominance of T_H1 cells and neutrophils in CRS patients without polyps (CRSsNP). However, recent studies using molecular methods have demonstrated up to 10 distinct endotypes of patients using markers of T_H1, T_H2, T_H17, eosinophil, and neutrophil activation (Tomassen et al., 2016). A separate study found four distinct groups of CRS patients characterized by distinct bacterial communities, each conferring a unique immune response (Cope et al., 2017). *Corynebacteriaceae*-dominated communities conferred increased *IL-5* gene expression and a higher risk for nasal polyps. While these findings are helpful, it is also important to understand how the sinus microbiota drive or exacerbate observed inflammatory endotypes so new therapeutics can target the initiation of disease.

Unified Airway Hypothesis

CRS often exists in the setting of concomitant lower airway disease, including asthma, and is frequently preceded by rhinitis. Upper respiratory infections can exacerbate asthma symptoms; bronchial hyperresponsiveness often occurs along with rhinitis (Leynaert et al., 2000; Passalacqua et al., 2001). Between 20 and 60% of CRS patients with nasal polyps have asthma (Larsen, 1996; Klossek et al., 2005). These observations have led to the "unified airway" hypothesis, which treats the respiratory tract as one organ rather than distinct organs affected by specific diseases. Supporting this hypothesis, medical and surgical treatment in the upper airways for CRS frequently results in reduced asthma symptoms. Therefore, respiratory inflammatory disease can be considered a disorder of the entire respiratory tract. Dysbiotic (altered) microbial communities have been described in the upper airways of CRS patients and in the lower airways of asthmatic patients, however, the microbiota of these two sites have not yet been examined in parallel. Below, we will discuss recent findings in CRS and asthmatic microbiota and potential interactions between these important microbial populations.

AIRWAY BACTERIAL MICROBIOME

The upper and lower airways harbor niche-specific microbial communities that relate to health status (Lemon et al., 2010; Huang and Boushey, 2014). Recent studies of the sinonasal microbiome generally demonstrate a loss of bacterial diversity and concomitant enrichment of pathobionts, resident microbes with pathogenic potential, in the sinuses of patients with CRS (Biswas et al., 2015; Cope et al., 2017; Lal et al., 2017; Wagner Mackenzie et al., 2017). Abreu and colleagues found that the sinus microbiome was characterized by reduced microbial richness and depletion of taxa associated with healthy individuals, such as lactic acid bacteria (Abreu et al., 2012). This depleted CRS microbiome was enriched with the pathobiont *Corynebacterium tuberculostearicum.* In a murine model, *C. tuberculostearicum* increased host mucin secretion, potentially increasing factors such as nutrient availability and adherence. Larger studies of upper airway microbiota in CRS patients demonstrate a heterogeneous, compositionally distinct microbiota which relate to distinct host immune responses (Cope et al., 2017; Lal et al., 2017). Studies comparing the lung microbiome of asthmatics and healthy controls found increased bacterial burden and diversity among asthmatics and a compositional shift in the bacterial microbiome composition characterized by enriched *Proteobacteria* (Huang et al., 2011; Han et al., 2012; Durack et al., 2017). Polymicrobial biofilms consisting of bacteria and fungi have been observed on sinonasal mucosa using fluorescence *in situ* hybridization, demonstrating that these taxa exist in mixed-species mucosal communities (Sanderson et al., 2006; Doble et al., 2007). More research is required on the upper and lower airway microbiota, and, in particular, on the relationship between sinonasal and pulmonary microbiota and mycobiota. The function and composition of these communities may shed light on the heterogeneity observed in CRS.

AIRWAY FUNGAL MICROBIOME

While the bacterial communities in airway inflammatory disease have been more extensively studied, fungal microbiota are still poorly characterized. This is due, in part, to the challenges faced by researchers who study the airway mycobiome. Low fungal biomass in the airways and databases that are curated from GI or environmental sources affect sample collection, DNA extraction, library preparation, and sequence analysis. In other body sites such as the GI tract, fungal dysbiosis has been implicated in diseases including inflammatory bowel disease. Although outside the scope of this perspective, gut mycobiome dysbiosis and interaction with the host immune response has been extensively reviewed (Underhill and Iliev, 2014; Mukherjee et al., 2015). In the airways, similarities between allergic fungal rhinosinusitis and ABPA illustrate the ability of fungi to act as antigens and invade the respiratory tract.

Studies on the fungal microbiome in CRS and asthma suggest that fungal communities in the airways likely influence host health. *Malassezia*, a known pathobiont on skin (Findley et al., 2013), is the predominant fungal genus in the sinonasal cavity in healthy individuals, CRS patients, and patients with allergic rhinitis (Cleland et al., 2014; Jung et al., 2015; Gelber et al., 2016). Lower abundance fungi in the sinuses of CRS patients include *Aspergillus*, *Alternaria*, *Fusarium*, and *Saccharomyces* (Cleland et al., 2014). A recent study using ITS gene sequencing found fungal presence in the sinuses of 63% of CRS patients and the predominant fungal genus in this study was *Aspergillus* (Zhao et al., 2017). PCR evaluations of *Aspergillus*, however, have demonstrated the presence of this fungus only in the sinuses of CRS patients with a known fungal subtype of CRS (Gelber et al., 2016). These discrepancies may be due to challenges related to fungal databases for human-associated ITS sequences. Analysis of the nasal vestibule of patients with AR and healthy individuals demonstrated increased diversity of fungal communities in AR characterized by preponderance of *Malassezia* (91–99% of 18S rRNA gene sequences) with lower abundance *Aspergillus* and *Alternaria* (Jung et al., 2015). In asthma, the lung mycobiome composition is altered between patients with severe asthma, ABPA, asthma with fungal sensitization (SAFS), and mild asthmatics. In ABPA, *Pseudomonas* abundance and fungal diversity both increased. Severe asthmatics were characterized by enrichment of *Aspergillus;* the relative abundance of *Aspergillus* increased approximately 15-fold compared to mild asthmatics (Chishimba et al., 2015). These results suggest that bacterial and fungal communities change in parallel in subsets of human disease. Future studies examining paired upper and lower airway microbiota (bacterial and fungal) in individuals with CRS and asthma are warranted.

Aspergillus fumigatus is often involved in the development of allergic fungal rhinosinusitis (AFRS) specific subtype CRS that accounts for an estimated 6–9% of all CRS patients undergoing surgery (Schubert, 2004). Patients with AFRS present with hyperplastic sinus disease, characterized by chronic eosinophilic-lymphocytic inflammation and nasal polyps. In

AFRS, fungal hyphae are found within allergic mucin. Sinus cultures obtained in surgical patients reveal the presence of fungal taxa such as *Bipolaris spicifera*, *Alternaria*, and *Aspergillus*. AFRS bears many similarities to ABPA in its immunopathology, treatment, and outcomes. ABPA and AFS patients have elevated levels of serum IgE, and ABPA patients have elevated *A. fumigatus*-specific IgE and IgG. Allergic mucin in AFRS patients is histologically identical to the bronchial mucus plugs in ABPA patients (Schubert, 2004). In ABPA, immunological hypersensitivity is higher relative to AFS, possibly due to differences in the diseased organ or etiological fungus.

Concurrent upper and lower airway disease may be a singular inflammatory process caused or mediated by microbial communities. Microbes or their metabolites might translocate between the upper and lower airways. Interactions between microbes, the environment, and the host inflammatory response could alter the mucus-associated microbial communities, causing dysbiosis. Fungi involved in these allergic fungal disorders, along with associated bacteria such as *S. aureus*, viruses, or other microbes, may act as the source of microbial T-cell superantigens. Fungal contributions to chronic inflammatory airway diseases are an emerging research interest, one which may advance the understanding of such unified airway disorders and treatment for patients with allergic airway disease.

FUNGAL-HOST AND FUNGAL-BACTERIAL INTERACTIONS AND HOST HEALTH

Bacteria and fungi co-inhabit the human body and mounting evidence suggests that microbial interactions in a given niche can influence human health and disease. Bacterial co-colonizers can directly affect fungal morphology (Peters et al., 2012), survival (Romano and Kolter, 2005; Harriott and Noverr, 2009), growth (Kerr et al., 1999), virulence (Schlecht et al., 2015), and attachment to host epithelia or other microbes (El-Azizi et al., 2004; Morales and Hogan, 2010). Current investigations have focused on bacterial interactions with the opportunistic pathogen *Candida albicans*. In the oral cavity, *C. albicans* selectively attaches to the bacterium *Streptococcus gordonii* to increase attachment and growth of the fungi (Holmes et al., 1996). *In vitro* experimental studies have demonstrated that *S. aureus* and *P. aeruginosa* selectively attach to *C. albicans* hyphae but not yeast (Peters et al., 2012; Schlecht et al., 2015), and these interactions have distinct outcomes. *S. aureus* attachment to *Candida* hyphae results in increased invasion of tissue (Schlecht et al., 2015), while *P. aeruginosa* forms biofilms on the hyphal cells, killing the fungus (Hogan and Kolter, 2002). Physical sensing and Quorum sensing (QS) through small molecules that allow microbes to respond to environmental stimuli play important roles in mediating these interactions. For example, the *C. albicans* QS molecule farnesol inhibits *Pseudomonas* quinolone signal (PQS) and virulence factor production (Cugini et al., 2010). *C. albicans* can respond to *P. aeruginosa* QS molecules, specifically 3-3-oxo-C12 homoserine lactone, which prevents hyphal formation in the fungus (Hogan et al., 2004). These species-specific interactions through physical contact and QS molecules likely affect microbial virulence or biofilm formation in host upper or lower airways. Ongoing research in our lab seeks to explore these interactions.

We are interested in interactions between sinonasal-associated fungi and bacterial co-colonizers, including *Malassezia* sp. (Mowat et al., 2010; Jung et al., 2015; Gelber et al., 2016). *Malassezia*, a dimorphic yeast implicated in a variety of conditions including dandruff and atopic dermatitis, is present in the upper airways (Cleland et al., 2014; Gelber et al., 2016). *Malassezia* species are known to be immunomodulatory and encode at least 13 different allergens; these fungi can exist as commensals and can be immunosuppressive (Ashbee and Evans, 2002) or as pathobionts and can become immunostimulatory (Gaitanis et al., 2012). Similar to CRS, *Malassezia dermatis* often co-occurs with *S. aureus* in atopic dermatitis, perhaps interacting with bacteria in driving skin inflammation. Further research is required to identify these *Malassezia* at the species or strain level and to compare populations between healthy subjects and those with allergic airway disease. Mechanistic studies are needed to reveal the interactions between *Malassezia*, bacterial community members, and the host immune system.

Airway epithelial cells and phagocytes express pattern recognition receptors, including C-type lectins, that sense and respond to components of the fungal cell wall. Dectin-1, a pattern recognition receptor expressed by dendritic cells, neutrophils, and macrophages, recognizes the fungal polysaccharide β-1,3 glucan motif found on fungal cell walls. Dectin-1 may mediate host immune responses to these fungi. In mice, loss of dectin-1 resulted in more severe colitis compared to wild-type mice. DSS-induced colitis was associated with increased *Candida* and *Trichosporon* and decreased *Saccharomyces*. Fungi were also found to invade inflamed tissues in dectin-1 KO mice, but not in wild-type mice. When given the antifungal fluconazole, dectin-1 KO mice had milder symptoms (Iliev et al., 2012). These results indicate that fungi contribute to the aggravation of inflammatory responses in colitis, and dectin-1 is an immune mechanism that regulates fungal community composition. Another C-type lectin that specifically recognizes *Malassezia* has recently been described (Yamasaki et al., 2009). When expressed on macrophages, Mincle (also called Clec4e and Clecsf9) sensing of *Malassezia* induces pro-inflammatory responses (Yamasaki et al., 2009), although this receptor has also been implicated in T_H2-polarization upon activation (Geijtenbeek and Gringhuis, 2016). Mincle-deficient mice showed a 2-4 fold reduction of IL-10, TNF-alpha, and MIP1 when challenged with *Malassezia* but not *A. fumigatus* (Yamasaki et al., 2009) Host-associated commensal fungi, therefore, specifically interact with the mucosal immune system, maintaining host and microbial homeostasis.

Fungal microbial communities in the GI tract can also affect the development of respiratory disease. Mice treated with antifungals exhibited increased development of allergic airway disease along with increased disease severity in models of colitis (Wheeler et al., 2016). The microbiomes of these mice revealed a restructuring of fungal and bacterial communities characterized by decreased *Candida* (2.25-fold) and increased *Aspergillus* (2.5-fold), *Wallemia* (6-fold), and *Epicoccum* (4-fold). Oral supplements of *A. amstelodami*, *W. sebi*, and *E. nigrum* recapitulated the effects of antifungal drugs to exacerbate allergic airway disease as measured by increased pulmonary eosinophils and lymphocytes and elevated serum IgE (Wheeler et al., 2016). Colonization of the gut by *Candida albicans* has also previously been shown to influence allergic airway disease and asthma in a mouse model, perhaps through secretion of prostaglandin-like immunomodulatory molecules (Noverr et al., 2005; Huffnagle, 2010; Marsland and Salami, 2015). Thus, fungal dysbiosis in the gut can affect disease development and exacerbation in the respiratory tract, indicating that a healthy mycobiome mediates immune responses throughout the body. Whether these effects reflect a direct action of fungi on the host or occur indirectly by altering the bacterial microbiome composition is unknown. Further studies are needed to elucidate the mechanisms for these responses and to determine whether fungal communities in other organs can influence allergic airway disease.

FUTURE DIRECTIONS

The evidence for the influence of dysbiotic fungal and bacterial microbiota interactions to drive or exacerbate chronic airway inflammatory disease is compelling. These findings open up the potential for targeted manipulation of the airway or gastrointestinal tract microbiota to improve airway health and manage airway disease in patients with CRS, asthma, and cystic fibrosis, among others. Current mycobiome studies focus on comparisons in healthy vs. diseased models, but more research needs to occur to understand the interactions between fungal and bacterial communities with the host immune system in host body sites. Fungi produce diverse secondary metabolites that can affect bacteria, while bacteria can keep the mycobiome in check by producing substances that inhibit the yeast to hyphae transition of fungal pathobionts such as in *C. albicans*.

Additional research is needed to define the functional effects of the mycobiome (**Figure 1**). Rather than focusing on taxonomic diversity, studies comparing microbial community function are potentially more useful. Further understanding of the contribution of the mycobiome and bacterial-fungal interactions should move beyond the GI tract into respiratory tract, where fungal dysbiosis has been shown to influence disease development and exacerbate symptoms. A variety of methods, combining metagenomics and sequencing approaches, experimental models, and functional studies should be used to clarify mechanisms by which the mycobiome impacts health and disease. In particular, focusing on microbial interactions with mucosal surfaces and their influence on local immune response

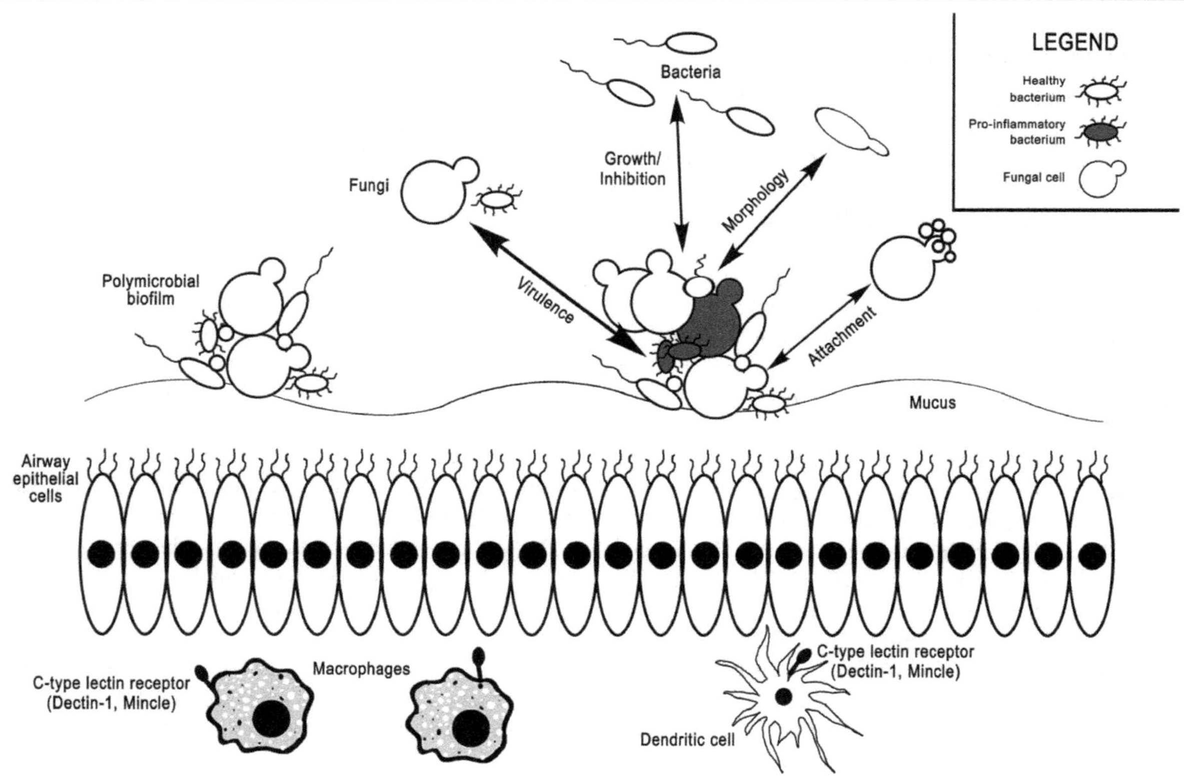

FIGURE 1 | Interactions between the bacterial microbiome, mycobiome, and host immune system in the airways. Bacteria and fungi coexist in mucosa-attached polymicrobial biofilms in the airways. Fungi can selectively induce or inhibit growth of various bacterial taxa, increase bacterial virulence factors, alter bacterial morphology, or act as attachment sites for bacteria. Vice versa, bacteria can also alter fungal growth, virulence, morphology, and attachment. C-type lectin receptors on macrophages and dendritic cells, such as dectin-1 and Mincle, can sense fungi and mediate host inflammatory responses.

likely will provide additional insights into the role of microbes in inflammatory disease of the upper and lower airways. Because fungal dysbiosis can have as great an effect on the host as bacterial dysbiosis, we cannot overlook the contributions of these largely unexplored fungal communities in our search to understand the microbial mechanisms behind health and disease.

AUTHOR CONTRIBUTIONS

IZ drafted and revised the manuscript. EC conceived, wrote, and revised the manuscript. BB conceived, wrote and revised the manuscript. AG conceived and revised the manuscript. SP conceived, wrote, and revised the manuscript.

REFERENCES

1. Abreu, N. A., Nagalingam, N. A., Song, Y., Roediger, F. C., Pletcher, S. D., Goldberg, A. N., et al. (2012). Sinus microbiome diversity depletion and *Corynebacterium tuberculostearicum* enrichment mediates rhinosinusitis. *Sci. Transl. Med.* 4:151ra124. doi: 10.1126/scitranslmed.3003783
2. Akdis, C. A., Bachert, C., Cingi, C., Dykewicz, M. S., Hellings, P. W., Naclerio, R. M., et al. (2013). Endotypes and phenotypes of chronic rhinosinusitis: a PRACTALL document of the european academy of allergy and clinical immunology and the american academy of allergy, asthma & immunology. *J. Allergy Clin. Immunol.* 131, 1479–1490. doi: 10.1016/j.jaci.2013.02.036
3. Ashbee, H. R., and Evans, E. G. V. (2002). Immunology of diseases associated with *Malassezia* species. *Clin. Microbiol. Rev.* 15, 21–57. doi: 10.1128/CMR.15.1.21-57.2002
4. Bassis, C. M., Erb-Downward, J. R., Dickson, R. P., Freeman, C. M., Schmidt, T. M., Young, V. B., et al. (2015). Analysis of the upper respiratory tract microbiotas as the source of the lung and gastric microbiotas in healthy individuals. *MBio* 6:e00037-15. doi: 10.1128/mBio.00037-15
5. Biswas, K., Hoggard, M., Jain, R., Taylor, M. W., and Douglas, R. G. (2015). The nasal microbiota in health and disease: variation within and between subjects. *Front. Microbiol.* 9:134. doi: 10.3389/fmicb.2015.00134
6. Caulley, L., Thavorn, K., Rudmik, L., Cameron, C., and Kilty, S. J. (2015). Direct costs of adult chronic rhinosinusitis by using 4 methods of estimation: results of the US medical expenditure panel survey. *J. Allergy Clin. Immunol.* 136, 1517–1522. doi: 10.1016/j.jaci.2015.08.037
7. Chaudhary, N., and Marr, K. A. (2011). Impact of *Aspergillus fumigatus* in allergic airway diseases. *Clin. Transl. Allergy* 1:4.doi: 10.1186/2045-7022-1-4
8. Chen, J.-C., Chuang, J.-G., Su, Y.-Y., Chiang, B.-L., Lin, Y.-S., and Chow, L.-P. (2011). The protease allergen Pen c 13 induces allergic airway inflammation and changes in epithelial barrier integrity and function in a murine model. *J. Biol. Chem.* 286, 26667–26679. doi: 10.1074/jbc.M110.193987
9. Chishimba, L., Niven, R., Fraczek, M., Bowyer, P., Smyth, L., Simpson, A., et al. (2015). Lung microbiome is associated with asthma severity in fungal associated asthma. *Eur. Respir. J.* 46:OA1462. doi: 10.1183/13993003.congress-2015.OA1462
10. Ciprandi, G., Caimmi, D., Miraglia Del Giudice, M., La Rosa, M., Salpietro, C., and Marseglia, G. L. (2012). Recent developments in United airways disease. *Allergy Asthma Immunol. Res.* 4, 171–177. doi: 10.4168/aair.2012.4.4.171
11. Cleland, E. J., Bassiouni, A., Bassioni, A., Boase, S., Dowd, S., Vreugde, S., et al. (2014). The fungal microbiome in chronic rhinosinusitis: richness, diversity, postoperative changes and patient outcomes. *Int. Forum Allergy Rhinol.* 4, 259–265. doi: 10.1002/alr.21297
12. Cohen, N. A. (2006). Sinonasal mucociliary clearance in health and disease. *Ann. Otol. Rhinol. Laryngol. Suppl.* 196, 20–26. doi: 10.1177/00034894061150S904 Cohen, R., Roth, F. J., Delgado, E., Ahearn, D. G., and Kalser, M. H. (1969). Fungal flora of the normal human small and large intestine. *N. Engl. J. Med.* 280, 638–641.doi: 10.1056/NEJM196903202801204
13. Cope, E. K., Goldberg, A. N., Pletcher, S. D., and Lynch, S. V. (2017). Compositionally and functionally distinct sinus microbiota in chronic rhinosinusitis patients have immunological and clinically divergent consequences. *Microbiome* 5:53. doi: 10.1186/s40168-017-0266-6
14. Cugini, C., Morales, D. K., and Hogan, D. A. (2010). *Candida albicans*-produced farnesol stimulates *Pseudomonas* quinolone signal production in LasR-defective *Pseudomonas aeruginosa* strains. *Microbiology* 156, 3096–3107.doi: 10.1099/mic.0.037911-0
15. Doble, P., Kern, R. C., Healy, D. Y., Leid, J., Sanderson, A. R., and Hunsaker, D. H. (2007). 11:06: biofilms with fungi in chronic rhinosinusitis. *Otolaryngol. Head Neck Surg.* 137, P40–P40. doi: 10.1016/j.otohns.2007.06.033
16. Durack, J., Lynch, S. V., Nariya, S., Bhakta, N. R., Beigelman, A., Castro, M., Dyer, A.-M., Israel, E., Kraft, M., Martin, R. J., et al. (2017). Features of the bronchial bacterial microbiome associated with atopy, asthma, and responsiveness to inhaled corticosteroid treatment. *J. Allergy Clin. Immunol.* 140, 63–75. doi: 10.1016/j.jaci.2016.08.055
17. El-Azizi, M. A., Starks, S. E., and Khardori, N. (2004). Interactions of *Candida albicans* with other *Candida* spp. and bacteria in the biofilms. *J. Appl. Microbiol.* 96, 1067–1073. doi: 10.1111/j.1365-2672.2004.02213.x
18. Findley, K., Oh, J., Yang, J., Conlan, S., Deming, C., Meyer, J. A., Schoenfeld, D., Nomicos, E., and Park, M. (2013). NIH intramural sequencing center comparative sequencing program, et al. topographic diversity of fungal and bacterial communities in human skin. *Nature* 498, 367–370. doi: 10.1038/nature12171
19. Fokkens, W. J., van Drunen, C., Georgalas, C., and Ebbens, F. (2012). Role of fungi in pathogenesis of chronic rhinosinusitis: the hypothesis rejected. *Curr. Opin. Otolaryngol. Head Neck Surg.* 20, 19–23. doi: 10.1097/MOO.0b013e32834 e9084
20. Foreman, A., Boase, S., Psaltis, A., and Wormald, P.-J. (2012). Role of bacterial and fungal biofilms in chronic rhinosinusitis. *Curr. Allergy Asthma Rep.* 12, 127–135. doi: 10.1007/s11882-012-0246-7
21. Gaitanis, G., Magiatis, P., Hantschke, M., Bassukas, I. D., and Velegraki, A. (2012). The Malassezia genus in skin and systemic diseases. *Clin. Microbiol. Rev.* 25, 106–141. doi: 10.1128/CMR.00021-11
22. Geijtenbeek, T. B. H., and Gringhuis, S. I. (2016). C-type lectin receptors in the control of T helper cell differentiation. *Nat. Rev. Immunol.* 16, 433–448. doi: 10.1038/nri.2016.55
23. Gelber, J. T., Cope, E. K., Goldberg, A. N., and Pletcher, S. D. (2016). Evaluation of *Malassezia* and common fungal pathogens in subtypes of chronic rhinosinusitis. *Int. Forum Allergy Rhinol.* 6, 950–955. doi: 10.1002/alr. 21777
24. Han, M. K., Huang, Y. J., Lipuma, J. J., Boushey, H. A., Boucher, R. C., Cookson, W. O., et al. (2012). Significance of the microbiome in obstructive lung disease. *Thorax* 67, 456–463. doi: 10.1136/thoraxjnl-2011-201183
25. Harriott, M. M., and Noverr, M. C. (2009). *Candida albicans* and *Staphylococcus aureus* form polymicrobial biofilms: effects on antimicrobial resistance. *Antimicrob. Agents Chemother.* 53, 3914–3922. doi: 10.1128/AAC.00657-09
26. Hogan, D. A., and Kolter, R. (2002). *Pseudomonas-candida* interactions: an ecological role for virulence factors. *Science* 296, 2229–2232.doi: 10.1126/science.1070784
27. Hogan, D. A., Vik, Å., and Kolter, R. (2004). A *Pseudomonas aeruginosa* quorum- sensing molecule influences *Candida albicans* morphology. *Mol. Microbiol.* 54, 1212–1223.doi: 10.1111/j.1365-2958.2004.04349.x
28. Holmes, A. R., McNab, R., and Jenkinson, H. F. (1996). *Candida albicans* binding to the oral bacterium *Streptococcus gordonii* involves multiple adhesin-receptor interactions. *Infect. Immun.* 64, 4680–4685.
29. Huang, Y. J., and Boushey, H. A. (2014). The microbiome and asthma. *Ann. Am. Thorac. Soc.* 11, S48–S51. doi: 10.1513/AnnalsATS.201306-187MG
30. Huang, Y. J., Nelson, C. E., Brodie, E. L., Desantis, T. Z., Baek, M. S., Liu, J., et al. (2011). Airway microbiota and bronchial hyperresponsiveness in patients with suboptimally controlled asthma. *J. Allergy Clin. Immunol.* 127, 372–381.e1–3. doi: 10.1016/j.jaci.2010.10.048
31. Huffnagle, G. B. (2010). The microbiota and allergies/asthma. *PLoS Pathog.* 6:e1000549. doi: 10.1371/journal.ppat.1000549
32. Huffnagle, G. B., and Noverr, M. C. (2013). The emerging world of the fungal microbiome. *Trends Microbiol.* 21, 334–341. doi: 10.1016/j.tim.2013.04.002 Human Microbiome Project Consortium (2012). Structure, function and diversity of the healthy human microbiome. *Nature* 486, 207–214. doi: 10.1038/nature11234
33. Iliev, I. D., Funari, V. A., Taylor, K. D., Nguyen, Q., Reyes, C. N., Strom, S. P., et al. (2012). Interactions between commensal fungi and the C- type lectin receptor Dectin-1 influence colitis. *Science* 336, 1314–1317. doi: 10.1126/science.1221789

34. Jeraldo, P., Sipos, M., Chia, N., Brulc, J. M., Dhillon, A. S., Konkel, M. E., et al. (2012). Quantification of the relative roles of niche and neutral processes in structuring gastrointestinal microbiomes. *Proc. Natl. Acad. Sci. U.S.A.* 109, 9692–9698. doi: 10.1073/pnas.1206721109
35. Jung, W. H., Croll, D., Cho, J. H., Kim, Y. R., and Lee, Y. W. (2015). Analysis of the nasal vestibule mycobiome in patients with allergic rhinitis. *Mycoses* 58, 167–172. doi: 10.1111/myc.12296
36. Kerr, J. R., Taylor, G. W., Rutman, A., Høiby, N., Cole, P. J., and Wilson, R. (1999). *Pseudomonas aeruginosa* pyocyanin and 1-hydroxyphenazine inhibit fungal growth. *J. Clin. Pathol.* 52, 385–387.
37. Klossek, J. M., Neukirch, F., Pribil, C., Jankowski, R., Serrano, E., Chanal, I., et al. (2005). Prevalence of nasal polyposis in France: a cross-sectional, case-control study. *Allergy* 60, 233–237. doi: 10.1111/j.1398-9995.2005.00688.x
38. Kumamoto, C. A. (2016). The fungal mycobiota: small numbers, large impacts. *Cell Host Microbe* 19, 750–751. doi: 10.1016/j.chom.2016.05.018
39. Lal, D., Keim, P., Delisle, J., Barker, B., Rank, M. A., Chia, N., et al. (2017). Mapping and comparing bacterial microbiota in the sinonasal cavity of healthy, allergic rhinitis, and chronic rhinosinusitis subjects. *Int. Forum Allergy Rhinol.* 7, 561–569. doi: 10.1002/alr.21934
40. Lambrecht, B. N., and Hammad, H. (2012). The airway epithelium in asthma. *Nat. Med.* 18, 684–692. doi: 10.1038/nm.2737
41. Larsen, K. (1996). The clinical relationship of nasal polyps to asthma. *Allergy Asthma Proc.* 17, 243–249. doi: 10.2500/108854196778662255
42. Lemon, K. P., Klepac-Ceraj, V., Schiffer, H. K., Brodie, E. L., Lynch, S. V., and Kolter, R. (2010). Comparative analyses of the bacterial microbiota of the human nostril and oropharynx. *MBio* 1. doi: 10.1128/mBio.00129-10
43. Leynaert, B., Neukirch, F., Demoly, P., and Bousquet, J. (2000). Epidemiologic evidence for asthma and rhinitis comorbidity. *J. Allergy Clin. Immunol.* 106, S201–S205. doi: 10.1067/mai.2000.110151
44. Licari, A., Caimmi, S., Bosa, L., Marseglia, A., Marseglia, G. L., and Caimmi, D. (2014). Rhinosinusitis and asthma: a very long engagement. *Int. J. Immunopathol. Pharmacol.* 27, 499–508. doi: 10.1177/039463201402700405
45. Liu, Z.-C., Wang, M., Sun, W.-K., Xia, D., Tan, M.-M., Ding, Y., et al. (2015). Up-regulation of Dectin-1 in airway epithelial cells promotes mice defense against invasive pulmonary aspergillosis. *Int. J. Clin. Exp. Med.* 8, 17489–17497.
46. Marsland, B. J., and Salami, O. (2015). Microbiome influences on allergy in mice and humans. *Curr. Opin. Immunol.* 36, 94–100. doi: 10.1016/j.coi.2015.07.005
47. Morales, D. K., and Hogan, D. A. (2010). *Candida albicans* interactions with bacteria in the context of human health and disease. *PLoS Pathog.* 6:e1000886. doi: 10.1371/journal.ppat.1000886
48. Mowat, E., Rajendran, R., Williams, C., McCulloch, E., Jones, B., Lang, S., et al. (2010). *Pseudomonas aeruginosa* and their small diffusible extracellular molecules inhibit *Aspergillus fumigatus* biofilm formation. *FEMS Microbiol. Lett.* 313, 96–102. doi: 10.1111/j.1574-6968.2010.02130.x
49. Mukherjee, P. K., Sendid, B., Hoarau, G., Colombel, J.-F., Poulain, D., and Ghannoum, M. A. (2015). Mycobiota in gastrointestinal diseases. *Nat. Rev. Gastroenterol. Hepatol.* 12, 77–87. doi: 10.1038/nrgastro.2014.188
50. Noverr, M. C., Falkowski, N. R., McDonald, R. A., McKenzie, A. N., and Huffnagle,G. B. (2005). Development of allergic airway disease in mice following antibiotic therapy and fungal microbiota increase: role of host genetics, antigen, and interleukin-13. *Infect. Immun.* 73, 30–38. doi: 10.1128/IAI.73.1.30-38.2005 Orlandi, R. R., Kingdom, T. T., and Hwang, P. H. (2016). International consensus statement on allergy and rhinology: rhinosinusitis executive summary. *Int. Forum Allergy Rhinol.* 6(Suppl. 1), S3–S21. doi: 10.1002/alr.21694
51. Orlandi, R. R., and Marple, B. F. (2010). Fungus and chronic rhinosinusitis: weighing the evidence. *Otolaryngol. Head Neck Surg.* 143, 611–613. doi: 10.1016/j.otohns.2010.07.002
52. Passalacqua, G., Ciprandi, G., and Canonica, G. W. (2001). The nose-lung interaction in allergic rhinitis and asthma: united airways disease. *Curr. Opin. Allergy Clin. Immunol.* 1, 7–13.
53. Peleg, A. Y., Hogan, D. A., and Mylonakis, E. (2010). Medically important bacterial–fungal interactions. *Nat. Rev. Microbiol.* 8, 340–349. doi: 10.1038/nrmicro2313
54. Peters, B. M., Ovchinnikova, E. S., Krom, B. P., Schlecht, L. M., Zhou, H., Hoyer, L. L., et al. (2012). *Staphylococcus aureus* adherence to *Candida albicans* hyphae is mediated by the hyphal adhesin Als3p. *Microbiology* 158, 2975–2986. doi: 10.1099/mic.0.062109-0
55. Ponikau, J. U., Sherris, D. A., and Kern, E. B. (2000). Role of fungi in allergic fungal sinusitis and chronic Rhinosinusitis: in response. *Mayo Clin. Proc.* 75, 540–541. doi: 10.4065/75.5.540-a
56. Ponikau, J. U., Sherris, D. A., Kern, E. B., Homburger, H. A., Frigas, E., Gaffey, T. A., et al. (1999). The diagnosis and incidence of allergic fungal sinusitis. *Mayo Clin. Proc.* 74, 877–884. doi: 10.4065/74.9.877
57. deShazo, R. D., Chapin, K., and Swain, R. E. (1997). Fungal sinusitis. *N. Engl. J. Med.* 337, 254–259. doi: 10.1056/NEJM199707243370407
58. Romano, J. D., and Kolter, R. (2005). *Pseudomonas-Saccharomyces* interactions: influence of fungal metabolism on bacterial physiology and survival. *J. Bacteriol.* 187, 940–948. doi: 10.1128/JB.187.3.940-948.2005
59. Roy, R. M., and Klein, B. S. (2013). Fungal glycan interactions with epithelial cells in allergic airway disease. *Curr. Opin. Microbiol.* 16, 404–408. doi: 10.1016/j.mib.2013.03.004
60. Ryu, J.-H., Yoo, J.-Y., Kim, M.-J., Hwang, S.-G., Ahn, K. C., Ryu, J.-C., et al. (2013). Distinct TLR-mediated pathways regulate house dust mite-induced allergic disease in the upper and lower airways. *J. Allergy Clin. Immunol.* 131, 549–561. doi: 10.1016/j.jaci.2012.07.050
61. Sanderson, A. R., Leid, J. G., and Hunsaker, D. (2006). Bacterial biofilms on the sinus mucosa of human subjects with chronic rhinosinusitis. *Laryngoscope* 116, 1121–1126. doi: 10.1097/01.mlg.0000221954.05467.54
62. Schlecht, L. M., Peters, B. M., Krom, B. P., Freiberg, J. A., Hänsch, G. M., Filler, S. G., et al. (2015). Systemic *Staphylococcus aureus* infection mediated by *Candida albicans* hyphal invasion of mucosal tissue. *Microbiology* 161, 168–181. doi: 10.1099/mic.0.083485-0
63. Schubert, M. S. (2004). Allergic fungal sinusitis: pathogenesis and management strategies. *Drugs* 64, 363–374. doi: 10.2165/00003495-200464040-00002
64. Tieu, D. D., Kern, R. C., and Schleimer, R. P. (2009). Alterations in epithelial barrier function and host defense responses in chronic rhinosinusitis. *J. Allergy Clin. Immunol.* 124, 37–42. doi: 10.1016/j.jaci.2009.04.045
65. Tomassen, P., Vandeplas, G., Van Zele, T., Cardell, L.-O., Arebro, J., Olze, H., et al. (2016). Inflammatory endotypes of chronic rhinosinusitis based on cluster analysis of biomarkers. *J Allergy Clin Immunol* 137, 1449–1456. e4. doi: 10.1016/j.jaci.2015.12.1324
66. Underhill, D. M., and Iliev, I. D. (2014). The mycobiota: interactions between commensal fungi and the host immune system. *Nat. Rev. Immunol.* 14, 405–416. doi: 10.1038/nri3684
67. Wagner Mackenzie, B., Waite, D. W., Hoggard, M., Douglas, R. G., Taylor, M. W., and Biswas, K. (2017). Bacterial community collapse: a meta-analysis of the sinonasal microbiota in chronic rhinosinusitis. *Environ. Microbiol.* 19, 381–392. doi: 10.1111/1462-2920.13632
68. Wheeler, M. L., Limon, J. J., Bar, A. S., Leal, C. A., Gargus, M., Tang, J., et al. (2016). Immunological consequences of intestinal fungal dysbiosis. *Cell Host Microbe* 19, 865–873. doi: 10.1016/j.chom.2016.05.003
69. Yamasaki, S., Matsumoto, M., Takeuchi, O., Matsuzawa, T., Ishikawa, E., Sakuma, M., et al. (2009). C-type lectin mincle is an activating receptor for pathogenic fungus, Malassezia. *Proc. Natl. Acad. Sci. U.S.A.* 106, 1897–1902. doi: 10.1073/pnas.0805177106
70. Zhao, Y. C., Bassiouni, A., Tanjararak, K., Vreugde, S., Wormald, P.-J., and Psaltis, A. J. (2017). Role of fungi in chronic rhinosinusitis through ITS sequencing. *Laryngoscope*. doi: 10.1002/lary.26702. [Epub ahead of print].

9

Methods of Controlling Invasive Fungal Infections using CD8$^+$ T Cells

Pappanaicken R. Kumaresan[1], Thiago Aparecido da Silva[1] and Dimitrios P. Kontoyiannis[2]**

[1] Department of Pediatrics, The University of Texas MD Anderson Cancer Center, Houston, TX, United States, [2] Department of Infectious Diseases, The University of Texas MD Anderson Cancer Center, Houston, TX, United States

***Correspondence:**
Pappanaicken R. Kumaresan
pkumaresan@mdanderson.org;
Dimitrios P. Kontoyiannis
dkontoyi@mdanderson.org

Invasive fungal infections (IFIs) cause high rates of morbidity and mortality in immunocompromised patients. Pattern-recognition receptors present on the surfaces of innate immune cells recognize fungal pathogens and activate the first line of defense against fungal infection. The second line of defense is the adaptive immune system which involves mainly CD4$^+$ T cells, while CD8$^+$ T cells also play a role. CD8$^+$ T cell-based vaccines designed to prevent IFIs are currently being investigated in clinical trials, their use could play an especially important role in acquired immune deficiency syndrome patients. So far, none of the vaccines used to treat IFI have been approved by the FDA. Here, we review current and future antifungal immunotherapy strategies involving CD8$^+$ T cells. We highlight recent advances in the use of T cells engineered using a Sleeping Beauty vector to treat IFIs. Recent clinical trials using chimeric antigen receptor (CAR) T-cell therapy to treat patients with leukemia have shown very promising results. We hypothesized that CAR T cells could also be used to control IFI. Therefore, we designed a CAR that targets β-glucan, a sugar molecule found in most of the fungal cell walls, using the extracellular domain of Dectin-1, which binds to β-glucan. Mice treated with D-CAR$^+$ T cells displayed reductions in hyphal growth of *Aspergillus* compared to the untreated group. Patients suffering from IFIs due to primary immunodeficiency, secondary immunodeficiency (e.g., HIV), or hematopoietic transplant patients may benefit from bioengineered CAR T cell therapy.

Keywords: fungal infection, immunotherapy, chimeric antigen receptor, D-CAR$^+$ T cells, cell therapy, Sleeping Beauty, CD8$^+$ T cells, adoptive T cell therapy

INTRODUCTION

Opportunistic invasive fungal infections (IFIs) are a major threat to the immunocompromised individual; neutropenia is a major risk factor for these infections (1, 2). Patients who require prolonged immunosuppressive therapy, for example, those who have undergone solid organ transplantation or hematopoietic stem-cell transplantation (HSCT) and those who have severe autoimmune diseases are also highly susceptible to IFIs (3–7). Other risk factors include long-term stays in an intensive care unit, the use of indwelling catheters, chemotherapy, or broad-spectrum antibiotics. The main causative agents of IFI are *Aspergillus* spp., *Candida* spp., and *Cryptococcus* spp. The incidence of IFI is increasing worldwide (2, 8, 9) (**Table 1**), and the worldwide crude mortality rate of invasive aspergillosis and invasive candidiasis has been estimated to be 0.4 deaths per 100,000 people. However, mortality rates associated with IFIs in immunocompromised patients are considerably higher, reaching 60–85% for invasive aspergillosis. The emergence of fungal strains that are resistant to currently available antifungal drugs such as polyenes, triazoles, and echinocandins poses a dangerous

TABLE 1 | Incidence and patterns of fungal infections worldwide.

Fungal infection	Incidence per year	Reference	Main routes	Comments
Invasive aspergillosis (*Aspergillus*)	>300,000	(15)	Pulmonary	
Invasive candidiasis (*Candida*)	8–10 cases/100,000	(16, 17)	Cutaneous Oropharyngeal Gastrointestinal Genitourinary	
Mucormycosis (*Mucorales*)	1.7 cases/1000,000	(18)	Sinopulmonary Disseminated	
Cryptococcus	~1,000,000	(19)	Pulmonary	
Pneumocystis pneumonia	In the US, 9% among hospitalized HIV/ acquired immune deficiency syndrome patients and 1% among solid organ transplant recipients	(20, 21)	Pulmonary	In immunocompromised patients, the mortality rate ranges from 5 to 40% in those who receive treatment. The mortality rate approaches 100% without therapy

CNS, central nervous system.
Since there is a lack of epidemiological data in many countries, the world incidence rate may be overestimated.

problem (10) and immune-based treatments are giving new hope to combat these deadly fungal infections (11–14).

The host response to fungal infection depends on several factors, including the host immune status, site of infection, fungal morphotype (yeast or hyphae), cell wall complexity, and virulence traits, such as the production of fungal exotoxins (22–25). The routes of various fungal infections are listed in **Table 1**; the majority occurs *via* the sinopulmonary and gastrointestinal routes (22). The host immune response to fungal infection occurs in a coordinated way *via* both innate and adaptive immune cells. Innate effector cells, mainly macrophages and neutrophils, are the first line of defense against inhaled fungal spores (11, 26). As a result, most initial fungal encounters go unnoticed (27). Pattern-recognition receptors (PRRs) are a family of receptors that is composed of the C-type lectin receptors (CLRs), toll-like receptors (TLRs), Nod-like receptors, and other receptors that initiate immune responses against invading fungal pathogens. Cellular expression and signaling mechanism of the PRRs have been reviewed previously (28–30).

Most of the sugars present on the fungal cell wall are recognized by the receptors from the CLR family, underscoring the constant vigil of the host innate immune system against invading fungal pathogens (28, 31–33). CLRs recognize the various carbohydrate glycoprotein components of the fungal cell wall, such as β-glucan or α-mannan, which trigger downstream signaling cascades that are essential for inducing protective immunity against fungi (34–37). When the fungal insult cannot be quickly controlled, adaptive immune cells, mainly $CD4^+$ T cells, activate other cellular responses and antibody production. Adaptive immune cells produce cytokines to activate B cells, which in turn secrete antibodies against fungal antigens and activate the release of antimicrobial peptides from endothelial cells. Recent comprehensive reviews have already detailed the mechanisms of $CD4^+$ T cells and surveyed current immunotherapeutic strategies to control fungal diseases (12, 38, 39). Despite having intact innate immune systems, patients with acquired immune deficiency syndrome (AIDS) are highly susceptible to fungal infections, highlighting the importance of the adaptive immune system. When $CD4^+$ T cell counts are low, as in patients with AIDS, $CD8^+$ T cells have a heightened role in controlling fungal infections (40). In this review, we focus on the functional role of $CD8^+$ T cells in the immune response to fungal infections. We then discuss a new method of combating fungal infections, engineering T cells with the "Sleeping Beauty" (SB) vector system.

CURRENT AND FUTURE STRATEGIES TO CONTROL FUNGAL INFECTIONS

Drug Therapy

Antifungal drugs have had only modest success in reducing the high mortality rates associated with IFIs. In large part, this is because diagnosis of fungal infection and identification of the responsible organism is often delayed, leading to a delay in the administration of directed antifungal therapy. The use of available antifungal drugs is also restricted by their route of administration, spectrum of activity, and bioavailability in target tissues such as the brain (41). Additional issues include toxicity, undesirable drug interactions, and drug resistance. Use of the triazoles, for example, is limited by their interactions with statins, corticosteroids, and other drugs (42).

Despite tremendous improvements in the response rates of aspergillosis to modern antifungal agents, fatality rates of 40% are common in contemporary real life cohorts of unselected patients with leukemia and transplant recipients (43). The high rate of mortality following *A. fumigatus* infection is a result of the suboptimal diagnostic tools available, leading to late diagnosis. Other factors include rising *Aspergillus* resistance, and even more importantly, the relative ineffectiveness of existing antifungal drugs against established *Aspergillus* infections (44).

The development of effective and safe immune enhancement therapies is a major unmet need. Some patients with candidiasis struggle with poor outcomes, although this is less common in the era of widespread azole prophylaxis given to high-risk patients. Randomized controlled studies typically exclude high-risk immunosuppressed patients by use of their inclusion criteria (45).

Immunotherapy

Innate Immune Cells

Immunotherapy, which comprises cell-based therapies, such as the adoptive transfer of T cells, dendritic cells (DCs), or neutrophils, and other humoral approaches, such as antibodies and recombinant pentraxins, is a viable option for control of IFIs. Among immunocompetent individuals, the innate immunity efficiently prevents and clears IFIs (26). Alveolar macrophages are the first line of fungal defense; they recognize, phagocytize, and destroy fungal spores (46). Neutrophils also play a key role in killing fungal hyphae. They eliminate fungal hyphae by inducing an oxidative burst and by forming neutrophil extracellular traps (NETs) (47). Neutrophils utilize NETs to trap the invading pathogens by releasing chromatin fibers to form a meshwork adorned with cytoplasmic granules containing the antimicrobial enzymes myeloperoxidase, cathepsin G and neutrophil elastase that destroy trapped pathogens. The whole process is called NETosis (48).

To date, immunotherapeutic strategies to combat IFIs have primarily focused on augmenting the number of granulocytes, since these cells are known to have fungicidal activity. Granulocyte-focused immunotherapies include granulocyte transfusions (49), infusion of growth factors [granulocyte colony-stimulating factor (G-CSF), or granulocyte macrophage colony-stimulating factor (GM-CSF)] to increase granulocyte numbers (50), and the administration of cytokines such as interferon (IFN)-γ (51) and/or interleukin (IL)-15, the latter of which promotes the production of IL-8 (52), to augment phagocytic and cytotoxic function. However, the reconstitution of granulocytes is hampered by an inability to numerically expand large numbers of cells *ex vivo*. Moreover, after infusion, reconstituted granulocytes exhibit poor persistence owing to increased apoptosis, weak potency, and a propensity to become trapped in the pulmonary vasculature (53).

Natural Killer (NK) Cells

Natural killer cells are another type of innate immune cell reported to be involved in controlling fungal infections. NK cells make up from 5 to 15% of the peripheral blood mononuclear cells (PBMCs) of healthy individuals; the NK cell population is made up of $CD56^+$ and $CD3^-$ cells (54). NK cells are activated when signals from activating receptors outweigh signals from inhibitory receptors, leading to cytotoxicity directed against tumor cells and virus-infected cells. NK cells also recognize infectious fungal pathogens, including *A. fumigatus*, *C. albicans*, *C. neoformans*, and *Mucorales* species (31, 55–58). Recently, CD56 has been identified as a PRR that can bind directly to both germ tubes and hyphae of *Aspergillus fumigatus* (59). Upon recognition, NK cells either induce lysis of these pathogens by secreting perforin and granulysin or trigger activation of other immune cells by releasing IFN-γ (60). Fungal pathogen-specific NK cell receptors and their mechanism of action has been reviewed (61, 62).

Dendritic Cells

Dendritic cells are professional antigen presenting cells (APCs) that can recognize and phagocytize fungal conidia and hyphae through PRRs and degrade them by fusing with lysosome vesicles (63). PRRs activate DCs to secrete cytokines, such as IL-12, IL-6, IL-4, and IL-1β that induce T-cell differentiation in the lymph nodes. During *A. fumigatus* infection, pulmonary DCs secrete IL-12 upon exposure to conidia, while IL-4 and IL-10 are secreted after exposure to hyphae. Therefore, IL-12 signaling generates a T helper (Th) 1 cell response, while IL-4 and IL-10 signaling generates a Th2 response. DCs also secrete tumor necrosis factor (TNF)-α and the chemokine CXCL8 which recruit neutrophils to the infection site (64).

In conventional DCs, β-glucan-induced Dectin-1-mediated signaling promotes secretion of the cytokines IL-2 and IL-23. The release of IL-23 induces Th17 differentiation but it is tightly regulated by IL-2 (65, 66). These data suggest that DCs direct naïve T cells to mature into functional T-cell sub types by secreting specific cytokines in the microenvironment based upon stimuli received by PRRs from different forms of fungi.

$CD4^+$ T Cells

Even though DCs help to reduce the fungal burden to some extent through fusion with lysosome vesicles, the major function of DCs is to present fungal antigens to naive T-cells. DCs present processed antigens *via* major histocompatibility complex (MHC) class I or class II molecules and interact with naive T cells through formation of an immunological synapse. T cells are broadly classified into helper $CD4^+$ T cells and cytotoxic $CD8^+$ T cells. In fungal infections, both $CD4^+$ and $CD8^+$ T cells participate in the elimination of fungal pathogens (67, 68). On the basis of their function and cytokine secretion profile, $CD4^+$ T cells are classified into several subsets: Th1, Th2, Th9, Th17, Th22, regulatory T cells, and follicular helper T cells. The activity of $CD4^+$ T cells against fungal infection in immunocompetent individuals has been very well characterized. The most important $CD4^+$ T cells in the antifungal immune response are the Th1 and Th17 helper T cells. After priming by DCs, $CD4^+$ T cells differentiate into Th1 and Th17 helper T cells. Th1 helper T cells secrete the cytokines IFN-γ and TNF-α which activate innate immune cells, such as neutrophils, macrophages, DCs, and inflammatory monocytes, to fight against invading fungi and bacteria (12, 27). The cytokines secreted by Th1 cells also activate B cells, leading to the secretion of antigen-specific antibodies against fungi. IL-17 secreted by Th17 cells controls fungal infection by mobilizing neutrophils and protecting mucosal body sites by inducing epithelial cells to secrete defensin (69). IL-17 deficiency has been shown to enhance susceptibility to *Candida albicans* infections at mucosal sites (70).

$CD8^+$ T Cells

Like $CD4^+$ T cells, $CD8^+$ T cells also have sub types, namely Tc1, Tc2, and Tc17 (**Figure 1**). APCs, mainly DCs, cross-present fungal antigens to $CD8^+$ T cells. $CD8^+$ T cells can be primed to recognize fungi by utilizing a "cross-presentation" and "cross-priming" approach, in which exogenous or fungal antigens are presented on MHC-I molecules (71). DCs internalize exogenous fungal products by CLRs and scavenger receptors for processing and presenting to MHC-I, and this process is called cross-presentation. Along with cross-presentation, some of the CLRs, for example, Dectin-1 activates DCs *via* Syk kinase signaling to produce IL-12, which favors Tc1 differentiation (72). Curdlan

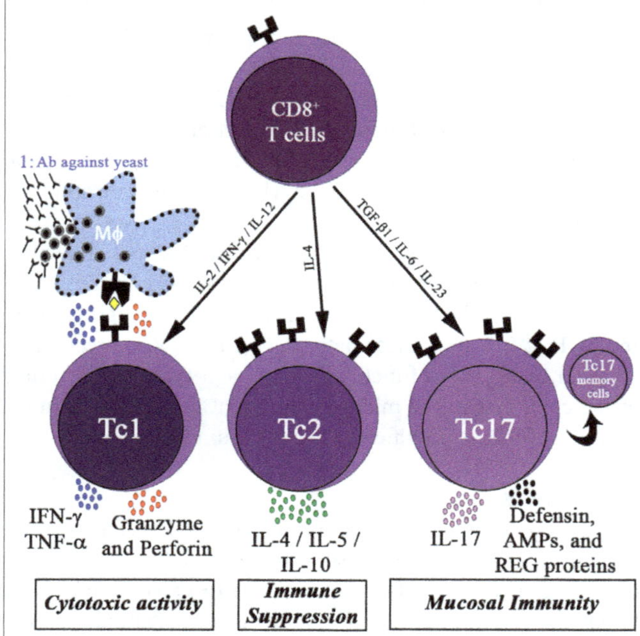

FIGURE 1 | CD8+ T cells activity in the immune response. Differentiation of CD8+ T cells into three functional subsets: the cytotoxic cells (Tc1) cells, producing high levels of interferon (IFN)-γ, tumor necrosis factor (TNF)-α, granzyme, and perforin, which contribute to the killing of yeast infected host cells; Tc1 kills fungal infected macrophages and allows the participation of humoral immunity (marked as 1); Tc2 cells, release high amounts of interleukin (IL)-4 and IL-10, promoting immune suppression; Tc17 cells secrete IL-17, which activates mucosal immunity by inducing epithelial cells to secrete defensin, antimicrobial peptides (AMPs), and regenerating proteins (REG). Some of the activated Tc17 cells may differentiate into memory Tc17 cells.

has been demonstrated to stimulate the Dectin-1-syk-CARD pathway, producing IL-23 to boost the differentiation of Th17 cells (73).

Upon recognition of fungal peptides presented by APCs, CD8+ T cells differentiate into Tc1 cells and Tc17 cells (CD8+ T cells that secrete IL-17A), depending on the cytokines present in the environment. Several reports highlighted the role of Tc1 and Tc17 cells in protecting humans from fungal infection (74, 75). Tc1 cells act indirectly by secreting cytokines such as IFN-γ, TNF-α, and GM-CSF to activate innate immune cells such as neutrophils and macrophages involved in antifungal defense. Furthermore, Tc1 cells directly kill unresponsive fungal-infected macrophages by secreting cytotoxic factors such as perforin, granulysin, and granzyme K (76). DCs uptake fungal breakdown products from apoptotic macrophages by endocytosis to cross prime CD8+ Tc1 cells. CD8+ Tc17 cells, like CD4+ Th17 cells, secrete IL-17A cytokines to activate epithelial cells (mucosal immunity) to secrete antimicrobial products such as defensin to fight against fungal infections (**Figure 1**).

The Tc1 and Tc17 subtypes can be divided into effector T cells and effector memory T cells on the basis of their expression of surface receptors. The cell-surface markers used for phenotypical characterization of Tc1 and Tc17 cells are shown in **Table 2**. Tc1 cells that express C–X–C motif chemokine receptor 3 migrate

TABLE 2 | Phenotypic characterization of CD8+ Tc1 and Tc17 cell subtypes.

Cell-surface markers/ cytokines	Expression in Tc1	Expression in Tc17	Transcription factors/ secretory granules	Expression in Tc1	Expression in Tc17
CCR6	N.I.	High	TCF-1	High	High
CXCR3	High	Low	T-bet	Low/High	Low
CD62L	Low/high	Low	Eomes	High	
CD27	High	Low/ intermediate	Ror (γ)t	Low	High
KLRG-1	Low	Low			
IFN-γ	+	–	Perforin	+	–
TNF-α	+	–	Granulysin	+	–
IL-2	+	–			
GM-CSF	+	–			
IL-17	–	+			

CCR6, C–C motif chemokine receptor 6; CXCR3, C–X–C motif chemokine receptor 3; CD62L, cluster of differentiation 62L;CD27, cluster of differentiation 27; KLRG-1, co-inhibitory receptor killer-cell lectin-like receptor G1; IFN, interferon; TNF, tumor necrosis factor; IL-2, interleukin-2; GM-CSF, Granulocyte macrophage colony-stimulating factor; IL-17, interleukin-17; TCF-1, T cell factor 1; T-bet, T-box transcription factor; ROR-γt, RAR-related orphan receptor gamma; N.I., no information.

to the lungs during pulmonary infections, such as pneumocystis (77). Tc17 cells have increased levels of effector memory phenotype markers on the cell surface ($CD62L^{lo}$ and $CD27^{it/lo}$) as compared with Tc1 cells, suggesting that Tc17 cells may play a role in preserving long-term antifungal immunity in the host. Cytokines secreted by CD8+ Tc1 and Tc17 cells boost the innate immune system as well as the mucosal immune system to give protection from IFI (**Figure 1**).

CD8+ T cells share many cellular functional mechanisms with NK cells, such as releasing cytolytic granules and cytokines. As indicated above, in the absence of CD4+ T cells, CD8+ T cells also play a major role in controlling fungal infection. Immunotherapy could take advantage of several properties of CD8+ T cells, they can kill pathogen infected cells, be propagated in large numbers *ex vivo*, be genetically modified to recognize particular antigens, and contribute to immunologic memory.

THE ROLE OF CD8+ T CELLS IN THE ANTIFUNGAL IMMUNE RESPONSE

The immune response to fungi elicited by CD8+ T cells can broadly classified into two processes: (1) T-cell receptor (TCR) mediated and (2) TLR and scavenger receptor mediated.

TCR-Mediated CD8+ T Cell Activation

Vaccines are a promising avenue for the treatment and prevention of IFIs (78–84) mediated through TCR receptors. The vaccine candidates developed against fungal antigens typically activate CD4+ T cells and Th17 cells. Several highly immunogenic and protective vaccine formulations for candidiasis are currently undergoing clinical trials (84, 85). It is worth noting that vaccination against fungi has mainly focused on yeast pathogens, such as *Candida* spp. and *Cryptococcus* spp. (80), and endemic mycoses that infect immunocompetent individuals, such as *Coccidioides* spp. (86), *Blastomyces* spp., and *Histoplasma* spp. It remains to be seen whether similar strategies will be as effective against opportunistic fungi, such as *Aspergillus*. In preclinical studies, vaccination using both crude and recombinant *Aspergillus* antigens improved the survival of immunocompromised mice following inhalation and intravenous administration of *Aspergillus fumigatus* (87).

Direct killing by CD8+ T cells has not been widely explored in the development of an immunotherapy against fungi, even though studies demonstrated the essential role of the CD8+ T-cell response in controlling fungal infections after vaccination (74, 75, 88–92) (**Table 3**). However, the presence of *Aspergillus*-specific CD8+ T cells has been shown in both mice and humans (93–96). Moreover, *Mucorales* (97) and *Fusarium*-specific T-cells (98) were reported in hematologic patients with IFI. Type I CD8+ T-cells (Tc1) were shown to provide protection against pneumocystis in mice (99). Preclinical studies demonstrated that the direct effect of CD8+ T cell-mediated cytotoxic activity and TNF-α and IFN-γ production were necessary to clear infected macrophages containing *H. capsulatum* (76), and provided full protection against coccidioidomycosis (88, 89). The activation of CD8+ T cells also contributed a protective response during *Cryptococcus neoformans* infection; involvement of Type 1 CD8+ T (Tc1) cells was triggered through immunization with the cytosolic proteins of the pathogens (90). Moreover, CD8+ T-cells secrete IL-17A to give protection against lethal fungal diseases, such as *Blasotomyces dermatitidis* and *Histoplasma capsulatum*, by supporting neutrophil activity (74) (**Figure 1**).

However, there are limitations for vaccine therapy. Currently, no FDA-approved vaccines are available to prevent the major opportunistic fungal infections, specifically candidiasis, aspergillosis, and cryptococcosis. Several reasons underlie the paucity of viable candidates. First, these infections are relatively uncommon, compared to viral and bacterial infections and typically occur in severely ill patients. Thus, finding sizable niche patient population who can benefit from a cost-effective vaccine strategy is difficult and not an area of priority for development by the pharmaceutical industry. Second, high-risk patients have pleiotropic and ever-evolving defects in both innate and adaptive immunity. As responses to fungi depend on both arms of the immune response, and because such responses are complex, depending on the site of infection (mucosal vs systemic infection) and the type of fungus (e.g., *Candida* or *Cryptococcus* vs a mold), much more groundwork needs to be done to decipher the key elements of a successful vaccine. In addition, there are questions regarding the efficacy and feasibility of using a vaccine in immunocompromised patients, since they are incapable of mounting a complete immune response (105).

TLR-Mediated CD8+ T Cell Activation

Toll-like receptors of the innate immune system play a major role in recognizing fungal cell wall carbohydrates, cell wall breakdown products, RNA, and DNA (13, 106–108) and thereby activate immune cells. One possible mechanism TLRs use to augment T-cell activation is when DCs activate fungal-specific CD8+ T cells by cross-presenting fungal antigens. TLR3 plays a crucial role in this process by sensing fungal RNA derived from necrotic cells and activating CD8+ memory T cells along with DCs. Indeed, TLR3$^{-/-}$ mice are more susceptible to *Aspergillus* infection than are control mice (101), and people with mutations in key TLR3 and TLR4 signaling components are susceptible to various fungal infections (109–111).

T-Cell Activation Mediated by Scavenger Receptors and Other Receptors

The scavenger receptor proteins are a highly heterogeneous set of proteins expressed on the cell surface that are involved in the uptake of modified low-density lipoproteins and a variety of microbes. One of the scavenger receptors, CD5, has been shown to bind β-glucan, a fungal cell wall sugar moiety, as well as many strains of yeast cells (112). CD5 is expressed on T cells and a small subset of mature B cells, where it associates with antigen receptors. Upon stimulation with zymosan (a protein-sugar moiety derived from the yeast cell wall), a CD5-transfected cell line produces IL-8, suggesting that CD5 has a pro-inflammatory role in fungal infection (112).

Besides TLRs, T cells have other receptors such as CD23 and CD56 for direct recognition of fungal antigens. CD23 is an

TABLE 3 | Fungal vaccine candidates and their CD8+ T-cell mechanisms of action.

Fungal infection	Candidate	CD8+ T cell responses	Model	Reference
Aspergillosis	Recombinant fungal antigens Pep1p, Gel1p, and Crh1p	Cytotoxic activity	Murine	(100)
	Live *A. fumigatus* conidia or *A. fumigatus* cell wall glucanase Crf1p	Cytotoxic activity	Murine	(101)
Blastomycosis	Attenuated mutant lacking BAD1	Tc17 cells	Murine	(74, 91)
	Attenuated mutant lacking WI-1 adhesin	TNF-α, IFN-γ, and GM-CSF production; CD8+ T cell memory	Murine	(75, 92)
Coccidioidomycosis	Arthroconidia of the 95–291 strain	Cytotoxic activity; TNF-α production	Murine	(88)
	Live spores of the Δcts2/ard1/cts3 strain	IFN-γ production	Murine	(89)
Candidiasis	Candidal adhesin (rAls3p-N) plus aluminum hydroxide adjuvant	Cytotoxic activity	Murine	(102)
	Candida dubliniensis mannan–human serum albumin conjugate	Upregulation	Rabbit	(103)
	Cytosolic antigens entrapped in liposomes	Upregulation	Murine	(104)
Paracoccidioidomycosis	N.I.	N.I.	N.I.	N.I.
Cryptococcosis	Cytosolic proteins	Tc1 cells	Murine	(90)

TNF, tumor necrosis factor; IFN, interferon; GM-CSF, granulocyte macrophage colony-stimulating factor; N.I., no information.

inducible low-affinity receptor for immunoglobulin (Ig)E (113). It can recognize both β-glucan and α-mannan sugar moieties and thereby targets both yeast and hyphae forms of *Candida* (114), and upon activation, it upregulates nitric oxide production to destroy invading *Candida.* C-Jun N-terminal kinases (JNK1) activation suppresses the expression of CD23, which increases susceptibility of fungal infection. This was verified in JNK1 KO mice which showed resistance to *Candida* infection when compared to control mice (114). CD56 is a NK cell receptor that has been shown to bind to *Aspergillus* hyphae in a concentration-dependent manner. Blocking of CD56 reduced fungal-mediated NK cell activation (59). Activated T-cells expresses high levels of CD56 and its expression level directly correlates with T-cell effector functions (115). However, additional studies are warranted to verify that the CD56 mediated $CD8^+$ T cells are activated during fungal infection.

ADOPTIVE T-CELL THERAPY

Over the years, several immunotherapies have been used to treat fungal infection. One such immunotherapy, adoptive T-cell therapy (ACT), is a promising therapeutic strategy not only for cancer but also for treating viral and fungal infections (38, 116–118). ACT involves the isolation and *ex vivo* expansion of autologous T cells in an antigen-specific manner; these expanded T cells are later infused into the patient. ACT has been shown to be effective in controlling viral infections, such as cytomegalovirus (119) and fungal infections, such as *Aspergillus* in HSCT patients (120). Immunocompromised patients, especially patients undergoing allogeneic HSCT, are highly susceptible to IFIs (7). The mortality rate from IFI in this patient population remains unacceptably high, partly due to the long-lasting immunosuppression in patients after HSCT (3, 121). Most IFIs in these patients occur after engraftment of the innate immune system, which suggests that adaptive cellular immunity plays a major role in controlling IFIs. In fact, adoptive transfer of $CD4^+$ Th1 cells elicited significant protection against invasive aspergillosis in haploidentical HSCT settings (120). These findings have generated a growing interest in restoring adaptive immunity against fungal pathogens by infusing donor-derived antifungal T cells and in various *ex vivo* methods of propagating clinical-grade *Aspergillus*-specific T cells (38, 122). Recently, the FDA-approved chimeric antigen receptor (CAR) T-cell therapy to treat B-cell malignancies. CAR T cell technology can be applied to redirect T cell specificity to target fungal pathogens.

Three approaches are used to redirect T-cell specificity against a particular antigen (123).

1. Gene modification with antigen-specific TCRs in which the α and β chains of the TCR are cloned from tumor-associated antigen-specific T-cell clones (124, 125).
2. Gene modification using natural receptors other than TCRs, such as the Dectin-1 receptor (126).
3. Introduction of a CAR that recognizes tumor-associated antigens through a single-chain variable region (scFv) derived from the corresponding monoclonal antibody (127). Currently, there are no reports in the literature of scFv-derived CAR T-cells targeting fungal antigens. Therefore, in the current review, it was added for comparison analysis with D-CAR^+ T cells (**Figure 2**).

Gene Modification Using Pathogen-Specific TCRs

T-cell receptors are found on the surface of T cells as heterodimers of α and β chains and they recognize antigens presented by the MHC receptors of the APCs. For ACT, genes of tumor

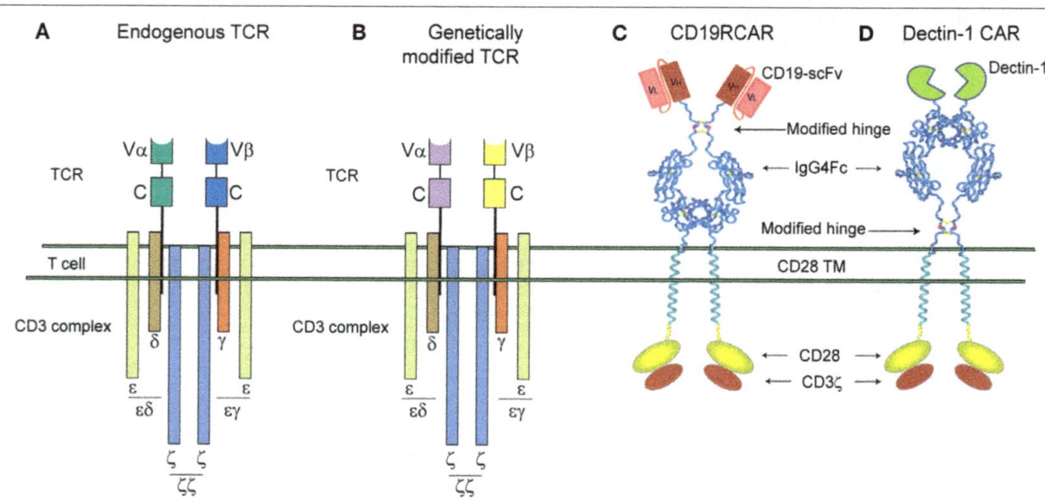

FIGURE 2 | Schematic representations of the T-cell receptor (TCR) complex and second-generation single-chain variable region (scFv)-chimeric antigen receptor (CAR) and D-CAR. **(A)** Unmodified endogenous TCR complex and **(B)** genetically modified TCR complex. The α and β chains are highlighted in different colors. **(C)** The CD19R-CAR derived from a scFv region of a CD19 antigen-specific mouse monoclonal antibody and **(D)** the D-CAR^+ derived from an extracellular domain of the Dectin-1 receptor. CD19R-CAR and D-CARs shown here have the same signaling domains, derived from costimulatory molecules, such as CD28 and CD3-ζ.

antigen-specific TCRs are isolated from patients and engineered into T cells using a viral or non-viral-based vector system (128, 129). These T cells are expanded *ex vivo* to generate large numbers for infusion. Improvements in vector design have recently increased the efficiency of this approach, and the avidity of the TCR has been improved by substituting amino acids in its complementarity determining region and introducing cysteines to form disulfide bonds, thereby preventing α- and β-chain mispairing. TCR-mediated T-cell responses to fungal antigens have been documented in both *in vitro* and *in vivo* studies (130, 131).

In colon cancer studies, TCR-driven ACT was effective in reducing tumor volume (132), but a high incidence of toxicity was reported, especially when high-avidity TCRs were used (133). Moreover, TCR-specific therapy is MHC restricted; if tumor cells lose antigen expression by downregulating MHC, they can evade the T cell attack (134). Hence, TCR-specific T-cell recognition is restricted to a single type of MHC molecule that presents the targeted antigen (135). In order to circumvent this problem, CAR-based therapy was developed. With CAR-based T-cell therapy, the tumor recognition of the CAR is not dependent on MHC (136).

Engineered CAR T-Cell Therapy

Engineered T-cell therapy involving the introduction of a CAR, which recognizes tumor-associated antigens through its scFv, is derived from the corresponding monoclonal antibody. CAR-based therapy involves the genetically engineered fusion of a variable light chain and a variable heavy chain that are specific for a cell-surface antigen and are tailored to produce an activating signal to host immune cells upon antigen engagement (137). CAR-based T-cell therapy has several key advantages. First, a CAR-based approach can be used in all tumor conditions expressing the antigen and is not MHC restricted. Second, tumor cells have no protection against CAR-based immunoediting. Finally, a varied range of tumor antigens can be targeted using this system, including glycoproteins and lipids (138, 139).

The CAR has been structurally refined over three generations of development. CAR's structure consists of four elements: an antigen-targeting domain, an extracellular linker/spacer, a membrane-spanning (transmembrane) domain, and an intracellular signaling domain. The antigen-specific domain is usually derived from the scFv of the monoclonal antibody targeting the antigen. The linker makes the CAR flexible so that it can reach the antigen. A mutated IgG derived Fc sequence incapable of activating innate immune cells is commonly used because of its stability in expressing the CAR on the cell surface (140). In first-generation CARs, the transmembrane domain used CD4 or CD8, while CD28 was used in second-generation CARs (141). The intracellular signaling domain of second-generation CARs used CD3-ζ along with the costimulatory signaling domain CD28 (142). Tumor clearance and persistence is better in second- and third-generation CAR$^+$ T cells than in the first generation (140, 141, 143, 144).

SB, a Non-Viral-Based Vector

Several vector systems have been used to introduce the CAR transgene into T cells. Of these, mammalian transposon/transposase-based vectors produce the most robust integration, have low immunogenicity, and allow for easy manipulation of plasmids. Multiple vectors have been studied in mammalian systems, including the SB transposon (derived from the fish, *Tanichthys albonubes*), the PiggyBac element (from the moth, *Trichoplusia ni*), Frog Prince (from the frog, *Rana pipiens*), Himar1 (from the horn fly, *Haematobia irritans*), Tol2 (from the fish, *Oryzias latipes*), and Passport (from the flatfish, *Pleuronectes platessa*) (145, 146).

Among all of the elements with activity in mammalian cells, the SB transposon is one of the most widely studied for use in gene transfer (147, 148). The SB transposase was derived by combining inactive transposase sequences from the genome of salmonid fish and then reversing the termination codon to activate transposase activity. A typical SB vector consists of 230-base pair (bp) regions containing long inverted and direct repeats (IR/DR) flanking the target gene sequence. These IR/DR sites bind with SB transposase to transfer the target gene to the host genome. In addition to the IR/DR sites, SB transposase also contains a DNA-recognition site, a nuclear localization signal, and a catalytic domain. Gene transfer using the SB transposon/transposase involves a cut-and-paste mechanism. The SB transposase protein is translated and accumulated in the cytoplasm, which is then imported into the nucleus using the nuclear localization signal. The SB transposase protein binds to the IR/DR sequence of the transposon, causing DNA breaks around the gene of interest.

The integration site of the gene cut from the SB transposon in the T cell genome depends on the presence of a TA dinucleotide site, DNA flexibility, and proximity of the donor and receiver (local hopping). More than 25% of integrations occur within a 200-bp region between the donor and receiver sites of the gene, and more than 75% of integrations are located in a single chromosome. CD19R-CAR T cells were developed using the SB vector and successfully used in clinical trials to treat acute myeloid leukemia and chronic lymphocytic leukemia (127, 149).

Dectin-1 CAR T-Cells to Target β-Glucan-Expressing Fungi

We modified this prototypical CAR to recognize carbohydrates by utilizing the pattern-recognition properties of Dectin-1 (126, 150, 151). It is specific for β-glucan, a glucose polymer consisting of β-1, 3-glucan and β-1, 6-glucan that is expressed on the cell wall of all known fungi (152–155). We hypothesized that the extracellular portion of Dectin-1 could be adapted as the specificity domain for a CAR (D-CAR) on T cells to redirect their specificity to β-glucan expressing fungi such as *Aspergillus*. Using the extracellular domain of Dectin-1, we engineered a CAR with specificity for the fungal cell wall sugar moiety β-glucan. This CAR was fused in frame to a modified human IgG4 hinge/Fc stalk (156), CD28 transmembrane domain, and a combination of CD28 and CD3ζ intracellular domains. This design is similar to that of our second-generation CD19-specific CAR (designated CD19RCD28), which is currently being employed in clinical trials to treat B-cell leukemia (157) (**Figure 2**).

Since this was the first time that a PRR was adapted to redirect T-cell specificity, we employed multiple assays [cell viability assay (XTT), cytokine production, upregulation of CD107a, and microscopy] to compare the ability of D-CAR+ T cells to target germinating *Aspergillus* hyphae with that of CD19-specific T cells. All of these assays demonstrated that the D-CAR activated the cytolytic machinery of the genetically modified T cells and probably their perforin/granzyme pathway as well (126). The production of IFN-γ by the D-CAR+ T cells may further augment innate immunity to IFIs. Treatment with recombinant IFN-γ or IFN-γ derived from CD4+ helper T cells or NK cells has been shown to augment anti-*Aspergillus* activity (55, 122). It remains to be determined whether the D-CAR-dependent production of IFN-γ contributes directly to the clearance of fungal infections or whether this activity works indirectly through the activation of granulocytes. Other cytokines, such as IL-17, may also participate in antifungal immunity. Reports indicate that IL-17 can activate neutrophils in a similar manner to IFN-γ (158), though increased levels of IL-17 are associated with mortality (153, 159–162) (**Figure 3**).

The use of combination therapies may supplement the antifungal efficacy of D-CAR+ T cells. For example, in *Aspergillus* pre-exposed to Caspofungin, β-glucan residues in the cell wall were unmasked, enhancing antifungal activity mediated by neutrophils (163). In HSCT patients, innate immune cells are present in the blood within 2 weeks of the stem-cell transplant, whereas it takes, on average, 7–12 months for NK, B, and T cells to be produced. Most IFIs occur during this period because no cellular immune system exists to support the innate immune system (7, 164). One clinical application for the add-back of donor-derived D-CAR+-T cells after allogeneic HSCT is to provide protection from IFIs by the recognition of β-glucan moieties present in all opportunistic fungi.

Bioengineered Dual CAR T Cells to Target B-Cell Leukemia and IFI

CD19-specific CAR T-cells have been used successfully to treat acute lymphoblastic leukemia (ALL) by eliminating both malignant and normal B-cells, since CD19 is also expressed on normal cells. However, total elimination of B-cells resulted in B-cell aplasia as a side effect (165). Therefore, patients undergoing CAR T cell therapy are typically given intravenous Ig to control bacterial and fungal infections. Also, high incidences

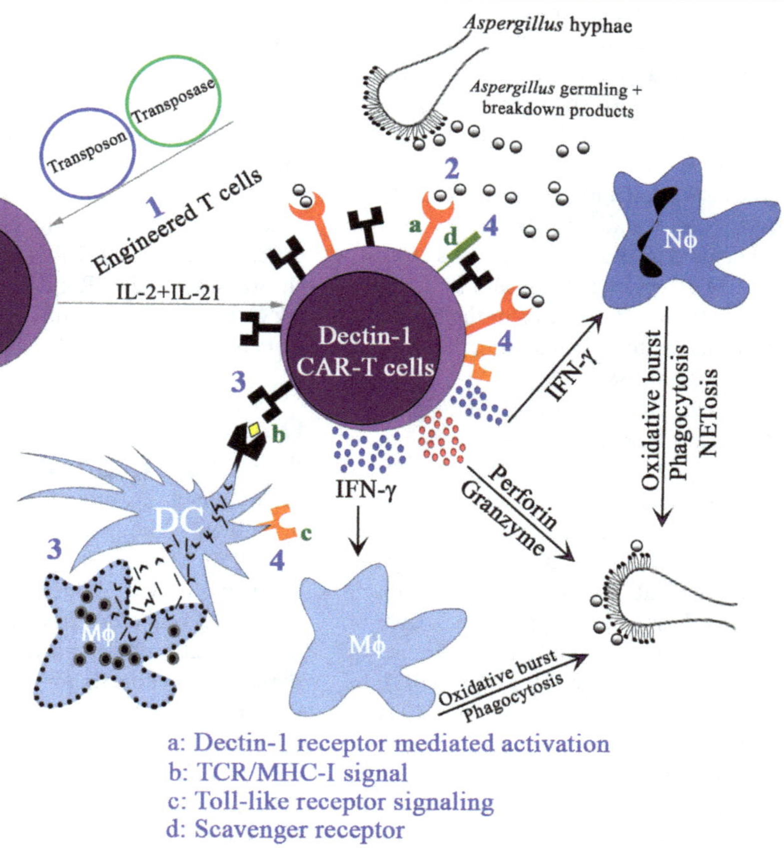

FIGURE 3 | The proposed activation pathways of Engineered T cells. (1) Expression of D-CAR+ using Sleeping Beauty system, namely, D-CAR+ T cells; (2) the β-glucan expressing *Aspergillus* germlings are recognized by D-CAR+-T cells and induce the production of interferon (IFN)-γ, which favors the microbicidal activity of macrophages and neutrophils. Activated D-CAR+-T cells also secrete granzyme and perforin to degrade fungal cell walls. (3) The activation of the D-CAR+-T cells can also occur by cross-presentation of dendritic cells (DCs) and recognition by specific T-cell receptor (TCR), and (4) direct interaction of fungal breakdown products with toll-like receptors and scavenger receptor-ligands.

of IFI are found in patients diagnosed with pediatric ALL (166, 167). These patients will gain additional benefits if CAR T cells can be engineered to destroy both IFI and tumor cells. With this goal in mind, we developed a novel gene therapy approach using dual CAR T cells to prevent IFIs such as *Aspergillus* and *Candida* and also treat B-cell leukemia. To target fungal infections, we adapted the PRR Dectin-1 to activate T cells *via* chimeric CD28 and CD3-ζ upon binding with β-1,3-gucan carbohydrate present in the fungal cell wall. The D-CAR$^+$ T cells exhibited specificity for β-1,3-gucan and led to damage to fungal hyphae and inhibition of hyphal growth of *Aspergillus* and *Candida* upon testing in both *in vitro* and mouse models. To target B-cell leukemia, we adapted chimeric CD19R-CD28-CD3-ζ T-cells that are currently being used in clinical trials (149). The D-CAR$^+$ T cells do not kill the yeast form of *Candida* so there should not be any reactions to normal commensals that live in the gut microbiota. Also, D-CAR T cells can control *Aspergillus* infections in the presence of immunosuppressive drugs at physiological concentrations (168). Thus, we propose utilizing the clinically appealing dual CAR T cells to control both leukemia and IFIs.

Future Directions for CAR T-Cell Therapy

The recent breakthroughs in bioengineered CAR T-cell therapy for cancer have opened up new horizons for targeting infectious disease-causing organisms, such as viruses, bacteria, and fungi. This approach promises to be especially useful in immunocompromised patients or those requiring long-term immunosuppressive drug therapy, such as solid organ transplant recipients. CAR T-cell therapy offers not only an immediate cure of the disease, but also long-term benefits because memory CAR T cells will protect the host from future attack by foreign invaders. This therapy will also give new hope to patients suffering from drug-resistant IFIs such as aspergillosis. An advantage of D-CAR T cell therapy is that it can be used with antifungal therapy such as Caspofungin and Amphotericin-B, thereby reducing drug-related toxicity such as nephrotoxicity associated with Amphotericin-B.

Several factors limit the immediate clinical applications of CAR T-cell therapy. Cytokine storm and neurotoxicity are the major side effects of CAR T-cell therapy and the good news is that now clinicians are successfully addressing these symptoms (169). Since D-CAR$^+$-T cells are activated by the β-glucan sugar moiety which is not present in the mammalian system, off-target related toxicities may be minimal. At present, we cannot rule out the possibility of other toxicities such as macrophage activation syndrome or GvHD that are observed in CAR T therapy to treat cancers (170).

At present, CAR T-cell therapy is a personalized therapy; more CAR T-cell manufacturing centers are needed to produce clinical-grade T cells in a cost-effective way. The therapeutic success of any form of ACT depends on infusing sufficient numbers of T cells that lack replicative senescence and terminal differentiation and have the desired specificity (171). The CAR T-cell therapy used in current clinical trials requires the use of a Good Manufacturing Practice (GMP)-compliant facility to generate the T cells; it takes 2–4 weeks to propagate enough CAR T cells for infusion into the patient. However, the length of time T cells spend in culture, especially if they are propagated under non-physiological conditions, may erode the quality of the product despite increasing its quantity. Thus, a technique to generate T cells that can be harvested from peripheral blood, minimally manipulated, and infused within a few days of collection is appealing. Pharmaceutical and biotechnology companies are actively evaluating methods for generating CAR T cells in less than a week by automating the cell culture process. Automation has immediate appeal, as it avoids the expense and risk of contamination associated with prolonged culture and reduces human labor-associated error. Rapid production may in fact improve the therapeutic potential of the manufactured T cells by allowing them to avoid the replication senescence and terminal differentiation that causes them to lose *in vivo* persistence.

Approaches to generating T cells in compliance with GMPs are based on the *ex vivo* use of reagents to identify antigen-specific T cells. One approach is to use fluorescence-labeled or paramagnetic-labeled probes that bind TCRs to identify T cells with the desired specificity. The labeled T cells are subsequently subjected to fluorescence-activated cell sorting or magnetic selection to generate a homogeneously tagged product that can be immediately infused upon meeting release criteria (172, 173). The success of this approach is measured in terms of the time needed to identify antigen-specific T cells and the specificity of the harvested product. In another approach, antigen-specific T cells are isolated from donor PBMCs using a cytokine-capture system. In this process, donor PBMCs are incubated with a peptide antigen for 4 h; the activated T cells secrete IFN-γ, which is captured by a magnetic bead-conjugated bi-specific antibody. One arm of the bi-specific antibody is specific to IFN-γ, and the other arm is specific for the cell-surface CD45. T cells that secrete IFN-γ are then separated by passing them through a column. However, this approach is limited by the number of antigen-specific T cells in the donor. If more than one donor is available, prescreening of T cells (obtained from potential donors by simple venipuncture) for antigen-specific secretion of IFN-γ will determine the most suitable donor.

Despite these limitations, adoptive transfer of viral-antigen-specific T cells that have been modified for minimal manipulation and immediate infusion has been successful in clinical trials (174). TCR sequences can be identified from these antigen-specific T cells and can be used to generate TCR CAR T cells (175). Some clinical applications, such as infusion of allogeneic antigen-specific T cells after HSCT, are not possible because the initial donor may be unavailable or anonymous. In these cases, however, potential recipients may benefit from infusion of "captured" T cells from third-party donors that can recognize antigens *via* a human leukocyte antigen molecule shared by the recipient and the donor. These "off-the-shelf" T cells could be premanufactured and cryopreserved for infusion on demand. This approach might be better for prophylaxis in high-risk patients than for treatment in patients with recalcitrant IFIs. A precedent for this approach was reported, in which third-party Epstein–Barr virus-specific T cells and multivirus-specific T cells were infused (119, 176, 177). This approach could be adapted to treat or prevent IFIs.

CONCLUSION

Adoptive T-cell therapy could play a key role in controlling IFI. The GMP grade protocols for isolation of fungal-specific T cells are well characterized. Fungal-specific CD8+ T-cells protect the host by activating the host innate immune system (Tc1 mediated) and mucosal immune system (Tc17 mediated) against IFI. For direct control, D-CAR+ T-cells have been developed by fusing the extra cellular domain of Dectin-1 and cytoplasmic domains of CD3 and CD28 receptors. It can target various fungi, such as *Aspergillus* and *Candida* (126), and such treatment is highly warranted to combat IFI infections in immunocompromised patients. We have also developed Bi-specific CARs to target both B-cell malignancies and IFI by expressing CD19R-CAR and D-CAR in the same T-cell. The high costs involved with providing CAR T-cell therapy may prohibit many patients from receiving this potentially life-saving therapy, especially those located in the developing world, where fungal infections are highly prevalent. To reduce the manufacturing costs, off-the-shelf products are being developed which can be adapted for treating IFI in near future.

AUTHOR CONTRIBUTIONS

DPK wrote the introduction and back ground. TAS wrote CD8+ T cell vaccines and PRK wrote engineered T-cells and Dectin1-CAR T cells. All authors have equally contributed for the tables. TAS and PRK contributed equally for figures.

ACKNOWLEDGMENTS

The authors thank Dr. Paul Hauser in the Pediatric Department and Dr. Amy Ninetto, Scientific Editor, Department of Scientific Publications for their assistance with proofreading this article.

FUNDING

National Institute of Allergy and Infectious Diseases grants R21 (AI127381-01), R33 (CA116127), P01 (CA148600); Burroughs Wellcome Fund; Cancer Prevention and Research Institute of Texas; CLL Global Research Foundation; DARPA (Defense Sciences Office); Department of Defense; Estate of Noelan L. Bibler; Gillson Longenbaugh Foundation; Harry T. Mangurian, Jr., Fund for Leukemia Immunotherapy; Institute of Personalized Cancer Therapy; Leukemia and Lymphoma Society; Lymphoma Research Foundation; MD Anderson Cancer Center's Sister Institution Network Fund; Miller Foundation; Mr. Herb Simons; Mr. and Mrs. Joe H. Scales; Mr. Thomas Scott; National Foundation for Cancer Research; Pediatric Cancer Research Foundation; William Lawrence and Blanche Hughes Children's Foundation. DK acknowledges the Texas 4000 Endowed Professorship for Cancer Research. Visiting scientist salary, Thiago Aparecido da Silva, was supported by funds received from Fundação de Amparo a Pesquisa do Estado de São Paulo (2016/23044-1).

REFERENCES

1. Perfect JR, Pfaller MA, Ostrosky-Zeichner L, Kontoyiannis DP. Invasive mycoses: evolving challenges and opportunities in antifungal therapy (multimedia activity). *Am J Med* (2011) 124:S2–3. doi:10.1016/j.amjmed.2011.08.001
2. Pfaller MA, Diekema DJ. Epidemiology of invasive mycoses in North America. *Crit Rev Microbiol* (2010) 36:1–53. doi:10.3109/10408410903241444
3. Neofytos D, Horn D, Anaissie E, Steinbach W, Olyaei A, Fishman J, et al. Epidemiology and outcome of invasive fungal infection in adult hematopoietic stem cell transplant recipients: analysis of multicenter prospective antifungal therapy (PATH) alliance registry. *Clin Infect Dis* (2009) 48:265–73. doi:10.1086/595846
4. Wirk B, Wingard JR. Current approaches in antifungal prophylaxis in high risk hematologic malignancy and hematopoietic stem cell transplant patients. *Mycopathologia* (2009) 168:299–311. doi:10.1007/s11046-009-9188-6
5. Person AK, Kontoyiannis DP, Alexander BD. Fungal infections in transplant and oncology patients. *Infect Dis Clin North Am* (2010) 24:439–59. doi:10.1016/j.idc.2010.01.002
6. Martino R, Subira M, Rovira M, Solano C, Vazquez L, Sanz GF, et al. Invasive fungal infections after allogeneic peripheral blood stem cell transplantation: incidence and risk factors in 395 patients. *Br J Haematol* (2002) 116:475–82. doi:10.1046/j.1365-2141.2002.03259.x
7. Kontoyiannis DP, Marr KA, Park BJ, Alexander BD, Anaissie EJ, Walsh TJ, et al. Prospective surveillance for invasive fungal infections in hematopoietic stem cell transplant recipients, 2001–2006: overview of the transplant-associated infection surveillance network (TRANSNET) database. *Clin Infect Dis* (2010) 50:1091–100. doi:10.1086/651263
8. Groll AH, Schrey D, Tragiannidis A, Bochennek K, Lehrnbecher T. Invasive aspergillosis in children and adolescents. *Curr Pharm Des* (2013) 19:3545–68. doi:10.2174/13816128113199990311
9. Tragiannidis A, Roilides E, Walsh TJ, Groll AH. Invasive aspergillosis in children with acquired immunodeficiencies. *Clin Infect Dis* (2012) 54:258–67. doi:10.1093/cid/cir786
10. Kontoyiannis DP. Antifungal prophylaxis in hematopoietic stem cell transplant recipients: the unfinished tale of imperfect success. *Bone Marrow Transplant* (2011) 46:165–73. doi:10.1038/bmt.2010.256
11. Lionakis MS, Iliev ID, Hohl TM. Immunity against fungi. *JCI Insight* (2017) 2:93156. doi:10.1172/jci.insight.93156
12. Wuthrich M, Deepe GS, Klein B. Adaptive immunity to fungi. *Annu Rev Immunol* (2012) 30:115–48. doi:10.1146/annurev-immunol-020711-074958
13. Romani L. Immunity to fungal infections. *Nat Rev Immunol* (2011) 11:275–88. doi:10.1038/nri2939
14. Armstrong-James D, Brown GD, Netea MG, Zelante T, Gresnigt MS, van de Veerdonk FL, et al. Immunotherapeutic approaches to treatment of fungal diseases. *Lancet Infect Dis* (2017) 17:e393–e402. doi:10.1016/S1473-3099(17)30442-5
15. Global Action Found for Fugal Disease. *Burden of Common Life-Threatening Fungal Infections*. Global Action Found for Fugal Disease (2017). Available from: https://www.gaffi.org/why/fungal-disease-frequency/
16. Gudlaugsson O, Gillespie S, Lee K, Berg JV, Hu JF, Messer S, et al. Attributable mortality of nosocomial candidemia, revisited. *Clin Infect Dis* (2003) 37:1172–7. doi:10.1086/378745
17. Hajjeh RA, Sofair AN, Harrison LH, Lyon GM, Arthington-Skaggs BA, Mirza SA, et al. Incidence of bloodstream infections due to *Candida* species and in vitro susceptibilities of isolates collected from 1998 to 2000 in a population-based active surveillance program. *J Clin Microbiol* (2004) 42:1519–27. doi:10.1128/JCM.42.4.1519-1527.2004
18. Gomes MZR, Lewis RE, Kontoyiannis DP. Mucormycosis caused by unusual mucormycetes, non-*Rhizopus*, -*Mucor*, and -*Lichtheimia* species. *Clin Microbiol Rev* (2011) 24:411. doi:10.1128/CMR.00056-10
19. Park BJ, Wannemuehler KA, Marston BJ, Govender N, Pappas PG, Chiller TA. Estimation of the current global burden of cryptococcal meningitis among persons living with HIV/AIDS. *AIDS* (2009) 23:525–30. doi:10.1097/QAD.0b013e328322ffac
20. Harris K, Maroun R, Chalhoub M, Elsayegh D. Unusual presentation of pneumocystis pneumonia in an immunocompetent patient diagnosed by open lung biopsy. *Heart Lung Circ* (2012) 21:221–4. doi:10.1016/j.hlc.2011.10.006

21. Harris JR, Marston BJ, Sangrujee N, DuPlessis D, Park B. Cost-effectiveness analysis of diagnostic options for pneumocystis pneumonia (PCP). *PLoS One* (2011) 6:e23158. doi:10.1371/journal.pone.0023158
22. Cottier F, Pavelka N. Complexity and dynamics of host-fungal interactions. *Immunol Res* (2012) 53:127–35. doi:10.1007/s12026-012-8265-y
23. Brunke S, Mogavero S, Kasper L, Hube B. Virulence factors in fungal pathogens of man. *Curr Opin Microbiol* (2016) 32:89–95. doi:10.1016/j.mib.2016.05.010
24. Gow NAR, Latge JP, Munro CA. The fungal cell wall: structure, biosynthesis, and function. *Microbiol Spectr* (2017) 5:1–25. doi:10.1128/microbiolspec.FUNK-0035-2016
25. Ibrahim AS. Host cell invasion in mucormycosis: role of iron. *Curr Opin Microbiol* (2011) 14:406–11. doi:10.1016/j.mib.2011.07.004
26. Espinosa V, Rivera A. First line of defense: innate cell-mediated control of pulmonary aspergillosis. *Front Microbiol* (2016) 7:272. doi:10.3389/fmicb.2016.00272
27. LeibundGut-Landmann S, Wuthrich M, Hohl TM. Immunity to fungi. *Curr Opin Immunol* (2012) 24:449–58. doi:10.1016/j.coi.2012.04.007
28. Plato A, Hardison SE, Brown GD. Pattern recognition receptors in antifungal immunity. *Semin Immunopathol* (2015) 37:97–106. doi:10.1007/s00281-014-0462-4
29. Claes AK, Zhou JY, Philpott DJ. NOD-like receptors: guardians of intestinal mucosal barriers. *Physiology (Bethesda)* (2015) 30:241–50. doi:10.1152/physiol.00025.2014
30. Kawai T, Akira S. The roles of TLRs, RLRs and NLRs in pathogen recognition. *Int Immunol* (2009) 21:317–37. doi:10.1093/intimm/dxp017
31. Hoving JC, Wilson GJ, Brown GD. Signalling C-type lectin receptors, microbial recognition and immunity. *Cell Microbiol* (2014) 16:185–94. doi:10.1111/cmi.12249
32. Underhill DM, Pearlman E. Immune interactions with pathogenic and commensal fungi: a two-way street. *Immunity* (2015) 43:845–58. doi:10.1016/j.immuni.2015.10.023
33. Naglik RJ. *Candida* immunity. *N J Sci* (2014) 2014:27. doi:10.1155/2014/390241
34. Brown GD, Taylor PR, Reid DM, Willment JA, Williams DL, Martinez-Pomares L, et al. Dectin-1 is a major beta-glucan receptor on macrophages. *J Exp Med* (2002) 196:407–12. doi:10.1084/jem.20020470
35. Robinson MJ, Osorio F, Rosas M, Freitas RP, Schweighoffer E, Gross O, et al. Dectin-2 is a Syk-coupled pattern recognition receptor crucial for Th17 responses to fungal infection. *J Exp Med* (2009) 206:2037–51. doi:10.1084/jem.20082818
36. Saijo S, Ikeda S, Yamabe K, Kakuta S, Ishigame H, Akitsu A, et al. Dectin-2 recognition of alpha-mannans and induction of Th17 cell differentiation is essential for host defense against *Candida albicans*. *Immunity* (2010) 32:681–91. doi:10.1016/j.immuni.2010.05.001
37. Zhu LL, Zhao XQ, Jiang C, You Y, Chen XP, Jiang YY, et al. C-type lectin receptors Dectin-3 and Dectin-2 form a heterodimeric pattern-recognition receptor for host defense against fungal infection. *Immunity* (2013) 39:324–34. doi:10.1016/j.immuni.2013.05.017
38. Papadopoulou A, Kaloyannidis P, Yannaki E, Cruz CR. Adoptive transfer of *Aspergillus*-specific T cells as a novel anti-fungal therapy for hematopoietic stem cell transplant recipients: progress and challenges. *Crit Rev Oncol Hematol* (2016) 98:62–72. doi:10.1016/j.critrevonc.2015.10.005
39. Castellano-Gonzalez G, Clancy LE, Gottlieb D. Prospects for adoptive T-cell therapy for invasive fungal disease. *Curr Opin Infect Dis* (2017) 30(6):518–27. doi:10.1097/QCO.0000000000000403
40. van de Veerdonk FL, Netea MG. T-cell subsets and antifungal host defenses. *Curr Fungal Infect Rep* (2010) 4:238–43. doi:10.1007/s12281-010-0034-6
41. Roemer T, Krysan DJ. Antifungal drug development: challenges, unmet clinical needs, and new approaches. *Cold Spring Harb Perspect Med* (2014) 4:a019703. doi:10.1101/cshperspect.a019703
42. Vermeulen E, Lagrou K, Verweij PE. Azole resistance in *Aspergillus fumigatus*: a growing public health concern. *Curr Opin Infect Dis* (2013) 26:493–500. doi:10.1097/QCO.0000000000000005
43. Heo ST, Tatara AM, Jimenez-Ortigosa C, Jiang Y, Lewis RE, Tarrand J, et al. Changes in in vitro susceptibility patterns of *Aspergillus* to triazoles and correlation with aspergillosis outcome in a tertiary care cancer center, 1999–2015. *Clin Infect Dis* (2017) 65:216–25. doi:10.1093/cid/cix297
44. Kontoyiannis DP, Selleslag D, Mullane K, Cornely OA, Hope W, Lortholary O, et al. Impact of unresolved neutropenia in patients with neutropenia and invasive aspergillosis: a post hoc analysis of the SECURE trial. *J Antimicrob Chemother* (2017) 1–17. doi:10.1093/jac/dkx423
45. Kontoyiannis DP. Echinocandin-based initial therapy in fungemic patients with cancer: a focus on recent guidelines of the infectious diseases society of America. *Clin Infect Dis* (2009) 49:638–39; author reply 639–40. doi:10.1086/603585
46. Waldorf AR, Levitz SM, Diamond RD. In vivo bronchoalveolar macrophage defense against *Rhizopus oryzae* and *Aspergillus fumigatus*. *J Infect Dis* (1984) 150:752–60. doi:10.1093/infdis/150.5.752
47. Amulic B, Cazalet C, Hayes GL, Metzler KD, Zychlinsky A. Neutrophil function: from mechanisms to disease. *Annu Rev Immunol* (2012) 30:459–89. doi:10.1146/annurev-immunol-020711-074942
48. Kaplan MJ, Radic M. Neutrophil extracellular traps: double-edged swords of innate immunity. *J Immunol* (2012) 189:2689–95. doi:10.4049/jimmunol.1201719
49. Mousset S, Hermann S, Klein SA, Bialleck H, Duchscherer M, Bomke B, et al. Prophylactic and interventional granulocyte transfusions in patients with haematological malignancies and life-threatening infections during neutropenia. *Ann Hematol* (2005) 84:734–41. doi:10.1007/s00277-005-1055-z
50. Grigull L, Beilken A, Schmid H, Kirschner P, Sykora KW, Linderkamp C, et al. Secondary prophylaxis of invasive fungal infections with combination antifungal therapy and G-CSF-mobilized granulocyte transfusions in three children with hematological malignancies. *Support Care Cancer* (2006) 14:783–6. doi:10.1007/s00520-005-0910-8
51. Gil-Lamaignere C, Winn RM, Simitsopoulou M, Maloukou A, Walsh TJ, Roilides E. Inteferon gamma and granulocyte-macrophage colony-stimulating factor augment the antifungal activity of human polymorphonuclear leukocytes against *Scedosporium* spp.: comparison with *Aspergillus* spp. *Med Mycol* (2005) 43:253–60. doi:10.1080/13693780412331271072
52. Winn RM, Gil-Lamaignere C, Roilides E, Simitsopoulou M, Lyman CA, Maloukou A, et al. Effects of interleukin-15 on antifungal responses of human polymorphonuclear leukocytes against *Fusarium* spp. and *Scedosporium* spp. *Cytokine* (2005) 31:1–8. doi:10.1016/j.cyto.2004.07.016
53. Valentini CG, Farina F, Pagano L, Teofili L. Granulocyte transfusions: a critical reappraisal. *Biol Blood Marrow Transplant* (2017) 23:2034–41. doi:10.1016/j.bbmt.2017.07.029
54. Robertson MJ, Ritz J. Biology and clinical relevance of human natural killer cells. *Blood* (1990) 76:2421–38.
55. Bouzani M, Ok M, McCormick A, Ebel F, Kurzai O, Morton CO, et al. Human NK cells display important antifungal activity against *Aspergillus fumigatus*, which is directly mediated by IFN-gamma release. *J Immunol* (2011) 187:1369–76. doi:10.4049/jimmunol.1003593
56. Schmidt S, Schneider A, Demir A, Lass-Florl C, Lehrnbecher T. Natural killer cell-mediated damage of clinical isolates of mucormycetes. *Mycoses* (2016) 59:34–8. doi:10.1111/myc.12431
57. Schmidt S, Tramsen L, Hanisch M, Latge JP, Huenecke S, Koehl U, et al. Human natural killer cells exhibit direct activity against *Aspergillus fumigatus* hyphae, but not against resting conidia. *J Infect Dis* (2011) 203:430–5. doi:10.1093/infdis/jiq062
58. Voigt J, Hunniger K, Bouzani M, Jacobsen ID, Barz D, Hube B, et al. Human natural killer cells acting as phagocytes against *Candida albicans* and mounting an inflammatory response that modulates neutrophil antifungal activity. *J Infect Dis* (2014) 209:616–26. doi:10.1093/infdis/jit574
59. Ziegler S, Weiss E, Schmitt AL, Schlegel J, Burgert A, Terpitz U, et al. CD56 is a pathogen recognition receptor on human natural killer cells. *Sci Rep* (2017) 7:6138. doi:10.1038/s41598-017-06238-4
60. Li SS, Kyei SK, Timm-McCann M, Ogbomo H, Jones GJ, Shi M, et al. The NK receptor NKp30 mediates direct fungal recognition and killing and is diminished in NK cells from HIV-infected patients. *Cell Host Microbe* (2013) 14:387–97. doi:10.1016/j.chom.2013.09.007
61. Ogbomo H, Mody CH. Granule-dependent natural killer cell cytotoxicity to fungal pathogens. *Front Immunol* (2016) 7:692. doi:10.3389/fimmu.2016.00692
62. Schmidt S, Condorelli A, Koltze A, Lehrnbecher T. NK cells and their role in invasive mold infection. *J Fungi* (2017) 8:1623. doi:10.3389/fimmu.2017.01623

63. Fidan I, Kalkanci A, Yesilyurt E, Erdal B. In vitro effects of *Candida albicans* and *Aspergillus fumigatus* on dendritic cells and the role of beta glucan in this effect. *Adv Clin Exp Med* (2014) 23:17–24. doi:10.17219/acem/37016
64. Bozza S, Gaziano R, Spreca A, Bacci A, Montagnoli C, di Francesco P, et al. Dendritic cells transport conidia and hyphae of *Aspergillus fumigatus* from the airways to the draining lymph nodes and initiate disparate Th responses to the fungus. *J Immunol* (2002) 168:1362–71. doi:10.4049/jimmunol.168.3.1362
65. Thakur R, Anand R, Tiwari S, Singh AP, Tiwary BN, Shankar J. Cytokines induce effector T-helper cells during invasive aspergillosis; what we have learned about T-helper cells? *Front Microbiol* (2015) 6:429. doi:10.3389/fmicb.2015.00429
66. Zelante T, Wong AY, Ping TJ, Chen J, Sumatoh HR, Vigano E, et al. CD103(+) dendritic cells control Th17 cell function in the lung. *Cell Rep* (2015) 12:1789–801. doi:10.1016/j.celrep.2015.08.030
67. Cutler JE, Deepe GS Jr, Klein BS. Advances in combating fungal diseases: vaccines on the threshold. *Nat Rev Microbiol* (2007) 5:13–28. doi:10.1038/nrmicro1537
68. Chen K, Kolls JK. T cell-mediated host immune defenses in the lung. *Annu Rev Immunol* (2013) 31:605–33. doi:10.1146/annurev-immunol-032712-100019
69. Hernandez-Santos N, Gaffen SL. Th17 cells in immunity to *Candida albicans*. *Cell Host Microbe* (2012) 11:425–35. doi:10.1016/j.chom.2012.04.008
70. Huppler AR, Bishu S, Gaffen SL. Mucocutaneous candidiasis: the IL-17 pathway and implications for targeted immunotherapy. *Arthritis Res Ther* (2012) 14:217. doi:10.1186/ar3893
71. Backer R, van Leeuwen F, Kraal G, den Haan JM. CD8-dendritic cells preferentially cross-present *Saccharomyces cerevisiae* antigens. *Eur J Immunol* (2008) 38:370–80. doi:10.1002/eji.200737647
72. Leibundgut-Landmann S, Osorio F, Brown GD, Reis e Sousa C. Stimulation of dendritic cells via the dectin-1/Syk pathway allows priming of cytotoxic T-cell responses. *Blood* (2008) 112:4971–80. doi:10.1182/blood-2008-05-158469
73. LeibundGut-Landmann S, Gross O, Robinson MJ, Osorio F, Slack EC, Tsoni SV, et al. Syk- and CARD9-dependent coupling of innate immunity to the induction of T helper cells that produce interleukin 17. *Nat Immunol* (2007) 8:630–8. doi:10.1038/ni1460
74. Nanjappa SG, Heninger E, Wuthrich M, Gasper DJ, Klein BS. Tc17 cells mediate vaccine immunity against lethal fungal pneumonia in immune deficient hosts lacking $CD4^+$ T cells. *PLoS Pathog* (2012) 8:e1002771. doi:10.1371/journal.ppat.1002771
75. Nanjappa SG, Heninger E, Wuthrich M, Sullivan T, Klein B. Protective antifungal memory CD8(+) T cells are maintained in the absence of CD4(+) T cell help and cognate antigen in mice. *J Clin Invest* (2012) 122:987–99. doi:10.1172/JCI58762
76. Lin JS, Yang CW, Wang DW, Wu-Hsieh BA. Dendritic cells cross-present exogenous fungal antigens to stimulate a protective CD8 T cell response in infection by *Histoplasma capsulatum*. *J Immunol* (2005) 174:6282–91. doi:10.4049/jimmunol.174.10.6282
77. McAllister F, Ruan S, Steele C, Zheng M, McKinley L, Ulrich L, et al. CXCR3 and IFN protein-10 in pneumocystis pneumonia. *J Immunol* (2006) 177:1846–54. doi:10.4049/jimmunol.177.3.1846
78. Nanjappa SG, Klein BS. Vaccine immunity against fungal infections. *Curr Opin Immunol* (2014) 28:27–33. doi:10.1016/j.coi.2014.01.014
79. Reichard U, Herrmann S, Asif AR. Vaccination approaches against opportunistic fungal infections caused by *Aspergillus fumigatus*. *Curr Protein Pept Sci* (2014) 15:424–9. doi:10.2174/1389203715666140512122037
80. Chaturvedi AK, Wormley FL. *Cryptococcus* antigens and immune responses: implications for a vaccine. *Expert Rev Vaccines* (2013) 12:1261–72. doi:10.1586/14760584.2013.840094
81. Bocca AL, Amaral AC, Teixeira MM, Sato P, Shikanai-Yasuda MA, Soares Felipe MS. Paracoccidioidomycosis: eco-epidemiology, taxonomy and clinical and therapeutic issues. *Future Microbiol* (2013) 8:1177–91. doi:10.2217/fmb.13.68
82. Nguyen C, Barker BM, Hoover S, Nix DE, Ampel NM, Frelinger JA, et al. Recent advances in our understanding of the environmental, epidemiological, immunological, and clinical dimensions of coccidioidomycosis. *Clin Microbiol Rev* (2013) 26:505–25. doi:10.1128/CMR.00005-13
83. Levitz SM, Huang H, Ostroff GR, Specht CA. Exploiting fungal cell wall components in vaccines. *Semin Immunopathol* (2015) 37:199–207. doi:10.1007/s00281-014-0460-6
84. Edwards JE Jr. Fungal cell wall vaccines: an update. *J Med Microbiol* (2012) 61:895–903. doi:10.1099/jmm.0.041665-0
85. Cassone A, Casadevall A. Recent progress in vaccines against fungal diseases. *Curr Opin Microbiol* (2012) 15:427–33. doi:10.1016/j.mib.2012.04.004
86. Yoon HJ, Clemons KV. Vaccines against *Coccidioides*. *Korean J Intern Med* (2013) 28:403–7. doi:10.3904/kjim.2013.28.4.403
87. Ito JI, Lyons JM, Hong TB, Tamae D, Liu YK, Wilczynski SP, et al. Vaccinations with recombinant variants of *Aspergillus fumigatus* allergen Asp f 3 protect mice against invasive aspergillosis. *Infect Immun* (2006) 74:5075–84. doi:10.1128/IAI.00815-06
88. Fierer J, Waters C, Walls L. Both $CD4^+$ and $CD8^+$ T cells can mediate vaccine-induced protection against *Coccidioides immitis* infection in mice. *J Infect Dis* (2006) 193:1323–31. doi:10.1086/502972
89. Hung CY, Gonzalez A, Wuthrich M, Klein BS, Cole GT. Vaccine immunity to coccidioidomycosis occurs by early activation of three signal pathways of T helper cell response (Th1, Th2, and Th17). *Infect Immun* (2011) 79:4511–22. doi:10.1128/IAI.05726-11
90. Khan AA, Jabeen M, Chauhan A, Owais M. Vaccine potential of cytosolic proteins loaded fibrin microspheres of *Cryptococcus neoformans* in BALB/c mice. *J Drug Target* (2012) 20:453–66. doi:10.3109/1061186X.2012.685474
91. Nanjappa SG, Hernandez-Santos N, Galles K, Wuthrich M, Suresh M, Klein BS. Intrinsic MyD88-Akt1-mTOR signaling coordinates disparate Tc17 and Tc1 responses during vaccine immunity against fungal pneumonia. *PLoS Pathog* (2015) 11:e1005161. doi:10.1371/journal.ppat.1005161
92. Wuthrich M, Filutowicz HI, Warner T, Deepe GS Jr, Klein BS. Vaccine immunity to pathogenic fungi overcomes the requirement for CD4 help in exogenous antigen presentation to $CD8^+$ T cells: implications for vaccine development in immune-deficient hosts. *J Exp Med* (2003) 197:1405–16. doi:10.1084/jem.20030109
93. Templeton SP, Buskirk AD, Law B, Green BJ, Beezhold DH. Role of germination in murine airway $CD8^+$ T-cell responses to *Aspergillus* conidia. *PLoS One* (2011) 6:e18777. doi:10.1371/journal.pone.0018777
94. Chaudhary N, Staab JF, Marr KA. Healthy human T-cell responses to *Aspergillus fumigatus* antigens. *PLoS One* (2010) 5:e9036. doi:10.1371/journal.pone.0009036
95. Cui N, Wang H, Long Y, Liu D. CD8(+) T-cell counts: an early predictor of risk and mortality in critically ill immunocompromised patients with invasive pulmonary aspergillosis. *Crit Care* (2013) 17:R157. doi:10.1186/cc12836
96. Sun Z, Zhu P, Li L, Wan Z, Zhao Z, Li R. Adoptive immunity mediated by HLA-A*0201 restricted Asp f16 peptides-specific $CD8^+$ T cells against *Aspergillus fumigatus* infection. *Eur J Clin Microbiol Infect Dis* (2012) 31:3089–96. doi:10.1007/s10096-012-1670-2
97. Potenza L, Vallerini D, Barozzi P, Riva G, Forghieri F, Zanetti E, et al. Mucorales-specific T cells emerge in the course of invasive mucormycosis and may be used as a surrogate diagnostic marker in high-risk patients. *Blood* (2011) 118:5416–9. doi:10.1182/blood-2011-07-366526
98. Vallerini D, Forghieri F, Lagreca I, Riva G, Barozzi P, Quadrelli C, et al. Detection of *Fusarium*-specific T cells in hematologic patients with invasive fusariosis. *J Infect* (2017) 74:314–8. doi:10.1016/j.jinf.2016.12.008
99. McAllister F, Steele C, Zheng M, Young E, Shellito JE, Marrero L, et al. T cytotoxic-1 $CD8^+$ T cells are effector cells against pneumocystis in mice. *J Immunol* (2004) 172:1132–8. doi:10.4049/jimmunol.172.2.1132
100. De Luca A, Iannitti RG, Bozza S, Beau R, Casagrande A, D'Angelo C, et al. CD4(+) T cell vaccination overcomes defective cross-presentation of fungal antigens in a mouse model of chronic granulomatous disease. *J Clin Invest* (2012) 122:1816–31. doi:10.1172/JCI60862
101. Carvalho A, De Luca A, Bozza S, Cunha C, D'Angelo C, Moretti S, et al. TLR3 essentially promotes protective class I-restricted memory CD8(+) T-cell responses to *Aspergillus fumigatus* in hematopoietic transplanted patients. *Blood* (2012) 119:967–77. doi:10.1182/blood-2011-06-362582
102. Spellberg B, Ibrahim AS, Lin L, Avanesian V, Fu Y, Lipke P, et al. Antibody titer threshold predicts anti-candidal vaccine efficacy even though the mechanism of protection is induction of cell-mediated immunity. *J Infect Dis* (2008) 197:967–71. doi:10.1086/529204
103. Paulovicova E, Machova E, Tulinska J, Bystricky S. Cell and antibody mediated immunity induced by vaccination with novel *Candida dubliniensis* mannan immunogenic conjugate. *Int Immunopharmacol* (2007) 7:1325–33. doi:10.1016/j.intimp.2007.05.014

104. Chauhan A, Swaleha Z, Ahmad N, Farazuddin M, Vasco A, Abida M, et al. Escheriosome mediated cytosolic delivery of *Candida albicans* cytosolic proteins induces enhanced cytotoxic T lymphocyte response and protective immunity. *Vaccine* (2011) 29:5424–33. doi:10.1016/j.vaccine.2011.05.066
105. Hamad M. Universal fungal vaccines: could there be light at the end of the tunnel? *Hum Vaccin Immunother* (2012) 8:1758–63. doi:10.4161/hv.21838
106. Meier A, Kirschning CJ, Nikolaus T, Wagner H, Heesemann J, Ebel F. Toll-like receptor (TLR) 2 and TLR4 are essential for *Aspergillus*-induced activation of murine macrophages. *Cell Microbiol* (2003) 5:561–70. doi:10.1046/j.1462-5822.2003.00301.x
107. Shoham S, Huang C, Chen JM, Golenbock DT, Levitz SM. Toll-like receptor 4 mediates intracellular signaling without TNF-alpha release in response to *Cryptococcus neoformans* polysaccharide capsule. *J Immunol* (2001) 166:4620–6. doi:10.4049/jimmunol.166.7.4620
108. Calich VL, Pina A, Felonato M, Bernardino S, Costa TA, Loures FV. Toll-like receptors and fungal infections: the role of TLR2, TLR4 and MyD88 in paracoccidioidomycosis. *FEMS Immunol Med Microbiol* (2008) 53:1–7. doi:10.1111/j.1574-695X.2008.00378.x
109. Firinu D, Pisanu M, Piras B, Meleddu R, Lorrai MM, Manconi PE, et al. Genetic susceptibility to *Candida* infection: a new look at an old entity. *Chin Med J (Engl)* (2013) 126:378–81.
110. Nahum A, Dadi H, Bates A, Roifman CM. The biological significance of TLR3 variant, L412F, in conferring susceptibility to cutaneous candidiasis, CMV and autoimmunity. *Autoimmun Rev* (2012) 11:341–7. doi:10.1016/j.autrev.2011.10.007
111. Bochud PY, Chien JW, Marr KA, Leisenring WM, Upton A, Janer M, et al. Toll-like receptor 4 polymorphisms and aspergillosis in stem-cell transplantation. *N Engl J Med* (2008) 359:1766–77. doi:10.1056/NEJMoa0802629
112. Vera J, Fenutria R, Canadas O, Figueras M, Mota R, Sarrias MR, et al. The CD5 ectodomain interacts with conserved fungal cell wall components and protects from zymosan-induced septic shock-like syndrome. *Proc Natl Acad Sci U S A* (2009) 106:1506–11. doi:10.1073/pnas.0805846106
113. Acharya M, Borland G, Edkins AL, Maclellan LM, Matheson J, Ozanne BW, et al. CD23/FcepsilonRII: molecular multi-tasking. *Clin Exp Immunol* (2010) 162:12–23. doi:10.1111/j.1365-2249.2010.04210.x
114. Zhao X, Guo Y, Jiang C, Chang Q, Zhang S, Luo T, et al. JNK1 negatively controls antifungal innate immunity by suppressing CD23 expression. *Nat Med* (2017) 23:337–46. doi:10.1038/nm.4260
115. Van Acker HH, Capsomidis A, Smits EL, Van Tendeloo VF. CD56 in the immune system: more than a marker for cytotoxicity? *Front Immunol* (2017) 8:892. doi:10.3389/fimmu.2017.00892
116. McLaughlin LP, Gottschalk S, Rooney CM, Bollard CM. EBV-directed T cell therapeutics for EBV-associated lymphomas. *Methods Mol Biol* (2017) 1532:255–65. doi:10.1007/978-1-4939-6655-4_19
117. Kumaresan P, Figliola M, Moyes JS, Huls MH, Tewari P, Shpall EJ, et al. Automated cell enrichment of cytomegalovirus-specific T cells for clinical applications using the cytokine-capture system. *J Vis Exp* (2015) 104:e52808. doi:10.3791/52808
118. Krishnamurthy J, Rabinovich BA, Mi T, Switzer KC, Olivares S, Maiti SN, et al. Genetic engineering of T cells to target HERV-K, an ancient retrovirus on melanoma. *Clin Cancer Res* (2015) 21:3241–51. doi:10.1158/1078-0432.CCR-14-3197
119. Tzannou I, Papadopoulou A, Naik S, Leung K, Martinez CA, Ramos CA, et al. Off-the-shelf virus-specific T cells to treat BK virus, human herpesvirus 6, cytomegalovirus, Epstein-Barr virus, and adenovirus infections after allogeneic hematopoietic stem-cell transplantation. *J Clin Oncol* (2017) 35:3547–57. doi:10.1200/JCO.2017.73.0655
120. Perruccio K, Tosti A, Burchielli E, Topini F, Ruggeri L, Carotti A, et al. Transferring functional immune responses to pathogens after haploidentical hematopoietic transplantation. *Blood* (2005) 106:4397–406. doi:10.1182/blood-2005-05-1775
121. Pagano L, Caira M, Nosari A, Van Lint MT, Candoni A, Offidani M, et al. Fungal infections in recipients of hematopoietic stem cell transplants: results of the SEIFEM B-2004 study – Sorveglianza Epidemiologica Infezioni Fungine Nelle Emopatie Maligne. *Clin Infect Dis* (2007) 45:1161–70. doi:10.1086/522189
122. Beck O, Topp MS, Koehl U, Roilides E, Simitsopoulou M, Hanisch M, et al. Generation of highly purified and functionally active human TH1 cells against *Aspergillus fumigatus*. *Blood* (2006) 107:2562–9. doi:10.1182/blood-2005-04-1660
123. Shi H, Liu L, Wang Z. Improving the efficacy and safety of engineered T cell therapy for cancer. *Cancer Lett* (2013) 328:191–7. doi:10.1016/j.canlet.2012.09.015
124. Stauss HJ, Morris EC, Abken H. Cancer gene therapy with T cell receptors and chimeric antigen receptors. *Curr Opin Pharmacol* (2015) 24:113–8. doi:10.1016/j.coph.2015.08.006
125. Klebanoff CA, Rosenberg SA, Restifo NP. Prospects for gene-engineered T cell immunotherapy for solid cancers. *Nat Med* (2016) 22:26–36. doi:10.1038/nm.4015
126. Kumaresan PR, Manuri PR, Albert ND, Maiti S, Singh H, Mi T, et al. Bioengineering T cells to target carbohydrate to treat opportunistic fungal infection. *Proc Natl Acad Sci U S A* (2014) 111:10660–5. doi:10.1073/pnas.1312789111
127. Singh H, Huls H, Kebriaei P, Cooper LJ. A new approach to gene therapy using Sleeping Beauty to genetically modify clinical-grade T cells to target CD19. *Immunol Rev* (2014) 257:181–90. doi:10.1111/imr.12137
128. Dembic Z, Haas W, Weiss S, McCubrey J, Kiefer H, von Boehmer H, et al. Transfer of specificity by murine alpha and beta T-cell receptor genes. *Nature* (1986) 320:232–8. doi:10.1038/320232a0
129. Zhao Y, Zheng Z, Cohen CJ, Gattinoni L, Palmer DC, Restifo NP, et al. High-efficiency transfection of primary human and mouse T lymphocytes using RNA electroporation. *Mol Ther* (2006) 13:151–9. doi:10.1016/j.ymthe.2005.07.688
130. Hohl TM. Immune responses to invasive aspergillosis: new understanding and therapeutic opportunities. *Curr Opin Infect Dis* (2017) 30:364–71. doi:10.1097/QCO.0000000000000381
131. Grazziutti M, Przepiorka D, Rex JH, Braunschweig I, Vadhan-Raj S, Savary CA. Dendritic cell-mediated stimulation of the in vitro lymphocyte response to *Aspergillus*. *Bone Marrow Transplant* (2001) 27:647–52. doi:10.1038/sj.bmt.1702832
132. Robbins PF, Morgan RA, Feldman SA, Yang JC, Sherry RM, Dudley ME, et al. Tumor regression in patients with metastatic synovial cell sarcoma and melanoma using genetically engineered lymphocytes reactive with NY-ESO-1. *J Clin Oncol* (2011) 29:917–24. doi:10.1200/JCO.2010.32.2537
133. Parkhurst MR, Yang JC, Langan RC, Dudley ME, Nathan DA, Feldman SA, et al. T cells targeting carcinoembryonic antigen can mediate regression of metastatic colorectal cancer but induce severe transient colitis. *Mol Ther* (2011) 19:620–6. doi:10.1038/mt.2010.272
134. Garrido F, Ruiz-Cabello F, Cabrera T, Perez-Villar JJ, Lopez-Botet M, Duggan-Keen M, et al. Implications for immunosurveillance of altered HLA class I phenotypes in human tumours. *Immunol Today* (1997) 18:89–95. doi:10.1016/S0167-5699(96)10075-X
135. Kershaw MH, Teng MW, Smyth MJ, Darcy PK. Supernatural T cells: genetic modification of T cells for cancer therapy. *Nat Rev Immunol* (2005) 5:928–40. doi:10.1038/nri1729
136. Eshhar Z. Tumor-specific T-bodies: towards clinical application. *Cancer Immunol Immunother* (1997) 45:131–6. doi:10.1007/s002620050415
137. Kuwana Y, Asakura Y, Utsunomiya N, Nakanishi M, Arata Y, Itoh S, et al. Expression of chimeric receptor composed of immunoglobulin-derived V regions and T-cell receptor-derived C regions. *Biochem Biophys Res Commun* (1987) 149:960–8. doi:10.1016/0006-291X(87)90502-X
138. Dotti G, Gottschalk S, Savoldo B, Brenner MK. Design and development of therapies using chimeric antigen receptor-expressing T cells. *Immunol Rev* (2014) 257:107–26. doi:10.1111/imr.12131
139. Johnson LA, June CH. Driving gene-engineered T cell immunotherapy of cancer. *Cell Res* (2017) 27:38–58. doi:10.1038/cr.2016.154
140. Hombach A, Hombach AA, Abken H. Adoptive immunotherapy with genetically engineered T cells: modification of the IgG1 Fc 'spacer' domain in the extracellular moiety of chimeric antigen receptors avoids 'off-target' activation and unintended initiation of an innate immune response. *Gene Ther* (2010) 17:1206–13. doi:10.1038/gt.2010.91
141. Bridgeman JS, Hawkins RE, Bagley S, Blaylock M, Holland M, Gilham DE. The optimal antigen response of chimeric antigen receptors harboring the CD3zeta transmembrane domain is dependent upon incorporation of the receptor into the endogenous TCR/CD3 complex. *J Immunol* (2010) 184:6938–49. doi:10.4049/jimmunol.0901766

142. Finney HM, Lawson AD, Bebbington CR, Weir AN. Chimeric receptors providing both primary and costimulatory signaling in T cells from a single gene product. *J Immunol* (1998) 161:2791–7.
143. Savoldo B, Ramos CA, Liu E, Mims MP, Keating MJ, Carrum G, et al. CD28 costimulation improves expansion and persistence of chimeric antigen receptor-modified T cells in lymphoma patients. *J Clin Invest* (2011) 121:1822–6. doi:10.1172/JCI46110
144. Porter DL, Levine BL, Kalos M, Bagg A, June CH. Chimeric antigen receptor-modified T cells in chronic lymphoid leukemia. *N Engl J Med* (2011) 365:725–33. doi:10.1056/NEJMoa1103849
145. Hackett PB, Largaespada DA, Cooper LJN. A transposon and transposase system for human application. *Mol Ther* (2010) 18:674–83. doi:10.1038/mt.2010.2
146. Hackett PB, Ekker SC, Largaespada DA, McIvor RS. Sleeping Beauty transposon-mediated gene therapy for prolonged expression. *Adv Genet* (2005) 54:189–232. doi:10.1016/S0065-2660(05)54009-4
147. Hudecek M, Izsvak Z, Johnen S, Renner M, Thumann G, Ivics Z. Going non-viral: the Sleeping Beauty transposon system breaks on through to the clinical side. *Crit Rev Biochem Mol Biol* (2017) 52:355–80. doi:10.1080/10409238.2017.1304354
148. Kebriaei P, Izsvak Z, Narayanavari SA, Singh H, Ivics Z. Gene therapy with the Sleeping Beauty transposon system. *Trends Genet* (2017) 33:852–70. doi:10.1016/j.tig.2017.08.008
149. Kebriaei P, Singh H, Huls MH, Figliola MJ, Bassett R, Olivares S, et al. Phase I trials using Sleeping Beauty to generate CD19-specific CAR T cells. *J Clin Invest* (2016) 126:3363–76. doi:10.1172/JCI86721
150. Brown GD, Herre J, Williams DL, Willment JA, Marshall AS, Gordon S. Dectin-1 mediates the biological effects of beta-glucans. *J Exp Med* (2003) 197:1119–24. doi:10.1084/jem.20021890
151. Gringhuis SI, den Dunnen J, Litjens M, van der Vlist M, Wevers B, Bruijns SC, et al. Dectin-1 directs T helper cell differentiation by controlling noncanonical NF-kappaB activation through Raf-1 and Syk. *Nat Immunol* (2009) 10:203–13. doi:10.1038/ni.1692
152. Bowman SM, Free SJ. The structure and synthesis of the fungal cell wall. *Bioessays* (2006) 28:799–808. doi:10.1002/bies.20441
153. Werner JL, Metz AE, Horn D, Schoeb TR, Hewitt MM, Schwiebert LM, et al. Requisite role for the dectin-1 beta-glucan receptor in pulmonary defense against *Aspergillus fumigatus*. *J Immunol* (2009) 182:4938–46. doi:10.4049/jimmunol.0804250
154. Taylor PR, Brown GD, Reid DM, Willment JA, Martinez-Pomares L, Gordon S, et al. The beta-glucan receptor, dectin-1, is predominantly expressed on the surface of cells of the monocyte/macrophage and neutrophil lineages. *J Immunol* (2002) 169:3876–82. doi:10.4049/jimmunol.169.7.3876
155. Willment JA, Marshall AS, Reid DM, Williams DL, Wong SY, Gordon S, et al. The human beta-glucan receptor is widely expressed and functionally equivalent to murine Dectin-1 on primary cells. *Eur J Immunol* (2005) 35:1539–47. doi:10.1002/eji.200425725
156. Cooper LJ, Topp MS, Serrano LM, Gonzalez S, Chang WC, Naranjo A, et al. T-cell clones can be rendered specific for CD19: toward the selective augmentation of the graft-versus-B-lineage leukemia effect. *Blood* (2003) 101:1637–44. doi:10.1182/blood-2002-07-1989
157. Singh H, Manuri PR, Olivares S, Dara N, Dawson MJ, Huls H, et al. Redirecting specificity of T-cell populations for CD19 using the Sleeping Beauty system. *Cancer Res* (2008) 68:2961–71. doi:10.1158/0008-5472.CAN-07-5600
158. Ouyang W, Kolls JK, Zheng Y. The biological functions of T helper 17 cell effector cytokines in inflammation. *Immunity* (2008) 28:454–67. doi:10.1016/j.immuni.2008.03.004
159. Zelante T, De Luca A, Bonifazi P, Montagnoli C, Bozza S, Moretti S, et al. IL-23 and the Th17 pathway promote inflammation and impair antifungal immune resistance. *Eur J Immunol* (2007) 37:2695–706. doi:10.1002/eji.200737409
160. Ivanov II, McKenzie BS, Zhou L, Tadokoro CE, Lepelley A, Lafaille JJ, et al. The orphan nuclear receptor RORgammat directs the differentiation program of proinflammatory IL-17^{+} T helper cells. *Cell* (2006) 126:1121–33. doi:10.1016/j.cell.2006.07.035
161. Ivanov II, Zhou L, Littman DR. Transcriptional regulation of Th17 cell differentiation. *Semin Immunol* (2007) 19:409–17. doi:10.1016/j.smim.2007.10.011
162. Yang XO, Pappu BP, Nurieva R, Akimzhanov A, Kang HS, Chung Y, et al. T helper 17 lineage differentiation is programmed by orphan nuclear receptors ROR alpha and ROR gamma. *Immunity* (2008) 28:29–39. doi:10.1016/j.immuni.2007.11.016
163. Lamaris GA, Lewis RE, Chamilos G, May GS, Safdar A, Walsh TJ, et al. Caspofungin-mediated beta-glucan unmasking and enhancement of human polymorphonuclear neutrophil activity against *Aspergillus* and non-*Aspergillus* hyphae. *J Infect Dis* (2008) 198:186–92. doi:10.1086/589305
164. Ogonek J, Kralj Juric M, Ghimire S, Varanasi PR, Holler E, Greinix H, et al. Immune reconstitution after allogeneic hematopoietic stem cell transplantation. *Front Immunol* (2016) 7:507. doi:10.3389/fimmu.2016.00507
165. Bonifant CL, Jackson HJ, Brentjens RJ, Curran KJ. Toxicity and management in CAR T-cell therapy. *Mol Ther Oncolytics* (2016) 3:16011. doi:10.1038/mto.2016.11
166. Pana ZD, Roilides E, Warris A, Groll AH, Zaoutis T. Epidemiology of invasive fungal disease in children. *J Pediatric Infect Dis Soc* (2017) 6:S3–11. doi:10.1093/jpids/pix046
167. Bochennek K, Hassler A, Perner C, Gilfert J, Schoning S, Klingebiel T, et al. Infectious complications in children with acute myeloid leukemia: decreased mortality in multicenter trial AML-BFM 2004. *Blood Cancer J* (2016) 6:e382. doi:10.1038/bcj.2015.110
168. Pappanaicken RK, Nathaniel A, Harjeet S, Simon O, Sourindra NM, Tiejuan M, et al. Bioengineered Dectin-1 CAR^{+} T cells to control invasive fungal infection. *Inaugural International Cancer Immunotherapy Conference*. New York (2016). Abstract A193.
169. Brudno JN, Kochenderfer JN. Toxicities of chimeric antigen receptor T cells: recognition and management. *Blood* (2016) 127:3321–30. doi:10.1182/blood-2016-04-703751
170. Neelapu SS, Tummala S, Kebriaei P, Wierda W, Gutierrez C, Locke FL, et al. Chimeric antigen receptor T-cell therapy – assessment and management of toxicities. *Nat Rev Clin Oncol* (2018) 15:47–62. doi:10.1038/nrclinonc.2017.148
171. DiGiusto D, Cooper L. Preparing clinical grade Ag-specific T cells for adoptive immunotherapy trials. *Cytotherapy* (2007) 9:613–29. doi:10.1080/14653240701650320
172. Yee C. Adoptive T-cell therapy for cancer: boutique therapy or treatment modality? *Clin Cancer Res* (2013) 19:4550–2. doi:10.1158/1078-0432.CCR-13-1367
173. Casati A, Varghaei-Nahvi A, Feldman SA, Assenmacher M, Rosenberg SA, Dudley ME, et al. Clinical-scale selection and viral transduction of human na < ve and central memory CD8($^{+}$) T cells for adoptive cell therapy of cancer patients. *Cancer Immunol Immunother* (2013) 62:1563–73. doi:10.1007/s00262-013-1459-x
174. Leen AM, Bollard CM, Mendizabal AM, Shpall EJ, Szabolcs P, Antin JH, et al. Multicenter study of banked third-party virus-specific T cells to treat severe viral infections after hematopoietic stem cell transplantation. *Blood* (2013) 121:5113–23. doi:10.1182/blood-2013-02-486324
175. Park TS, Rosenberg SA, Morgan RA. Treating cancer with genetically engineered T cells. *Trends Biotechnol* (2011) 29:550–7. doi:10.1016/j.tibtech.2011.04.009
176. Cruz CRY, Micklethwaite KP, Savoldo B, Ramos CA, Lam S, Ku S, et al. Infusion of donor-derived CD19-redirected virus-specific T cells for B-cell malignancies relapsed after allogeneic stem cell transplant: a phase 1 study. *Blood* (2013) 122:2965–73. doi:10.1182/blood-2013-06-506741
177. Barker JN, Doubrovina E, Sauter C, Jaroscak JJ, Perales MA, Doubrovin M, et al. Successful treatment of EBV-associated posttransplantation lymphoma after cord blood transplantation using third-party EBV-specific cytotoxic T lymphocytes. *Blood* (2010) 116:5045–9. doi:10.1182/blood-2010-04-281873

10

MicroRNA Regulation of Host Immune Responses Following Fungal Exposure

Tara L. Croston[1], Angela R. Lemons[1], Donald H. Beezhold[2] and Brett J. Green[1]*

[1] *Allergy and Clinical Immunology Branch, Health Effects Laboratory Division, National Institute for Occupational Safety and Health, Centers for Disease Control and Prevention, Morgantown, WV, United States,* [2] *Health Effects Laboratory Division, National Institute for Occupational Safety and Health, Centers for Disease Control and Prevention, Morgantown, WV, United States*

***Correspondence:**
Tara L. Croston
xzu9@cdc.gov

Fungal bioaerosols are ubiquitous in the environment and human exposure can result in a variety of health effects ranging from systemic, subcutaneous, and cutaneous infections to respiratory morbidity including allergy, asthma, and hypersensitivity pneumonitis. Recent research has focused on the role of microRNAs (miRNAs) following fungal exposure and is overlooked, yet important, group of regulators capable of influencing fungal immune responses through a variety of cellular mechanisms. These small non-coding ribose nucleic acids function to regulate gene expression at the post-transcriptional level and have been shown to participate in multiple disease pathways including cancer, heart disease, apoptosis, as well as immune responses to microbial hazards and occupational allergens. Recent animal model studies have characterized miRNAs following the exposure to inflammatory stimuli. Studies focused on microbial exposure, including bacterial infections, as well as exposure to different allergens have shown miRNAs, such as miR-21, miR-146, miR-132, miR-155, and the let-7 family members, to be involved in immune and inflammatory responses. Interestingly, the few studies have assessed that the miRNA profiles following fungal exposure have identified the same critical miRNAs that have been characterized in other inflammatory-mediated and allergy-induced experimental models. Review of available *in vitro*, animal and human studies of exposures to *Aspergillus fumigatus*, *Candida albicans*, *Cryptococcus neoformans*, *Paracoccidioides brasiliensis*, and *Stachybotrys chartarum* identified several miRNAs that were shared between responses to these species including miR-125 a/b (macrophage polarization/activation), miR-132 [toll-like receptor (TLR)2-mediated signaling], miR-146a (TLR mediated signaling, alternative macrophage activation), and miR-29a/b (natural killer cell function, C-leptin signaling, inhibition of Th1 immune response). Although these datasets provide preliminary insight into the role of miRNAs in fungal exposed models, interpretation of miRNA datasets can be challenging for researchers. To assist in navigating this rapidly evolving field, the aim of this review is to describe miRNAs in the framework of host recognition mechanisms and provide initial insight into the regulatory pathways in response to fungal exposure.

Keywords: fungal exposure, microRNA, fungi, immune response, inflammatory response

---

Abbreviations: Clec7a, C-type lectin domain family 7 member A; IL, interleukin; LPS, lipopolysaccharide; RNA, ribose nucleic acid; RNAi, RNA interference; mRNA, messenger RNA; miRNA, microRNA; TLRs, Toll-like receptors.

INTRODUCTION

Fungi are ubiquitous eukaryotic microorganisms that can be prevalent in indoor, outdoor, and occupational environments. A small portion of 1.5 million fungal species estimated to exist (1) are primary or opportunistic pathogens, whereas the vast majority is ubiquitous saprophytes that obtain nutrients from organic matter. Fungi are composed of membrane bound organelles that are encased by a rigid cell wall but do not contain chlorophyll. The cell wall is composed of ergosterol, chitin, glucans, such as (1 → 3)-β-d-glucan, and mannose proteins (2). Fungal lifeforms broadly vary from unicellular yeasts to multicellular filamentous hyphae that include the production of mitotic or meiotically produced spores. In some cases, fungi are dimorphic and share both lifecycles. Upon disturbance, fungal spores can be aerosolized and in some occupational environments the airborne concentrations may exceed 1×10^5 spores/m^3 (3).

Personal exposure to fungal species has been associated with a broad variety of adverse health effects that range from pulmonary, sinus, and subcutaneous infections to respiratory morbidities that may include hypersensitivity pneumonitis, allergy, and asthma (3). Each of these health effects is dependent on the host's immune responsiveness and fungal species exposed (4). In specific geographical regions, dimorphic fungi that cause endemic mycoses exist as either a filamentous fungus in the environment or as a pathogenic yeast in the host. In the environment, the filamentous hyphae grow in soil at ambient temperatures and produce infectious spores (5, 6). Soil disturbance can aerosolize spores that can be inhaled by a mammalian host. In a process that is thermoregulated, the spores can germinate into a pathogenic yeast phase that helps these fungi avoid the hosts' immune responses. For example, *Blastomyces dermatitidis* can proliferate on the respiratory mucosa, *Histoplasma capsulatum* modulates the monocyte phagolysosome compartment, and *Coccidioides immitis* develops a large spherule containing endospores that is resistant to phagocytosis. These dimorphic fungal species as well as others, including *Paracoccidioides brasiliensis* (paracoccidioidomycosis) and *Talaromyces* (*Penicillium*) *marneffei* (talaromycosis), affect the lungs, although the latter can also affect the liver and mouth.

By contrast, opportunistic fungal pathogens consist of fungi that are environmentally ubiquitous and affect those who are immunocompromised, especially patients who have received a transplant or undergoing chemotherapy or corticosteroid therapy. Examples of fungi that are commonly implicated in opportunistic infections include, *Candida albicans* (candidiasis), *Pneumocystis jirovecii* (*Pneumocystis* pneumonia), *Cryptococcus neoformans/gattii* (cryptococcosis), and *Aspergillus fumigatus* (aspergillosis). Infections can be acquired through the inhalation of conidia or yeast depending on the species and can result in systemic mycoses. With the increase in broad-spectrum antibiotic usage and other medical and therapeutic strategies, invasive opportunistic fungal infections are of particular concern in the hospital setting, as nosocomial infections may be life-threatening for critically ill individuals (7).

The World Health Organization and the Institute of Medicine have published consensus documents that report respiratory morbidities are associated with damp indoor environments (8, 9). Recent epidemiological evidence has further built on these consensus findings and shown exposure to mold in damp indoor environments to be associated with adverse respiratory health effects (10, 11). Following recent natural disasters and flooding events associated with Hurricanes Harvey, Irma, and Maria, water-infiltrated occupational, and residential buildings are environments where mold can grow and proliferate on water damaged building materials. Returning to these environments and disturbing contaminated building materials can result in the aerosolization of fungal spores (12) that can pose a significant health risk especially if the person is immunocompromised. Fungi associated with colonizing wet building materials include, *Aspergillus versicolor*, *Ulocladium chartarum*, *Chaetomium globosum*, and *Stachybotrys chartarum* that are hydrophilic and require a high water activity for growth and proliferation. Of these hydrophilic fungal contaminants, *S. chartarum* is the most widely studied and many reports have identified exposure to contribute to negative health effects (12–15).

Due to increased community concern regarding personal exposure to these pathogenic fungi and the potential result of life-threatening health outcomes, it is important to characterize the mechanisms that contribute to the host innate and adaptive immune responses. Previous research has focused on host responses in fungal exposure models by analyzing functional, histological, and immunological endpoints; however, research examining the molecular mechanisms that underlie these responses remains unclear for many clinically relevant fungal species. Although many studies have been published that have explored pulmonary immunological responses to acute and chronic fungal spores exposures, the microRNAs (miRNAs) that regulate these deficiencies have not been fully characterized. In this review, the state-of-knowledge of miRNAs characterized in various animal models, including those that have evaluated fungal exposures, will be reviewed with emphasis placed on the mechanistic insights that these studies have provided in relation to the host response following fungal exposure.

MICRORNAs

MicroRNAs are an important group of regulators capable of influencing gene expression through different mechanisms (16–20). Consisting of short, single stranded noncoding ribose nucleic acids (RNAs), miRNAs bind to target messenger RNA (mRNA) to downregulate gene expression post-transcriptionally through RNA silencing or RNA degradation (21, 22). Depending on the complementarity of base pairing, gene expression is repressed, as observed in humans and animals, or mRNA is cleaved, as observed in plants (23–25). More recently, studies have shown that miRNA can also activate the translation of certain target mRNA (17, 18, 26). Providing insight into how altered miRNA profiles affect upstream processes can be methodologically challenging. For example, a single miRNA can regulate from one to multiple genes, whereas studies have also shown that multiple miRNAs can regulate the same gene (27–30). Several miRNAs, as well as miRNA families, have been extensively studied and have been characterized in models of cancer, heart disease, aging, apoptosis, and immune responses to inflammatory stimuli (19, 20, 31–34).

Influence of miRNAs on IL-13-Mediated Allergic Responses

The *let-7* family is the most abundant pulmonary miRNAs and has been identified in cancer, diabetes, and aging studies (35–39). The *let-7* family members have been shown to target interleukin (IL)-13 in *in vivo* and *in vitro* models, although the regulatory *in vivo* mechanisms of *let-7* are complex (20, 40). miR-21 is another widely studied miRNA and has been shown to participate in the inflammatory response elicited by different stimuli, including doxycycline-induced allergic airway inflammation (41), as well as viral, bacterial and protozoan infections (42–44). One of the most upregulated miRNAs in human patients with allergic eosinophilic esophagitis is miR-21 (45, 46), which correlates with studies that reported miR-21 and miR-223 as regulators of eosinophilic development in an *ex vivo* model of bone-derived eosinophils (47, 48). miR-375 has also been reported to be downregulated in epithelial cells derived from patients with eosinophilic esophagitis, as well as in IL-13 stimulated epithelial cells indicating the role of miR-375 as a regulator of IL-13-mediated responses (21).

miRNA Involvement in Allergy-Induced Asthma

In rodent models exposed to house dust mite allergen, increased miR-126, miR-106a, and miR-145 expression have been shown to contribute to allergic inflammation (49–51). Studies involving airborne pollutants, such as cigarette smoke, reported a down-regulation in let-7c, let-7f, miR-34b, miR-34c, and miR-222, all of which contributed to pulmonary inflammation in rodent models (52–54). Research examining aberrant miRNA profiles in asthmatic hosts has also revealed novel miRNAs that contribute to allergic airway disease. Examination of $CD4^+$ T cells isolated from the bronchoalveolar lavage fluid from asthmatic human patients revealed that miR-19a had the highest expression (55), which promoted a Th2-mediated cell response, a known response contributing to allergic asthma. In another study, miR-221 and miR-485-3p were upregulated in peripheral blood from pediatric asthmatic human patients compared with controls, suggesting that these miRNAs contribute to the development of asthma (56). In a chemical allergen model examining the murine miRNA profile following dermal exposure to toluene 2,4-diisocyanate, miR-21, miR-22, miR-27b, miR-31, miR-126, miR-155, miR-210, and miR-301a expression were increased (57). While this study identified miRNAs that are known to participate in the immune response associated with asthma (miR-21, miR-31, miR-126, and miR-155), new miRNAs were proposed as potential biomarkers for allergic sensitization to toluene 2,4-diisocyanate (miR-22, miR-27b, miR-301a, and miR-210).

miRNA Regulation on Adaptive Immunity

MicroRNAs critically influence the development and responses of the immune system, but the contributing biological mechanisms are poorly characterized (22, 58–60). Overexpression of the miR-17-92 cluster and miR-181 enhanced B-cell proliferation, while miR-150 regulated B-cell differentiation (61–64). When overexpressed, miR-181 has been shown to decrease T-cell numbers (61), but enhance T-cell receptor signaling (65). When T cells are activated, the miRNA expression profiles are altered (66–68). T-cell activation has additionally been found to induce the miR-17~92 family members (69), as well as the gene that encodes miR-155 (70). miR-155 has also been reported to regulate antigen presentation (71) and to negatively regulate toll-like receptor (TLR) and cytokine signaling (72). The miR-17-92 cluster promotes Th1 type immune responses along with inhibiting regulatory T-cell differentiation (69). Rodriquez et al. showed that miR-155 is required for normal functioning of B and T lymphocytes as well as dendritic cells (73).

Macrophage Development and TLR Signaling

In human macrophages, miR-155 targets and subsequently decreases IL-13Rα1, modulating the IL-13 pathway and the switching between classic and alternatively activated macrophages (74). Macrophage polarization is transcriptionally controlled by either miR-146b or miR-34a, directing an M1 macrophage polarization, whereas miR-18a/miR-34a, miR-130b, or miR-125-5p dictates an M2 macrophage phenotype (75). miR-21 has also been reported to direct macrophage polarization from an M1 phenotype toward an M2 phenotype (75). Alveolar macrophages isolated from a fibrotic mouse model showed significantly increased miR-let-7c levels compared with control and that overexpression of this miRNA regulated macrophage polarization (76). Expression of miR-124 and miR-223 in macrophages has also been reported to contribute to macrophage polarization (77, 78).

Located on the surface of sentinel cells, such as macrophages, TLRs play a critical role in the innate immune system by recognizing pathogen-associated molecular patterns expressed on pathogens and signaling for the production of cytokine to elicit an immune response. These TLRs participate in macrophage activation and have been shown to induce miR-155, miR-146, miR-147, miR-9, and miR-21 (79, 80). An upregulation of miR-21 has been observed in both primary human airway epithelial cells (41) and in an IL-13 transgenic mouse model with the latter study identifying that the observed miR-21 upregulation was through an IL-13Rα1-dependent mechanism (81). This increase in miR-21 was also associated with inhibited Th1 cytokine signaling (41). Using an ovalbumin-induced miR-21 deficient mouse model, Th1 cytokines were found to be increased (82), supporting the contribution of miR-21 in Th2 type immune responses. One study confirmed that miR-21 expression inhibited murine pulmonary inflammation by suppressing TLR2 signaling (81). When secreted from tumor cells, miR-21 and miR-29a have also been reported to interact with TLRs, specifically TLR7 and TLR8, respectively (83). Upon lipopolysaccharide (LPS) stimulation, miR-146a/b was shown to be induced and predicted to negatively regulate TLR and cytokine signaling (72).

Influence of miRNA on T-Helper Cell Responses

Macrophage surface activation is induced in the presence of over-expressed miR-125b (84) and in an eosinophilic rhinosinusitis animal model, miR-125b is increased resulting in increased interferon gamma and a Th1 type immune response (85). miR-19a has also been shown to be critical in regulating Th1 type responses through the production of interferon gamma following antigen stimulation in a mouse model (69). Upregulation of miR-19a also caused increased inflammation and promoted a Th2 type response (55). miR-19a is a member of the miR-17–92 cluster, which has been reported to be upregulated during T-cell activation (69, 86). Th17 cell differentiation has also been shown to be regulated by the miR-106-363 cluster (87) and in an experimental autoimmune encephalomyelitis model, Th17 cell-mediated inflammation was shown to be induced by both miR-326 and miR-21 (88, 89).

In summary, multiple studies have characterized the role of miRNAs on immune processes in a variety of diverse animal models of inflammation, but few studies have evaluated the miRNA profiles following fungal exposures (**Figure 1**). Investigation of these miRNA profiles could provide insight into the immune mechanisms and regulatory pathways involved in the host response to fungal exposure.

FUNGAL EXPOSURE: ROLE OF miRNAs

Several research studies have focused on the miRNA profiles following acute or chronic fungal exposures (**Figure 1**). **Table 1** shows 10 studies that have preliminarily characterized differentially expressed miRNAs following exposure to five clinically relevant fungal species including *A. fumigatus* (22, 90, 91), *C. albicans* (22, 92–94), *C. neoformans* (95), *P. brasiliensis* (96, 97), and *S. chartarum* (98). The paucity of research investigating the regulation of miRNAs on pulmonary and systemic responses to fungal exposure highlights the need for research examining the role miRNAs play in the immunological mechanisms associated with endemic, opportunistic, and environmental fungal exposures.

miRNA Profiles following *P. brasiliensis* Infection

Paracoccidioidomycosis, caused by the dimorphic fungus *P. brasiliensis*, is a public health burden in Latin America (117). This fungus can be isolated in the form of yeast from infected individuals and armadillos, and has also been sporadically isolated from soil, dog food, and bat feces (118, 119). The disease begins with the inhalation of spores into the lungs that germinate into yeast and cause a primary lung infection or disseminate systemically resulting in oral and cutaneous lesions. To date, two studies have evaluated differentially expressed miRNAs following *P. brasiliensis* exposure in a murine model and in a human model. Turini Gonzales Marioto et al. (97) evaluated the miRNA profiles in mice intravenously administered *P. brasiliensis* and showed that the most upregulated miRNAs at 28 days included miR-126a-5p, miR-340-5p, miR-30b-5p, miR-19b-3p, miR-221-3p, miR-20a-5p, miR-130a-3p, and miR-301a-3p, whereas after 56 days, miRNAs from the let-7 family, as well as miR-26b-5p, and miR-369-3p were the greatest upregulated miRNAs (97). The only miRNA that was upregulated at both time points was miR-466k (**Table 1**). This study identified differentially expressed miRNAs that are known to contribute to the immune response through T cell function and proliferation, as well as monocyte and erythrocyte differentiation. The contribution of miR-466k on the immune response is unknown; however, this miRNA has been identified in prostate cancer and graft rejection studies (120, 121). Another study examined the miRNA profile in the serum of human patients infected with *P. brasiliensis* and found that of the 752 miRNAs analyzed, 8 were differentially expressed (96). The upregulated miRNAs included miR-132-3p, miR-604, miR-186-5p, miR-29b-3p, miR-125b-5p, miR-376c-3p, and miR-30b-5p, where the only downregulated miRNA was miR-423-3p

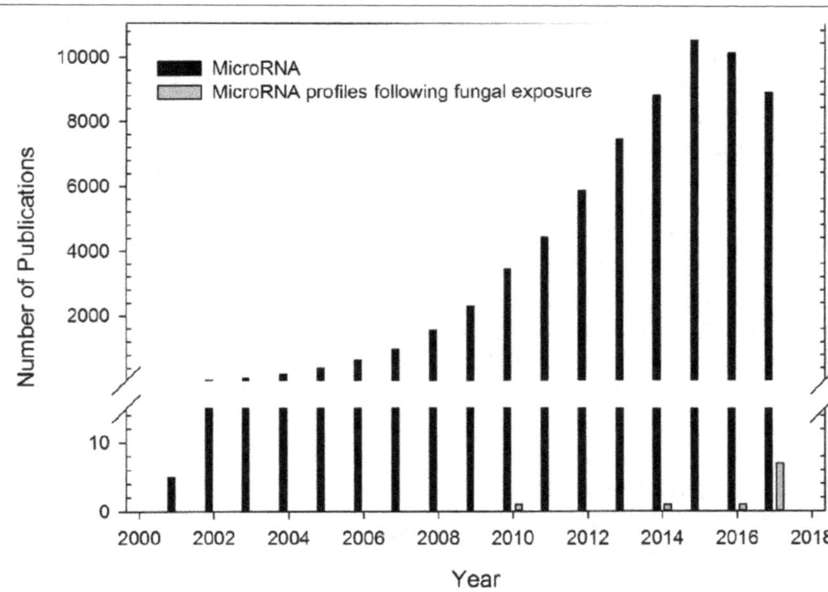

FIGURE 1 | Number of microRNA (miRNA) publications. Number of publications examining miRNA profiles from different diseased models (black bars) or following fungal exposure (gray bars). Data obtained from PubMed search using term "microRNA" for publications using different diseased models. One study in 1972 was not included. Number of publications using a fungal exposed model were obtained from studies included in this review.

TABLE 1 | Differentially expressed microRNAs following fungal exposure.

miRNA	Regulation	Fungal species	Exposure model	Function in immune response[a]	Reference
let-7 family	↑	*Paracoccidioides brasiliensis*	Inv	Regulator of TLR mediated signaling; Involved in IL-13 production	(97, 99)
miR-125a	↑	*Candida albicans*	Cc	Regulator of macrophage polarization; Enhances classical activation of macrophages	(92, 100)
miR-125b	↑	*P. brasiliensis*	Hu		(96, 101)
miR-126a	↑	*P. brasiliensis*	Inv	Promoter of Th2 immune response	(49, 97)
miR-129	↑	*C. albicans* with *Aspergillus fumigatus*	Cc	Regulation of cell cycle	(22, 102)
miR-130a	↑	*P. brasiliensis*	Inv	Involved in CD8⁺T cell activation	(97, 103)
miR-132	↑	*P. brasiliensis*	Hu	Regulator of TLR2 mediated signaling; induces alternative activation of macrophages	(96, 104)
	↑	*A. fumigatus*	Cc		(90, 101)
	↑	*C. albicans* with *A. fumigatus*	Cc		(22)
miR-146a	↑	*C. albicans*	Cc	Negative regulator of TLR mediated signaling; Induces alternative activation of macrophages	(92, 94)
	↑	*Cryptococcus neoformans*	Cc		(95, 101, 105)
miR-155	↑	*C. albicans*	Cc	Regulator of TLR mediated signaling; Enhances classical activation of macrophages; Promoter of Th2 immune response	(92, 93, 100, 101, 105)
miR-186	↑	*P. brasiliensis*	Hu	Involved in TLR2 mediated signaling	(96, 106)
miR-19b	↑	*P. brasiliensis*	Inv	Promoter of Th17 immune response; T cell proliferation	(97, 107)
miR-20a	↑	*P. brasiliensis*	Inv	Negatively regulates monocyte differentiation	(97, 108)
miR-210	↑	*C. albicans*	Cc	Involved in Th1 and Th17 cell differentiation	(94, 109)
miR-21a	↑	*Stachybotrys chartarum*	Inh	Regulator of TLR mediated signaling; Involved in monocyte, dendritic, macrophage, and Th2 cell differentiation; Involved in macrophage activation	(98, 109, 110)
miR-212	↑	*C. albicans* with *A. fumigatus*	Cc	Involved in B-cell development and Th17 cell differentiation	(22, 111)
miR-23	↓	*A. fumigatus*	Inh	Involved in cell differentiation of B cells; Mediate macrophage polarization	(91, 109)
miR-26b	↑	*P. brasiliensis*	Inv	Increases regulatory T cells	(97, 107)
miR-29a	↓	*A. fumigatus*	Inh	Regulator of natural killer cell function; Involved in C-leptin signaling; Inhibition of Th1 immune response	(91, 112)
miR-29b	↑	*P. brasiliensis*	Hu		(96, 100)
miR-30	↑	*C. albicans*	Cc	Involved in natural killer cell function and B cell activation; Involved in cell function of cytotoxic T cells	(94, 109)
miR-30b	↑	*P. brasiliensis*	Inv, Hu		(96, 97)
miR-301a	↑	*P. brasiliensis*	Inv	Promoter of Th17 immune response; Regulator of TLR mediated signaling	(97, 100, 113)
miR-340	↑	*P. brasiliensis*	Inv	Unknown	(97)
miR-369	↑	*P. brasiliensis*	Inv	Activation of protein translation	(26, 97)
miR-376c	↑	*P. brasiliensis*	Hu	Unknown	(96)
miR-423	↓	*P. brasiliensis*	Hu	Regulator of chemokine expression	(96, 114)
miR-455	↑	*C. albicans*	Cc	Involved in nuclear factor-kappaB signaling	(92, 115)
miR-466k	↑	*P. brasiliensis*	Inv	Unknown	(97)
miR-604	↑	*P. brasiliensis*	Hu	Unknown	(96)
miR-706	↑	*S. chartarum*	Inh	Promotes granulocyte production	(98, 116)

[a]*Immune responses are compiled from different studies utilizing a variety of diseased models and not necessarily from fungal exposure studies. Exposure model abbreviations are Inv, Intravenous administration or through; Inh, Inhalation using a mouse model; Cc, cell culture; Hu, Human patients.*

(**Table 1**). These miRNAs are known to mediate macrophage polarization or are involved in TLR2 signaling, indicative of a Th1 immune response. Interestingly, both studies reported an upregulation in miR-30b-5p, suggesting a possible biomarker for *P. brasiliensis* infection.

miR-132 Is Induced by *A. fumigatus* Exposure

Aspergillus fumigatus is a commonly encountered pathogenic fungal species and is often found in the soil, occupational environments [i.e., biowaste containment facilities (122, 123)] or indoor environments [i.e., hospitals (124, 125)]. Inhalation of *A. fumigatus* unicellular spores can result in varying degrees of infection, known as aspergillosis, depending on the preexisting conditions of the host. miR-132 has been shown to be induced in human monocytes and dendritic cells following stimulation with *A. fumigatus* compared with control, LPS (90). These datasets suggest a Th2-mediated response, which is further supported by other recent animal models of inhalation exposure to *A. fumigatus* (91).

Upregulation of miR-146 in *Candida* and *C. neoformans*

Candidiasis, an infection caused by several endogenous *Candida* species, results in varying symptoms depending on the site of infection (126, 127). Candidiasis is among the most common opportunistic fungal infections localized in the gastrointestinal tract (thrush), occluded regions of the hands, feet, and groin, or can develop into invasive candidiasis and disseminate systemically in the blood (candidemia), heart, brain, eyes, and bones. Invasive candidiasis is the most common type of fungal infection in critically ill patients, with approximately 46,000 healthcare-associated cases occurring each year in the United

States (126–128). Although *Candida* infections are typically resolved by antifungal therapy, some *Candida* species are resistant or are becoming resistant, such as *C. auris* (129). In murine macrophages stimulated by 10^6 cells/mL heat killed *C. albicans,* miR-146a and miR-155, as well as miR-455 and miR-125a were upregulated (92), indicative of the involvement of these miRNAs in macrophage polarization.

Similar to the findings of Monk et al. (92), miR-146a was also shown to be upregulated in human monocytic THP-1 cells exposed to *C. neoformans*, inhibiting nuclear factor-κB activation and the release of inflammatory cytokines (95). *C. neoformans* is one of two pathogenic *Cryptococcus* species, and along with *C. gatti,* typically live in the soil surrounding trees and are capable of causing infection following inhalation. Exposure to *Cryptococcus* species usually causes adverse respiratory health effects; however, it can affect other parts of the body such as the brain, known as cryptococcal meningitis (130, 131).

miRNA Profiles following a Mixed Fungal Exposure

In a model of mixed fungal exposure, Dix et al. co-cultured human monocyte-derived dendritic cells with *A. fumigatus, C. albicans,* and LPS and showed that differentially expressed miRNAs were increased after 6 and 12 h, with a stronger regulation observed after 12 h (22). Twenty six miRNAs were identified to be differently expressed in response to the exposure. The authors also showed that strongly modified miRNAs after exposure to fungi clustered separately from the strongly modified miRNAs exposed to LPS. This clustering pattern suggests that examination of miRNA profiles could distinguish between fungal and bacterial exposure. For example, miR-132 and miR-212-5p were specific to fungal exposure at 6 h time point, whereas miR-132, miR-212, and miR-129-5p were specific to fungal exposure at 12 h time point.

C-Leptin Receptors and Associated miRNAs

Critical to antifungal innate immunity, Dectin-1 is a surface receptor that recognizes (1,3)-β-D-glucan found on the cell wall of germinating conidia (91, 132). To date, several studies have attempted to explore the regulatory mechanisms involving miRNAs that underlie Dectin-1 associated immune responses. In a murine model of subchronic *A. fumigatus* inhalation exposure, Croston et al. showed that significantly downregulated miR-29a-3p was predicted to regulate C-type lectin domain family 7 member A, the gene that codes for Dectin-1 (91). A recent study found that following exposure to *C. albicans*, Dectin-1 is required for the upregulation of miR-155 in murine macrophages (93). Along with an increase in miR-30-5p and miR-210-3p in THP-1 cells treated with β-glucan isolated from *C. albicans*, Du et al. found that miR-146a was increased upon Dectin-1 stimulation and negatively regulated the resultant inflammatory response (94). Results from the Croston et al. study using a murine model of subchronic *A. fumigatus* inhalation exposure also determined that significantly downregulated miR-23b-3p and miR-145a-5p was predicted to regulate the mannose receptor gene, *Mrc1* (data not reported).

miRNA Profiles following Exposure to Occupationally Relevant Fungal Species

In order to elucidate the influence of germination on the ensuing immune response, Croston et al. utilized an acoustical generator system to deliver dry fungal spores to mice housed in a multi-animal nose-only inhalation chamber (91). This murine inhalation model reproduces exposures that could be encountered in contaminated indoor or occupational environments (133). Furthermore, this study included a heat inactivated conidia group that was used as a biological control to examine the influence of germination on the miRNA profiles. Along with a downregulation of miR-29a-3p, miR-23b-3p, a miRNA predicted to target *SMAD2,* as well as genes involved in IL-13 and IL-33 responses, was also downregulated following subchronic exposure to a dry aerosol containing viable *A. fumigatus* conidia (91). Furthermore, out of 415 miRNAs detected, approximately 50% were altered in mice exposed to viable versus heat inactivated conidia 48 h post fungal exposure. Taken together, these results demonstrate that *A. fumigatus* germination is an important variable that can lead to the induction of allergic inflammation in the lungs, potentially through an IL-13/IL-33-driven mechanism.

Stachybotrys chartarum is a hydrophilic fungal species prevalent in water infiltrated occupational and residential environments. Exposure to *S. chartarum* is currently of heightened public health interest following recent natural disasters, such as floods and hurricanes, that can lead to the contamination of indoor building materials. Following the consensus reports published by the IOM and WHO, the immunological mechanisms that contribute to the host response to fungal contaminant exposure require further elucidation. Recent studies have attempted to characterize critical interactions that influence these pulmonary immunological responses. Croston et al. found that miR-21a was the only miRNA upregulated in murine whole lung homogenate 48 h following subchronic exposure to *S. chartarum* (98) (**Figure 2**). Although miR-21a is known to promote a Th2 phenotype, a more dominant Th1 phenotype was evident. Since then, miR-706 was also discovered to be upregulated at the same time point. Interestingly, out of 468 miRNA evaluated, only 2 were upregulated with no downregulated miRNAs. These preliminary results suggest that miRNA regulation mechanisms induced by *S. chartarum* vary from *A. fumigatus* in these studies using the same exposure system and the pulmonary immunological responses to this species require further evaluation.

Figure 2 depicts a disease pathway generated by Ingenuity Pathway Analysis (IPA) that includes predicted miRNAs involved in the inflammation of organs. Once miRNA data are uploaded into IPA, the integrated knowledge base predicts associations between miRNAs from the dataset and different disease pathways or biological functions. These predictions are primarily based on previously publish datasets derived from a broad diversity of animal models. The miRNA dataset included in **Figure 2** was obtained from murine whole lung homogenate 48 h following subchronic exposure to *S. chartarum* (98). When analyzing the top diseases and biological functions of the miRNAs included in the dataset, a handful were predicted to be involved in an inflammatory response, specifically in organ inflammation,

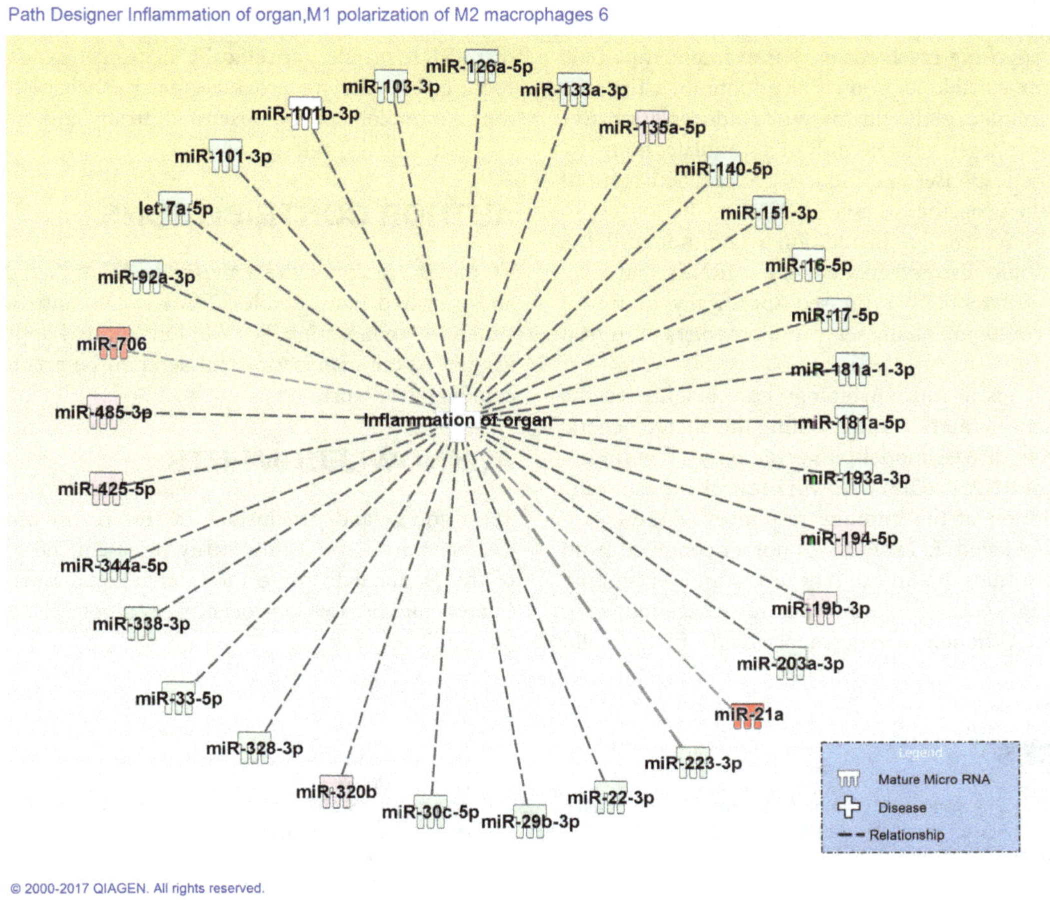

FIGURE 2 | Disease network map generated by Ingenuity Pathway Analysis depicting miRNAs involved in the inflammation of organs. miRNAs are color-coded (red or green for up- and downregulation, respectively) for the expression of miRNAs in *Stachybotrys chartarum* exposed versus control at 13 weeks, 48 h post-exposure. Gray dotted lines represent predicted regulated relationship.

illustrated in **Figure 2**. The miRNAs are color-coded depending on the respective expression level (red or green for up- and down-regulation, respectively). The absence of confirmed associations between miRNAs, evidenced by gray dotted lines, supports the lack of miRNA profile studies following fungal exposure.

To date, only a handful of studies have examined the altered miRNA profiles following fungal exposure; therefore, more researches are required to fully understand the mechanistic influence miRNAs have on the immune response. With the increased interest in studying miRNAs, methodological approaches are becoming more advanced by using next-generation sequencing methods that examine miRNA profiles in more depth and at a higher precision compared with miRNA arrays. Once the more influential miRNAs are identified, strategies can be developed in order to manipulate the host response. With the use of transgenic or knockout animal models, the functionality of miRNAs or genes can be elucidated; however, the manipulation of the genome may in fact alter normal miRNA production or function, contributing to the phenotype of the disease (134). Targeting miRNAs that are upregulated or replacing the expression of miRNAs that are downregulated are potential strategies that could be tested in animal models as a new therapeutic strategy to treat fungal infections and diseases. This targeting strategy could be completed through the use of an anti-miR or a miRNA mimic (135), and may allow for the manipulation of a group of genes or proteins that participate in the progression of the infection or disease. Ultimately, these targeting strategies will help bridge the knowledge gap between the identification of miRNAs and the host responses to fungal exposure, potentially leading to advanced therapeutics to combat adverse effects resulting from exposure to pathogenic fungi.

CONCLUSION

In this review, the identification and influence of miRNAs on the host immune responses following fungal exposure were examined. Compared with existing datasets examining miRNA profiles in allergic and inflammatory models, some common differentially expressed miRNAs were identified in fungal exposed models. Influential miRNAs altered in different disease models, such as miR-132, functions to maintain a normal hematopoietic output during an immune response and regulates genes at the beginning of an immune response to regain homeostasis of the immune system. Other common miRNAs identified in multiple inflammatory disease models, including miR-155 and miR-146a, regulates critical genes involved in the host defense system through opposing mechanisms. For example, miR-146a is known to decrease cytokine production and inhibits Th1 cells following an inflammatory stimulus, as well as induces alternative activation of macrophages, whereas miR-155 stimulates both Th1 and Th17 immune responses and induces classical activation of macrophages. Taken together, these miRNAs act in concert to defend the host from infection. In addition to common miRNAs identified in multiple diseased models, the miRNAs that were observed to be differentially expressed specifically in fungal exposed models could potentially serve as biomarkers for fungal exposures.

Recent discoveries in miRNA biology have heightened the research community's interest in examining the altered genetic profiles in different disease models; however, only a few studies have examined miRNA profiles following fungal exposure. As such, the description of the immune responses to the corresponding miRNAs listed in **Table 1** was not all compiled from fungal exposure studies due to the lack of research examining the influence of miRNAs on the immune responses following fungal exposure. Although advancements made in this field have helped elucidate mechanisms underlying host responses to a variety of infections and diseases, further examination of miRNA profiles, specifically in fungal exposed models, is required in order to provide greater mechanistic insight into the immunological response to clinically and environmentally relevant fungal species.

AUTHOR CONTRIBUTIONS

TC and BG designed the manuscript; TC drafted the manuscript and prepared figures/tables; TC, AL, DB, and BG revised and edited the manuscript; TC, AL, DB, and BG approved the final version of the manuscript and agree to be accountable for the content of the work.

ACKNOWLEDGMENTS

The findings and conclusions in this report are those of the author(s) and do not necessarily represent the official position of the National Institute for Occupational Safety and Health, Centers for Disease Control and Prevention. The author declares no conflict of interest.

REFERENCES

1. Hawksworth DL. The magnitude of fungal diversity: the 1.5 million species estimate revised. *Mycol Res* (2001) 105:1422–32. doi:10.1017/S0953756201004725
2. Cannon R, Firth N. Fungi and fungal infections of the oral cavity. In: Lamont R, Burne R, Lantz M, LeBlanc D, editors. *Oral Microbiology and Immunology* Washington, DC: ASM Press (2006). p. 333–48.
3. Eduard W. Fungal spores: a critical review of the toxicological and epidemiological evidence as a basis for occupational exposure limit setting. *Crit Rev Toxicol* (2009) 39(10):799–864. doi:10.3109/10408440903307333
4. Benedict K, Park BJ. Invasive fungal infections after natural disasters. *Emerg Infect Dis* (2014) 20(3):349–55. doi:10.3201/eid2003.131230
5. Klein BS, Tebbets B. Dimorphism and virulence in fungi. *Curr Opin Microbiol* (2007) 10(4):314–9. doi:10.1016/j.mib.2007.04.002
6. Gauthier GM. Dimorphism in fungal pathogens of mammals, plants, and insects. *PLoS Pathog* (2015) 11(2):e1004608. doi:10.1371/journal.ppat.1004608
7. Bajwa SJ, Kulshrestha A. Fungal infections in intensive care unit: challenges in diagnosis and management. *Ann Med Health Sci Res* (2013) 3(2):238–44. doi:10.4103/2141-9248.113669
8. IoM. *Damp Indoor Spaces and Health.* Washington, DC, USA: National Academies Press (2004).
9. WHO. *WHO Guidelines for Indoor Air Quality: Dampness and Mould.* Geneva: WHO Regional Office for Europe (2009).
10. Mendell MJ, Mirer AG, Cheung K, Tong M, Douwes J. Respiratory and allergic health effects of dampness, mold, and dampness-related agents: a review of the epidemiologic evidence. *Environ Health Perspect* (2011) 119(6):748–56. doi:10.1289/ehp.1002410
11. Quansah R, Jaakkola MS, Hugg TT, Heikkinen SA, Jaakkola JJ. Residential dampness and molds and the risk of developing asthma: a systematic review and meta-analysis. *PLoS One* (2012) 7(11):e47526. doi:10.1371/journal.pone.0047526
12. Kuhn DM, Ghannoum MA. Indoor mold, toxigenic fungi, and *Stachybotrys chartarum*: infectious disease perspective. *Clin Microbiol Rev* (2003) 16(1):144–72. doi:10.1128/CMR.16.1.144-172.2003
13. Johanning E, Biagini R, Hull D, Morey P, Jarvis B, Landsbergis P. Health and immunology study following exposure to toxigenic fungi (*Stachybotrys chartarum*) in a water-damaged office environment. *Int Arch Occup Environ Health* (1996) 68(4):207–18. doi:10.1007/s004200050052
14. Andersson MA, Nikulin M, Koljalg U, Andersson MC, Rainey F, Reijula K, et al. Bacteria, molds, and toxins in water-damaged building materials. *Appl Environ Microbiol* (1997) 63(2):387–93.
15. Pestka JJ, Yike I, Dearborn DG, Ward MD, Harkema JR. *Stachybotrys chartarum*, trichothecene mycotoxins, and damp building-related illness: new insights into a public health enigma. *Toxicol Sci* (2008) 104(1):4–26. doi:10.1093/toxsci/kfm284
16. Bartel DP. MicroRNAs: genomics, biogenesis, mechanism, and function. *Cell* (2004) 116(2):281–97. doi:10.1016/S0092-8674(04)00045-5
17. Li L-C, Okino ST, Zhao H, Pookot D, Place RF, Urakami S, et al. Small dsRNAs induce transcriptional activation in human cells. *Proc Natl Acad Sci U S A* (2006) 103(46):17337–42. doi:10.1073/pnas.0607015103
18. Janowski BA, Younger ST, Hardy DB, Ram R, Huffman KE, Corey DR. Activating gene expression in mammalian cells with promoter-targeted duplex RNAs. *Nat Chem Biol* (2007) 3(3):166–73. doi:10.1038/nchembio860
19. Place RF, Li L-C, Pookot D, Noonan EJ, Dahiya R. MicroRNA-373 induces expression of genes with complementary promoter sequences. *Proc Natl Acad Sci U S A* (2008) 105(5):1608–13. doi:10.1073/pnas.0707594105
20. Lu TX, Rothenberg ME. Diagnostic, functional, and therapeutic roles of microRNA in allergic diseases. *J Allergy Clin Immunol* (2013) 132(1):3–13; quiz 14. doi:10.1016/j.jaci.2013.04.039
21. Lu TX, Lim E-J, Wen T, Plassard AJ, Martin LJ, Aronow BJ, et al. miR-375 is down-regulated in epithelial cells after IL-13 stimulation and regulates an IL-13 induced epithelial transcriptome. *Mucosal Immunol* (2012) 5(4):388–96. doi:10.1038/mi.2012.16
22. Dix A, Czakai K, Leonhardt I, Schäferhoff K, Bonin M, Guthke R, et al. Specific and novel microRNAs are regulated as response to fungal infection in human dendritic cells. *Front Microbiol* (2017) 8:270. doi:10.3389/fmicb.2017.00270
23. Lewis BP, Shih IH, Jones-Rhoades MW, Bartel DP, Burge CB. Prediction of mammalian microRNA targets. *Cell* (2003) 115(7):787–98. doi:10.1016/S0092-8674(03)01018-3
24. Lewis BP, Burge CB, Bartel DP. Conserved seed pairing, often flanked by adenosines, indicates that thousands of human genes are microRNA targets. *Cell* (2005) 120(1):15–20. doi:10.1016/j.cell.2004.12.035
25. Carthew RW, Sontheimer EJ. Origins and mechanisms of miRNAs and siRNAs. *Cell* (2009) 136(4):642–55. doi:10.1016/j.cell.2009.01.035

26. Vasudevan S, Tong Y, Steitz JA. Switching from repression to activation: microRNAs can up-regulate translation. *Science* (2007) 318(5858):1931–4. doi:10.1126/science.1149460
27. Krek A, Grun D, Poy MN, Wolf R, Rosenberg L, Epstein EJ, et al. Combinatorial microRNA target predictions. *Nat Genet* (2005) 37(5):495–500. doi:10.1038/ng1536
28. Rajewsky N. MicroRNA target predictions in animals. *Nat Genet* (2006) 38:S8–13. doi:10.1038/ng1798
29. Bartel DP. MicroRNA target recognition and regulatory functions. *Cell* (2009) 136(2):215–33. doi:10.1016/j.cell.2009.01.002
30. Friedman RC, Farh KK-H, Burge CB, Bartel DP. Most mammalian mRNAs are conserved targets of microRNAs. *Genome Res* (2009) 19(1):92–105. doi:10.1101/gr.082701.108
31. van Rooij E, Olson EN. MicroRNAs: powerful new regulators of heart disease and provocative therapeutic targets. *J Clin Invest* (2007) 117(9):2369–76. doi:10.1172/JCI33099
32. Hackl M, Brunner S, Fortschegger K, Schreiner C, Micutkova L, Muck C, et al. miR-17, miR-19b, miR-20a, and miR-106a are down-regulated in human aging. *Aging Cell* (2010) 9(2):291–6. doi:10.1111/j.1474-9726.2010.00549.x
33. Leopold JA, Maron BA. Molecular mechanisms of pulmonary vascular remodeling in pulmonary arterial hypertension. *Int J Mol Sci* (2016) 17(5):761. doi:10.3390/ijms17050761
34. Posch W, Heimdörfer D, Wilflingseder D, Lass-Flörl C. Invasive candidiasis: future directions in non-culture based diagnosis. *Expert Rev Anti Infect Ther* (2017) 15:829–38. doi:10.1080/14787210.2017.1370373
35. Frost RJA, Olson EN. Control of glucose homeostasis and insulin sensitivity by the Let-7 family of microRNAs. *Proc Natl Acad Sci U S A* (2011) 108(52):21075–80. doi:10.1073/pnas.1118922109
36. Jen-Liang S, Pai-Sheng C, Gunnar J, Min-Liang K. Function and regulation of let-7 family microRNAs. *Microrna* (2012) 1(1):34–9. doi:10.2174/2211536611201010034
37. Jun-Hao ET, Gupta RR, Shyh-Chang N. Lin28 and let-7 in the metabolic physiology of aging. *Trends Endocrinol Metab* (2016) 27(3):132–41. doi:10.1016/j.tem.2015.12.006
38. Brennan E, Wang B, McClelland A, Mohan M, Marai M, Beuscart O, et al. Protective effect of let-7 miRNA family in regulating inflammation in diabetes-associated atherosclerosis. *Diabetes* (2017) 66(8):2266. doi:10.2337/db16-1405
39. Pal AS, Kasinski AL. Animal models to study microRNA function. *Adv Cancer Res* (2017) 135:53–118. doi:10.1016/bs.acr.2017.06.006
40. Polikepahad S, Knight JM, Naghavi AO, Oplt T, Creighton CJ, Shaw C, et al. Proinflammatory role for let-7 microRNAS in experimental asthma. *J Biol Chem* (2010) 285(39):30139–49. doi:10.1074/jbc.M110.145698
41. Lu TX, Munitz A, Rothenberg ME. MicroRNA-21 is up-regulated in allergic airway inflammation and regulates IL-12p35 expression. *J Immunol* (2009) 182(8):4994–5002. doi:10.4049/jimmunol.0803560
42. Zhou R, Hu G, Liu J, Gong AY, Drescher KM, Chen XM. NF-kappaB p65-dependent transactivation of miRNA genes following *Cryptosporidium parvum* infection stimulates epithelial cell immune responses. *PLoS Pathog* (2009) 5(12):e1000681. doi:10.1371/journal.ppat.1000681
43. Eulalio A, Schulte L, Vogel J. The mammalian microRNA response to bacterial infections. *RNA Biol* (2012) 9(6):742–50. doi:10.4161/rna.20018
44. Chen Y, Chen J, Wang H, Shi J, Wu K, Liu S, et al. HCV-induced miR-21 contributes to evasion of host immune system by targeting MyD88 and IRAK1. *PLoS Pathog* (2013) 9(4):e1003248. doi:10.1371/journal.ppat 1003248
45. Lu S, Mukkada VA, Mangray S, Cleveland K, Shillingford N, Schorl C, et al. MicroRNA profiling in mucosal biopsies of eosinophilic esophagitis patients pre and post treatment with steroids and relationship with mRNA targets. *PLoS One* (2012) 7(7):e40676. doi:10.1371/journal.pone.0040676
46. Lu TX, Sherrill JD, Wen T, Plassard AJ, Besse JA, Abonia JP, et al. MicroRNA signature in patients with eosinophilic esophagitis, reversibility with glucocorticoids, and assessment as disease biomarkers. *J Allergy Clin Immunol* (2012) 129(4):1064–75.e9. doi:10.1016/j.jaci.2012.01.060
47. Lu TX, Lim E-J, Besse JA, Itskovich S, Plassard AJ, Fulkerson PC, et al. miR-223 deficiency increases eosinophil progenitor proliferation. *J Immunol* (2013) 190(4):1576–82. doi:10.4049/jimmunol.1202897
48. Lu TX, Lim E-J, Itskovich S, Besse JA, Plassard AJ, Mingler MK, et al. Targeted ablation of miR-21 decreases murine eosinophil progenitor cell growth. *PLoS One* (2013) 8(3):e59397. doi:10.1371/journal.pone.0059397
49. Mattes J, Collison A, Plank M, Phipps S, Foster PS. Antagonism of microRNA-126 suppresses the effector function of T(H)2 cells and the development of allergic airways disease. *Proc Natl Acad Sci U S A* (2009) 106(44):18704–9. doi:10.1073/pnas.0905063106
50. Collison A, Mattes J, Plank M, Foster PS. Inhibition of house dust mite-induced allergic airways disease by antagonism of microRNA-145 is comparable to glucocorticoid treatment. *J Allergy Clin Immunol* (2011) 128(1):160–167.e4. doi:10.1016/j.jaci.2011.04.005
51. Sharma A, Kumar M, Ahmad T, Mabalirajan U, Aich J, Agrawal A, et al. Antagonism of mmu-mir-106a attenuates asthma features in allergic murine model. *J Appl Physiol (1985)* (2012) 113(3):459–64. doi:10.1152/japplphysiol.00001.2012
52. Izzotti A, Bagnasco M, Cartiglia C, Longobardi M, Balansky RM, Merello A, et al. Chemoprevention of genome, transcriptome, and proteome alterations induced by cigarette smoke in rat lung. *Eur J Cancer* (2005) 41(13):1864–74. doi:10.1016/j.ejca.2005.04.011
53. Izzotti A, Calin GA, Arrigo P, Steele VE, Croce CM, De Flora S. Downregulation of microRNA expression in the lungs of rats exposed to cigarette smoke. *FASEB J* (2009) 23(3):806–12. doi:10.1096/fj.08-121384
54. Izzotti A, Calin GA, Steele VE, Croce CM, De Flora S. Relationships of microRNA expression in mouse lung with age and exposure to cigarette smoke and light. *FASEB J* (2009) 23(9):3243–50. doi:10.1096/fj.09-135251
55. Simpson LJ, Patel S, Bhakta NR, Choy DF, Brightbill HD, Ren X, et al. A microRNA upregulated in asthma airway T cells promotes TH2 cytokine production. *Nat Immunol* (2014) 15(12):1162–70. doi:10.1038/ni.3026
56. Liu F, Qin HB, Xu B, Zhou H, Zhao DY. Profiling of miRNAs in pediatric asthma: upregulation of miRNA-221 and miRNA-485-3p. *Mol Med Rep* (2012) 6(5):1178–82. doi:10.3892/mmr.2012.1030
57. Anderson SE, Beezhold K, Lukomska E, Richardson J, Long C, Anderson K, et al. Expression kinetics of miRNA involved in dermal toluene 2,4-diisocyanate sensitization. *J Immunotoxicol* (2014) 11(3):250–9. doi:10.3109/1547691X.2013.835891
58. Carissimi C, Fulci V, Macino G. MicroRNAs: novel regulators of immunity. *Autoimmun Rev* (2009) 8(6):520–4. doi:10.1016/j.autrev.2009.01.008
59. Xiao C, Rajewsky K. MicroRNA control in the immune system: basic principles. *Cell* (2009) 136(1):26–36. doi:10.1016/j.cell.2008.12.027
60. Gracias DT, Katsikis PD. MicroRNAs: key components of immune regulation. In: Pulendran B, Katsikis PD, Schoenberger SP, editors. *Crossroads between Innate and Adaptive Immunity III*. New York, NY: Springer (2011). p. 15–26.
61. Chen CZ, Li L, Lodish HF, Bartel DP. MicroRNAs modulate hematopoietic lineage differentiation. *Science* (2004) 303(5654):83–6. doi:10.1126/science.1091903
62. Xiao C, Calado DP, Galler G, Thai TH, Patterson HC, Wang J, et al. miR-150 controls B cell differentiation by targeting the transcription factor c-Myb. *Cell* (2007) 131(1):146–59. doi:10.1016/j.cell.2007.07.021
63. Xiao C, Srinivasan L, Calado DP, Patterson HC, Zhang B, Wang J, et al. Lymphoproliferative disease and autoimmunity in mice with elevated miR-17-92 expression in lymphocytes. *Nat Immunol* (2008) 9(4):405–14. doi:10.1038/ni1575
64. Lu L-F, Liston A. MicroRNA in the immune system, microRNA as an immune system. *Immunology* (2009) 127(3):291–8. doi:10.1111/j.1365-2567.2009.03092.x
65. Li Q-J, Chau J, Ebert PJR, Sylvester G, Min H, Liu G, et al. miR-181a is an intrinsic modulator of T cell sensitivity and selection. *Cell* (2007) 129(1):147–61. doi:10.1016/j.cell.2007.03.008
66. Monticelli S, Ansel KM, Xiao C, Socci ND, Krichevsky AM, Thai T-H, et al. MicroRNA profiling of the murine hematopoietic system. *Genome Biol* (2005) 6(8):R71–71. doi:10.1186/gb-2005-6-8-r71
67. Wu H, Neilson JR, Kumar P, Manocha M, Shankar P, Sharp PA, et al. miRNA profiling of naïve, effector and memory CD8 T cells. *PLoS One* (2007) 2(10):e1020. doi:10.1371/journal.pone.0001020
68. Jeker LT, Bluestone JA. MicroRNA regulation of T-cell differentiation and function. *Immunol Rev* (2013) 253(1):65–81. doi:10.1111/imr.12061
69. Jiang S, Li C, Olive V, Lykken E, Feng F, Sevilla J, et al. Molecular dissection of the miR-17-92 cluster's critical dual roles in promoting Th1 responses and preventing inducible Treg differentiation. *Blood* (2011) 118(20):5487–97. doi:10.1182/blood-2011-05-355644

70. Haasch D, Chen Y-W, Reilly RM, Grace Chiou X, Koterski S, Smith ML, et al. T cell activation induces a noncoding RNA transcript sensitive to inhibition by immunosuppressant drugs and encoded by the proto-oncogene, BIC. *Cell Immunol* (2002) 217(1):78–86. doi:10.1016/S0008-8749(02)00506-3
71. Pedersen I, David M. MicroRNAs in the immune response. *Cytokine* (2008) 43(3):391–4. doi:10.1016/j.cyto.2008.07.016
72. Taganov KD, Boldin MP, Chang K-J, Baltimore D. NF-κB-dependent induction of microRNA miR-146, an inhibitor targeted to signaling proteins of innate immune responses. *Proc Natl Acad Sci U S A* (2006) 103(33):12481–6. doi:10.1073/pnas.0605298103
73. Rodriguez A, Vigorito E, Clare S, Warren MV, Couttet P, Soond DR, et al. Requirement of bic/microRNA-155 for normal immune function. *Science* (2007) 316(5824):608–11. doi:10.1126/science.1139253
74. Martinez-Nunez RT, Louafi F, Sanchez-Elsner T. The interleukin 13 (IL-13) pathway in human macrophages is modulated by microRNA-155 via direct targeting of interleukin 13 receptor α1 (IL13Rα1). *J Biol Chem* (2011) 286(3):1786–94. doi:10.1074/jbc.M110.169367
75. Roy S. miRNA in macrophage development and function. *Antioxid Redox Signal* (2016) 25(15):795–804. doi:10.1089/ars.2016.6728
76. Banerjee S, Xie N, Cui H, Tan Z, Yang S, Icyuz M, et al. MicroRNA let-7c regulates macrophage polarization. *J Immunol* (2013) 190(12):6542–9. doi:10.4049/jimmunol.1202496
77. Zhuang G, Meng C, Guo X, Cheruku PS, Shi L, Xu H, et al. A novel regulator of macrophage activation clinical perspective miR-223 in obesity-associated adipose tissue inflammation. *Circulation* (2012) 125(23):2892–903. doi:10.1161/CIRCULATIONAHA.111.087817
78. Veremeyko T, Siddiqui S, Sotnikov I, Yung A, Ponomarev ED. IL-4/IL-13-dependent and independent expression of miR-124 and its contribution to M2 phenotype of monocytic cells in normal conditions and during allergic inflammation. *PLoS One* (2013) 8(12):e81774. doi:10.1371/journal.pone.0081774
79. O'Connell RM, Rao DS, Chaudhuri AA, Baltimore D. Physiological and pathological roles for microRNAs in the immune system. *Nat Rev Immunol* (2010) 10(2):111–22. doi:10.1038/nri2708
80. O'Neill LA, Sheedy FJ, McCoy CE. MicroRNAs: the fine-tuners of toll-like receptor signalling. *Nat Rev Immunol* (2011) 11(3):163–75. doi:10.1038/nri2957
81. Case SR, Martin RJ, Jiang D, Minor MN, Chu HW. MicroRNA-21 inhibits toll-like receptor 2 agonist-induced lung inflammation in mice. *Exp Lung Res* (2011) 37(8):500–8. doi:10.3109/01902148.2011.596895
82. Lu TX, Hartner J, Lim EJ, Fabry V, Mingler MK, Cole ET, et al. MicroRNA-21 limits in vivo immune response-mediated activation of the IL-12/IFN-gamma pathway, Th1 polarization, and the severity of delayed-type hypersensitivity. *J Immunol* (2011) 187(6):3362–73. doi:10.4049/jimmunol.1101235
83. Chen X, Liang H, Zhang J, Zen K, Zhang CY. MicroRNAs are ligands of toll-like receptors. *RNA* (2013) 19(6):737–9. doi:10.1261/rna.036319.112
84. Chaudhuri AA, So AY, Sinha N, Gibson WS, Taganov KD, O'Connell RM, et al. MicroRNA-125b potentiates macrophage activation. *J Immunol* (2011) 187(10):5062–8. doi:10.4049/jimmunol.1102001
85. Zhang XH, Zhang YN, Li HB, Hu CY, Wang N, Cao PP, et al. Overexpression of miR-125b, a novel regulator of innate immunity, in eosinophilic chronic rhinosinusitis with nasal polyps. *Am J Respir Crit Care Med* (2012) 185(2):140–51. doi:10.1164/rccm.201103-0456OC
86. Podshivalova K, Salomon DR. MicroRNA regulation of T lymphocyte immunity: modulation of molecular networks responsible for T cell activation, differentiation and development. *Crit Rev Immunol* (2013) 33(5):435–76. doi:10.1615/CritRevImmunol.2013006858
87. Kastle M, Bartel S, Geillinger-Kastle K, Irmler M, Beckers J, Ryffel B, et al. MicroRNA cluster 106a~363 is involved in T helper 17 cell differentiation. *Immunology* (2017) 152:402–13. doi:10.1111/imm.12775
88. Liston A, Lu LF, O'Carroll D, Tarakhovsky A, Rudensky AY. Dicer-dependent microRNA pathway safeguards regulatory T cell function. *J Exp Med* (2008) 205(9):1993–2004. doi:10.1084/jem.20081062
89. Murugaiyan G, da Cunha AP, Ajay AK, Joller N, Garo LP, Kumaradevan S, et al. MicroRNA-21 promotes Th17 differentiation and mediates experimental autoimmune encephalomyelitis. *J Clin Invest* (2015) 125(3):1069–80. doi:10.1172/JCI74347
90. Das Gupta M, Fliesser M, Springer J, Breitschopf T, Schlossnagel H, Schmitt AL, et al. *Aspergillus fumigatus* induces microRNA-132 in human monocytes and dendritic cells. *Int J Med Microbiol* (2014) 304(5–6):592–6. doi:10.1016/j.ijmm.2014.04.005
91. Croston TL, Nayak AP, Lemons AR, Goldsmith WT, Gu JK, Germolec DR, et al. Influence of *Aspergillus fumigatus* conidia viability on murine pulmonary microRNA and mRNA expression following subchronic inhalation exposure. *Clin Exp Allergy* (2016) 46(10):1315–27. doi:10.1111/cea.12783
92. Monk CE, Hutvagner G, Arthur JSC. Regulation of miRNA transcription in macrophages in response to *Candida albicans*. *PLoS One* (2010) 5(10):e13669. doi:10.1371/journal.pone.0013669
93. Agustinho DP, de Oliveira MA, Tavares AH, Derengowski L, Stolz V, Guilhelmelli F, et al. Dectin-1 is required for miR155 upregulation in murine macrophages in response to *Candida albicans*. *Virulence* (2017) 8(1):41–52. doi:10.1080/21505594.2016.1200215
94. Du L, Chen X, Duan Z, Liu C, Zeng R, Chen Q, et al. miR-146a negatively regulates dectin-1-induced inflammatory responses. *Oncotarget* (2017) 8(23):37355–66. doi:10.18632/oncotarget.16958
95. Chen H, Jin Y, Chen H, Liao N, Wang Y, Chen J. MicroRNA-mediated inflammatory responses induced by *Cryptococcus neoformans* are dependent on the NF-kappaB pathway in human monocytes. *Int J Mol Med* (2017) 39(6):1525–32. doi:10.3892/ijmm.2017.2951
96. De Lacorte Singulani J, De Fátima Da Silva J, Gullo FP, Costa MC, Fusco-Almeida AM, Enguita FJ, et al. Preliminary evaluation of circulating microRNAs as potential biomarkers in paracoccidioidomycosis. *Biomed Rep* (2017) 6(3):353–7. doi:10.3892/br.2017.849
97. Turini Gonzales Marioto D, Navarro Dos Santos Ferraro AC, Goulart de Andrade F, Barros Oliveira M, Itano EN, Petrofeza S, et al. Study of differential expression of miRNAs in lung tissue of mice submitted to experimental infection by *Paracoccidioides brasiliensis*. *Med Mycol* (2017) 55:774–84. doi:10.1093/mmy/myw135
98. Croston TL, Nayak AP, Lemons AR, Goldsmith WT, Germolec DM, Beezhold DH, et al. Pulmonary immune response following subchronic *Stachybotrys chartarum* exposure. *J Allergy ClinImmunol* (2017) 139(2):AB75. doi:10.1016/j.jaci.2016.12.291
99. Kumar M, Ahmad T, Sharma A, Mabalirajan U, Kulshreshtha A, Agrawal A, et al. Let-7 microRNA-mediated regulation of IL-13 and allergic airway inflammation. *J Allergy ClinImmunol* (2011) 128(5):1077–85.e1–10. doi:10.1016/j.jaci.2011.04.034
100. Liu G, Abraham E. MicroRNAs in immune response and macrophage polarization. *Arterioscler Thromb Vasc Biol* (2013) 33(2):170–7. doi:10.1161/ATVBAHA.112.300068
101. Essandoh K, Li Y, Huo J, Fan G-C. miRNA-mediated macrophage polarization and its potential role in the regulation of inflammatory response. *Shock* (2016) 46(2):122–31. doi:10.1097/SHK.0000000000000604
102. Rivera A, Barr T, Rais M, Engelmann F, Messaoudi I. MicroRNAs regulate host immune response and pathogenesis during influenza infection in rhesus macaques. *Viral Immunol* (2016) 29(4):212–27. doi:10.1089/vim.2015.0074
103. Zhang N, Bevan MJ. Dicer controls CD8(+) T-cell activation, migration, and survival. *Proc Natl Acad Sci U S A* (2010) 107(50):21629–34. doi:10.1073/pnas.1016299107
104. Nahid MA, Yao B, Dominguez-Gutierrez PR, Kesavalu L, Satoh M, Chan EK. Regulation of TLR2-mediated tolerance and cross-tolerance through IRAK4 modulation by miR-132 and miR-212. *J Immunol* (2013) 190(3):1250–63. doi:10.4049/jimmunol.1103060
105. Testa U, Pelosi E, Castelli G, Labbaye C. miR-146 and miR-155: two key modulators of immune response and tumor development. *Noncoding RNA* (2017) 3(3):22. doi:10.3390/ncrna3030022
106. Suman S, Jones-Reed DZZ, Schmidt ML, Clark GJ, Klinge C, Barve S, et al. Alteration of miR-186 expression modifies inflammatory markers in normal epithelial and prostate cancer cell models. *FASEB J* (2017) 31(1 Suppl):757.716. doi:10.1096/fasebj.31.1_supplement.757.16
107. Liu S-Q, Jiang S, Li C, Zhang B, Li Q-J. miR-17-92 cluster targets phosphatase and tensin homology and Ikaros family zinc finger 4 to promote T(H)17-mediated inflammation. *J Biol Chem* (2014) 289(18):12446–56. doi:10.1074/jbc.M114.550723
108. Fontana L, Pelosi E, Greco P, Racanicchi S, Testa U, Liuzzi F, et al. MicroRNAs 17-5p-20a-106a control monocytopoiesis through AML1 targeting and

M-CSF receptor upregulation. *Nat Cell Biol* (2007) 9(7):775–87. doi:10.1038/ncb1613

109. Paladini L, Fabris L, Bottai G, Raschioni C, Calin GA, Santarpia L. Targeting microRNAs as key modulators of tumor immune response. *J Exp Clin Cancer Res* (2016) 35(1):103. doi:10.1186/s13046-016-0375-2
110. Quinn SR, O'Neill LA. A trio of microRNAs that control toll-like receptor signalling. *Int Immunol* (2011) 23(7):421–5. doi:10.1093/intimm/dxr034
111. Mehta A, Baltimore D. MicroRNAs as regulatory elements in immune system logic. *Nat Rev Immunol* (2016) 16(5):279–94. doi:10.1038/nri.2016.40
112. Ma F, Xu S, Liu X, Zhang Q, Xu X, Liu M, et al. The microRNA miR-29 controls innate and adaptive immune responses to intracellular bacterial infection by targeting interferon-[gamma]. *Nat Immunol* (2011) 12(9):861–9. doi:10.1038/ni.2073
113. Huang L, Liu Y, Wang L, Chen R, Ge W, Lin Z, et al. Down-regulation of miR-301a suppresses pro-inflammatory cytokines in toll-like receptor-triggered macrophages. *Immunology* (2013) 140(3):314–22. doi:10.1111/imm.12139
114. Vegh P, Magee DA, Nalpas NC, Bryan K, McCabe MS, Browne JA, et al. MicroRNA profiling of the bovine alveolar macrophage response to *Mycobacterium bovis* infection suggests pathogen survival is enhanced by microRNA regulation of endocytosis and lysosome trafficking. *Tuberculosis* (2015) 95(1):60–7. doi:10.1016/j.tube.2014.10.011
115. Warg LA, Oakes JL, Burton R, Neidermyer AJ, Rutledge HR, Groshong S, et al. The role of the E2F1 transcription factor in the innate immune response to systemic LPS. *Am J Physiol Lung Cell Mol Physiol* (2012) 303(5):L391–400. doi:10.1152/ajplung.00369.2011
116. Wong JJL, Ritchie W, Gao D, Lau KA, Gonzalez M, Choudhary A, et al. Identification of nuclear-enriched miRNAs during mouse granulopoiesis. *J Hematol Oncol* (2014) 7:42. doi:10.1186/1756-8722-7-42
117. Goldani LZ, Wirth F. Animal models and antifungal agents in paracoccidioidomycosis: an overview. *Mycopathologia* (2017) 182(7):633–43. doi:10.1007/s11046-017-0130-z
118. Terçarioli GR, Bagagli E, Reis GM, Theodoro RC, Bosco Sde M, Macoris SA, et al. Ecological study of *Paracoccidioides brasiliensis* in soil: growth ability, conidia production and molecular detection. *BMC Microbiol* (2007) 7(1):92. doi:10.1186/1471-2180-7-92
119. Bagagli E, Theodoro RC, Bosco SM, McEwen JG. *Paracoccidioides brasiliensis*: phylogenetic and ecological aspects. *Mycopathologia* (2008) 165(4–5):197–207. doi:10.1007/s11046-007-9050-7
120. Seo M, Choi J-S, Rho CR, Joo C-K, Lee SK. MicroRNA miR-466 inhibits lymphangiogenesis by targeting prospero-related homeobox 1 in the alkali burn corneal injury model. *J Biomed Sci* (2015) 22(1):3. doi:10.1186/s12929-014-0104-0
121. Colden M, Dar AA, Saini S, Dahiya PV, Shahryari V, Yamamura S, et al. MicroRNA-466 inhibits tumor growth and bone metastasis in prostate cancer by direct regulation of osteogenic transcription factor RUNX2. *Cell Death Dis* (2017) 8(1):e2572. doi:10.1038/cddis.2017.15
122. Wéry N. Bioaerosols from composting facilities—a review. *Front Cell Infect Microbiol* (2014) 4:42. doi:10.3389/fcimb.2014.00042
123. Pearson C, Littlewood E, Douglas P, Robertson S, Gant TW, Hansell AL. Exposures and health outcomes in relation to bioaerosol emissions from composting facilities: a systematic review of occupational and community studies. *J Toxicol Environ Health B Crit Rev* (2015) 18(1):43–69. doi:10.1080/10937404.2015.1009961
124. Symoens F, Burnod J, Lebeau B, Viviani MA, Piens MA, Tortorano AM, et al. Hospital-acquired *Aspergillus fumigatus* infection: can molecular typing methods identify an environmental source? *J Hosp Infect* (2002) 52(1):60–7. doi:10.1053/jhin.2002.1263
125. Kauffman CA, Nicolasora NP. Epidemiology of invasive pulmonary aspergillosis. In: Comarú Pasqualotto A, editor. *Aspergillosis: From Diagnosis to Prevention*. Dordrecht, Netherlands: Springer (2010). p. 329–44.
126. CDC. *Fungal Diseases: Candidiasis*. (2017). Available from: https://www.cdc.gov/fungal/diseases/candidiasis/index.html
127. CDC. *Fungal Diseases: Invasive Candidiasis Statistics*. (2017). Available from: https://www.cdc.gov/fungal/diseases/candidiasis/invasive/statistics.html
128. Epelbaum O, Chasan R. Candidemia in the intensive care unit. *Clin Chest Med* (2017) 38(3):493–509. doi:10.1016/j.ccm.2017.04.010
129. CDC. *Fungal Diseases: General Information about Candida auris*. (2016). Available from: https://www.cdc.gov/fungal/diseases/candidiasis/candida-auris-qanda.html
130. CDC. *Fungal Diseases: C. gattii Infection*. (2015). Available from: https://www.cdc.gov/fungal/diseases/cryptococcosis-gattii/index.html
131. CDC. *Fungal Diseases: C. neoformans Infection*. (2015). Available from https://www.cdc.gov/fungal/diseases/cryptococcosis-neoformans/index.html
132. Brown GD. Dectin-1: a signalling non-TLR pattern-recognition receptor. *Nat Rev Immunol* (2006) 6(1):33–43. doi:10.1038/nri1745
133. Buskirk AD, Green BJ, Lemons AR, Nayak AP, Goldsmith WT, Kashon ML, et al. A murine inhalation model to characterize pulmonary exposure to dry *Aspergillus fumigatus* conidia. *PLoS One* (2014) 9(10):e109855. doi:10.1371/journal.pone.0109855
134. Bhaskaran M, Mohan M. MicroRNAs: history, biogenesis, and their evolving role in animal development and disease. *Vet Pathol* (2014) 51(4):759–74. doi:10.1177/0300985813502820
135. Gibson NW. Engineered microRNA therapeutics. *J R Coll Physicians Edinb* (2014) 44(3):196–200. doi:10.4997/JRCPE.2014.302

11

Mechanisms of *Cryptococcus neoformans*-Mediated Host Damage

Arturo Casadevall[1]*, Carolina Coelho[1] and Alexandre Alanio[1,2,3]

[1] Department of Molecular Microbiology and Immunology, Johns Hopkins School of Public Health, Baltimore, MD, United States, [2] Institut Pasteur, Molecular Mycology Unit, CNRS UMR2000, Paris, France, [3] Laboratoire de Parasitologie-Mycologie, Hôpital Saint-Louis, Groupe Hospitalier Lariboisière, Saint-Louis, Fernand Widal, Assistance Publique-Hôpitaux de Paris (AP-HP), Université Paris Diderot, Sorbonne Paris Cité, Paris, France

***Correspondence:**
Arturo Casadevall
acasade1@jhu.edu

Cryptococcus neoformans is not usually considered a cytotoxic fungal pathogen but there is considerable evidence that this microbe can damage host cells and tissues. In this essay, we review the evidence that *C. neoformans* damages host cells and note that the mechanisms involved are diverse. We consider *C. neoformans*-mediated host damage at the molecular, cellular, tissue, and organism level. Direct mechanisms of cytotoxicity include lytic exocytosis, organelle dysfunction, phagolysosomal membrane damage, and cytoskeletal alterations. Cytotoxicity contributes to pathogenesis by interfering with immune effector cell function and disrupting endothelial barriers thus allowing dissemination. When *C. neoformans*-mediated and immune-mediated host damage is sufficient to affect homeostasis, cryptococcosis occurs at the organism level.

Keywords: *Cryptococcus*, cryptococcosis, disease, damage, macrophage, cytotoxicity

Anyone with expertise in routine laboratory tissue culture knows that fungal contamination rapidly kills mammalian cells *in vitro*. However, *Cryptococcus neoformans* is unusual among fungi in that it has minimal toxicity for animal cells in tissue culture, such that it is possible to maintain yeast cells and macrophages for days without major cytotoxicity for the latter. Macrophage-like cells that have phagocytosed *C. neoformans* are capable of replicating and divide their yeast cargo among daughter cells (1). This implies that *C. neoformans* does not release major cytotoxic products, at least *in vitro*. Consistent with this notion, cryptococcal infections are not associated with tissue necrosis as seen in infections caused by other fungal pathogens, such as those caused by *Aspergillus* or Mucorales spp. In fact, cryptococcosis often shows many features of a chronic infection, and host death is often due to the effects of physical compression of tissue, and defects in resorption of cerebrospinal fluid (CSF) (possibly due to increased viscosity from fungal polysaccharide shedding into CSF) and overwhelming brain edema (2). While these observations might lead to the conclusion that *C. neoformans* infections are associated with minimal host damage, a review of available knowledge reveals otherwise. In this essay, we survey the available evidence that *C. neoformans* is capable of inflicting direct damage on host cells and tissues. We note that host damage following cryptococcal infection can come from microbe and the host (3, 4), with the latter culminating in a dramatic pathology known as Immune Reconstitution Inflammatory Syndrome (IRIS).

We consider damage at four levels of scale: molecular, cellular, tissue, and organism level. Molecular damage is that caused by enzymes or molecules produced by *C. neoformans,* which induces modifications of host molecules and cells and manifests itself at the molecular or organelle level. Cellular damages are those causing modifications of the architecture and structure of the host cells due to the toxic action of *C. neoformans*. Tissue damages cause anatomical and functional disorganization beyond cellular injury. Together, these combine to produce the disease of cryptococcosis at the organism level. We recognize that these are not independent, since molecular damage leads to cellular damage, cellular damage leads to tissue damage, and all three combine to produce organismal damage. Furthermore, we note that a process of *C. neoformans* can damage the host at more than one level. For example, cryptococcal phospholipase can cause molecular damage by destroying surfactant molecules (5) while also being a potential cause of cellular damage at the level of macrophage (6). Despite these important caveats, these mechanisms are sufficiently distinct that it is possible to discuss them separately. Our goal is integrating them to produce a holistic view of *C. neoformans*-mediated host damage into a new synthesis for approaching cryptococcal pathogenesis (**Figure 1**).

MOLECULAR DAMAGE

In the section for molecular damage, we consider how cryptococcal products damage host molecules (**Figure 1**). *C. neoformans*-mediated molecular damage enhances its likelihood of survival in tissues. As a soil-dwelling organism that obtains its nutrition from digesting material in the environment, *C. neoformans* secretes a large suite of enzymes with the potential to degrade host molecules (7). Among all enzymes produced by the fungus, the major candidates as mediators of host toxicity at the molecular level are proteases, urease, phospholipase, and nuclease (7). *C. neoformans* can metabolize immunoglobulins and complement proteins for growth as these compounds are presumably degraded by released proteases (8). Hence, proteases may interfere with host defense mechanisms by cleaving immunologically important molecules and directly damaging effector cells. Cryptococcal serine proteases promote increased blood–brain barrier (BBB) permeability (9), which may help in the process of brain infection. Although not directly related to host damage *C. neoformans* releases a protease that cleaves a peptide, which functions as quorum-sensing molecule to increase virulence (10). Urease is a virulence factor for *C. neoformans* (11), which is important for brain invasion (12). The mechanism by which urease promotes brain invasion could involve catalyzing the hydrolysis of urea to ammonia to locally damage endothelial cells in the brain vasculature. Another group of enzymes involved in the pathogenesis of *C. neoformans*

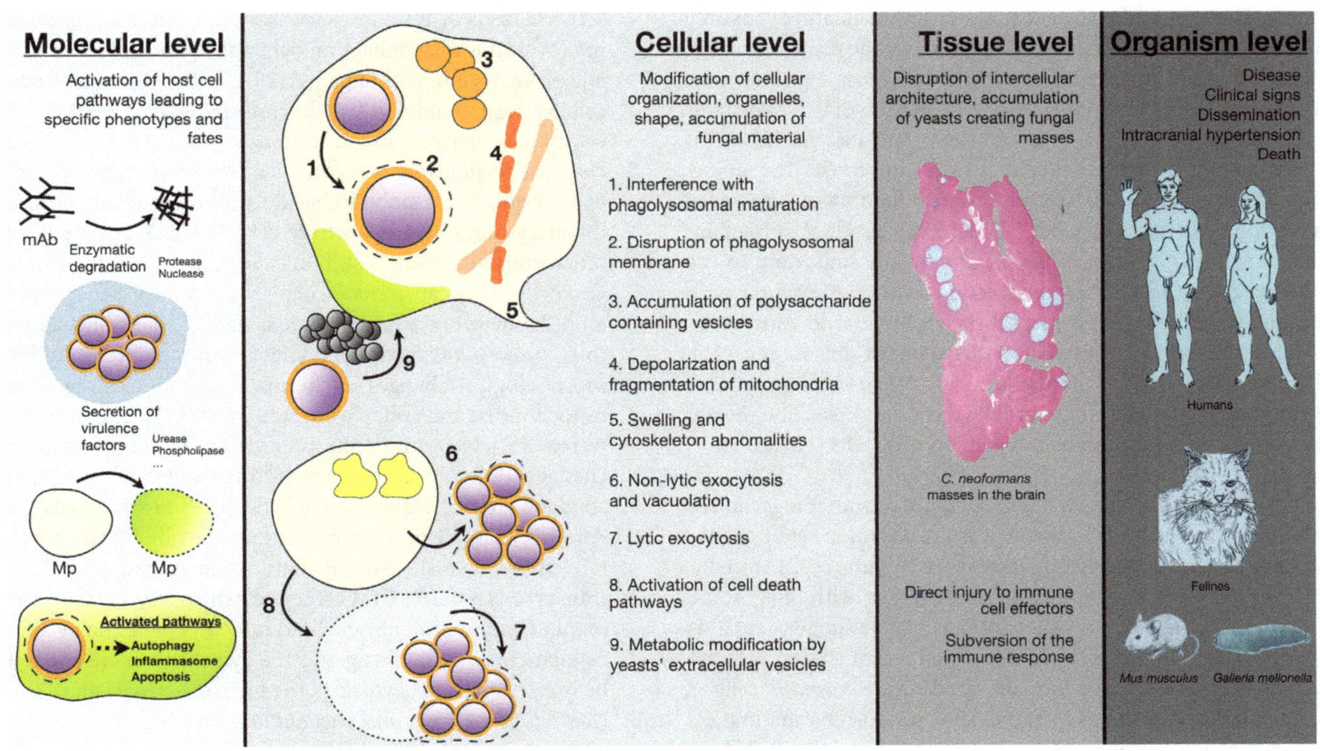

FIGURE 1 | Schematic representation of the different *Cryptococcus neoformans*-mediated cell host damages are various scales. Damage at the molecular level results from the secretion of various enzyme by *C. neoformans* (proteases, nuclease, urease, phospholipase) that degrade host molecules such as antibodies and/or modify cells membranes. *C. neoformans* ingestion is also able to trigger autophagy, apoptosis, and cell death in the host (mAb, monoclonal antibodies; Mp, macrophages). Damage at the cellular level involves modification of cellular compartments such as accumulation of polysaccharide vacuoles (1), inhibition of phagolysosomal maturation (2), phagolysosomal leakage (3), mitochondrial fission and depolarization (4), swelling and cytoskeleton abonomalities (5) or metabolic modification due to *C. neoformans* vesicles secretions (6), *C. neoformans* engulfment resulted also in non-lytic (7), or lytic (8) exocytosis. Damage at the tissue level consisted typical cryptococcal lesions in the brain parenchyma after intravenous inoculation of *C. neoformans* to outbred mice (sacrifice seven days after inoculation). No granuloma and accumulation of yeast masses without inflammatory cells can be observed engendering tissue disorganization. Coloration Alcian Blue (magnification 4×). Damage at the organism level combines to produce the clinical signs associated with cryptococcal diseases in humans with dissemination and neurological abnormalities as the most severe clinical presentation leading to death. Felines are also naturally susceptible to cryptococcosis with localized to disseminate infections. *Mus musculus* and *Galleria mellonella* are well established organisms for experimental models of infection that help understanding the pathophysiology of the disease and the biology of the yeast in relation to the host.

are phospholipases. *C. neoformans* produces both phospholipase B and C (6, 13–17). Phospholipases cleave phospholipids, which in turn allow them to damage membranes. Phospholipase-deficient *C. neoformans* manifest delayed intracellular replication, which could result in better maintenance of phagosomal membrane integrity and subsequent enhanced fungal control (6). *In vitro* phospholipase-mediated cleaves surfactant and promotes the attachment of *C. neoformans* to human lung epithelial cells, a process *in vivo* could promote pulmonary infection (5). Ingestion of *C. neoformans* results in the activation of autophagy initiation complex pathways, which results in a global reprogramming of host kinase signaling (18).

CELLULAR DAMAGE

By cellular damage, we consider mechanisms for cytotoxicity. *C. neoformans*-mediated cytotoxicity contributes to establishment of disease *via* at least two major mechanisms. First, damage to host immune system in tissue, which inflicts damage to immune system, the surrounding tissues and may cause symptoms to the host while ultimately allowing persistence of infection. Second, damage to the endothelial cells in the brain vasculature (possibly in other organs as well), precedes invasion of the central nervous system to cause meningoencephalitis, the most common life-threatening form of cryptococcosis. Interaction of *C. neoformans* with the epithelial barriers is transient, and internalization of *C. neoformans* by epithelial cells is rarely observed. The airway epithelium is critical to trigger initial inflammatory response to the inhaled spores or yeast (19) and can produce surfactant, which agglutinates yeast cells (20). Potentially important to the pathogenesis of disease are interactions of *C. neoformans* with neurons or (micro)glial cells and their potential to cause neurological dysfunction but so far this remains an enigma (21). Therefore, we focus our discussion on cytotoxic damage primarily on phagocytic cells and endothelial cells and describe several forms of damage that can be inflicted on host cells (**Figure 1**).

1. *Interference with phagolysosomal maturation:* after ingestion of *C. neoformans* by macrophage, yeasts resides in an acidic phagosome that has maturation markers such as Lamp1 (22). Initially, *C. neoformans* was thought to not interfere with phagosome maturation similar to other intracellular pathogens such as *Mycobacterium tuberculosis*. However, recent studies indicate that ingestion of live but not dead *C. neoformans* cells is associated with the early removal of phagosome maturation markers Rab5 and Rab11 and interference with proteolytic activity and calcium fluxes (23). Interference with phagolysosome maturation would have the effect of interfering with its microbicidal activity, which in turn would promote fungal intracellular replication.
2. *Phagolysosomal leakage:* electron microscopic studies revealed damaged membranes in phagosomes containing *C. neoformans* (24). Phagolysosomal leakage was confirmed with fluorescence labeled microscopy (25). Leakage of phagosome components into the cytoplasm interferes with microbicidal activity of this organelle but may also trigger the inflammasome (26). The amount of phagosomal leakage is modulated by IFN gamma (27), and likely other cytokines, with consequences to the amount of host cell damage and death.
3. *Accumulation of polysaccharide-containing vesicles in macrophage: C. neoformans* infection in macrophage is accompanied by the accumulation of vesicles, which contain fungal polysaccharide (24). The presence of large numbers of vesicles in the macrophage cytoplasm can give the impression of "holes" and these cells were named "hueco cells" after the Spanish term for hole (24). These intracytoplasmic vesicles appear to bud from the cryptococcal phagolysosome (24). Whether vesicle accumulation causes direct damage to the host cell is not known but their presence crowds out the cytoplasm and could impair cellular function. In addition, the generation of such large number of vesicles must tax lipid reserves (as composed of lipid bilayer), which could put a stress on membrane generation and repair.
4. *Interference with organelle function:* upon detection of intruder pathogenic microbes, macrophage undergo a series of concerted metabolic changes ranging from shift to a glycolytic metabolism to adaptation of the rate of protein synthesis or autophagy. Cellular quality control is a method to detect pathogen interference with organelle and cellular function and, therefore, is a mechanism of immune surveillance. When *C. neoformans* infects murine macrophage mitochondria potential is decreased and protein synthesis is impaired (28, 29). The presence of extracellular yeast could also lead to chromosomal aberrations and cell cycle impairment in murine macrophage (28).
5. *Host cell swelling and cytoskeleton alterations:* several decades ago, cryptococcal polysaccharide was reported to produce direct swelling effects on host cells (30–32). Although the biochemical pathway is still unknown, there is evidence that *C. neoformans* can induce changes in the cellular cytoskeleton. Human brain microvascular endothelial cells exposed to *C. neoformans* manifested changes in membrane ruffling, morphological changes in the nucleus, and swelling of the mitochondria and endoplasmic reticulum (33). These changes were associated with dephosphorylation of cofilin and actin changes (33). Similarly, changes to the cell cytoskeleton are apparent after *C. neoformans* infection of *Drosophila* cells (34). Another example of close interaction, and possible manipulation of host cytoskeleton, is that in human endothelial cells transcytosis requires CD44 receptor, which triggers the movement of actin to membrane lipid rafts that presumably aid in cellular entry (35). Changes to the cytoskeleton are apparent before non-lytic exocytosis in the form of actin flashes (36).
6. *Non-lytic exocytosis and vacuolation:* non-lytic exocytosis is generally considered a relatively benign process on the host cell given that cells from which *C. neoformans* has exited can replicate (37, 38). However, it is likely that any interference in the process of non-lytic exocytosis will trigger lytic exocytosis (discussed below). In this regard, macrophage deficient in annexin A2, a protein that is involved in a myriad of cellular functions including membrane fusion and plasma membrane repair, manifested increased lytic exocytosis, possibly due to problems in membrane fusion during the process of exocytosis (39). Non-lytic exocytosis results in the formation of large vacuoles in the cytoplasm of macrophage from which *C. neoformans* exited (40). Although the mechanism and

etiology of the formation of these large vacuoles is not known they represent an anatomic cellular abnormality that could interfere with cell homeostasis.

7. *Lytic exocytosis*: unchecked replication of *C. neoformans* inside macrophage can lead to rupture of the host cell with the release of fungal cells. This process, known as lytic exocytosis results in the death of the host cell. Host cell lysis by *C. neoformans* is also triggered by a mechanism that is dependent on cell wall mannosylation of cell wall and does not require the viability of the fungal cell (41).
8. *Activation of cell death pathways*: the toxic presence of *C. neoformans* may culminate in host cellular death. However, a study of death pathways activated in murine macrophage upon *C. neoformans* infection did not find a specific pathway predominantly activated. In contrast, in dendritic cells *C. neoformans* caused robust inflammasome activation and cell death, which was critically dependent of caspase-1, or caspase-8 in the absence of caspase-1 (28, 29). *C. neoformans* polysaccharides glucuronoxylomannan and galactoxylomannan can mediate direct cytoxicity to macrophage by activation of Fas ligand and triggering apoptosis (42, 43). Galatoxylomannan trigger B cell death leading to lymphocyte depletion and abrogating effective antibody responses (44).
9. *Extracellular vesicle (EV) effects. C. neoformans* like all other fungi that have been examined produces EVs (45). In the case of *C. neoformans*, these vesicles contain many fungal products associated with virulence, including fungal secreted polysaccharides and laccase (46). EVs are not directly toxic to host cells but they can have powerful stimulatory effects on host macrophage (47). In fact, incubation of macrophage with *C. neoformans* EVs led to activation changes that increased the antifungal activity (47). Although at this time EVs are not linked directly to cryptococcal pathogenesis there is considerable indirect evidence for an important role in virulence. For example, capsular polysaccharides have been implicated in negative effects on immune function, and this material is exported across the cryptococcal cell wall in EVs (5, 48). In addition to capsular polysaccharides, EVs have been shown to carry numerous short RNA molecules raising the possibility that these microbial RNAs are involved in modulating host cell function (49).

TISSUE AND ORGAN DAMAGE

Immune System Damage

The immune system is damaged during *C. neoformans* infection by direct injury to its effector cells and by interference with effective immunity. The outcome of the *C. neoformans*–macrophage interaction is a critical determinant for the fate of the microbe and host during infection. The ability of *C. neoformans* to replicate inside macrophage correlated with mice and rat susceptibility to infection (50, 51). In humans, the capacity of *C. neoformans* strains to replicate in macrophage to higher intracellular burden correlated with worse clinical outcomes (52, 53). Hence, the available evidence suggests that factors and interventions that modulate macrophage function, in particular when T-cell function is impaired, could control cryptococcal disease, whereas the capacity of the yeast to efficiently replicate intracellularly is associated with progression of infection. In this light, it is apparent that mechanisms that damage macrophage are likely to impair the antifungal capacity of these cells, which in turn facilitates intracellular growth. Hence, mitochondrial damage, phagosomal damage, and induction of programmed cellular pathways can be expected to directly aid in fungal survival *in vivo* through impairment of monocyte mononuclear macrophage as well as other innate immune cells. Damage to other immune cells has been less studied, but for example direct effects of shed polysaccharides on adaptive cellular responses (54) will magnify the impairment to macrophage function by providing inadequate activation of microbicidal capacity. A last point is that shear physical force of capsule and cell body growth, to dimensions surpassing 10 μm, may exhaust intracellular membranes of the host and that this fungal gigantism could physically damage host cell (55, 56), as seen with capsule growth and titan cell formation.

The second form of damage to the immune system is interference with its ability to organize an effective response. Here, the damage is multifaceted and originates from the cellular damage described above as well direct effects of cryptococcal components that affect the response of immune cells, which in turn interfere with effective immunity. The major cryptococcal polysaccharides have protean effects on the function of immune cells, which contribute to dysregulated process [reviewed in Ref. (57)]. In addition, to polysaccharide-mediated effects, the presence of the *C. neoformans* urease in the lung promotes the accumulation of immature dendritic cells and the emergence of a non-protective T2 polarized inflammatory response (58). *C. neoformans* produces a variety of prostaglandins and leukotrienes, which have direct effects on inflammatory cells and thus may have a major effect in altering the local immune response to infection (3, 41). Synthesis of eicosanoids is dependent on phospholipase activity thus implicating this enzyme in several different possible mechanism of virulence (47). Cryptococcal polysaccharides interfere with leukocyte migration toward chemoattractants (59). The mechanism for this effect includes induction of L-selectin shedding from neutrophils (60). Interference with leukocyte migration could account for the notoriously poor inflammatory responses observed in many individuals (61).

The immune response to *C. neoformans* can also mediate host damage. This phenomenon was first when HIV-infected individuals successfully treated for cryptococcosis manifested a worsening of symptoms after the initiation of antiretroviral therapy (62, 63). What came to be known as "immune reconstitution inflammatory syndrome" was the result of immune system recovery reacting to residual cryptococcal antigens in tissue, which resulted in inflammation and local organ damage (63). Recently, T cells have been associated with immune injury in experimental murine cryptococcosis establishing a mechanism by which immune dysregulation in response to infection can produce host damage (3). The contribution of immune-mediated damage to the pathogenesis of cryptococcosis could help explain the paradoxical observation that the prognosis of cryptococcal meningitis if more favorable in patients with HIV infection and severe immunodeficiency than in those without obvious immune impairment (4). There is also some evidence that cryptococcal infection in the lung predisposes the host to develop allergic inflammation that could progress to hyperreactive airway diseases, such as asthma (64–66).

Damage to BBB

The major cause of mortality and morbidity during cryptococcosis is meningoencephalitis. For *C. neoformans* to invade the central nervous system yeast cells must cross the BBB. *C. neoformans* crosses the BBB by two mechanisms: transcytosis, whereby yeast cells transit directly through endothelial cells and a Trojan Horse-like mechanism involving carriage inside an infected macrophage (67–69). The former mechanism involves undermining the integrity of the BBB (70) and is enhanced by brain inositol (71). The yeasts are trapped in the brain capillaries because of their size, allowing for active transcytosis (12). For the efficient Trojan horse crossing, *C. neoformans* must survive inside macrophage and, as noted above, fungal-mediated damage to the phagocytic cell enhances intracellularly cryptococcal survival. However, it is still poorly understood if the crossing of the BBB merely causes a transient disruption in integrity of the BBB or whether it has more pernicious consequences. One could hypothesize that entrapment of yeast in brain capillaries cause ischemia to surrounding tissues but this issue has not been formally addressed.

Tissue Masses

A distinctive feature of many cases of cryptococcal meningoencephalitis is the formation of masses of yeast cells in the brain with little or no inflammation (**Figure 1**). This feature distinctively distinguishes these structures from granuloma where inflammation and immune response are well organized. These structures are so distinctive that they have been referred as "soap bubbles" as they are composed of gelatinous pseudocysts composed of packed *C. neoformans* cells with a particular appearance in magnetic resonance imaging (72). For masses of *C. neoformans* to form in the brain, they must grow in a manner that displaces or destroys brain tissue to create the space for the fungal mass. Given the propensity of *C. neoformans* to replicate inside cells and trigger host cell death such soap bubble anatomic lesions could be the result of progressive lysis of host cells at the fungal–brain interface. In this regard, *C. neoformans* can replicate inside microglial cells, the brain resident macrophage population (73). Alternatively, it is possible that such lesions represent fungal replication that creates spaces in the brain through compression of brain tissue through the force generated by fungal replication. Hence, irrespective of the mechanism of formation, soap bubble lesions represent *prima facie* evidence of direct fungal damage to brain tissue.

ORGANISM DAMAGE

At the organism level the combination of molecular, cellular, and tissue damage leads to cryptococcosis (**Figure 1**). The damage–response framework of microbial pathogenesis posits that disease occurs when host damage is sufficient to affect hemostasis, which in turn produces clinical symptoms (74). For *C. neoformans* infections, host damage can come from both the microbe, as reviewed in this essay, and from the immune response (3, 4, 75). Although a discussion of how tissue damage results in clinical signs and symptoms that can ultimately lead to death is beyond the scope of this review, there is a clear connection between the types of damage discussed here and the disease.

RELATION OF CYTOTOXICITY TO ENVIRONMENTAL SELECTION PRESSURES—AMOEBA

Evolution of *C. neoformans* virulence, virulence being defined as capacity to survive or to cause disease in mammalian hosts, was proposed to arise from selection pressures in the environment by phagocytic predators such as amoeba (76, 77). According to this view, *C. neoformans* virulence factors needed for animal pathogenicity function emerged as characteristics that protect fungal cells against phagocytic predators. For example, the capsule, melanin, and phospholipase each contribute to fungal cell survival when preyed upon by amoeba (76). The outcome of amoeba–*C. neoformans* interactions is highly dependent on the conditions of the experiment. In conditions where there are minimal nutrients such as phosphate-buffered saline, *C. neoformans* is ascendant but the reverse occurs when there are nutrients for amoeba (48). The presence of extracellular Ca^{2+} and Mg^{2+} is enough to tilt the balance of the host–*C. neoformans* and allow amoeba to kill a significant portion of *C. neoformans* (78). Although far less is known about how *C. neoformans* damages amoeba than for mammalian cells, it is likely to have parallels in the mechanisms for cytotoxicity. In this regard, accumulation of polysaccharide-containing vesicles was observed in the cytoplasm of amoeba that ingested *C. neoformans* (76).

A SYNTHESIS FOR *C. neoformans*-MEDIATED HOST DAMAGE

For the purposes of this essay, we have considered host damage as a function of size scales but it is important to stress that damage is continuous from the molecular to organism level (**Figure 1**). Disseminated cryptococcosis is a rare disease in hosts with intact immunity, which means that host defense mechanisms are highly effective at confiding damage form inhaled *C. neoformans* to the molecular and cellular level in the lungs, such that damage does not rise to the level where homeostasis is affected and clinical symptoms ensue. Since cryptococcal infection is common and diseases is rare, and *C. neoformans* are common in the environment, it is likely that repeated cycles of macrophage infection occur in the lives of human hosts. Although we do not know the sequence of events that follow these interactions, the fact that these are asymptomatic suggests fungal control in the lung with minimal tissue damage. However, once there is impairment to the immune system, commonly following immunosuppression, HIV infection or iatrogenic, cryptococcal infection transforms from silent or latent, to a slow but inexorable progressive condition that invariably kills the host without aggressive therapy. However, more than a half century after the introduction of the first antifungal agent in the form of amphotericin B, the mortality and morbidity of cryptococcosis remains stubbornly high. Improvements in therapy may require a better understanding of the mechanisms of host damage that will allow the development of new therapeutic interventions. A critical synthesis of how the various types of host damage synergize to impair tissue function is an important next step for understanding the pathogenesis of cryptococcosis.

AUTHOR CONTRIBUTIONS

All authors listed have made a substantial, direct, and intellectual contribution to the work and approved it for publication.

ACKNOWLEDGMENTS

The authors thank Françoise Dromer, for providing the picture of the whole brain of infected mouse. The authors also credit the histopathology and models unit at the Institute Pasteur, Paris. AC was supported by grants 5R01HL059842, 5R01AI033774, 5R37AI033142, and 5R01AI052733. AA was supported by a Fullbright and Monahan fellowship and a grant from the Philippe Foundation.

REFERENCES

1. Luo Y, Alvarez M, Xia L, Casadevall A. The outcome of phagocytic cell division with infectious cargo depends on single phagosome formation. *PLoS One* (2008) 3:e3219. doi:10.1371/journal.pone.0003219
2. Robertson EJ, Najjuka G, Rolfes MA, Akampurira A, Jain N, Anantharanjit J, et al. *Cryptococcus neoformans* ex vivo capsule size is associated with intracranial pressure and host immune response in HIV-associated cryptococcal meningitis. *J Infect Dis* (2014) 209:74–82. doi:10.1093/infdis/jit435
3. Neal LM, Xing E, Xu J, Kolbe JL, Osterholzer JJ, Segal BM, et al. CD4(+) T cells orchestrate lethal immune pathology despite fungal clearance during *Cryptococcus neoformans* meningoencephalitis. *MBio* (2017) 8:e01415-7. doi:10.1128/mBio.01415-17
4. Pirofski LA, Casadevall A. Immune-mediated damage completes the parabola: *Cryptococcus neoformans* pathogenesis can reflect the outcome of a weak or strong immune response. *MBio* (2017) 8:e02063-17. doi:10.1128/mBio.02063-17
5. Ganendren R, Carter E, Sorrell T, Widmer F, Wright L. Phospholipase B activity enhances adhesion of *Cryptococcus neoformans* to a human lung epithelial cell line. *Microbes Infect* (2006) 8:1006–15. doi:10.1016/j.micinf.2005.10.018
6. Cox GM, Mcdade HC, Chen SC, Tucker SC, Gottfredsson M, Wright LC, et al. Extracellular phospholipase activity is a virulence factor for *Cryptococcus neoformans*. *Mol Microbiol* (2001) 39:166–75. doi:10.1046/j.1365-2958.2001.02236.x
7. Almeida F, Wolf JM, Casadevall A. Virulence-associated enzymes of *Cryptococcus neoformans*. *Eukaryot Cell* (2015) 14:1173–85. doi:10.1128/EC.00103-15
8. Chen L-C, Blank E, Casadevall A. Extracellular proteinase activity of *Cryptococcus neoformans*. *Clin Diagn Lab Immunol* (1996) 3:570–4.
9. Xu CY, Zhu HM, Wu JH, Wen H, Liu CJ. Increased permeability of blood-brain barrier is mediated by serine protease during *Cryptococcus* meningitis. *J Int Med Res* (2014) 42:85–92. doi:10.1177/0300060513504365
10. Homer CM, Summers DK, Goranov AI, Clarke SC, Wiesner DL, Diedrich JK, et al. Intracellular action of a secreted peptide required for fungal virulence. *Cell Host Microbe* (2016) 19:849–64. doi:10.1016/j.chom.2016.05.001
11. Cox GM, Mukherjee J, Cole GT, Casadevall A, Perfect JR. Urease as a virulence factor in experimental cryptococcosis. *Infect Immun* (2000) 68:443–8. doi:10.1128/IAI.68.2.443-448.2000
12. Shi M, Li SS, Zheng C, Kim KS, Zhou H, Kubes P, et al. Real-time imaging of trapping and urease-dependent transmigration of *Cryptococcus* in the brain. *J Clin Invest* (2010) 120:1683–93. doi:10.1172/JCI41963
13. Chen SC, Wright LC, Santangelo RT, Muller M, Moran VR, Kuchel PW, et al. Identification of extracellular phospholipase B, lysophospholipase, and acyltransferase produced by *Cryptococcus neoformans*. *Infect Immun* (1997) 65:405–11.
14. Santangelo R, Zoellner H, Sorrell T, Wilson C, Donald C, Djordjevic J, et al. Role of extracellular phospholipases and mononuclear phagocytes in dissemination of cryptococcosis in a murine model. *Infect Immun* (2004) 72:2229–39. doi:10.1128/IAI.72.4.2229-2239.2004
15. Shea JM, Kechichian TB, Luberto C, Del Poeta M. The cryptococcal enzyme inositol phosphosphingolipid-phospholipase C confers resistance to the antifungal effects of macrophages and promotes fungal dissemination to the central nervous system. *Infect Immun* (2006) 74:5977–88. doi:10.1128/IAI.00768-06
16. Chayakulkeeree M, Sorrell TC, Siafakas AR, Wilson CF, Pantarat N, Gerik KJ, et al. Role and mechanism of phosphatidylinositol-specific phospholipase C in survival and virulence of *Cryptococcus neoformans*. *Mol Microbiol* (2008) 69:809–26. doi:10.1111/j.1365-2958.2008.06310.x
17. Djordjevic JT. Role of phospholipases in fungal fitness, pathogenicity, and drug development – lessons from *Cryptococcus neoformans*. *Front Microbiol* (2010) 1:125. doi:10.3389/fmicb.2010.00125
18. Pandey A, Ding SL, Qin QM, Gupta R, Gomez G, Lin F, et al. Global reprogramming of host kinase signaling in response to fungal infection. *Cell Host Microbe* (2017) 21:637–49.e636. doi:10.1016/j.chom.2017.04.008
19. Taylor-Smith LM. *Cryptococcus*-epithelial interactions. *J Fungi (Basel)* (2017) 3:E53. doi:10.3390/jof3040053
20. Schelenz S, Malhotra R, Sim RB, Holmskov U, Bancroft GJ. Binding of host collectins to the pathogenic yeast *Cryptococcus neoformans*: human surfactant D acts as an agglutinin for acapsular cells. *Infect Immun* (1995) 63:3360–6.
21. Drummond RA. Neuro-immune mechanisms of anti-cryptococcal protection. *J Fungi (Basel)* (2017) 4:E4. doi:10.3390/jof4010004
22. Levitz SM, Nong SH, Seetoo KF, Harrison TS, Speizer RA, Simons ER. *Cryptococcus neoformans* resides in an acidic phagolysosome of human macrophages. *Infect Immun* (1999) 67:885–90.
23. Smith LM, Dixon EF, May RC. The fungal pathogen *Cryptococcus neoformans* manipulates macrophage phagosome maturation. *Cell Microbiol* (2015) 17:702–13. doi:10.1111/cmi.12394
24. Feldmesser M, Kress Y, Novikoff P, Casadevall A. *Cryptococcus neoformans* is a facultative intracellular pathogen in murine pulmonary infection. *Infect Immun* (2000) 68:4225–37. doi:10.1128/IAI.68.7.4225-4237.2000
25. Tucker SC, Casadevall A. Replication of *Cryptococcus neoformans* in macrophages is accompanied by phagosomal permeabilization and accumulation of vesicles containing polysaccharide in the cytoplasm. *Proc Natl Acad Sci U S A* (2002) 99:3165–70. doi:10.1073/pnas.052702799
26. Chen M, Xing Y, Lu A, Fang W, Sun B, Chen C, et al. Internalized *Cryptococcus neoformans* activates the canonical caspase-1 and the noncanonical caspase-8 inflammasomes. *J Immunol* (2015) 195:4962–72. doi:10.4049/jimmunol.1500865
27. Davis MJ, Eastman AJ, Qiu Y, Gregorka B, Kozel TR, Osterholzer JJ, et al. *Cryptococcus neoformans*-induced macrophage lysosome damage crucially contributes to fungal virulence. *J Immunol* (2015) 194:2219–31. doi:10.4049/jimmunol.1402376
28. Ben-Abdallah M, Sturny-Leclere A, Ave P, Louise A, Moyrand F, Weih F, et al. Fungal-induced cell cycle impairment, chromosome instability and apoptosis via differential activation of NF-kappaB. *PLoS Pathog* (2012) 8:e1002555. doi:10.1371/journal.ppat.1002555
29. Coelho C, Souza AC, Derengowski Lda S, De Leon-Rodriguez C, Wang B, Leon-Rivera R, et al. Macrophage mitochondrial and stress response to ingestion of *Cryptococcus neoformans*. *J Immunol* (2015) 194:2345–57. doi:10.4049/jimmunol.1402350
30. Hirano A, Zimmerman HM, Levine S. Fine structure of cerebral fluid accumulation. V. Transfer of fluid from extracellular compartments in acute phase of cryptococcal polysaccharide lesions. *Arch Neurol* (1964) 11:632–41. doi:10.1001/archneur.1964.00460240064009
31. Hirano A, Zimmerman HM, Levine S. The fine structure of cerebral fluid accumulation. III. Extracellular spread of cryptococcal polysaccharides in the acute stage. *Am J Pathol* (1964) 45:1–19.
32. Hirano A, Zimmerman HM, Levine S. The fine structure of cerebral fluid accumulation. VII. Reactions of astrocytes to cryptococcal polysaccharide implantation. *J Neuropathol Exp Neurol* (1965) 24:386–96. doi:10.1097/00005072-196507000-00002

33. Chen SH, Stins MF, Huang SH, Chen YH, Kwon-Chung KJ, Chang Y, et al. *Cryptococcus neoformans* induces alterations in the cytoskeleton of human brain microvascular endothelial cells. *J Med Microbiol* (2003) 52:961–70. doi:10.1099/jmm.0.05230-0
34. Qin QM, Luo J, Lin X, Pei J, Li L, Ficht TA, et al. Functional analysis of host factors that mediate the intracellular lifestyle of *Cryptococcus neoformans*. *PLoS Pathog* (2011) 7:e1002078. doi:10.1371/journal.ppat.1002078
35. Jong A, Wu CH, Shackleford GM, Kwon-Chung KJ, Chang YC, Chen HM, et al. Involvement of human CD44 during *Cryptococcus neoformans* infection of brain microvascular endothelial cells. *Cell Microbiol* (2008) 10:1313–26. doi:10.1111/j.1462-5822.2008.01128.x
36. Johnston SA, May RC. The human fungal pathogen *Cryptococcus neoformans* escapes macrophages by a phagosome emptying mechanism that is inhibited by Arp2/3 complex-mediated actin polymerisation. *PLoS Pathog* (2010) 6:e1001041. doi:10.1371/journal.ppat.1001041
37. Alvarez M, Casadevall A. Phagosome fusion and extrusion, and host cell survival following *Cryptococcus neoformans* phagocytosis by macrophages. *Curr Biol* (2006) 16:2161–5. doi:10.1016/j.cub.2006.09.061
38. Ma H, Croudace JE, Lammas DA, May RC. Expulsion of live pathogenic yeast by macrophages. *Curr Biol* (2006) 16:2156–60. doi:10.1016/j.cub.2006.09.032
39. Stukes S, Coelho C, Rivera J, Jedlicka AE, Hajjar KA, Casadevall A. The membrane phospholipid binding protein annexin A2 promotes phagocytosis and nonlytic exocytosis of *Cryptococcus neoformans* and impacts survival in fungal infection. *J Immunol* (2016) 197(4):1252–61. doi:10.4049/jimmunol.1501855
40. Alvarez M, Casadevall A. Cell-to-cell spread and massive vacuole formation after *Cryptococcus neoformans* infection of murine macrophages. *BMC Immunol* (2007) 8:16. doi:10.1186/1471-2172-8-16
41. O'Meara TR, Veri AO, Ketela T, Jiang B, Roemer T, Cowen LE. Global analysis of fungal morphology exposes mechanisms of host cell escape. *Nat Commun* (2015) 6:6741. doi:10.1038/ncomms7741
42. Pericolini E, Cenci E, Monari C, De Jesus M, Bistoni F, Casadevall A, et al. *Cryptococcus neoformans* capsular polysaccharide component galactoxylomannan induces apoptosis of human T-cells through activation of caspase-8. *Cell Microbiol* (2006) 8:267–75. doi:10.1111/j.1462-5822.2005.00619.x
43. Monari C, Paganelli F, Bistoni F, Kozel TR, Vecchiarelli A. Capsular polysaccharide induction of apoptosis by intrinsic and extrinsic mechanisms. *Cell Microbiol* (2008) 10:2129–37. doi:10.1111/j.1462-5822.2008.01196.x
44. De JM, Nicola AM, Frases S, Lee IR, Mieses S, Casadevall A. Galactoxylomannan-mediated immunological paralysis results from specific B cell depletion in the context of widespread immune system damage. *J Immunol* (2009) 183:3885–95. doi:10.4049/jimmunol.0900449
45. Rodrigues ML, Nimrichter L, Oliveira DL, Frases S, Miranda K, Zaragoza O, et al. Vesicular polysaccharide export in *Cryptococcus neoformans* is a eukaryotic solution to the problem of fungal trans-cell wall transport. *Eukaryot Cell* (2007) 6:48–59. doi:10.1128/EC.00318-06
46. Rodrigues ML, Nakayasu ES, Oliveira DL, Nimrichter L, Nosanchuk JD, Almeida IC, et al. Extracellular vesicles produced by *Cryptococcus neoformans* contain protein components associated with virulence. *Eukaryot Cell* (2008) 7:58–67. doi:10.1128/EC.00370-07
47. Oliveira DL, Freire-De-Lima CG, Nosanchuk JD, Casadevall A, Rodrigues ML, Nimrichter L. Extracellular vesicles from *Cryptococcus neoformans* modulate macrophage functions. *Infect Immun* (2010) 78:1601–9. doi:10.1128/IAI.01171-09
48. Garcia-Solache MA, Izquierdo-Garcia D, Smith C, Bergman A, Casadevall A. Fungal virulence in a lepidopteran model is an emergent property with deterministic features. *MBio* (2013) 4:e00100-13. doi:10.1128/mBio.00100-13
49. Peres da Silva R, Puccia R, Rodrigues ML, Oliveira DL, Joffe LS, Cesar GV, et al. Extracellular vesicle-mediated export of fungal RNA. *Sci Rep* (2015) 5:7763. doi:10.1038/srep07763
50. Shao X, Mednick A, Alvarez M, Van Rooijen N, Casadevall A, Goldman DL. An innate immune system cell is a major determinant of species-related susceptibility differences to fungal pneumonia. *J Immunol* (2005) 175:3244–51. doi:10.4049/jimmunol.175.5.3244
51. Zaragoza O, Alvarez M, Telzak A, Rivera J, Casadevall A. The relative susceptibility of mouse strains to pulmonary *Cryptococcus neoformans* infection is associated with pleiotropic differences in the immune response. *Infect Immun* (2007) 75:2729–39. doi:10.1128/IAI.00094-07
52. Alanio A, Desnos-Ollivier M, Dromer F. Dynamics of *Cryptococcus neoformans*-macrophage interactions reveal that fungal background influences outcome during cryptococcal meningoencephalitis in humans. *MBio* (2011) 2:e00158-111. doi:10.1128/mBio.00158-11
53. Sabiiti W, Robertson E, Beale MA, Johnston SA, Brouwer AE, Loyse A, et al. Efficient phagocytosis and laccase activity affect the outcome of HIV-associated cryptococcosis. *J Clin Invest* (2014) 124:2000–8. doi:10.1172/JCI72950
54. Yauch LE, Lam JS, Levitz SM. Direct inhibition of T-cell responses by the *Cryptococcus* capsular polysaccharide glucuronoxylomannan. *PLoS Pathog* (2006) 2:e120. doi:10.1371/journal.ppat.0020120
55. Zaragoza O, García-Rodas R, Nosanchuk JD, Cuenca-Estrella M, Rodriguez-Tudela JL, Casadevall A. Fungal cell gigantism during mammalian infection. *PLoS Pathog* (2010) 6:e1000945. doi:10.1371/journal.ppat.1000945
56. Hommel B, Mukaremera L, Cordero RJB, Coelho C, Desjardins CA, Sturny-Leclère A, et al. Titan cells formation in *Cryptococcus neoformans* is finely tuned by environmental conditions and modulated by positive and negative genetic regulators. *PLoS Pathog* (2018) (in press).
57. Vecchiarelli A, Pericolini E, Gabrielli E, Kenno S, Perito S, Cenci E, et al. Elucidating the immunological function of the *Cryptococcus neoformans* capsule. *Future Microbiol* (2013) 8:1107–16. doi:10.2217/fmb.13.84
58. Osterholzer JJ, Surana R, Milam JE, Montano GT, Chen GH, Sonstein J, et al. Cryptococcal urease promotes the accumulation of immature dendritic cells and a non-protective T2 immune response within the lung. *Am J Pathol* (2009) 174:932–43. doi:10.2353/ajpath.2009.080673
59. Ellerbroek PM, Walenkamp AME, Hoepelman AIM, Coenjaerts FEJ. Effects of the capsular polysaccharides of Cryptococcus neoformans on phagocyte migration and inflammatory mediators. *Curr Med Chem* (2004) 11(2):253–66.
60. Dong ZM, Murphy JW. Cryptococcal polysaccharides induce L-selectin shedding and tumor necrosis factor receptor loss from the surface of human neutrophils. *J Clin Invest* (1996) 97:689–98. doi:10.1172/JCI118466
61. Casadevall A, Perfect JR. *Cryptococcus Neoformans*. Washington, DC: ASM Press (1998).
62. Lortholary O, Fontanet A, Memain N, Martin A, Sitbon K, Dromer F, et al. Incidence and risk factors of immune reconstitution inflammatory syndrome complicating HIV-associated cryptococcosis in France. *AIDS* (2005) 19:1043–9. doi:10.1097/01.aids.0000174450.70874.30
63. Beatty GW. Immune reconstitution inflammatory syndrome. *Emerg Med Clin North Am* (2010) 28:393–407. doi:10.1016/j.emc.2010.01.004
64. Goldman DL, Davis J, Bommartio F, Shao X, Casadevall A. Enhanced allergic inflammation and airway responsiveness in rats with chronic *Cryptococcus neoformans* infection suggests a potential role for fungal pulmonary infection in the pathogenesis of asthma. *J Infect Dis* (2006) 193:1178–86. doi:10.1086/501363
65. Goldman DL, Vicencio AG. The chitin connection. *MBio* (2012) 3:e00056-12. doi:10.1128/mBio.00056-12
66. Grahnert A, Muller U, Von Buttlar H, Treudler R, Alber G. Analysis of asthma patients for cryptococcal seroreactivity in an urban German area. *Med Mycol* (2015) 53:576–86. doi:10.1093/mmy/myv024
67. Charlier C, Nielsen K, Daou S, Brigitte M, Chretien F, Dromer F. Evidence of a role for monocytes in dissemination and brain invasion by *Cryptococcus neoformans*. *Infect Immun* (2009) 77:120–7. doi:10.1128/IAI.01065-08
68. Sorrell TC, Juillard PG, Djordjevic JT, Kaufman-Francis K, Dietmann A, Milonig A, et al. Cryptococcal transmigration across a model brain blood-barrier: evidence of the Trojan horse mechanism and differences between *Cryptococcus neoformans* var. grubii strain H99 and *Cryptococcus gattii* strain R265. *Microbes Infect* (2016) 18:57–67. doi:10.1016/j.micinf.2015.08.017
69. Santiago-Tirado FH, Onken MD, Cooper JA, Klein RS, Doering TL. Trojan horse transit contributes to blood-brain barrier crossing of a eukaryotic pathogen. *MBio* (2017) 8:e2183–2116. doi:10.1128/mBio.02183-16
70. Chang YC, Stins MF, Mccaffery MJ, Miller GF, Pare DR, Dam T, et al. Cryptococcal yeast cells invade the central nervous system via transcellular penetration of the blood-brain barrier. *Infect Immun* (2004) 72:4985–95. doi:10.1128/IAI.72.11.6753.2004
71. Liu TB, Kim JC, Wang Y, Toffaletti DL, Eugenin E, Perfect JR, et al. Brain inositol is a novel stimulator for promoting *Cryptococcus* penetration of the blood-brain barrier. *PLoS Pathog* (2013) 9:e1003247. doi:10.1371/journal.ppat.1003247

72. Vieira MA, Costa CH, Ribeiro JC, Nunes-Filho LP, Rabelo MG, Almeida-Neto WS. Soap bubble appearance in brain magnetic resonance imaging: cryptococcal meningoencephalitis. *Rev Soc Bras Med Trop* (2013) 46:658–9. doi:10.1590/0037-8682-0142-2013
73. Lee SC, Kress Y, Zhao M-L, Dickson DW, Casadevall A. *Cryptococcus neoformans* survive and replicate in spacious phagosomes in human microglia. *Lab Invest* (1995) 73:871–9.
74. Casadevall A, Pirofski L. The damage-response framework of microbial pathogenesis. *Nat Microbiol Rev* (2003) 1:17–24. doi:10.1038/nrmicro732
75. Panackal AA, Williamson KC, Van De Beek D, Boulware DR, Williamson PR. Fighting the monster: applying the host damage framework to human central nervous system infections. *MBio* (2016) 7:e1906–15. doi:10.1128/mBio.01906-15
76. Steenbergen JN, Shuman HA, Casadevall A. *Cryptococcus neoformans* interactions with amoebae suggest an explanation for its virulence and intracellular pathogenic strategy in macrophages. *Proc Natl Acad Sci U S A* (2001) 18:15245–50. doi:10.1073/pnas.261418798
77. Casadevall A. Amoeba provide insight into the origin of virulence in pathogenic fungi. *Adv Exp Med Biol* (2012) 710:1–10. doi:10.1007/978-1-4419-5638-5_1
78. Fu MS, Casadevall A. Divalent metal cations potentiate the predatory capacity of amoeba for *Cryptococcus neoformans*. *Appl Environ Microbiol* (2017) 84(3):e01717. doi:10.1128/AEM.01717-17

12

Disease Tolerance Mediated by Phosphorylated Indoleamine-2,3 Dioxygenase Confers Resistance to a Primary Fungal Pathogen

*Eliseu Frank de Araújo[1], Flávio Vieira Loures[1], Cláudia Feriotti[1], Tania Costa[1], Carmine Vacca[2], Paolo Puccetti[2], Luigina Romani[2] and Vera Lúcia Garcia Calich[1]**

[1] *Department of Immunology, Institute of Biomedical Sciences, University of São Paulo, São Paulo, Brazil,*
[2] *Department of Experimental Medicine, University of Perugia, Perugia, Italy*

Correspondence:
Vera Lúcia Garcia Calich
vlcalich@icb.usp.br

Resistance to primary fungal pathogens is usually attributed to the proinflammatory mechanisms of immunity conferred by interferon-γ activation of phagocytes that control microbial growth, whereas susceptibility is attributed to anti-inflammatory responses that deactivate immunity. This study challenges this paradigm by demonstrating that resistance to a primary fungal pathogen such as *Paracoccidiodes brasiliensis* can be mediated by disease tolerance, a mechanism that preserves host fitness instead of pathogen clearance. Among the mechanisms of disease tolerance described, a crucial role has been ascribed to the enzyme indoleamine-2,3 dioxygenase (IDO) that concomitantly controls pathogen growth by limiting tryptophan availability and reduces tissue damage by decreasing the inflammatory process. Here, we demonstrated in a pulmonary model of paracoccidioidomycosis that IDO exerts a dual function depending on the resistant pattern of hosts. IDO activity is predominantly enzymatic and induced by IFN-γ signaling in the pulmonary dendritic cells (DCs) from infected susceptible (B10.A) mice, whereas phosphorylated IDO (pIDO) triggered by TGF-β activation of DCs functions as a signaling molecule in resistant mice. IFN-γ signaling activates the canonical pathway of NF-κB that promotes a proinflammatory phenotype in B10.A DCs that control fungal growth but ultimately suppress T cell responses. In contrast, in A/J DCs IDO promotes a tolerogenic phenotype that conditions a sustained synthesis of TGF-β and expansion of regulatory T cells that avoid excessive inflammation and tissue damage contributing to host fitness. Therefore, susceptibility is unexpectedly mediated by mechanisms of proinflammatory immunity that are usually associated with resistance, whereas genetic resistance is based on mechanisms of disease tolerance mediated by pIDO, a phenomenon never described in the protective immunity against primary fungal pathogens.

Keywords: paracoccidioidomycosis, indoleamine-2,3 dioxygenase, pulmonary dendritic cells, disease tolerance, phosphorylated indoleamine-2,3 dioxygenase, signaling function

INTRODUCTION

Indolamine-2,3 dioxygenase (IDO) is an intracellular enzyme of crucial importance in the tryptophan (trp) catabolism acting along the kynurenine (Kyn) pathway (1, 2). Although the cytokine IFN-γ is considered the major inducer of IDO expression, TGF-β, TNF-α, prostaglandins, and toll-like receptor ligands, among others, have also the ability of promoting the expression of the

IDO gene (3–5). Among the cells of myeloid origin, dendritic cells (DCs) express the highest levels of IDO that is, however, also present in macrophages, neutrophils, endothelial cells, and fibroblasts (6–8).

Indolamine-2,3 dioxygenase was primarily associated with host defense against pathogens due to trp starvation caused by its enzymatic activity (9–11). However, the recent recognition of the immunoregulatory function of IDO and Kyn, put IDO to another level of knowledge and relevance (11, 12). Indeed, IDO has been recently described as an intracellular signaling molecule involved in the sustained tolerogenic activity of DCs. This signaling function is mediated by TGF-β signaling that promotes IDO phosphorylation that ultimately activates the non-canonical NF-κB pathway that triggers genes encoding IDO, TGF-β, IFN-α, and IFN-β, that maintain the tolerogenic activity of murine plasmacytoid DCs (12). This likely accounts for the ability of IDO to favor microbial persistence and concomitant control of inflammation during chronic infections (13–15). Therefore, acting *via* trp depletion and Kyn production, IDO can inhibit the proliferation of T cells and induce their apoptosis. Furthermore, through the aryl hydrocarbon receptor (AhR), IDO directs the conversion of naive $CD4^+$ T cells into regulatory $Foxp3^+$ T cells (Tregs) (13, 16, 17).

Paracoccidioidomycosis (PCM), is a granulomatous disease caused by two species of the dimorphic fungus *Paracoccidioides*, *P. brasilienis*, and *P. lutzii* (18). Immunoprotection to PCM is mediated by prevalent Th1/Th17 immunity, whereas Th2 and Th9 responses are associated with severe forms of the disease (19–21). In a murine model of pulmonary infection, A/J and B10.A mice were described, respectively, as resistant and susceptible to PCM. The A/J mouse strain develops a chronic and regressive PCM restricted to the lungs. Its adaptive immunity is mainly mediated by Th1 and Th17 cells that are tightly controlled by elevated numbers and activity of Treg cells. In contrast, B10.A mice develop a progressive disseminated disease associated with large amounts of fungi in non-organized lesions that mimic the severe forms of PCM (22–27).

The immunoregulatory mechanisms that control resistance to PCM are complex and not yet completely solved. In this aspect, our previous studies have shown that in pulmonary PCM, IDO is an important immunoregulatory enzyme that promotes fungal clearance and inhibits T cell immunity but only in susceptible mice IDO inhibition by 1-methyl-DL-tryptophan (1MT) caused progressive tissue pathology and increased mortality rates (28). This difference appeared to be mediated by the opposite innate immunity of resistant and susceptible mice. Indeed, in B10.A mice the innate immunity is preferentially proinflammatory with elevated production of IL-12 and IFN-γ, whereas in A/J mice the predominant TGF-β secretion provides an anti-inflammatory innate response (26–30). These findings led us to hypothesize that IDO has distinct immunoregulatory roles in pulmonary PCM. Here, we could demonstrate that in susceptible mice IDO activity is IFN-γ-dependent and mediated by its catalytic activity, whereas in resistant mice a prevalent TGF-β signaling triggered IDO phosphorylation imparting a signaling and tolerogenic function. The TGF-β-IDO-Treg interplay generates an early pathogen tolerance that allows A/J mice to interact with a primary fungal pathogen as a commensal microbe. Thus, the signaling function of pIDO may lead to an unusual fungus–host interaction that efficiently balances tolerance and resistance mechanisms to the benefit both the pathogen and the host.

MATERIALS AND METHODS

Mice

Susceptible (B10.A) and resistant (A/J) mouse strains to *Paracoccidiodes brasiliensis* infection were obtained from our Isogenic Unit (Immunology Department of Institute of Biomedical Sciences of University of São Paulo, Brazil) and used at 8–11 weeks of age. Specific pathogen free mice were fed with sterilized laboratory chow and water *ad libitum*.

Fungus

Paracoccidiodes brasiliensis 18 (Pb 18) was used throughout this investigation. To ensure the maintenance of its virulence, the isolate was used after three serial animal passages. Pb18 yeast cells were maintained by weekly sub cultivation in semisolid Fava Netto culture medium at 37°C and used on the seventh day of culture. Phosphate-buffered saline (PBS)-washed yeast cells were adjusted to 20×10^6 cells/mL based on hemocytometer counts. Viability was determined with Janus Green B vital dye (Merck) and was always higher than 85%.

1MT Treatment and *In Vivo* Fungal Infection

Mice were anesthetized and submitted to intratracheal (i.t.) *P. brasiliensis* infection as previously described. Briefly, after intraperitoneal (i.p.) anesthesia the animals were infected with 1×10^6 Pb18 yeast cells, contained in 50 μL of PBS, by surgical i.t. inoculation, which allowed dispensing of the fungal cells directly into the lungs. The skins of the animals were then sutured, and the mice were allowed to recover under a heat lamp. Groups of infected B10.A and A/J mice were treated with daily i.p. injections of 5 mg/mL of 1MT or 1 mg/mL of rice starch (Sigma-Aldrich) as control. Mice were sacrificed after 96 h or 2 weeks of infection.

Isolation of Pulmonary $CD11c^+$ Cells

Cell suspensions of the lungs were prepared as previously described (31). The lungs were removed and digested for 60 min in RPMI-1640 medium (Sigma) containing collagenase (2 mg/mL) and DNAse (30 mg/mL). The erythrocytes were lysed with lysis buffer (TRIS + ammonium chloride) and viability determined by the trypan blue exclusion test with viabilities at least to 95%. $CD11c^+$ cells were isolated from total pulmonary cells by magnetic microbeads (Miltenyi Biotec, Cologne, Germany). Pellets of pulmonary cells were counted, washed and the pellet resuspended in 400 μL of buffer (PBS + 0.5% BSA + 2 mM EDTA). The Trypan Blue dye exclusion test was used to determine the number of viable cells present in the DCs cell suspension with viabilities at least to 95%. Then, 100 μL of anti-CD11c coated microspheres were added for each 10^8 cells and incubated for 15 min at 2–8°C. The cells were washed, the supernatant removed and the pellet resuspended in 500 μL of buffer. This suspension was fractionated using a magnetic separation column, and purified CD11c$^+$ DCs obtained. This fractionation methods resulted in 95% CD11c$^+$ DCs.

Culture of DCs

The viability of the CD11c$^+$ cells was evaluated by the trypan blue exclusion test, and was always higher than 85%. The cell suspensions were then centrifuged at 1,200 rpm at 4°C for 10 min and resuspended in 1.0 mL of culture medium (RPMI-1640 medium) supplemented with 10% fetal bovine serum. The cells were adjusted to 1×10^6/mL and 500 µL were dispensed into each well of 24-well culture plates. The cultures were incubated at 37°C in an incubator containing 5% CO_2–95% air for 24 h. After this period, the cells were removed and used in different assays.

CFU Assay

After 24 h of culture, DCs suspensions were lysed and 100 µL of pellets plated onto BHI-agar medium containing 5% "fungus growth factor" and 4% horse serum (32). The plates were incubated at 35°C and colonies counted daily until no increase in CFU was observed. The numbers (log10) of viable *P. brasiliensis* colonies are expressed as the mean ± SE. Two to three experiments were performed separately.

NO and Cytokines Measurement

Supernatants from DC cultures and lymphoproliferation assays were separated and stored at −70°C. The levels of IL-12, IL-1β, TGF-β, IL-6, IFN-γ, and IL-12 were measured by a capture enzyme-linked immunosorbent assay (ELISA) with antibody pairs purchased from eBioscience. The ELISA procedure was performed according to the manufacturer's protocol, and absorbance was measured with a Versa Max Microplate Reader (Molecular Devices). The concentrations of cytokines were determined based on a standard curve of serial twofold dilutions of murine recombinant cytokines. Nitric oxide production was quantified by the accumulation of nitrite in the supernatants from *in vitro* protocols by a standard Griess reaction (33). All determinations were performed in duplicate, and results were expressed as micromolar concentration of NO.

Determination of IDO Enzymatic Activity

To monitor IDO enzymatic activity, Kyn were measured using a modified spectrophotometric assay (5). The amount of 50 µL of 30% trichloroacetic acid was added to 100 µL of DCs supernatants, vortexed, and centrifuged at 800 *g* for 5 min. A volume of 75 µL of the supernatant was then added to an equal volume of Ehrlich reagent (100 mg P-dimethylbenzaldehyde, 5 mL glacial acetic acid) in a 96-well microtiter plate. Optical density was measured at 492 nm, using a Multiskan MS (Labsystems) microplate reader. A standard curve of defined L-Kyn concentrations (0–100 mM) was used to determine unknown Kyn concentrations.

Flow Cytometry

Pulmonary CD11c$^+$ cells were obtained from 1MT treated and untreated B10.A and A/J mice 96 h and 2 weeks after *P. brasiliensis* infection. For cell surface staining, pulmonary CD11c$^+$ cells were washed and resuspended at a concentration of 1×10^6 cells/mL in staining buffer (PBS 1×, 2% FBS, 0.5% NaN_3). Fc receptors were blocked by the addition of unlabeled anti-CD16/32 (Fc block; BD Biosciences). The DCs cells were then stained for 30 min at 4°C with the optimal dilution of each labeled antibody [phycoerythrin tandem Cy7 (PE-Cy7)-labeled anti-CD11c, phycoerythrin tandem Cy5 (PE-Cy5)-labeled anti-B220, pacific blue (PB)-labeled anti-CD8, phycoerythrin (PE)-labeled anti-CD86, allophycocyanin (APC)-labeled anti-CD40, fluorescein isothiocyanate (FITC)-labeled MHCII (IA^K), and allophycocyanin tandem Cy7 (APC-Cy7)-labeled anti-CD11b] monoclonal antibodies (mAbs) from BD Biociences. Cells were washed twice with staining buffer, fixed with 2% paraformaldehyde (Sigma), and acquired using a FACSCanto II equipment and FACSDiva® software (BD Biosciences) and analyzed by the FlowJo software (Tree-Star, Ashland, OR, USA).

Lymphocyte Proliferation Assay and Lymphocyte Phenotyping

Spleen lymphocytes from normal B10.A and A/J mice were resuspended (1×10^7) in 1 mL PBS-BSA (0.1%) and 1 µL of carboxyfluorescein diacetate, succinimidly ester (CFSE) at a concentration of 5 mM (Molecular Probes, USA). The cells were incubated at room temperature in the dark for 15 min (shaking the tube constantly). The cells were washed two times with 10 mL of supplemented RPMI, counted and resuspended at a concentration of 1×10^6/mL. In parallel, purified DCs from A/J and B10.A of uninfected and infected mice, treated and untreated with 1MT were centrifuged and adjusted to 1×10^6/mL in supplemented medium. Aliquots of 100 µL of DCs were placed in a 96-well plate (U-bottom) in the presence of lymphocytes previously stained with CFSE at a ratio 1:10 (100 µL of DCs + 100 µL of lymphocytes). CFSE stained lymphocytes (treated or not with 1MT), purified DCs and Concanavalin A (Sigma) stimulated lymphocytes were used as controls. The cells were incubated at 37°C in 5% CO_2 for 3 days. After the incubation period the cells were washed and resuspended at a concentration of 1×10^6 cells/mL in staining buffer (PBS 1×, 2% FBS, 0.5% NaN_3). Fc receptors were blocked by the addition of unlabeled anti-CD16/32 (Fc block; BD Biosciences). The leukocytes were then stained for 30 min at 4°C with the optimal dilution of each antibody labeled with the adequate fluorochrome (BD Biosciences). PE-labeled anti-CD44, PE-Cy7-labeled anti-CD8, and PerCP Cy5.5-labeled anti-CD4, and APC-Cy7-labeled anti-CD62L mAbs from BD Biosciences were used. Cells were washed twice with staining buffer, resuspended in 100 µL, and an equal volume of 2% paraformaldehyde was added to fix the cells. A minimum of 50,000 events was acquired on a FACSCanto II flow cytometer using FACSDiva® and FlowJo softwares. Lymphocytes were identified on forward-scatter (FSC) and side-scatter (SSC) analysis. Gated cells were measured for CD4$^+$ and CD8$^+$ expression followed by CD44 expression, and cells expressing high and low levels of this molecule were gated. Gated CD44^{high} cells were then measured for expression of low levels of CD62L identifying the effector/memory CD4$^+$CD44^{high}CD62L^{low} and CD8$^+$CD44^{high}CD62L^{low} subpopulations. Gated CD44^{low} cells were then measured for the expression of high levels of CD62L identifying the naive

CD4$^+$CD44^{low}CD62L^{high} and CD8$^+$CD44^{low}CD62L^{high} subpopulations (Figure S2A,B in Supplementary Material). Fifty thousand cells were acquired and the data expressed as the frequency of positive cells. The cell division index was calculated and was based on the number of CFSE$^+$CD4$^+$ or CFSE$^+$CD8$^+$ T cells found in the stimulated culture/number of CFSE$^+$CD4$^+$ or CFSE$^+$CD8$^+$ T cells in the unstimulated culture (34).

Assay to Determine Regulatory CD4$^+$CD25$^+$FoxP3$^+$ T Cells

Culture supernatants from lymphoproliferation assays were removed and stored at 70°C for the determination of cytokines according to the manufacturer's protocol. The pellet cells were stained with PE-Cy7-labeled anti-CD4 and PerCP Cy5.5-labeled anti-CD25 for 30 min 4°C. Next, the cells were washed with staining buffer containing 2% fetal bovine serum, 0.1% azide, 100 µL of PBS. The permeabilization of the plasma membrane was made with the addition of 200 µL/well of Cytofix/Cytoperm (Fixation/permeabilization kit BD Biosciences®) at room temperature for 30 min, protected from light. After washing the cells, the nuclear membrane permeabilization was performed using 130 µL/well of solution containing 750 µL of 1× PBS, 250 µL of 4% paraformaldehyde, 5 µL Tween 20 Sigma® and the cells then were incubated for 30 min at room temperature in the dark.

After washing the cells with cold PBS (1×), the nuclear transcription factor FoxP3 was stained with an anti-FoxP3 labeled specific antibody (APC-labeled anti-FoxP3) diluted 1/100 in staining buffer. The cells were incubated for 30 min at 4°C, then washed with ice-cold PBS, fixed with 2% paraformaldehyde and stored at 4°C in the dark until analyzed by flow cytometry. Gates for total lymphocytes, CD4$^+$, CD4$^+$CD25$^+$, and finally for TCD4$^+$CD25$^+$FoxP3$^+$ T lymphocytes were made (Figure S2C in Supplementary Material). Cells were analyzed as above described.

Intracellular Cytokines Measurement

Dendritic cells obtained from the lungs were adjusted to 1×10^6 cells and stimulated for 6 h in complete medium in the presence of 100 ng/mL phorbol 12-myristate 13-acetate, 500 ng/mL ionomycin (both from Sigma-Aldrich; Germany), and monensin (3 mM, eBioscience). Cells were subsequently labeled for surface molecules (PerCP-labeled anti-CD11c) and then treated according to the manufacturer's protocol for intracellular staining using the Cytofix/Cytoperm kit (BD Biosciences). Cells were then stained with Alexa Fluor 488-labeled anti-IDO, PB-labeled anti-IL-10, PE-Cy7-labeled anti-TNF-α, APC-CY7-labeled anti-IL-12, and PE-labeled anti-TGF-β. Cells were washed twice with staining buffer, resuspended in 100 µL, and fixed with an equal volume of 2% paraformaldehyde. The flow cytometry data were acquired and analyzed as above described.

Quantitative Analysis of mRNA Expression

RNA was extracted from DC cultures and the RNA concentrations were determined by spectrophotometer readings at an absorbance of 260 nm. First-strand cDNAs were synthesized from 2 µg RNA using the High Capacity RNA-to-cDNA kit (Applied Biosystems) according to the manufacturer's instructions. Gene expression for the genes of IDO, IFN-γ, janus kinase 1 (JAK1), signal transducer and activator of transcription (STAT1), IL-6, suppressor of cytokine signaling 3 (SOCS3), nitric oxide synthase (NOS2), TGF-β, small mothers against decapentaplegic 2 and 3 (SMAD2 and SMAD3), inositol polyphosphate phosphatase (SHIP), Shp1, Shp2 (tyrosine phosphatases 1 and 2), IFN-α, IFN-β, and the endogenous gene glyceraldehyde 3-phosphate dehydrogenase (GAPDH) was characterized by the real time polymerase chain reaction technique (qPCR) (Applied Biosystems, Foster City, CA, USA). The samples were placed in microcentrifuge tubes for specific qPCR testing of samples containing 2 µL cDNA + 10 µL TaqMan PCR Master Mix (Applied Biosystems, Foster City, CA, USA) according to the following thermal profile: denaturation at 95°C for 15 s, annealing at 60°C for 15 s, and extension at 72°C for 15 s. The sequences of the primers for Indo (Mm00492506_m1), IFN-γ (Mm01168134_m1), Jak1 (Mm00600614_m1), Stat1 (Mm00439531_m1), IL-6 (Mm00446190_m1), SOCS3 (Mm00545913_s1), NOS2 (Mm00440502_m1), TGF-β1 (Mm01178820_m1), Smad2 (Mm00487530_m1), Smad3 (Mm01170760_m1), Inpp5d (Ship—Mm00494987_m1), Nr0b2 (Shp1—Mm00442278_m1), PTPN11 (Shp2—Mm00448434_m1), Ifna1 (Mm03030145_gH), Ifnb1 (Mm00439552_s1), and the gene endogenous transcripts in all cell types, GAPDH (Mm99999915_g1) were obtained from Applied Biosystems ready for testing. Probes were labeled with 6-carboxyfluorescein (FAM) at their 5′-terminal end. Data were normalized to GAPDH gene expression. *Taq*Man PCR assays were performed on a Stratagene MxP3005P QPCR System and data were developed using the MxPro QPCR software (Agilent Technologies, USA).

NF-κB DNA-Binding ELISA-Based Assay

Nuclear extracts from DCs were obtained using the Nuclear Extract Kit (Active Motif, USA). DNA binding activity of NF-κB was measured by an ELISA-based assay using a TransAM NF-κB kit (Active Motif, USA) following manufacturer's instructions. This assay detects binding of p65, p50, p52, and Rel-B proteins to oligonucleotides containing an NF-κB consensus-binding site immobilized onto 96-well plates by specific primary antibodies that recognize an epitope that is accessible only when NF-κB is activated and bound to its target DNA. The binding activity was quantified by spectrophotometry, the OD at 450 nm.

Western Blot Analysis

Dendritic cells were lysed in ice-cold NP40 buffer [1% NP40, 50 mmol/L Tris-HCl (pH 7.4), 150 mmol/L NaCl, 5 mmol/L EDTA containing 5 mmol/L NaF, 2 mmol/L Na_3VO_4, 1 mmol/L phenylmethylsulfonyl fluoride, 5 µg/mL leupeptine, and 5 µg/mL aprotinine], supplemented with a protease inhibitor (Sigma Aldrich). After incubating for 30 min at 4°C, the samples were centrifuged, and the supernatants were kept as the NP40-soluble fraction. The pellets were resuspended in SDS buffer [2% SDS, 80 mmol/L Tris (pH 6.8), 100 mmol/L DTT, and 10% glycerol]. Protein concentration was determined using a BCA assay (Pierce), and 40–60 µg of protein was run on a polyacrylamide gel and transferred to a 0.45 µm nitrocellulose membrane

(Millipore, Bedford, MA, USA). Western blotting was carried out according to a standard procedure using horseradish peroxidase-conjugated secondary antibodies (Thermo Scientific and Sigma Aldrich). mAbs used were as follows: Jak1 (1:500; Sigma-Aldrich), Stat1 (1:500; Sigma-Aldrich), IKK-β (1:1,500; Sigma-Aldrich), IDO1 (1:200; Santa Cruz Biotechnology), Smad2 (1:2,500; Sigma-Aldrich), Ship (1:2,000; Sigma-Aldrich), Shp-2 (1:1,500; Sigma-Aldrich), and IKK-α (1:500; Sigma-Aldrich). Mouse monoclonal antimouse β-tubulin antibody (1:2,000; Sigma-Aldrich) was used as an internal control. The immunoreactive proteins were visualized with ECL plus reagents (ECL Western blotting Detection Reagents; Amersham) and enhanced chemiluminescence apparatus (ImageQuant LAS 4000). Densitometry was performed using ImageQuant TL 8.1 software.

Detection of Phosphorylated IDO1 (pIDO1) by Western Blotting

Lung extracts from susceptible and resistant mice infected with *P. brasiliensis*, treated or not with 1MT were obtained after 96 h and 2 weeks of infection and analyzed by Western Blot for detection of pIDO1. Protein concentration was determined using a BCA assay (Pierce), and 20 µg of protein in Ripa Buffer and Sample buffer 4× were heated at 100°C for 4 min, then on ice for 2 min and loaded 60 µL per well in 10% SDS PAGE, 1.5 mm. Electrophoresis was then performed by transferring the gel to the nitrocellulose membrane with the membranes incubated in blocking solution [5% non-fat dry milk, 0.1% Tween 20 in Tris buffer saline (TBS) for 1 h at room temperature on an orbital shaker]. After two washings steps with TBS-Tween 20 0.1% buffer, the membrane was incubated overnight with polyclonal rabbit antimouse pIDO (CV223 AP8), 2 µg/mL in 5% non-fat dry milk, TBS at 0.1%. The membrane was washed three times for 5 min and incubated with HRP-labeled secondary antibody (anti-rabbit HRP–Pierce) for 1 h at room temperature. The revelation was performed using the ECL chemiluminescent substrate for enzymatic activity detection of peroxidase (HRP) on photographic film. The same method was used with a monoclonal rabbit anti-mouse IDO1 (CV152) and monoclonal antimouse β-tubulin antibody (1:1,000).

Statistical Analysis

Data are expressed as the mean ± SEM. Differences between groups were analyzed by analysis of variance followed by the Bonferroni test. Differences between survival times were determined with the LogRank test. Data were analyzed using GraphPad Prism 6.2 software for Windows (GraphPad). A p value of ≤0.05 was considered statistically significant.

RESULTS

IDO Inhibition Increases the Fungal Loads of DCs from Susceptible and Resistant Mice and Down Regulates NO, Kyn, and IDO mRNA Production

Initially, we sought to investigate the effects of IDO inhibition on the interaction of DCs, from susceptible and resistant mice, with *P. brasiliensis*. To explore the effect of IDO inhibition *in vivo*, A/J and B10.A mice were treated or not with 1MT and subsequently subjected to i.t. infection with *P. brasiliensis* yeasts cells. Pulmonary DCs from both mouse strains were isolated 96 h and 2 weeks after infection and *in vitro* cultivated for 24 h. Using a CFU assay it was verified that in both postinfection periods 1MT-treated DCs presented increased fungal loads than untreated cells (**Figures 1A,B**). The levels of NO and Kyn in the supernatants of 1MT-treated DCs from both mouse strains was significantly lower than those detected in untreated groups (**Figures 1A,B**). In agreement with the reduced levels of Kyn observed, a reduced expression of IDO mRNA was found in 1MT-treated groups of both mouse strains (**Figures 1A,B**). The reduced IDO mRNA expression was further

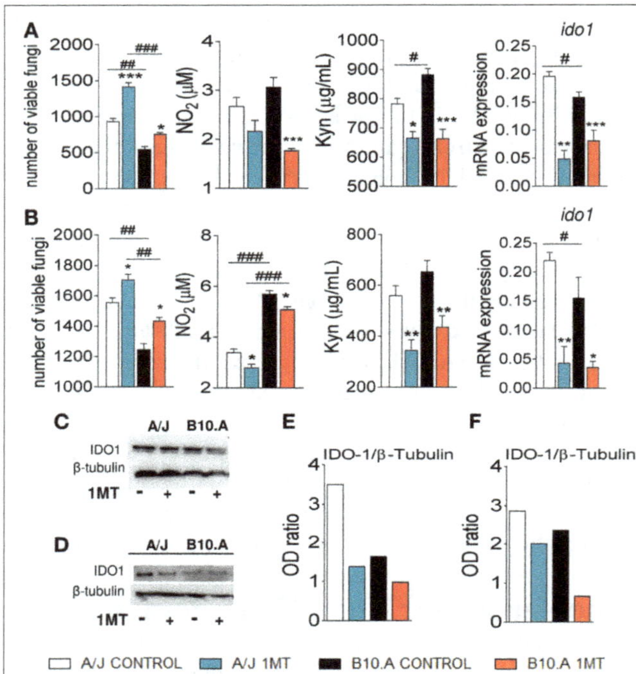

FIGURE 1 | 1-Methyl-DL-tryptophan (1MT) treatment reduces fungicidal activity, NO, kynurenine (Kyn), and indoleamine-2,3 dioxygenase (IDO) production by pulmonary dendritic cells (DCs) of *Paracoccidiodes Brasiliensis*-infected mice. Pulmonary DCs were obtained from 1MT treated or untreated A/J and B10.A mice at 96 h **(A,C,E)** and 2 weeks **(B,D,F)** after infection with 1 × 10^6 yeasts and cultivated for 24 h at 37°C in 5% CO_2. The cells were centrifuged, resuspended in 100 µL of culture medium and assayed for the presence of viable yeasts by a CFU assay. Supernatants from DCs cultures were used to determine the levels of nitrite and Kyn. In the same experimental conditions, DCs of A/J and B10.A mice were mixed with TRizol reagent for RNA extraction. IDO mRNA was measured using TaqMan real-time PCR assay. IDO was also characterized by immunoblotting. DCs were lysed, supernatants supplemented with a protease inhibitor and protein concentration determined by a BCA assay. After electrophoresis in polyacrylamide gel, proteins were transferred to nitrocellulose membranes and stained with anti-IDO antisera. Densitometry of bands was performed using ImageQuant TL 8.1 software. NO production was measured by Griess reagent, and Kyn were evaluated using Ehrlich's reagent. The data represent the mean ± SEM of three independent experiments where the asterisk (*) represents a statistically significant difference between treatments (*$p < 0.05$, **$p < 0.01$, and ***$p < 0.001$) and the hashtag (#) represents the difference between the strains (#$p < 0.05$, ##$p < 0.01$, and ###$p < 0.001$).

confirmed by the decreased protein production as detected by Western Blotting (**Figures 1C–F**).

IDO Inhibition Promotes Increased Influx and Activation of Pulmonary DCs

We have then asked whether the *in vivo* treatment with 1MT could interfere with the number and activation of pulmonary DCs. We verified that 96 h and 2 weeks after infection, 1MT-treated A/J and B10.A mice presented a higher number of myeloid, lymphoid and plasmacytoid DCs (CD11c^{high}CD11b^{+}, CD11c^{high}CD8^{+}, and CD11c^{low}B220^{+}, respectively) in their lungs than control mice (**Figures 2A,B**; Figure S1 in Supplementary Material). In addition, 1MT treatment increased the expression of MHC class II (IA^{K}) and costimulatory molecules (CD86, CD40, CD11b) in pulmonary DCs obtained at both periods and mouse strains studied (**Figures 2C,D**). Thus, the increased fungal loads induced by 1MT treatment were associated with increased migration and activation of pulmonary DCs.

IDO Inhibition Reduces the Secretion of DCs Cytokines

To better evaluate the function of pulmonary DCs from 1MT-treated and untreated A/J and B10.A mice, the levels of cytokines present in the supernatants of cultivated cells were measured by ELISA. First, we could confirm our previous findings (28, 30) demonstrating that DCs from A/J mice produce higher levels of TGF-β and IL-1β than DCs from B10.A mice that, in contrast, are better producer of IL-12 and IL-6. We have found that 1MT treatment reduced the levels of TGF-β and IL-1β produced by A/J DCs at both postinfection periods. In contrast, reduced levels of IL-12, IL-6, and IL-1β were observed in 1MT-treated B10.A mice at both time points (**Figures 3A,B**).

IDO Inhibition Reduces the Levels of Intracellular IDO and TGF-β in A/J DCs, whereas in B10.A DCs Diminishes the Levels of IDO, IL-12, and IL-6

We have also characterized the effect of IDO inhibition on intracellular cytokine (TGF-β, IL-12, and IL-6) content and IDO. Purified DCs from 1MT-treated and -untreated A/J and B10.A mice obtained at two postinfection periods were cultivated and intracellular proteins analyzed by flow cytometry. Compared with untreated controls, decreased numbers of TGF-β^{+} DCs were observed in the lungs of 1MT treated A/J mice. In contrast, reduced numbers of IL-12^{+} and IL-6^{+} DCs were detected in 1MT treated B10A mice (**Figures 4A,B**). In addition, reduced numbers of IDO positive DCs were observed in 1MT-treated DCs from both mouse strains and postinfection periods analyzed (**Figures 4A,B**).

IDO Inhibition Enhances the Immunogenic Activity of DCs

Dendritic cells were isolated from 1MT-treated and untreated A/J and B10.A mice 96 h and 2 weeks after *P. brasiliensis* infection. In parallel, splenic lymphocytes from uninfected A/J and B10. A mice were obtained, labeled with CFSE, and cocultured with purified DCs. After 3 days, the proliferation of lymphocytes was analyzed by flow cytometry and the levels of IFN-γ and IL-2 were measured in culture supernatants.

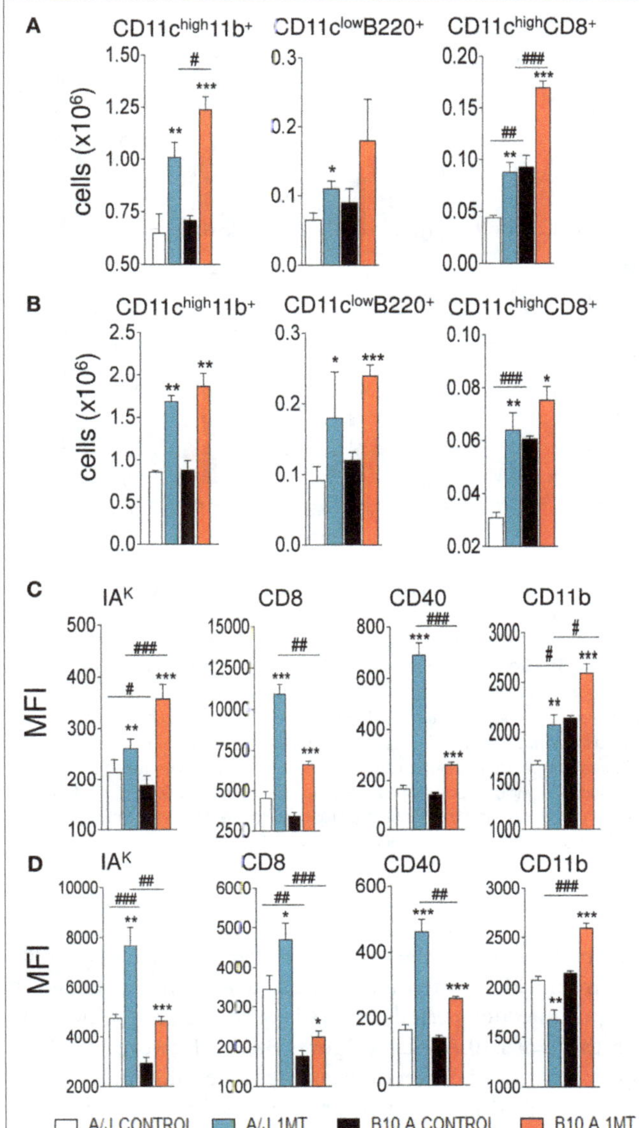

FIGURE 2 | 1-Methyl-DL-tryptophan (1MT) treatment increases the influx of activated dendritic cells (DCs) to the lungs. Number of myeloid (CD11c^{high}11b^{+}), plasmacytoid (CD11c^{low}B220^{+}), and lymphoid (CD11c^{high}CD8^{+}) DCs obtained 96 h **(A)** and 2 weeks **(B)** after infection of 1MT-treated and -untreated A/J and B10.A mice with 1 × 10^{6} viable *Paracoccidiodes brasiliensis* yeasts. Median fluorescence intensity (MFI) of IA^{K}, CD86, CD11b, and CD40 expressed by DCs obtained from lungs of A/J and B10.A mice 96 h **(C)** and 2 weeks **(D)** after infection. DCs were obtained by fractionation with anti-CD11c magnetic beads and DCs phenotyping determined by flow cytometry. The data represent the mean ± SEM of three independent experiments where the asterisk (*) represents a statistically significant difference between treatments (*$p < 0.05$, **$p < 0.01$, and ***$p < 0.001$) and the hashtag (#) represents the difference between the strains (#$p < 0.05$, ##$p < 0.01$, and ###$p < 0.001$).

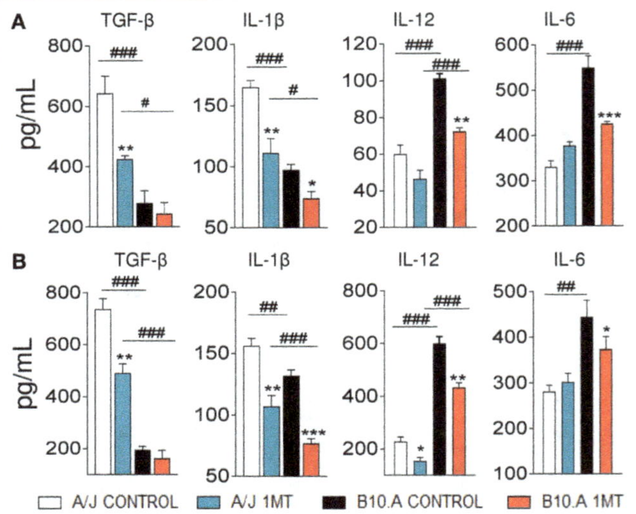

FIGURE 3 | 1-Methyl-DL-tryptophan (1MT) reduces the levels of TGF-β and IL-1β secreted by A/J dendritic cells (DCs), whereas in B10.A DCs 1MT reduces the secretion of IL-12, IL-6, and IL-1β. Pulmonary DCs were purified from lungs of 1MT treated and untreated A/J and B10.A mice 96 h **(A)** and 2 weeks **(B)** after i.t. infection with 1×10^6 *Paracoccidiodes brasiliensis* yeasts cells. DCs were cultivated by 24 h, supernatants collected and analyzed for cytokine content (TGF-β, IL-1β, IL-12, and IL-6) by enzyme-linked immunosorbent assay (ELISA). The data represent the mean ± SEM of three independent determinations where the asterisk (*) represents a statistically significant difference between treatments ($^{*}p < 0.05$, $^{**}p < 0.01$, and $^{***}p < 0.001$) and the hashtag (#) represents the difference between the strains ($^{\#}p < 0.05$, $^{\#\#}p < 0.01$, and $^{\#\#\#}p < 0.001$).

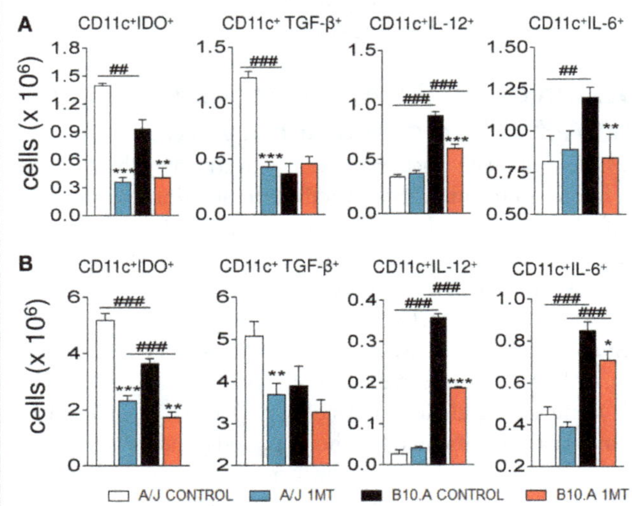

FIGURE 4 | 1-Methyl-DL-tryptophan (1MT) treatment reduces the number of TGF-β+ dendritic cells (DCs) in the lungs of A/J mice but reduces IL-12+ and IL-6+ DCs in B10.A mice. Number of pulmonary DCs expressing intracellular indoleamine-2,3 dioxygenase (IDO), TGF-β, IL-12, and IL-6, 96 h **(A)** and 2 weeks **(B)** after *Paracoccidiodes brasiliensis* infection. Lungs of 1MT treated and untreated A/J and B10.A-infected mice were excised and DCs cells purified by magnetic beads. The expression of intracellular cytokines was analyzed by flow cytometry. The data represent the mean ± SEM of three independent determinations where the asterisk (*) represents a statistically significant difference between treatments ($^{*}p < 0.05$, $^{**}p < 0.01$, and $^{***}p < 0.001$) and the hashtag (#) represents the difference between the strains ($^{\#\#}p < 0.01$ and $^{\#\#\#}p < 0.001$).

As previously described with untreated pulmonary DCs (30), DCs from 1MT-treated and untreated A/J mice obtained at both postinfection periods induced higher lymphoproliferation than those of B10.A mice (**Figures 5A,B**). However, for both mouse strains at both periods assayed, the inhibition of IDO activity resulted in increased proliferation of lymphocytes. In cultures of 1MT treated or untreated naive lymphocytes no cell proliferation was observed. In the same way, no observe lymphoproliferation were detected when 1MT treated or untreated DCs from uninfected mice were used as APC. The levels of IFN-γ and IL-2 were measured in the supernatants of lymphoproliferation assays. Compared with untreated controls, decreased levels of IFN-γ were observed in 1MT-treated DCs from B10.A mice, whereas increased levels of IL-2 were detected in 1MT-treated DCs of both mouse strains and postinfection periods (**Figures 5A,B**).

IDO Inhibition Expands CD4+ and CD8+ T Cells and Reduces Treg Cells

Treatment of B10.A and A/J mice with 1MT increased the differentiation of T cells induced by isolated DCs. Indeed, increased frequency of naive CD4+ and CD8+ ($CD4^+CD44^{low}CD62L^{high}$ and $CD8^+CD44^{low}CD62L^{high}$, respectively) as well as memory/effector CD4+ and CD8+ ($CD4^+CD44^{high}CD62L^{low}$ and $CD8^+CD44^{high}CD62L^{low}$) T cells were detected when 1MT-treated DCs were cocultivated with naive lymphocytes (**Figures 6A,B**). Accordingly, reduced frequencies of Treg ($CD4^+CD25^+FoxP3^+$) cells were observed when DCs were obtained from 1MT-treated resistant and susceptible mice (**Figures 6C,D**).

IDO Is Regulated by IFN-γ in B10.A Mice and TGF-β in A/J Mice

Two opposite pathways were shown to regulate the production and activity of IDO. The IDO-IFN-γ axis is associated with the catalyst function of this enzyme, leads to trp starvation and Kyn production that in turn control pathogen growth and the inflammatory response. In contrast, the TGF-β-IDO axis has been linked with a signaling function of pIDO that promotes the differentiation of tolerogenic DCs and increases the expansion of Treg cells (12, 35, 36).

We speculated that in B10.A mice IDO is mainly catalyst and IFN-γ-regulated, whereas in A/J mice IDO has a predominant signaling function induced by TGF-β. To explore this hypothesis, the expression of genes and proteins related to IFN-γ and TGF-β signaling pathways was characterized in DCs of 1MT-treated and untreated B10.A and A/J mice obtained at 96 h (Figures S3 and S4 in Supplementary Material) and 2 weeks of infection. As shown in **Figure 7A**, DCs from B10.A mice expressed higher levels of mRNA from genes associated with IFN-γ signaling. Therefore, 1MT treatment led to a marked reduction in the expression of IFN-γ, Jak1, Stat1, IL-6, SOCS3, and NOS2 mRNA by DCs from B10.A mice but had a minor effect on DCs of A/J mice (**Figure 7A**).

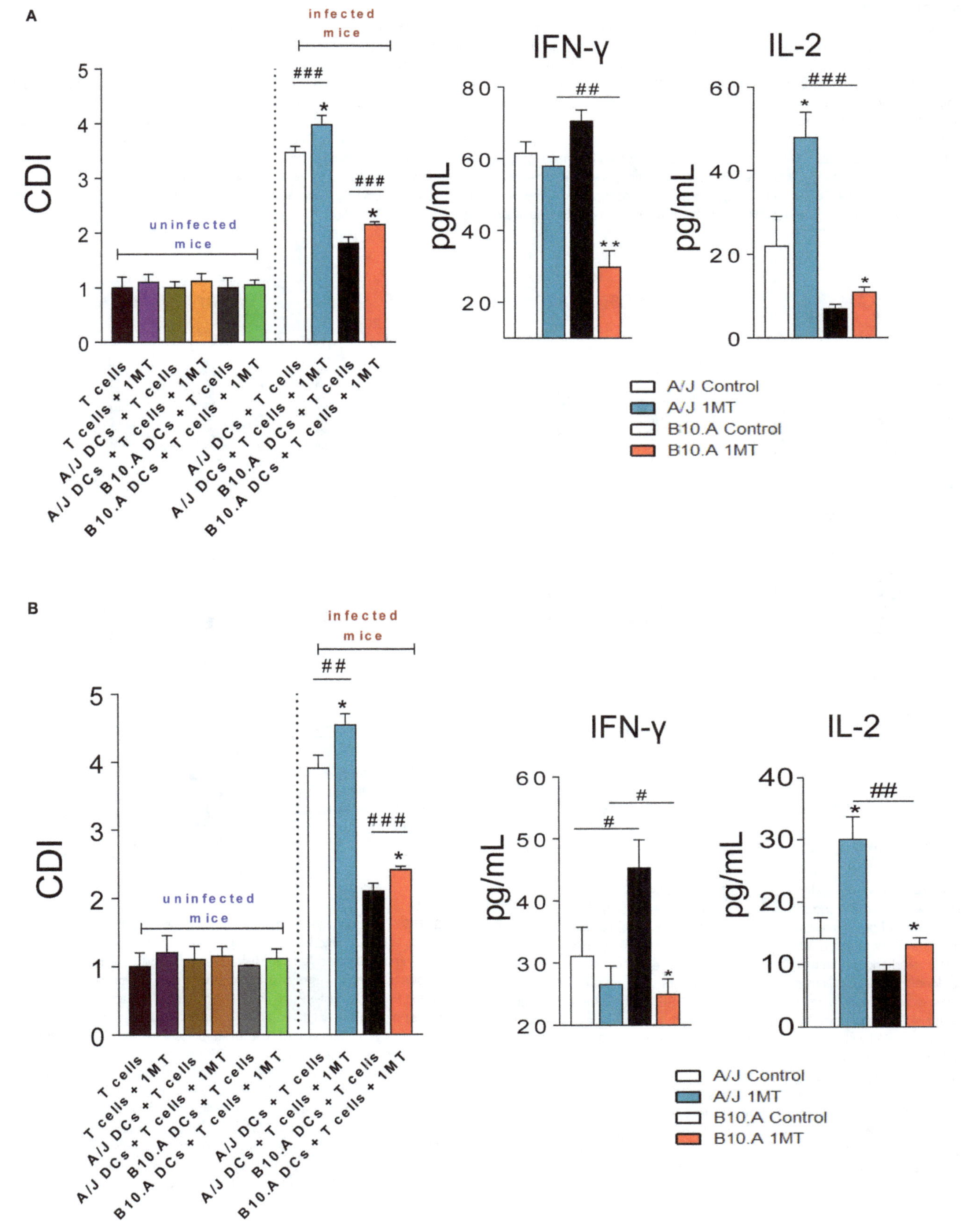

FIGURE 5 | Continued

FIGURE 5 | Continued
1-Methyl-DL-tryptophan (1MT) treatment increases the immunogenicity of dendritic cells (DCs). Pulmonary DCs were obtained from lungs of 1MT-treated and -untreated A/J and B10.A mice 96 h **(A)** and 2 weeks **(B)** after infection with 1 × 10^6 *Paracoccidiodes brasiliensis* yeasts using anti-DC11c magnetic beads. Spleen lymphocytes from non-infected A/J and B10.A mice were obtained, labeled with CFSE (5 mM) and cocultivated with DCs, in a ratio of DC:lymphocyte of 1:10. In parallel, pulmonary DCs from uninfected A/J and B10.A mice were obtained using anti-DC11c magnetic beads and cultured with 1MT treated or untreated naive lymphocytes from non-infected A/J and B10.A mice. After 3 days of cocultivation, the cells were adjusted to 1 × 10^6, labeled with specific anti-CD4 **(A)** and anti-CD8 **(B)** antibodies and analyzed by flow cytometry. The cell division index (CDI) was calculated as previously described by Mannering et al. (34) and was based on the number of CFSE$^+$CD4$^+$ or CFSE$^+$CD8$^+$ T cells found in the stimulated culture/number of CFSE$^+$CD4$^+$ or CFSE$^+$CD8$^+$ T cells in the unstimulated culture. The lymphocyte population was gated by FSC/SSC analysis. Fifty thousand cells were counted, and the data expressed as frequency of positive cells. Results are representative of three independent experiments. Cytokines were measured in culture supernatants by enzyme-linked immunosorbent assay (ELISA). The data represent the mean ± SEM of three independent determinations where the asterisk (*) represents a statistically significant difference between treatments (*$p < 0.05$ and **$p < 0.01$) and the hashtag (#) represents the difference between the strains (##$p < 0.01$ and ###$p < 0.001$).

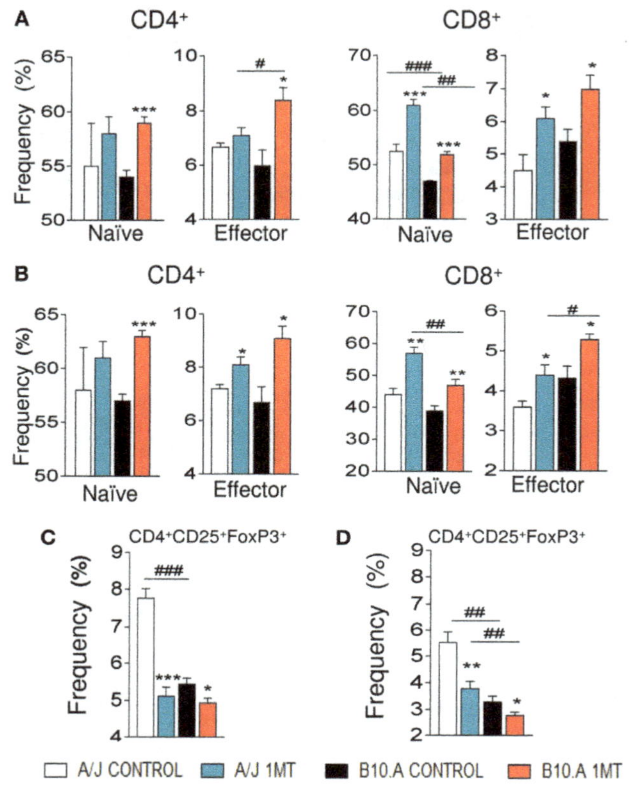

FIGURE 6 | 1-Methyl-DL-tryptophan (1MT) increases proliferation of CD4$^+$ and CD8$^+$ T cells and reduces regulatory T cells (Tregs). Spleen lymphocytes from uninfected A/J and B10.A mice were labeled with CFSE (5 mM) and cocultivated with purified dendritic cells (DCs) obtained from 1MT treated and untreated A/J and B10.A mice infected i.t. with 1 × 10^6 of *Paracoccidiodes brasiliensis* yeasts. DCs were obtained at 96 h and 2 weeks postinfection, and cocultivated with CFSE labeled lymphocytes at a ratio of DC:lymphocyte of 1:10. After 3 days, the cells were adjusted to 1 × 10^6, labeled with specific anti-CD4, -CD8, -CD44, -CD62L, -CD25, and -FoxP3 antibodies. The frequency of naive (CD62L^{high}CD44^{low}) and effector/memory (CD62L^{low}CD44^{high}) CD4$^+$ and CD8$^+$ T cells as well as CD4$^+$CD25$^+$FoxP3 regulatory T cells (Tregs) was measured by flow cytometry. Values are the mean of three independent experiments. The asterisks represent statistically significant differences between treatments (*$p < 0.05$, **$p < 0.01$, and ***$p < 0.05$). The hashtag marks represent statistically significant differences between strains (#$p < 0.05$, ##$p < 0.01$, ###$p < 0.001$).

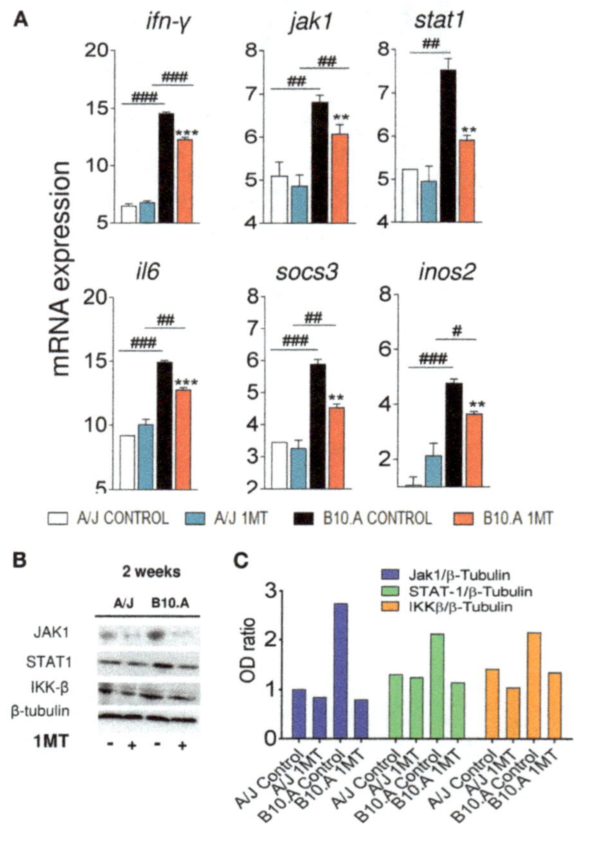

FIGURE 7 | Indoleamine-2,3 dioxygenase (IDO) is regulated by IFN-γ signaling in B10.A dendritic cells (DCs). 1-Methyl-DL-tryptophan (1MT) treated or untreated B10.A and A/J mice were infected i.t. with 1 × 10^6 yeast cells of *P. brasiliensis*, and 2 weeks after infection total lung inflammatory cells were obtained and DCs purified with anti-CD11c magnetic beads. **(A)** The relative expression of mRNA of IFN-γ, Janus kinase 1 (Jak1), signal transducer and activator of transcription (STAT1), IL-6, suppressor of cytokine signaling 3 (SOCS3), and nitric oxide synthase (NOS2) was measured by real-time PCR. **(B)** Presence of Jak1, Stat1, IKK-β, and IDO1 proteins was assessed by western blot in supernatants of lysed DCs. Proteins were estimated by analyzing the intensity of each band normalized by β-tubulin, used as control. **(C)** Densitometry of bands was performed using ImageQuant TL 8.1 software. The asterisks represent statistically significant differences between treatments (*$p < 0.05$, **$p < 0.01$, ***$p < 0.001$). The hash marks represent statistically significant differences between strains (#$p < 0.05$, ##$p < 0.01$, ###$p < 0.001$).

These findings indicate that in B10.A mice the IDO function is mainly catalyst and regulated by IFN-γ. Immunoblotting assays confirmed these results, and increased amounts of JAK1, STAT-1, and IKK-β proteins were found in DCs extracts of B10.A DCs (**Figures 7B,C**). In contrast to B10.A mice, DCs from A/J mice express higher levels of mRNA from genes associated with TGF-β signaling (**Figure 8**). Compared with B10.A DCs, cells from A/J mice expressed increased levels of TGF-β, SMAD2, SMAD3, SHIP1, SHP1, SHP2, IFN-α, and IFN-β mRNA. 1MT treatment significantly reduced these levels in A/J DCs but had no effect on B10.A cells (**Figure 8A**). Assessing protein production, increased amounts of SMAD2, SHIP1, SHP2, and IKK-α were observed in extracts of A/J DCs (**Figures 8B,C**).

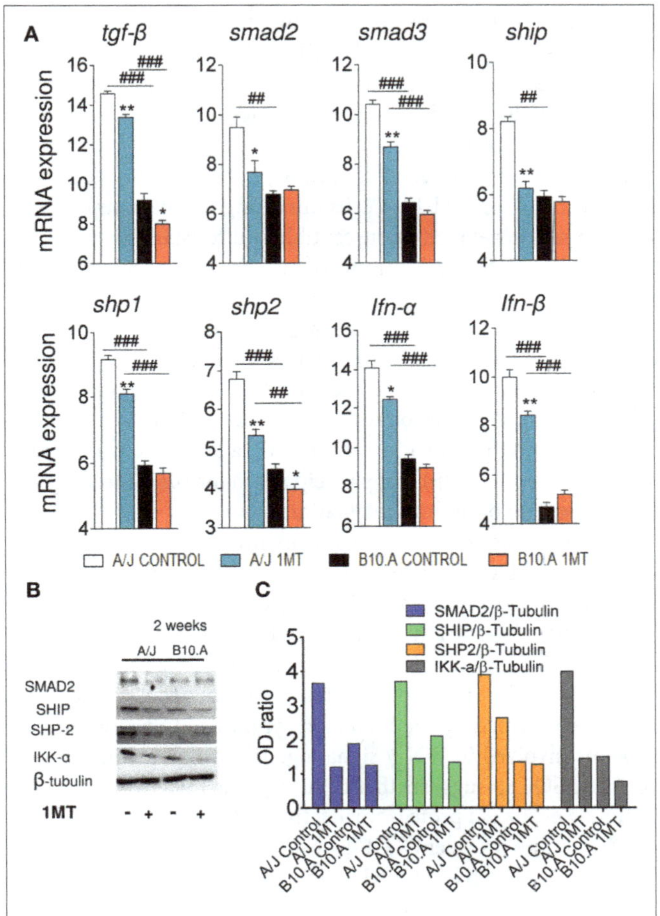

FIGURE 8 | Indoleamine-2,3 dioxygenase (IDO) is regulated by TGF-β signaling in A/J dendritic cells (DCs). 1-Methyl-DL-tryptophan (1MT)-treated or -untreated B10.A and A/J mice were infected i.t. with 1 × 10[6] *Paracoccidiodes brasiliensis* yeasts and 2 weeks after infection total lung inflammatory cells were obtained and DCs purified by anti-CD11c magnetic beads. **(A)** The relative expression of mRNA of TGF-β, Smad2, Smad 3, SHIP, SHP-1, SHP-2, IFN-α, and IFN-β was measured by real-time PCR. **(B)** Smad2, Ship, Shp-2, IKK-α and IDO1, and pIDO protein expression was assessed by western blot in supernatants of lysed DCs. **(C)** Proteins were estimated by analyzing the intensity of each band normalized by β-tubulin, used as control. Densitometry of bands was performed using ImageQuant TL 8.1 software. Values are the mean ± SEM of three independent experiments; the asterisks represent statistically significant differences between treatments (*$p < 0.001$, **$p < 0.01$, ***$p < 0.05$). The hash marks represent statistically significant differences between strains (#$p < 0.05$, ##$p < 0.01$, ###$p < 0.001$).

A/J Mice pIDO and Use the Non-Canonical Pathway of NF-κB Activation, whereas B10.A Mice Express Non-pIDO and Use the Canonical Pathway of NF-κB Activation

Because IFN-γ signaling is mainly associated with the canonical pathway of NF-κB activation, the transcriptional activity of p65 and p50 were measured by ELISA in the nuclear extracts of 1MT-treated and -untreated DCs. As depicted in **Figure 9A**, DCs from B10.A mice showed increased expression of nuclear

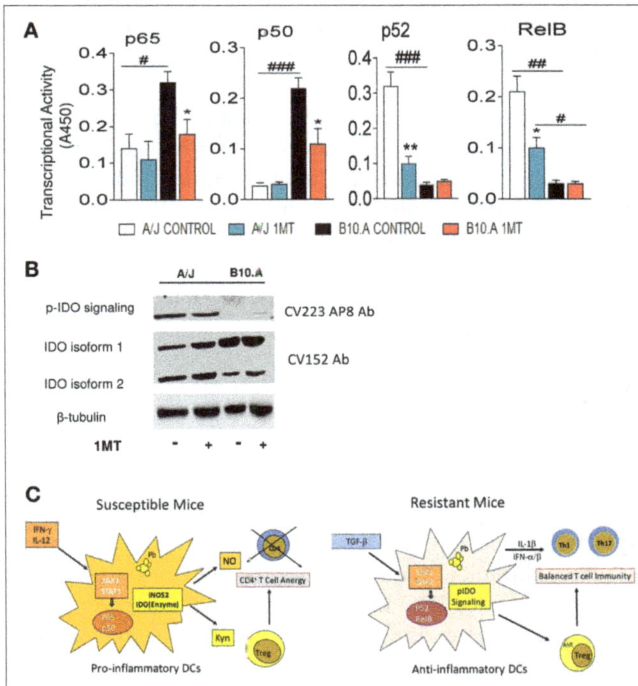

FIGURE 9 | Phosphorylated indoleamine-2,3 dioxygenase (pIDO) of A/J mice activate the non-canonical pathway of NF-κB, whereas in B10.A dendritic cells (DCs) non-phosphorylated IDO use the canonical pathway of NF-κB activation. **(A,B)** Quantitation of p65, p50, p52, and RelB by enzyme-linked immunosorbent assay (ELISA) in nuclear extracts of DCs. Results are presented as absorbance at 450 nm (A450). Values are the mean ± SE of three independent experiments. The asterisks represent statistically significant differences between treatments (*$p < 0.05$, **$p < 0.01$, ***$p < 0.001$). The hash marks represent statistically significant differences between strains (#$p < 0.05$, ##$p < 0.01$, ###$p < 0.001$). **(C)** Expression of IDO and phosphorylated IDO was assessed by Western blot in lung homogenates of 1MT treated and untreated B10.A and A/J mice i.t. infected with 1 × 10[6] *Paracoccidiodes brasiliensis* yeasts. pIDO was revealed by a specific antibody that recognizes the phosphorylated ITIM moiety of IDO (CV223 AP8) and the enzyme IDO by a specific monoclonal antibody (CV152). **(D)** In susceptible B10.A mice, IDO is induced by IFN-γ that activates the canonical pathway of NF-κB and enhanced iNOS and IDO expression resulting in anergy of T cell immunity. In resistant A/J mice, TGF-β signaling induces the non-canonical pathway of NF-κB activation and continuous synthesis of TGF-β that confers an stable tolerogenic behavior to DCs that control immunity and tissue pathology.

p65 and p50 than those of A/J mice. Furthermore, 1MT treatment reduced this production by B10.A DCs (**Figure 9A**). An equivalent result was observed with DCs obtained at 96 h of infection (Figure S1 in Supplementary Material). The prevalent non-canonical activation of NF-κB by A/J DCs was confirmed by the increased levels of nuclear p52 and RelB produced in comparison with B10.A DCs (**Figure 9A**). Treatment with 1MT caused a significant reduction of p52 and RelB only in A/J DCs. Again, an equivalent result was observed in pulmonary DCs obtained at 96 h of infection (Figure S2 in Supplementary Material). To further confirm the signaling function of IDO, lung extracts from 1MT treated and untreated B10.A and A/J mice were obtained 2 weeks after infection and the expression of pIDO was assessed by immunoblotting using a specific mAb (CV223 AP8 Ab) that recognizes the phosphorylated moiety of ITIM. As depicted in **Figure 9B**, only A/J mice express pIDO and 1MT treatment did not affect the production of this molecule. Using a polyclonal antibody to IDO (CV152 Ab) two isoforms of IDO, and the elevated expression of the isoform 1 by B10.A DCs were seen. The **Figure 9C** summarizes the main findings obtained in this study. In susceptible B10.A mice IDO is induced by IFN-γ that activates the canonical pathway of NF-κB and enhanced iNOS and IDO expression resulting in anergy of T cell immunity. In resistant A/J mice, TGF-β signaling induces the non-canonical pathway of NF-κB activation and continuous synthesis of TGF-β that confers an stable tolerogenic behavior to DCs that control immunity and tissue pathology.

DISCUSSION

The antimicrobial and immunoregulatory functions of IDO have been recognized as central mechanisms used by infected hosts to control pathogen growth and excessive inflammatory responses (37, 38). Both IDO activities were used by resistant and susceptible mice to control pulmonary PCM. However, only in susceptible mice IDO inhibition caused a sustained fungal growth, exacerbated tissue pathology and increased mortality rates. This is the worst scenario for host fitness because it allies uncontrolled pathogen growth with exacerbated, but inefficient, inflammation resulting in warful tissue destruction. These findings led us to further investigate the role of IDO in the resistance mechanisms to pulmonary PCM. We found the IDO inhibitor 1-MT as two stereoisomers, 1-D-MT and 1-L-MT. Most of the studies using the IDO enzyme model employed the racemic mixture 1-DL-MT to inhibit the enzime, independent of the known IDO isoforms, IDO1 and IDO2, as we did in this work. Indeed, recent studies have shown that IDO1 is the preferentially inhibited by1-L-MT, while 1-D-MT inhibits IDO2 (39, 40). These findings directed our subsequent studies on the inhibition of IDO using deficient mice, as we have shown in recent work with IDO1-deficient (IDO1$^{-/-}$) C57BL/6 mice (41). Using isolated pulmonary DCs, we verified that IDO inhibition led to increased recovery of yeasts from both A/J and B10.A cells. This finding could be attributed to the increased trp availability (42–44), but also to the reduced NO production, an important fungicidal mechanism of macrophages and DCs (26, 27, 30).

Interestingly, IDO inhibition by 1MT reduced the typical cytokines that characterize the innate response of B10.A and A/J mice (26, 27, 30). Furthermore, the assessment of intracellular cytokines has further emphasized the higher expression of IDO and TGF-β by A/J DCs, whereas the proinflammatory cytokines IL-12 and IL-6 were more expressed by B10.A DCs. This led us to suppose that IDO activity in B10.A mice was mainly catalytic, whereas in A/J mice was predominantly signaling and involved in signal transduction (12). These investigators clearly demonstrated that TGF-β signaling activates the tyrosine kinase Fyn that phosphorylates ITIM domains of IDO. This promotes the association of IDO with tyrosine phosphatases (SHP1 and SHP2) that results in the activation of the non-canonical NF-κB pathway (p52/RelB), which translocates to the nucleus and induces the production of type I IFNs (IFNα/β), TGF-β, and IDO (12).

An increased number of activated DCs with enhanced immunogenic activity was detected in the lungs of 1MT treated mice. The increased proliferation of naive lymphocytes was concomitant with increased levels of IL-2 in A/J cocultures and decreased IFN-γ production by B10.A cells mirroring the main suppressive mechanisms developed by susceptible and resistant mice (25, 26, 30, 45). This enhanced immunogenic activity was further confirmed by the increased proliferation of naive and effector/memory CD8$^+$ and CD4$^+$ T cells concomitant with a reduced expansion of CD4$^+$CD25$^+$Foxp3$^+$ Treg cells, a finding that may be also associated with the reduced levels of suppressive Kyn produced (46, 47). This tolerogenic function of Kyn on T cells has been linked to the activation of the AhR, a ligand activated transcription factor (46, 48) that participates in the pIDO-tolerogenic DCs-Treg loop of immunoregulation.

Further assessing the mechanisms involved in the divergent activities of IDO, a preferential IFN-γ signaling pathway was observed in B10.A DCs which expressed increased levels of IFN-γ and Jak1-Stat1 mRNA. In addition, elevated expression of IL-6, SOCS3, and iNOS2 were also observed indicating a prevalent IFN-γ-mediated catalytic activity. These data were further confirmed by the increased synthesis of Stat1 and IKKβ proteins involved in IFN-γ signaling. The striking expression of IL-6 and SOCS3 suggests that in susceptible mice IDO degradation may occur *via* proteasomal activity (49). Our data have also confirmed the prevalent TGF-β-mediated activation of A/J DCs which express high levels of TGF-β, Smad2, Smad3, Ship1, Shp1, Shp2, IFN-α, and IFN-β mRNA. The high levels of IKK-α, Smad2, Ship1, and Sph-2 proteins expressed by these cells have further confirmed the use of the TGF-β signaling by DCs from resistant mice. These data agree with the signaling cascade that promotes a tolerogenic activity in plasmacytoid DCs previously described (12) and explain the continuous production of TGF-β by A/J mice that favors the conversion of naive CD4$^+$ T cells into Treg cells (25). Interestingly, the inhibition of IDO by 1MT showed little interference with the TGF-β-mediated pathway used by A/J DCs, demonstrating that the signaling function of the IDO has predominance over the catalytic activity.

The predominant expression of p65 and p50 by B10.A DCs and the higher levels of p52 and RelB by A/J DCs showed that B10.A and A/J mice predominantly use, respectively, the canonical and non-canonical pathways of NF-κB activation by their pulmonary

DCs. More importantly, we could demonstrate the presence of pIDO in the lungs of A/J mice, whereas the non-phosphorylated forms of IDO were observed in B10.A mice. Therefore, we could clearly demonstrate that in B10.A mice IDO has a catalytic function, whereas in A/J mice, besides its enzymatic activity, IDO has a signaling function.

Similarly with other infection models, our studies in experimental PCM showed that the balanced activation of the immune system is fundamental for the protection of hosts, and the excessive activation of pro- or anti-inflammatory mechanisms leads to severe disease (23, 29, 50, 51). In susceptible mice, the suppression of adaptive immunity is a consequence of excessive proinflammatory activity of innate immunity, which, however, is able to control the initial fungal burden. In resistant mice, in contrast, an anti-inflammatory, TGF-β dominated activity is observed impairing the initial control of fungal growth. This activity, however, is balanced by the marked dectin-1 and NLRP3 activation and synthesis of TNF-α and IL-1β by A/J macrophages. This pattern of response leads to the subsequent differentiation of Th1/Th17 mixed responses, always strongly controlled by elevated numbers and activity of Treg cells that control tissue pathology (24–27, 29). This is an unusual but interesting model of resistance and susceptibility to a primary fungal pathogen not previously described.

Host defenses are governed by resistance and tolerance mechanisms, the first involved in reducing pathogen burden during infection and the latter associated with mechanisms that protected the hosts from pathogen- or immune-induced damage. These mechanisms are not mutually exclusive and both can contribute to host fitness (52, 53). Our studies clearly demonstrate that the resistance mechanisms used by A/J mice are governed by resistance to disease and not resistance to pathogen growth. Although not intensely explored in infectious diseases, tolerance to disease is well known and described in plant pathology where host protection is not achieved by the elimination of the pathogen, but it is based on the control of host fitness. This response allows pathogen persistence, but its excessive growth is partially controlled by immune mechanisms that are tightly regulated to preserve homeostasis (52). This is also observed in the host responses to commensal fungi such as *Aspergillus fumigatus* and *Candida albicans* that concomitantly control fungal burden and tissue inflammation (50). The important influence of the IDO-AhR-Treg-Th17 axis on the control of this well-balanced response to commensal microorganisms has been well described in the literature (44, 54, 55). The IFN-γ-mediated activation of IDO and Kyn synthesis by tissue DCs confers a tolerogenic profile to these cells that promote the differentiation of naive T cells into Foxp3$^+$ Treg cells *via* activation of AhR that prevents the excessive expansion of Th17 lymphocytes (44, 56). This tightly regulated immunity confers immune homeostasis that results in microbial resistance but avoids tissue pathology. The signaling property of pIDO induced by TGF-β activation of DCs was more recently described and has been viewed as potent immunoregulatory mechanism that confer stable tolerogenic activity to plasmacytoid DCs (12, 36). This tolerogenic mechanism was validated in an *in vivo* model of delayed-type hypersensitivity response and in the protective activity of endotoxin-tolerant state of mice against Gram-positive and Gram-negative infections (12, 54). However, to our knowledge, this mechanism of disease tolerance was never described in the host responses to primary fungal pathogens. Indeed, pIDO-mediated tolerance was poorly explored in infectious pathologies. Therefore, we believe that the findings here described open new perspectives to understand resistance against PCM and other primary fungal pathogens where excessive inflammation has been associated with severe disease (57, 58). Previous results obtained in our experimental model, not well understood when originally obtained, can now be re-evaluated in face of the IDO mediated mechanism here described. The subcutaneous immunization with viable yeasts confers sterile immunity to i.p. challenged B10.A mice but partial immunoprotection in A/J mice that remain with persistent fungal loads (59, 60). We have also previously described that *P. brasiliensis* infection induces an elevated production of TGF-β and M2 differentiation of A/J macrophages which express high levels of arginase (27), another finding which is not in line with the mechanisms of pathogen resistance usually described. However, this finding may now be better appreciated and understood because arginase expression by innate immune cells was reported to be essential for the differentiation of tolerogenic DCs, whose signaling activity is mediated by pIDO (61).

In conclusion, the studies here reported clearly demonstrate that susceptibility to PCM is mainly mediated by the proinflammatory IFN-γ-IDO axis of innate responses, whereas resistance relies on the initial anti-inflammatory TGF-β-pIDO-Treg axis that confers a sustained tolerogenic phenotype in pulmonary DCs allowing A/J mice to interact with a primary fungal pathogen as a commensal microbe. Therefore, the signaling function of pIDO leads to an efficient fungus--host interaction that conditions advantages to both partners, and may be explored in further innovative anti-fungal therapy.

ETHICS STATEMENT

The experiments were performed in strict accordance with the Brazilian Federal Law 11,794 establishing procedures for the scientific use of animals, and the State Law establishing the Animal Protection Code of the State of São Paulo. All efforts were made to minimize animal suffering. The procedures were approved by the Ethics Committee on Animal Experiments of the Institute of Biomedical Sciences of University of São Paulo (Proc.180/11/CEEA).

AUTHOR CONTRIBUTIONS

EA designed, conducted, and analyzed all experiments. FL, CF, and TC performed and analyzed experiments. CV characterized IDO and IDO phosphorylation. LR provided conceptual help and wrote the article. PP provided conceptual help and precious reagents for IDO analysis. VC planned experiments, supervised the overall study, and wrote the manuscript.

FUNDING

This work was supported by grants from the Fundação de Amparo à Pesquisa do Estado de São Paulo (FAPESP) research grants no. 2011/51258-2 (VLGC) and 2016/04783-2 (FVL); FAPESP postdoctoral felowships no. 2013/02396-9 (CF) and 2014/18668-0 (EFA); and Conselho Nacional de Pesquisas (CNPq) research grant no. 4713172012 (VLGC).

REFERENCES

1. Mellor AL, Munn DH. IDO expression by dendritic cells: tolerance and tryptophan catabolism. *Nat Rev Immunol* (2004) 4:762–74. doi:10.1038/nri1457
2. Grohmann U, Fallarino F, Puccetti P. Tolerance, DCs and tryptophan: much ado about IDO. *Trends Immunol* (2003) 24:242. doi:10.1016/S1471-4906(03)00072-3
3. von Bergwelt-Baildon MS, Popov A, Saric T, Chemnitz J, Classen S, Stoffel MS, et al. CD25 and indoleamine 2,3-dioxygenase are up-regulated by prostaglandin E2 and expressed by tumor-associated dendritic cells *in vivo*: additional mechanisms of T-cell inhibition. *Blood* (2006) 108:228–37. doi:10.1182/blood-2005-08-3507
4. Popov A, Schultze JL. IDO-expressing regulatory dendritic cells in cancer and chronic infection. *J Mol Med* (2008) 86:145–60. doi:10.1007/s00109-007-0262-6
5. Braun D, Longman RS, Albert ML. A two-step induction of indoleamine 2,3 dioxygenase (IDO) activity during dendritic-cell maturation. *Blood* (2005) 106:2375–81. doi:10.1182/blood-2005-03-0979
6. Beutelspacher SC, Tan PH, McClure MO, Larkin DF, Lechler RI, George AJ. Expression of indoleamine 2,3-dioxygenase (IDO) by endothelial cells: implications for the control of alloresponses. *Am J Transplant* (2006) 6:1320–30. doi:10.1111/j.1600-6143.2006.01324.x
7. Uyttenhove C, Pilotte L, Théate I, Stroobant V, Colau D, Parmentier N, et al. Evidence for a tumoral immune resistance mechanism based on tryptophan degradation by indoleamine 2,3-dioxygenase. *Nat Med* (2003) 9:1269–74. doi:10.1038/nm934
8. Wang D, Feng X, Lu L, Konkel JE, Zhang H, Chen Z, et al. A CD8 T cell/indoleamine 2,3-dioxygenase axis is required for mesenchymal stem cell suppression of human systemic lupus erythematosus. *Arthritis Rheumatol* (2014) 66:2234–45. doi:10.1002/art.38674
9. Taylor MW, Feng GS. Relationship between interferon-gamma, indoleamine 2,3-dioxygenase, and tryptophan catabolism. *FASEB J* (1991) 5:2516–22.
10. Sedlmayr P, Blaschitz A, Wintersteiger R, Semlitsch M, Hammer A, MacKenzie CR, et al. Localization of indoleamine 2,3-dioxygenase in human female reproductive organs and the placenta. *Mol Hum Reprod* (2002) 8:385–91. doi:10.1093/molehr/8.4.385
11. Munn DH, Mellor AL. Indoleamine 2,3 dioxygenase and metabolic control of immune responses. *Trends Immunol* (2013) 34:137–43. doi:10.1016/j.it.2012.10.001
12. Pallotta MT, Orabona C, Volpi C, Vacca C, Belladonna ML, Bianchi R, et al. Indoleamine 2,3-dioxygenase is a signaling protein in long-term tolerance by dendritic cells. *Nat Immunol* (2011) 12:870–8. doi:10.1038/ni.2077
13. Barth H, Raghuraman S. Persistent infectious diseases say – IDO. Role of indoleamine-2,3-dioxygenase in disease pathogenesis and implications for therapy. *Crit Rev Microbiol* (2014) 40:360–8. doi:10.3109/1040841X.2012.742037
14. Baban B, Penberthy WT, Mozaffari MS. The potential role of indoleamine 2, 3 dioxygenase (IDO) as a predictive and therapeutic target for diabetes treatment: a mythical truth. *EPMA J* (2010) 1:46–55. doi:10.1007/s13167-010-0009-2
15. Zelante T, Fallarino F, Bistoni F, Puccetti P, Romani L. Indoleamine 2,3-dioxygenase in infection: the paradox of an evasive strategy that benefits the host. *Microbes Infect* (2009) 11:133–41. doi:10.1016/j.micinf.2008.10.007
16. Fallarino F, Grohmann U, You S, McGrath BC, Cavener DR, Vacca C, et al. The combined effects of tryptophan starvation and tryptophan catabolites down-regulate T cell receptor zeta-chain and induce a regulatory phenotype in naive T cells. *J Immunol* (2006) 176:6752–61. doi:10.4049/jimmunol.176.11.6752
17. Sharma MD, Baban B, Chandler P, Hou DY, Singh N, Yagita H, et al. Plasmacytoid dendritic cells from mouse tumor-draining lymph nodes directly activate mature Tregs via indoleamine 2,3-dioxygenase. *J Clin Invest* (2007) 117:2570–82. doi:10.1172/JCI31911
18. Teixeira MM, Theodoro RC, de Carvalho MJ, Fernandes L, Paes HC, Hahn RC, et al. Phylogenetic analysis reveals a high level of speciation in the *Paracoccidioides* genus. *Mol Phylogenet Evol* (2009) 52:273–83. doi:10.1016/j.ympev.2009.04.005
19. De Castro LF, Ferreira MC, da Silva RM, Blotta MH, Longhi LN, Mamoni RL. Characterization of the immune response in human paracoccidioidomycosis. *J Infect* (2013) 67:470–85. doi:10.1016/j.jinf.2013.07.019
20. Mamoni RL, Blotta MH. Kinetics of cytokines and chemokines gene expression distinguishes *Paracoccidioides brasiliensis* infection from disease. *Cytokine* (2005) 32:20–9. doi:10.1016/j.cyto.2005.07.006
21. Benard G, Romano CC, Cacere CR, Juvenale M, Mendes-Giannini MJ, Duarte AJ. Imbalance of IL-2, IFN-gamma and IL-10 secretion in the immunosuppression associated with human paracoccidioidomycosis. *Cytokine* (2001) 13:248–52. doi:10.1006/cyto.2000.0824
22. Calich VL, Singer-Vermes LM, Siqueira AM, Burger E. Susceptibility and resistance of inbred mice to *Paracoccidioides brasiliensis*. *Br J Exp Pathol* (1985) 66:585–94.
23. Calich VLG, Blotta MHSL. Pulmonary paracoccidioidomycosis. 1st ed. In: Fidel P, Huffnagle G, editors. *Fungal Immunology, From an Organ Perspective*. (Vol. 1), New York: Springer (2005). p. 201–28.
24. Cano LE, Singer-Vermes LM, Vaz CAC, Russo M, Calich VL. Pulmonary paracoccidioidomycosis in resistant and susceptible mice: relationship among progression of infection, bronchoalveolar cell activation, cellular immune response, and specific isotype patterns. *Infect Immun* (1995) 63:1777–83.
25. Felonato M, Pina A, de Araujo EF, Loures FV, Bazan SB, Feriotti C, et al. Anti-CD25 treatment depletes Treg cells and decreases disease severity in susceptible and resistant mice infected with *Paracoccidioides brasiliensis*. *PLoS One* (2012) 7:e51071. doi:10.1371/journal.pone.0051071
26. Pina A, Bernardino S, Calich VL. Alveolar macrophages from susceptible mice are more competent than those of resistant mice to control initial *Paracoccidioides brasiliensis* infection. *J Leukoc Biol* (2008) 83:1088–99. doi:10.1189/jlb.1107738
27. Feriotti C, Loures FV, Frank de Araújo E, da Costa TA, Calich VL. Mannosyl-recognizing receptors induce an M1-like phenotype in macrophages of susceptible mice but an M2-like phenotype in mice resistant to a fungal infection. *PLoS One* (2013) 8:e54845. doi:10.1371/journal.pone.0054845
28. Araújo EF, Loures FV, Bazan SB, Feriotti C, Pina A, Schanoski AS, et al. Indoleamine 2,3-dioxygenase controls fungal loads and immunity in paracoccidioidomicosis but is more important to susceptible than resistant hosts. *PLoS Negl Trop Dis* (2014) 8:e3330. doi:10.1371/journal.pntd.0003330
29. Calich VLG, da Costa TA, Felonato M, Arruda C, Bernardino S, Loures FV, et al. Innate immunity to *Paracoccidioides brasiliensis* infection. *Mycopathologia* (2008) 165:223–36. doi:10.1007/s11046-007-9048-1
30. Pina A, de Araujo EF, Felonato M, Loures FV, Feriotti C, Bernardino S, et al. Myeloid dendritic cells (DCs) of mice susceptible to paracoccidioidomycosis suppress T cell responses whereas myeloid and plasmacytoid DCs from resistant mice induce effector and regulatory T cells. *Infect Immun* (2013) 81:1064–77. doi:10.1128/IAI.00736-12
31. Felonato M, Pina A, Bernardino S, Loures FV, de Araujo EF, Calich VLG. CD28 exerts protective and detrimental effects in a pulmonary model of paracoccidioidomycosis. *Infect Immun* (2010) 78:4922–35. doi:10.1128/IAI.00297-10
32. Singer-Vermes LM, Caldeira CB, Burger E, Calich VLG. Experimental murine paracoccidioidomycosis: relationship among dissemination of the infection, humoral and cellular responses. *Clin Exp Immunol* (1993) 94:75–9. doi:10.1111/j.1365-2249.1993.tb05980.x
33. Ding AH, Nathan CF, Stuehr DJ. Release of reactive nitrogen intermediates and reactive oxygen intermediates from mouse peritoneal macrophages. Comparison of activating cytokines and evidence for independent production. *J Immunol* (1988) 141:2407–12.

34. Mannering SI, Morris JS, Jensen KP, Purcell AW, Honeyman MC, van Endert PM, et al. A sensitive method for detecting proliferation of rare autoantigen-specific human T cells. *J Immunol Methods* (2003) 283:173–83. doi:10.1016/j.jim.2003.09.004
35. Orabona C, Pallotta MT, Volpi C, Fallarino F, Vacca C, Bianchi R, et al. SOCS3 drives proteasomal degradation of indoleamine 2,3-dioxygenase (IDO) and antagonizes IDO-dependent tolerogenesis. *Proc Natl Acad Sci U S A* (2008) 105:20828–33. doi:10.1073/pnas.0810278105
36. Orabona C, Pallotta MT, Grohmann U. Different partners, opposite outcomes: a new perspective of the immunobiology of indoleamine 2,3-dioxygenase. *Mol Med* (2012) 18:834–42. doi:10.2119/molmed.2012.00029
37. Bonifazi P, Zelante T, D'Angelo C, De Luca A, Moretti S, Bozza S, et al. Balancing inflammation and tolerance *in vivo* through dendritic cells by the commensal *Candida albicans*. *Mucosal Immunol* (2009) 2:362–74. doi:10.1038/mi.2009.17
38. Araújo EF, Medeiros DH, Galdino NA, Condino-Neto A, Calich VL, Loures FV. Tolerogenic plasmacytoid dendritic cells control *Paracoccidioides brasiliensis* infection by inducting regulatory T cells in an IDO-dependent manner. *PLoS Pathog* (2016) 12:e1006115. doi:10.1371/journal.ppat.1006115
39. Lob S, Konigsrainer A, Zieker D, Brucher BL, Rammensee HG, Opelz G, et al. IDO1 and IDO2 are expressed in human tumors: levo- but not dextro-1-methyl tryptophan inhibits tryptophan catabolism. *Cancer Immunol Immunother* (2009) 58:153–7. doi:10.1007/s00262-008-0513-6
40. Metz R, Duhadaway JB, Kamasani U, Laury-Kleintop L, Muller AJ, Prendergast GC. Novel tryptophan catabolic enzyme IDO2 is the preferred biochemical target of the antitumor indoleamine 2,3-dioxygenase inhibitory compound D-1-methyl-tryptophan. *Cancer Res* (2007) 67:7082–7. doi:10.1158/0008-5472.CAN-07-1872
41. De Araújo EF, Feriotti C, Galdino NAL, Preite NW, Calich VLG, Loures FV. The IDO-AhR axis controls Th17/Treg immunity in a pulmonary model of fungal infection. *Front Immunol* (2017) 8:880. doi:10.3389/fimmu.2017.00880
42. Montagnoli C, Bozza S, Gaziano R, Zelante T, Bonifazi P, Moretti S, et al. Immunity and tolerance to *Aspergillus fumigatus*. *Novartis Found Symp* (2006) 279:66–77. doi:10.1002/9780470035399.ch6
43. Schmidt SV, Schultze JL. New insights into IDO biology in bacterial and viral infections. *Front Immunol* (2014) 5:384–93. doi:10.3389/fimmu.2014.00384
44. Zelante T, Iannitti RG, Fallarino F, Gargaro M, De Luca A, Moretti S, et al. Tryptophan feeding of the IDO1-AhR axis in host-microbial symbiosis. *Front Immunol* (2014) 5:640. doi:10.3389/fimmu.2014.00640
45. Cano LE, Kashino SS, Arruda C, André D, Xidieh CF, Singer-Vermes LM, et al. Protective role of gamma interferon in experimental pulmonary paracoccidioidomycosis. *Infect Immun* (1998) 66:800–6.
46. Mezrich JD, Fechner JH, Zhang X, Johnson BP, Burlingham WJ, Bradfield CA. An interaction between kynurenine and the aryl hydrocarbon receptor can generate regulatory T cells. *J Immunol* (2010) 185:3190–8. doi:10.4049/jimmunol.0903670
47. Puccetti P, Grohmann U. IDO and regulatory T cells: a role for reverse signalling and non-canonical Nf-κB activation. *Nat Rev Immunol* (2007) 7:817–23. doi:10.1038/nri2163
48. Fallarino F, Vacca C, Orabona C, Belladonna ML, Bianchi R, Marshall B, et al. Functional expression of indoleamine 2,3-dioxygenase by murine CD8a+ dendritic cells. *Int Immunol* (2002) 14:1206–12. doi:10.1093/intimm/14.1.65
49. Grohmann U, Fallarino F, Bianchi R, Belladonna ML, Vacca C, Orabona C, et al. IL-6 inhibits the tolerogenic function of CD8α+ dendritic cells expressing indoleamine 2,3-dioxygenase. *J Immunol* (2001) 167:708–14. doi:10.4049/jimmunol.167.2.708
50. Romani L. Immunity to fungal infections. *Nat Rev Immunol* (2011) 11:275–88. doi:10.1038/nri2939
51. Loures FV, Pina A, Felonato M, Araújo EF, Leite KR, Calich VL. Toll-like receptor 4 signaling leads to severe fungal infection associated with enhanced proinflammatory immunity and impaired expansion of regulatory T cells. *Infect Immun* (2010) 78:1078–88. doi:10.1128/IAI.01198-09
52. Read AF, Graham AL, Raberg L. Animal defenses against infectious agents: is damage control more important than pathogen control. *PLoS Biol* (2008) 6:e4. doi:10.1371/journal.pbio.1000004
53. Schneider DS, Ayres JS. Two ways to survive infection: what resistance and tolerance can teach us about treating infectious diseases. *Nat Rev Immunol* (2008) 8:889–95. doi:10.1038/nri2432
54. Bessede A, Gargaro M, Pallotta MT, Matino D, Servillo G, Brunacci C, et al. Aryl hydrocarbon receptor control of a disease tolerance defence pathway. *Nature* (2014) 511:184–90. doi:10.1038/nature13323
55. Romani L, Zelante T, De Luca A, Iannitti RG, Moretti S, Bartoli A, et al. Microbiota control of a tryptophan-AhR pathway in disease tolerance to fungi. *Eur J Immunol* (2014) 44:3192–200. doi:10.1002/eji.201344406
56. Nguyen NT, Kimura A, Nakahama T, Chinen I, Masuda K, Nohara K, et al. Aryl hydrocarbon receptor negatively regulates dendritic cell immunogenicity via a kynurenine-dependent mechanism. *Proc Natl Acad Sci U S A* (2010) 107:19961–6. doi:10.1073/pnas.1014465107
57. Benard G, Costa AN, Ravanini J, Goulart S, Nicodemo EL, Barbas CS, et al. Fatal acute respiratory distress syndrome in a patient with paracoccidioidomycosis: first case report. *Med Mycol* (2010) 48:542–5. doi:10.3109/13693780903330563
58. Gryschek RC, Pereira RM, Kono A, Patzina RA, Tresoldi AT, Shikanai-Yasuda MA, et al. Paradoxical reaction to treatment in 2 patients with severe acute paracoccidioidomycosis: a previously unreported complication and its management with corticosteroids. *Clin Infect Dis* (2010) 50:e56–8. doi:10.1086/652290
59. Arruda C, Kashino SS, Fazioli RA, Calich VL. A primary subcutaneous infection with *Paracoccidioides brasiliensis* leads to immunoprotection or exacerbated disease depending on the route of challenge. *Microbes Infect* (2007) 9:308–16. doi:10.1016/j.micinf.2006.12.005
60. Arruda C, Vaz CA, Calich VL. Aseptic cure of pulmonary paracoccidioidomycosis can be achieved after a previous subcutaneous immunization of susceptible but not resistant mice. *Microbes Infect* (2007) 9:704–13. doi:10.1016/j.micinf.2007.02.014
61. Mondanelli G, Bianchi R, Pallotta MT, Orabona C, Albini E, Iacono A, et al. A relay pathway between arginine and tryptophan metabolism confers immunosuppressive properties on dendritic cells. *Immunity* (2017) 46:233–44. doi:10.1016/j.immuni.2017.01.005

13

RIPK3/Fas-Associated Death Domain Axis Regulates Pulmonary Immunopathology to Cryptococcal Infection Independent of Necroptosis

Zhenzong Fa[1,2,3†], Qun Xie[4,5†], Wei Fang[2,6†], Haibing Zhang[5], Haiwei Zhang[5], Jintao Xu[3], Weihua Pan[1,2], Jinhua Xu[6], Michal A. Olszewski[2,3], Xiaoming Deng[4*] and Wanqing Liao[1,2*]*

[1] PLA Key Laboratory of Mycosis, Department of Dermatology and Venereology, Changzheng Hospital, Shanghai, China, [2] Shanghai Key Laboratory of Molecular Medical Mycology, Shanghai Institute of Medical Mycology, Second Military Medical University, Shanghai, China, [3] Division of Pulmonary and Critical Care Medicine, Department of Internal Medicine, University of Michigan Health System, Ann Arbor, MI, United States, [4] Department of Anesthesiology and Intensive Care, Changhai Hospital, Second Military Medical University, Shanghai, China, [5] Key Laboratory of Nutrition and Metabolism, Institute for Nutritional Sciences, Shanghai Institutes for Biological Sciences, Chinese Academy of Sciences, Shanghai, China, [6] Department of Dermatology, Huashan Hospital, Fudan University, Shanghai, China

***Correspondence:**
Michal A. Olszewski
olszewsm@med.umich.edu;
Xiaoming Deng
deng_x@yahoo.com;
Wanqing Liao
liaowanqing@sohu.com

†These authors have contributed equally to this work.

Fas-associated death domain (FADD) and receptor interacting protein kinase 3 (RIPK3) are multifunctional regulators of cell death and immune response. Using a mouse model of cryptococcal infection, the roles of FADD and RIPK3 in anti-cryptococcal defense were investigated. Deletion of RIPK3 alone led to increased inflammatory cytokine production in the *Cryptococcus neoformans*-infected lungs, but in combination with FADD deletion, it led to a robust Th1-biased response with M1-biased macrophage activation. Rather than being protective, these responses led to paradoxical *C. neoformans* expansion and rapid clinical deterioration in *Ripk3*$^{-/-}$ and *Ripk3*$^{-/-}$*Fadd*$^{-/-}$ mice. The increased mortality of *Ripk3*$^{-/-}$ and even more accelerated mortality in *Ripk3*$^{-/-}$*Fadd*$^{-/-}$ mice was attributed to profound pulmonary damage due to neutrophil-dominant infiltration with prominent upregulation of pro-inflammatory cytokines. This phenomenon was partially associated with selective alterations in the apoptotic frequency of some leukocyte subsets, such as eosinophils and neutrophils, in infected *Ripk3*$^{-/-}$*Fadd*$^{-/-}$ mice. In conclusion, our study shows that RIPK3 in concert with FADD serve as physiological "brakes," preventing the development of excessive inflammation and Th1 bias, which in turn contributes to pulmonary damage and defective fungal clearance. This novel link between the protective effect of FADD and RIPK3 in antifungal defense and sustenance of immune homeostasis may be important for the development of novel immunomodulatory therapies against invasive fungal infections.

Keywords: *Cryptococcus neoformans*, immune responses, inflammation, Fas-associated death domain, receptor-interacting serine/threonine kinase 3

INTRODUCTION

Invasive fungal infections have become an increasingly significant challenge to public health due to the ever-increasing population of immunosuppressed patients, associated with aging of the global population, immunosuppressive infections such as HIV, and the growing use of immunosuppressive therapies. Among the major fungal pathogens, *Cryptococcus neoformans* causes life-threatening invasive infections in both immunocompromised and immunocompetent hosts (1, 2). As supported by evidence from both the clinic and animal infection models, an insufficiency of Th1 and Th17 responses and subsequent classical activation of macrophages are major triggers of cryptococcal infection. However, emerging evidence supports the view that the excessive inflammation and pathology is frequently derived by a Th1 response, which is often initiated during highly active antiretroviral therapy (HAART) in HIV$^+$ patients with cryptococcosis. This paradoxical response, known as immune reconstitution inflammatory syndrome (IRIS), contributes to worsening symptoms and patient mortality despite ongoing antifungal and HAART treatments (3). This unique clinical problem underscores the importance of immunoregulatory processes during opportunistic fungal infections of which many aspects remain to be elucidated.

Fas-associated death domain protein (FADD) is known as a critical mediator of death receptor-triggered extrinsic apoptosis, which plays a role in removing "no longer needed" inflammatory cells, thereby serving as a crucial immune-regulatory pathway at the site of infection, preventing excessive inflammation (4, 5). Besides its role in apoptosis, FADD also has been shown to function in regulating cell cycle progression (6, 7), cytokine signaling (8, 9), and T-cell activation (10), which are all involved in regulation of immune responses. Targeted deletion of FADD in mice causes embryonic lethality due to spontaneous activation of another programmed cell death (PCD) pathway, necroptosis (11). However, co-deletion of receptor interacting protein kinase 3 (RIPK3), which is an essential serine/threonine kinase for necroptosis, rescues these mice (12). In addition to necroptosis, RIPK3 has also been reported as an important inflammatory signal adaptor because it functions in NF-κB activation, inflammasome activation, and cytokine signaling (13–15) and participates in the pathogenesis of several inflammatory diseases. However, it remains unknown whether these molecules play important roles during immune responses to fungal infections.

Here, we explored the roles of FADD and RIPK3 in a mouse model of cryptococcal infection and identified previously unknown, critical contributions of these molecules in pulmonary immune responses to cryptococcal infection. Deletion of FADD and RIPK3 induced robust Th1 responses, which, paradoxically, led to *C. neoformans* expansion and increased mortality in the infected mice. These effects were attributed to an excessive accumulation of neutrophils, over exuberant inflammatory cytokine production, and development of severe lung pathology. Collectively, these findings establish a novel link between these PCD components and immune response to cryptococcal challenge, demonstrating the crucial importance of FADD and RIPK3 in maintaining immune homeostasis during invasive fungal infection.

MATERIALS AND METHODS

Mice

Female wild-type (WT) C57BL/6 mice were housed in a specific pathogen-free facility. *Ripk3*$^{-/-}$ mice have been previously described (16). *Fadd*$^{+/-}$ mice were generated using the CRISPR–Cas9 mutation system (Shanghai Bioray Laboratory, Inc.). A 100-bp deletion was introduced into exon 1 of the *Fadd* gene (Figure S1 in Supplementary Material). Because ablation of *Fadd* in mice causes embryonic death, we crossed the *Ripk3*$^{-/-}$ mice with *Fadd*$^{+/-}$ mice to obtain *Ripk3*$^{-/-}$*Fadd*$^{-/-}$ mice. All mice genotypes were confirmed by PCR (Figure S1 in Supplementary Material). Mice were 8–10 weeks old at the time of infection and were humanely euthanized by CO_2 inhalation at the time of data collection. Animal experiments were conducted in accordance with the National Institutes of Health Guide for the Care and Use of Laboratory Animals with the approval of the Scientific Investigation Board of Second Military Medical University.

Cryptococcus neoformans

Encapsulated *C. neoformans* strain H99 (serotype A) was recovered from 10% glycerol-frozen stocks stored at −80°C. The strains were cultured on yeast extract-peptone-dextrose agar plates at 30°C. Liquid cultures were grown in Sabouraud dextrose broth at 30°C for 20–24 h in a shaking incubator at 180 rpm. Fungal cells were centrifuged at 2,000 × *g* for 3 min, washed three times, and resuspended in sterile PBS.

Inoculation

Mice were anesthetized by intraperitoneal injection of ketamine (100 mg/kg, Sigma, St. Louis, MO, USA) and secured onto a clean foam board (17). Next, 50 µl [10^5 colony-forming units (CFU)] of the washed yeast (2×10^6 yeast cells/ml in sterile PBS) were used for intranasal infection as previously described (18). After inoculation, the mice were kept warm and monitored during recovery from anesthesia.

Tissue Collection and Lung Leukocyte Isolation

The procedures were performed as previously described with modifications (19). At the time of data collection, the mice were sacrificed and perfused with 5 ml sterile PBS. The lungs were removed, minced with scissors, and added to homogenization gentleMACS C tubes containing proprietary catalysts for mechanical and enzymatic digestion (Miltenyi Biotec, Auburn, CA, USA). This process was followed by lung tissue homogenization using the gentle MACS dissociator (Miltenyi Biotec) and incubation at 37°C for 30 min in 4 ml/mouse digestion buffer (RPMI 1640, 5% fetal bovine serum, penicillin, and streptomycin, Invitrogen, Grand Island, NY, USA), 1 mg/ml collagenase A (Roche Diagnostics, Indianapolis, IN, USA), and 30 g/ml DNase I (Sigma, St. Louis, MO, USA). Erythrocytes were removed using 1× RBC lysis buffer (eBioscience, San Diego, CA, USA). Homogenized tissue was then passed through a 70 µm cell strainer (BD Falcon, Bedford, MA, USA) and centrifuged at 300 × *g* for 10 min to pellet the cells. The filtrate was centrifuged for 25 min at 1,500 × *g* in the presence of 20% Percoll (Sigma, St. Louis, MO, USA) in complete RPMI 1640 medium (RPMI 1640, 5% FBS, penicillin, and streptomycin) with no brake to separate leukocytes from cell debris and epithelial cells. Leukocyte pellets were resuspended in 5 ml complete RPMI 1640 medium and counted in a hemocytometer using trypan blue staining to exclude dead cells.

Tissue CFU Assay

For determining the fungal burden in the lungs, brains, and spleens, tissues were removed and homogenized in 1 ml sterile PBS. Ten-fold dilutions of the samples were plated in duplicate on Sabouraud dextrose agar plates. Colonies were counted after 48 h of growth at 30°C, and CFU were calculated on a per-gram basis.

Cytokine Analysis

Mouse serum was obtained from blood samples collected by heart puncture before lung excision and centrifugation at 10,000 × *g* for 10 min. Homogenates of lungs were centrifuged, and supernatants were diluted for cytokine analysis. Leukocytes isolated from mice lung were plated at 10^7 cells/ml, and supernatants were collected by centrifugation of the culture medium. Mouse TNFα, IFNγ, IL-1α, IL-1β, IL-4, IL-6, IL-12, IL-17A, and IL-33 ELISA kits were from eBioscience. CXCL1 was from Raybio (Norcross, GA, USA).

Flow Cytometry Analysis of Leukocyte Populations

For the flow cytometry experiments, antibodies were purchased from eBioscience, BioLegend, or BD Biosciences, including anti-murine CD16/CD32; CD45 conjugated to PerCP-Cy5.5; CD3, CD193, CD80, IFNγ, and annexin V conjugated to FITC; CD4, Ly6G, and Siglec F conjugated to APC; CD8, CD11c, CD19, and CD80 conjugated to PE-Cy7; CD19, CD40, F4/80, and IL-4 conjugated to PE; CD11b, CD326, and MHC II conjugated to APC-Cy7; and CD19, CD80, and Ly6C conjugated to BV421.

Leukocytes were isolated from the lung and lymph nodes of mice. Cell surface immunofluorescence staining involved the addition of a fluorochrome-conjugated antibody mixture containing antibodies specific to various leukocyte subpopulations to the staining buffer. Cells were incubated on ice for 30 min in the dark and washed twice with PBS. For intracellular cytokine staining, cells were fixed in IC fixation buffer and stimulated with cell stimulation cocktail (plus protein transportation inhibitors) from eBioscience. Cells were resuspended in permeabilization buffer (eBioscience) and stained with intracellular antibody cocktail. After staining, cells were immediately analyzed by flow cytometry (FACSAria III, BD Biosciences). FlowJo (For Mac OS X, version X 10.0.7r2, Tree Star, San Carlos, CA, USA) was used for data analysis. Leukocyte populations were identified using the following markers as previously described (20, 21): neutrophils ($CD45^+$ $Ly6G^+$ $CD11b^+$), dendritic cells (DCs, $CD45^+$ $CD11c^+$ MHC II high), resident macrophage ($CD45^+$ $CD11b^-$ Siglec F^+), eosinophils ($CD45^+$ $CD11b^+$ Siglec F^+), monocyte-derived DCs or macrophages ($CD45^+$ $Ly6C^+$ $CD11c^-$), CD4 T cells ($CD45^+$ $CD3^+$ $CD4^+$), CD8 T cells ($CD45^+$ $CD3^+$ $CD8^+$), and B cells ($CD45^+$ $CD19^+$). DCs in lymph nodes were stained with extracellular CD45, CD11c, and CD80, and intracellular TNFα, IFNγ, and IL-4. Total numbers of each cell population were calculated by multiplying the frequency of the population by the total number of leukocytes (the percentage of $CD45^+$ cells multiplied by the original hemocytometer counts for total cells).

Lung-Associated Lymph Node (LALN) Leukocyte Isolation

Lung-associated lymph node leukocytes were collected as previously described with modifications (22). Lymph nodes were removed from the mediastinum and then mechanically dispersed using an 1-ml sterile syringe plunger to press them through a 70 μm cell strainer (BD Falcon, Bedford, MA, USA) in complete medium. After centrifugation at 2,500 × *g* for 5 min, the supernatant was removed and the cell pellets saved for further use.

Immunoblot Analysis

Lungs were ground up in liquid nitrogen and suspended in lysis buffer containing Tris–HCl (50 mM; pH 8.0), NaCl (150 mM), EDTA (1 mM), NP-40 (1%), PMSF (1 mM; Sigma), phosphatase inhibitor (Sigma), and a protease inhibitor cocktail (Roche Biochemical Laboratories). After incubation on ice for 30 min, the cell lysates were collected after centrifugation (14,000 × *g* for 10 min) at 4°C, and protein concentrations were determined using the Pierce BCA Protein Assay Kit (Thermo Scientific). A total of 30 μg protein was loaded for western blot analysis using the following antibodies: RIPK3 (Prosci) and caspase-3 (Cell Signaling Technology).

Real-time PCR Analysis

Total RNA (10–100 ng, depending on the abundance of the target gene) was purified using TRIzol reagent (Ambion by Life Technologies) for RT-qPCR in a one-step reaction with Reverse Transcriptase (Takara) and SYBR green master mix (Takara) using a 7900 Real-Time PCR system (Applied Biosystems). All primers used for RT-qPCR are listed in Table S1 in Supplementary Material. The qPCR analysis was performed using the $2^{-\Delta Ct}$ method, and target genes were normalized to the housekeeping genes in each strain.

Histology, Immunohistochemistry, and Immunofluorescence

Lungs were instilled with 1 ml 10% neutral-buffered formalin, excised, immersed in 10% neutral-buffered formalin, and embedded in paraffin as described previously (19). Five-micrometer sections were cut and stained with hematoxylin & eosin. Immunohistochemical and immunofluorescent staining was performed using formalin-fixed, paraffin-embedded tissue sections with rabbit anti-RIPK3 (Prosci) antibodies. Sections were photographed by Zeiss light microscopy (ZEISS, AXIO) and Olympus confocal microscopy (FV1000). Lung tissue inflammation and injury score were performed by three different pathologists in a blinded fashion. The quantify criteria is referenced to previous studies with modification (23).

Fungal Killing and Cell Viability Assays

Bone marrow-derived macrophages (BMDMs) were generated as previously described (24, 25). Briefly, marrow was flushed from the C57Bl/6 mouse femurs and tibias and dispersed into a single-cell suspension. The cells were cultured for 7 days in RPMI medium supplemented with 10% FBS and 50 ng/ml M-CSF. The cultures were additionally nourished with M-CSF-containing medium on the third day of culture. All *in vitro* experiments were performed in RPMI 1640 containing 10% FCS and 5 ng/ml M-CSF.

For the fungal killing assay, freshly isolated BMDMs were diluted to 10^6 cells/ml and plated on a 96-well cell culture plate. *C. neoformans* were washed twice with PBS, resuspended in RPMI medium, and adjusted to 10^5 cells/ml. The yeast cells were further opsonized with anti-GXM antibody for 1 h at 37°C, followed by the addition of 100 µl opsonized *C. neoformans* to each well of the BMDM culture plate and incubation at 37°C with 5% CO_2 for 24 h. BMDM cells were lysed in sterile water for 20 min, mixed with the supernatant, and then diluted and plated on Sabouraud agar plates. CFU were counted after 2 days at 30°C.

Statistical Analysis

All data are expressed as means ± SEMs. The data obtained for the animal survival assays were plotted as Kaplan–Meier survival curves and analyzed with the log-rank test using GraphPad Prism version 6.00 for Windows (GraphPad Software, San Diego, CA, USA). The remaining statistical analyses were conducted with the ANOVA, Student's *t*-test, and Kruskal–Wallis test as well as Dunn's test for non-parametric measures. The results were considered statistically significant when the *P* value was less than 0.05.

RESULTS

RIPK3 and FADD Critically Contribute to Host Defenses against *C. neoformans* Infection

To gain insight into the roles of RIPK3 and FADD during host responses to *C. neoformans* infection, we first evaluated kinetics of RIPK3 and FADD protein expression in the lungs of the infected C57BL6 mice. Immunoblot study results showed that RIPK3 displayed extensive upregulation in the lung tissue at 10 days postinfection (dpi) (**Figure 1A**), further verified by immunochemistry and immunofluorescence assays (**Figures 1B,C**). In contrast, FADD, albeit abundant, appeared to be expressed constitutively throughout the studied time points of infection. Next, to elucidate the relative importance of FADD and RIPK3-related signaling in anti-cryptococcal defense, we compared the survival of infected *Ripk3*$^{-/-}$, *Ripk3*$^{-/-}$*Fadd*$^{-/-}$, and WT mice. As shown in **Figure 1D**, compared with WT mice, infected *Ripk3*$^{-/-}$ mice showed accelerated onset of mortality (8 vs 20 dpi) and reduced median survival (median: 16.5 ± 5.8 vs 21.5 ± 1.4 days). *Ripk3*$^{-/-}$*Fadd*$^{-/-}$ mice showed even greater susceptibility to *C. neoformans* infection compared with either *Ripk3*$^{-/-}$ ($P < 0.05$) or WT mice ($P < 0.001$), exhibiting a drastically shortened median survival

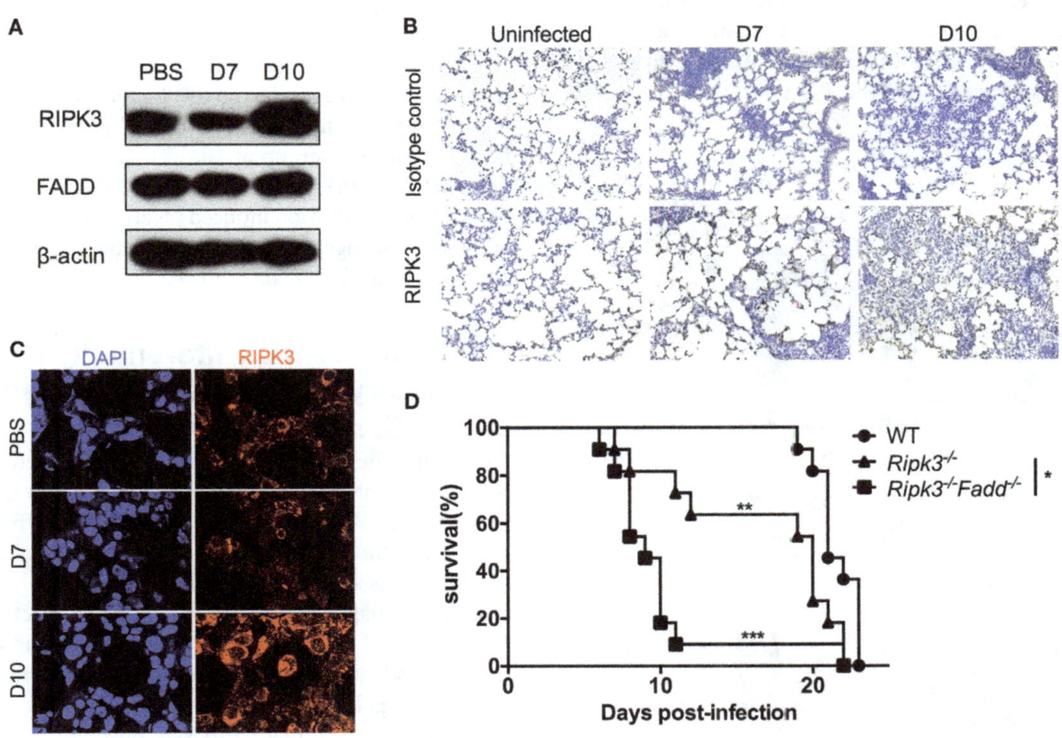

FIGURE 1 | RIPK3 and Fas-associated death domain (FADD) critically contribute to host defenses against *Cryptococcus neoformans* infection. C57BL/6 mice were inoculated intranasally with 10^5 *C. neoformans* strain H99 and sacrificed at 7 or 10 days postinfection. PBS-treated mice were utilized as control. Pulmonary expression of RIPK3 and FADD were examined by western blot analysis in total protein form lung homogenates **(A)**. RIPK3 displayed enhanced expression, while FADD remains unchanged in the lung of infected mice. Fungal infection significantly enhanced the local recruitment of RIPK3 in the lung as shown by immunohistochemistry **(B)** and immunofluorescence **(C)** assays. Wild-type (WT), *Ripk3*$^{-/-}$, *Ripk3*$^{-/-}$*Fadd*$^{-/-}$ mice (n = 11 for each group) were infected *via* inhalation to determine effects of RIPK3 and FADD in host defenses against *C. neoformans*. Survival studies showed that *Ripk3*$^{-/-}$ and *Ripk3*$^{-/-}$*Fadd*$^{-/-}$ mice were more susceptible to fungal infection **(D)**. *$P < 0.05$, **$P < 0.01$, ***$P < 0.001$. The survival study was repeated three times independently.

(9.9 ± 4.3 days). These results demonstrate that FADD and RIPK3 signaling critically contribute to host defense against cryptococcal infection, most likely in a synergistic fashion.

RIPK3 or RIPK3/FADD Deletion Reduced Host Ability to Control Fungal Growth and Dissemination

Next, we examined fungal burden in the lung, brain, and spleen in infected mice to assess the effects of RIPK3 and FADD in the control of fungal growth and systemic dissemination. Compared with WT, *Ripk3*$^{-/-}$ mice had a significantly higher pulmonary fungal load at 10 dpi ($P < 0.05$), and RIPK3/FADD double deletions further enhanced the fungal burden in the lungs at both 7 dpi ($P < 0.05$) and 10 dpi (**Figure 2A**, $P < 0.01$). Similar trends in fungal burden were noted in the spleens and brains of *Ripk3*$^{-/-}$ and *Ripk3*$^{-/-}$*Fadd*$^{-/-}$ mice at 10 dpi (**Figures 2B,C**), with frequencies of positive spleen and brain cultures in *Ripk3*$^{-/-}$*Fadd*$^{-/-}$ mice doubling those in the WT mice. Thus, both RIPK3 and FADD significantly contributed fungal containment during pulmonary cryptococcal infection.

RIPK3/FADD Deletions Lead to Severe Lung Pathologies in *C. neoformans*-Infected Mice

Rapid clinical deterioration and the accelerated mortality, especially in the infected *Ripk3*$^{-/-}$*Fadd*$^{-/-}$ mice could not be solely explained by relatively modest increases in fungal burdens at 7 and 10 dpi. Thus, we examined the effects of RIPK3 or RIPK3/FADD on the development of lung inflammation and pathologies post-*C. neoformans* infection. We examined lung sections obtained from sham-infected (PBS) and *C. neoformans*-infected WT and *Ripk3*$^{-/-}$ or *Ripk3*$^{-/-}$*Fadd*$^{-/-}$ mice. Deletion of RIPK3 or RIPK3/FADD induced no visible alterations in uninfected lungs (data not shown), demonstrating that genetic defects in RIPK3 or RIPK3/FADD did not affect baseline pulmonary morphology. Comparative histopathological assessments of lung sections from each group at 10 dpi (**Figure 3**) demonstrated only subtle enhancement of pulmonary leukocyte infiltration in *Ripk3*$^{-/-}$ mice relative to the control mice with largely similar pattern of inflammatory lesions (**Figures 3A–C** vs **Figures 3D–F**). The borders between inflamed regions and normal alveoli remained distinct at 10 dpi in WT and *Ripk3*$^{-/-}$ mice (**Figures 3A–F**). Consistently, blinded pathology score, albeit showing an increasing trend, has not increased significantly (Figure S2 in Supplementary Material). In contrast, *Ripk3*$^{-/-}$*Fadd*$^{-/-}$ mice exhibited progressive pulmonary inflammation with severe tissue damage (**Figures 3G–I**). Less organized inflammatory infiltrates (predominantly neutrophils and lymphocytes) were spread diffusely through bilateral lung fields at 10 dpi. The margins of inflamed regions from uninvolved alveoli were less distinct. Features of suppurative bronchopneumonia, such as airway plugins with polymorphic neutrophil and dense cellular exudate, were observed throughout the lung (**Figures 3H,I**). The blinded pathology score showed that inflammation/pathology score was significantly greater in lungs of *Ripk3*$^{-/-}$*Fadd*$^{-/-}$ mice compared to both WT and *Ripk3*$^{-/-}$ mice (Figure S2 in Supplementary Material), which corroborated well with survival data on day 10 (80% mortality in *Ripk3*$^{-/-}$*Fadd*$^{-/-}$ group, only 20% mortality in *Ripk3*$^{-/-}$ group, and all mice surviving in the WT group at 10 dpi). Collectively, these findings demonstrated that concurrent RIPK3/FADD deletion during *C. neoformans* infection induced severe pulmonary inflammation and tissue damage, potentially explaining the highly accelerated mortality in *Ripk3*$^{-/-}$*Fadd*$^{-/-}$ mice.

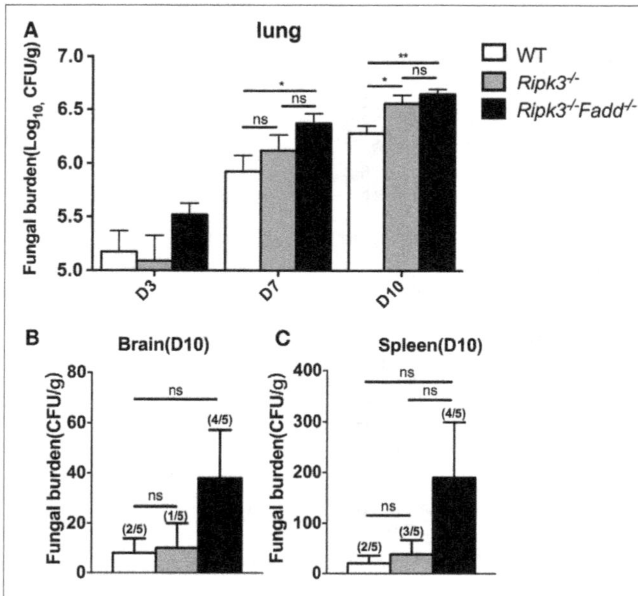

FIGURE 2 | *Ripk3*$^{-/-}$ and *Ripk3*$^{-/-}$*Fadd*$^{-/-}$ mice showed diminished fungal clearance. Pulmonary fungal burden assay revealed diminished fungal clearance of *Ripk3*$^{-/-}$ and *Ripk3*$^{-/-}$*Fadd*$^{-/-}$ mice **(A)** and similar but not significant trends in extra-pulmonary dissemination **(B,C)**. Values represent means and SEMs ($n = 5$ mice for each group); ns, no significant difference, $^{*}P < 0.05$, $^{**}P < 0.01$, between compared groups. The colony-forming units (CFU) assay was repeated three times independently.

RIPK3 and FADD Differentially Modulated Pulmonary Leukocyte Accumulation during *C. neoformans* Infection

To quantify the effects of RIPK3 and FADD deletions on cellular components of the inflammatory response to *C. neoformans*, we compared leukocyte populations isolated from uninfected or infected lungs of WT, *Ripk3*$^{-/-}$, or *Ripk3*$^{-/-}$*Fadd*$^{-/-}$ mice at 10 dpi. Consistent with the histopathological findings, flow cytometric analysis revealed no significant effect of *Ripk3*$^{-/-}$ or *Ripk3*$^{-/-}$*Fadd*$^{-/-}$ mutations in uninfected mice for any of the leukocyte subsets (data not shown). While only a borderline increase in total lung leukocytes counts (CD45^{+} cells) was observed in *Ripk3*$^{-/-}$ mice compared to WT mice at 10 dpi (**Figure 4A**, $P < 0.06$), significant increases in two pulmonary leukocyte subsets were observed. *Ripk3*$^{-/-}$ mice showed increased numbers of neutrophils (**Figure 4B**), increasing trend in macrophages (**Figure 4C**), and elevated CD4^{+} T cells (**Figure 4D**) compared to the infected WT mice.

Ripk3$^{-/-}$*Fadd*$^{-/-}$ mice infected by *C. neoformans* showed more profound alterations in both the magnitude of inflammation

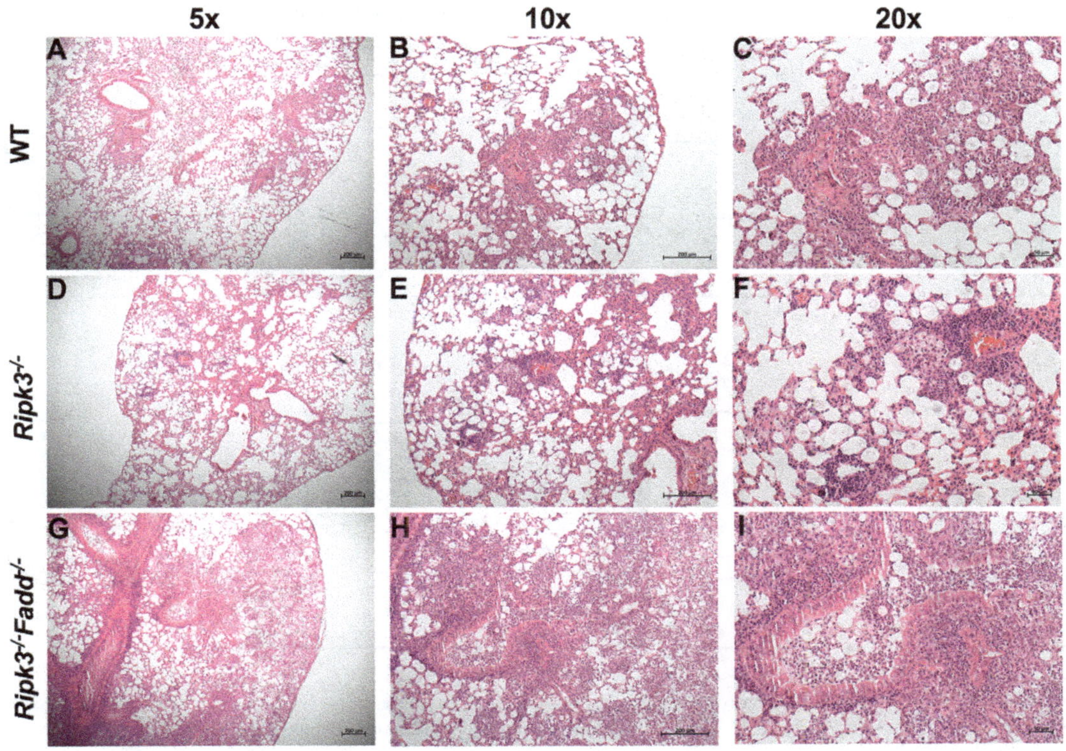

FIGURE 3 | Deletion of RIPK3 and Fas-associated death domain (FADD) resulted in enhanced inflammatory infiltration and tissue injury following pulmonary infection with *Cryptococcus neoformans*. Wild-type (WT) **(A–C)**, *Ripk3*$^{-/-}$ **(D–F)**, or *Ripk3*$^{-/-}$*Fadd*$^{-/-}$ **(G–I)** mice were infected with 10^5 *C. neoformans* strain H99. At 10 days postinfection, lungs were harvested, and the morphological pattern of inflammation was evaluated by light microscopy. Pulmonary representative photomicrographs (hematoxylin and eosin stained) taken from each group were shown [**(A,D,G)** magnification of ×50; **(B,E,H)** magnification of ×100; **(C,F,I)** magnification of ×200]. Compared with WT group, *Ripk3*$^{-/-}$ mice displayed similar pulmonary histological pattern but a slight enhancement in inflammatory infiltration. However, extensive pulmonary inflammation and severe tissue damage were observed in *Ripk3*$^{-/-}$*Fadd*$^{-/-}$ mice, indicating distinct patterns of lung pathology.

and leukocyte composition. There was a significant increase in total leukocyte counts in *Ripk3*$^{-/-}$*Fadd*$^{-/-}$ mice compared with WT group at 10 dpi (**Figure 4A**), mostly driven by a remarkable increase in neutrophil numbers (more than threefold) relative to infected WT (**Figure 4B**) and significantly greater than in the *Ripk3*$^{-/-}$ mice. Significant increases in other myeloid cell subsets (monocytes, **Figure 4E**; DCs, **Figure 4F**) were also detected relative to both WT and *Ripk3*$^{-/-}$ mice. However, the numbers of pulmonary eosinophils observed in abundance in the WT and *Ripk3*$^{-/-}$ mice at 10 dpi were suppressed in *Ripk3*$^{-/-}$*Fadd*$^{-/-}$ mice (**Figure 4G**), suggesting a shift away from a Th2 response in the WT following RIPK3/FADD double deletion. Collectively, these data further support that RIPK3 and more so the RIPK3/FADD double deletion results in an exuberant accumulation of inflammatory cells and immunomodulation that alters the course of inflammatory response implicated in lung injury we observe during *C. neoformans* infection in the absence of these factors.

RIPK3/FADD Deletion Reshaped the Cytokine Responses during Pulmonary *C. neoformans* Infection

Having demonstrated that RIPK3 and RIPK3/FADD deletions altered inflammatory infiltrate compositions, we further examined the roles of RIPK3 and FADD in pulmonary and systemic cytokine levels during cryptococcal infection. Lung homogenates and serum isolated at 10 dpi from *Ripk3*$^{-/-}$, *Ripk3*$^{-/-}$*Fadd*$^{-/-}$, and WT mice were analyzed by ELISA (**Figures 5A–N**). We found that depletion of RIPK3 alone significantly increased pro-inflammatory cytokine production: TNF-α, IL-1α, and IL-1β (**Figures 5A–C**). Non-significant increasing trend in IFN-γ level and decreasing trend in IL-4 production suggested cytokine profile drifting away from Th2 to Th1 pattern (**Figure 5J**), which corresponded to diminished serum IgE accumulation (**Figure 5N**). However, no difference in IL-12p40 or IL-33 have been observed (**Figures 5E,H**), suggesting that the full switch from Th2 to Th1 has not occurred as a result of RIPK3 deletion. Finally, while we observe significant increase in IL-17A, which might have suggested shift toward Th17 response as a result of RIPK3 deletion, there was no concurrent increase in IL-6 or IL-12p40, which would be associated with Th17 response. Most of these trends were reproduced in pulmonary leukocyte cell culture supernatants (Figure S4 in Supplementary Material). Consistently, with lung cytokines, serum cytokine analysis showed somewhat elevated TNFα (**Figure 5K**), suggesting more pronounced inflammatory response levels in infected *Ripk3*$^{-/-}$ mice, but no increase in serum IL-6 or IFNγ (**Figures 5L,M**).

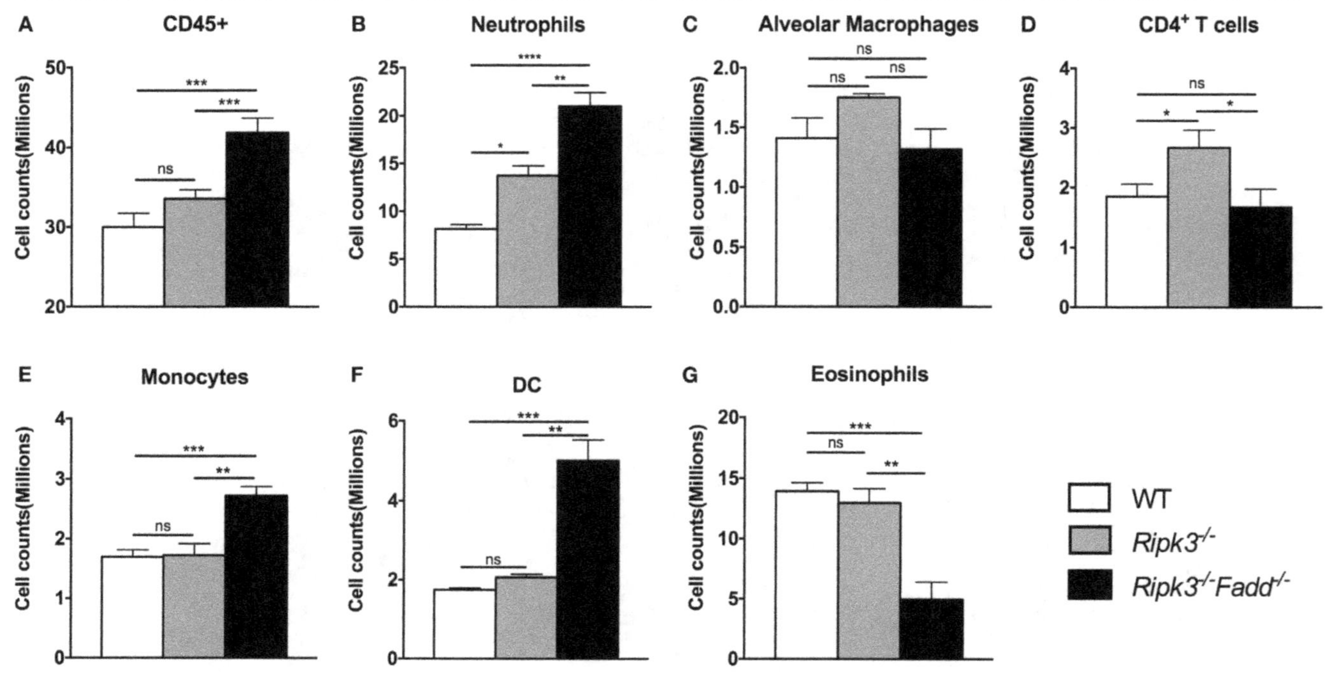

FIGURE 4 | RIPK3 or RIPK3/Fas-associated death domain (FADD) deletions resulted in alteration in pulmonary leukocyte recruitments. Mice were inhalationally infected with 10^5 *Cryptococcus neoformans* strain H99. Lungs were harvested at 10 days postinfection, and total pulmonary leukocytes were isolated from infected lungs after enzymatic digestion. Magnitude and/or leukocyte composition were then analyzed by flow cytometry. Results are illustrated as absolute numbers of each lymphocyte subset **(A–G)** in the total lung sample. Values represent means and SEMs ($n = 6$ mice per group). ns, no significant difference, $^*P < 0.05$, $^{**}P < 0.01$, $^{***}P < 0.001$, $^{****}P < 0.0001$, between compared groups. This experiment was repeated twice.

RIPK3/FADD double deletion further enhanced pro-inflammatory TNFα and IL-6 (**Figures 5A,D**) and showed sustained elevation of IL-1α, IL-1β compared to the WT mice. Furthermore, Th1 cytokines IL-12p40 and IFNγ were significantly elevated relative to WT and RIPK3 mice levels of these cytokines (**Figures 5E,F**), suggesting not only strongly intensified inflammatory response but also shift to a Th1. Consistently, Th2 cytokines (IL-4 and IL-33) were profoundly suppressed (**Figures 5G,H**), and the Th1/Th2 ratio increased more than 300-fold in RIPK3/FADD-depleted relative to WT mice (**Figure 5J**), with corresponding absence of serum IgE accumulation (**Figure 5N**) strongly suggesting the development of a robust Th1 bias in these mice. Interestingly, IL-17 was not elevated as in the *Ripk3*$^{-/-}$ mice but showed level similar to that in the infected WT mice (**Figure 5I**).

To further investigate effects of RIPK3/FADD on T cell polarization, we performed intracellular flow analysis on pulmonary CD4$^+$ T cells from infected *Ripk3*$^{-/-}$ *and Ripk3*$^{-/-}$*Fadd*$^{-/-}$ mice at 10 dpi. Consistent with the cytokine data, percentage of IFNγ$^+$ CD4$^+$ T cells was not different in *Ripk3*$^{-/-}$ mice compared to the WT, but increased in *Ripk3*$^{-/-}$*Fadd*$^{-/-}$ mice compared to WT and *Ripk3*$^{-/-}$ mice (**Figures 6A,B**). Moreover, we found there was a borderline reduction in the frequencies of GATA3 positive CD4$^+$ T cells in *Ripk3*$^{-/-}$ ($P = 0.07$) mice and significant decrease in frequencies of GATA3$^+$ CD4$^+$ T cells in *Ripk3*$^{-/-}$*Fadd*$^{-/-}$ mice relative to WT mice at 10 dpi (**Figures 6C,D**). Interestingly, no increase in IL-17A or RorγT$^+$ CD4 or CD8 T-cells was observed in either *Ripk3*$^{-/-}$ or *Ripk3*$^{-/-}$*Fadd*$^{-/-}$ mice compared to the WT mice (data not shown). Collectively, analysis of cytokine responses and T-cell polarization profile showed that RIPK3 single deletion led to enhanced pro-inflammatory responses, while that additional deletion of FADD further potentiated these effects, leading to a very strong Th1 bias systemic and "cytokine storm" at 10 dpi with *C. neoformans*.

RIPK3 and FADD Deletions Potentiated Classical Activation of Macrophages and Their Fungicidal Responses *In Vitro*

Macrophages are distal effector cells that execute anti-*C. neoformans*-based cytokine responses in infected organs (26, 27). Having determined that RIPK3 increased inflammatory cytokines, including pro-M1 cytokine TNFα and RIPK3/FADD additionally promoted Th1 responses, we assessed the M1/M2 polarization patterns of macrophages in the infected lungs and the fungicidal ability of BMDMs isolated from uninfected mice from each strain. The expression of M1- and M2-associated genes was evaluated by real-time PCR (**Figures 7A–C**). Consistent with absence of major increase in IFNγ production, *Ripk3*$^{-/-}$ mice macrophages did not show significant upregulation of the M1 activation marker iNOS (**Figure 7A**). However, in concert with less pronounced Th2 and more pro-inflammatory environment in the lungs, we observed diminished upregulation of M2 markers arginase Arg1 and Fizz1

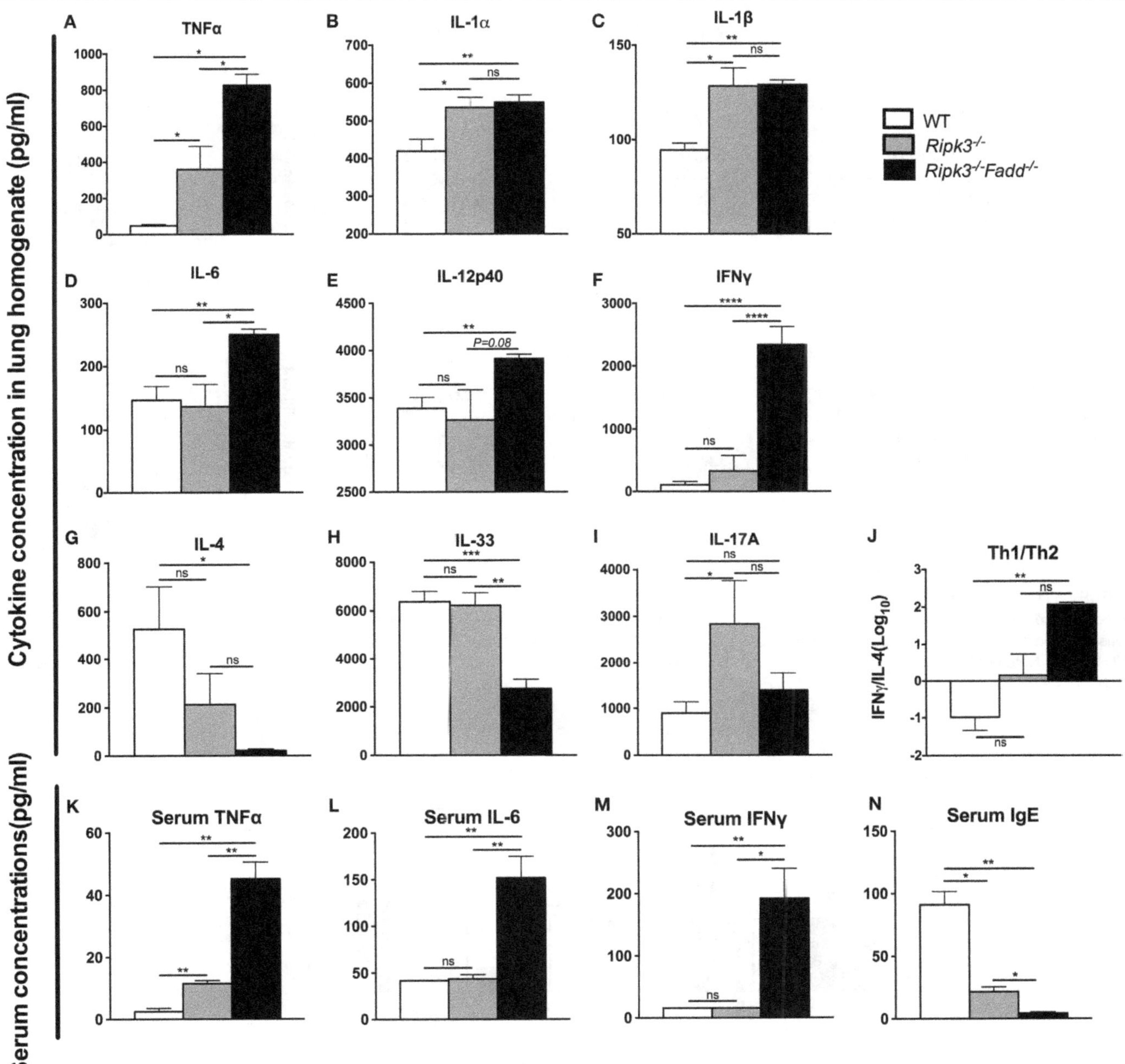

FIGURE 5 | RIPK3 or RIPK3/Fas-associated death domain (FADD) deletions altered Th-polarizing and pro-inflammatory cytokine profiles in local (pulmonary) and systemic response to cryptococcal infection. Total lung tissue **(A–I)** and serum **(K–N)** were harvested from wild-type (WT), *Ripk3*$^{-/-}$, or *Ripk3*$^{-/-}$*Fadd*$^{-/-}$ mice at 10 days postinfection with *Cryptococcus neoformans*. Supernatants of lung tissue homogenates were collected by centrifuging. Cytokine levels were detected in the supernatants and serum by ELISA. Total serum IgE concentrations were also measured by ELISA. Values represent mean cytokine or IgE concentrations [**(A–I)**, **(K–N)**, pg/ml] or **(J)**, the IFNγ/IL-33 ratio (n = 5 mice per group). ns, no statistical significance. $^{*}P < 0.05$, $^{**}P < 0.01$, $^{***}P < 0.001$. This experiment was repeated twice.

(**Figures 7B,C**), suggesting that macrophages in the infected lung of *Ripk3*$^{-/-}$ mice were less M2 biased. Consistent with the strong Th1-type polarization, expression of the iNOS was significantly upregulated in *Ripk3*$^{-/-}$*Fadd*$^{-/-}$ mice (**Figure 7A**), while Fizz1 expression was diminished (**Figure 7C**) compared to both WT and *Ripk3*$^{-/-}$ mice. Compared with WT mice, arginase1 expression in *Ripk3*$^{-/-}$ *Fadd*$^{-/-}$ mice also showed downward effects (**Figures 7B,C**). To determine whether the fungicidal potential (typically linked to M1/M2 activation) was affected in *Ripk3*$^{-/-}$ and *Ripk3*$^{-/-}$*Fadd*$^{-/-}$ mice, we evaluated the fungicidal effect of the BMDMs after co-incubation with 10^6 CFU of *C. neoformans* (**Figure 7D**). Significantly reduced cryptococcal survival was observed in *Ripk3*$^{-/-}$ and, to an even greater extent, in *Ripk3*$^{-/-}$ *Fadd*$^{-/-}$ macrophages, further demonstrating

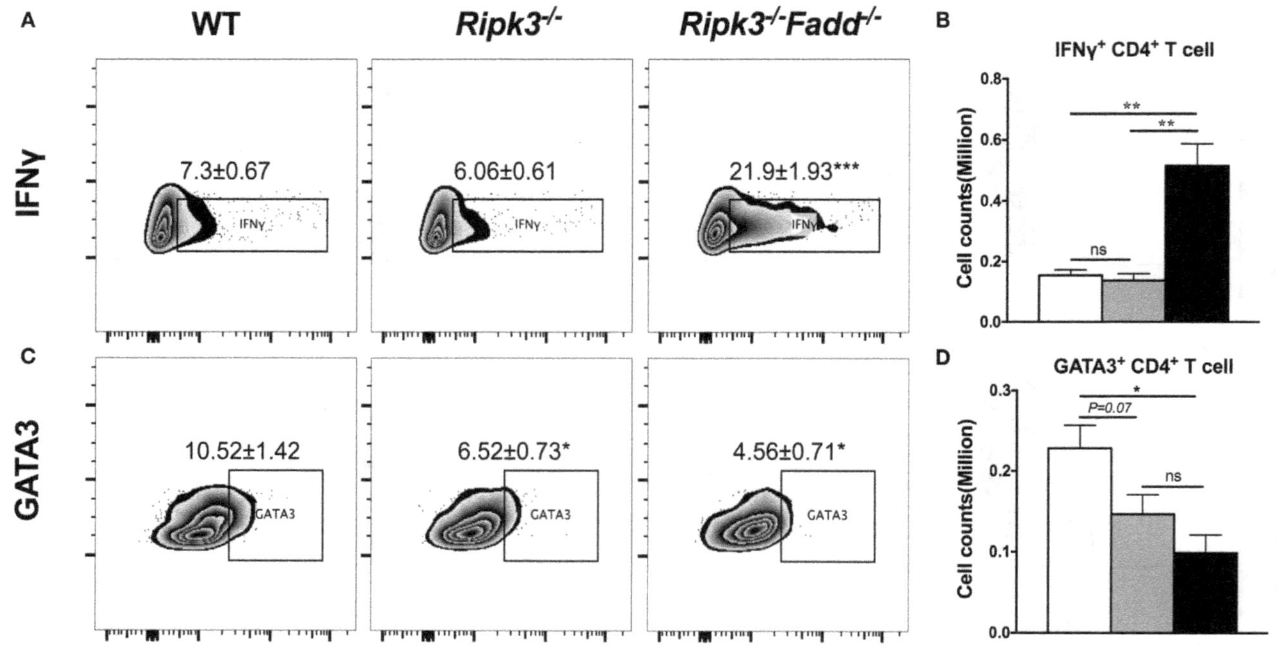

FIGURE 6 | RIPK3 or RIPK3/Fas-associated death domain (FADD) deletions affects Th1/Th2 polarization of CD4 T cells in the lungs of *Cryptococcus neoformans*-infected mice. Leukocytes were isolated from the lungs of infected mice on 9 days postinfection. Intracellular expression of IFNγ, GATA3 in CD4 T cell population was analyzed by flow cytometry. **(A,C)** show the percentage of IFNγ or GATA positive populations in CD4$^+$ cells. **(B,D)** show the cell counts of IFNγ or GATA positive CD4 T cells. Note that there was a significant increase of Th1 markers in the *Ripk3*$^{-/-}$*Fadd*$^{-/-}$ mice compared to the control mice **(A,B)**. Results represent means and SEMs (n = 4 mice for each time point). $^*P < 0.05$, $^{**}P < 0.01$.

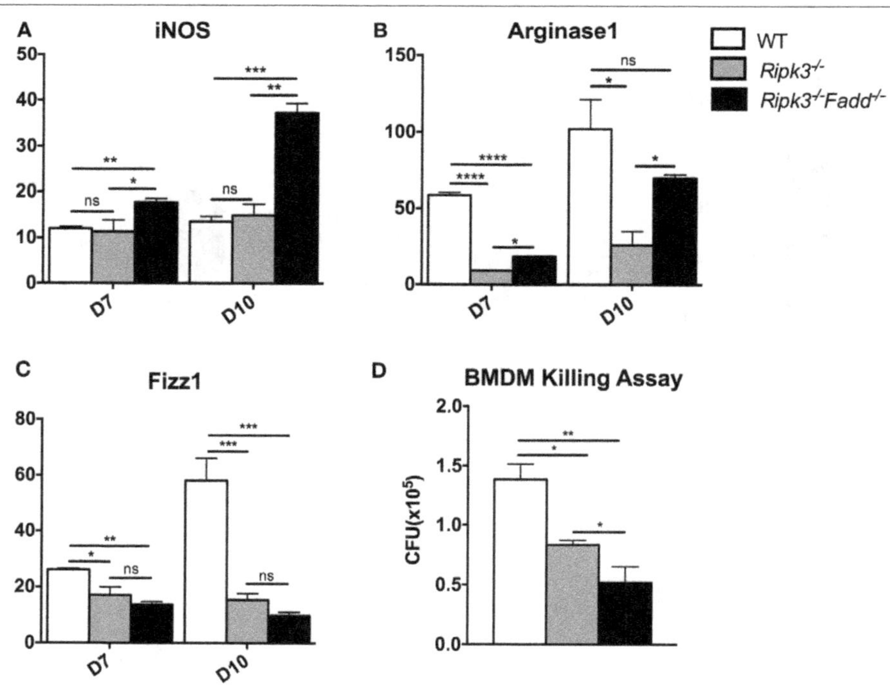

FIGURE 7 | *Ripk3*$^{-/-}$ and *Ripk3*$^{-/-}$*Fadd*$^{-/-}$ mice displayed progressively classical activation and increased fungicidal ability of macrophages. mRNA was isolated from pulmonary macrophages of WT, *Ripk3*$^{-/-}$, or *Ripk3*$^{-/-}$*Fadd*$^{-/-}$ mice at 10 days postinfection. Real-time PCR was performed to determine the expression levels of iNOS [nitric oxide synthase, **(A)**], Arg1 [arginase 1, **(B)**], and Fizz1 **(C)**. Bone marrow-derived macrophages (BMDMs) isolated from uninfected mice were utilized to detect intracellular fungicidal ability **(D)**. ns, no significant difference, $^*P < 0.05$, $^{**}P < 0.01$, $^{***}P < 0.001$, $^{****}P < 0.0001$, between compared groups. These experiments were repeated three times.

that, at the cellular level, RIPK3 and FADD deletions promoted rather than suppressed fungicidal M1 macrophage polarization. Thus, the effects of RIPK3 and FADD deletion on M1/M2 gene expression were consistent with the cytokine profiles in the *C. neoformans*-infected lungs.

RIPK3 and FADD Deletions Promoted DC Differentiation and Activation in LALNs after Cryptococcal Infection

Having determined that joint RIPK3/FADD deletion promoted a strong shift toward the Th1/M1 response in contrast with single RIPK3 deletion, we analyzed the phenotype of DCs, the central regulators of Th-immune polarization. Flow cytometric analysis of LALN DC was performed to examine the frequency and intensity of CD80, a major co-stimulatory molecule expressed during DC co-stimulatory maturation, and the DC1 phenotypic marker and predictor of Th1 response development (22, 28, 29). Consistent with strong Th1 bias, the frequency and intensity of CD80 expression was significantly enhanced in LALN DC of *Ripk3*$^{-/-}$*Fadd*$^{-/-}$ mice at both 7 and 10 dpi (**Figures 8A–D**) compared to those in the infected WT mice. *Ripk3*$^{-/-}$ mice also did not show this effect on day 7 but displayed some increase of CD80 surface expression at 10 dpi (**Figures 8C,D**). Cytokine expression in LALN DCs was also analyzed by intracellular staining. Again, notable enhancement of IFNγ expression was observed in LALN DCs of *Ripk3*$^{-/-}$*Fadd*$^{-/-}$ mice at day 7 but without significant alteration in IL-4 production (**Figures 8E,F**). However, RIPK3 deletion had no effect on LALN DC cytokine expression. Taken together, these results suggested that RIPK3/FADD deletions synergistically promoted LALN DC maturation and DC1 activation during *C. neoformans* infection, which was consistent with strongly enhanced Th1 responses in these mice, while changes in *Ripk3*$^{-/-}$ DC activation profile was subtle, consistent with subtle

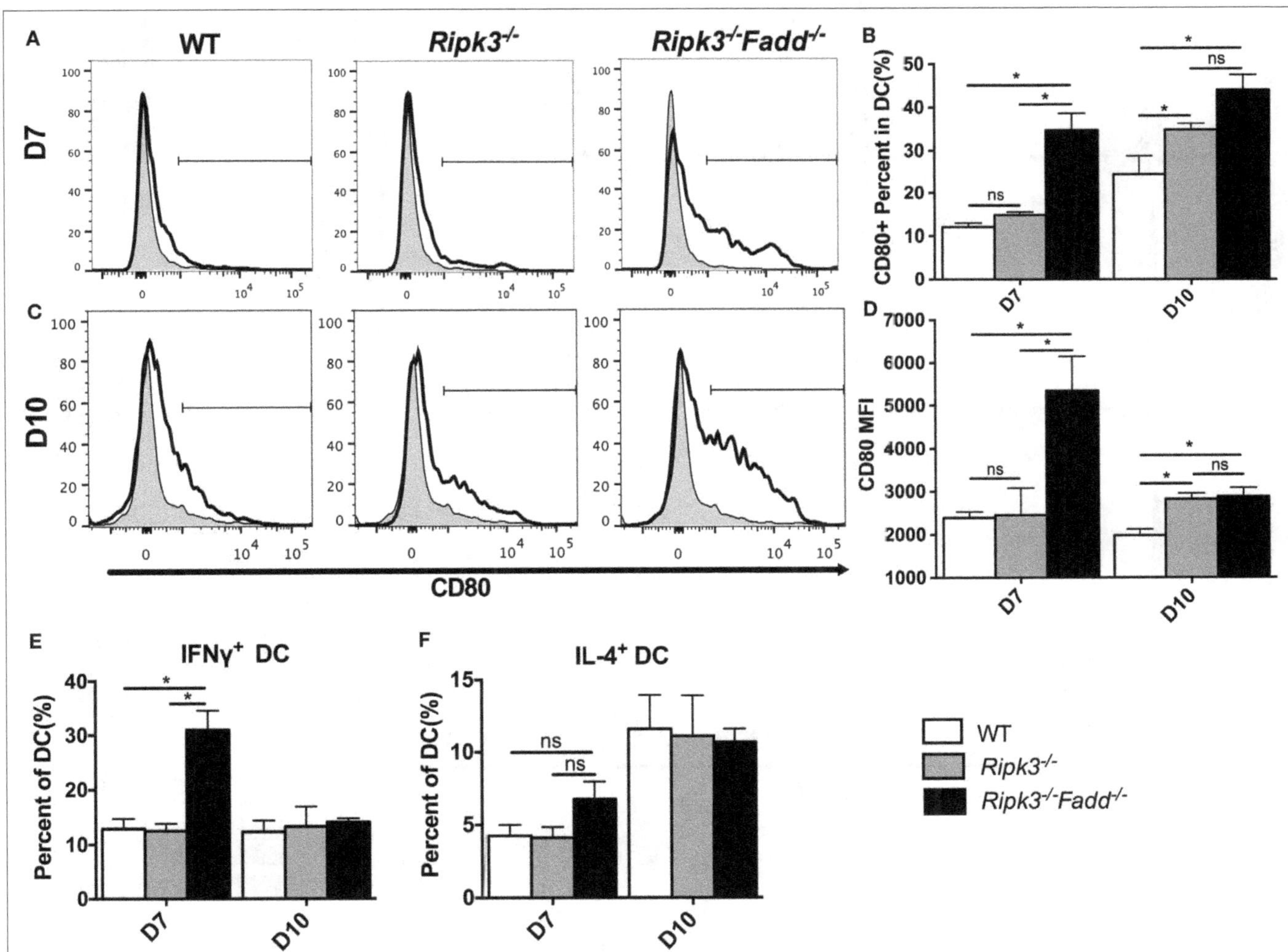

FIGURE 8 | RIPK3 and Fas-associated death domain (FADD) deletion promoted dendritic cells differentiation and activation in lung-associated lymph nodes (LALNs) after cryptococcal infection. LALNs were harvested from wild-type (WT), *Ripk3*$^{-/-}$, and *Ripk3*$^{-/-}$*Fadd*$^{-/-}$ mice at 7 and 10 days postinfection. Dendritic cells (DCs) were stained with CD80 **(A–D)**, IFNγ **(E)**, and IL-4 **(F)** antibodies and then analyzed by flow cytometry. Data are expressed as means and SEMs (n = 3 or more mice per group). ns, no significant difference, $^{*}P < 0.05$, between compared groups. This experiment was repeated twice.

effects of *Ripk3*$^{-/-}$ Th polarization profile during *C. neoformans* infection.

RIPK3 and FADD Deletions Differentially Affected Apoptosis Rate in Leukocyte Subsets during *C. neoformans* Infection

To further explore the potential mechanisms underlying how RIPK3 and RIPK3/FADD pathways prevent excessive accumulation of inflammatory cells, we next evaluated the apoptotic frequency (APF) of various pulmonary cell subsets in *C. neoformans*-infected lungs. Lungs were dissociated gently, and single-cell suspensions were analyzed by flow cytometry with annexin V. Since different leukocyte subsets have distinct life spans and exploit different cell death pathways (30, 31), we first tested their APF in each group without fungal infection to exclude direct effects of FADD and/or RIPK3 deletions. All three groups exhibited similar rates of apoptosis in all subsets, indicating that FADD and RIPK3 were dispensable for homeostatic survival of these immune cells under physiological condition (data not shown).

Similar APFs in each cell subset at 10 dpi in *Ripk3*$^{-/-}$ compared with WT mouse infected lungs (**Figure 9** and Figure S4 in Supplementary Material) suggested that RIPK3 alone had no effect on inflammatory cell apoptosis during *C. neoformans* infection. However, *C. neoformans* infection significantly reduced eosinophil cell death in WT and *Ripk3*$^{-/-}$ mice (eosinophil APF from uninfected mice: 60.9% ± 1.4% decreasing to 6.6 ± 1.2%

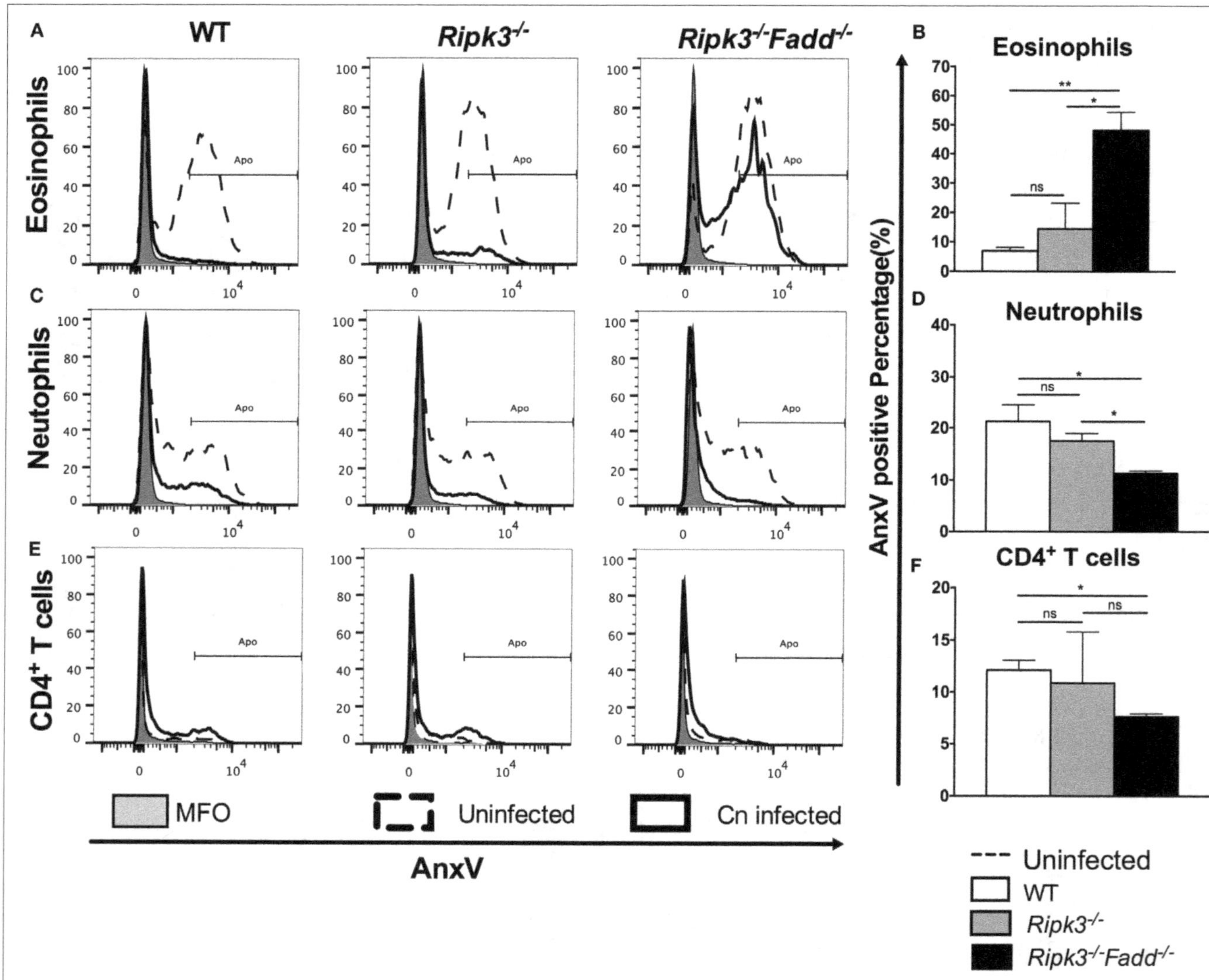

FIGURE 9 | Fas-associated death domain (FADD) deletion differentially regulated the apoptosis of pulmonary leukocyte subsets during *Cryptococcus neoformans* infection. Total pulmonary cells were gently harvested from wild-type (WT), *Ripk3*$^{-/-}$, and *Ripk3*$^{-/-}$*Fadd*$^{-/-}$ mice at 0 or 10 days postinfection. Isolated cells were stained with antibodies specific for annexin V and then analyzed by flow cytometry. After *C. neoformans* infection, eosinophils **(A,B)**, neutrophils **(C,D)**, and CD4^{+} T cells **(E,F)** displayed significant alteration in apoptosis frequency in the lungs from *Ripk3*$^{-/-}$*Fadd*$^{-/-}$ mice but not *Ripk3*$^{-/-}$ mice. Other pulmonary subsets had no significant difference in apoptosis frequency between WT and genetic-defect mice (see Figure S2 in Supplementary Material). ns, no significant difference, *$P < 0.05$, **$P < 0.01$, between compared groups. This experiment was repeated twice.

and 16.3 ± 8.0, respectively), whereas the absence of FADD completely abolished this phenomenon (**Figures 9A,B**). A considerable reduction of neutrophil cell death was also detected in each group after fungal infection. Furthermore, consistent with the greatest increase in neutrophil accumulation, the *Ripk3*$^{-/-}$*Fadd*$^{-/-}$ group exhibited significant decrease in the APF of neutrophils (11.0% ± 0.9%) compared with WT (21.5 ± 3.2%, $P < 0.05$) or *Ripk3*$^{-/-}$ mice (17.6 ± 1.5%, **Figures 9C,D**). Among lymphocyte subsets, only CD4$^+$ T cells displayed a significant reduction in cell death frequency in *Ripk3*$^{-/-}$*Fadd*$^{-/-}$ mice (12.1 ± 1.7 vs 7.6 ± 0.5%, $P < 0.05$), while other subpopulations did not differ in each group (**Figures 9E,F**). Collectively, these data indicated that *C. neoformans* infection affected apoptotic rates of certain leukocyte subsets, but only the joint FADD/RIPK3 deletion led to detectable alterations in the APF in these leukocyte subsets during fungal infection.

The Effects of RIPK3 and FADD Deletions on Anti-Cryptococcal Defenses Appear to Be Unrelated to Their Role Necroptosis Pathway

RIPK3 and FADD, among other functions, are major intracellular upstream mediators of necroptosis cell death pathway, a PCD resulting in cell lysis, reported to influence mycobacterial-infected macrophages (32) but also to regulate type-I IFN signaling in macrophages independent of necroptosis in influenza-infected macrophages (33). We first asked a question about the global role of necroptosis pathway in anti-cryptococcal host defenses. To asses this, mice with deletion of a distal effector kinase in necroptosis pathway, MLKL along with the WT mice, were infected with *C. neoformans* and mouse survival was monitored. Results show that MLKL deletion had no significant effect on *C. neoformans*-infected mouse survival (**Figure 10A**) demonstrating that the necroptosis cell death pathway is dispensable for mouse resistance to *C. neoformans*.

We next asked, whether *C. neoformans* infection triggers significant level of lytic cell death in macrophages and whether this was affected by RIPK3 and FADD deletion. To address this question, we used LDH release assay on WT, *Ripk3*$^{-/-}$, *Ripk3*$^{-/-}$*Fadd*$^{-/-}$ BMDM following *in vitro* H99-challenge. Results show that *C. neoformans* infection triggered lytic death in a small subset of BMDM (14.3 ± 1.7%, **Figure 10B**), The cell death rate was potentiated when the cells were treated with *C. neoformans* opsonized with M18B7 antibody, which increases cryptococcal uptake by macrophages (34) reaching (21.7 ± 1.6%, **Figure 10B**). However, neither RIPK3 deletion alone nor combined with FADD deletion had effect on the rate of BMDM death. This was in contrast with the positive control, in which lytic cell death triggered by the combined simulation of lipopolysaccharide with pan-caspase inhibition was profoundly reduced *Ripk3*$^{-/-}$, *Ripk3*$^{-/-}$*Fadd*$^{-/-}$ BMDM (**Figure 10B**). Thus, RIPK3 and FADD are not directly involved in regulation lytic cell death of BMDM population infected with *C. neoformans in vitro*, further supporting that the susceptibility of *Ripk3*$^{-/-}$ mice and *Ripk3*$^{-/-}$*Fadd*$^{-/-}$ mice to cryptococcal infection was mechanistically unrelated to the role RIPK3 and FADD in necroptosis pathway.

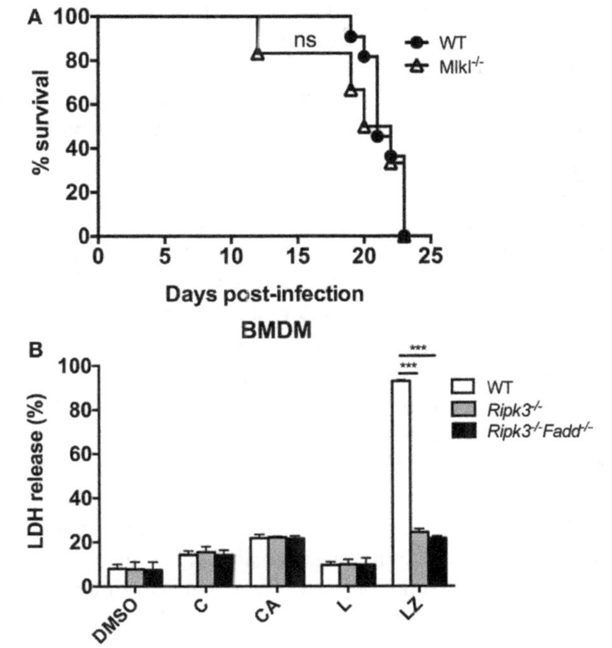

FIGURE 10 | Necroptosis process is not likely a mechanism for RIPK3- and Fas-associated death domain (FADD)-mediated effects on immune responses to *Cryptococcus neoformans*. **(A)** Wild-type (WT) mice and mice with targeted deletion of MLKL gene (distal effector kinase in necroptosis pathway) were infected with *C. neoformans* and analyzed as in **Figure 1D**. MLKL deletion had no effects on survival, unlike RIPK3 and FADD deletions (per **Figure 1D**) (n = 10 for each group). **(B)** Bone marrow-derived macrophage (BMDM) cells isolated form WT, *Ripk3*$^{-/-}$, *Ripk3*$^{-/-}$*Fadd*$^{-/-}$ mice were challenged with H99 at MOI = 5:1 and co-incubated for 24 h. Supernatants were collected followed by LDH release detection. In RIPK3 or FADD, deletions had no effect on magnitude of BMDM lytic death in contrast with their effects in positive control (LZ). Study was repeated three times independently. Symbols signify C, *C. neoformans*; A, anti-cryptococcal antibody M18B7; Z, zVAD (pan-caspase inhibitor); L, lipopolysaccharide; ns, no significant difference, ***$P < 0.001$, between compared groups.

DISCUSSION

While PCD pathway components were shown to be important regulators of the immune responses, the role of FADD and RIPK3 in host defenses against fungal infection remained unknown up to this point. This report provides novel data demonstrating that RIPK3 and FADD are crucial for fungal containment and survival of the infected host during *C. neoformans* infection. Here, we show that these factors serve jointly as physiological "brake" that prevents the development of over exuberant inflammation and profound Th1 bias, which in their absence leads to pulmonary damage and rapid deterioration of the infected host.

Our first group of studies elucidated the involvement of both RIPK3 and FADD during cryptococcal infection. Upon *C. neoformans* infection, RIPK3 expression in infected lungs was strongly upregulated and deletion of RIPK3 significantly shortened the survival of the infected mice and impaired fungal clearance. Furthermore, the absence of RIPK3 resulted in upregulation of pro-inflammatory components including, increased T cell numbers and neutrophils and increased inflammatory cytokines in response to *C. neoformans* (**Figures 5** and **6**). These data provide evidence that RIPK3 exerts an important role in controlling inflammatory responses with some minor effect on Th polarization during *C. neoformans* infection that is at least in part independent of its well-established function in apoptosis and necroptosis pathways (**Figures 9** and **10**).

While FADD expression appeared to be constitutive, RIPK3/FADD double deletion resulted in further enhanced susceptibility to *C. neoformans* infection, demonstrating that joint absence of these factors further potentiated the effects of single RIPK3 deletion at both severity of pathology and upregulation of inflammatory components in the infected lungs. In *Ripk3*$^{-/-}$*Fadd*$^{-/-}$ mice, lung pathology and the effect on mouse survival effects were more severe (**Figures 1–3**; Figure S2 in Supplementary Material) compared to *Ripk3*$^{-/-}$ mice. Our pathology studies further highlighted that in *C. neoformans*-infected lungs RIPK3 and FADD play important roles in protecting infected lungs against the rapid development of severe pathology associated with excessive accumulation and activation of leukocytes. One caveat here could be that the enhanced inflammatory pathology was driven by the increased fungal burden; however, the lung CFU burdens even in the most profoundly affected *Ripk3*$^{-/-}$*Fadd*$^{-/-}$ mice were still relatively modest and could not explain the profound pathological changes and 80% mortality at day 10 in these mice. This, together with very selective amplification of Th1 response, clearly indicates that rather small differences in fungal burdens were not the major driver of the profound changes in the immunophenotype and increase in mouse mortality.

Inflammation is a double-edged sword in the pathogenesis of infectious disease. While suboptimal production of pro-inflammatory cytokines hinders control of *C. neoformans* infection (35–37), highly elevated production of these factors promotes severe inflammatory tissue damage (38). In at least some aspects of host–fungus interactions, a preservation immune strategy denoted "protective tolerance" may be optimal to limit immunopathology while controlling fungal infection (39). This is the first report that RIPK3 and FADD may play jointly important roles by critically fine-tuning "protective tolerance" mechanisms during anti-cryptococcal defense. We show that both of these two molecules are required for optimal control of fungal growth and host protection against the severe inflammatory pathology that develops in lungs in the absence of these factors. Interestingly, the phenotypes that developed in *Ripk3*$^{-/-}$*Fadd*$^{-/-}$ mice mirror many findings in cryptococcosis patients who suffer from IRIS. The development of IRIS (pre-IRIS phase) is specifically characterized by increasing pro-inflammatory responses without efficient clearance of the fungal pathogen (40, 41). Several cohort studies highlighted the relationship of IL-6 signaling and the risk of developing IRIS. For every twofold increase in IL-6 or C-reactive protein, the hazard of IRIS increased by 1.6 and 1.5, respectively (42). Another recent study proposed TNFα, IL-1β, and IL-12 to be predictors of IRIS (43), and these factors found to be elevated as a result of RIPK3 and FADD deletion in *C. neoformans*-infected lungs. Finally, a central IRIS characteristic reproduced in *Ripk3*$^{-/-}$*Fadd*$^{-/-}$ mice is massive systemic elevation of IFNγ, TNFα, and IL-6 (42) (**Figure 5**).

Besides the findings that mimic cytokine profiles of IRIS patients in our models, we also found the elevated induction of IL-1α and neutrophil recruitment in the lungs of both *Ripk3*$^{-/-}$ and *Ripk3*$^{-/-}$*Fadd*$^{-/-}$ mice. Neutrophil recruitment linked to IL-1α upregulation have been shown to contribute to the development of lethal lung pathology during fungal infection with *Aspergillus* (44). Together, these findings document that responses "designed" to be protective in fungal infections can become highly detrimental to the host, and our data demonstrate such detrimental outcomes following RIPK3 and FADD deletion. Collectively, our study shows that RIPK3 and FADD factors are crucial elements of regulatory network that allows protecting the host from pathological effects of inflammation while supporting clearance of the invasive fungal infection.

Another novel finding was that the expression of RIPK3 and FADD proteins was required for the development of Th2 polarization during *C. neoformans* infection. While the deletion of RIPK3 had subtle effect on Th2 polarization, joint deletion of RIPK3 and FADD resulted in a complete switch from strong Th2 to strong Th1 response. Cytokine profile analysis revealed massive reduction in Th2 cytokines (IL-4 and IL-33) in infected lungs of *Ripk3*$^{-/-}$*Fadd*$^{-/-}$ mice. This, together with diminished GATA3 expression by CD4 T cells and the absence of serum IgE accumulation (**Figures 5** and **6**), indicates the loss of Th2 polarization in response to RIPK3 and FADD deletion. Th1 and Th2 responses are known to counterbalance each other. Therefore, it remains to be determined if the primary effect of RIPK3/FADD deletion was the lack of the Th2 development with its subsequent "replacement by Th1," other way around, or else both pathways were subjected to concurrent regulation by joint action of RIPK3 and FADD. While future studies are needed to address these points, our data show that the net effect of RIPK3 and FADD is directly or indirectly counter-regulate Th1 and support Th2 polarization during *C. neoformans* infection. Another highly unexpected finding was that strong Th1 bias and the absence of Th2 rather than being protective (19, 45) resulted in non-protective response in the infected *Ripk3*$^{-/-}$*Fadd*$^{-/-}$ mice. This absence of improved clearance following RIPK3/FADD deletions, despite the robust Th1 polarization, was not due to a defect in macrophage M1 polarization or loss of their intrinsic fungicidal function downstream of robust M1 polarization. Our studies of macrophages from infected lungs and BMDMs revealed neither a defect in macrophage polarization nor a defect in their effector (killing) functions resulting from RIPK3 or FADD deletion (**Figure 7**). In spite of this, we observed paradoxical impairment of pulmonary fungal clearance in both *Ripk3*$^{-/-}$ and *Ripk3*$^{-/-}$*Fadd*$^{-/-}$ mice (**Figure 2**). One explanation for this observation is that the excessive systemic activation of microbicidal factors in macrophages resulted in exhaustion of these effectors cells even before they reached the infection site, which seems to be consistent with our data (**Figures 4** and **7**). Additionally, the excessive accumulation of neutrophils at the infection site interfered with macrophage and T-cell fungicidal functions, since an excessive accumulation of neutrophils have been reported to contribute to tissue damage and defects in the clearance of other fungal organisms (46, 47).

The final observation in our study is quite strong independence of immunoregulatory effects exerted by RIPK3 and FADD from apoptotic or necroptotic cell death. The diminished apoptosis likely contributed to the excessive accumulation of neutrophils in infected *Ripk3*$^{-/-}$*Fadd*$^{-/-}$ mice (**Figures 3**, **4B** and **9D**) since, granulocytes are typically eliminated *via* extrinsic apoptosis after a brief period of activation (48). However, the enhanced neutrophil accumulation was also observed in *Ripk3*$^{-/-}$ mice without significant effect on neutrophil APF. Likewise, APF in eosinophil population was greatest in the infected *Ripk3*$^{-/-}$*Fadd*$^{-/-}$ mice, demonstrating that eosinophil apoptosis did not require or was positively regulated by FADD or RIPK3. Furthermore, RIPK3 and FADD deletion appeared not to affect APF in T and B cells, monocytes, DCs, or epithelial cells in the infected lungs. Finally, our outcomes do not favor necroptosis as an important pathway defining host–pathogen interactions during *C. neoformans* infection (**Figure 10**). Thus, while future studies are needed to provide definitive answers, our data favor the hypothesis that RIPK3 and FADD can induce immunoregulatory effects during fungal infection in a manner independent of their role in PCD responses, but chiefly by regulating cytokine responses.

In summary, our results demonstrate, for the first time, that RIPK3 and FADD are vital components of the immune responses to fungal pathogens. These molecules are required for fine-tuning of inflammatory responses during infection, acting as powerful regulators of Th1 and Th2 polarization. They contribute to optimal fungal clearance and serve as indispensable "nodes" in immunoregulatory network supporting tissue damage control in fungal-infected host.

AUTHOR CONTRIBUTIONS

ZF, QX, and WF contributed to study concept and design, performing the experiments, and drafting of the manuscript. WL, MO, XD, and HZ (Haibing Zhang) contributed to study concept and design, analysis, and interpretation of data, critical revision of the manuscript, obtained funding, and provided study supervision, administrative, and technical support. HZ (Haiwei Zhang), JX (Jintao Xu), WP, and JX (Jinhua Xu) contributed to data analysis and revision of the manuscript. All listed authors gave final approval of the manuscript.

FUNDING

This study was supported by the National Key Basic Research Programs of China (2013CB531601 and 2013CB531606), the National Natural Science Foundation of China (81401651, 81471926, 81271799, and 81501728), China Postdoctoral Science Foundation Grant (2016M600286), and Shanghai Key Laboratory of Molecular Medical Mycology (14DZ2272900). MO's work is supported by Merit Review Grant 1I01BX000656 and VA Research Career Scientist Award 1K6BX003615 from the US Department of Veterans' Affairs.

REFERENCES

1. Fang W, Fa Z, Liao W. Epidemiology of *Cryptococcus* and cryptococcosis in China. *Fungal Genet Biol* (2015) 78:7–15. doi:10.1016/j.fgb.2014.10.017
2. Heitman J, Kozel TR, Kwon-Chung KJ, Perfect JR, Casadevall A. *Cryptococcus: From Human Pathogen to Model Yeast*. Washington, DC: ASM Press (2010). 275 p.
3. Maziarz EK, Perfect JR. Cryptococcosis. *Infect Dis Clin North Am* (2016) 30(1):179–206. doi:10.1016/j.idc.2015.10.006
4. Greenlee-Wacker MC. Clearance of apoptotic neutrophils and resolution of inflammation. *Immunol Rev* (2016) 273(1):357–70. doi:10.1111/imr.12453
5. Ortega-Gomez A, Perretti M, Soehnlein O. Resolution of inflammation: an integrated view. *EMBO Mol Med* (2013) 5(5):661–74. doi:10.1002/emmm.201202382
6. Scaffidi C, Volkland J, Blomberg I, Hoffmann I, Krammer PH, Peter ME. Phosphorylation of FADD/MORT1 at serine 194 and association with a 70-kDa cell cycle-regulated protein kinase. *J Immunol* (2000) 164(3):1236–42. doi:10.4049/jimmunol.164.3.1236
7. Zhang J, Kabra NH, Cado D, Kang C, Winoto A. FADD-deficient T cells exhibit a disaccord in regulation of the cell cycle machinery. *J Biol Chem* (2001) 276(32):29815–8. doi:10.1074/jbc.M103838200
8. Balachandran S, Thomas E, Barber GN. A FADD-dependent innate immune mechanism in mammalian cells. *Nature* (2004) 432(7015):401–5. doi:10.1038/nature03124
9. Ma Y, Liu H, Tu-Rapp H, Thiesen HJ, Ibrahim SM, Cole SM, et al. Fas ligation on macrophages enhances IL-1R1-toll-like receptor 4 signaling and promotes chronic inflammation. *Nat Immunol* (2004) 5(4):380–7. doi:10.1038/ni1054
10. Zhang J, Cado D, Chen A, Kabra NH, Winoto A. Fas-mediated apoptosis and activation-induced T-cell proliferation are defective in mice lacking FADD/Mort1. *Nature* (1998) 392(6673):296–300. doi:10.1038/32681
11. Varfolomeev EE, Schuchmann M, Luria V, Chiannilkulchai N, Beckmann JS, Mett IL, et al. Targeted disruption of the mouse caspase 8 gene ablates cell death induction by the TNF receptors, Fas/Apo1, and DR3 and is lethal prenatally. *Immunity* (1998) 9(2):267–76. doi:10.1016/S1074-7613(00)80609-3
12. He S, Wang L, Miao L, Wang T, Du F, Zhao L, et al. Receptor interacting protein kinase-3 determines cellular necrotic response to TNF-alpha. *Cell* (2009) 137(6):1100–11. doi:10.1016/j.cell.2009.05.021
13. Kasof GM, Prosser JC, Liu D, Lorenzi MV, Gomes BC. The RIP-like kinase, RIP3, induces apoptosis and NF-kappaB nuclear translocation and localizes to mitochondria. *FEBS Lett* (2000) 473(3):285–91. doi:10.1016/S0014-5793(00)01473-3
14. Moriwaki K, Balaji S, McQuade T, Malhotra N, Kang J, Chan FK. The necroptosis adaptor RIPK3 promotes injury-induced cytokine expression and tissue repair. *Immunity* (2014) 41(4):567–78. doi:10.1016/j.immuni.2014.09.016
15. Shlomovitz I, Zargrian S, Gerlic M. Mechanisms of RIPK3-induced inflammation. *Immunol Cell Biol* (2017) 95(2):166–72. doi:10.1038/icb.2016.124
16. Zhang X, Fan C, Zhang H, Zhao Q, Liu Y, Xu C, et al. MLKL and FADD are critical for suppressing progressive lymphoproliferative disease and activating the NLRP3 inflammasome. *Cell Rep* (2016) 16(12):3247–59. doi:10.1016/j.celrep.2016.06.103

17. Semighini CP, Averette AF, Perfect JR, Heitman J. Deletion of *Cryptococcus neoformans* AIF ortholog promotes chromosome aneuploidy and fluconazole-resistance in a metacaspase-independent manner. *PLoS Pathog* (2011) 7(11):e1002364. doi:10.1371/journal.ppat.1002364
18. Meng Y, Zhang C, Yi J, Zhou Z, Fa Z, Zhao J, et al. Deubiquitinase Ubp5 is required for the growth and pathogenicity of *Cryptococcus gattii*. *PLoS One* (2016) 11(4):e0153219. doi:10.1371/journal.pone.0153219
19. Zhang Y, Wang F, Tompkins KC, McNamara A, Jain AV, Moore BB, et al. Robust Th1 and Th17 immunity supports pulmonary clearance but cannot prevent systemic dissemination of highly virulent *Cryptococcus neoformans* H99. *Am J Pathol* (2009) 175(6):2489–500. doi:10.2353/ajpath.2009.090530
20. Qiu Y, Dayrit JK, Davis MJ, Carolan JF, Osterholzer JJ, Curtis JL, et al. Scavenger receptor A modulates the immune response to pulmonary *Cryptococcus neoformans* infection. *J Immunol* (2013) 191(1):238–48. doi:10.4049/jimmunol.1203435
21. Stevens WW, Kim TS, Pujanauski LM, Hao X, Braciale TJ. Detection and quantitation of eosinophils in the murine respiratory tract by flow cytometry. *J Immunol Methods* (2007) 327(1–2):63–74. doi:10.1016/j.jim.2007.07.011
22. Xu J, Eastman AJ, Flaczyk A, Neal LM, Zhao G, Carolan J, et al. Disruption of early tumor necrosis factor alpha signaling prevents classical activation of dendritic cells in lung-associated lymph nodes and development of protective immunity against cryptococcal infection. *MBio* (2016) 7(4):e00510–16. doi:10.1128/mBio.00510-16
23. Szarka RJ, Wang N, Gordon L, Nation PN, Smith RH. A murine model of pulmonary damage induced by lipopolysaccharide via intranasal instillation. *J Immunol Methods* (1997) 202(1):49–57. doi:10.1016/S0022-1759(96)00236-0
24. Davis MJ, Tsang TM, Qiu Y, Dayrit JK, Freij JB, Huffnagle GB, et al. Macrophage M1/M2 polarization dynamically adapts to changes in cytokine microenvironments in *Cryptococcus neoformans* infection. *MBio* (2013) 4(3):e00264–13. doi:10.1128/mBio.00264-13
25. Davis MJ, Eastman AJ, Qiu Y, Gregorka B, Kozel TR, Osterholzer JJ, et al. *Cryptococcus neoformans*-induced macrophage lysosome damage crucially contributes to fungal virulence. *J Immunol* (2015) 194(5):2219–31. doi:10.4049/jimmunol.1402376
26. Alanio A, Desnos-Ollivier M, Dromer F. Dynamics of *Cryptococcus neoformans*-macrophage interactions reveal that fungal background influences outcome during cryptococcal meningoencephalitis in humans. *MBio* (2011) 2(4):e00158-11. doi:10.1128/mBio.00158-11
27. Arora S, Olszewski MA, Tsang TM, McDonald RA, Toews GB, Huffnagle GB. Effect of cytokine interplay on macrophage polarization during chronic pulmonary infection with *Cryptococcus neoformans*. *Infect Immun* (2011) 79(5):1915–26. doi:10.1128/IAI.01270-10
28. Herring AC, Falkowski NR, Chen GH, McDonald RA, Toews GB, Huffnagle GB. Transient neutralization of tumor necrosis factor alpha can produce a chronic fungal infection in an immunocompetent host: potential role of immature dendritic cells. *Infect Immun* (2005) 73(1):39–49. doi:10.1128/IAI.73.1.39-49.2005
29. Eastman AJ, Osterholzer JJ, Olszewski MA. Role of dendritic cell-pathogen interactions in the immune response to pulmonary cryptococcal infection. *Future Microbiol* (2015) 10(11):1837–57. doi:10.2217/fmb.15.92
30. Hampson P, Hazeldine J, Lord JM. Neutrophil apoptosis and its induction as a potential treatment for chronic inflammatory disease. *Curr Opin Hematol* (2013) 20(1):10–5. doi:10.1097/MOH.0b013e32835b06be
31. Shen ZJ, Malter JS. Determinants of eosinophil survival and apoptotic cell death. *Apoptosis* (2015) 20(2):224–34. doi:10.1007/s10495-014-1072-2
32. Zhao X, Khan N, Gan H, Tzelepis F, Nishimura T, Park SY, et al. Bcl-xL mediates RIPK3-dependent necrosis in *M. tuberculosis*-infected macrophages. *Mucosal Immunol* (2017). doi:10.1038/mi.2017.12
33. Downey J, Pernet E, Coulombe F, Allard B, Meunier I, Jaworska J, et al. RIPK3 interacts with MAVS to regulate type I IFN-mediated immunity to influenza A virus infection. *PLoS Pathog* (2017) 13(4):e1006326. doi:10.1371/journal.ppat.1006326
34. Zebedee SL, Koduri RK, Mukherjee J, Mukherjee S, Lee S, Sauer DF, et al. Mouse-human immunoglobulin G1 chimeric antibodies with activities against *Cryptococcus neoformans*. *Antimicrob Agents Chemother* (1994) 38(7):1507–14. doi:10.1128/AAC.38.7.1507
35. Bauman SK, Huffnagle GB, Murphy JW. Effects of tumor necrosis factor alpha on dendritic cell accumulation in lymph nodes draining the immunization site and the impact on the anticryptococcal cell-mediated immune response. *Infect Immun* (2003) 71(1):68–74. doi:10.1128/IAI.71.1.68-74.2003
36. Siddiqui AA, Shattock RJ, Harrison TS. Role of capsule and interleukin-6 in long-term immune control of *Cryptococcus neoformans* infection by specifically activated human peripheral blood mononuclear cells. *Infect Immun* (2006) 74(9):5302–10. doi:10.1128/IAI.00661-06
37. Axton PJ, Bancroft GJ. In vivo analysis of immune responses to *Cryptococcus neoformans* – role of interferon-gamma in host resistance. *Biochem Soc Trans* (1997) 25(2):276S. doi:10.1042/bst025276s
38. Osterholzer JJ, Milam JE, Chen GH, Toews GB, Huffnagle GB, Olszewski MA. Role of dendritic cells and alveolar macrophages in regulating early host defense against pulmonary infection with *Cryptococcus neoformans*. *Infect Immun* (2009) 77(9):3749–58. doi:10.1128/IAI.00454-09
39. Roussey JA, Olszewski MA, Osterholzer JJ. Immunoregulation in fungal diseases. *Microorganisms* (2016) 4(4):E47. doi:10.3390/microorganisms4040047
40. Skiest DJ, Hester LJ, Hardy RD. Cryptococcal immune reconstitution inflammatory syndrome: report of four cases in three patients and review of the literature. *J Infect* (2005) 51(5):e289–97. doi:10.1016/j.jinf.2005.02.031
41. Wiesner DL, Boulware DR. *Cryptococcus*-related immune reconstitution inflammatory syndrome (IRIS): pathogenesis and its clinical implications. *Curr Fungal Infect Rep* (2011) 5(4):252–61. doi:10.1007/s12281-011-0064-8
42. Boulware DR, Meya DB, Bergemann TL, Wiesner DL, Rhein J, Musubire A, et al. Clinical features and serum biomarkers in HIV immune reconstitution inflammatory syndrome after cryptococcal meningitis: a prospective cohort study. *PLoS Med* (2010) 7(12):e1000384. doi:10.1371/journal.pmed.1000384
43. Sun J, Chen H, Xie Y, Su J, Huang Y, Xu L, et al. Nuclear factor of activated T cells and cytokines gene expression of the T cells in AIDS patients with immune reconstitution inflammatory syndrome during highly active antiretroviral therapy. *Mediators Inflamm* (2017) 2017:1754741. doi:10.1155/2017/1754741
44. Caffrey AK, Lehmann MM, Zickovich JM, Espinosa V, Shepardson KM, Watschke CP, et al. IL-1alpha signaling is critical for leukocyte recruitment after pulmonary *Aspergillus fumigatus* challenge. *PLoS Pathog* (2015) 11(1): e1004625. doi:10.1371/journal.ppat.1004625
45. Holmer SM, Evans KS, Asfaw YG, Saini D, Schell WA, Ledford JG, et al. Impact of surfactant protein D, interleukin-5, and eosinophilia on cryptococcosis. *Infect Immun* (2014) 82(2):683–93. doi:10.1128/IAI.00855-13
46. Zelante T, De Luca A, D'Angelo C, Moretti S, Romani L. IL-17/Th17 in anti-fungal immunity: what's new? *Eur J Immunol* (2009) 39(3):645–8. doi:10.1002/eji.200839102
47. Zelante T, De Luca A, Bonifazi P, Montagnoli C, Bozza S, Moretti S, et al. IL-23 and the Th17 pathway promote inflammation and impair antifungal immune resistance. *Eur J Immunol* (2007) 37(10):2695–706. doi:10.1002/eji.200737409
48. Fox S, Leitch AE, Duffin R, Haslett C, Rossi AG. Neutrophil apoptosis: relevance to the innate immune response and inflammatory disease. *J Innate Immun* (2010) 2(3):216–27. doi:10.1159/000284367

14

Host Responses to *Malassezia* spp. in the Mammalian Skin

Florian Sparber* and Salomé LeibundGut-Landmann*

Section of Immunology, Vetsuisse Faculty, University of Zürich, Zürich, Switzerland

****Correspondence:***
Florian Sparber
florian.sparber@uzh.ch;
Salomé LeibundGut-Landmann
salome.leibundgut-landmann@uzh.ch

The skin of mammalian organisms is home for a myriad of microbes. Many of these commensals are thought to have beneficial effects on the host by critically contributing to immune homeostasis. Consequently, dysbiosis can have detrimental effects for the host that may manifest with inflammatory diseases at the barrier tissue. Besides bacteria, fungi make an important contribution to the microbiota and among these, the yeast *Malassezia* widely dominates in most areas of the skin in healthy individuals. There is accumulating evidence that *Malassezia* spp. are involved in a variety of skin disorders in humans ranging from non- or mildly inflammatory conditions such as dandruff and pityriasis versicolor to more severe inflammatory skin diseases like seborrheic eczema and atopic dermatitis. In addition, *Malassezia* is strongly linked to the development of dermatitis and otitis externa in dogs. However, the association of *Malassezia* spp. with such diseases remains poorly characterized. Until now, studies on the fungus–host interaction remain sparse and they are mostly limited to experiments with isolated host cells *in vitro*. They suggest a multifaceted crosstalk of *Malassezia* spp. with the skin by direct activation of the host *via* conserved pattern recognition receptors and indirectly *via* the release of fungus-derived metabolites that can modulate the function of hematopoietic and/or non-hematopoietic cells in the barrier tissue. In this review, we discuss our current understanding of the host response to *Malassezia* spp. in the mammalian skin.

Keywords: *Malassezia*, commensalism, opportunistic pathogenic fungi, skin disorders, innate immunity, adaptive immunity, allergic response, indoles

INTRODUCTION

Malassezia spp. are lipophilic yeasts, which are part of the skin microbiota of many mammals and birds. In fact, the genus *Malassezia* is by far the most abundant eukaryotic member of the microbial flora of the skin in these organisms (1). Most *Malassezia* spp. have a predilection for seborrheic skin sites such as the scalp and the trunk. They rely on exogenous fatty acid sources for their nutritive requirements because of their lack of genes encoding for the fatty acid synthase and genes involved in carbohydrate metabolism (2–4). In agreement, the cell wall of *Malassezia* spp. is particularly rich in lipids (5).

The genus *Malassezia* currently comprises 17 species, three of which have only recently been proposed (6–8). *Malassezia globosa*, *Malassezia restricta*, and *Malassezia sympodialis* are most frequently isolated from the healthy human skin with distinct relative frequencies at specific body sites (1, 9). The age of the host and geographic factors also influence their distribution (10). *Malassezia pachydermatis*, *Malassezia nana*, and *Malassezia caprae* are found predominantly in non-human hosts (6). Surprisingly, the microbial communities of the skin are astonishingly stable and maintained over time, despite the skin's exposure to the external environment (11). It is currently unknown whether *Malassezia* spp. play a mutualistic role and may thus contribute to immune homeostasis of the host.

Apart from their commensal nature, *Malassezia* spp. are also associated with common skin disorders such as pityriasis versicolor and seborrheic dermatitis as well as more severe inflammatory skin pathologies including atopic eczema and atopic dermatitis in humans (10) and dermatitis and otitis externa in animals, most frequently in dogs (12). The composition of the skin mycobiome can vary under pathological conditions and some species of *Malassezia* such as *M. sympodialis* and *Malassezia furfur* are found to be enriched in certain skin disorders (10). To date, a causative link between *Malassezia* and disease development has only been made for Pityriasis versicolor, while the role of the yeast in other pathologies remains correlative (10, 13, 14). Changes in the degree of colonization in diseased compared to healthy skin have been documented in dogs (15) but remain uncertain in humans (16).

The pathophysiology of *Malassezia*-associated skin conditions is largely unknown. The lack of knowledge on the cellular and molecular interactions between *Malassezia* spp. and the host preclude a better understanding of the factors determining commensalism versus disease. Herein, we review the current knowledge with regard to how the host recognizes *Malassezia* spp. and responds to it (**Figure 1**).

SENSING OF *Malassezia* spp. BY THE HOST

Through their localization in the skin, *Malassezia* spp. interact primarily with keratinocytes, tissue-resident dendritic cells (DCs), and macrophages, as well as with myeloid cells that are recruited to the skin under inflammatory conditions. Activation of DCs is key for induction of adaptive immunity and memory formation. The fungus is recognized by the host either directly through interaction of fungal cell wall components with membrane bound pattern recognition receptors (PRRs) or indirectly through soluble metabolites that are released by *Malassezia* spp. The set of receptors expressed by the hematopoietic and the non-hematopoietic compartment are largely distinct.

Direct Recognition of *Malassezia* spp. by Surface-Bound Receptors

The fungal cell wall is rich in carbohydrates and glycoproteins that are recognized by PRRs of the family of Syk-coupled C-type lectin receptor (CLR), which are expressed primarily by myeloid cells (17, 18). Binding to these receptors results in ligand internalization and activation of multiple signaling pathways, including the MAPK, NF-κB, and NFAT pathways as well as the inflammasome.

The polysaccharides of the *Malassezia* cell wall are organized differently than in other fungal species analyzed to date (19, 20).

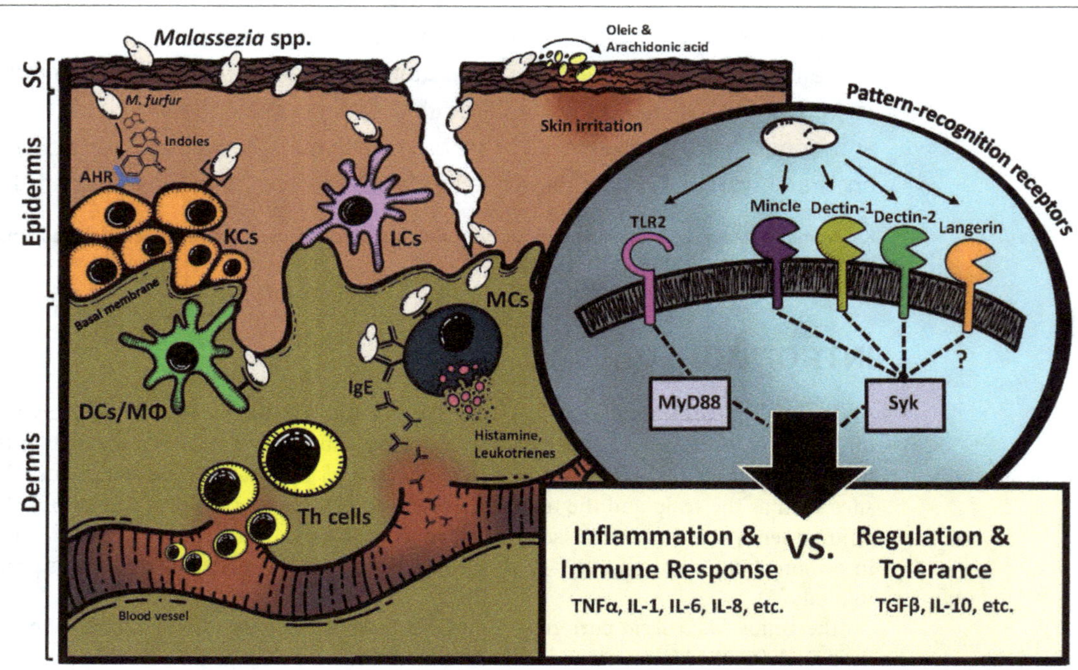

FIGURE 1 | Interaction of *Malassezia* spp. with the mammalian skin. Direct interactions involve various PRRs, which recognize fungal cell wall constituents and are distinctly expressed on the surface of non-hematopoietic (i.e., keratinocytes) and hematopoietic cells (i.e., Langerhans cells, mast cells) of the skin. Spatial and temporal signal integration of different PRR signals results in the induction of inflammation and immunity or, alternatively, in the regulation and tolerance of the host toward *Malassezia* spp. Indirect interactions of *Malassezia* spp. with the skin include fungus-derived metabolites such as irritant fatty acids on the one hand and indoles that are potent agonists for the AhR, which is expressed by various skin cells, on the other hand. KCs, keratinocytes; LCs, Langerhans cells; DCs, dendritic cells; Mϕ, macrophages; MCs, mast cells; SC, stratum corneum; AhR, aryl hydrocarbon receptor.

Moreover, the cell wall is surrounded by a lipid-rich outer layer (21). Several CLRs have been shown to respond to *Malassezia* spp. *in vitro*. The two FcRγ-associated receptors Dectin-2 and Mincle sense *Malassezia* spp., albeit through recognition of distinct ligands (22). While Mincle binds to two distinct glycolipids in *Malassezia*, Dectin-2 recognizes the fungus through α-1,2-linked mannose. High-mannose binding is a general feature of Dectin-2, which is reported to recognize a variety of fungi, including *Candida albicans, Saccharomyces cerevisiae, Blastomyces dermatitidis, Aspergillus fumigatus, Cryptococcus neoformans*, and *Fonsecaea pedrosoi* (23). In contrast, *Malassezia* spp. were initially found to be unique agonists of Mincle when a large panel of 50 different fungi was tested in a glycoconjugate microarray (24). More recently, other fungi such as *Pneumocystis carinii, F. pedrosoi*, and *Fonsecaea monomorpha* were also reported to engage Mincle (25–27), in addition to bacterial ligands (28–32), mammalian alarmins released from damaged cells (33, 34) and even cholesterol crystals (35, 36). Mincle is thus a highly pleiotropic receptor, which can bind chemically and structurally distinct ligands through at least two complementary binding sites (37–40). The β-glucan receptor Dectin-1, which was the first member of the family of Syk-coupled CLRs to be identified (41), was also found to sense *Malassezia* and was linked to the activation of the NLRP3 inflammasome (42). Finally, Langerin was suggested to act as a receptor for *Malassezia* in the skin due to its prominent expression by epidermal Langerhans cells and by a subset of dermal DCs. Direct binding of the fungus to recombinant Langerin was indeed observed (43, 44).

Activation of myeloid cells by *Malassezia* spp. *via* these different CLRs was shown to induce the secretion of proinflammatory cytokines. However, the relative contribution of individual receptors to fungal control *in vivo* during commensalism and in infectious settings remains to be determined. At least partial redundancy of receptors that signal *via* the same pathway may occur, similarly to what was found for other fungi (45, 46). Dissecting the role of Mincle in the context of *Malassezia* spp. in more detail will also be interesting in light of its reported antagonizing activity, e.g., in response to *Fonsacaea* spp. (27), and thus this receptor may also mediate regulatory or inhibitory responses to *Malassezia* spp.

In addition to CLRs, Toll-like receptors (TLRs), and in particular TLR2, also contribute to fungal recognition by the host. TLR2 was implicated in sensing of *Malassezia* spp. and inducing a proinflammatory response characterized by the release of cytokines, chemokines and antimicrobial peptides by keratinocytes (47–50).

The proinflammatory response is generally enhanced by lipid removal from the yeast to enhance exposure of fungal cell wall carbohydrates (51, 52). In contrast, thymic stromal lymphopoietin secretion from keratinocytes was found to be induced specifically by the lipid layer components of *M. restricta* and *M. globosa* but not by yeasts that were depleted of lipids (53).

Indirect Interaction

Specific products of *Malassezia* metabolic pathways are thought to act as virulence factors promoting inflammation and pathology, while others downregulate the production of inflammatory mediators and thereby contribute to immune regulation. Fungal strains with altered production of such factors have been linked to *Malassezia*-associated skin disorders (54–56).

Malassezia-derived lipases and phospholipases, which are required to assimilate host-derived lipids, can initiate an inflammatory response in the skin by releasing unsaturated free fatty acids from the sebum lipids (57–60). Oleic acid has irritant and desquamative effects on keratinocytes (61–63), whereas arachidonic acid produces proinflammatory eicosanoids and leads to inflammation and damage to the stratum corneum, thereby contributing to the disruption of the epithelial barrier function and induction of abnormal keratinization (64).

Malassezia furfur is able to convert tryptophan into a variety of indole alkaloids. This pathway is mainly active if tryptophan is the sole source of nitrogen (65). *M. furfur*-derived indoles including malassezin, indirubin, and indolo [3,2-b] carbazole (ICZ) serve as potent ligands for the host aryl hydrocarbon receptor (AhR) and thereby potentially modify the function of all cells in the epidermis expressing this receptor (54, 55, 66, 67). For example, some tryptophan metabolites can promote apoptosis of melanocytes (68) or inhibit the respiratory burst in neutrophils (69). Given the broad spectrum of biological responses that are influenced by AhR activity, *M. furfur* may engage this pathway to modulate inflammation and/or promote skin immune homeostasis (70) but may also promote skin pathology (71) or even contribute to carcinogenesis (72). The significance of yeast-derived indoles in each of these contexts remains to be demonstrated *in vivo*.

INNATE IMMUNITY TO *Malassezia* spp.

The majority of what is currently known about the host response to *Malassezia* spp. is based on *in vitro* studies with isolated myeloid cells or keratinocyte cell lines. Stimulation of these cells with *Malassezia* yeast leads to the induction of mainly proinflammatory cytokines, chemokines, and antimicrobial peptides (22, 24, 47–52, 73–76). In line with an inflammatory character of the innate response to the fungus, the intraperitoneal injection of *Malassezia* into mice results in the recruitment of neutrophils to the peritoneum (24). Only few studies have examined regulatory cytokines such as IL-10 and TGF-β by the yeast (24, 49, 51, 74, 77), but these may be relevant with regard to the role of *Malassezia* spp. as a skin commensal.

Given the association of *Malassezia* spp. with inflammatory skin disorders and allergic responses, the fungus may also interact with mast cells. Progenitor cell-derived mast cells from atopic patients show increased release of proinflammatory cytokines upon stimulation with *Malassezia* (76) and are enriched in the skin of atopic eczema patients where they are positioned in the superficial dermis and can interact with the fungus (78). Mast cell activation in response to *Malassezia* spp. has also been reported in studies with bone-marrow-derived mast cells. These cells are directly activated by the fungus in a TLR2-dependent manner and release inflammatory mediators and cytokines (79). Moreover, the crosslinking of the high-affinity IgE receptor (FcεRI) by antigen-bound IgE can induce mast cell degranulation (79). Therefore, mast cells may contribute to further barrier disruption and thereby amplify the inflammatory response.

The access of *Malassezia* to immune cells in the skin may be facilitated by disruption of the epithelial barrier as it frequently occurs during chronic inflammation. Moreover, *Malassezia* spp. were reported to release nanovesicles/exosomes that contain immunogenic proteins and trigger increased release of cytokines by DCs (80).

ADAPTIVE IMMUNITY TO *Malassezia* spp.

As a commensal, *Malassezia* interacts continuously with the immune system. Therefore, cellular and humoral immune memory to the fungus can be evidenced in healthy individuals (81). Although there are fewer studies related to dogs when compared with humans, dogs also develop cellular and humoral immune responses to their commensal yeast, *M. pachydermatis* (82–84). Generally, the adaptive immune responses are heightened and qualitatively distinct in patients with *Malassezia*-associated diseases.

Humoral Responses

During steady state, *Malassezia*-specific antibodies are predominantly of the IgG and IgM isotypes (81). In contrast, although *Malassezia*-specific IgE is not usually detected in healthy individuals, it is common in atopic patients (85). A positive correlation was found between the sensitization to *Malassezia*-specific IgE and the severity of atopic dermatitis (86, 87). Similar observations were made in atopic dogs (83, 84). However, whether the IgE response plays a pathogenic role in atopic and other *Malassezia*-associated inflammatory disorders or rather serves as a marker for the severity of disease remains unclear.

T Cell Responses

Patients with atopic dermatitis often show positive skin prick test and atopic patch test reactions to *Malassezia* (85). T cell-responsiveness to *Malassezia* in such patients was associated with a Th2 response (88), in line with the classical paradigm of Th2-polarized allergic T cells. GATA3$^+$ T cells were identified in pityriasis versicolor lesions (89) and likewise *Malassezia*-specific T cell in allergic dogs were found to be strongly polarized toward a type 2 response (82). More recently, other T helper cell subsets such as Th17 and Th22 cells have been found enriched in allergic individuals (90, 91) as well as in non-allergic immune-mediated skin diseases such as psoriasis (92). Consistent with this notion, *Malassezia*-reactive skin homing T cells from *Malassezia*-sensitized atopic dermatitis patients comprise not only Th1 and Th2 subsets but also IL-17- and IL-22-secreting cells (93). Of note, IL-4/IL-17 coproducers have also been described in the context of atopic eczema especially in children (94). Importantly, Th17 differentiation is a hallmark of T cell responses induced by CLR signaling (46) and T cells directed against other fungi, in particular *Candida* spp., belong predominantly to the Th17 subset (95). Whether and how IL-17 and/or IL-22 may contribute to pathogenicity in atopic dermatitis remains to be determined. It is also unknown to which subset *Malassezia*-specific T cells belong in healthy individuals and to what extent T cell plasticity contributes to sensitization.

Malassezia Allergens

To date, 13 *Malassezia*-derived allergens have been identified from *M. furfur* and *M. sympodialis* (3, 96). Interestingly, more allergens are released from *M. sympodialis* when cultured at the increased pH conditions of atopic skin compared with culture at the pH of healthy skin (97). Several of the known allergens belong to a class of phylogenetically highly conserved proteins and display a high degree of homology with the corresponding mammalian proteins. Cross-reactivity between *Malassezia*-derived allergens and endogenous human proteins (e.g., thioredoxin, manganese-dependent superoxide dismutase) has been indeed demonstrated (93, 98, 99). Therefore, the induction of autoreactive T cells by *Malassezia* allergens may play a role in sustained inflammation.

CONCLUSION

Malassezia spp. have been implicated in various pathologies. Yet, direct evidence for a causal relationship between *Malassezia* spp. and the mammalian host remains elusive. For instance, it is unclear whether *Malassezia* actively promotes atopic dermatitis or whether the inflammatory environment in the atopic skin triggers a dysregulated immune response toward the fungus.

At the basis of this is the key question of what determines the balance between commensalism and pathogenicity of *Malassezia* spp. The answer likely relates to changes occurring in both the fungus (55) (e.g., variable secretion of AhR agonists) and in the host (e.g., barrier defects, changes in immune polarization) which are responsible for promoting the development of pathology. Changes in the environment such as seasonal variations in sebum production have also been linked to altered disease prevalence (100).

Inter-species variations in the skin mycobiome may further contribute as different species of *Malassezia* can induce variable inflammatory responses (51, 75, 101). Moreover, *Malassezia* spp. have been shown to display a large intra-species diversity (73) similarly to what is known for other opportunistic fungal pathogens (102), and thus the exact composition of *Malassezia* strains and species present in an individual at a given time may contribute to different outcomes in the interaction between the fungus and the host. The recently completed assembly and detailed annotation of the genome of *M. sympodialis* makes an important contribution to approach this complexity (103). Future research will help fill the important gaps in our knowledge on the pathophysiology of and the host response to *Malassezia in vivo*. Enhanced understanding of host-*Malassezia* interactions may contribute to improved diagnostic and therapeutic options for patients affected by *Malassezia*-associated pathologies.

AUTHOR CONTRIBUTIONS

Both authors listed have made a substantial, direct, and intellectual contribution to the work and approved it for publication.

FUNDING

This work was supported by the University of Zürich.

REFERENCES

1. Findley K, Oh J, Yang J, Conlan S, Deming C, Meyer JA, et al. Topographic diversity of fungal and bacterial communities in human skin. *Nature* (2013) 498:367–70. doi:10.1038/nature12171
2. Wu G, Zhao H, Li C, Rajapakse MP, Wong WC, Xu J, et al. Genus-wide comparative genomics of *Malassezia* delineates its phylogeny, physiology, and niche adaptation on human skin. *PLoS Genet* (2015) 11:e1005614. doi:10.1371/journal.pgen.1005614
3. Gioti A, Nystedt B, Li W, Xu J, Andersson A, Averette AF, et al. Genomic insights into the atopic eczema-associated skin commensal yeast *Malassezia sympodialis*. *mBio* (2013) 4: e00572-12. doi:10.1128/mBio.00572-12
4. Park M, Cho Y-J, Lee YW, Jung WH. Whole genome sequencing analysis of the cutaneous pathogenic yeast *Malassezia restricta* and identification of the major lipase expressed on the scalp of patients with dandruff. *Mycoses* (2017) 60:188–97. doi:10.1111/myc.12586
5. Thompson E, Colvin JR. Composition of the cell wall of *Pityrosporum ovale* (Bizzozero) Castellani and Chalmers. *Can J Microbiol* (1970) 16:263–5. doi:10.1139/m70-048
6. Cabanes FJ. *Malassezia* yeasts: how many species infect humans and animals? *PLoS Pathog* (2014) 10:e1003892. doi:10.1371/journal.ppat.1003892
7. Honnavar P, Prasad GS, Ghosh A, Dogra S, Handa S, Rudramurthy SM. *Malassezia arunalokei* sp. nov., a novel yeast species isolated from seborrheic dermatitis patients and healthy individuals from India. *J Clin Microbiol* (2016) 54:1826–34. doi:10.1128/JCM.00683-16
8. Cabanes FJ, Coutinho SDA, Puig L, Bragulat MR, Castella G. New lipid-dependent *Malassezia* species from parrots. *Rev Iberoam Micol* (2016) 33: 92–9. doi:10.1016/j.riam.2016.03.003
9. Jo J-H, Kennedy EA, Kong HH. Topographical and physiological differences of the skin mycobiome in health and disease. *Virulence* (2017) 8:324–33. doi:10.1080/21505594.2016.1249093
10. Prohic A, Jovovic Sadikovic T, Krupalija-Fazlic M, Kuskunovic-Vlahovljak S. *Malassezia* species in healthy skin and in dermatological conditions. *Int J Dermatol* (2016) 55:494–504. doi:10.1111/ijd.13116
11. Oh J, Byrd AL, Park M, Kong HH, Segre JA. Temporal stability of the human skin microbiome. *Cell* (2016) 165:854–66. doi:10.1016/j.cell.2016.04.008
12. Bond R, Guillot J, Cabañes FJ. *Malassezia* yeasts in animal disease. In: Boekhout T, Mayser P, Guého-Kellermann E, Velegraki A, editors. *Malassezia and the Skin: Science and Clinical Practice*. Berlin, Heidelberg: Springer (2010). p. 271–99.
13. Johansson C, Eshaghi H, Linder MT, Jakobson E, Scheynius A. Positive atopy patch test reaction to *Malassezia furfur* in atopic dermatitis correlates with a T helper 2-like peripheral blood mononuclear cells response. *J Invest Dermatol* (2002) 118:1044–51. doi:10.1046/j.1523-1747.2002.01758.x
14. Casagrande BF, Fluckiger S, Linder MT, Johansson C, Scheynius A, Crameri R, et al. Sensitization to the yeast *Malassezia sympodialis* is specific for extrinsic and intrinsic atopic eczema. *J Invest Dermatol* (2006) 126:2414–21. doi:10.1038/sj.jid.5700431
15. Cafarchia C, Gallo S, Romito D, Capelli G, Chermette R, Guillot J, et al. Frequency, body distribution, and population size of *Malassezia* species in healthy dogs and in dogs with localized cutaneous lesions. *J Vet Diagn Invest* (2005) 17:316–22. doi:10.1177/104063870501700403
16. Sandstrom Falk MH, Tengvall Linder M, Johansson C, Bartosik J, Back O, Sarnhult T, et al. The prevalence of *Malassezia* yeasts in patients with atopic dermatitis, seborrhoeic dermatitis and healthy controls. *Acta Derm Venereol* (2005) 85:17–23. doi:10.1080/00015550410022276
17. Underhill DM, Pearlman E. Immune interactions with pathogenic and commensal fungi: a two-way street. *Immunity* (2015) 43:845–58. doi:10.1016/j.immuni.2015.10.023
18. Dambuza IM, Brown GD. C-type lectins in immunity: recent developments. *Curr Opin Immunol* (2015) 32:21–7. doi:10.1016/j.coi.2014.12.002
19. Stalhberger T, Simenel C, Clavaud C, Eijsink VGH, Jourdain R, Delepierre M, et al. Chemical organization of the cell wall polysaccharide core of *Malassezia restricta*. *J Biol Chem* (2014) 289:12647–12656.
20. Kruppa MD, Lowman DW, Chen YH, Selander C, Scheynius A, Monteiro MA, et al. Identification of (1 – >6)-beta-D-glucan as the major carbohydrate component of the *Malassezia sympodialis* cell wall. *Carbohydr Res* (2009) 344:2474–9. doi:10.1016/j.carres.2009.09.029
21. Mittag H. Fine structural investigation of *Malassezia furfur*. II. The envelope of the yeast cells. *Mycoses* (1995) 38:13–21. doi:10.1111/j.1439-0507.1995.tb00003.x
22. Ishikawa T, Itoh F, Yoshida S, Saijo S, Matsuzawa T, Gonoi T, et al. Identification of distinct ligands for the C-type lectin receptors Mincle and Dectin-2 in the pathogenic fungus *Malassezia*. *Cell Host Microbe* (2013) 13:477–88. doi:10.1016/j.chom.2013.03.008
23. Brown GD, Crocker PR. Lectin receptors expressed on myeloid cells. *Microbiol Spectr* (2016) 4(5):MCHD-0036-2016. doi:10.1128/microbiolspec.MCHD-0036-2016
24. Yamasaki S, Matsumoto M, Takeuchi O, Matsuzawa T, Ishikawa E, Sakuma M, et al. C-type lectin Mincle is an activating receptor for pathogenic fungus, *Malassezia*. *Proc Natl Acad Sci U S A* (2009) 106:1897–902. doi:10.1073/pnas.0805177106
25. Kottom TJ, Hebrink DM, Jenson PE, Nandakumar V, Wuthrich M, Wang H, et al. The interaction of pneumocystis with the C-type lectin receptor Mincle exerts a significant role in host defense against infection. *J Immunol* (2017) 198:3515–25. doi:10.4049/jimmunol.1600744
26. Sousa MdG., Reid DM, Schweighoffer E, Tybulewicz V, Ruland J, Langhorne J, et al. Restoration of pattern recognition receptor costimulation to treat chromoblastomycosis, a chronic fungal infection of the skin. *Cell Host Microbe* (2011) 9:436–43. doi:10.1016/j.chom.2011.04.005
27. Wevers BA, Kaptein TM, Zijlstra-Willems EM, Theelen B, Boekhout T, Geijtenbeek TBH, et al. Fungal engagement of the C-type lectin mincle suppresses dectin-1-induced antifungal immunity. *Cell Host Microbe* (2014) 15:494–505. doi:10.1016/j.chom.2014.03.008
28. Iborra S, Martinez-Lopez M, Cueto FJ, Conde-Garrosa R, Del Fresno C, Izquierdo HM, et al. *Leishmania* uses Mincle to target an inhibitory ITAM signaling pathway in dendritic cells that dampens adaptive immunity to infection. *Immunity* (2016) 45:788–801. doi:10.1016/j.immuni.2016.09.012
29. Rabes A, Zimmermann S, Reppe K, Lang R, Seeberger PH, Suttorp N, et al. The C-type lectin receptor Mincle binds to Streptococcus pneumoniae but plays a limited role in the anti-pneumococcal innate immune response. *PLoS One* (2015) 10:e0117022. doi:10.1371/journal.pone.0117022
30. Devi S, Rajakumara E, Ahmed N. Induction of Mincle by *Helicobacter pylori* and consequent anti-inflammatory signaling denote a bacterial survival strategy. *Sci Rep* (2015) 5:15049. doi:10.1038/srep15049
31. Schick J, Etschel P, Bailo R, Ott L, Bhatt A, Lepenies B, et al. Toll-like receptor 2 and Mincle cooperatively sense corynebacterial cell wall glycolipids. *Infect Immun* (2017) 85:e00075-17. doi:10.1128/IAI.00075-17
32. Matsunaga I, Moody DB. Mincle is a long sought receptor for mycobacterial cord factor. *J Exp Med* (2009) 206:2865–8. doi:10.1084/jem.20092533
33. Yamasaki S, Ishikawa E, Sakuma M, Hara H, Ogata K, Saito T. Mincle is an ITAM-coupled activating receptor that senses damaged cells. *Nat Immunol* (2008) 9:1179–88. doi:10 1038/ni.1651
34. Nagata M, Izumi Y, Ishikawa E, Kiyotake R, Doi R, Iwai S, et al. Intracellular metabolite beta-glucosylceramide is an endogenous Mincle ligand possessing immunostimulatory activity. *Proc Natl Acad Sci U S A* (2017) 114:E3285–94. doi:10.1073/pnas.1618133114
35. Kiyotake R, Oh-Hora M, Ishikawa E, Miyamoto T, Ishibashi T, Yamasaki S. Human Mincle binds to cholesterol crystals and triggers innate immune responses. *J Biol Chem* (2015) 290:25322–32. doi:10.1074/jbc.M115.645234
36. Kostarnoy AV, Gancheva PG, Lepenies B, Tukhvatulin AI, Dzharullaeva AS, Polyakov NB, et al. Receptor Mincle promotes skin allergies and is capable of recognizing cholesterol sulfate. *Proc Natl Acad Sci U S A* (2017) 114:E2758–65. doi:10.1073/pnas.1611665114
37. Feinberg H, Jegouzo SAF, Rowntree TJW, Guan Y, Brash MA, Taylor ME, et al. Mechanism for recognition of an unusual mycobacterial glycolipid by the macrophage receptor mincle. *J Biol Chem* (2013) 288:28457–65. doi:10.1074/jbc.M113.497149
38. Jegouzo SAF, Harding EC, Acton O, Rex MJ, Fadden AJ, Taylor ME, et al. Defining the conformation of human mincle that interacts with mycobacterial trehalose dimycolate. *Glycobiology* (2014) 24:1291–300. doi:10.1093/glycob/cwu072
39. Feinberg H, Rambaruth NDS, Jegouzo SAF, Jacobsen KM, Djurhuus R, Poulsen TB, et al. Binding sites for acylated trehalose analogs of glycolipid

ligands on an extended carbohydrate recognition domain of the macrophage receptor Mincle. *J Biol Chem* (2016) 291:21222–33.

40. Furukawa A, Kamishikiryo J, Mori D, Toyonaga K, Okabe Y, Toji A, et al. Structural analysis for glycolipid recognition by the C-type lectins Mincle and MCL. *Proc Natl Acad Sci U S A* (2013) 110:17438–43. doi:10.1073/pnas.1312649110
41. Brown GD, Gordon S. Immune recognition. A new receptor for beta-glucans. *Nature* (2001) 413:36–7. doi:10.1038/35092620
42. Kistowska M, Fenini G, Jankovic D, Feldmeyer L, Kerl K, Bosshard P, et al. *Malassezia* yeasts activate the NLRP3 inflammasome in antigen-presenting cells via Syk-kinase signalling. *Exp Dermatol* (2014) 23:884–9. doi:10.1111/exd.12552
43. de Jong MA, Vriend LE, Theelen B, Taylor ME, Fluitsma D, Boekhout T, et al. C-type lectin Langerin is a beta-glucan receptor on human Langerhans cells that recognizes opportunistic and pathogenic fungi. *Mol Immunol* (2010) 47: 1216–25. doi:10.1016/j.molimm.2009.12.016
44. Tateno H, Ohnishi K, Yabe R, Hayatsu N, Sato T, Takeya M, et al. Dual specificity of Langerin to sulfated and mannosylated glycans via a single C-type carbohydrate recognition domain. *J Biol Chem* (2010) 285:6390–400. doi:10.1074/jbc.M109.041863
45. Robinson MJ, Osorio F, Rosas M, Freitas RP, Schweighoffer E, Gross O, et al. Dectin-2 is a Syk-coupled pattern recognition receptor crucial for Th17 responses to fungal infection. *J Exp Med* (2009) 206:2037–51. doi:10.1084/jem.20082818
46. LeibundGut-Landmann S, Gross O, Robinson MJ, Osorio F, Slack EC, Tsoni SV, et al. Syk- and CARD9-dependent coupling of innate immunity to the induction of T helper cells that produce interleukin 17. *Nat Immunol* (2007) 8:630–8. doi:10.1038/ni1460
47. Angrisano T, Pero R, Paoletti I, Keller S, Lembo L, Baroni A, et al. Epigenetic regulation of IL-8 and beta-defensin genes in human keratinocytes in response to *Malassezia furfur*. *J Invest Dermatol* (2013) 133:2104–2104. doi:10.1038/jid.2013.143
48. Buommino E, De Filippis A, Parisi A, Nizza S, Martano M, Iovane G, et al. Innate immune response in human keratinocytes infected by a feline isolate of *Malassezia pachydermatis*. *Vet Microbiol* (2013) 163:90–6. doi:10.1016/j.vetmic.2012.12.001
49. Donnarumma G, Perfetto B, Paoletti I, Oliviero G, Clavaud C, Del Bufalo A, et al. Analysis of the response of human keratinocytes to *Malassezia globosa* and restricta strains. *Arch Dermatol Res* (2014) 306:763–8. doi:10.1007/s00403-014-1479-1
50. Baroni A, Orlando M, Donnarumma G, Farro P, Iovene MR, Tufano MA, et al. Toll-like receptor 2 (TLR2) mediates intracellular signalling in human keratinocytes in response to *Malassezia furfur*. *Arch Dermatol Res* (2006) 297:280–8. doi:10.1007/s00403-005-0594-4
51. Thomas DS, Ingham E, Bojar RA, Holland KT. In vitro modulation of human keratinocyte pro- and anti-inflammatory cytokine production by the capsule of *Malassezia* species. *FEMS Immunol Med Microbiol* (2008) 54:203–14. doi:10.1111/j.1574-695X.2008.00468.x
52. Kesavan S, Holland KT, Ingham E. The effects of lipid extraction on the immunomodulatory activity of *Malassezia* species in vitro. *Med Mycol* (2000) 38:239–47. doi:10.1080/mmy.38.3.239.247
53. Ishibashi Y, Sugawara K, Sugita T, Nishikawa A. Secretion of thymic stromal lymphopoietin from human keratinocytes in response to *Malassezia* yeasts. *J Dermatol Sci* (2011) 62:124–8. doi:10.1016/j.jdermsci.2011.02.012
54. Magiatis P, Pappas P, Gaitanis G, Mexia N, Melliou E, Galanou M, et al. *Malassezia* yeasts produce a collection of exceptionally potent activators of the Ah (dioxin) receptor detected in diseased human skin. *J Invest Dermatol* (2013) 133:2023–30. doi:10.1038/jid.2013.92
55. Gaitanis G, Magiatis P, Stathopoulou K, Bassukas ID, Alexopoulos EC, Velegraki A, et al. AhR ligands, malassezin, and indolo[3,2-b]carbazole are selectively produced by *Malassezia furfur* strains isolated from seborrheic dermatitis. *J Invest Dermatol* (2008) 128:1620–5. doi:10.1038/sj.jid.5701252
56. Cafarchia C, Otranto D. Association between phospholipase production by *Malassezia pachydermatis* and skin lesions. *J Clin Microbiol* (2004) 42:4868–9. doi:10.1128/JCM.42.10.4868-4869.2004
57. Riciputo RM, Oliveri S, Micali G, Sapuppo A. Phospholipase activity in *Malassezia furfur* pathogenic strains. *Mycoses* (1996) 39:233–5. doi:10.1111/j.1439-0507.1996.tb00131.x
58. Plotkin LI, Squiquera L, Mathov I, Galimberti R, Leoni J. Characterization of the lipase activity of *Malassezia furfur*. *J Med Vet Mycol* (1996) 34:43–8. doi:10.1080/02681219680000071
59. Lee YW, Lee SY, Lee Y, Jung WH. Evaluation of expression of lipases and phospholipases of *Malassezia restricta* in patients with seborrheic dermatitis. *Ann Dermatol* (2013) 25:310–4. doi:10.5021/ad.2013.25.3.310
60. Ortiz G, Martin MC, Carrillo-Muñoz AJ, Paya MJ. [Phospholipase and proteinase production by *Malassezia pachydermatis* isolated in dogs with and without otitis]. *Rev Iberoam Micol* (2013) 30:235–8. doi:10.1016/j.riam.2013.01.006
61. DeAngelis YM, Gemmer CM, Kaczvinsky JR, Kenneally DC, Schwartz JR, Dawson TL Jr. Three etiologic facets of dandruff and seborrheic dermatitis: *Malassezia* fungi, sebaceous lipids, and individual sensitivity. *J Investig Dermatol Symp Proc* (2005) 10:295–7. doi:10.1111/j.1087-0024.2005.10119.x
62. Katsuta Y, Iida T, Hasegawa K, Inomata S, Denda M. Function of oleic acid on epidermal barrier and calcium influx into keratinocytes is associated with N-methyl D-aspartate-type glutamate receptors. *Br J Dermatol* (2009) 160:69–74. doi:10.1111/j.1365-2133.2008.08860.x
63. Katsuta Y, Iida S, Fau-Inomata T, Inomata M, Fau-Denda S, Denda M. Unsaturated fatty acids induce calcium influx into keratinocytes and cause abnormal differentiation of epidermis. *J Investig Dermatol* (2005) 124:1008–13. doi:10.1111/j.0022-202X.2005.23682.x
64. Plotkin LI, Mathov I, Squiquera L, Leoni J. Arachidonic acid released from epithelial cells by *Malassezia furfur* phospholipase A2: a potential pathophysiologic mechanism. *Mycologia* (1998) 90:163–9. doi:10.2307/3761291
65. Mayser P, Wille G, Imkampe A, Thoma W, Arnold N, Monsees T. Synthesis of fluorochromes and pigments in *Malassezia furfur* by use of tryptophan as the single nitrogen source. *Mycoses* (1998) 41:265–71. doi:10.1111/j.1439-0507.1998.tb00336.x
66. Mexia N, Gaitanis G, Velegraki A, Soshilov A, Denison MS, Magiatis P. Pityriazepin and other potent AhR ligands isolated from *Malassezia furfur* yeast. *Arch Biochem Biophys* (2015) 571:16–20. doi:10.1016/j.abb.2015.02.023
67. Mayser P, Schafer U, Kramer HJ, Irlinger B, Steglich W. Pityriacitrin – an ultraviolet-absorbing indole alkaloid from the yeast *Malassezia furfur*. *Arch Dermatol Res* (2002) 294:131–4. doi:10.1007/s00403-002-0294-2
68. Kramer HJ, Podobinska M, Bartsch A, Battmann A, Thoma W, Bernd A, et al. Malassezin, a novel agonist of the aryl hydrocarbon receptor from the yeast *Malassezia furfur*, induces apoptosis in primary human melanocytes. *Chembiochem* (2005) 6:860–5. doi:10.1002/cbic.200400247
69. Kramer HJ, Kessler D, Hipler UC, Irlinger B, Hort W, Bodeker RH, et al. Pityriarubins, novel highly selective inhibitors of respiratory burst from cultures of the yeast *Malassezia furfur*: comparison with the bisindolylmaleimide arcyriarubin A. *Chembiochem* (2005) 6:2290–7. doi:10.1002/cbic.200500163
70. Vlachos C, Schulte BM, Magiatis P, Adema GJ, Gaitanis G. *Malassezia*-derived indoles activate the aryl hydrocarbon receptor and inhibit toll-like receptor-induced maturation in monocyte-derived dendritic cells. *Br J Dermatol* (2012) 167:496–505. doi:10.1111/j.1365-2133.2012.11014.x
71. Tauchi M, Hida A, Negishi T, Katsuoka F, Noda S, Mimura J, et al. Constitutive expression of aryl hydrocarbon receptor in keratinocytes causes inflammatory skin lesions. *Mol Cell Biol* (2005) 25:9360–8. doi:10.1128/MCB.25.21.9360-9368.2005
72. Gaitanis G, Velegraki A, Magiatis P, Pappas P, Bassukas ID. Could *Malassezia* yeasts be implicated in skin carcinogenesis through the production of aryl-hydrocarbon receptor ligands? *Med Hypotheses* (2011) 77:47–51. doi:10.1016/j.mehy.2011.03.020
73. Buommino E, Nocera FP, Parisi A, Rizzo A, Donnarumma G, Mallardo K, et al. Correlation between genetic variability and virulence factors in clinical strains of *Malassezia pachydermatis* of animal origin. *New Microbiol* (2016) 39:216–23.
74. Donnarumma G, Paoletti I, Buommino E, Orlando M, Tufano MA, Baroni A. *Malassezia furfur* induces the expression of beta-defensin-2 in human keratinocytes in a protein kinase C-dependent manner. *Arch Dermatol Res* (2004) 295:474–81. doi:10.1007/s00403-003-0445-0
75. Watanabe S, Kano R, Sato H, Nakamura Y, Hasegawa A. The effects of *Malassezia yeasts* on cytokine production by human keratinocytes. *J Invest Dermatol* (2001) 116:769–73. doi:10.1046/j.1523-1747.2001.01321.x
76. Ribbing C, Engblom C, Lappalainen J, Lindstedt K, Kovanen PT, Karlsson MA, et al. Mast cells generated from patients with atopic eczema have enhanced

levels of granule mediators and an impaired Dectin-1 expression. *Allergy* (2011) 66:110–9. doi:10.1111/j.1398-9995.2010.02437.x

77. Baroni A, Paoletti I, Ruocco E, Agozzino M, Tufano MA, Donnarumma G. Possible role of *Malassezia furfur* in psoriasis: modulation of TGF-beta1, integrin, and HSP70 expression in human keratinocytes and in the skin of psoriasis-affected patients. *J Cutan Pathol* (2004) 31:35–42. doi:10.1046/j.0303-6987.2004.0135.x
78. Jarvikallio A, Naukkarinen A, Harvima IT, Aalto ML, Horsmanheimo M. Quantitative analysis of tryptase- and chymase-containing mast cells in atopic dermatitis and nummular eczema. *Br J Dermatol* (1997) 136:871–7. doi:10.1111/j.1365-2133.1997.tb03927.x
79. Selander C, Engblom C, Nilsson G, Scheynius A, Andersson CL. TLR2/MyD88-dependent and -independent activation of mast cell IgE responses by the skin commensal yeast *Malassezia sympodialis*. *J Immunol* (2009) 182:4208–16. doi:10.4049/jimmunol.0800885
80. Gehrmann U, Qazi KR, Johansson C, Hultenby K, Karlsson M, Lundeberg L, et al. Nanovesicles from *Malassezia sympodialis* and host exosomes induce cytokine responses – novel mechanisms for host-microbe interactions in atopic eczema. *PLoS One* (2011) 6:e21480. doi:10.1371/journal.pone.0021480
81. Ashbee HR, Bond R. *Malassezia* species and immunity: host–pathogen interactions. In: Boekhout T, Mayser P, Guého-Kellermann E, Velegraki A, editors. *Malassezia and the Skin: Science and Clinical Practice*. Berlin, Heidelberg: Springer (2010). p. 139–73.
82. Valli JL, Williamson A, Sharif S, Rice J, Shewen PE. In vitro cytokine responses of peripheral blood mononuclear cells from healthy dogs to distemper virus, *Malassezia* and Toxocara. *Vet Immunol Immunopathol* (2010) 134:218–29. doi:10.1016/j.vetimm.2009.09.023
83. Kang MH, Kim HJ, Jang HJ, Park HM. Sensitization rates of causative allergens for dogs with atopic dermatitis: detection of canine allergen-specific IgE. *J Vet Sci* (2014) 15:545–50. doi:10.4142/jvs.2014.15.4.545
84. Chen TA, Halliwell RE, Pemberton AD, Hill PB. Identification of major allergens of *Malassezia pachydermatis* in dogs with atopic dermatitis and *Malassezia* overgrowth. *Vet Dermatol* (2002) 13:141–50. doi:10.1046/j.1365-3164.2002.00291.x
85. Johansson C, Sandstrom MH, Bartosik J, Sarnhult T, Christiansen J, Zargari A, et al. Atopy patch test reactions to *Malassezia* allergens differentiate subgroups of atopic dermatitis patients. *Br J Dermatol* (2003) 148:479–88. doi:10.1046/j.1365-2133.2003.05093.x
86. Glatz M, Buchner M, von Bartenwerffer W, Schmid-Grendelmeier P, Worm M, Hedderich J, et al. *Malassezia* spp.-specific immunoglobulin E level is a marker for severity of atopic dermatitis in adults. *Acta Derm Venereol* (2015) 95:191–6. doi:10.2340/00015555-1864
87. Zhang E, Tanaka T, Tajima M, Tsuboi R, Kato H, Nishikawa A, et al. Anti-*Malassezia*-specific IgE antibodies production in Japanese patients with head and neck atopic dermatitis: relationship between the level of specific ige antibody and the colonization frequency of cutaneous *Malassezia* species and clinical severity. *J Allergy (Cairo)* (2011) 2011:645670. doi:10.1155/2011/645670
88. Johansson C, Ahlborg N, Andersson A, Lundeberg L, Karlsson MA, Scheynius A, et al. Elevated peripheral allergen-specific T cell response is crucial for a positive atopy patch test reaction. *Int Arch Allergy Immunol* (2009) 150:51–8. doi:10.1159/000210380
89. Levy JMS, Magro C. Atrophying pityriasis versicolor as an idiosyncratic T cell-mediated response to *Malassezia*: a case series. *J Am Acad Dermatol* (2017) 76:730–5. doi:10.1016/j.jaad.2016.08.062
90. Eyerich S, Eyerich K, Pennino D, Carbone T, Nasorri F, Pallotta S, et al. Th22 cells represent a distinct human T cell subset involved in epidermal immunity and remodeling. *J Clin Invest* (2009) 119:3573–85. doi:10.1172/JCI40202
91. Cavani A, Pennino D, Eyerich K. Th17 and Th22 in skin allergy. *Chem Immunol Allergy* (2012) 96:39–44. doi:10.1159/000331870
92. Di Cesare A, Di Meglio P, Nestle FO. The IL-23/Th17 axis in the immunopathogenesis of psoriasis. *J Invest Dermatol* (2009) 129:1339–50. doi:10.1038/jid.2009.59
93. Balaji H, Heratizadeh A, Wichmann K, Niebuhr M, Crameri R, Scheynius A, et al. *Malassezia* sympodialis thioredoxin-specific T cells are highly cross-reactive to human thioredoxin in atopic dermatitis. *J Allergy Clin Immunol* (2011) 128: 92–9.e4. doi:10.1016/j.jaci.2011.02.043
94. Esaki H, Brunner PM, Renert-Yuval Y, Czarnowicki T, Huynh T, Tran G, et al. Early-onset pediatric atopic dermatitis is TH2 but also TH17 polarized in skin. *J Allergy Clin Immunol* (2016) 138:1639–51. doi:10.1016/j.jaci.2016.07.013
95. Sparber F, LeibundGut-Landmann S. Interleukin 17-mediated host defense against *Candida albicans*. *Pathogens* (2015) 4:606–19. doi:10.3390/pathogens4030606
96. Gaitanis G, Magiatis P, Hantschke M, Bassukas ID, Velegraki A. The *Malassezia* genus in skin and systemic diseases. *Clin Microbiol Rev* (2012) 25:106–41. doi:10.1128/CMR.00021-11
97. Selander C, Zargari A, Mollby R, Rasool O, Scheynius A. Higher pH level, corresponding to that on the skin of patients with atopic eczema, stimulates the release of *Malassezia sympodialis* allergens. *Allergy* (2006) 61:1002–8. doi:10.1111/j.1398-9995.2006.01108.x
98. Schmid-Grendelmeier P, Fluckiger S, Disch R, Trautmann A, Wuthrich B, Blaser K, et al. IgE-mediated and T cell-mediated autoimmunity against manganese superoxide dismutase in atopic dermatitis. *J Allergy Clin Immunol* (2005) 115:1068–75. doi:10.1016/j.jaci.2005.01.065
99. Glaser AG, Limacher A, Fluckiger S, Scheynius A, Scapozza L, Crameri R. Analysis of the cross-reactivity and of the 1.5 A crystal structure of the *Malassezia sympodialis* Mala s 6 allergen, a member of the cyclophilin pan-allergen family. *Biochem J* (2006) 396:41–9. doi:10.1042/BJ20051708
100. Youn SW, Na JI, Choi SY, Huh CH, Park KC. Regional and seasonal variations in facial sebum secretions: a proposal for the definition of combination skin type. *Skin Res Technol* (2005) 11:189–95. doi:10.1111/j.1600-0846.2005.00119.x
101. Kim SY, Kim SH, Kim SN, Kim AR, Kim YR, Kim MJ, et al. Isolation and identification of *Malassezia* species from Chinese and Korean patients with seborrheic dermatitis and in vitro studies on their bioactivity on sebaceous lipids and IL-8 production. *Mycoses* (2016) 59:274–80. doi:10.1111/myc.12456
102. Schonherr FA, Sparber F, Kirchner FR, Guiducci E, Trautwein-Weidner K, Gladiator A, et al. The intraspecies diversity of *C. albicans* triggers qualitatively and temporally distinct host responses that determine the balance between commensalism and pathogenicity. *Mucosal Immunol* (2017) 10:1335–50. doi:10.1038/mi.2017.2
103. Zhu Y, Engstrom PG, Tellgren-Roth C, Baudo CD, Kennell JC, Sun S, et al. Proteogenomics produces comprehensive and highly accurate protein-coding gene annotation in a complete genome assembly of *Malassezia sympodialis*. *Nucleic Acids Res* (2017) 45:2629–43.

15

The IDO–AhR Axis Controls Th17/Treg Immunity in a Pulmonary Model of Fungal Infection

Eliseu Frank de Araújo, Claudia Feriotti, Nayane Alves de Lima Galdino, Nycolas Willian Preite, Vera Lúcia Garcia Calich† and Flávio Vieira Loures†*

Departamento de Imunologia, Instituto de Ciências Biomédicas, Universidade de São Paulo, São Paulo, Brazil

***Correspondence:**
Flávio Vieira Loures
loures@icb.usp.br

†These authors have contributed equally to this work.

In infectious diseases, the enzyme indoleamine 2,3 dioxygenase-1 (IDO1) that catalyzes the tryptophan (Trp) degradation along the kynurenines (Kyn) pathway has two main functions, the control of pathogen growth by reducing available Trp and immune regulation mediated by the Kyn-mediated expansion of regulatory T (Treg) cells *via* aryl hydrocarbon receptor (AhR). In pulmonary paracoccidioidomycosis (PCM) caused by the dimorphic fungus *Paracoccidioides brasiliensis*, IDO1 was shown to control the disease severity of both resistant and susceptible mice to the infection; however, only in resistant mice, IDO1 is induced by TGF-β signaling that confers a stable tolerogenic phenotype to dendritic cells (DCs). In addition, in pulmonary PCM, the tolerogenic function of plasmacytoid dendritic cells was linked to the IDO1 activity. To further evaluate the function of IDO1 in pulmonary PCM, IDO1-deficient (IDO1$^{-/-}$) C57BL/6 mice were intratracheally infected with *P. brasiliensis* yeasts and the infection analyzed at three postinfection periods regarding several parameters of disease severity and immune response. The fungal loads and tissue pathology of IDO1$^{-/-}$ mice were higher than their wild-type controls resulting in increased mortality rates. The evaluation of innate lymphoid cells showed an upregulated differentiation of the innate lymphoid cell 3 phenotype accompanied by a decreased expansion of ILC1 and NK cells in the lungs of infected IDO1$^{-/-}$ mice. DCs from these mice expressed elevated levels of costimulatory molecules and cytokine IL-6 associated with reduced production of IL-12, TNF-α, IL-1β, TGF-β, and IL-10. This response was concomitant with a marked reduction in AhR production. The absence of IDO1 expression caused an increased influx of activated Th17 cells to the lungs with a simultaneous reduction in Th1 and Treg cells. Accordingly, the suppressive cytokines IL-10, TGF-β, IL-27, and IL-35 appeared in reduced levels in the lungs of IDO1$^{-/-}$ mice. In conclusion, the immunological balance mediated by the axis IDO/AhR is fundamental to determine the balance between Th17/Treg cells and control the severity of pulmonary PCM.

Keywords: paracoccidioidomycosis, indolemine 2,3 dioxygenase, aryl hydrocarbon receptor, Th17 cells, regulatory T cells, innate lymphoid cells

INTRODUCTION

Indoleamine 2,3 dioxygenase-1 (IDO1) is an enzyme that catalyzes the degradation of tryptophan (Trp) along the kynurenines (Kyn) pathway. This enzyme plays a critical role to host defenses against a wide range of pathogens by inducing Trp starvation and controlling inflammatory and immunological reactions. Several types of cells, such as macrophages, epithelial, and dendritic cells (DCs), express IDO1 that can be induced by proinflammatory cytokines (e.g., IFN-γ), toll-like receptor ligands (e.g., lipopolysaccharide), and interactions between immune cells through the engagement of costimulatory molecules (CD80, CD86) on antigen-presenting cells by cytotoxic T-lymphocyte antigen-4 on T cells (1–3). It is known that IDO1 can affect immunity through two non-exclusive mechanisms: (a) establishment of a local response with "amino acid deprivation" that inhibits cell and pathogen proliferation (2); (b) cascade generation of Trp metabolites with specific immunomodulatory or cytotoxic functions that inhibit T cell activation and modulate the differentiation of naïve T cells into regulatory T (Treg) cells (1, 4, 5). Furthermore, Kyn has immunomodulatory effects in the absence of Trp starvation, *via* activation of the transcription factor aryl hydrocarbon receptor (AhR) (6, 7).

Indeed, in fungal infections, IDO1 modulates the mechanisms of host resistance and disease tolerance by inducing Treg cells and inhibiting Th17 differentiation (1, 8–10). Importantly, Treg and Th17 cells share similar developmental pathways (11) and may arise from a common progenitor depending on the balance of inflammatory and anti-inflammatory cytokines (12), presence of retinoic acid (13), and the activation of the AhR, a ligand-activated transcription factor (14, 15).

Paracoccidioidomycosis (PCM), a systemic granulomatous disease caused either by the dimorphic fungi *Paracoccidioides brasiliensis* or *P. lutzii*, is considered the most prevalent deep mycosis in Latin America (16). The inhalation of conidia usually leads to an asymptomatic infection but a few infected individuals evolve to overt disease. The diverse patterns of T cell responses of *P. brasiliensis*-infected individuals are associated with different clinical manifestations. The resistance to infection observed in asymptomatic individuals is mediated by a predominant Th1 response, which is responsible for macrophage activation and fungal killing. The most severe form of the disease, the juvenile form, presents a prevalent Th2/Th9 response and enhanced antibody production. In the chronic inflammatory response characteristic of the adult form of the disease, a prominent Th17 immunity with important participation of Th1 cells was described (17, 18). The polar patterns of the disease could be reproduced in a murine model of pulmonary PCM (19). Resistant mice (A/Sn) develop a chronic regressive disease allied with prevalent Th1/Th17 immunity concomitant with protective $CD8^+$ T cells that synthesize large amounts of IFN-γ. The relative protection of susceptible mice (B10.A) is mediated by $CD8^+$ T cells that, however, are not able to compensate the $CD4^+$ T cell anergy induced by excessive proinflammatory innate response (20–23).

Recent studies of our group have demonstrated that in pulmonary PCM, IDO1 is an important immunoregulatory enzyme that promotes fungal clearance and inhibits T cell immunity and inflammation, with prominent importance to susceptible hosts. Actually, only in the susceptible mouse strain IDO1 inhibition by 1-methyl-dl-tryptophan (1MT) caused progressive tissue pathology and increased mortality rates. In addition, as previously reported in candidiasis and aspergillosis (24, 25), an IDO1-mediated immunomodulatory function of plasmacytoid dendritic cells (pDCs) has been described in pulmonary PCM. The *in vivo* depletion of IDO1 expressing pDCs resulted in less severe disease and increased T cell immunity (26), demonstrating that in pulmonary PCM pDCs have a tolerogenic function as previously described by Pallotta et al. (27). These findings led us to further elucidate the role of IDO1 in pulmonary PCM by comparing the disease developed by IDO1-deficient ($IDO1^{-/-}$) C57BL/6 mice with their normal wild-type (WT) controls. $IDO1^{-/-}$ mice developed a more severe infection with elevated fungal burdens, accompanied by increased inflammatory reactions mediated by prevalent Th17 responses insufficiently controlled by Treg cells. The absence of IDO1 expression has also influenced the expression of AhR and the differentiation of pulmonary innate lymphoid cells (ILCs) by increasing the presence of innate lymphoid cell 3 (ILC3) and reducing NK cells. Both the uncontrolled fungal growth and the exuberant inflammatory reactions appear to have contributed to the exacerbated tissue pathology that resulted in increased mortality rates of $IDO1^{-/-}$ mice.

MATERIALS AND METHODS

Ethics Statement

The experiments were performed in strict accordance with the Brazilian Federal Law 11,794 establishing procedures for the scientific use of animals, and the State Law establishing the Animal Protection Code of the State of São Paulo. All efforts were made to minimize animal suffering. The procedures were approved by the Ethics Committee on Animal Experiments of the Institute of Biomedical Sciences of University of São Paulo (Proc.180/11/CEEA).

Mice

Eight- to twelve-week-old male C57Bl/6 WT and C57Bl/6 $IDO1^{-/-}$ mice originally obtained from Jackson Laboratories and bred as the specific pathogen-free mice at the Isogenic Breeding Unit of the Department of Immunology, Institute of Biomedical Sciences, University of São Paulo were used throughout this study.

Fungus and Infection

The highly virulent *P. brasiliensis* 18 isolate (Pb18) was used throughout this study. Yeast cells were maintained by weekly cultivation in Fava Netto culture medium at 36°C and used on days 6–8 of culture. The viability of fungal suspensions, determined by Janus Green B vital dye (Merck), was always higher than 95%. Mice were anesthetized and submitted to intra-tracheal (i.t.) infection as previously described (19). Briefly, after intraperitoneal (i.p.) injection of ketamine and xylazine, animals were infected with 1×10^6 yeast cells, contained in 50 μL of PBS, by surgical i.t. inoculation, which allowed dispensing of the fungal cells directly into the lungs.

Colony-Forming Unit (CFU) Assays, Mortality Rates, and Histological Analysis

The numbers of viable yeasts in infected organs (lung and liver) were determined by counting the number of CFUs as previously described (28). Mortality studies were done with groups of 10–12 mice. Deaths were registered daily. For histological examinations, the left lung of infected mice was removed and fixed in 10% formalin. Five-micrometer sections were stained by hematoxylin-eosin for an analysis of the lesions and were silver stained (Grocott stain) for fungal evaluation. Morphometrical analysis was performed using a Nikon DXM 1200c camera and Nikon NIS AR 2.30 software. The areas of lesions were measured (in square micrometers) in 10 microscopic fields per slide in 5 mice per group as previously described (29). Results are expressed as the mean ± SEM of total area of lesions for each mouse.

Assessment of Leukocyte Subpopulations and Flow Cytometric Analysis

The lungs from *P. brasiliensis*-infected WT and IDO1$^{-/-}$ mice were collected after 96 h, 2, and 10 weeks of infection and digested enzymatically for 45 min with collagenase (1 mg/mL; Sigma) in RPMI culture medium (Sigma). Total lung leukocyte numbers were assessed with trypan blue, and viability was always >95%. For cell-surface staining, lung cells were washed and suspended at 1×10^6 cells/mL in staining buffer (PBS, 2% fetal calf serum and 0.1% NaN_3). Fc receptors were blocked by the addition of unlabeled anti-CD16/32 (eBioscience). The cells were then stained in the dark for 20 min at 4°C with the optimal dilution of each monoclonal antibody. To myeloid cells: anti-CD11b, CD11c, CD40, CD80, CD86, and MHC-II; lymphocytes: CD4, CD25, CD8, CD44, and CD62L; NK: NK1.1, CD49b, NPK46, and Eomes; ILC1: CD127, CD49a, IL-12Rβ1, and Tbet; ILC2: CD127, ICOS, IL-17RB, IL-33, and GATA3; ILC3: CD127, IL-23R, IL-22, and RORC (eBiosciences or BioLegend). Cells were washed twice with staining buffer, fixed with 2% paraformaldehyde (PFA; Sigma). For intracellular detection of cytokines, leukocytes obtained from lungs were stimulated for 6 h in complete RPMI medium containing 50 ng/mL phorbol 12-myristate 13-acetate, 500 ng/mL ionomycin (Sigma), and 3 mM monensin (eBioscience). Next, cells were labeled for surface molecules and then treated according to the manufacturer's protocol for intracellular staining using the Cytofix/Cytoperm kit (BD Biosciences) and specifics antibodies anti-17, IL-4, IFN-γ, IL, 22, IL-1β, IL-12, TNF-α, IL-6, TGF-β, FoxP3, IDO1, and AhR. Cells were washed twice with staining buffer, suspended in 100 µL, and an equal volume of PFA was added to fix the cells. A minimum of 50,000 events was acquired on FACScanto II flow cytometer (BD Biosciences) using the FACSDiva software (BD Biosciences). Lymphocytes, myeloid cells, NK cells, and ILCs were gated as judged from forward and side light scatter. For Treg cell characterization, FACS plots or histograms were gated on live $CD4^+$ $CD25^+$ cells and the expression of $FoxP3^+$ were determined. The cell-surface expression of leukocyte markers as well as intracellular cytokine expression was analyzed using the FlowJo software (Tree Star).

RNA Isolation and cDNA Synthesis

Lungs were homogenized in TRIzol reagent using tissue grinders. Phase separation was realized following addition of 0.2 mL chloroform per mL of TRIzol and centrifugation at 12,000 × *g* for 15 min at 4°C. The upper aqueous RNA phase was removed to a new tube and further purified using Ultraclean Tissue and Cells RNA Isolation Kit (MO BIO Laboratories) according to the manufacturer's protocol. RNA purity and concentration were assessed on a NanoDrop ND-1000 spectrophotometer. An amount of 1 µg total RNA was reverse transcribed in a 20 µL reaction mixture using the High Capacity RNA-to-cDNA kit (Applied Biosystems) following the manufacturer's directions.

Real-time Quantitative Polymerase Chain Reaction (RT-PCR)

The cDNA was amplified using TaqMan Universal PCR Master Mix (Applied Biosystems) and pre-developed TaqMan assay primers and probes (*Ahr*, Mm00478932_m1, *Ifng*, Mm001168134_m1, *Tnf*, Mm99999068_m1, *Il-6*, Mm00446190_m1, *Il-10*, Mm00439614_m1, *Tgfb1*, Mm00117882_m1, *Il-17*, Mm00439618_m1, *Il-22*, Mm01226722_m1, *Tbx21*, Mm00450960_m1; *GATA3*, Mm00484683_m1; *Rorc*, Mm01261022_m1; *Foxp3*, Mm 00475162_m1; all from Applied Biosystems). PCR assays were performed on an MxP3000P qPCR System and data were developed using the MxPro qPCR software (Stratagene). The average threshold cycle (C_T) values of samples were normalized to C_T value of *Gapdh* gene. The relative expression was determined by the $2^{-\Delta\Delta C_T}$ method.

Cytokines and Nitric Oxide (NO) Detection

The lungs from *P. brasiliensis*-infected WT and IDO1$^{-/-}$ mice were collected after 96 h, 2, and 10 weeks of infection with one million *P. brasiliensis* yeasts. The organs were aseptically removed and individually disrupted in 7 mL of PBS. Supernatants were separated from cell debris by centrifugation at 3,000 × *g* for 10 min and stored at −80°C. The levels of IL-1β, IL-2, IL-4, IL-6, IL-10, IL-12, IL-17, IL-22, IL-27, IL-35, IL-23, TNF-α, IFN-γ, and TGF-β were measured by capture enzyme-linked immunosorbent assay (ELISA) with antibody pairs purchased from eBioscience or PBL. NO production was quantified by the accumulation of nitrite in the supernatants from *in vitro* protocols by a standard Griess reaction. All determinations were performed in duplicate, and results were expressed as micro molar concentration of NO. Plates were read using a spectrophotometric plate reader (VersaMax, Molecular Devices).

Statistical Analysis

Differences between groups were analyzed by the Student's *t*-test or analysis of variance (ANOVA) followed by the Tukey test. For comparisons of greater than two groups, significance was determined using the one- or two-way ANOVA with Tukey's multiple correction. Differences between survival times were determined with the log-rank test. Calculations were performed using statistical software (GraphPad Prism 7.03). Data are expressed as the mean ± SEM. *p*-Values of ≤0.05 were considered statistically significant.

RESULTS

Absence of IDO1 Increases Mortality Rates Associated With Increased Fungal Loads and Tissue Damage

C57Bl/6 WT and C57Bl/6 IDO1$^{-/-}$ mice were infected with one million yeasts of *P. brasiliensis* by the i.t. route. After 96 h, 2, and 10 weeks, the severity of fungal infection was assessed by a CFU assay. As shown in the **Figure 1**, pulmonary (**Figure 1A**) fungal burdens were increased in IDO1$^{-/-}$ mice at 96 h, 2, and 10 weeks after infection, whereas significant increases in the liver (**Figure 1B**) after 10 weeks of infection showed that the lung of both IDO1$^{-/-}$ mice were concomitant with increased levels of pulmonary NO (**Figure 1C**), considered a potent fungicidal mediator in pulmonary PCM. The histopathology analysis (**Figures 1D–G**) after 10 weeks of infection showed that the lung of both IDO1$^{-/-}$ and WT control mice exhibited large granulomas containing many yeasts and large damaged areas. Comparatively, lung lesions from IDO1$^{-/-}$ were more severe than those of WT mice, involved extensive areas of the lung parenchyma and were composed of isolated or confluent granulomas containing a huge number of fungi (**Figures 1F,G**). To assess the influence of IDO1 deficiency in the disease outcome, the area of lung lesions and mortality of infected mice were evaluated. The absence of IDO1 expression led to increased lung pathology (**Figure 1H**) and mortality rates (**Figure 1I**). At day 127 of infection, all IDO1$^{-/-}$ mice were dead; whereas in the same period, 5 of 10 WT mice were still alive and apparently healthy.

IDO-1 Expression Increases the Expansion of ILC3 but Reduces ILC1 and NK Cells in the Lungs of *P. brasiliensis*-Infected Mice

Innate lymphoid cells are a new family of lymphocytes that lack specific antigen receptors, produce significant amounts of cytokines, and may be cytotoxic upon activation. The distinct ILC subsets exhibit transcription factors and cytokine signatures found in the CD4^{+} T helper (Th) subpopulations in their responses to specific antigens (30, 31). These features include the shared expression of Tbet and IFN-γ by ILC1 and Th1 cells; GATA-3, IL-5, and IL-13 by Th2 and ILC2 cells; RORC, IL-17, and IL-22 by ILC3 and Th17/Th22 cells, as well as Eomes, IFN-γ, and cytolytic molecules by CD8^{+} T cells and conventional NK cells (30). The previously described influence of IDO/AhR in the differentiation of T and ILC cells (30–33) led us to investigate the influence of IDO1 in the differentiation of pulmonary ILC subsets after *P. brasiliensis* infection of WT and IDO1$^{-/-}$ mice. Therefore, ILC subpopulations were characterized by their expression of specific surface molecules and transcription factors. The NK cells are classified as NK11^{+} CD49b^{+} NKp46^{+} Eomes^{+}, the ILC1 are CD127^{+} CD49a^{+} IL-12Rβ1^{+} Tbet^{+}, the ILC2 are CD127^{+} ICOS^{+} IL-17RB^{+} IL-33^{+} GATA3^{+}, and the ILC3 are CD127^{+} IL-23R^{+} RORC^{+}, or the so-called NCR^{+} ILC3^{+}, producing IL-22 that are CD127^{+} IL-23R^{+} RORC^{+} IL-22^{+}. After 96 h and 2 weeks of infection, a significant decrease in NK cells was detected in the lung infiltrating lymphocytes of IDO1$^{-/-}$ mice. However, at all periods of infection studied, a significant increase in the number of ILC3 and NRC IL-22 was observed in the lungs of IDO1$^{-/-}$ mice. Differences in the influx of ILC1 and ILC2 to the lungs were only observed at week 10 postinfection, when ILC1 appeared in reduced numbers whereas ILC2 appeared in increased numbers in the lungs of IDO1$^{-/-}$ mice (**Figure 2**).

IDO1 Controls the Expression of Activation Markers and Intracellular Cytokines of Pulmonary CD11b^{+} and CD11c^{+} Cells

IDO1$^{-/-}$ and WT C57BL/6 mice were infected with one million yeasts of *P. brasiliensis* by the i.t. route. After 96 h, 2, and 10 weeks of infection, the lung infiltrating leukocytes were obtained and analyzed for the expression of surface molecules by flow cytometry. We analyzed the expression of some activation molecules (IA^{b}, CD40, CD80, and CD86) expressed by CD11b^{+} and CD11c^{+} (**Figure 3**). When compared with the WT control group, CD11b^{+} (**Figure 3A**) and CD11c^{+} (**Figure 3B**) cells from IDO1$^{-/-}$ mice expressed increased levels of IA^{b}, CD80, and CD86 at all postinfection periods studied. CD40 was also upregulated in CD11b^{+} cells of IDO1$^{-/-}$ mice.

To better define the activation profile of the lung infiltrating leukocytes of both infected mouse strains, the expression of intracellular cytokines (IL-12, TNF-α, IL-1β, IL-6, TGF-β, and IL-10) was characterized by flow cytometry. At all periods after infection analyzed, a decreased number of CD11b^{+} cells expressing intracellular IL-12, TNF-α, IL-1β, TGF-β, and IL-10 was detected in the lungs of IDO1$^{-/-}$ mice when compared to the WT counterparts. By contrast, increased numbers of CD11b^{+} IL-6^{+} cells were found at all postinfection periods in the lungs of IDO$^{-/-}$ mice (**Figure 4A**). When CD11c^{+} cells were studied, similar results were obtained (**Figure 4B**), but the differences between IDO1 sufficient and deficient mice was much more evident.

IDO1 Controls the Expression of AhR in CD11b^{+} and CD11c^{+} Cells

The influence of IDO1 in the expression of the transcription factor AhR was also investigated in our experimental model. Thus, after 96 h, 2, and 10 weeks of infection with yeasts, the lung infiltrating leukocytes from the IDO1$^{-/-}$ and WT mice were obtained, and the presence of intracellular IDO1 and AhR in CD11b^{+} and CD11c^{+} cells detected by flow cytometry. As depicted in **Figure 5**, only WT mice expressed intracellular IDO1. The AhR was detected in both mouse strains, but this transcription factor was markedly downregulated in IDO1$^{-/-}$ mice. These results clearly showed that the expression of AhR is dependent on the enzyme IDO1 (**Figure 5**).

The Absence of IDO1 Increases the Influx of Naïve and Activated TCD4^{+} and TCD8^{+} Lymphocytes to the Lungs

We have also assessed whether the absence IDO1 modulates the influx of T cell subpopulations to the lung of *P. brasiliensis*-infected mice. As showed in the **Figure 6**, after 2 and 10 weeks of infection, a higher number of both naive (CD4^{+} CD44^{low} CD62L^{high}) and effector/activated (CD4^{+} CD44^{high} CD62L^{low}) CD4^{+} T cells were found in the lungs of IDO1$^{-/-}$ mice when compared with WT counterparts. Similarly, a larger number of naïve and activated CD8^{+} T cells (CD8^{+} CD44^{low} CD62L^{high} and

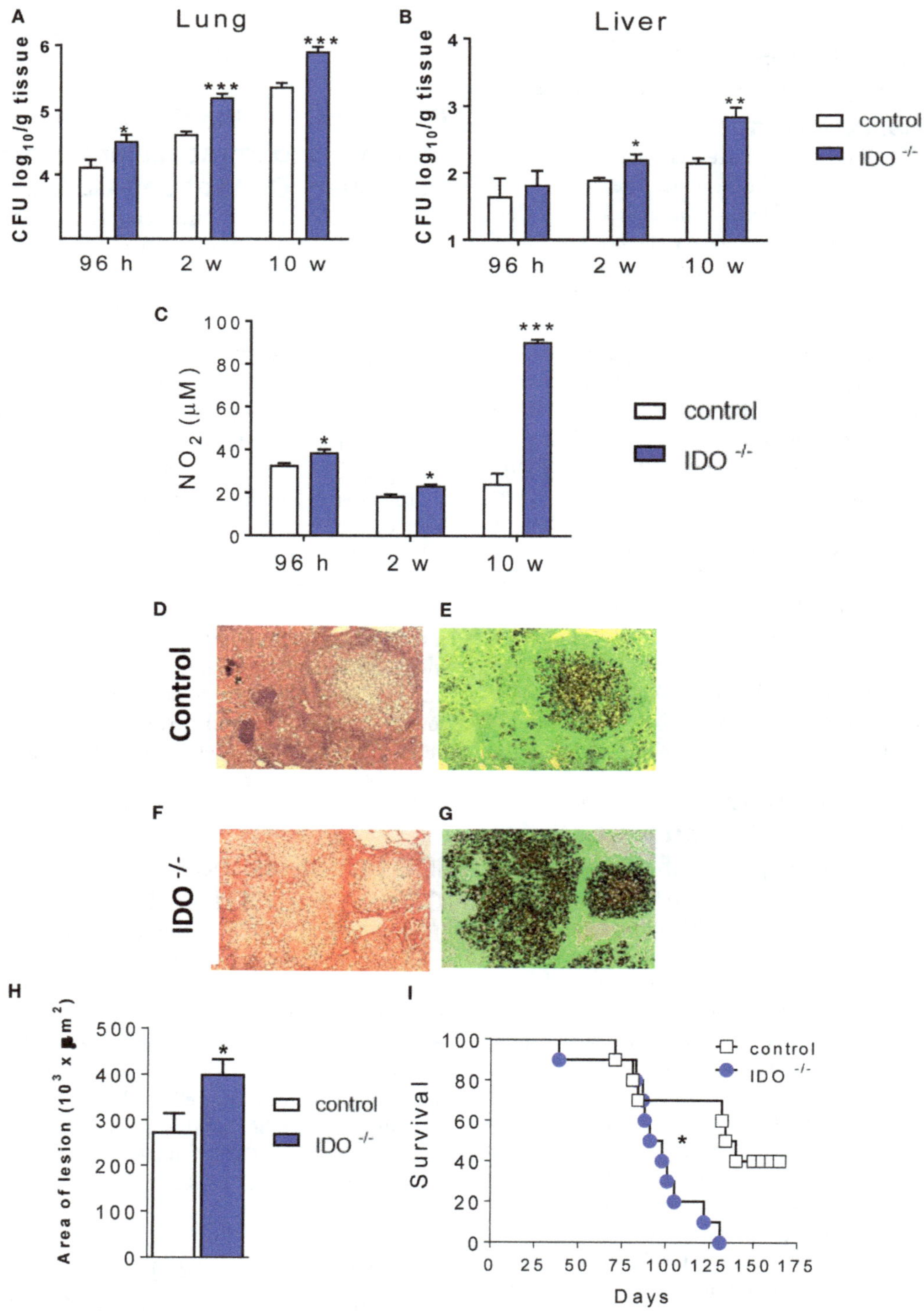

FIGURE 1 | Indoleamine 2,3 dioxygenase-1 (IDO1) controls fungal loads, tissue pathology, and mortality rates. Groups of IDO1$^{-/-}$ and wild-type (WT) C57Bl/6 mice were infected i.t. with 1×10^6 of *Paracoccidioides brasiliensis* yeasts. The colony-forming units (CFUs) from lungs **(A)** and liver **(B)** were determined 96 h, 2, and 8 weeks after infection. The bars represent means ± SEs of the mean (SEM) of $\log_{10}$ CFU counts obtained from groups of 4 to 5 mice. **(C)** Supernatants of lung homogenates were obtained from infected mice (n = 4–5/time point) and used to determine the levels of nitrite by the Griess reagent. **(D–G)** Photomicrographs of lung lesions of WT control mice **(D,E)** and IDO1$^{-/-}$ mice **(F,G)** at week 10 of infection. Lesions were stained with hematoxylin-eosin (left panels) and Grocott (right panels). **(H)** Morphometrical analysis of lung lesions. The areas of lesions were measured (in square micrometers) in 10 microscopic fields per slide in five mice per group and results expressed as the mean ± SEM of total area of lesions for each mouse. **(I)** Survival curves of WT and IDO1$^{-/-}$ infected mice (n = 10) were determined in a period of 175 days. Data represent the means ± SEM and are representative of two independent experiments with equivalent results ($^*p < 0.05$; $^{**}p < 0.01$, and $^{***}p < 0.001$).

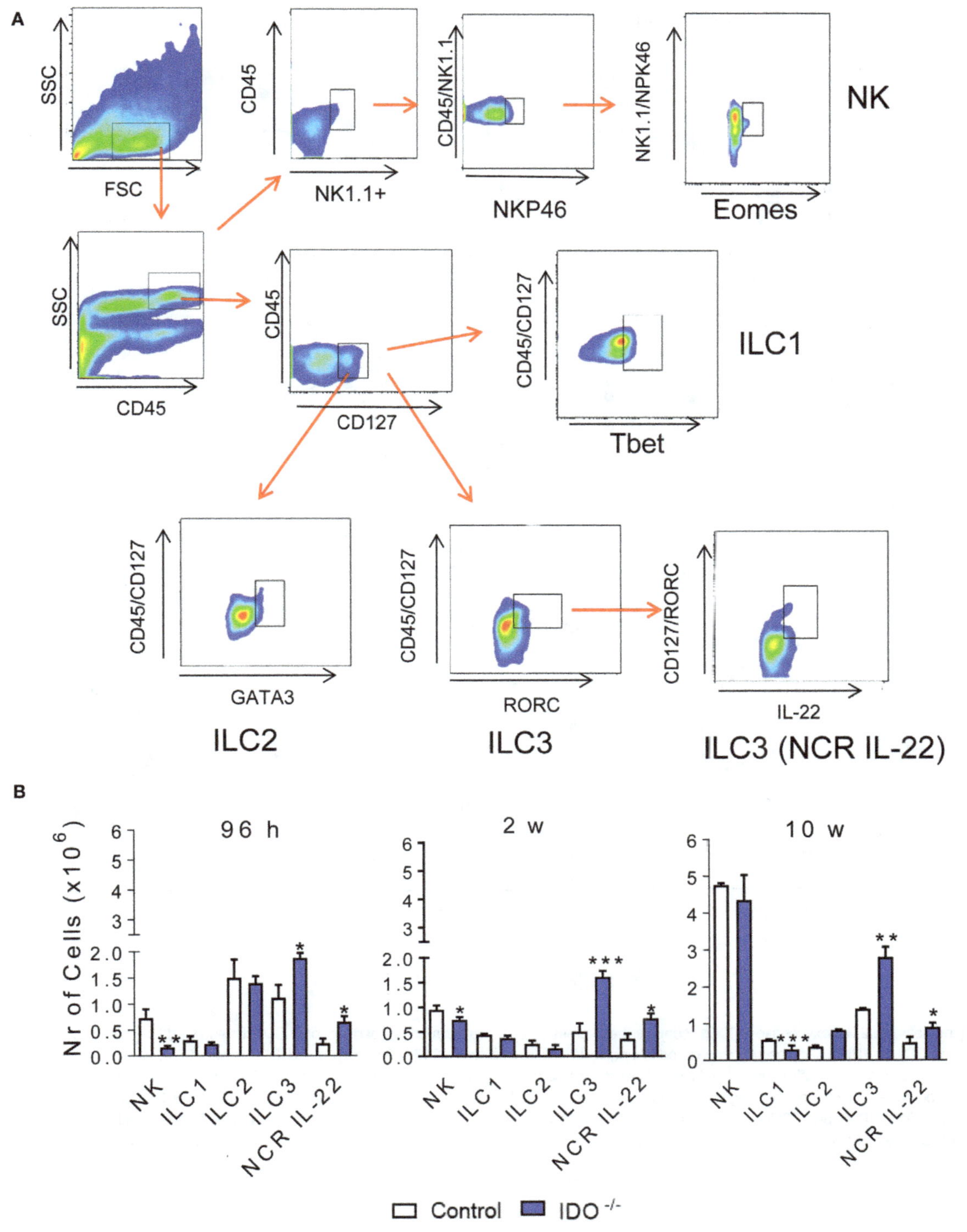

FIGURE 2 | Influence of indoleamine 2,3 dioxygenase-1 (IDO1) expression on the presence of innate lymphoid cells (ILCs) in the lungs of *Paracoccidioides brasiliensis*-infected mice. The phenotypic analysis of ILCs in the lungs of C57Bl/6 wild-type and IDO1$^{-/-}$ mice was performed after 96 h, 2, and 10 weeks after *P. brasiliensis* infection. The lung infiltrating leukocytes were labeled with specific antibodies and analyzed for the phenotypes of NK, ILC1, ILC2, innate lymphoid cell 3 (ILC3), and ILC3 (NCR IL-22) subsets according to the respective markers: for NK: NK1.1, CD49b, NKp46, and Eomes; ILC1: CD127, CD49a, IL-12Rβ1, and Tbet; ILC2: CD127, ICOS, IL-17RB, IL-33, and GATA3; ILC3: CD127, IL-23R, IL-22, and RORC **(A)**. The cell surface and intracellular markers were measured by flow cytometry. One hundred thousand cells were counted and the data expressed by number of positive cells **(B)**. Data are expressed as means ± SEM and are representative of three independent experiments using five mice of each mouse strain per group (*$p < 0.05$; **$p < 0.01$, and ***$p < 0.001$).

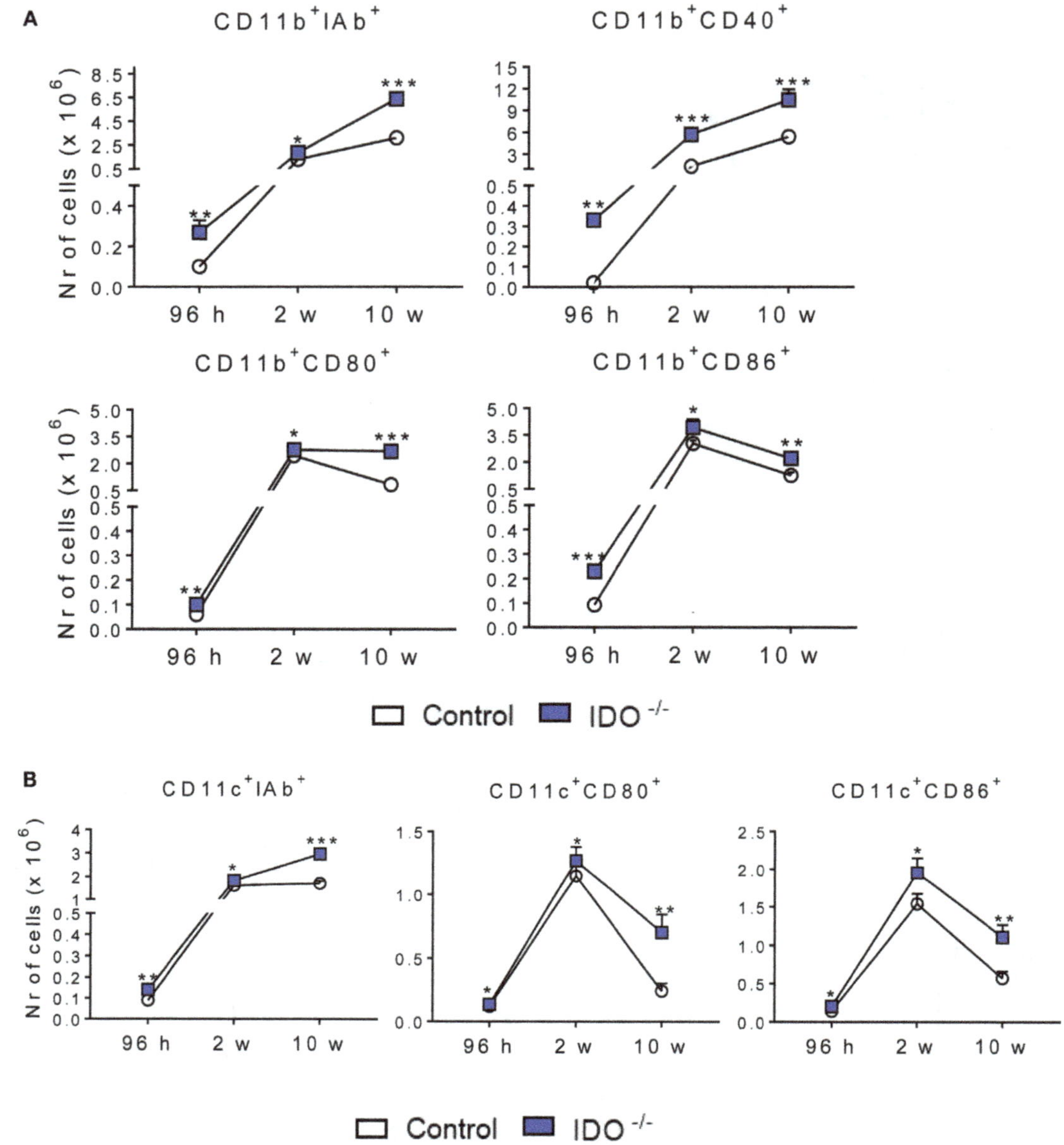

FIGURE 3 | Indoleamine 2,3 dioxygenase-1 (IDO1) controls the expression of activation markers of pulmonary CD11b⁺ and CD11c⁺ cells. The activation markers of CD11b⁺ **(A)** and CD11c⁺ **(B)** leukocytes were measured in lung infiltrating leukocytes of C57Bl/6 wild-type and IDO1$^{-/-}$ after 96 h, 2, and 10 weeks of infection with 1×10^6 *Paracoccidioides brasiliensis* yeasts. The lung cells were obtained as described in Section "Materials and Methods" and labeled with antibodies conjugated to different fluorochromes. The lung infiltrating leukocytes were gated by FSC/SSC analysis. The cells were gated for CD11b⁺ or CD11c⁺ expression and then for the presence of IA^b, CD40, CD80, and CD86. One hundred thousand cells were acquired on FACS CANTO II and, subsequently, analyzed by FlowJo software. Data are expressed as means ± SEM and are representative of three independent experiments using five mice of each mouse strain per group (*$p < 0.05$; **$p < 0.01$, and ***$p < 0.001$).

CD8⁺ CD44^{high} CD62L^{low}, respectively) were detected in the lungs of IDO1$^{-/-}$ mice when compared with WT controls (**Figure 6**).

IDO1 Controls the Secretion of Pulmonary Cytokines

Lung homogenates were obtained from infected IDO1$^{-/-}$ and control WT mice at 96 h, 2, and 10 weeks of infection. Cytokines were measured by ELISA. Similar patterns of cytokines production were seen at all postinfection periods studied. Most cytokines (TNF-α, IL-1β, IFN-γ, TGF-β, IL-27, IL-10, and IL-35) appeared in decreased levels, whereas IL-2, IL-6, IL-17, and IL-22 were detected in higher concentrations than those detected in WT mice (**Figure 7**).

IDO1 Influences the Expression of Cytokines and Transcription Factors Genes

We have also assayed the gene expression of transcription factors and cytokines involved in Th1, Th2, Th17, Th22, and Treg differentiation as well as the transcription of AhR gene

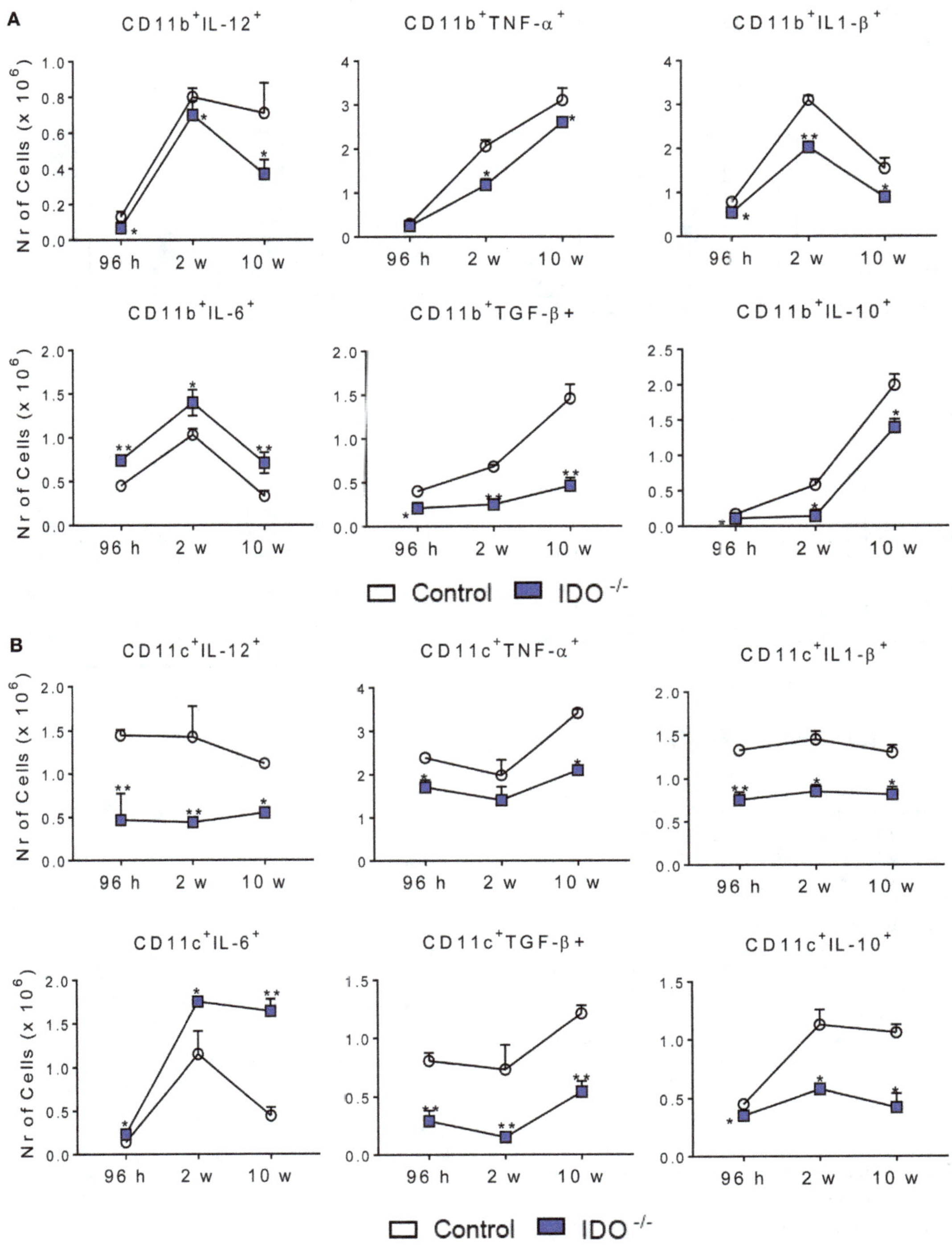

FIGURE 4 | Indoleamine 2,3 dioxygenase-1 (IDO1) expression controls the presence of intracellular cytokines in pulmonary CD11b+ and CD11c+ cells. The intracellular cytokines were determined in CD11b+ **(A)** and CD11c+ **(B)** lung infiltrating leukocytes of wild-type and IDO1−/− C57BL/6 mice after 96 h, 2, and 10 weeks of infection with 1 × 10^6 *Paracoccidioides brasiliensis* yeasts. The lung cells were obtained as described in Section "Materials and Methods" and labeled with antibodies conjugated to different fluorochromes. The lung infiltrating leukocytes were gated by FSC/SSC analysis. The cells were gated for CD11b+ or CD11c+ expression and then for the presence of cytokines IL-12, TNF-α, IL-1β, IL-10, IL-6, and TGF-β. One hundred thousand cells were acquired on FACS CANTO II and subsequently analyzed by FlowJo software. Data are expressed as means ± SEM and are representative of three independent experiments using five mice of each mouse strain per group (*$p < 0.05$; **$p < 0.01$, and ***$p < 0.001$).

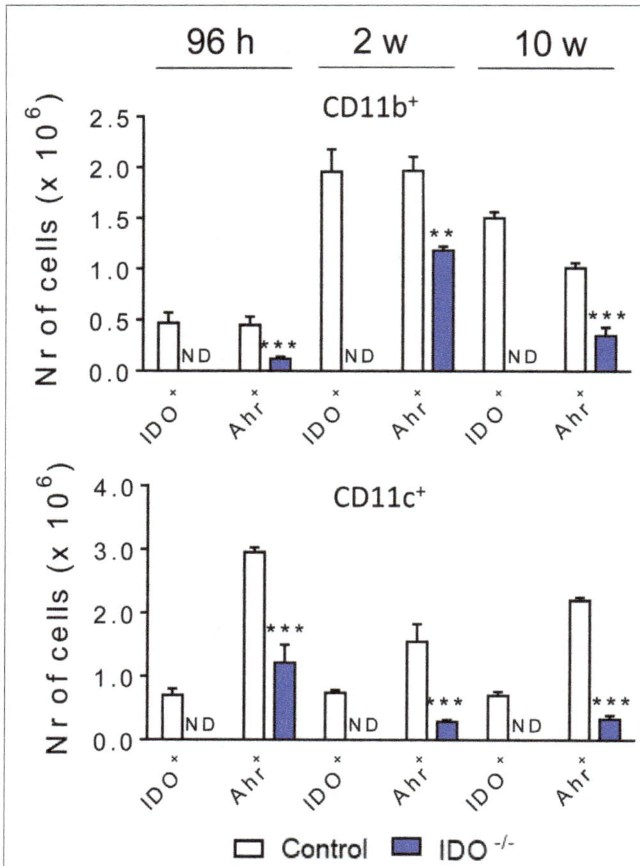

FIGURE 5 | Indoleamine 2,3 dioxygenase-1 (IDO1) expression controls the levels of aryl hydrocarbon receptor (AhR) present in CD11b+ e CD11c+ cells. The number of IDO1 and AhR expressing cells was determined in lung infiltrating leukocytes of wild-type and IDO1$^{-/-}$ mice 96 h, 2, and 10 weeks after infection with 1 × 10^6 viable *Paracoccidioides brasiliensis* yeasts. The lung cells were obtained as described in Section "Materials and Methods" and labeled with antibodies conjugated to different fluorochromes. The lung infiltrating leukocytes were gated by FSC/SSC analysis. The cells were gated for CD11b+ or CD11c+ and then for IDO1 and AhR expression. One hundred thousand cells were acquired on FACS CANTO II and subsequently analyzed by FlowJo software. Data are expressed as means ± SEM and are representative of three independent experiments using five mice of each mouse strain per group (**$p < 0.01$, and ***$p < 0.001$).

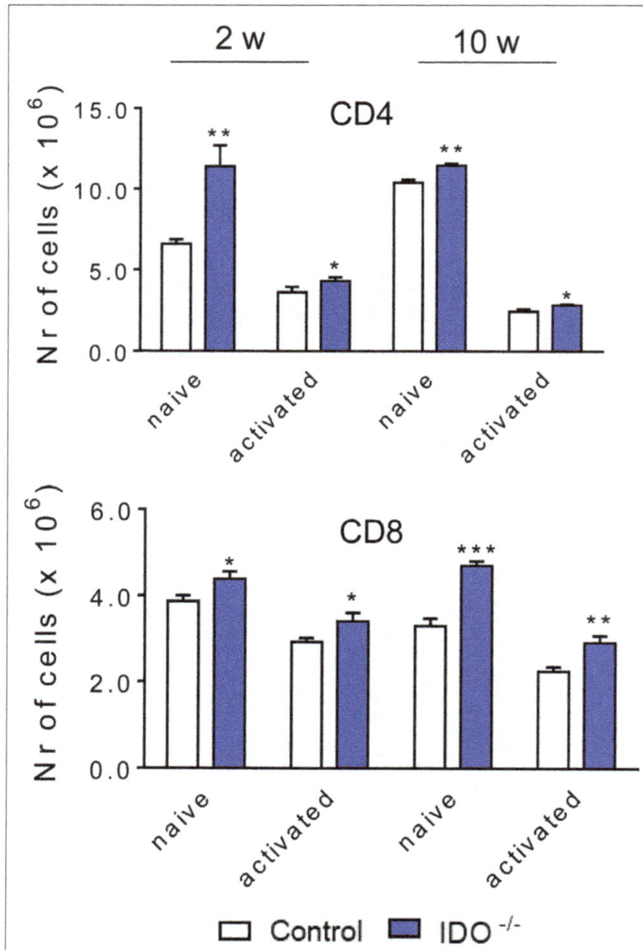

FIGURE 6 | The absence of indoleamine 2,3 dioxygenase-1 (IDO1) induces an increased influx of naïve and activated TCD4+ and TCD8+ lymphocytes to the lungs. The phenotypic analysis of lung infiltrating lymphocytes from wild-type and IDO1$^{-/-}$ mice was performed at 2 and 10 weeks of *Paracoccidioides brasiliensis* infection. The lung cells were obtained as described in Section "Materials and Methods" and labeled with antibodies conjugated to different fluorochromes. The lung infiltrating leukocytes were gated by FSC/SSC analysis. The cells were then gated for CD4+ or CD8+ expression and for membrane markers that characterize naïve and effector/activated cells (CD44^{low} CD62L^{high} and CD44^{high} CD62L^{low}, respectively). For gate strategy, see Figures S1 and S2 in Supplementary Material. One hundred thousand cells were acquired on FACS CANTO II and subsequently analyzed by FlowJo software. Data are expressed as means ± SEM and are representative of three independent experiments using five mice of each mouse strain per group (*$p < 0.05$; **$p < 0.01$, and ***$p < 0.001$).

associated with IDO1 activity. Thus, the mRNA expression in whole lung cells of WT and IDO1$^{-/-}$ mice was determined at weeks 2 and 10 after infection by RT-PCR using appropriate primers. As shown in **Figure 8**, the absence of the IDO1 led to a marked decrease in the relative expression of IFN-γ mRNA, a cytokine associated with the induction and catalytic function of the enzyme. The mRNA expression for IFN-γ, TNF-α, TGF-β, and IL-10 also appeared in decreased levels, whereas IL-6, IL-17 and IL-22 mRNAs were upregulated in IDO1$^{-/-}$ mice. In both postinfection periods, a concomitant decrease in mRNA for AhR and Foxp3 transcription factors were observed in IDO1 deficient mice. By contrast, RORC and GATA3 mRNAs were significantly upregulated IDO1$^{-/-}$ mice, but minor differences were seen in Tbet expression.

The Absence of IDO1 Increases the Number of Th17 but Reduces the Presence of Th1 and Treg Cells

It has been demonstrated in fungal infections the fundamental role of IDO1 in promoting the balance between T cell subpopulations (8, 9). IDO1 and Kyn, have the ability of inducing Treg and inhibit Th17 cells, contributing to the regulatory mechanisms that govern immunity or tolerance to pathogens (1, 2, 10). Therefore, we have also characterized the Th subpopulations in the course of

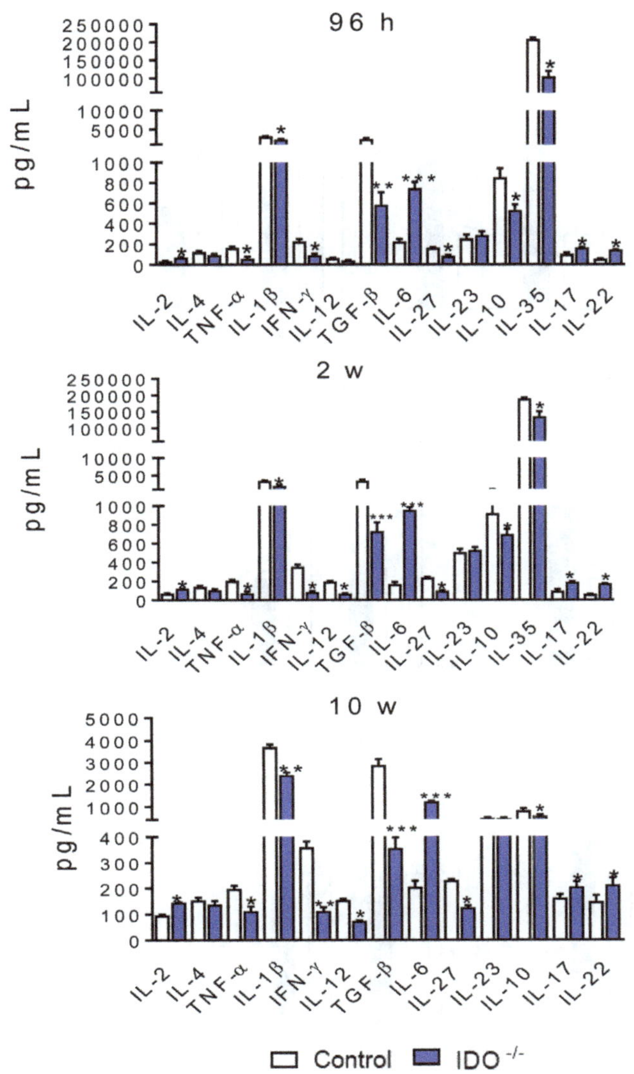

FIGURE 7 | Indoleamine 2,3 dioxygenase-1 (IDO1) regulates the levels of pulmonary cytokines. Quantitation of cytokines in lung homogenates of wild-type and IDO1$^{-/-}$ mice infected with 1 × 10^6 viable *Paracoccidioides brasiliensis* yeasts. The levels of cytokines were measured by enzyme-linked immunosorbent assay in lung homogenates obtained at 96 h, 2, and 10 weeks of infection. Bars show mean ± SEM from at least four mice per group and are representative of two independent experiments (*$p < 0.05$, **$p < 0.01$, and ***$p < 0.001$).

P. brasiliensis infection of IDO1-deficient and -sufficient mice. As depicted in **Figure 9A**, the IDO1 enzyme consistently regulates the differentiation of Th lymphocytes. In the absence of IDO1, decreased numbers of CD4$^+$ IFN-γ$^+$ T cells, which define the Th1 subset, were found. Regarding the Th2 subpopulation, only at the 10th week postinfection, a significant increase in the number of CD4$^+$ IL-4$^+$ cells were detected in the lungs of IDO1$^{-/-}$ mice. After 2 and 10 weeks of infection, CD4$^+$ IL-17$^+$ cells appeared in higher numbers in the lungs of IDO1$^{-/-}$ mice when compared with their WT controls. Furthermore, a marked reduction in pulmonary Treg (CD4$^+$ CD25$^+$ Foxp3$^+$) cells was observed in IDO1$^{-/-}$ mice at weeks 2 and 10 of infection (**Figure 9B**).

DISCUSSION

The characterization of the infectious environment is particularly important to define how the innate immunity influences the subsequent development of adaptive immunity. The regulatory mechanisms that operate at both phases of immunity are also fundamental to determine how the immune response will dictate the severity of an infectious process. In this context, we have explored the regulatory mechanism exerted by IDO1 in both phases of immunity against *P. brasiliensis* infection. It was quite interesting to verify the close association between T cell phenotypes and those of ILCs, a class of innate cells present at the site of infection at the early phase of infection that plays an important role in the further definition of adaptive immunity (34). We could demonstrate for the first time that ILC participate in the immune response against pulmonary PCM. The expansion of these ILCs was influenced by the IDO1 activity that controls the initial cytokines microenvironment and the induction of transcription factors that determine ILC and T cell phenotypes. Reduced Eomes and Tbet led, respectively, to diminished NKp46 and ILC1, whereas increased RORC resulted in increased ILC3, including those associated with increased production of IL-22, despite the reduced AhR mRNA and protein, which were previously shown to be involved in the synthesis of IL-22 (15).

Previous studies demonstrated that IDO1 plays an important role in the immunity against commensal fungi such as *Aspergillus fumigatus* and *Candida albicans* by regulating fungal tolerance mediated by prevalent activity of IDO and Treg cells and fungal immunity mediated by Th1/Th17 cells properly controlled by Treg cells (24, 35, 36). Here, we demonstrated that in pulmonary PCM, IDO1 controls both, fungal burdens and inflammatory reactions, functions that were concomitant with the expression of AhR. The absence of IDO1 expression increases fungal burdens potentiating the expression of activation markers in macrophages and DCs resulting in exacerbated Th17 immunity poorly controlled by Treg cells. This is the worst scenario for a host response where the uncontrolled fungal growth is accompanied by intense but inefficient inflammatory reactions causing exaggerated tissue pathology that ultimately leads to the precocious death of infected mice.

In vaginal candidiasis, the selective deficiency of IDO1 significantly increased fungal loads and tissue damage mediated by inflammatory cells (37) and here a similar result was observed. Our finding of elevated fungal loads may be attributed to the inability of IDO1$^{-/-}$ mice to deplete Trp, which possibly contributed to the enhanced *P. brasiliensis* growth. However, the more severe infection of IDO1$^{-/-}$ mice was accompanied by increased levels of NO, a well-recognized fungicidal mediator in *P. brasiliensis* infection (38–44). Actually, a dual role regarding NO activity has been observed in PCM. It can enhance the killing ability of IFN-γ activated macrophages exerting a protective effect (40–42) but is also suppressive as previously demonstrated (39, 43, 44). Interestingly, Bernardino et al. using iNOS$^{-/-}$ mice clearly showed that absence of NO production is accompanied by an early reduction in fungal loads and increased T cell responses. At the chronic phase of the disease, iNOS$^{-/-}$ mice showed increased numbers of viable yeasts circumscribed, however, by elevated presence of activated

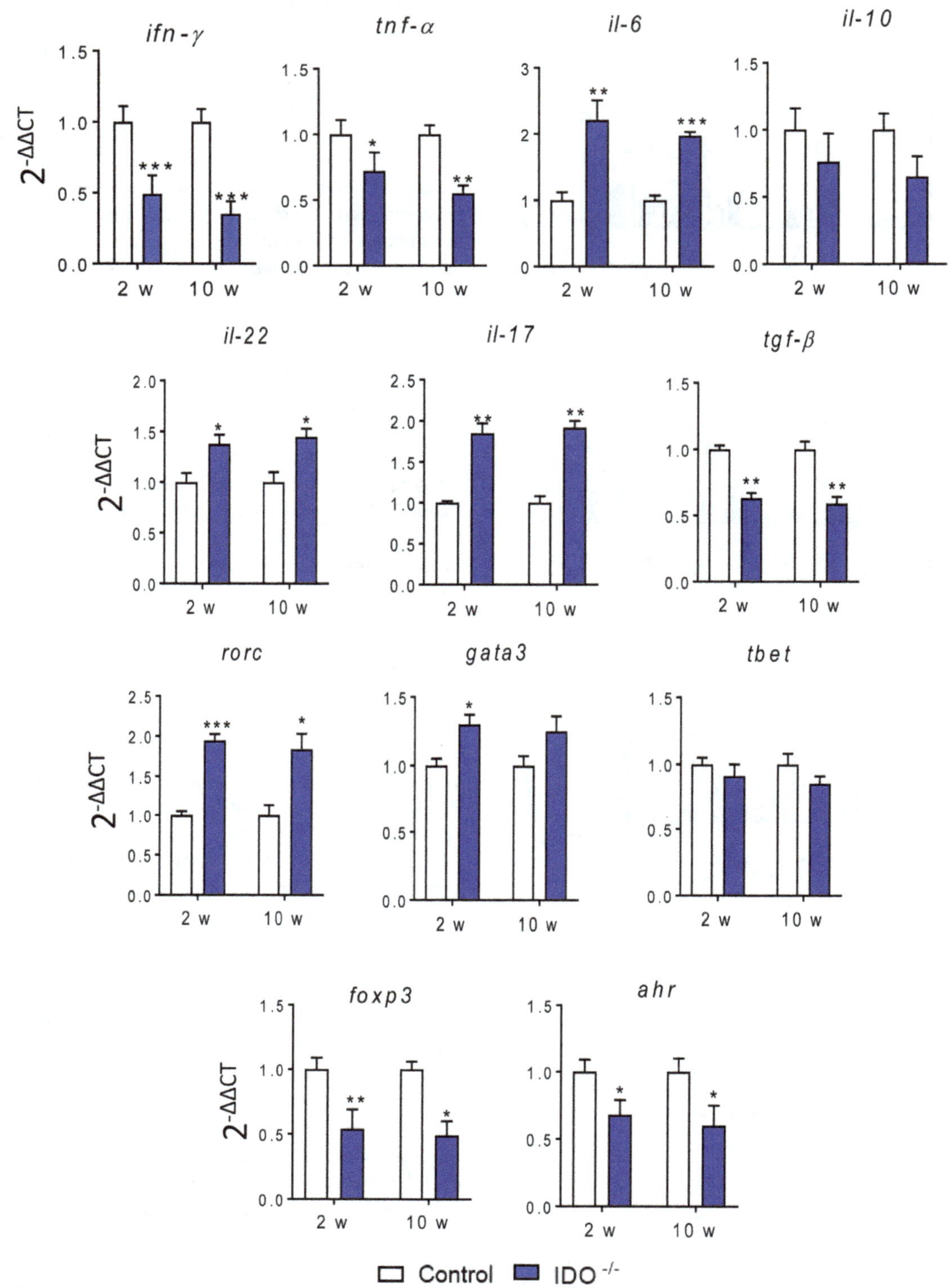

FIGURE 8 | Indoleamine 2,3 dioxygenase-1 (IDO1) influences the gene expression of cytokines and transcription factors. Relative expression of mRNA for aryl hydrocarbon receptor (AhR), IFN-γ, TNF-α, IL-6, RORC, Tbet, GATA3, FoxP3, IL-10, TGF-β, IL-17, and IL-22 in whole lung cells of wild-type and IDO1$^{-/-}$ mice after 10 weeks of *Paracoccidioides brasiliensis* infection. The level of gene transcription was determined by real-time PCR. Bars show mean ± SEM from at least four mice per group and are representative of three independent experiments (*$p < 0.05$; **$p < 0.01$, and ***$p < 0.001$).

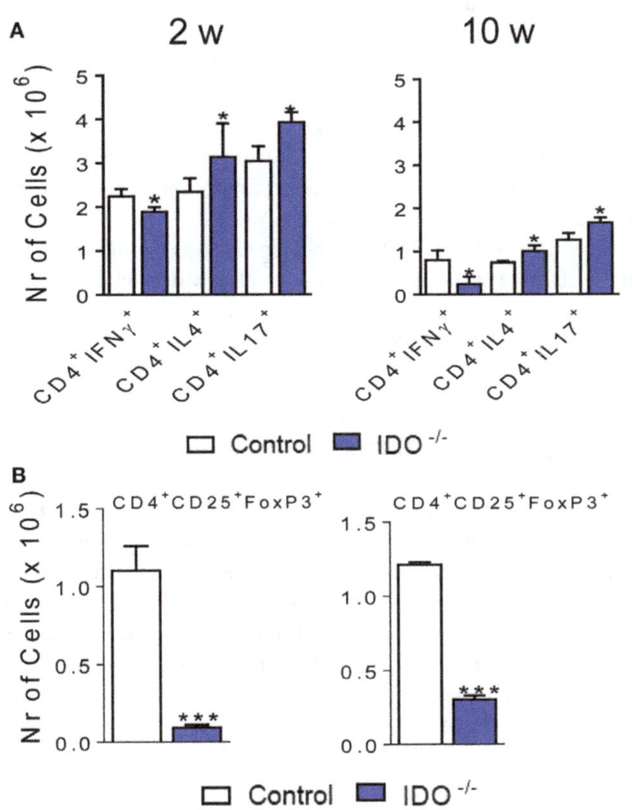

FIGURE 9 | The absence of indoleamine 2,3 dioxygenase-1 (IDO1) increases the number of pulmonary Th17 but reduces the presence of Th1 and regulatory T (Treg) cells. The phenotypic analysis of lung infiltrating lymphocytes from wild-type and IDO1$^{-/-}$ mice was performed at 2 and 10 weeks of *Paracoccidioides brasiliensis* infection. The lung cells were obtained as described in Section "Materials and Methods" and labeled with antibodies conjugated to different fluorochromes. The lung infiltrating leukocytes were gated by FSC/SSC analysis. **(A)** The cells were gated for CD4$^+$ and then for intracellular expression of IL-4, IL-17, IL-22, or IFN-γ. **(B)** For Treg cells characterization, the FoxP3$^+$ cells were counted on CD4$^+$ CD25$^+$ double-positive cells. For gate strategy, see Figure S3 in Supplementary Material. One hundred thousand cells were acquired on FACS CANTO II and subsequently analyzed by FlowJo software. Data are expressed as means ± SEM and are representative of three independent experiments using five mice of each mouse strain per group (*$p < 0.05$ and ***$p < 0.001$).

macrophages and T cells (39). Importantly, the inducible isoform of nitric oxide synthase (iNOS) responsible for NO production and the enzyme IDO1 are induced by IFN-γ (45, 46), but NO and IDO1 are mutually inhibitory (47–49). This reciprocal regulation was here detected, since the absence of IDO1 expression enhanced NO production that, however, was insufficient to control the fungal growth. These results were consistent with our previous studies in resistant and susceptible mice where increased NO synthesis paralleled the IDO1 inhibition by 1MT (50). More than that, these results demonstrate that, contrary to the well accepted paradigm, in murine PCM the control of fungal growth by IDO1 activity is more relevant than that mediated by NO production.

At all postinfection periods analyzed, cells from IDO1$^{-/-}$ mice showed increased levels of activation molecules that can explain the increased expansion of T cell that migrate to the lungs of IDO1$^{-/-}$ mice. Despite its statistical significance, it is difficult to point out the biological relevance of the increased expression of costimulatory molecules by DC11b$^+$ cells. However, our previous report (26) showed that CD11c$^+$ cells from IDO$^{-/-}$ mice have an enhanced ability of inducing T cell proliferation. DCs from IDO$^{-/-}$ mice have an increased ability of expanding CD4$^+$ and CD8$^+$ T lymphocytes in parallel with decreased proliferation of Treg cells (CD4$^+$ CD25$^+$ Foxp3$^+$ T cells). This study has also shown that the CD4$^+$ and CD8$^+$ T cells exhibited an activated phenotype as revealed by the expression of CD25 and DC69. Interestingly, both CD11b$^+$ and CD11c$^+$ cells showed significantly reduced levels of pro- and anti-inflammatory cytokines, in parallel with increased presence of IL-6$^+$ cells. These effects were more prominent in CD11c$^+$ than in CD11b$^+$ cells as expected by the major expression and activity of IDO1 by the former leukocyte subpopulation (2, 51). Furthermore, the new cytokines balance in CD11b$^+$ and CD11c$^+$ cells of IDO1$^{-/-}$ mice led to increased proliferation of ILC3 and Th17 lymphocytes that are known to be expanded by IL-6 and TGF-β signaling, STAT3 phosphorylation, and expression of RORγτ, the molecular signature of IL-17 secreting cells.

It has been reported that IDO1 and AhR expression are mutually regulated (1, 52), and here we could demonstrate the profound impairment of AhR production caused by IDO1 deficiency. These findings can explain the increased inflammatory reactions developed by IDO1$^{-/-}$ mice and the phenotype of T cells and ILC developed by *P. brasiliensis*-infected IDO1$^{-/-}$ mice: increased presence of ILC3 and Th17 cells associated with reduced numbers of Treg cells.

The elevated numbers of naïve and activated lymphocytes found in the lung of IDO1$^{-/-}$ mice were consistent with the higher levels of IL-2 produced by IDO1$^{-/-}$ mice, a fact also observed in our studies with IDO1 inhibition by 1MT (50). In addition, the decreased levels of IL-12, TNF-α, and IFN-γ reflect the reduced expansion of Th1 cells as shown by the low numbers of CD4$^+$ IFN-γ$^+$ lymphocytes present in the lungs of IDO$^{-/-}$ mice. The reduced levels of IL-10, TGF-β, and IL-35 are also concordant with the drastic reduction in the expansion and migration of Treg cells to the lungs of IDO1$^{-/-}$ mice. Furthermore, the blockade of the IDO/AhR axis increased the levels of IL-17 and IL-22 secreted into the lungs parenchyma and reflected the major development of Th17 cells that have the ability of secreting both cytokines. The same shift for the IL-17/IL-22 production was reported by De Luca et al. (37) studying the effect of genetic IDO1 deficiency in vaginal candidiasis.

It is apparently conflicting the elevated levels of IL-22 and reduced presence of AhR observed in the lungs of IDO1$^{-/-}$-infected mice. IL-22 belongs to the IL-10 family of cytokines, is produced by many cell subpopulations, including Th22, Th1, and Th17, classical and non-classical NK cells (NK-22), NKT cells, and ILC3 (15, 53–58). IL-22 plays an important role in fungal infections due to its ability of epithelial tissue repair and the induction of antimicrobial peptides (37, 59–61). Several transcription factors, including STAT3, RORC, and AhR, have been described as essential regulators of IL-22 (15) and AhR$^{-/-}$ mice are highly compromised in terms of IL-22 production (15, 62). By inducing

IL-22-producing ILC, AhR can promote initial fungal clearance, and by expanding Treg cells contributes with disease tolerance but both mechanisms cooperate to maintain host homeostasis (37). The increase in IL-22 here observed could not be ascribed to the effect of AhR on T cells but to the increased influx of IL-22-secreting Th17 and ILC3 into the lungs of IDO1$^{-/-}$ mice. Indeed, besides Th17 cells, an increased presence of ILC3 positive for intracellular IL-22 were seen in the whole course of infection of IDO1$^{-/-}$ mice.

The increased number of effector T cells in the lung cell infiltrates of IDO1-deficient mice was concurrent with a vigorous reduction of CD4^{+} CD25^{+} FoxP3^{+} Treg cells. This decreased expansion can be attributed to the impaired production of Kyn and the reduced activity of AhR, a transcription factor that needs its ligand to upregulate the transcription of FoxP3 necessary to Treg cells differentiation (63, 64). Indeed, the IDO1 enzyme and its catalytic products can act as a bridge between DCs and Treg cells that can use reverse signaling and non-canonical NFκB activation of DCs for its suppressive effector function and self-propagation (3). The low numbers of Treg cells were concomitant with the reduced levels of several suppressive cytokines detected in the lung cell infiltrates of IDO1$^{-/-}$ mice. IL-27 has a wide range of immunomodulatory activities. Although it may promote the development of Th1, IL-27 can suppress T cell responses and the development of pathogenic Th17 cells (65, 66), functions that were possibly impaired in the immune response developed by IDO1$^{-/-}$ mice. Besides IL-27, the low levels of typical Treg cytokines (TGF-β, IL-35, and IL-10) possibly contributed to the enhanced Th17 inflammatory reactions developed by IDO1-deficient mice.

The IDO1 activity has been associated with the tolerogenic function of plasmacytoid DCs (26), a phenomenon also described in our pulmonary model of PCM. We verified that *P. brasiliensis*-infected pDCs enhanced the expansion of Treg cells and reduce the expansion and activation of Th1 and Th17 cells. This finding was not detected when pDCs from IDO1$^{-/-}$ mice were used. More importantly, the *in vivo* depletion of pDC by a specific antibody resulted in reduced levels of IDO1 and regulatory cytokines in the lungs of pDC-depleted mice, once more highlighting the close connection between IDO1 and expression of immunosuppressive cytokines in PCM (26).

Although no studies have directly addressed the role of IL-17 in PCM, several indirect findings have clearly demonstrated that the well-balanced production of pro- and anti-inflammatory cytokines is crucial for establishing protective immunity. Indeed, the absence of TLR2 signaling induces an excessive tissue pathology associated with increased Th17 differentiation and impaired Treg development (67). By contrast, in the case of TLR4 deficiency, the excessive proliferation of Treg cells and reduced Th17 differentiation led to more severe disease due to uncontrolled fungal burden (68). In addition, the immunity mediated by the adoptive transfer of effector T cells without the concomitant presence of Treg cells led to increased Th17 differentiation and tissue pathology (69). These previous findings and those reported here clearly indicate that the balanced differentiation of Th1/Th17/Treg cells is fundamental for achieving protection in pulmonary PCM.

In conclusion, the IDO1–Ahr–Treg axis was strongly impaired in the absence of the IDO1 expression clearly demonstrating that this regulatory loop plays an important role in the control of immunity and severity of pulmonary PCM.

ETHICS STATEMENT

The experiments were performed in strict accordance with the Brazilian Federal Law 11,794 establishing procedures for the scientific use of animals, and the State Law establishing the Animal Protection Code of the State of São Paulo. All efforts were made to minimize animal suffering. The procedures were approved by the Ethics Committee on Animal Experiments of the Institute of Biomedical Sciences of University of São Paulo (Proc.180/11/CEEA).

AUTHOR CONTRIBUTIONS

Conceived and designed experiments: EA, VC, and FL. Contributed with reagent: FL and VC. Performed the experiments: EA, CF, NG, NP, and FL. Analyzed the data: CF, VC, FL, and EA. Wrote the paper: EA, NP, VC, and FL.

ACKNOWLEDGMENTS

We are grateful to Tânia A. Costa for her invaluable technical assistance. We thank Paulo Albe for processing the histological samples.

FUNDING

This work was supported by a grant from the Fundação de Amparo à Pesquisa do Estado de São Paulo (FAPESP-grant to VC 2011/51258-2 and fellowship to EA 2014/18668-2; grant to FL 2014/04783-2) and Conselho Nacional de Pesquisas (CNPq).

REFERENCES

1. Grohmann U, Fallarino F, Puccetti F, Tolerance P. DCs and tryptophan: much ado about IDO. *Trends Immunol* (2003) 24:242–8. doi:10.1016/S1471-4906(03)00072-3
2. Mellor AL, Munn DH. IDO expression by dendritic cells: tolerance and tryptophan catabolism. *Nat Rev Immunol* (2004) 4:762–74. doi:10.1038/nri1457
3. Puccetti P, Grohmann U. IDO and regulatory T cells: a role for reverse signalling and non-canonical NF-kappaB activation. *Nat Rev Immunol* (2007) 7:817–23. doi:10.1038/nri2163
4. Terness P, Bauer TM, Röse L, Dufter C, Watzlik A, Simon H, et al. Inhibition of allogeneic T cell proliferation by indoleamine 2,3-dioxygenase-expressing dendritic cells: mediation of suppression by tryptophan metabolites. *J Exp Med* (2002) 196:447–57. doi:10.1084/jem.20020052

5. Hayashi T, Mo JH, Gong X, Rossetto C, Jang A, Beck L, et al. 3-Hydroxyanthranilic acid inhibits PDK1 activation and suppresses experimental asthma by inducing T cell apoptosis. *Proc Natl Acad Sci U S A* (2007) 104:18619–24. doi:10.1073/pnas.0709261104
6. Grohmann U, Puccetti P. The coevolution of IDO1 and AhR in the emergence of regulatory T-cells in mammals. *Front Immunol* (2015) 6:58. doi:10.3389/fimmu.2015.00058
7. Platten M, Litzenburger U, Wick W. The aryl hydrocarbon receptor in tumor immunity. *Oncoimmunology* (2012) 1:396–7. doi:10.4161/onci.19071
8. Romani L, Puccetti P. Protective tolerance to fungi: the role of IL-10 and tryptophan catabolism. *Trends Microbiol* (2006) 14:183–9. doi:10.1016/j.tim.2006.02.003
9. Romani L, Puccetti P. Controlling pathogenic inflammation to fungi. *Expert Rev Anti Infect Ther* (2007) 5:1007–17. doi:10.1586/14787210.5.6.1007
10. Popov A, Schultze JL. IDO-expressing regulatory dendritic cells in cancer and chronic infection. *J Mol Med* (2008) 86:145–60. doi:10.1007/s00109-007-0262-6
11. Bettelli E, Carrier Y, Gao W, Korn T, Strom TB, Oukka M, et al. Reciprocal developmental pathways for the generation of pathogenic effector TH17 and regulatory T cells. *Nature* (2006) 441:235–8. doi:10.1038/nature04753
12. Mangan PR, Harrington LE, O'Quinn DB, Helms WS, Bullard DC, Elson CO, et al. Transforming growth factor-b induces development of the Th17 lineage. *Nature* (2006) 441:231–4. doi:10.1038/nature04754
13. Mucida D, Park Y, Kim G, Turovskaya O, Scott I, Kronenberg M, et al. Reciprocal Th17 and regulatory T cell differentiation mediated by retinoic acid. *Science* (2007) 317:256–60. doi:10.1126/science.1145697
14. Quintana FJ, Basso AS, Iglesias AH, Korn T, Farez MF, Bettelli E, et al. Control of Treg and TH17 cell differentiation by the aryl hydrocarbon receptor. *Nature* (2008) 453:65–71. doi:10.1038/nature06880
15. Veldhoen M, Hirota K, Westendorf AM, Buer J, Dumoutier L, Renauld JC, et al. The aryl hydrocarbon receptor links TH17-cell-mediated autoimmunity to environmental toxins. *Nature* (2008) 453:106–9. doi:10.1038/nature06881
16. Marques AS. Paracoccidioidomycosis: epidemiological, clinical, diagnostic and treatment up-dating. *An Bras Dermatol* (2013) 88:700–11. doi:10.1590/abd1806-4841.20132463
17. De Castro LF, Ferreira MC, da Silva RM, Blotta MH, Longhi LN, Mamoni RL. Characterization of the immune response in human paracoccidioidomycosis. *J Infect* (2013) 67:470–85. doi:10.1016/j.jinf.2013.07.019
18. Shikanai-Yasuda MA, Telles FQ, Mendes RP, Colombo AL, Moretti ML. Guidelines in paracoccidioidomycosis. *Rev Soc Bras Med Trop* (2006) 39:297–310. doi:10.1590/S0037-86822006000300017
19. Cano LE, Singer-Vermes LM, Vaz CA, Russo M, Calich VL. Pulmonary paracoccidioidomycosis in resistant and susceptible mice: relationship among progression of infection, bronchoalveolar cell activation, cellular immune response, and specific isotype patterns. *Infect Immun* (1995) 63:1777–83.
20. Cano LE, Kashino SS, Arruda C, André D, Xidieh CF, Singer-Vermes LM, et al. Protective role of gamma interferon in experimental pulmonary paracoccidioidomycosis. *Infect Immun* (1998) 66:800–6.
21. Cano LE, Singer-Vermes LM, Costa TA, Mengel JO, Xidieh CF, Arruda C, et al. Depletion of CD8(+) T cells in vivo impairs host defense of mice resistant and susceptible to pulmonary paracoccidioidomycosis. *Infect Immun* (2000) 68:352–9. doi:10.1128/IAI.68.1.352-359.2000
22. Chiarella AP, Arruda C, Pina A, Costa TA, Ferreira RC, Calich VL. The relative importance of CD4+ and CD8+T cells in immunity to pulmonary paracoccidioidomycosis. *Microbes Infect* (2007) 9:1078–88. doi:10.1016/j.micinf.2007.04.016
23. Calich VL, Costa TA, Felonato M, Arruda C, Bernardino S, Loures FV, et al. Innate immunity to *Paracoccidioides brasiliensis* infection. *Mycopathologia* (2008) 165:223–36. doi:10.1007/s11046-007-9048-1
24. Romani L. Immunity to fungal infections. *Nat Rev Immunol* (2011) 11:275–88. doi:10.1038/nri2939
25. Zelante T, Fallarino F, Bistoni F, Puccetti P, Romani L. Indoleamine 2,3-dioxygenase in infection: the paradox of an evasive strategy that benefits the host. *Microbes Infect* (2009) 11:133–41. doi:10.1016/j.micinf
26. Araújo EF, Medeiros DH, Galdino NA, Condino-Neto A, Calich VL, Loures FV. Tolerogenic plasmacytoid dendritic cells control *Paracoccidioides brasiliensis* infection by inducting regulatory T cells in an IDO-dependent manner. *PLoS Pathog* (2016) 12:e1006115. doi:10.1371/journal.ppat.1006115
27. Pallotta MT, Orabona C, Volpi C, Vacca C, Belladonna ML, Bianchi R, et al. Indoleamine 2,3-dioxygenase is a signaling protein in long-term tolerance by dendritic cells. *Nat Immunol* (2011) 12:870–8. doi:10.1038/ni.2077
28. Singer-Vermes LM, Caldeira CB, Burger E, Calich VLG. Experimental murine paracoccidioidomycosis: relationship among dissemination of the infection, humoral and cellular responses. *Clin Exp Immunol* (1993) 94:75–9. doi:10.1111/j.1365-2249.1993.tb05980.x
29. Loures FV, Pina A, Felonato M, Feriotti C, Araújo EF, Calich VL. MyD88 signaling is required for efficient innate and adaptive immune responses to *Paracoccidioides brasiliensis* infection. *Infect Immun* (2011) 79:2470–80. doi:10.1128/IAI.00375-10
30. Huntington ND, Carpentier S, Vivier E, Belz GT. Innate lymphoid cells: parallel checkpoints and coordinate interactions with T cells. *Curr Opin Immunol* (2016) 38:86–93. doi:10.1016/j.coi.2015.11.008
31. Spits H, Artis D, Colonna M, Diefenbach A, Di Santo JP. Innate lymphoid cells – a proposal for uniform nomenclature. *Nat Rev Immunol* (2013) 13:145–9. doi:10.1038/nri3365
32. Lee JS, Cella M, McDonald KG, Garlanda C, Kennedy GD, Nukaya M, et al. AHR drives the development of gut ILC22 cells and postnatal lymphoid tissues via pathways dependent on and independent of Notch. *Nat Immunol* (2011) 13:144–51. doi:10.1038/ni.2187
33. Stockinger B, Hirota K, Duarte J, Veldhoen M. External influences on the immune system via activation of the aryl hydrocarbon receptor. *Semin Immunol* (2011) 23:99–105. doi:10.1016/j.smim.2011.01.008
34. Gronke K, Kofoed-Nielsen M, Diefenbach A. Innate lymphoid cells, precursors and plasticity. *Immunol Lett* (2016) 9:9–18. doi:10.1016/j.imlet.2016.07.004
35. Romani L, Zelante T, De Luca A, Bozza S, Bonifazi P, Moretti S, et al. Indoleamine 2,3-dioxygenase (IDO) in inflammation and allergy to *Aspergillus*. *Med Mycol* (2009) 1:154–61. doi:10.1080/13693780802139867
36. Bonifazi P, Zelante T, D'Angelo C, De Luca A, Moretti S, Bozza S, et al. Balancing inflammation and tolerance in vivo through dendritic cells by the commensal *Candida albicans*. *Mucosal Immunol* (2009) 2:362–74. doi:10.1038/mi.2009.17
37. De Luca A, Carvalho A, Cunha C, Iannitti RG, Pitzurra L, Giovannini G, et al. IL-22 and IDO1 affect immunity and tolerance to murine and human vaginal candidiasis. *PLoS Pathog* (2013) 9:e1003486. doi:10.1371/journal.ppat.1003486
38. Brummer E, Hanson LH, Stevens DA. Gamma-interferon activation of macrophages for killing of *Paracoccidioides brasiliensis* and evidence for nonoxidative mechanisms. *Int J Immunopharmacol* (1988) 10:945–52. doi:10.1016/0192-0561(88)90041-0
39. Bernardino S, Pina A, Felonato M, Costa TA, Frank de Araújo E, Feriotti C, et al. TNF-α and CD8+ T cells mediate the beneficial effects of nitric oxide synthase-2 deficiency in pulmonary paracoccidioidomycosis. *PLoS Negl Trop Dis* (2013) 7:e2325. doi:10.1371/journal.pntd.0002325
40. Livonesi MC, Rossi MA, de Souto JT, Campanelli AP, de Sousa RL, Maffei CM, et al. Inducible nitric oxide synthase-deficient mice show exacerbated inflammatory process and high production of both Th1 and Th2 cytokines during paracoccidioidomycosis. *Microbes Infect* (2009) 11:123–32. doi:10.1016/j.micinf.2008.10.015
41. Batista AC, Soares CT, Lara VS. Failure of nitric oxide production by macrophages and decrease in CD4+ T cells in oral paracoccidioidomycosis: possible mechanisms that permit local fungal multiplication. *Rev Inst Med Trop Sao Paulo* (2005) 47:267–73. doi:10.1590/S0036-46652005000500006
42. Gonzalez A, de Gregori W, Velez D, Restrepo A, Cano LE. Nitric oxide participation in the fungicidal mechanism of gamma interferon-activated murine macrophages against *Paracoccidioides brasiliensis* conidia. *Infect Immun* (2000) 68:2546–52. doi:10.1128/IAI.68.5.2546-2552.2000
43. Bocca AL, Hayashi EE, Pinheiro AG, Furlanetto AB, Campanelli AP, Cunha FQ, et al. Treatment of *Paracoccidioides brasiliensis*-infected mice with a nitric oxide inhibitor prevents the failure of cell-mediated immune response. *J Immunol* (1998) 161:3056–63.
44. Nascimento FR, Calich VL, Rodríguez D, Russo M. Dual role for nitric oxide in paracoccidioidomycosis: essential for resistance, but overproduction associated with susceptibility. *J Immunol* (2002) 168:4593–600. doi:10.4049/jimmunol.168.9.4593
45. Geller DA, Billiar TR. Molecular biology of nitric oxide synthases. *Cancer Metastasis Rev* (1998) 17:7–23. doi:10.1023/A:1005940202801
46. Kleinert H, Schwarz PM, Förstermann U. Regulation of the expression of inducible nitric oxide synthase. *Biol Chem* (2003) 384:1343–64. doi:10.1515/BC.2003.152

47. Alberati-Giani D, Malherbe P, Ricciardi-Castagnoli P, Köhler C, Denis-Donini S, Cesura AM. Differential regulation of indoleamine 2,3-dioxygenase expression by nitric oxide and inflammatory mediators in IFN-gamma-activated murine macrophages and microglial cells. *J Immunol* (1997) 159:419–26.
48. Thomas SR, Mohr D, Stocker R. Nitric oxide inhibits indoleamine 2,3-dioxygenase activity in interferon-gamma primed mononuclear phagocytes. *J Biol Chem* (1994) 269:14457–64.
49. Hucke C, MacKenzie CR, Adjogble KD, Takikawa O, Däubener W. Nitric oxide-mediated regulation of gamma interferon-induced bacteriostasis: inhibition and degradation of human indoleamine 2,3-dioxygenase. *Infect Immun* (2004) 72:2723–30. doi:10.1128/IAI.72.5.2723-2730.2004
50. Araújo EF, Loures FV, Bazan SB, Feriotti C, Pina A, Schanoski AS, et al. Indoleamine 2,3-dioxygenase controls fungal loads and immunity in Paracoccidioidomicosis but is more important to susceptible than resistant hosts. *PLoS Negl Trop Dis* (2014) 8:e3330. doi:10.1371/journal.pntd.0003330
51. Harden JL, Egilmez NK. Indoleamine 2,3-dioxygenase and dendritic cell tolerogenicity. *Immunol Invest* (2012) 41:738–64. doi:10.3109/08820139.2012.676122
52. Vogel CF, Goth SR, Dong B, Pessah IN, Matsumura F. Aryl hydrocarbon receptor signaling mediates expression of indoleamine 2,3-dioxygenase. *Biochem Biophys Res Commun* (2008) 375:331–5. doi:10.1016/j.bbrc.2008.07.156
53. Witte E, Witte K, Warszawska K, Sabat R, Wolk K. Interleukin-22: a cytokine produced by T, NK and NKT cell subsets, with importance in the innate immune defense and tissue protection. *Cytokine Growth Factor Rev* (2010) 21:365–79. doi:10.1016/j.cytogfr.2010.08.002
54. Martin B, Hirota K, Cua DJ, Stockinger B, Veldhoen M. Interleukin-17-producing gammadelta T cells selectively expand in response to pathogen products and environmental signals. *Immunity* (2009) 31:321–30. doi:10.1016/j.immuni.2009.06.020
55. Hughes T, Becknell B, McClory S, Briercheck E, Freud AG, Zhang X, et al. Stage 3 immature human natural killer cells found in secondary lymphoid tissue constitutively and selectively express the TH 17 cytokine interleukin-22. *Blood* (2009) 113:4008–10. doi:10.1182/blood-2008-12-192443
56. Cella M, Fuchs A, Vermi W, Facchetti F, Otero K, Lennerz JK, et al. A human natural killer cell subset provides an innate source of IL-22 for mucosal immunity. *Nature* (2009) 457:722–5. doi:10.1038/nature07537
57. Crellin NK, Trifari S, Kaplan CD, Cupedo T, Spits H. Human NKp44+IL-22+ cells and LTi-like cells constitute a stable RORC+ lineage distinct from conventional natural killer cells. *J Exp Med* (2010) 207:281–90. doi:10.1084/jem.20091509
58. Trifari S, Kaplan CD, Tran EH, Crellin NK, Spits H. Identification of a human helper T cell population that has abundant production of interleukin 22 and is distinct from T(H)-17, T(H)1 and T(H)2 cells. *Nat Immunol* (2009) 10:864–71. doi:10.1038/ni.1770
59. Goupil M, Cousineau-Côté V, Aumont F, Sénéchal S, Gaboury L, Hanna Z, et al. Defective IL-17- and IL-22-dependent mucosal host response to *Candida albicans* determines susceptibility to oral candidiasis in mice expressing the HIV-1 transgene. *BMC Immunol* (2014) 15:49. doi:10.1186/s12865-014-0049-9
60. Delsing CE, Becker KL, Simon A, Kullberg BJ, Bleeker-Rovers CP, van de Veerdonk FL, et al. Th17 cytokine deficiency in patients with *Aspergillus* skull base osteomyelitis. *BMC Infect Dis* (2015) 15:140. doi:10.1186/s12879-015-0891-2
61. Borghi M, De Luca A, Puccetti M, Jaeger M, Mencacci A, Oikonomou V, et al. Pathogenic NLRP3 inflammasome activity during *Candida* infection is negatively regulated by IL-22 via activation of NLRC4 and IL-1Ra. *Cell Host Microbe* (2015) 18:198–209. doi:10.1016/j.chom.2015.07.004
62. Kiss EA, Vonarbourg C, Kopfmann S, Hobeika E, Finke D, Esser C, et al. Natural aryl hydrocarbon receptor ligands control organogenesis of intestinal lymphoid follicles. *Science* (2011) 334:1561–5. doi:10.1126/science.1214914
63. Mezrich JD, Fechner JH, Zhang X, Johnson BP, Burlingham WJ, Bradfield CA. An interaction between kynurenine and the aryl hydrocarbon receptor can generate regulatory T cells. *J Immunol* (2010) 185:3190–8. doi:10.4049/jimmunol.0903670
64. Munn DH, Mellor AL. Indoleamine 2,3 dioxygenase and metabolic control of immune responses. *Trends Immunol* (2013) 34:137–43. doi:10.1016/j.it.2012.10.001
65. Fitzgerald DC, Zhang GX, El-Behi M, Fonseca-Kelly Z, Li H, Yu S, et al. Suppression of autoimmune inflammation of the central nervous system by interleukin 10 secreted by interleukin 27-stimulated T cells. *Nat Immunol* (2007) 8:1372–9. doi:10.1038/ni1540
66. Awasthi A, Carrier Y, Peron JP, Bettelli E, Kamanaka M, Flavell RA, et al. A dominant function for interleukin 27 in generating interleukin 10-producing anti-inflammatory T cells. *Nat Immunol* (2007) 8:1380–9. doi:10.1038/ni1541
67. Loures FV, Pina A, Felonato M, Calich VL. TLR2 is a negative regulator of Th17 cells and tissue pathology in a pulmonary model of fungal infection. *J Immunol* (2009) 183:1279–90. doi:10.4049/jimmunol.0801599
68. Loures FV, Pina A, Felonato M, Araújo EF, Leite KR, Calich VL. Toll-like receptor 4 signaling leads to severe fungal infection associated with enhanced proinflammatory immunity and impaired expansion of regulatory T cells. *Infect Immun* (2010) 78:1078–88. doi:10.1128/IAI.01198-09
69. Bazan SB, Costa TA, de Araújo EF, Feriotti C, Loures FV, Pretel FD, et al. Loss- and gain-of-function approaches indicate a dual role exerted by regulatory T cells in pulmonary paracoccidioidomycosis. *PLoS Negl Trop Dis* (2015) 9:e0004189. doi:10.1371/journal.pntd.0004189

16

Immunology of Cryptococcal Infections: Developing a Rational Approach to Patient Therapy

***Waleed Elsegeiny*[1]*, Kieren A. Marr*[2] *and Peter R. Williamson*[1]*

[1] *Laboratory of Clinical Immunology and Microbiology (LCIM), National Institute of Allergy and Infectious Diseases (NIAID), National Institutes of Health (NIH), Bethesda, MD, United States,* [2] *Johns Hopkins University, Baltimore, MD, United States*

***Correspondence:**
Peter R. Williamson
williamsonpr@mail.nih.gov

Cryptococcal meningoencephalitis is responsible for upwards of 15% of HIV-related deaths worldwide and is currently the most common cause of non-viral meningitis in the US, affecting both previously healthy and people with immune suppression caused by cancer chemotherapy, transplantation, and biologic therapies. Despite a continued 30–50% attributable mortality, recommended therapeutic strategies have remained largely unchanged since the 1950s. Recent murine models and human studies examining the role of the immune system in both susceptibility to the infection as well as host damage have begun to influence patient care decisions. The Damage Framework Response, originally proposed in 1999, was recently used to discuss dichotomous etiologies of host damage in cryptococcal disease. These include patients suffering microbiological damage with low host immunity (especially those immunosuppressed with HIV) and those having low (live) microbiological burden but high immune-mediated damage (HIV-related immune reconstitution syndrome and non-HIV-related postinfectious inflammatory response syndrome). Cryptococcal disease in previously healthy hosts, albeit rare, has been known for a long time. Immunophenotyping and dendritic cell-T cell signaling studies on cerebral spinal fluid of these rare patients reveal immune capacity for recognition and T-cell activation pathways including increased levels of HLA-DR and CD56. However, despite effective T-cell signals, brain biopsy and autopsy specimens demonstrated an M2 alternative macrophage polarization and poor phagocytosis of fungal cells. These studies expand the paradigm for cryptococcal disease susceptibility to include a prominent role for immune-mediated damage and suggest a need for careful individual consideration of immune activation during therapy of cryptococcal disease in diverse hosts.

Keywords: *Cryptococcus*, cryptococcal, meningoencephalitis, meningitis, neurology, infection, fungus

INTRODUCTION

Cryptococcus is an opportunistic fungus, which most frequently presents as a pulmonary infection or meningoencephalitis. Cryptococcosis has a high impact on immunocompromised populations such as patients with HIV-AIDS and a wide array of non-HIV patients including those with hematopoietic malignancies, autoimmune diseases, or genetic immunodeficiency syndromes, as well as patients receiving immunosuppressive cancer-therapies, undergoing transplant conditioning, in combination with age-related immunosenescence (1–5). HIV-related cryptococcal meningitis (CM) is one the most common causes of adult meningitis with an estimated 223,100 cases and 181,100 deaths in 2014, globally (6, 7). In countries with access to optimal medical care, non-HIV CM accounts for at least 25% of all CM-related hospitalization and deaths and is currently the leading cause of non-viral meningitis in the U.S. (2, 8–10). Interestingly, there are rare reports that as much as 30% of non-HIV patients with cryptococcal infection were previously healthy with no known underlying condition (11).

Cryptococcus is a basidiomycete yeast with over 30 known species; however, the majority of human infections are caused by either *Cryptococcus neoformans* and *Cryptococcus gattii*. *C. neoformans* is the main source of infections in CM patients with CD4+ T-cell deficiency while *C. gattii* is a predominant species in the previously healthy (12, 13). *C. neoformans* and *C. gattii* both can both be found in the vicinity of a variety of trees, and *C. neoformans* can also be found in soils and bird feces (14). Although the life-cycle of *Cryptococcus* is not dependent on an animal host, *C. neoformans* has the potential to infect a wide range of warm- and cold-blooded species (15). Cryptococci are considered sapronotic due to their ability to cause an opportunistic infection without coevolution of a host–parasite virulence (16) although molecular optimization of virulence has been noted in environmental strains after mammalian residence (17). Key to infecting such a wide-ranging host population is its adaptation to environmental conditions and defenses against innate plant defenses as well as phagocytic predators such as parasites and insects (18–21).

CRYPTOCOCCAL INFECTION

Cryptococcal infection is believed to be transmitted by inhalation of infective particles such as yeast cells and/or spores from an environmental source. It is believed that humans encounter the organism early in life, evident by the gradual increase of cryptococcal-specific antibodies in humans with age (22) and isolation of infective strain types from the country of origin in immigrant patients later presenting with HIV/AIDS-related CM (23). As most immunocompetent humans are asymptomatic and resolve the infection, there are limited observations as to the mechanism in which infection is cleared. *Cryptococcus* spp. has a unique repertoire of immune reactivity from other fungi because of distinguishing attributes such as a large polysaccharide capsule that limits exposure to immune dominant carbohydrate epitopes (20), immunomodulatory enzymes such as a phenol oxidizing and cytokine-inducing laccase (24), and a robust tolerance to low nutrient conditions such as within brain tissues (25, 26).

INNATE RECOGNITION

Although no single pattern recognition receptor (PRR) has been shown to be required for binding of *Cryptococcus*, it is hypothesized that alveolar macrophages recognize *Cryptococcus* and initialize the immune response through multiple receptors such as Dectin-1, Mincle, mannose receptor, CD14, and toll-like receptors (TLRs) (**Figure 1**.1) (27–30). The role of AMs and phagocytosis is believed to be critical during early infection, as observed *in vivo* imaging studies and animal models demonstrating enhanced susceptibility after AM depletion (31, 32). The complexity of how PRRs impact the immune response to cryptococcal infection is still being studied as there may be redundancy among them. For example, TLRs 2, 4, and 9 individually appear to play only a minor role, although the use of agonists *in vitro* enhances the proinflammatory responses by microglial cells to *Cryptococcus* (33). NOD-like receptor family pyrin domain containing 3 (NLRP3) is another cryptococcal recognizing PRR and has been shown in mice to be involved with leukocyte infiltration in the lung during infection (34). Interestingly, *in vitro*, NLRP3 activation appears to be inhibited by capsulated cryptococcal cells further suggesting that PRRs may have a more important role during early infection (35). Some PRRs have been linked to specific immune responses such as scavenger receptors B1 and SR-B3, which appear to be important for the induction of IL-1β (36). Macrophage receptor with collagenous structure (MARCO) is also a scavenger receptor which plays a role in cellular recruitment to the lung, cytokine production, and pathogen uptake by mononuclear phagocytes during early stages of cryptococcal infection (37). However, during the adaptive phase of infection, mice deficient in MARCO have improved fungal clearance which is marker by a type-I skewed immune response. Although MARCO plays an important role initially during infection, it is believed that *Cryptococcus* is capable of exploiting MARCO to polarize toward a non-protective immune response (38). There are other recognition receptors such as scavenger receptor A (SR-A) that are also associated with poorer response to *Cryptococcus*. SR-A-deficient mice show enhanced fungal clearance, which was correlated with decreased production of IL-4 and IL-13 (39). This may be linked to the unique cell wall composition of *C. neoformans* compared to other fungi, which contains high levels of acetylated chitin and deacetylated chitosan polymers (40). Furthermore, the binding of PRRs will become hindered as the infectious propagule starts forming the polysaccharide capsule, thus, it is believed that the successful phagocytosis of *Cryptococcus* also requires antibody- or complement-type opsonins (41, 42).

INITIATION OF THE IMMUNE RESPONSE

Although recognition and phagocytosis are important in induction of the immune response to *Cryptococcus*, the processes and pathways involved in breaking down and clearing pathogens and their antigens are also critical steps to mounting an optimal immune response. One clinical study observed that *in vitro* macrophage phagocytosis was directly correlated with clinical outcome (43). This suggests that other factors including phagocyte polarization and lysosomal activity may also need to be regulated for successful clearance.

Unlike other intracellular pathogens such as *Mycobacterium tuberculosis*, cryptococci do not interfere with phagosome formation or maturation; however, they are capable of surviving within vesicles or escaping by phagosome permeabilization or vesicular release (44–47). *C. neoformans* is thus capable of using macrophages as a host for immune evasion, and can escape through expulsion, lysis, or rupture due to excessive intracellular proliferation (47, 48). Alveolar macrophages are required; however, to recruit monocytes and dendritic cells (DCs) primarily through

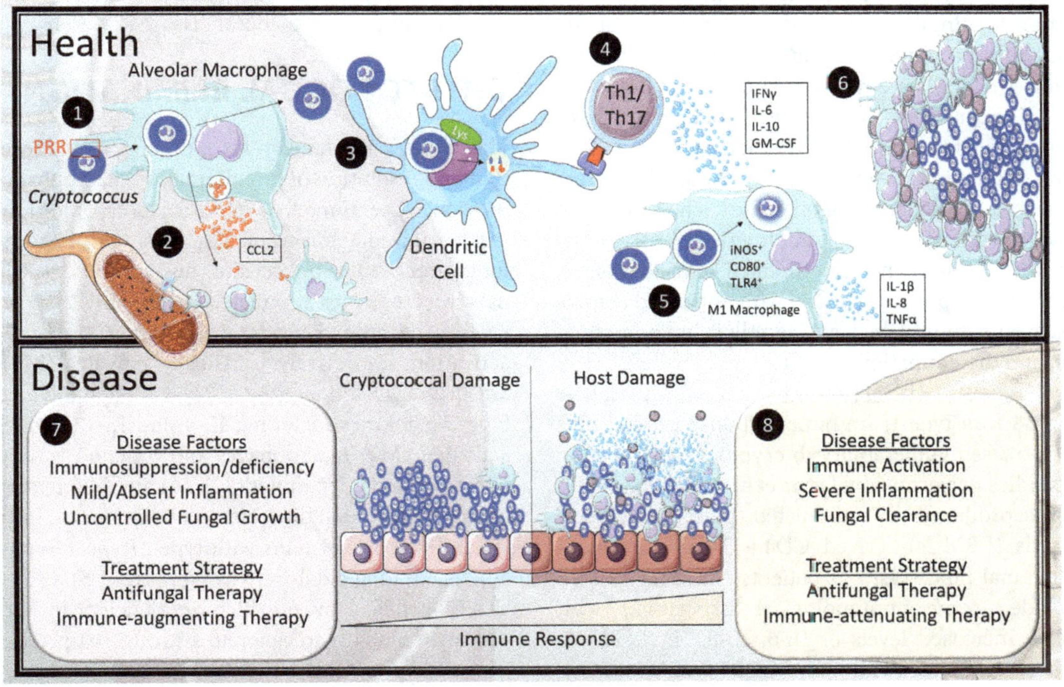

FIGURE 1 | Current model of the immune response to *Cryptococcus* in health and disease. In health (1) inhaled cryptococcal spores are recognized by alveolar macrophages through pattern recognition receptors (PRR). (2) This stimulates the macrophages to release CCL2 to recruit monocytes and dendritic cells (DCs) to the lung. (3) Recruited DCs are capable of breaking down cryptococcal lifeforms and present antigen to CD4+ T-cells. (4) Activated T helper 1 and Th17 T-cells secrete IFNγ, IL-6, IL-10, and granulocyte-macrophage colony stimulating factor (GM-CSF) to recruit and differentiate classical (M1) macrophages. (5) The exact mechanism of fungicidal activity by M1 macrophages against *Cryptococcus* is still unclear in humans; however, they are known to upregulate iNOS, CD80, and TLR4 as well as produce IL-1β, IL-8, and TNFα. (6) The cytokine and chemokine mileu organize the leukocytes to encapsulate and eliminate cryptococcal organisms within granulomas. The most common association with cryptococcal disease (7) is the absence or dysfunction of at least one aspect of the healthy immune response, which leads to uncontrolled fungal growth. (8) However, some patients develop a skewed hyper-immune response to the pathogen that causes inflammatory damage to the host tissue even with fungal clearance.

the production of macrophage chemotactic protein 1 (MCP1). In rats, MCP1 (also known as CCL2) and its receptor, CCR2, are essential for the recruitment of DCs, formation of granulomas, antigen presentation, and T-cell responses (**Figure 1**.2) (49, 50). Granulomas are a sign for control of infections and are composed of macrophages and giant multinucleated cells that contain cryptococcal cells, as well as CD4+ T-cells. These granulomas encompass the fungi and often resolve without additional medical assistance, but treatment with antifungal therapy or surgical removal of the lesions may expedite recovery. It has also been suggested that cryptococci may also be able to latently persist within granulomas and macrophages without degradation (51, 52). In patients with HIV-related pulmonary cryptococcosis, multinucleated giant cells are still present; however, the cryptococci are mainly extracellular and propagate within alveolar spaces (53). DCs are considered the primary antigen-presenting cell (APC) in the context of cryptococcal infection and have an advantage over macrophages in stimulating T-cell proliferation (54). Recruited DCs phagocytose cryptococcal bodies, which then are passaged through lysosomes to be degraded by both oxidative and non-oxidative mechanisms (**Figure 1**.3) (54, 55). For example, cathepsin-B has a non-enzymatic role to fracture the cell wall through osmotic lysis. Degraded components are then loaded onto major histocompatibility complex class II to initiate the adaptive immune response through CD4+ T-cell stimulation (**Figure 1**.4) (56). Eosinophils from a rat model of cryptococcosis have also been demonstrated to have the ability to phagocytose cryptococci and prime of CD4+ and CD8+ T-cells, *in vitro* (57) although their role in human infections is less clear. Their ability to function as APCs is associated with a decrease in nitric oxide and hydrogen peroxide production, followed by migration to the lymphatic system (58). However, their involvement in priming the adaptive immune response is associated with increased fungal burden and lung pathology by skewing immunity toward a type-II response (59, 60).

THE ADAPTIVE IMMUNE RESPONSE

Involvement of the adaptive immune compartment is critical for control of a cryptococcal infection; however, it may also have a detrimental affect depending on the type of response. In draining lymphoid tissues, APCs carrying cryptococcal antigens stimulate several types of lymphocytes including CD4+ T cells, CD8+ T cells, and natural killer T cells. Once activated, CD4+ T-cells can further differentiate into unique effector subsets with distinctive cytokine profiles including: T helper 1 (Th1), Th2, and

Th17 cells. Cxcr5+ T follicular helper cells are also induced and primarily function to stimulate B cell maturation and antibody production, as well as activate inflammatory macrophages (61, 62). Th1 and Th17 cells are recognized by their production of IFNγ and IL-17, respectively, and both help mediate the resolution of cryptococcal infection. On the other hand, Th2 cells, which are described as producers of IL-4, IL-5, and IL-13, are associated with more of a detrimental outcome such as increases in inflammation, worsened pathology, and increased risk of dissemination. In both humans and experimental murine models, deficiencies in type-II responses is linked with enhanced control of fungal burden and diminished eosinophilia, inflammation, airway damage, and dissemination (63, 64).

Interestingly, patients with HIV infection will gradually shift from a type-I to a type-II immune response profile, thus developing an increased vulnerability to cryptococcal infection (65). Profiling studies performed by Jarvis et al. on the cytokines and chemokines produced by stimulated peripheral blood mononuclear cells (PBMCs)-derived CD4+ T-cell as well as within cerebral spinal fluid (CSF) of patients with HIV-related CM have provided some immunological associations with survival (66, 67). Increased levels of IL-6, IL-8, IL-10, IL-17, IFN-γ, tumor necrosis factor (TNF), and CCL5 within the CSF correlated with high white cell counts, macrophage activation, reduced cryptococcal burden, and survival. A high proportion of IFN-γ and TNF double producing PBMC-derived CD4+ T-cells was also associated with survival (68). This study corroborates the critical importance of maintaining a Th1/Th17 profile in both cryptococcal pulmonary infection and meningoencephalitis. Most CM studies have been performed in the context of HIV patients and *C. neoformans*; however, little is known about the immune profile in patients with non-HIV CM, particularly those with *C. gattii* infection.

HUMORAL IMMUNITY TO *Cryptococcus*

As previously mentioned, serum antibodies to *Cryptococcus* can be detected in early life. However, immunocompromised patients at risk for cryptococcal infection appear to have a defect in antibody responses, such as loss of glucuronoxylomannan (GXM), a capsular component reactive B-cells, as well as overall lower levels of peripheral blood memory IgM B cells (69, 70). Lower serum GXM-IgM antibody levels in both HIV+ as well as HIV− solid organ transplant patients is also associated with increased risk for development of cryptococcosis (71–73). Antibody-mediated phagocytosis may be important as the increase in capsular size has been shown to reduce complement-mediated phagocytosis (74). Furthermore, murine studies have demonstrated that the murine equivalent of IgM memory B cells, B-1 cells, can dampen fungal growth *in vitro* and *in vivo*, by inducing an earlier T-cell response, reducing dissemination, and enhancing macrophage phagocytosis (75–78). Additionally, adoptive transfer of IgM-sufficient wild-type mouse serum into Rag1$^{-/-}$ mice demonstrated enhanced alveolar macrophage phagocytosis and a reduction in early dissemination compared to mice treated with IgM-deficient serum. The use of vaccines or antibody therapy to boost antifungal titers may thus provide protection against the development of cryptococcal disease (79–81).

CRYPTOCOCCAL ELIMINATION

The primary mechanism for pulmonary clearance is the formation and resolution of granulomas by macrophages. However, as previously mentioned, *Cryptococcus* is capable of surviving within resident alveolar macrophages, thus the macrophages required for clearance must be recruited and activated by CD4+ T-cell signals. Macrophages stimulated under Th1/Th17 or Th2 cytokine profiles become skewed toward either classical or alternative activation, respectively. Classically activated (M1) macrophages are primarily induced by IFNγ and lipopolysaccharide, while type-2 cytokines including IL-4 and IL-13 induce alternatively activated (M2) macrophages and function through production of proline and polyamines (82). M1 and M2 macrophages *in vitro* have demonstrated different outcomes during intracellular parasitism by *C. neoformans* with type-1/type-17 conditions having enhanced fungicidal activity (**Figure 1**.5) (83). Furthermore, STAT1-deficient mice, which are deficient in M1 macrophages due to an inability to generate a strong Th1 profile, have a defect in anti-cryptococcal activity, which correlated with a decrease in NO production (84). In the previously mentioned cohort of non-HIV patients with CM, although there were intact Th1 signaling found in the CSF, autopsy results revealed an overrepresentation of M2 macrophages within central nervous system (CNS) tissues (85). Similarly, patients with granulocyte-macrophage colony stimulating factor (GM-CSF) autoantibodies are also at risk for CM and have an abundance of Th1 CD4+ T-cells, but also have a skewed M2 macrophage phenotype (86, 87). Activated M1 macrophages, with CD4+ T-cells, resolve the infection by entrapping and degrading the cryptococcal propagules through the formation of granulomas (**Figure 1**.6).

BRAIN DISSEMINATION

Uncontrolled cryptococcal infection will inevitably disseminate into the (CNS) leading to a life-threatening CM. There are currently three known methods of cryptococcal dissemination from the lung: (1) the disruption of blood vesicle integrity allowing passive transport into the blood stream, (2) intact endothelial cells may phagocytose the spores from the lung and expulse them into the blood stream, (3) macrophages may act as a Trojan Horse by transporting phagocytosed spores to the brain, and regurgitating the spores in a process known as vomocytosis. Both microbial and host factors have been identified to be involved in CNS invasion, including cryptococcal matrix metalloprotease, production of a urease enzyme (88), and increases in host brain inositol levels (89, 90).

NON-HIV FACTORS OF SUSCEPTIBILITY

Over 1,000 cases of CM are reported to occur in previously healthy people in the U.S. annually. Studying this population reveals unique vulnerability risks, including previously undiagnosed, rare immune-associated monogenic disorders or autoimmune diseases. Patients with autoantibodies to (GM-CSF) and interferon-gamma (IFNγ) were recently demonstrated to be susceptible to CM, emphasizing the T-cell/monocyte signaling pathway that is required for a successful immune response (86, 87, 91). Interestingly, poor macrophage function was also demonstrated in a cohort of clinically refractory patients by a lack of iNOS expression and intact M2-related CD200R1 expression using immunohistochemistry of infected brain tissue (85). Further studies also demonstrated defective CSF activated macrophage TNF-α secretion, which may explain a lack of symptomatology and diagnostic delays in non-HIV related CM.

Historically, the most common syndrome associated with risk for CM is an idiopathic CD4 lymphopenia (ICL) that presents as a non-HIV-associated reduction or loss of CD4+ T-cells. The tremendous impact of CM on AIDS patients makes the importance of CD4 T-cells self-evident. However, ICL is a very heterogeneous disorder that has been implicated as a serious risk factor (92, 93) but many patients with ICL remain healthy. Recently, the concept of a "two hit" hypothesis was advanced by the finding of two ICL patients with CM who had additional autoantibodies to GMCSF or an otherwise benign, but functionally significant mutation in the IKBKG/NEMO gene, with reductions in NFKB T-cell signaling (94). Similarly, patients with monocytopenia, such as patients with a GATA2 deficiency, also have increased risk of developing CM (95–97). Monogenic disorders such as X-linked CD40L deficiency, chronic granulomatous disease, and Job syndrome are also associated with susceptibility to CM (98–100). T-cell suppressing biological therapy such as natalizumab or fingolimod is also a risk factor (101, 102).

CRYPTOCOCCAL DISEASE: A REFLECTION OF HOST AND MICROBIOLOGICAL FACTORS

Recently, there has been a greater appreciation that host damage can occur from either the toxic products of an overwhelming microbial infection or a pathological inflammatory response to the invading pathogen (**Figure 1**.7 and **Figure 1**.8), recently termed the damage-framework response (103). Cryptococcal disease is a classic example of this phenomenon, often occurring within the same patient during different stages of treatment (104, 105). For example, in the setting of HIV infection, clinical outcomes of primary therapy are related to clearance of the fungus (106, 107). However, a paradoxical immune reconstitution syndrome can also be seen in these same patients whereby, in the setting of microbiological control, reconstitution of the immune system after initiation of antiretroviral therapy (ART) results in a pathological central nervous system inflammatory response (7, 108–110). HIV-related CM further exemplifies differences in disease at these polar extremes of immune response—a recent study of 90 HIV patients with cryptococcal disease found that high levels of Th1-related cytokines INF-gamma and IL-6 were predictive of 2-week initial survival when pathogen load was high; whereas, the development of symptomatic cIRIS was associated with elevated activation with increased macrophage-related cytokines such as CCL2/MCP-1, CCL3/MIP1a, and GM-CSF (67).

Similar to HIV-related disease, in the initial stages of therapy of non-HIV patients, failure to achieve negative CSF fungal cultures at 2 weeks is associated with clinical failure (107). However, many more such patients develop refractory symptoms and/or clinical deterioration despite microbiological control, recently described in previously healthy patients as a postinfectious inflammatory response syndrome. Similar to that encountered with cIRIS in HIV, these patients have an activated CD4+ T-cell intrathecal compartment with minimal Th2 presence (85). Additionally, these patients have high levels of CD4+ T-cells in the CSF and within the intracranial Virchow–Robin channels, which displayed an activated phenotype, as measured by HLA-DR4 and CD56 positivity. Elevated Th1 CSF soluble cytokines such as INF-gamma and interferon-related CXCL10 confirmed the activated T-cell phenotype. In addition, a strong relationship between the T-cell activation marker, sCD27, and elevations of the axonal damage marker, neurofilament light chain protein, suggest that such inflammation is not a benign event, but pathological (85, 111). These observations were surprising as these conditions typically define a successful immune response, as understood from susceptibility studies in HIV-related disease. However, as described above, many of the previously healthy have defects in macrophage polarization, allowing disease susceptibility in the face of unrestrained T-cell-mediated host damage. Such findings also highlight the limitations of applying disease principles from one host to another without careful consideration.

TREATMENT: TOWARD A MORE RATIONAL APPROACH TO ADJUNCTIVE THERAPY

Treatment strategies for patients with CM should be developed to address damage caused by both the microbe and the pathological immune response; the disease framework is a useful guide (**Figure 1**.7 and **Figure 1**.8) (104). In the initial therapy of all patients, treatment with antifungals with a fungicidal agent such as amphotericin B is paramount and is typically continued with oral azole therapy to prevent relapse after initial negative CSF cultures (112–114). Those with low immunity (HIV prior to ART or patients with a skewed Th2 response to *Cryptococcus*), may benefit from potential immune adjuvants such as IFNγ to accelerate microbiological control (115). In these cases, rates of clearance of CSF fungal cultures (early fungicidal activity) may be an important parameter (67, 106, 107). However, in HIV patients who have developed cryptococcal immune reconstitution inflammatory syndrome (cIRIS) or in refractory non-HIV patients after microbiological control, attention needs to be drawn to the host damage side of the disease model—to minimize pathological effects of a dysregulated host response. In this setting, application of immune enhancers such as IFNγ therapy (116) may lead to exacerbated inflammation and potentially cause irreparable neurological damage (85). In these patients, use of adjunctive immunosuppression including corticosteroids is increasingly reported to suppress pathological inflammatory responses and control cerebral edema, improving clinical response (85, 115–120). In the previously healthy CM patient, for example, successful application of adjunctive corticosteroids in refractory disease or after clinical deterioration requires a personalized strategy. When resources are available, CSF culture negativity and measures of CSF inflammation such as sCD27, CSF glucose, or choroid plexitis or ependymitis by MRI imaging can provide specific biomarkers of the relative contributions of microbe and host toward understanding an individual patient's condition (85, 120, 121). Biomarkers may also be useful guides during corticosteroid tapers to prevent exacerbations.

However, applying this type of therapy successfully requires careful attention to concurrent microbial control, as corticosteroids suppress innate and acquired immune responses needed to maintain fungal clearance (122, 123). Indeed, corticosteroids can have a deleterious effect when applied without pre-established microbial control during primary therapy of HIV-related CM (**Figure 1**.7) (124). The complex heterogeneity of clinical pathologies that occur in various patients during HIV-related CM requires a thoughtful approach that considers the evolving damage caused by both the fungus and the immune response within defined sub-groups. In resource-limited regions, more studies are needed to understand the damage-response framework as it relates to CM in poorly controlled HIV infection, and to identify markers that can tailor resource appropriate therapies. Strategies to prevent CM in HIV (125) and to diagnose or empirically treat co-infections (126) may have tremendous impact on outcomes given the high prevalence of disease in some geographic regions (8). Risks for co-infections, such as tuberculosis and bacterial infections can be exacerbated without specific therapy in the presence of corticosteroids (127).

Other populations with risks for CM such as solid organ transplant recipients also require tailored approaches to prevention and treatment. In populations other than HIV, the feasibility of prevention strategies is limited due to a lower prevalence of disease. Much more attention is needed to better define appropriate therapeutic strategies. Since diagnosis typically occurs late, fungal burden can be high at diagnosis; at the same time, relatively intact and variable inflammatory responses can lead to exuberant inflammatory neurological damage similar to cIRIS (128). In these patients, attention needs to be focused on personalizing therapies according to which side of the host damage framework is most responsible for neurological pathology (128, 129).

AUTHOR CONTRIBUTIONS

The bulk of this manuscript was written by WE under the guidance of PW. PW and KM provided revisional comments and suggestions on both content and organization of this manuscript.

ACKNOWLEDGMENTS

The introduction diagram was designed using Servier Medical Art.

FUNDING

This research was supported by NIH/NIAID extramual grants AI109657 as well as funding from the NIAID intramural program, AI001123, and AI001124.

REFERENCES

1. Sloan DJ, Parris V. Cryptococcal meningitis: epidemiology and therapeutic options. *Clin Epidemiol* (2014) 6:169–82. doi:10.2147/CLEP.S38850
2. Pyrgos V, Seitz AE, Steiner CA, Prevots DR, Williamson PR. Epidemiology of cryptococcal meningitis in the US: 1997-2009. *PLoS One* (2013) 8(2):e56269. doi:10.1371/journal.pone.0056269
3. Durski KN, Kuntz KM, Yasukawa K, Virnig BA, Meya DB, Boulware DR. Cost-effective diagnostic checklists for meningitis in resource-limited settings. *J Acquir Immune Defic Syndr* (2013) 63(3):e101–8. doi:10.1097/QAI.0b013e31828e1e56
4. Rajasingham R, Rhein J, Klammer K, Musubire A, Nabeta H, Akampurira A, et al. Epidemiology of meningitis in an HIV-infected Ugandan cohort. *Am J Trop Med Hyg* (2015) 92(2):274–9. doi:10.4269/ajtmh.14-0452
5. Jarvis JN, Meintjes G, Williams A, Brown Y, Crede T, Harrison TS. Adult meningitis in a setting of high HIV and TB prevalence: findings from 4961 suspected cases. *BMC Infect Dis* (2010) 10:67. doi:10.1186/1471-2334-10-67
6. Park BJ, Wannemuehler KA, Marston BJ, Govender N, Pappas PG, Chiller TM. Estimation of the current global burden of cryptococcal meningitis among persons living with HIV/AIDS. *AIDS* (2009) 23(4):525–30. doi:10.1097/QAD.0b013e328322ffac
7. Williamson PR, Jarvis JN, Panackal AA, Fisher MC, Molloy SF, Loyse A, et al. Cryptococcal meningitis: epidemiology, immunology, diagnosis and therapy. *Nat Rev Neurol* (2017) 13(1):13–24. doi:10.1038/nrneurol.2016.167
8. Rajasingham R, Smith R, Park B, Jarvis J, Govender NP, Chiller T, et al. Global Burden of Disease of HIV-associated cryptococcal meningitis: an updated analysis. *Lancet Infect Dis* (2017) 17(8):873–81. doi:10.1016/S1473-3099(17)30243-8
9. Pappas PG, Perfect JR, Cloud GA, Larsen RA, Pankey GA, Lancaster DJ, et al. Cryptococcosis in human immunodeficiency virus-negative patients in the era of effective azole therapy. *Clin Infect Dis* (2001) 33(5):690–9. doi:10.1086/322597
10. Brizendine KD, Baddley JW, Pappas PG. Predictors of mortality and differences in clinical features among patients with cryptococcosis according to immune status. *PLoS One* (2013) 8(3):e60431. doi:10.1371/journal.pone.0060431
11. Zhu LP, Wu JQ, Xu B, Ou XT, Zhang QQ, Weng XH. Cryptococcal meningitis in non-HIV-infected patients in a Chinese tertiary care hospital, 1997-2007. *Med Mycol* (2010) 48(4):570–9. doi:10.3109/13693780903437876
12. Farrer RA, Desjardins CA, Sakthikumar S, Gujja S, Saif S, Zeng Q, et al. Genome evolution and innovation across the four major lineages of *Cryptococcus gattii*. *MBio* (2015) 6(5):e868–815. doi:10.1128/mBio.00868-15
13. Hagen F, Khayhan K, Theelen B, Kolecka A, Polacheck I, Sionov E, et al. Recognition of seven species in the *Cryptococcus gattii/Cryptococcus neoformans* species complex. *Fungal Genet Biol* (2015) 78:16–48. doi:10.1016/j.fgb.2015.02.009
14. Reimão JQ, Drummond ED, Terceti Mde S, Lyon JP, Franco MC, de Siqueira AM. Isolation of *Cryptococcus neoformans* from hollows of living trees in the city of Alfenas, MG, Brazil. *Mycoses* (2007) 50(4):261–4. doi:10.1111/j.1439-0507.2007.01374.x
15. Casadevall A, Steenbergen JN, Nosanchuk JD. 'Ready made' virulence and 'dual use' virulence factors in pathogenic environmental fungi – the *Cryptococcus neoformans* paradigm. *Curr Opin Microbiol* (2003) 6(4):332–7. doi:10.1016/S1369-5274(03)00082-1
16. Kuris AM, Lafferty KD, Sokolow SH. Sapronosis: a distinctive type of infectious agent. *Trends Parasitol* (2014) 30(8):386–93. doi:10.1016/j.pt.2014.06.006
17. Hu G, Chen SH, Qiu J, Bennett JE, Myers TG, Williamson PR. Microevolution during serial mouse passage demonstrates FRE3 as a virulence adaptation gene in *Cryptococcus neoformans*. *MBio* (2014) 5(2):e941–914. doi:10.1128/mBio.00941-14

18. Warpeha KM, Park YD, Williamson PR. Susceptibility of intact germinating *Arabidopsis thaliana* to human fungal pathogens *Cryptococcus neoformans and C. gattii. Appl Environ Microbiol* (2013) 79(9):2979–88. doi:10.1128/AEM.03697-12
19. McDonald T, Wiesner DL, Nielsen K. Cryptococcus. *Curr Biol* (2012) 22(14):R554–5. doi:10.1016/j.cub.2012.05.040
20. Zaragoza O, Rodrigues ML, De Jesus M, Frases S, Dadachova E, Casadevall A. The capsule of the fungal pathogen *Cryptococcus neoformans. Adv Appl Microbiol* (2009) 68:133–216. doi:10.1016/S0065-2164(09)01204-0
21. Steenbergen JN, Casadevall A. The origin and maintenance of virulence for the human pathogenic fungus *Cryptococcus neoformans. Microbes Infect* (2003) 5(7):667–75. doi:10.1016/S1286-4579(03)00092-3
22. Goldman DL, Khine H, Abadi J, Lindenberg DJ, Pirofski L, Niang R, et al. Serologic evidence for *Cryptococcus neoformans* infection in early childhood. *Pediatrics* (2001) 107(5):E66. doi:10.1542/peds.107.5.e66
23. Dromer F, Ronin O, Dupont B. Isolation of *Cryptococcus neoformans* var. gattii from an Asian patient in France: evidence for dormant infection in healthy subjects. *J Med Vet Mycol* (1992) 30(5):395–7. doi:10.1080/02681219280000511
24. Valdez PA, Vithayathil PJ, Janelsins BM, Shaffer AL, Williamson PR, Datta SK. Prostaglandin E2 suppresses antifungal immunity by inhibiting interferon regulatory factor 4 function and interleukin-17 expression in T cells. *Immunity* (2012) 36(4):668–79. doi:10.1016/j.immuni.2012.02.013
25. Waterman S, Hacham M, Hu G, Zhu X, Park Y, Shin S, et al. Role of a CUF1-CTR4 copper regulatory axis in the virulence of *Cryptococcus neoformans*. *J Clin Invest* (2007) 117:794–802. doi:10.1172/JCI30006
26. Hu G, Hacham M, Waterman SR, Panepinto J, Shin S, Liu X, et al. PI3K signaling of autophagy is required for starvation tolerance and virulenceof *Cryptococcus neoformans*. *J Clin Invest* (2008) 118(3):1186–97. doi:10.1172/JCI32053
27. Garcia-Rodas R, Zaragoza O. Catch me if you can: phagocytosis and killing avoidance by *Cryptococcus neoformans. FEMS Immunol Med Microbiol* (2012) 64(2):147–61. doi:10.1111/j.1574-695X.2011.00871.x
28. Levitz SM. Innate recognition of fungal cell walls. *PLoS Pathog* (2010) 6(4):e1000758. doi:10.1371/journal.ppat.1000758
29. Walsh NM, Wuthrich M, Wang H, Klein B, Hull CM. Characterization of C-type lectins reveals an unexpectedly limited interaction between *Cryptococcus neoformans* spores and Dectin-1. *PLoS One* (2017) 12(3): e0173866. doi:10.1371/journal.pone.0173866
30. Mansour MK, Reedy JL, Tam JM, Vyas JM. Macrophage *Cryptococcus* interactions: an update. *Curr Fungal Infect Rep* (2014) 8(1):109–15. doi:10.1007/s12281-013-0165-7
31. Shao X, Mednick A, Alvarez M, van Rooijen N, Casadevall A, Goldman DL. An innate immune system cell is a major determinant of species-related susceptibility differences to fungal pneumonia. *J Immunol* (2005) 175(5): 3244–51. doi:10.4049/jimmunol.175.5.3244
32. Bojarczuk A, Miller KA, Hotham R, Lewis A, Ogryzko NV, Kamuyango AA, et al. *Cryptococcus neoformans* intracellular proliferation and capsule size determines early macrophage control of infection. *Sci Rep* (2016) 6:21489. doi:10.1038/srep21489
33. Redlich S, Ribes S, Schütze S, Eiffert H, Nau R. Toll-like receptor stimulation increases phagocytosis of *Cryptococcus neoformans* by microglial cells. *J Neuroinflammation* (2013) 10:71. doi:10.1186/1742-2094-10-71
34. Lei G, Chen M, Li H, Niu JL, Wu S, Mao L, et al. Biofilm from a clinical strain of *Cryptococcus neoformans* activates the NLRP3 inflammasome. *Cell Res* (2013) 23(7):965–8. doi:10.1038/cr.2013.49
35. Guo C, Chen M, Fa Z, Lu A, Fang W, Sun B, et al. Acapsular *Cryptococcus neoformans* activates the NLRP3 inflammasome. *Microbes Infect* (2014) 16(10):845–54. doi:10.1016/j.micinf.2014.08.013
36. Means TK, Mylonakis E, Tampakakis E, Colvin RA, Seung E, Puckett L, et al. Evolutionarily conserved recognition and innate immunity to fungal pathogens by the scavenger receptors SCARF1 and CD36. *J Exp Med* (2009) 206(3):637–53. doi:10.1084/jem.20082109
37. Xu J, Flaczyk A, Neal LM, Fa Z, Eastman AJ, Malachowski AN, et al. Scavenger receptor MARCO orchestrates early defenses and contributes to fungal containment during cryptococcal infection. *J Immunol* (2017) 198(9):3548–57. doi:10.4049/jimmunol.1700057
38. Xu J, Flaczyk A, Neal LM, Fa Z, Cheng D, Ivey M, et al. Exploitation of scavenger receptor, macrophage receptor with collagenous structure, by *Cryptococcus neoformans* promotes alternative activation of pulmonary lymph node CD11b(+) conventional dendritic cells and non-protective Th2 bias. *Front Immunol* (2017) 8:1231. doi:10.3389/fimmu.2017.01231
39. Qiu Y, Dayrit JK, Davis MJ, Carolan JF, Osterholzer JJ, Curtis JL, et al. Scavenger receptor a modulates the immune response to pulmonary *Cryptococcus neoformans* infection. *J Immunol* (2013) 191(1):238–48. doi:10.4049/jimmunol.1203435
40. Wiesner DL, Specht CA, Lee CK, Smith KD, Mukaremera L, Lee ST, et al. Chitin recognition via chitotriosidase promotes pathologic type-2 helper T cell responses to cryptococcal infection. *PLoS Pathog* (2015) 11(3):e1004701. doi:10.1371/journal.ppat.1004701
41. Dan JM, Wang JP, Lee CK, Levitz SM. Cooperative stimulation of dendritic cells by *Cryptococcus neoformans* mannoproteins and CpG oligodeoxynucleotides. *PLoS One* (2008) 3(4):e2046. doi:10.1371/journal.pone.0002046
42. Voelz K, May RC. Cryptococcal interactions with the host immune system. *Eukaryot Cell* (2010) 9(6):835–46. doi:10.1128/EC.00039-10
43. Sabiiti W, Robertson E, Beale MA, Johnston SA, Brouwer AE, Loyse A, et al. Efficient phagocytosis and laccase activity affect the outcome of HIV-associated cryptococcosis. *J Clin Invest* (2014) 124(5):2000–8. doi:10.1172/JCI72950
44. Johnston SA, May RC. The human fungal pathogen *Cryptococcus neoformans* escapes macrophages by a phagosome emptying mechanism that is inhibited by Arp2/3 complex-mediated actin polymerisation. *PLoS Pathog* (2010) 6(8):e1001041. doi:10.1371/journal.ppat.1001041
45. Levitz SM, Nong SH, Seetoo KF, Harrison TS, Speizer RA, Simons ER. *Cryptococcus neoformans* resides in an acidic phagolysosome of human macrophages. *Infect Immun* (1999) 67(2):885–90.
46. Vandal OH, Nathan CF, Ehrt S. Acid resistance in mycobacterium tuberculosis. *J Bacteriol* (2009) 191(15):4714–21. doi:10.1128/JB.00305-09
47. Tucker SC, Casadevall A. Replication of *Cryptococcus neoformans* in macrophages is accompanied by phagosomal permeabilization and accumulation of vesicles containing polysaccharide in the cytoplasm. *Proc Natl Acad Sci U S A* (2002) 99(5):3165–70. doi:10.1073/pnas.052702799
48. Alvarez M, Casadevall A. Cell-to-cell spread and massive vacuole formation after *Cryptococcus neoformans* infection of murine macrophages. *BMC Immunol* (2007) 8:16. doi:10.1186/1471-2172-8-16
49. He W, Casadevall A, Lee SC, Goldman DL. Phagocytic activity and monocyte chemotactic protein expression by pulmonary macrophages in persistent pulmonary cryptococcosis. *Infect Immun* (2003) 71(2):930–6. doi:10.1128/IAI.71.2.930-936.2003
50. Osterholzer JJ, Curtis JL, Polak T, Ames T, Chen GH, McDonald R, et al. CCR2 mediates conventional dendritic cell recruitment and the formation of bronchovascular mononuclear cell infiltrates in the lungs of mice infected with *Cryptococcus neoformans*. *J Immunol* (2008) 181(1):610–20. doi:10.4049/jimmunol.181.1.610
51. Garcia-Hermoso D, Janbon G, Dromer F. Epidemiological evidence for dormant *Cryptococcus neoformans* infection. *J Clin Microbiol* (1999) 37(10): 3204–9.
52. Goldman DL, Lee SC, Mednick AJ, Montella L, Casadevall A. Persistent *Cryptococcus neoformans* pulmonary infection in the rat is associated with intracellular parasitism, decreased inducible nitric oxide synthase expression, and altered antibody responsiveness to cryptococcal polysaccharide. *Infect Immun* (2000) 68(2):832–8. doi:10.1128/IAI.68.2.832-838.2000
53. Shibuya K, Hirata A, Omuta J, Sugamata M, Katori S, Saito N, et al. Granuloma and cryptococcosis. *J Infect Chemother* (2005) 11(3):115–22. doi:10.1007/s10156-005-0387-X
54. Syme RM, Spurrell JC, Amankwah EK, Green FH, Mody CH. Primary dendritic cells phagocytose *Cryptococcus neoformans* via mannose receptors and Fcgamma receptor II for presentation to T lymphocytes. *Infect Immun* (2002) 70(11):5972–81. doi:10.1128/IAI.70.11.5972-5981.2002
55. Mansour MK, Latz E, Levitz SM. *Cryptococcus neoformans* glycoantigens are captured by multiple lectin receptors and presented by dendritic cells. *J Immunol* (2006) 176(5):3053–61. doi:10.4049/jimmunol.176.5.3053
56. Hole CR, Bui H, Wormley FL Jr, Wozniak KL. Mechanisms of dendritic cell lysosomal killing of *Cryptococcus*. *Sci Rep* (2012) 2:739. doi:10.1038/srep00739
57. Garro AP, Chiapello LS, Baronetti JL, Masih DT. Rat eosinophils stimulate the expansion of *Cryptococcus neoformans*-specific CD4(+) and CD8(+) T cells with a T-helper 1 profile. *Immunology* (2011) 132(2):174–87. doi:10.1111/j.1365-2567.2010.03351.x

58. Garro AP, Chiapello LS, Baronetti JL, Masih DT. Eosinophils elicit proliferation of naive and fungal-specific cells in vivo so enhancing a T helper type 1 cytokine profile in favour of a protective immune response against *Cryptococcus neoformans* infection. *Immunology* (2011) 134(2):198–213. doi:10.1111/j.1365-2567.2011.03479.x
59. Huffnagle GB, Boyd MB, Street NE, Lipscomb MF. IL-5 is required for eosinophil recruitment, crystal deposition, and mononuclear cell recruitment during a pulmonary *Cryptococcus neoformans* infection in genetically susceptible mice (C57BL/6). *J Immunol* (1998) 160(5):2393–400.
60. Holmer SM, Evans KS, Asfaw YG, Saini D, Schell WA, Ledford JG, et al. Impact of surfactant protein D, interleukin-5, and eosinophilia on cryptococcosis. *Infect Immun* (2014) 82(2):683–93. doi:10.1128/IAI.00855-13
61. Luckheeram RV, Zhou R, Verma AD, Xia B. CD4(+)T cells: differentiation and functions. *Clin Dev Immunol* (2012) 2012:925135. doi:10.1155/2012/925135
62. Gibson JF, Johnston SA. Immunity to *Cryptococcus neoformans and C. gattii* during cryptococcosis. *Fungal Genet Biol* (2015) 78:76–86. doi:10.1016/j.fgb.2014.11.006
63. Zhang Y, Wang F, Tompkins KC, McNamara A, Jain AV, Moore BB, et al. Robust Th1 and Th17 immunity supports pulmonary clearance but cannot prevent systemic dissemination of highly virulent *Cryptococcus neoformans* H99. *Am J Pathol* (2009) 175(6):2489–500. doi:10.2353/ajpath.2009.090530
64. Flaczyk A, Duerr CU, Shourian M, Lafferty EI, Fritz JH, Qureshi ST. IL-33 signaling regulates innate and adaptive immunity to *Cryptococcus neoformans*. *J Immunol* (2013) 191(5):2503–13. doi:10.4049/jimmunol.1300426
65. Altfeld M, Addo MM, Kreuzer KA, Rockstroh JK, Dumoulin FL, Schliefer K, et al. T(H)1 to T(H)2 shift of cytokines in peripheral blood of HIV-infected patients is detectable by reverse transcriptase polymerase chain reaction but not by enzyme-linked immunosorbent assay under non-stimulated conditions. *J Acquir Immune Defic Syndr* (2000) 23(4):287–94. doi:10.1097/00042560-200004010-00001
66. Jarvis JN, Casazza JP, Stone HH, Meintjes G, Lawn SD, Levitz SM, et al. The phenotype of the *Cryptococcus*-specific CD4 + memory T-cell response is associated with disease severity and outcome in HIV-associated cryptococcal meningitis. *J Infect Dis* (2013) 207(12):1817–28. doi:10.1093/infdis/jit099
67. Jarvis JN, Meintjes G, Bicanic T, Buffa V, Hogan L, Mo S, et al. Cerebrospinal fluid cytokine profiles predict risk of early mortality and immune reconstitution inflammatory syndrome in HIV-associated cryptococcal meningitis. *PLoS Pathog* (2015) 11(4):e1004754. doi:10.1371/journal.ppat.1004754
68. Scriven JE, Rhein J, Hullsiek KH, von Hohenberg M, Linder G, Rolfes MA, et al. Early ART after cryptococcal meningitis is associated with cerebrospinal fluid pleocytosis and macrophage activation in a multisite randomized trial. *J Infect Dis* (2015) 212(5):769–78. doi:10.1093/infdis/jiv067
69. Pirofski LA. Polysaccharides, mimotopes and vaccines for fungal and encapsulated pathogens. *Trends Microbiol* (2001) 9(9):445–51. doi:10.1016/S0966-842X(01)02134-5
70. Subramaniam K, Metzger B, Hanau LH, Guh A, Rucker L, Badri S, et al. IgM(+) memory B cell expression predicts HIV-associated cryptococcosis status. *J Infect Dis* (2009) 200(2):244–51. doi:10.1086/599318
71. Deshaw M, Pirofski LA. Antibodies to the *Cryptococcus neoformans* capsular glucuronoxylomannan are ubiquitous in serum from HIV + and HIV-individuals. *Clin Exp Immunol* (1995) 99(3):425–32. doi:10.1111/j.1365-2249.1995.tb05568.x
72. Subramaniam K, French N, Pirofski LA. *Cryptococcus neoformans*-reactive and total immunoglobulin profiles of human immunodeficiency virus-infected and uninfected Ugandans. *Clin Diagn Lab Immunol* (2005) 12(10):1168–76. doi:10.1128/CDLI.12.10.1168-1176.2005
73. Jalali Z, Ng L, Singh N, Pirofski LA. Antibody response to *Cryptococcus neoformans* capsular polysaccharide glucuronoxylomannan in patients after solid-organ transplantation. *Clin Vaccine Immunol* (2006) 13(7):740–6. doi:10.1128/CVI.00139-06
74. Zaragoza O, Taborda CP, Casadevall A. The efficacy of complement-mediated phagocytosis of *Cryptococcus neoformans* is dependent on the location of C3 in the polysaccharide capsule and involves both direct and indirect C3-mediated interactions. *Eur J Immunol* (2003) 33(7):1957–67. doi:10.1002/eji.200323848
75. Fleuridor R, Zhong Z, Pirofski L. A human IgM monoclonal antibody prolongs survival of mice with lethal cryptococcosis. *J Infect Dis* (1998) 178(4):1213–6. doi:10.1086/515688
76. Subramaniam KS, Datta K, Quintero E, Manix C, Marks MS, Pirofski LA. The absence of serum IgM enhances the susceptibility of mice to pulmonary challenge with *Cryptococcus neoformans*. *J Immunol* (2010) 184(10):5755–67. doi:10.4049/jimmunol.0901638
77. Szymczak WA, Davis MJ, Lundy SK, Dufaud C, Olszewski M, Pirofski LA. X-linked immunodeficient mice exhibit enhanced susceptibility to *Cryptococcus neoformans* Infection. *MBio* (2013) 4(4):e265–213. doi:10.1128/mBio.00265-13
78. Rohatgi S, Pirofski LA. Molecular characterization of the early B cell response to pulmonary *Cryptococcus neoformans* infection. *J Immunol* (2012) 189(12):5820–30. doi:10.4049/jimmunol.1201514
79. Chaturvedi AK, Hameed RS, Wozniak KL, Hole CR, Leopold Wager CM, Weintraub ST, et al. Vaccine-mediated immune responses to experimental pulmonary *Cryptococcus gattii* infection in mice. *PLoS One* (2014) 9(8):e104316. doi:10.1371/journal.pone.0104316
80. Mirza Z, Soto ER, Dikengil F, Levitz SM, Ostroff GR. Beta-glucan particles as vaccine adjuvant carriers. *Methods Mol Biol* (2017) 1625:143–57. doi:10.1007/978-1-4939-7104-6_11
81. Spero D, Levitz L, De Groot AS. Report from the field: overview of the sixth annual vaccine renaissance conference. *Hum Vaccin Immunother* (2013) 9(7):1555–7. doi:10.4161/hv.24833
82. Classen A, Lloberas J, Celada A. Macrophage activation: classical versus alternative. *Methods Mol Biol* (2009) 531:29–43. doi:10.1007/978-1-59745-396-7_3
83. Voelz K, Lammas DA, May RC. Cytokine signaling regulates the outcome of intracellular macrophage parasitism by *Cryptococcus neoformans*. *Infect Immun* (2009) 77(8):3450–7. doi:10.1128/IAI.00297-09
84. Leopold Wager CM, Hole CR, Wozniak KL, Olszewski MA, Wormley FL Jr. STAT1 signaling is essential for protection against *Cryptococcus neoformans* infection in mice. *J Immunol* (2014) 193(8):4060–71. doi:10.4049/jimmunol.1400318
85. Panackal AA, Wuest SC, Lin YC, Wu T, Zhang N, Kosa P, et al. Paradoxical immune responses in non-HIV cryptococcal meningitis. *PLoS Pathog* (2015) 11(5):e1004884. doi:10.1371/journal.ppat.1004884
86. Saijo T, Chen J, Chen SC, Rosen LB, Yi J, Sorrell TC, et al. Anti-granulocyte-macrophage colony-stimulating factor autoantibodies are a risk factor for central nervous system infection by *Cryptococcus gattii* in otherwise immunocompetent patients. *MBio* (2014) 5(2):e912–4. doi:10.1128/mBio.00912-14
87. Rosen LB, Freeman AF, Yang LM, Jutivorakool K, Olivier KN, Angkasekwinai N, et al. Anti-GM-CSF autoantibodies in patients with cryptococcal meningitis. *J Immunol* (2013) 190(8):3959–66. doi:10.4049/jimmunol.1202526
88. Cox G, Mukherjee J, Cole G, Casadeval lA, Perfect J. Urease as a virulence factor in experimental cryptococcosis. *Infect Immun* (2000) 68:443–8. doi:10.1128/IAI.68.2.443-448.2000
89. Liu TB, Kim JC, Wang Y, Toffaletti DL, Eugenin E, Perfect JR, et al. Brain inositol is a novel stimulator for promoting *Cryptococcus* penetration of the blood-brain barrier. *PLoS Pathog* (2013) 9(4):e1003247. doi:10.1371/journal.ppat.1003247
90. Santiago-Tirado FH, Onken MD, Cooper JA, Klein RS, Doering TL. Trojan horse transit contributes to blood-brain barrier crossing of a eukaryotic pathogen. *MBio* (2017) 8(1). doi:10.1128/mBio.02183-16
91. Browne SK, Burbelo PD, Chetchotisakd P, Suputtamongkol Y, Kiertiburanakul S, Shaw PA, et al. Adult-onset immunodeficiency in Thailand and Taiwan. *N Engl J Med* (2012) 367(8):725–34. doi:10.1056/NEJMoa1111160
92. Ahmad DS, Esmadi M, Steinmann WC. Idiopathic CD4 lymphocytopenia: spectrum of opportunistic infections, malignancies, and autoimmune diseases. *Avicenna J Med* (2013) 3(2):37–47. doi:10.4103/2231-0770.114121
93. Gorska MM, Alam R. A mutation in the human Uncoordinated 119 gene impairs TCR signaling and is associated with CD4 lymphopenia. *Blood* (2012) 119(6):1399–406. doi:10.1182/blood-2011-04-350686
94. Panackal AA, Rosen LB, Uzel G, Davis MJ, Hu G, Adeyemo A, et al. Susceptibility to cryptococcal meningoencephalitis associated with idiopathic CD4(+) lymphopenia and secondary germline or acquired defects. *Open Forum Infect Dis* (2017) 4(2):ofx082. doi:10.1093/ofid/ofx082
95. Vinh DC, Patel SY, Uzel G, Anderson VL, Freeman AF, Olivier KN, et al. Autosomal dominant and sporadic monocytopenia with susceptibility to mycobacteria, fungi, papillomaviruses, and myelodysplasia. *Blood* (2010) 115(8):1519–29. doi:10.1182/blood-2009-03-208629
96. Hsu AP, Sampaio EP, Khan J, Calvo KR, Lemieux JE, Patel SY, et al. Mutations in GATA2 are associated with the autosomal dominant and sporadic

monocytopenia and mycobacterial infection (MonoMAC) syndrome. *Blood* (2011) 118(10):2653–5. doi:10.1182/blood-2011-05-356352

97. Spinner MA, Sanchez LA, Hsu AP, Shaw PA, Zerbe CS, Calvo KR, et al. GATA2 deficiency: a protean disorder of hematopoiesis, lymphatics, and immunity. *Blood* (2014) 123(6):809–21. doi:10.1182/blood-2013-07-515528
98. Jacobs DH, Macher AM, Handler R, Bennett JE, Collen MJ, Gallin JI. Esophageal cryptococcosis in a patient with the hyperimmunoglobulin E-recurrent infection (Job's) syndrome. *Gastroenterology* (1984) 87(1):201–3.
99. Holland SM, DeLeo FR, Elloumi HZ, Hsu AP, Uzel G, Brodsky N, et al. STAT3 mutations in the hyper-IgE syndrome. *N Engl J Med* (2007) 357(16):1608–19. doi:10.1056/NEJMoa073687
100. Winkelstein JA, Marino MC, Ochs H, Fuleihan R, Scholl PR, Geha R, et al. The X-linked hyper-IgM syndrome: clinical and immunologic features of 79 patients. *Medicine (Baltimore)* (2003) 82(6):373–84. doi:10.1097/01.md.0000100046.06009.b0
101. Forrestel AK, Modi BG, Longworth S, Wilck MB, Micheletti RG. Primary cutaneous *Cryptococcus* in a patient with multiple sclerosis treated with fingolimod. *JAMA Neurol* (2016) 73(3):355–6. doi:10.1001/jamaneurol.2015.4259
102. Valenzuela RM, Pula JH, Garwacki D, Cotter J, Kattah JC. Cryptococcal meningitis in a multiple sclerosis patient taking natalizumab. *J Neurol Sci* (2014) 340(1–2):109–11. doi:10.1016/j.jns.2014.03.007
103. Casadevall A, Pirofski LA. Host-pathogen interactions: redefining the basic concepts of virulence and pathogenicity. *Infect Immun* (1999) 67(8):3703–13.
104. Pirofski LA, Casadevall A. Immune-mediated damage completes the parabola: *Cryptococcus neoformans* pathogenesis can reflect the outcome of a weak or strong immune response. *MBio* (2017) 8(6):e02063-17. doi:10.1128/mBio.02063-17
105. Panackal AA, Williamson KC, van de Beek D, Boulware DR, Williamson PR. Fighting the monster: applying the host damage framework to human central nervous system infections. *MBio* (2016) 7(1):e1906–15. doi:10.1128/mBio.01906-15
106. Bicanic T, Meintjes G, Wood R, Hayes M, Rebe K, Bekker L, et al. Fungal burden, early fungicidal activity, and outcome in cryptococcal meningitis in antiretroviral-naive or antiretroviral-experienced patients treated with amphotericin B or fluconazole. *Clin Infect Dis* (2007) 45:76–80. doi:10.1086/518607
107. Montezuma-Rusca JM, Powers JH, Follmann D, Wang J, Sullivan B, Williamson PR. Early fungicidal activity as a candidate surrogate endpoint for all-cause mortality in cryptococcal meningitis: a systematic review of the evidence. *PLoS One* (2016) 11(8):e0159727. doi:10.1371/journal.pone.0159727
108. Jenny-Avital E, Abadi M. Immune reconstitution cryptococcosis after initiation of successful highly active antiretroviral therapy. *Clin Infect Dis* (2002) 35:e128–33. doi:10.1086/344467
109. Longley N, Harrison TS, Jarvis JN. Cryptococcal immune reconstitution inflammatory syndrome. *Curr Opin Infect Dis* (2013) 26(1):26–34. doi:10.1097/QCO.0b013e32835c21d1
110. Bahr NC, Wallace J, Frosch AE, Boulware DR. Unmasking cryptococcal meningitis immune reconstitution inflammatory syndrome due to granulocyte colony-stimulating factor use in a patient with a poorly differentiated germ cell neoplasm. *Case Rep Oncol* (2014) 7(1):1–5. doi:10.1159/000357666
111. Bacioglu M, Maia LF, Preische O, Schelle J, Apel A, Kaeser SA, et al. Neurofilament light chain in blood and CSF as marker of disease progression in mouse models and in neurodegenerative diseases. *Neuron* (2016) 91(1):56–66. doi:10.1016/j.neuron.2016.05.018
112. Dismukes WE, Cloud G, Gallis HA, Kerkering TM, Medoff G, Craven PC, et al. Treatment of cryptococcal meningitis with combination amphotericin B and flucytosine for four as compared with six weeks. *N Engl J Med* (1987) 317(6):334–41. doi:10.1056/NEJM198708063170602
113. Sarosi GA, Parker JD, Doto IL, Tosh FE. Amphotericin B in cryptococcal meningitis. Long-term results of treatment. *Ann Intern Med* (1969) 71(6):1079–87. doi:10.7326/0003-4819-71-6-1079
114. Diamond RD, Bennett JE. Prognostic factors in cryptococcal meningitis. A study in 111 cases. *Ann Intern Med* (1974) 80(2):176–81. doi:10.7326/0003-4819-80-2-176
115. Jarvis JN, Meintjes G, Rebe K, Williams GN, Bicanic T, Williams A, et al. Adjunctive interferon-gamma immunotherapy for the treatment of HIV-associated cryptococcal meningitis: a randomized controlled trial. *AIDS* (2012) 26(9):1105–13. doi:10.1097/QAD.0b013e3283536a93
116. Perfect JR, Dismukes WE, Dromer F, Goldman DL, Graybill JR, Hamill RJ, et al. Clinical practice guidelines for the management of cryptococcal disease: 2010 update by the infectious diseases society of america. *Clin Infect Dis* (2010) 50(3):291–322. doi:10.1086/649858
117. Pappas PG, Bustamante B, Ticona E, Hamill RJ, Johnson PC, Reboli A, et al. Recombinant interferon-gamma 1b as adjunctive therapy for AIDS-related acute cryptococcal meningitis. *J Infect Dis* (2004) 189(12):2185–91. doi:10.1086/420829
118. Chen SC, Korman TM, Slavin MA, Marriott D, Byth K, Bak N, et al. Antifungal therapy and management of complications of cryptococcosis due to *Cryptococcus gattii*. *Clin Infect Dis* (2013) 57(4):543–51. doi:10.1093/cid/cit341
119. Mehta GU, Panackal AA, Murayi R, Bennett JE, Williamson PR, Chittiboina P. Corticosteroids for shunted previously healthy patients with non-HIV cryptococcal meningoencephalitis. *J Neurol Neurosurg Psychiatry* (2017). doi:10.1136/jnnp-2017-315830
120. Panackal AA, Komori M, Kosa P, Khan O, Hammoud DA, Rosen LB, et al. Spinal arachnoiditis as a complication of cryptococcal meningoencephalitis in non-HIV previously healthy adults. *Clin Infect Dis* (2017) 64(3):275–83. doi:10.1093/cid/ciw739
121. Hammoud DA, Mahdi E, Panackal AA, Wakim P, Sheikh V, Sereti I, et al. Choroid plexitis and ependymitis by magnetic resonance imaging are biomarkers of neuronal damage and inflammation in HIV-negative cryptococcal meningoencephalitis. *Sci Rep* (2017) 7(1):9184. doi:10.1038/s41598-017-09694-0
122. Collins JV, Tong D, Bucknall RG, Warin AP. Cryptococcal meningitis as a complication of systemic lupus erythematosus treated with systemic corticosteroids. *Postgrad Med J* (1972) 48(555):52–5. doi:10.1136/pgmj.48.555.52
123. Rinehart JJ, Balcerzak SP, Sagone AL, LoBuglio AF. Effects of corticosteroids on human monocyte function. *J Clin Invest* (1974) 54(6):1337–43. doi:10.1172/JCI107880
124. Beardsley J, Wolbers M, Kamali A, Cuc NT, Chierakul W, Chan AK, et al. Adjunctive Corticosteroids in HIV-associated cryptococcal meningitis: a randomised controlled trial in African and Southeast Asian Adults. *55th Interscience Conference on Antimicrobial Agents and Chemotherapy*. San Diego, CA (2015).
125. Meya DB, Manabe YC, Castelnuovo B, Cook BA, Elbireer AM, Kambugu A, et al. Cost-effectiveness of serum cryptococcal antigen screening to prevent deaths among HIV-infected persons with a CD4 + cell count < or = 100 cells/microL who start HIV therapy in resource-limited settings. *Clin Infect Dis* (2010) 51(4):448–55. doi:10.1086/655143
126. Hakim J, Musiime V, Szubert AJ, Mallewa J, Siika A, Agutu C, et al. Enhanced prophylaxis plus antiretroviral therapy for advanced HIV infection in Africa. *N Engl J Med* (2017) 377(3):233–45. doi:10.1056/NEJMoa1615822
127. Beale MA, Sabiiti W, Robertson EJ, Fuentes-Cabrejo KM, O'Hanlon SJ, Jarvis JN, et al. Genotypic diversity is associated with clinical outcome and phenotype in cryptococcal meningitis across Southern Africa. *PLoS Negl Trop Dis* (2015) 9(6):e0003847. doi:10.1371/journal.pntd.0003847
128. Sun HY, Singh N. Opportunistic infection-associated immune reconstitution syndrome in transplant recipients. *Clin Infect Dis* (2011) 53(2):168–76. doi:10.1093/cid/cir276
129. Sun HY, Wagener MM, Singh N. Cryptococcosis in solid-organ, hematopoietic stem cell, and tissue transplant recipients: evidence-based evolving trends. *Clin Infect Dis* (2009) 48(11):1566–76. doi:10.1086/598936

17

Evaluation of Simpler Criteria for Diagnosing Allergic Bronchopulmonary Aspergillosis Complicating Asthma

Ritesh Agarwal[1†], Puneet Saxena[1†], Valliappan Muthu[1], Inderpaul Singh Sehgal[1], Sahajal Dhooria[1], Kuruswamy Thurai Prasad[1], Ashutosh Nath Aggarwal[1] and Arunaloke Chakrabarti[2]*

[1] Department of Pulmonary Medicine, Postgraduate Institute of Medical Education and Research (PGIMER), Chandigarh, India, [2] Department of Medical Microbiology, Postgraduate Institute of Medical Education and Research (PGIMER), Chandigarh, India

***Correspondence:**
Ritesh Agarwal
agarwal.ritesh@outlook.in*

Background: The modified International Society for Human and Animal Mycology (ISHAM) allergic bronchopulmonary aspergillosis (ABPA) working group (AWG) criteria lists up to five components for diagnosing ABPA in asthmatics. Whether eliminating specific components of the existing criteria would have the same diagnostic utility as the original remains unknown.

Objective: To evaluate the performance of several simplified criteria for diagnosing ABPA.

Methods: We compared the performance of seven new criteria (after excluding or modifying one or more of the components of the ISHAM-AWG criteria) with the modified ISHAM-AWG criteria in asthmatic subjects using latent class analysis (LCA). We also tested the performance of the newer criteria using accuracy measures against a multidisciplinary team (MDT) diagnosis of ABPA. We considered the diagnostic accuracy of the newer criteria to be acceptable if the correct classification and false-negative rates were >95% and <5%, respectively, on an MDT evaluation.

Results: We analyzed data from 543 asthmatic subjects (58.8% women; mean age, 36.8 years). Using LCA, the sensitivity of the *A.fumigatus*-specific IgE-based criteria ranged from 92-99%, while the specificity varied between 92% and 100%. The MDT diagnosed ABPA in 106 (19.5%) subjects. Using MDT as the reference standard, the correct classification and false-negative rates were >95% for three of the seven and <5% for four of the seven newer criteria.

Conclusions: We found several of the newly developed criteria to perform, like the modified ISHAM-AWG criteria, for diagnosing ABPA complicating asthma. A prospective study in current clinical algorithms is required for validating our observations.

Keywords: allergic bronchopulmonary mycosis (ABPM), *Aspergillus*, fungal sensitization, allergy, latent class analysis (LCA), bronchial asthma, cystic fibrosis

INTRODUCTION

Allergic bronchopulmonary aspergillosis (ABPA) is a pulmonary disorder caused by complex hypersensitivity reactions targeted against *Aspergillus fumigatus* colonizing the airways of patients with asthma (and cystic fibrosis) (Agarwal et al., 2022). ABPA patients typically present with difficult-to-control asthma, recurrent pulmonary opacities, and bronchiectasis (Kosmidis and Denning, 2015). Several criteria based on a composite of imaging and immunological parameters are available for diagnosing ABPA (Rosenberg et al., 1977; Agarwal et al., 2010; Agarwal et al., 2013; Agarwal et al., 2013; Dhooria and Agarwal, 2014; Asano et al., 2021; Saxena et al., 2021). Rosenberg et al. were the first to suggest a criterion for diagnosing ABPA (Rosenberg et al., 1977). Subsequently, several modifications to these criteria have been proposed to incorporate the advances in immunological tests.

In 2013, the ABPA working group (AWG) of the International Society for Human and Animal Mycology (ISHAM) had framed a criterion for diagnosing ABPA in asthmatics (Agarwal et al., 2013). The ISHAM-AWG criterion relies on certain thresholds of a combination of several immunological and other tests. Although this criterion is a refinement over previously available diagnostic algorithms, it would be interesting to both clinicians and patients if the diagnostic standards could be made even easier. Recently, we proposed specific changes to the ISHAM-AWG criteria that made the criteria simpler and resulted in better sensitivity and specificity (Saxena et al., 2021). The current study aimed to evaluate the performance of several simpler criteria for diagnosing ABPA.

METHODS

We conducted a retrospective analysis of prospectively collected data (July 2017 and September 2018) at our institute's chest clinic. The Institute Ethics Committee approved the study protocol. We obtained written informed consent from all the participants.

Patients

We have published the detailed methodology and the patient details previously (Saxena et al., 2021). Briefly, we included consecutive cases of physician-diagnosed bronchial asthma. We classified the asthma severity based on the 2004 Global Initiative for Asthma (GINA) recommendations that adjusts for the effect of treatment on disease severity (Global Initiative for Asthma (GINA), 2004).

Study Procedure

We performed spirometry, *Aspergillus* skin test (AST), *A.fumigatus*-specific IgE and IgG, serum total IgE, peripheral blood total eosinophil count (TEC), chest radiograph, and computed tomography (CT) of the chest in all the asthmatic subjects. We performed AST intradermally using the crude *A.fumigatus* antigen (Alcit India, New Delhi, India), and an immediate cutaneous hyperreactivity was considered a positive AST (Agarwal et al., 2013). Serum total IgE, *A.fumigatus*-specific IgE, and IgG were assayed using the commercially available fluorescent enzyme immunoassay (Phadia 100, Thermofisher Scientific, Uppsala, Sweden) (Agarwal et al., 2016; Agarwal et al., 2017). We classified the chest radiographic findings as transient or permanent (Agarwal et al., 2016). We obtained CT of the thorax using a multidetector CT scanner and defined bronchiectasis according to previously described criteria (Agarwal et al., 2011). We performed spirometry on an ultrasonic flow-sensing spirometer and interpreted the spirometric values, namely the forced expiratory volume in the first second (FEV1) and the forced vital capacity (FVC), using the Indian guidelines (Aggarwal et al., 2019).

Newer, Simpler Criteria

The modified ISHAM-AWG criterion requires up to 5 components for diagnosing ABPA in asthmatics (**Table 1**). We tested five criteria (criteria 1, 2, 5, 6, and 7) with only four components (eliminating TEC or *A.fumigatus*-specific IgG or CT) and two criteria (criteria 3 and 4) with only three components (eliminating both TEC and *A.fumigatus*-specific IgG). All the criteria included *A.fumigatus*-specific IgE except criteria 4 that replaced it with AST. Criteria 1 and 2 were like criteria 5 and 6, respectively, except for a higher threshold for *A.fumigatus*-specific IgE (0.5 kUA/L instead of 0.35 kUA/L). Criteria 4 was similar to criteria 3, except it replaced *A.fumigatus*-specific IgE with AST. Finally, criteria 7 was solely an immunological criterion, and we excluded CT chest.

TABLE 1 | Various proposed simpler criteria for allergic bronchopulmonary aspergillosis (ABPA) evaluated in the current study.

Modified ISHAM-AWG criteria
Presence of both: (1) serum *Af*-specific IgE >0.35 kUA/L, and (2) serum total IgE >500 IU/mL AND two of the following: (1) serum *Af*-specific IgG >27 mgA/L; (2) bronchiectasis on CT chest; (3) TEC >500 cells/µL
New criteria
Criteria 1: Both: (1) serum *Af*-specific IgE >0.5 kUA/L, and (2) serum total IgE >500 IU/mL AND any of the following: (1) serum *Af*-specific IgG >27 mgA/L; (2) bronchiectasis on CT chest
Criteria 2: Both: (1) serum *Af*-specific IgE >0.5 kUA/L, and (2) serum total IgE >500 IU/mL AND any of the following: (1) bronchiectasis on CT chest; (2) TEC >500 cells/µL
Criteria 3: All the following: (1) serum *Af*-specific IgE >0.35 kUA/L; (2) serum total IgE >500 IU/mL; (3) bronchiectasis on CT chest
Criteria 4: All the following: (1) type 1 Aspergillus skin test positive; (2) serum total IgE >500 IU/mL; (3) bronchiectasis on CT chest
Criteria 5: Both: (1) serum *Af*-specific IgE >0.35 kUA/L, and (2) serum total IgE >500 IU/mL AND any of the following: (1) serum *Af*-specific IgG >27 mgA/L; (2) bronchiectasis on CT chest
Criteria 6: Both: (1) serum *Af*-specific IgE >0.35 kUA/L, and (2) serum total IgE >500 IU/mL AND any of the following: (1) bronchiectasis on CT chest; (2) TEC >500 cells/µL
Criteria 7: Both: (1) serum *Af*-specific IgE >0.35 kUA/L, and (2) serum total IgE >500 IU/mL AND any of the following: (1) serum *Af*-specific IgG >27 mgA/L; (2) TEC >500 cells/µL

Af, Aspergillus fumigatus; AWG, ABPA working group; CT, computed tomography; ISHAM, international society for human and animal mycology; TEC, peripheral blood total eosinophil count.

Objectives

The study objectives were to (1) compare the diagnostic performance of seven simpler standards (**Table 1**) with the recently published modified ISHAM-AWG criteria for diagnosing ABPA, using latent class analysis (LCA); (2) evaluate the diagnostic performance of the different newer benchmarks for detecting ABPA against a multidisciplinary team (MDT) diagnosis of ABPA.

Statistical Analysis

We analyzed data using the commercial statistical package SPSS version 22.0 (IBM SPSS Inc., Armonk NY, US) and present the data as number (percentage) or mean (standard deviation [SD]). We performed LCA using the R version of the TAGS software (Pouillot et al., 2002; Agarwal et al., 2013; Saxena et al., 2021). For performing LCA, we divided the asthma population into mild-to-moderate (consisting of mild intermittent, mild persistent, and moderate persistent asthma) or severe (comprising of severe persistent asthma), according to the 2004 GINA guidelines (Global Initiative for Asthma (GINA), 2004). We also analyzed the difference in LCA results when the study population is not divided into two groups (single group). We calculated the sensitivity and specificity (with corresponding 2.5-97.5% bootstrap confidence intervals [CI]), obtained by bootstrapping 5000 samples) using the maximum likelihood (ML) estimates. We used the Newton-Raphson and the expectation-maximization algorithms for deriving the ML estimates (Pouillot et al., 2002). We assumed conditional independence using the goodness-of-fit test and subsequently deduced the residual correlations between tests. A p-value >0.05 and the random distribution of the residuals near-zero indicated a good fit of the model. We used the DAG_stat application for calculating the sensitivity, specificity, correct classification rate (efficiency), and false-negative rate with 95% CI of the various criteria using MDT diagnosis as the reference standard (Mackinnon, 2000). We assumed the diagnostic accuracy of the newer criteria to be acceptable if the efficiency was >95% and the false-negative rate <5%.

RESULTS

We included 543 asthmatic subjects (58.8% women) with a mean (SD) age of 36.8 (13.9) years. There were 338 and 205 subjects with mild-to-moderate and severe asthma, respectively. The proportion of subjects with aspergillus skin test positivity, serum total IgE level >500 IU/mL, serum *A.fumigatus*-specific IgE values >0.35 and >0.5 kUA/L, serum *A.fumigatus*-specific IgG >27 mgA/L, and peripheral blood TEC >500 cells/μL were similar between the two groups (**Table 2**). We encountered bronchiectasis in 24.3% of the study subjects.

Objective 1

We compared seven newer criteria with the modified ISHAM-AWG criteria using LCA (**Table 3**). The sensitivity of the *A.fumigatus*-specific IgE-based criteria (criteria 1, 2, 3, 5, 6, and 7) ranged from 92-99%, while the specificity varied between 92% and 100%. The diagnostic performance of *A.fumigatus*-specific IgE dependent criteria was almost similar to the modified ISHAM-AWG criteria. The sensitivity was lowest for the new criteria 4 [also known as minimal diagnostic criteria for ABPA (Schwartz and Greenberger, 1991)]; however, the specificity was nearly 100%. When the skin test in the minimal essential criteria was replaced with *A.fumigatus*-specific IgE (**Table 3**, criteria 3), the sensitivity of the minimal criteria increased to 92% with no loss of specificity. The specificity was lowest for the new criteria 7 that excluded imaging, although the sensitivity was 93.5%. The ML estimate for the model was -857.8 (deviance, 631.8). The p-value for the goodness-of-fit test for conditional independence was 0.00001, indicating that the model does not fit. We found the results similar even if the population is kept as a single group (**Table 3**).

TABLE 2 | Select baseline characteristics of the study population.

	Total (n=543)	Mild-to-moderate asthma (n=338)	Severe asthma (n=205)	P value
Age (years)	36.8 ± 13.9	34.9 ± 13.5	39.9 ± 13.8	0.0001
Female sex	319 (58.8)	186 (55.0)	133 (64.9)	0.024
Body mass index (Kg/m^2)	23.6 ± 5.2	23.5 ± 5.2	23.8 ± 5.2	0.55
Asthma duration (years)	10.2 ± 9.5	9.0 ± 8.8	12.0 ± 10.3	0.0001
Spirometry				
FEV1 (% predicted)	73.5 ± 25.6	77.7 ± 24.0	66.6 ± 26.7	0.0001
FVC (% predicted)	82.7 ± 20.7	85.9 ± 19.8	77.2 ± 21.0	0.0001
Immunologic investigations				
Type 1 *Af* skin test positive	164 (30.2)	103 (30.5)	61 (29.8)	0.86
Serum *Af*-specific IgE >0.35 kUA/L	219 (40.5)	133 (39.5)	86 (42.2)	0.54
Serum *Af*-specific IgE >0.5 kUA/L	201 (37.0)	123 (36.4)	78 (33.0)	0.70
Serum *Af*-specific IgG >27 mgA/L	194 (35.7)	119 (35.2)	75 (36.6)	0.75
Peripheral blood eosinophil count >500 cells/μL	223 (41.1)	132 (39.1)	91 (44.4)	0.22
Serum total IgE level >500 IU/mL	300 (55.3)	186 (55)	114 (55.6)	0.90
Bronchiectasis on CT chest	132 (24.3)	73 (21.6)	59 (28.8)	0.059

All values are represented as mean ± SD or number (percentage) unless otherwise indicated.

Af, Aspergillus fumigatus; CT, computed tomography; FEV1, forced expiratory volume in the first second; FVC, forced vital capacity.

Mild-to-moderate asthma consisted of mild intermittent, mild persistent, and moderate persistent, while severe asthma comprised of severe persistent category as per the 2004 GINA guidelines.

TABLE 3 | Diagnostic performance of various criteria for diagnosing allergic bronchopulmonary aspergillosis (ABPA) using latent class analysis.

	Two populations		Single population	
	Sensitivity (95% CI)	**Specificity (95% CI)**	**Sensitivity (95% CI)**	**Specificity (95% CI)**
Modified ISHAM-AWG criteria	94.4 (89.7-98.3)	100 (100-100)	94.4 (89.7-98.3)	100 (100-100)
New criteria				
Criteria 1	96.3 (92.4-99.2)	97.7 (96.2-99.1)	93.5 (88.6-97.9)	97.7 (96.2-99.1)
Criteria 2	98.1 (95.2-100)	95.4 (93.3-97.3)	98.1 (95.0-100)	95.4 (93.4-97.3)
Criteria 3	91.7 (86.0-96.6)	100 (100-100)	91.6 (86.0-96.4)	100 (100-100)
Criteria 4	73.1 (64.4-81.3)	99.8 (99.3-100)	73.1 (64.7-81.3)	99.8 (99.3-100)
Criteria 5	99.1 (96.9-100)	97.7 (96.2-99.1)	99.1 (96.7-100)	97.7 (96.2-99.1)
Criteria 6	99.1 (96.9-100)	94.0 (91.7-96.1)	99.1 (96.7-100)	94.0 (91.7-96.2)
Criteria 7	93.5 (88.6-97.9)	91.7 (88.9-94.2)	93.5 (88.5-97.9)	91.7 (89.1-94.3)

The values in parentheses are confidence intervals that correspond to 2.5-97.5% bootstrap confidence intervals (CI) obtained by bootstrapping 5000 samples.
ISHAM-AWG, international society for human and animal mycology ABPA working group criteria
The p-value for goodness-of-fit for the models is <0.05, indicating that the model does not fit.
*The details of the various criteria are provided in **Table 1**.*

Objective 2

We further compared the diagnostic performance of various criteria using an MDT diagnosis as the reference standard. The MDT adjudicated 106 cases to have ABPA (**Table 4**). The correct classification rate was >95% for criteria 1, 3, and 5 (**Table 4**). The false-negative rate was <5% for all criteria except for criteria 3 (11.3%), 4 (26.5%), and 7 (26.3%).

DISCUSSION

We found several simpler criteria to perform reasonably well than the modified ISHAM-AWG criteria. An optimal diagnostic criterion should be valid, reliable, and straightforward. Moreover, the criterion should have a clearly defined diagnostic threshold. The modified ISHAM-AWG meets most of these attributes but involves several tests, and less complex criteria would be welcome. While minimal criteria have been proposed in the past (Schwartz and Greenberger, 1991), none have been evaluated for their performance. We designed seven criteria and compared them with the modified ISHAM-AWG criteria. We found no benefit on LCA and MDT evaluation by increasing the *A.fumigatus*-specific IgE threshold from 0.35 to 0.5 kUA/L (criteria 1 and 2 vs. 5 and 6). The minimal essential diagnostic criteria for ABPA (Aspergillus skin test positivity, raised serum total IgE, and bronchiectasis; criteria 4) (Schwartz and Greenberger, 1991) had a sensitivity of only about 74%, thus potentially missing many cases despite having a nearly 100% specificity. There was a considerable increase in the sensitivity of the minimal essential criteria once skin test positivity was replaced by *A.fumigatus*-specific IgE (criteria 3). Finally, criteria 7 based on immunological components alone had a reasonable performance on LCA but lower on the MDT evaluation due to the higher number of patients with bronchiectasis in the study.

We have previously used latent class analysis (LCA), a relatively modern computational approach for refining the diagnostic criteria for ABPA (Agarwal et al., 2013; Saxena et al., 2021). The LCA statistically calculates the relationship between certain user-defined input variables (for example, the various diagnostic criteria) that estimate a latent variable, namely the presence or absence of disease (ABPA) (Rindskopf and Rindskopf, 1986; Goetghebeur et al., 2000). Importantly, the LCA technique provides diagnostic accuracy measures without requiring a 2 x 2 matrix that is usually required when a reference standard is used. Conventionally, for performing LCA, we divide the asthma population into two groups. This procedure is complicated as

TABLE 4 | Performance of various diagnostic criteria for allergic bronchopulmonary aspergillosis (ABPA) using the multidisciplinary team (MDT) evaluation as the reference standard.

	MDT, Not ABPA (n=437)		MDT, ABPA (n=106)		Sensitivity (95% CI)	Specificity (95% CI)	Correct classification rate (95% CI)	False-negative rate (95% CI)
Criteria	**Present**	**Absent**	**Present**	**Absent**				
Modified ISHAM-AWG criteria	0	437	102	4	96.2 (90.6-98.9)	100 (99.2-100)	99.3 (98.1-99.8)	3.8 (1.0-9.4)
New criteria								
Criteria 1	14	423	102	4	96.2 (90.6-98.9)	96.8 (94.7-98.2)	96.7 (94.8-98.0)	3.8 (1.0-9.4)
Criteria 2	24	413	102	4	96.2 (90.6-98.9)	94.5 (91.9-96.5)	94.8 (92.6-96.6)	3.8 (1.0-9.4)
Criteria 3	05	432	94	12	88.7 (81.1-94.0)	98.9 (97.4-99.6)	96.9 (95.0-98.2)	11.3 (5.9-18.9)
Criteria 4	02	435	78	28	73.6 (64.1-81.7)	99.5 (98.4-99.9)	94.5 (92.2-96.2)	26.5 (18.3-35.9)
Criteria 5	15	422	102	4	96.2 (90.6-98.9)	96.6 (94.4-98.1)	96.5 (94.6-97.9)	3.8 (1.0-9.4)
Criteria 6	31	406	102	4	96.2 (90.6-98.9)	92.9 (90.1-95.1)	93.6 (91.2-95.5)	3.8 (1.0-9.4)
Criteria 7	36	401	101	5	73.7 (65.5-80.9)	98.8 (97.2-99.6)	92.5 (89.9-64.5)	26.3 (19.1-34.4)

CI, confidence intervals; ISHAM-AWG, international society for human and animal mycology ABPA working group criteria.
*The details of the various criteria are provided in **Table 1**.*

each criterion (n=8) has to be applied in every subject. Further, the presence or absence of the criteria is reported in a binary fashion (0 or 1), and we have 256 different permutations in two groups (**Table S1**). We evaluated whether the results would be different if the asthma population is not divided into two groups (**Table S2**). We found the results similar whether or not the population is divided into two groups (**Table 3**).

What is the clinical relevance of our study? We evaluated seven simpler models for diagnosing ABPA. The *A.fumigatus*-specific IgE-based criteria performed better than skin test-based criteria. Of the seven, the combination of IgE (total and *A.fumigatus*-specific) and either an elevated *A.fumigatus*-specific IgG or bronchiectasis (criteria 5) could be a reasonable alternative to the modified ISHAM-AWG criteria. Also, the specific IgE-based minimal essential criteria (criteria 3) might be another alternative, especially in resource-constrained settings, as it includes only three components. The skin test-based minimal essential criteria (criteria 4) may be used to confirm ABPA (specificity >99%) in settings without access to immunoassays. However, criteria 4 cannot be used to rule out ABPA, given its poor sensitivity (73%). Lastly, our study results provide clinicians with evidence regarding the certainty of diagnosing ABPA when one or more of the components of the existing criterion are missing.

Finally, our study has a few limitations. Our study is a single-center study conducted at a tertiary care hospital. Also, a large proportion of patients in our study had bronchiectasis. The diagnostic performance of the criteria might be different in the milder asthmatics seen in the community or those with serological ABPA. Another weakness can be attributed to the statistical technique as the probabilistic estimates cannot be perfect and depend on the positivity and negativity rates of the variables included in the model. Also, the goodness-of-fit for the model indicated that conditional independence was not met. Thus, the diagnostic performance may be fallible. While an MDT comparison of the various diagnostic criteria was also performed to circumvent this flaw, the criteria need to be validated prospectively. The study results are also not immediately applicable to other predisposing conditions for ABPA, including cystic fibrosis and others.

In conclusion, we found that specific IgE-based simpler criteria may be good alternatives for diagnosing ABPA. Because of the monocentric nature of our study, a prospective multicenter investigation is required for validating our study results.

AUTHOR CONTRIBUTIONS

RA conceived the idea, involved in patient management, statistical analysis, drafted, and revised the manuscript for intellectual content. PS: involved in patient management and data collection, revised the manuscript for intellectual content. VM: involved in patient management, statistical analysis, revised the manuscript. IS: involved in patient management, revised the manuscript. SD: involved in patient management, revised the manuscript. KP: involved in patient management, revised the manuscript. AA: involved in patient management, revised the manuscript for intellectual content. AC: involved in patient management, revised the manuscript for intellectual content. RA: guarantor of the paper. All authors contributed to the article and approved the submitted version.

REFERENCES

1. Agarwal, R., Aggarwal, A. N., Sehgal, I. S., Dhooria, S., Behera, D., and Chakrabarti, A. (2016). Utility of IgE (Total and Aspergillus Fumigatus Specific) in Monitoring for Response and Exacerbations in Allergic Bronchopulmonary Aspergillosis. *Mycoses* 59 (1), 1–6. doi: 10.1111/myc.12423
2. Agarwal, R., Chakrabarti, A., Shah, A., Gupta, D., Meis, J. F., Guleria, R., et al. (2013). Allergic Bronchopulmonary Aspergillosis: Review of Literature and Proposal of New Diagnostic and Classification Criteria. *Clin. Exp. Allergy* 43 (8), 850–873. doi: 10.1111/cea.12141
3. Agarwal, R., Dua, D., Choudhary, H., Aggarwal, A. N., Sehgal, I. S., Dhooria, S., et al. (2017). Role of Aspergillus Fumigatus-Specific IgG in Diagnosis and Monitoring Treatment Response in Allergic Bronchopulmonary Aspergillosis. *Mycoses* 60 (1), 33–39. doi: 10.1111/myc.12541
4. Agarwal, R., Khan, A., Garg, M., Aggarwal, A. N., and Gupta, D. (2011). Pictorial Essay: Allergic Bronchopulmonary Aspergillosis. *Indian J. Radiol. Imaging* 21 (4), 242–252. doi: 10.4103/0971-3026.90680
5. Agarwal, R., Khan, A., Gupta, D., Aggarwal, A. N., Saxena, A. K., and Chakrabarti, A. (2010). An Alternate Method of Classifying Allergic Bronchopulmonary Aspergillosis Based on High-Attenuation Mucus. *PloS One* 5 (12), e15346. doi: 10.1371/journal.pone.0015346
6. Agarwal, R., Maskey, D., Aggarwal, A. N., Saikia, B., Garg, M., Gupta, D., et al. (2013). Diagnostic Performance of Various Tests and Criteria Employed in Allergic Bronchopulmonary Aspergillosis: A Latent Class Analysis. *PloS One* 8 (4), e61105. doi: 10.1371/journal.pone.0061105
7. Agarwal, R., Muthu, V., Sehgal, I. S., Dhooria, S., Prasad, K. T., and Aggarwal, A. N. (2022). Allergic Bronchopulmonary Aspergillosis. *Clin. Chest Med.* 43 (1), 99–125. doi: 10.1016/j.ccm.2021.12.002
8. Agarwal, R., Sehgal, I. S., Dhooria, S., and Aggarwal, A. N. (2016). Developments in the Diagnosis and Treatment of Allergic Bronchopulmonary Aspergillosis. *Expert Rev. Respir. Med.* 10 (12), 1317–1334. doi: 10.1080/17476348.2016.1249853
9. Aggarwal, A. N., Agarwal, R., Dhooria, S., Prasad, K. T., Sehgal, I. S., Muthu, V., et al. (2019). Joint Indian Chest Society-National College of Chest Physicians (India) Guidelines for Spirometry. *Lung India* 36 (Supplement), S1–S35. doi: 10.4103/lungindia.lungindia_300_18
10. Asano, K., Hebisawa, A., Ishiguro, T., Takayanagi, N., Nakamura, Y., Suzuki, J., et al. (2021). New Clinical Diagnostic Criteria for Allergic Bronchopulmonary Aspergillosis/Mycosis and its Validation. *J. Allergy Clin. Immunol.* 147 (4), 1261–1268.e5. doi: 10.1016/j.jaci.2020.08.029
11. Dhooria, S., and Agarwal, R. (2014). Diagnosis of Allergic Bronchopulmonary Aspergillosis: A Case-Based Approach. *Future Microbiol.* 9 (10), 1195–1208. doi: 10.2217/fmb.14.74
12. Global Initiative for Asthma. (GINA). (2004). *Global Strategy for Asthma Management and Prevention.* Available at: https://ginasthma.org/wp-content/uploads/2019/01/2004-GINA.pdf.
13. Goetghebeur, E., Liinev, J., Boelaert, M., and van der Stuyft, P. (2000). Diagnostic Test Analyses in Search of Their Gold Standard: Latent Class Analyses With Random Effects. *Stat. Methods Med. Res.* 9 (3), 231–248. doi: 10.1177/ 096228020000900304
14. Kosmidis, C., and Denning, D. W. (2015). The Clinical Spectrum of Pulmonary Aspergillosis. *Thorax* 70 (3), 270–277. doi: 10.1136/thoraxjnl- 2014-206291
15. Mackinnon, A. (2000). A Spreadsheet for the Calculation of Comprehensive Statistics for the Assessment of Diagnostic Tests and Inter-Rater Agreement. *Comput. Biol. Med.* 30 (3), 127–134. doi: 10.1016/S0010-4825(00)00006-8
16. Pouillot, R., Gerbier, G., and Gardner, I. A. (2002). "TAGS", a Program for the Evaluation of Test Accuracy in the Absence of a Gold Standard. *Prev. Vet. Med.* 53 (1-2), 67–81. doi: 10.1016/S0167-5877(01)00272-0
17. Rindskopf, D., and Rindskopf, W. (1986). The Value of Latent Class Analysis in Medical Diagnosis. *Stat. Med.* 5 (1), 21–27. doi: 10.1002/sim.4780050105
18. Rosenberg, M., Patterson, R., Mintzer, R., Cooper, B. J., Roberts, M., and

Harris, K. E. (1977). Clinical and Immunologic Criteria for the Diagnosis of Allergic Bronchopulmonary Aspergillosis. *Ann. Intern. Med.* 86 (4), 405–414. doi: 10.7326/0003-4819-86-4-405

19. Saxena, P., Choudhary, H., Muthu, V., Sehgal, I. S., Dhooria, S., Prasad, K. T., et al. (2021). Which Are the Optimal Criteria for the Diagnosis of Allergic Bronchopulmonary Aspergillosis? A Latent Class Analysis. *J. Allergy Clin. Immunol. Pract.* 9 (1), 328–335.e1. doi: 10.1016/j.jaip.2020.08.043
20. Schwartz, H. J., and Greenberger, P. A. (1991). The Prevalence of Allergic Bronchopulmonary Aspergillosis in Patients With Asthma, Determined by Serologic and Radiologic Criteria in Patients at Risk. *J. Lab. Clin. Med.* 117 (2), 138–142.

18

Intestinal Infection of *Candida albicans*: Preventing the Formation of Biofilm by *C. albicans* and Protecting the Intestinal Epithelial Barrier

Ziyao Peng and Jianguo Tang*

Department of Trauma-Emergency and Critical Care Medicine, Shanghai Fifth People's Hospital, Fudan University, Shanghai, China

Correspondence:
Jianguo Tang
tangjianguo@5thhospital.com

The large mortality and morbidity rate of *C. albicans* infections is a crucial problem in medical mycology. Because the generation of biofilms and drug resistance are growing concerns, the growth of novel antifungal agents and the looking for newer objectives are necessary. In this review, inhibitors of *C. albicans* biofilm generation and molecular mechanisms of intestinal epithelial barrier protection are elucidated. Recent studies on various transcription elements; quorum-sensing molecules; host responses to adherence; and changes in efflux pumps, enzymes, bud to hyphal transition, and lipid profiles have increased the knowledge of the intricate mechanisms underlying biofilm resistance. In addition, the growth of novel biomaterials with anti-adhesive nature, natural products, drugs, bioactive compounds, proteins, lipids, and carbohydrates are being researched. Recently, more and more attention has been given to various metal nanoparticles that have also appeared as antibiofilm agents in *C. albicans*. The intestinal epithelial obstacle exerts an crucial effect on keeping intestinal homeostasis and is increasingly associated with various disorders associated with the intestine such as inflammatory bowel disease (IBD), irritable bowel syndrome, metabolic syndrome, allergies, hepatic inflammation, septic shock, etc. However, whether their involvement in the prevention of other intestinal disorders like IBD are useful in *C. albicans* remains unknown. Further studies must be carried out in order to validate their inhibition functions in intestinal *C. albicans*. This provides innovates ideas for intestinal *C. albicans* treatment.

Keywords: *Candida albicans*, antibiofilm, intestinal epithelial barrier, new combination therapy, fungal infection

INTRODUCTION

Fungal infections due to *Candida* species represent an important cause of nosocomial bloodstream infections and are especially usual among seriously ill and intensive care patients and the patients with a solid malignancy or in recovery from abdominal operation (Pongrácz et al., 2015; Li et al., 2016). As a normal element of human intestinal, oral, and vaginal microflora, *C. albicans* (*C. albicans*) is also the leading element causing nosocomial fungemia (Brown et al., 2012). Among several susceptible people, it is argued that *C. albicans* infections are disseminated by gastrointestinal transmission; data from researches applying both patients and animal models supports this assumption (Miranda et al., 2009; Maraki et al., 2015; Shankar et al., 2015). Life-threatening illness with obvious rates of mortality among immunocompromised patients and the patients undergoing immunosuppressive therapy can be caused by *C. albicans* (Pongrácz et al., 2015). *C. albicans* is a commensal fungus that asymptomatically adapts to the normal microflora of the host but is aggressive and virulent once transformed to its hyphal form and covered by an extracellular polymeric substance (EPS). This demonstrates that *C. albicans* is the most common fungal pathogen among humans. It can cause diseases varying from considerable mucosal infection to deadly invasive bloodstream infection, which has a 40% mortality rate (Gulati and Nobile, 2016; Lohse et al., 2018). The three phases of the growth of *C. albicans* include adherence of the yeast cells to medical equipment (early phase), differentiation of yeast cells and hyphal cells (intermediate phase), and a growth in the matrix, which is the mature stage (Alim et al., 2018).

Greatly structured biofilms consist of various cell kinds (i.e., round, budding yeast-form cells; oval pseudohyphal cells; and elongated, cylindrical hyphal cells) encased in an extracellular matrix (Chandra et al., 2001; Ramage et al., 2005, 2009; Fox and Nobile, 2012). Occupying 15% of all hospital-obtained sepsis cases, species within the CTG clade (mainly *C. albicans* and closely associated species) are the remarkable fungal species found in medical device infections as the fourth most usual cause of bloodstream infections in clinical context (Wenzel, 1995; Wisplinghoff et al., 2004; Pfaller and Diekema, 2007). Urinary and central venous catheters, mechanical heart valves, pacemakers, contact lenses, joint prostheses, and dentures are all impressionable to *C. albicans* biofilms (Donlan and Costerton, 2002; Kojic and Darouiche, 2004; Cauda, 2009; Seddiki et al., 2013). Once it generates on an implanted medical equipment, *Candida* biofilm is potential in seeding impregnated bloodstream infections and can result in aggressive systemic tissue and organ infections. Biofilms cause various infections by attaching to surfaces or interfaces and embedding in a matrix of extracellular polymeric substances (Costerton et al., 1999). Biofilm generation on biomaterials and medical equipment including catheters and heart valves, results in chronic infections with high morbidity and mortality rates (Uppuluri et al., 2010; Seddiki et al., 2013). Biofilms are remarkably resistant to drugs after implant-associated infections because matured biofilms form a protein- and carbohydrate-rich extracellular matrix. In virtue of the large resistance of fungal biofilms to present antifungal drugs, great antifungal doses and elimination of the colonized medical equipment are necessary for treating infections (Mermel et al., 2009; Lepak and Andes, 2011; Andes et al., 2012; Cornely et al., 2012; Lortholary et al., 2012). Therefore, novel antifungal agents are needed to stop biofilm generation (Zarnowski et al., 2014).

At the same time, intestinal epithelial cells (IECs) form the first physical and immunological protective wall against aggressive pathogens. They not only coexist with the intestinal microbiota resisting commensal bacteria, but also fight pathogens to maintain homeostasis (Sánchez de Medina et al., 2013). The intestinal mucosal obstacle is chiefly composed of a mechanical obstacle, a chemical obstacle, a microbial obstacle, and an immune obstacle, which exert a significant effect on blocking the attaching of pathogens including *Candida* species (Yan et al., 2013). The mucosal barrier can adequately contain luminal microorganisms and molecules while absorbing nutrients. Alterations of the intestinal epithelial barrier are increasingly being associated with various disorders related to the intestine such as irritable bowel syndrome, inflammatory bowel disease (IBD), metabolic syndrome, hepatic inflammation, allergy, septic shock, and others (Natividad and Verdu, 2013). Several studies have documented that farnesol promotes intestinal epithelial barrier transcriptional regulation by activating JAK/STAT3 signaling. The involved molecules may also represent a good potential target for the treatment of *C. albicans* invasion (Fang et al., 2019). Recent studies have shown that labetalol decreased TBI-caused sympathetic hyperactivity, and restrained histopathological intestinal injury combined with variations in gut permeability and gut TNF-α expression in a rat model of TBI (Lang et al., 2015).

Because broad-spectrum antifungal drugs are extensively applied, the appearance of resistant fungal strains in clinical cases has been a main issue in antifungal therapy. Here, we reviewed current research progress in inhibiting *C. albicans* biofilm formation and summarized current elements of intestinal epithelial barrier protection. These treatments may alter traditional antifungal drugs so that it becomes a novel therapeutic solution for *C. albicans* intestinal infection.

ANTIBIOFILM TREATMENT

Natural Products

Over the past several decades, natural compounds have become an important source of antimalarial, antibacterial, and chemotherapeutic agents. Currently, approximately 60% of drugs applied for treating cancers are obtained from natural sources. In addition, one of the most usual and productive methods of obtaining new therapeutic agents applying medicinal chemistry is to modify natural products (Zaki et al., 2019). Natural commodity screening has proven to be a hopeful strategy. Therefore, it is effective to target fungal biofilms with natural derivatives or synthetic analogs. Antibiofilm agents can make fungal biofilms more impressionable against traditional antibiotics and the hosts' immune systems but might not directly kill the bacteria. The search for *C. albicans* inhibitors has resulted in the identification of many compounds of potential therapeutic use.

Eucarobustol E (EE), a currently reported formyl-phloroglucinol meroterpenoid, displayed potent inhibitory roles against both *C. albicans* yeast cells and biofilms, but no poisonousness toward human cells. Observing an obvious increase in negative regulator genes (TUP1, NRG1), researchers assumed that eucarobustol E's suppression of carbon flow to ergosterol activated the mechanisms of negative hyphal development management and finally contributed to biofilm suppression *in vitro* model (Liu et al., 2017). The biofilm generation of *C. albicans* can be inhibited by CLEO, camphor, or fenchyl alcohol at 0.01% treatments. The treatments appear to prevent hyphal formation, which might be beneficial in controlling *C. albicans* infections *in vitro* (Manoharan et al., 2017). Several vitro studies have documented that purpurin suppressed *C. albicans* biofilm formation by blocking the yeast-to-hypha change under hypha-inducing conditions at sublethal concentration (3 mg/ml) and decreased the metabolic activity of mature biofilms in a way dependent of concentration (Tsang et al., 2012). Recently, researchers have tested 21 methylindoles and found that biofilm formation was effectively inhibited by 1-methylindole-2-carboxylic acid (5MI2CA) at 0.1 mM (17.5 μg\ml) and 5-methylindole-2-carboxylic acid (5MI2CA) at 0.1 mM with *C. albicans* DAY185 and ATCC10231 strains *in vitro* research (Lee et al., 2018). In another study, researchers investigated biofilm-inhibiting activity against *C. albicans* with used different indole derivatives. The vitro and vivo research showed that 7-benzyloxyindole, 4-fluoroindole and 5-iodoindole suppressed biofilm generation more effectively than antifungal agent fluconazole (FCZ). In particular, while reducing *C. albicans* biofilm formation,7-benzyloxyindole at 0.02 mM (4.5 lg m l μ1) did not have an additive effect on planktonic cells (Manoharan et al., 2018). A recently vitro and vivo study proposed a set of compounds, on basis of the Pseudomonas aeruginosa 2-heptyl-4(1H)-quinolone (HHQ) key quinolone interkingdom signal structure, that manifest non-cytotoxic antibiofilm activity in *C. albicans*' fungal pathogens (Reen et al., 2016). Interestingly, an *in vitro* study showed that riccardin D as a macrocyclic bisbibenzyl separated from Chinese liverwort Dumortiera hirsute plays an inhibitory role in *C. albicans* biofilm generation (Li et al., 2012). In addition, as a novel kind of antifungal agent, bisbibenzyls suppress morphogenesis switches and biofilm generation by upregulating DPP3 in *C. albicans in vitro* (Zhang et al., 2011). Perillaldehyde (PAE), a natural monoterpenoid agent extracted from *Perilla frutescens*, has been proved to have multiple physiological capabilities, which is available as an anti-inflammatory, anti-oxidative and antifungal agent. Study has demonstrated that PAE manifests powerful antifungal ability against *C. albicans* (*C. albicans*). Researchers found that PAE prevented NLRP3 inflammasome assembly, decreased the extreme accumulation of ROS and inhibited the p65 transfer in nuclear; all resulting in decreased inflammation in the host. Together, these evidences suggest using PAE to treat *C. albicans* infection *in vitro* and vivo (Chen et al., 2020). Shikonin (SK) is the president component of the red pigment extracts from the roots of the plant Lithospermum erythrorhizon. It not only could prevent the formation of biofilms but also break the maintenance of mature biofilms. In a mouse vulvovaginal candidiasis (VVC) model, the fungal burden was largely decreased after SK treatment. Another studies demonstrated that SK is able to prevent hyphae formation and decrease cellular surface hydrophobicity (CSH). Some hypha-and adhesion-specific genes were distinguishingly expressed in SK-treated biofilm, containing the downregulation of ECE1, HWP1, EFG1, CPH1, RAS1, ALS1, ALS3, and CSH1 and upregulation of TUP1, NRG1, and BCR1 *in vitro*. Furthermore, SK could induce the production of farnesol, a quorum sensing molecule, and an exogenous addition of farnesol improved the antibiofilm activity of SK (Yan et al., 2019).

Antifungal Agents

One challenge for clinicians is that there is a limited number of available antifungal agents. There are three classes of antifungals: azoles, polyenes, and echinocandins, which are primarily used for invasive infections. Although azoles including fluconazole, have been the major treatment method for *Candida* infections for nearly two decades, a decline in susceptibility to azoles, polyenes, and echinocandins by *Candida* species is an expanding problem. The limited quantity of antifungal drugs and the growth in resistance to present antifungals necessitate the discovery of some new antifungal agents (Vila et al., 2016). Presently, FCZ is the preferred therapy in systemic C. albicans infection (Parizkova et al., 1999). Recent research shows that a novel topical triazole PC945 has antifungal activity against emerging yeast Candida auris *in vitro* (Rudramurthy et al., 2019).

Miceli et al. (2009) have shown that doxycycline (DOX; 128 μg/mL) alone had similar role as FCZ (2–1,024 μg/mL) against *C. albicans* biofilms, which only result in a 22.9% reduction of biofilm metabolic activity. Nevertheless, when DOX (128 μg/mL) used in combination with FLC it had an important synergistic effect, resulting in that biofilm metabolic activity of the *C. albicans* biofilm was reduced 58.3%. Furthermore, when DOX was used alone at a higher concentration (2,048 μg/mL) it cause more significant effect, with a 85% increase in reduction. These results demonstrate that the combination of a high-dose DOX-based antimicrobial lock therapy and traditional antifungal agents may be more advantageous to the treatment of *C. albicans* biofilms *in vitro* (Miceli et al., 2009). In another vitro study, researchers have observed that the synergistic effects and mechanisms of the combination of FLC and DOX at a lower concentration (1–64 μg/mL) against *C. albicans* biofilms (Gao et al., 2013). Some vitro studies have reasearchedresearched the effects of NSAIDs on fungal growth inhibition, enzyme activation, and reduction in fungal prostaglandin E2 (PGE2) production, particularly focusing on inhibiting biofilm formation; For instance, aspirin could decrease the biofilm formation, with a 95% reduction. The inhibitory effects on *C. albicans* of COX inhibitors plus FLC block biofilm development through the PGE2-dependent mechanism, which suggests a new method of solving the biofilm resistance problem (Alem and Douglas, 2004; de Quadros et al., 2011; Ells et al., 2011). In addition, ibuprofen exhibited a synergistic effect with FLC against FLC-resistant strains (but not FLC-susceptible strains) *in vitro* (Arai et al., 2005; Ricardo et al., 2009). Vitro studies suggest that ambroxol (AMB) could easily penetrate the formed biofilm and exert antifungal effects, thereby blocking biofilm formation. The finding herein provide the first mode of action of the antifungal and antibiofilm activity of the mucolytic agent and its advantage to terbinafine as a commercially available antifungal that can prevent the fungal growth and biofilm formation (Rene et al., 2014). Researchers found that anidulafungin exerts an additive effect on immune cells which prevents *Candida* biofilms formation. Moreover, this additive interaction contributes to the release of the proinflammatory cytokine TNF-α and the chemokine IL-8 at different levels. The helpful Th1 response observed after therapy of biofilms with anidulafungin could provide new therapeutic ideas, including inhibiting the release of cytokines with harmful effects and the induction of others with beneficial effects *in vitro* (Katragkou et al., 2010). Besides, D,L-2-hydroxyisocaproic acid (HICA) may become a promising antifungal agent to prevent *C. albicans* cell growth and biofilm formation due to abnormal hyphae and collapsed hyphal structures in time of incubation with HICA at an acidic pH. It's necessary for treating bacterial-fungal biofilm infections to importantly decrease the mutagenic potential of *C. albicans* biofilms *in vitro* (Nieminen et al., 2014). Clioquinol [5-chloro-7-iodoquinolin-8-ol, (CQ)] manifested fungistatic and fungicidal activity against *C. albicans*. It blocked true hyphae formation in a way of concentration-dependent in a variety hyphae-inducing of conditions. CQ also has interferential effect on ion homeostasis in *C. albicans* to prevent the growth of fungi in the vitro model, which is different from the present antifungal agents (You et al., 2020). Farnesol as the first quorum-sensing molecule was discovered in a eukaryote. It prevented the development of biofilms formed by the resistant strain. Moreover, there were synergistic effects between farnesol and fluconazole/5-fluorocytosine, but there were antagonistic effects between farnesol and terbinafine/itraconazole, respectively, on the biofilms formed by the resistant strains *in vitro* (Xia et al., 2017).

The new bacterial quorum sensing quencher thiazolidinedione-8 (S-8) possessed given antibiofilm and antiadhesion activities against *C. albicans*. The expression extents of genes correlated with biofilm generation, adherence, and filamentation (HWP1, ALS3, and EAP1, respectively) were downregulated by S-8 dose-dependently. Therefore, S-8 presents a new antibiofilm therapeutic approach in treating and preventing biofilm-related *C. albicans* infections *in vitro* model (Feldman et al., 2014). Recently vitro and vivo studies revealed that phenylthiazole small molecules including compound 1, have become an important research subject as novel antifungal agents for drug-resistant *Candida* infections. These molecules manifest rapid fungicidal activity and reduce the metabolic activity of appendiculate *C. albicans* and *C. auris* biofilms by more than 66 and 50%, respectively (Mohammad et al., 2019). CD101 is a long-acting novel echinocandin with unique pharmacokinetic characteristics and effective stability and safety relative to the same drug class, acting powerfully against early and mature *C. albicans* biofilms *in vitro* (Chandra and Ghannoum, 2018). In addition, ionic liquids are a new category of molten salts. These compounds have been used as ingredients of active pharmaceutical ingredients and antimicrobials. Previous vitro studies have shown that imidazolium ionic liquid compounds have antifungal and antibiofilm activities by influencing different cellular processes (Reddy and Nancharaiah, 2020). In another study, oral management of the broad-spectrum antibiofilm compound toremifene suppresses *C. albicans* and staphylococcus aureus biofilm generation *in vivo*, exhibiting a promising possibility of toremifene use as a broad-spectrum oral antibiofilm compound (De Cremer et al., 2014). Nitric-oxide releasing aspirin (NO-ASA) has an antifungal/antibiofilm effect on *C. albicans* separates from denture stomatitis patients *in vitro*, which demonstrates NO-ASA's potential as a novel antibiofilm agent for treating fluconazole-resistant strains of *C. albicans* (Madariaga-Venegas et al., 2017). Miltefosine is an alkylphosphocholine showing potent antiparasitic activity. This compound has been proven to inhibit *C. albicans* and non-*albicans Candida* spp. biofilms and impair the interspersion of infectious cells *in vitro* (Vila et al., 2016). Another new discovery has identified that caspofungin as an antifungal agent acted effectively against biofilms by intensely reducing biofilm dispersion under flow conditions *in vitro* (Uppuluri et al., 2011).

Microorganisms

Microorganisms could synthesize different types of surface-active compounds that are effective as antifungal, antibacterial, anti-adhesive, and antibiofilm agents, which could make them useful as main immunomodulatory molecules or in vaccines and gene therapy (Barbosa et al., 2016).

A recent vitro study revealed that *Lactobacillus* strains acted against *Candida*, and the strains' biosurfactants were anti-adhesive and impeded biofilm activity against *C. albicans* (dos Santos et al., 2019). Given that *lactobacilli* and *C. albicans* are present in all regions of the human GI tract, including the low-biodiversity niches of the stomach and small intestine, *Lactobacillus* species may be central to preventing the outgrowth of *C. albicans* and other similarly resilient opportunistic pathogens (Zeise et al., 2021). *In vitro* study, butyrate isolated from *Lactobacillus* cultures can inhibit *C. albicans* hyphal morphogenesis (Noverr and Huffnagle, 2004). Another group found that butyrate inhibited C. albicans growth and filamentation but also enhanced the production of nitric oxide by macrophages and thus their ability to kill C. albicans cells *in vitro*. While this group did not implement any experiments to directly link the effects of butyrate on C. albicans virulence to its function as a HDACi, they hypothesized that that was the likely mechanism (Nguyen et al., 2011). In addition, researchers show that some *Lactobacillus* species produce a small molecule under laboratory conditions that blocks the C. albicans yeast-to-filament transition, an important virulence trait. Bioassay-guided fractionation of Lactobacillus-conditioned medium linked this activity to 1-acetyl-β-carboline (1-ABC). They use genetic approaches to show that filamentation inhibition by 1-ABC requires Yak1, a DYRK1-family kinase. Additional biochemical characterization of structurally related 1-ethoxycarbonyl-β-carboline confirms that it inhibits Yak1 and blocks *C. albicans* biofilm formation. Thus, our findings reveal Lactobacillus-produced 1-ABC can prevent the yeast-to-filament transition in *C. albicans* through inhibition of Yak1 *in vitro* and vivo (MacAlpine et al., 2021). Graham et al. (2017) reported that recognition of the *E. faecalis* bacteriocin, EntV [generated from the entV (ef1097) locus], is essential and adequate for decreasing C. albicans virulence and biofilm generation by inhibiting hyphal generation *in vitro* and vivo research (Graham et al., 2017). Phagocytic cells are crucial components of the innate immune system preventing *C. albicans* mucosal infections. *Streptococcus gordonii* and *Pseudomonas aeruginosa* often colonize mucosal sites, along with *C. albicans*. *S. gordonii* increased *C. albicans* survival and filamentation within macrophage phagosomes, while *P. aeruginosa* reduced fungal survival and filamentation (Salvatori et al., 2020). In another vitro and vivo study, *S. mutans* were capable of secreting subproducts that inhibited biofilm generation, morphogenesis and pathogenicity in

C. albicans, alleviating test candidiasis in the *G. mellonella* model (Barbosa et al., 2016).

Phenotypic Screening

Phenotypic screening means the nascent methodology for biological screening of chemical entities for the assessment of their therapeutic roles. Phenotypic screening was useful for identifying various small molecules exhibiting antibiofilm and anti-filamentation activity against *C. albicans*. Recent studies have shown that a new range of diazaspiro-decane structural analogs were often elements of bioactive compounds, which prevent processes connected with *C. albicans* virulence (most remarkably biofilm generation and filamentation) without affecting overall growth or generating resistance *in vitro* and vivo (Pierce et al., 2015). One study leads to the identification of about 2,293 compounds from the chemical library of the National Cancer Institute which were categorized into three sets- (i) NCI Natural set, Out of all the compounds present in this set six hits were confirmed against *C. albicans* biofilm formation. These include -Trichoderonin; Nanaomycin; Rapamycin; Anisomycin; V alinomycin and Bacitracin. Three of these molecules (Trichoderonin, Nanaomycin, Rapamycin) showed inhibition of both filamentation and biofilm formation while the rest three showed inhibition against biofilm formation. Furthermore in the next (ii)-NCI-Structural diversity set, in total, there were 12 hits out of which eight were identified as biofilm inhibitor. These compounds were-Phenanthroline Hydrochloride; 2-isoquinolin-2-ium-2-yl-1-phenanthren-3-ylethanone, Iodide; Metanilamide (3-aminobenzenesul fonamide); Mercury, (4-amino phenyl) (6-thioguano sinato-N7,S6)-; 2-[7-[3-(carboxymethyl)-5,10-dihydroxy-1-methyl-6,9-dioxo-3,4-dihydro-1H-benzo[g] isochr omen-7-yl]-5,10-dihydroxy-1-methyl-6,9-dioxo-3,4-dihydro-1H -benzo [g]isochromen-3-yl]acetic acid 37 are only biofilm inhibitors while Mercury,(2-aminio-1,9-dihydro-6H-purine-6 -thionato-N7,S6) hexyl-,2-benzo[a]phenothiazin-12-yl-N,N diet hylethanamine; 17-[1-[2(dimethylamino)ethylamino]ethyl]-13-methyl-6,7,8, 9,11,12,14,15,16,17 decahydrocyclopenta[a]phen anthren-3-ol. Next and last was (iii)-NCI-Challenge Set. In this, there were total 11 hits, of which 10 showed inhibition against *C. albicans* biofilm formation whereas only one compound inhibited filamentation transformation. Ten hits which were identified from these compounds displayed common inhibition against both biofilm and filament formation. These include Biofilm Inhibitor- Trichopolyn-B, Vengicide (Unamycin B, Toyocamycin), 4Z-4-[[4-(dimethylamino)phenyl]methylidene]-1-methyl-2-phenylpyrazolo[1,5-a]indol-1-ium-6-ol; trifluorome thanesulfonate, Anisomycin43, Azetidinecarbo thioic acid, [1-(2-pyridinyl) ethylidene] hydrazide4. Additionally, compounds with both antifilamentation and antibiofilm activity are—6-Hydroxy-3-[(methanesulfonyloxy) Methyl]-1-[(5,6,7-tri methoxyindol-2-yl) carbonylindoline, Hydrazineca rbothioamide, N,N-dipropyl-2-(2-pyridinemethylene)-,(N,N,S) copper(II)chloridecomplex(SP-4-3)3;3-Azabicyclo[3.22]nonane-3-carboselenoicacid,[1-(2pyridinyl)ethyidene] hydrazide, 2-hyd roxyethyl-[(2R)-2-hydroxyheptadecyl]-dimethylazaniumiodide, 1H-Azepine-1-carbothioic acid, hexa hydro-, [1-(2-pyridinyl) ethylidene]hydrazide (Pierce et al., 2014).

Protein and Peptide Inhibitors

Researchers are trying to develop effective and potent therapies designed to eradicate biofilm-associated infections. Among these, antimicrobial peptides (AMPs), cytokines, and various proteins have been examined extensively as new therapeutic agents. Repetitive Lysine-Tryptophan Peptide scans are capable of inhibiting cellular functions by binding to RNA and DNA after it has been translocated into the cell, contributing to the inhibition of biofilm formation in a fluconazole-resistant *C. albicans* strain and the eradication of *C. albicans* (Ramamourthy et al., 2020). Researchers selected four of peptides (cathelicidin-BF, Pc-CATH1, Cc-CATH2, Cc-CATH3) and human cathelicidin LL-37 to carefully examine their anti-*C. albicans* and antibiofilm activities *in vitro* and *in vivo*. Antimicrobial assay suggesting that Pc-CATH1, Cc-CATH2, Cc-CATH3 and cathelicidin-BF have valid antifungal activities against the eight tested *C. albicans* strains, containing standard and clinically isolated amphotericin B-resistant strains. Furthermore, cathelicidin-BF importantly prevented the formation of *C. albicans* biofilms at sub-antimicrobial concentrations, and also manifested powerful activity of killing *C. albicans* in preformed biofilms (Yu et al., 2016). In another study, the naturally happening host defense peptide, LL-37, and its truncated mimetics KE-18 and KR-12 were biocidal and antibiofilm against *C. albicans*, Escherichia coli and staphylococcus aureus *in vitro* (Luo et al., 2017). Adopting an *in vitro C. albicans* biofilm model, research suggests that TNF dose-dependently suppresses biofilm growth stopped by cultivating TNF with N,N′-diacetylchitobiose, a main carbohydrate ingredient of the *C. albicans*' cell wall (Rocha et al., 2017). Weiland-Brauer et al. (2019) assessed the capacity of a multitude of metagenome-derived bacterial quorum quenching (QQ) proteins to block biofilm development in *C. albicans* and S. epidermidis. Here, proteins QQ-5 and QQ-7 obstructed the morphogenesis of *C. albicans* by suppressing a yeast-to-hyphae conversion and impairing biofilm formation *in vitro* model (Weiland-Brauer et al., 2019). Along with this, the hLF1-11 peptide notably inhibited *C. albicans* biofilm formation primarily at early stages, disturbing biofilm cellular density and metabolic activity, and influenced morphogenesis in the Ras1-cAMP-Efg1 path *in vitro* (Morici et al., 2016). By binding the most hopeful amino acid substitutions, researchers observed that the double-substituted OSIP108 analog Q6R/G7K displayed eight-fold-grown antibiofilm activities (Delattin et al., 2014).

Lipid Inhibitors

Based cluded that cinnamaldehyde (CNMA) could potentially be used in multilamellar liposomes (ML) as an antifungal and antibiofilm agent. According to the outcomes of a vitro research, ML-CNMA blocks the proliferation of *C. albicans* and accelerates apoptosis (Khan et al., 2017). In another study, it has been proved that sophorolipid (SL), a glycolipid biosurfactant, has antimicrobial and anticancer characteristics. It was found that SL retards *C. albicans* biofilm formation and decreases the survival of conducted biofilms *in vitro* (Haque et al., 2016). Researchers have found that SMOFlipid did not damage *C. albicans* development, but it did notably suppress hypha generation and hyphal prolongation *in vitro*. Furthermore, development suppression could occur in intralipid when replenished with capric acid, there is a fatty acid in SMOFlipid but no in intralipid. *C. albicans* biofilm generation in PN solutions was also found to be dependently inhibited by capric acid (Willems et al., 2019).

Polysaccharide Inhibitors

Another newly discovered chitosan has been identified as a polysaccharide that inhibited *C. albicans* planktonic development (HMW, 1 mg/mL; LMW, 3 mg/mL) *in vitro*. With respect to biofilm development, chitosan suppressed *C. albicans* adherence (ca. 95%) and biofilm generation (> 90%) and decreased mature biofilms by ca. 65% and dual species biofilms (*C. albicans* and *S. mutans*) by ca. 70%. The outcomes demonstrate that this molecule is potential to be an anti-*Candida* agent working with *C. albicans* infections (Costa et al., 2014). Further, chitosan resistance, aggressive development, biofilm generation, and virulence in *C. albicans* require that MSS2 maintain mitochondrial function *in vitro* (Ke et al., 2021).

Nanoparticles

Due to the weak penetration and non-specificity of antifungal and non-antifungal drugs, it is hard to treat biofilm generation. For addressing this issue, researchers are striving to augment the penetration of drugs into the extracellular matrix of biofilm. Over the past few years, several metal nanoparticles have become potential candidates in treating microbial infections because of their potential as effective antimicrobial agents. Hernandez-Delgadillo et al. (2013) found that aqueous colloidal bismuth oxide nanoparticles exhibited antimicrobial activity against *C. albicans* development (decreasing colony scale by 85%) and inhibited biofilm generation *in vitro* (Hernandez-Delgadillo et al., 2013). In another vitro and vivo study, sustained nitric oxide-releasing nanoparticles resulted in cell death in *C. albicans* yeast and hyphal cells, inhibiting biofilm generation *in vitro* and in a rodent central venous catheter model (Ahmadi et al., 2016).

Other Inhibitors

RNA aptamers chosen against yeast cells suppress *C. albicans* biofilm generation *in vitro* (Bachtiar et al., 2019). Researchers screened three protease inhibitor libraries including 80 compounds for their inhibitive abilities against *C. albicans* biofilm formation *in vitro* and vivo and interference in mature biofilms. The outcomes demonstrate that through integrating normative antifungal agents with given protease inhibitors, it may be a therapy method to prevent and treat *C. albicans* biofilm infections (Lohse et al., 2020). Besides, it was observed that human serum weakens biofilm generation by preventing the adherence of *C. albicans* cells *in vitro*. This answer may relate to the downregulation of adherence-associated genes ALS1, ALS3, and BCR1. The administrative serum ingredient is protease-resistant and heat stable (Ding et al., 2014) (**Table 1**).

PROTECTING THE INTESTINAL EPITHELIAL BARRIER

Mucus Layer

The mucus layer covering gastrointestinal mucosa is the first line of defense against invasions generating from luminal content. It chiefly consists of high molecular weight glycoproteins called mucins (MUC).

Current researches have identified butyric acid as a main source of energy for intestinal epithelial cells that can raise the mucus-layer supplement rate *in vitro* models and butyrate is able to upregulate colonic mucins at the transcriptional level and even better when it is the major energy source of the cells (Gaudier et al., 2004). Inflammatory responses in the gut can be retarded by this short-chain fatty acid through reducing the expression of INF-γ, TLR2, and TNF-α *in vitro* (Elce et al., 2017). Some researches have disclosed that there are more A. muciniphila-derived extracellular vesicles (AmEVs) in the fecal samples of healthy controls compared with those of patients with T2D. In addition, AmEV administration enhanced tight junction function, reduced body weight gain and improved glucose tolerance in high-fat diet (HFD)-induced diabetic mice. To test the direct effect of AmEVs on human epithelial cells, cultured Caco-2 cells were treated with these vesicles. AmEVs decreased the gut permeability of lipopolysaccharide-treated Caco-2 cells, whereas Escherichia coli-derived EVs had no significant effect. Interestingly, the expression of occludin was increased by AmEV treatment. Thus, *A. muciniphila* is capable of restoring mucus layer thickness, relieving intestinal inflammation reactions from pili-like proteins, and improving inflammation-caused obstacle integrity damage, thus decreasing gut barrier demolition (Chelakkot et al., 2018). *In vivo* model showed that ILC3-derived IL-22 can also induce expression of tissue protective mucins and antimicrobial peptides including RegIIIβ, RegIIIγ, S100A8 and S100A9 by acting on the intestinal epithelium (Sonnenberg et al., 2011) (**Figure 1**).

Toll-Like Receptors

Toll-like receptors (TLRs), innate immune sensors, exert a significant role in molding intestinal microbiota. TLRs may be deemed to an interface among the intestinal epithelial barrier, microbiota, and the immune system (Frosali et al., 2015). Furthermore, intestinal flora can mediate TLRs' expression to maintain immune balance (Goto and Kiyono, 2012). TLR pathways are also closely associated with gastrointestinal barrier integrity and function; TLR2 and TLR4 paths are necessary for intestinal protection against acute mucosal injury by maintaining epithelial barrier integrity (Cario, 2008). The study suggests that the intestinal barrier offers places for intestinal flora that regulate TLRs' immune reactions, particularly those in the intestine, because there are several immune cells and non-immune cells that contain TLRs in the intestinal barrier (de Medina et al., 2014). According to vitro, animal models and human results, TLR signaling in the intestinal epithelial cells notably increased the generation of IgA in the intestine. This role was regulated by TLR-caused expression of a given series of

TABLE 1 | Inhibitors involved in *C. albicans* biofilm formation.

	Inhibitors	Biofilm-related processes inhibited	Research type	References
Natural product	Eucarobustol E	Hyphal formation	vitro	Liu et al., 2017
	CLEO, camphor, fenchyl alcohol	Hyphal formation	vitro	Manoharan et al., 2017
	purpurin	Hyphal formation and metabolic activity	vitro	Tsang et al., 2012
	5MI2CA and 1MI2CA	Hyphal formation	vitro and vivo	Lee et al., 2018
	7-benzyloxyindole	Hyphal formation	vitro and vivo	Manoharan et al., 2018
	HHQ	Unknown	vitro and vivo	Reen et al., 2016
	Riccardin D	Hyphal formation	vitro	Li et al., 2012
	Bisbibenzyls	Hyphal formation	vitro	Zhang et al., 2011
	Perillaldehyde	Hyphal formation	vitro and vivo	Chen et al., 2020
	Shikonin	Hyphal formation	vitro and vivo	Yan et al., 2019
Antifungal agents	doxycycline	Unknown	vitro	Miceli et al., 2009; Gao et al., 2013
	NSAIDs	Unknown	vitro	Alem and Douglas, 2004; de Quadros et al., 2011; Ells et al., 2011
	ibuprofen	Unknown	vitro	Arai et al., 2005; Ricardo et al., 2009
	Ambroxol	Unknown	vitro	Rene et al., 2014
	anidulafungin	Unknown	vitro	Katragkou et al., 2010
	D,L-2-hydroxyisocaproic acid	Hyphal formation	vitro	Nieminen et al., 2014
	clioquinol	Hyphal formation	vitro	You et al., 2020
	Farnesol	Hyphal formation and Decreased thickness	vitro	Xia et al., 2017
	thiazolidinedione-8	Adherence, Hyphal formation and metabolic activity	vitro	Feldman et al., 2014
	Phenylthiazole Small Molecule	Adherence and metabolic activity	vitro and vivo	Mohammad et al., 2019
	CD101	Adherence	vitro	Chandra and Ghannoum, 2018
	[C16MIM]Cl	Killing of biofilm cells	vitro	Reddy and Nancharaiah, 2020
	Toremifene	Unknown	vivo	De Cremer et al., 2014
	nitric-oxide releasing aspirin	Adherence and hyphal formation	vitro	Madariaga-Venegas et al., 2017
	Miltefosine	Dispersion	vitro	Vila et al., 2016
	caspofungin	Dispersion	vitro	Uppuluri et al., 2011
Microorganisms	Lactobacillus	Adherence and hyphal formation	vitro and vivo	Noverr and Huffnagle, 2004; Nguyen et al., 2011; dos Santos et al., 2019; MacAlpine et al., 2021; Zeise et al., 2021
	E. faecalis bacteriocin	Hyphal formation	Vitro and vivo	Graham et al., 2017
	Pseudomonas aeruginosa	Hyphal formation	vitro	Salvatori et al., 2020

(Continued)

TABLE 1 | (Continued)

	Inhibitors	Biofilm-related processes inhibited	Research type	References
	Streptococcus mutans	Hyphal formation	vitro and vivo	Barbosa et al., 2016
	diazaspiro-decane structural analogs	Hyphal formation	vitro and vivo	Pierce et al., 2015
Phenotypic screening				
	Trichoderonin, Nanaomycin, Rapamycin et al.	Hyphal formation	vitro	Pierce et al., 2014
	Anisomycin, V alinomycin, Bacitracin et al.	Unknown	vitro	Pierce et al., 2014
Protein and peptides inhibitors	Pc-CATH1, Cc-CATH2, Cc-CATH3 and cathelicidin-BF	Unknown	vitro and vivo	Yu et al., 2016
	LL-37, KE-18 and KR-12	Unknown	vitro	Luo et al., 2017
	TNF	metabolic activity and yeast morphology	vitro	Rocha et al., 2017
	QQ	Hyphal formation	vitro	Weiland-Brauer et al., 2019
	hLF1-11 peptide	metabolic activity and yeast morphology	vitro	Morici et al., 2016
	Decapeptide OSIP108	Unknown	vitro	Delattin et al., 2014
Lipid inhibitors	Cinnamaldehyde	Unknown	vitro	Khan et al., 2017
	Sophorolipid	Hyphal formation	vitro	Haque et al., 2016
	Smoflipid	Hyphal formation	vitro	Willems et al., 2019
Polysaccharide inhibitors	Chitosan	Adherence	vitro	Costa et al., 2014
	MSS2	Adherence	vitro	Ke et al., 2021
Nanoparticles	Bismuth oxide aqueous colloidal nanoparticles	Unknown	vitro and vivo	Hernandez-Delgadillo et al., 2013
	Nitric Oxide-Releasing Nanoparticles	metabolic activity	vitro and vivo	Ahmadi et al., 2016
Other inhibitors	RNA aptamers	Hyphal formation	vitro	Bachtiar et al., 2019
	Combination of Antifungal Drugs and Protease Inhibitors	Unknown	vitro	Lohse et al., 2020
	Human serum	Adherence	vitro	Ding et al., 2014

chemokines and cytokines that facilitated both the employment of B cells to the lamina propria and IgA class shift of B cells, thus decreasing intestinal inflammation (Shang et al., 2008). In a previous study, *B. fragilis,* through TLR2, induced cytokine generation and T cell differentiation. Unlike pathogens that trigger inflammatory responses through TLRs that result in immune responses to clear infections, symbiotic colonization by *B. fragilis* is actually enhanced by signaling *via* the TLR pathway promoting suppression of Th17 immunity. Thus, PSA evolved to engender host-bacterial mutualism by inducing mucosal tolerance through TLR2 activation of Treg cells (Round et al., 2011). Vitro and animal models also suggested the favorable impacts of *Yupingfeng* (YPF) are likely associated with stimulation of cytokines synthesis by triggering TLR2 and TLR4 paths, improving intestinal community structure and intestinal barrier integrity and functionality (Sun et al., 2016). A model of synergy was suggested whereby infection with *C. albicans* increases both the biomass of *S. oralis* and the TLR2 expression to critical levels required for mucosal proinflammatory signaling by this otherwise commensal organism. Whole mouse genome tongue microarray analysis showed that when compared with animals infected with one organism, the doubly infected animals had genes in the major categories of neutrophilic response/chemotaxis/inflammation significantly upregulated, indicative of an exaggerated inflammatory response. This response was dependent on TLR2 signaling since oral lesions, transcription of pro-inflammatory genes and neutrophil infiltration, were attenuated in TLR2-/-animals. Furthermore,

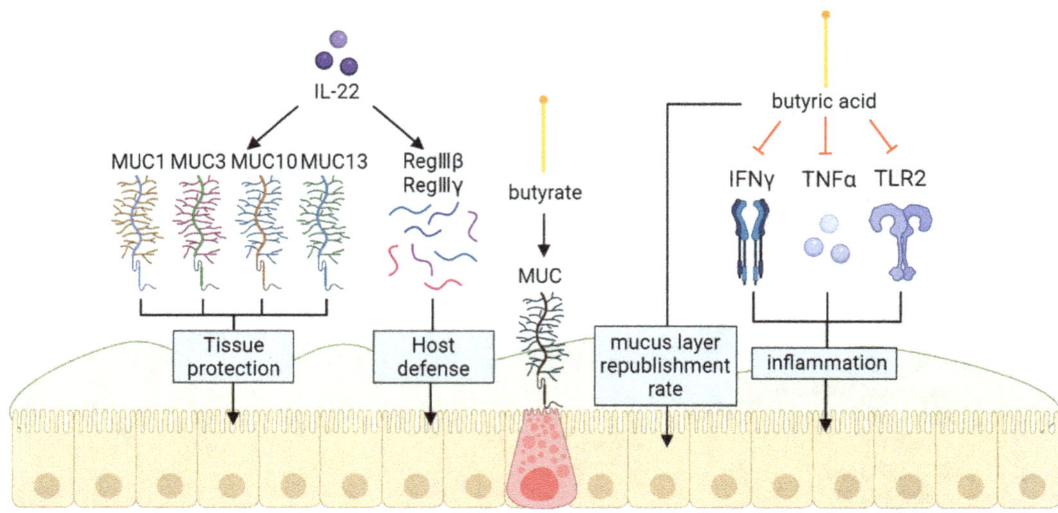

FIGURE 1 | Mucins (MUC) protect the intestinal epithelial barrier. Exogenous IL-22 involved in producing protective mucus (MUC1, MUC3, MUC10 and MUC13) and induce expression of antimicrobial peptides including RegIIIβ, RegIIIγ. Butyrate specifically modulates MUC gene expression in intestinal epithelial goblet cells deprived of glucose Inflammatory responses in the gut can be retarded by butyrate through reducing the expression of INF-γ, TLR2, and TNF-α *in vitro*.

S. oralis activated neutrophils in a TLR2-dependent manner *in vitro*. Thus, this study identifies a previously unrecognized pathogenic synergy between oral commensal bacteria (Xu et al., 2014). In intestinal epithelial cells, TLR2 stimulation efficiently preserves zonula occludens-1 (ZO-1)-associated barrier integrity against stress-induced damage, which is controlled by positive signaling crosstalk between PI3K-Akt and conventional protein kinase C (PKC) isoforms *via* MyD88. In parallel, the PI3K/Akt pathway limits proinflammatory TLR2-signaling through the Mapk-NFkB pathway (Cario, 2008) (**Figure 2**).

Aryl-Hydrocarbon Receptor

An aryl-hydrocarbon receptor (AHR) as a kind of toxin sensor binds to various endogenous and exogenous chemicals. In a previous study, AHR expression was increased in gut resident innate lymphoid cells (ILCs) (Jacquelot et al., 2019). The lack of AHR was related to decreased ILCs numbers and reduced ILC3-derived IL-22 *in vitro* and vivo (Kiss et al., 2012; Lee et al., 2012; Jacquelot et al., 2019). In another study, AHR directly bound to the Il22 promoter and had a synergistic effect on RORγt to promote Il22 expression in ILC3 (Qiu et al., 2012). *In vitro* and vivo studies have shown that the cytochrome P4501 (CYP1) family enzymes mediated the metabolic elimination of AHR ligands. Moreover, structural expression of CYP1 enzymes greatly decreased the usability of AHR ligands and resulted in the loss of gut ILC3 and Th17 cells (Schiering et al., 2017). Therefore, elementary Cyp1a expression or entire loss of AHR in mice improves their sensitivity to *C. rodentium* related to hindered IL-22 generation (Kiss et al., 2012; Qiu et al., 2012; Schiering et al., 2017). There are studies that emphasize how constant control of the availability of AHR ligands *via* intestinal epithelial cells offer vital feedback to immune cells, which thereby forms mucosal protections (Jacquelot et al., 2019) (**Figure 3**).

ATG16L1

The autophagy gene ATG16L1 is related to protection from intestinal epithelial infection. A study showed that there were fewer Paneth cells and unusual granule morphology in Atg16l1f/f × Villin-cre mice, resulting in decreased AMP expression. Congruent with these defective immune responses, Atg16l1f/f × Villin-cre mice had improved inflammation and comprehensive bacterial translocation by comparing with control mice. The limitations of such in mouse model was that there is a compelling need for new therapeutic approaches to modulate specific pathways important in autoinflammatory and infectious diseases. Researchers tried to screen Modulators of autophagy to evaluate their effects on antibacterial responses in human epithelial cells. But few studies have been done on humans (Conway et al., 2013) (**Figure 3**).

Cold-Inducible RNA-Binding Protein

Intestinal injuries occurring during deep hypothermic circulatory arrest (DHCA) are hazardous for clinical outcomes. Recent vitro and animals studies revealed that cold-inducible RNA-binding Protein (CIRBP) provides a protective effect in cases of hypothermia. These findings demonstrated the possibility of utilizing innate mechanisms of CIRBP to sustain the intestinal epithelial barrier during DHCA for the first time. This utilization will probably become a targeted treatment to prevent or relieve intestinal injury and relevant complications (Li et al., 2019) (**Figure 4**).

Epidermal Growth Factor Receptors

Vitro research results demonstrate that M3 receptor-induced activation of p38 MAPK might maintain epithelial barrier function by downregulating TNF-α signaling and activation of Epidermal growth factor receptors (EGFR) instead of H1 (Uwada et al., 2017). Both P40 and P75 proteins were shown to

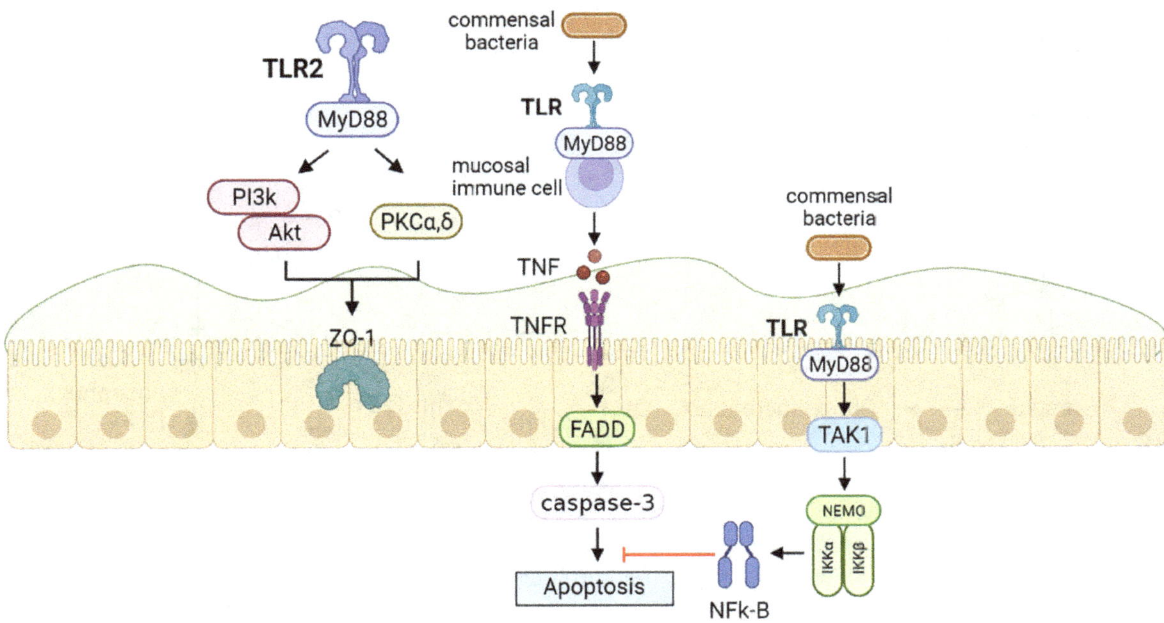

FIGURE 2 | Toll-like receptors (TLRs) protect the intestinal epithelial barrier. In intestinal epithelial cells, TLR2 stimulation efficiently preserves zonula occludens-1 (ZO-1)-associated barrier integrity against stress-induced damage, which is controlled by positive signaling crosstalk between PI3K-Akt and conventional protein kinase C (PKC) isoforms *via* MyD88. In the steady-state conditio, commensal bacteria induce the production of physiologically optimal concentration of inflammatory cytokines [tumor necrosis factor (TNF)] in mucosal immune cells. Epithelial nuclear factor-κB (NFκB) has an important role in the anti-apoptotic function that is mediated by TAK1 and the inhibitor of NF-jB kinase (IKK) complex including NF-jB essential modulator (NEMO), IKKa, and IKKb.

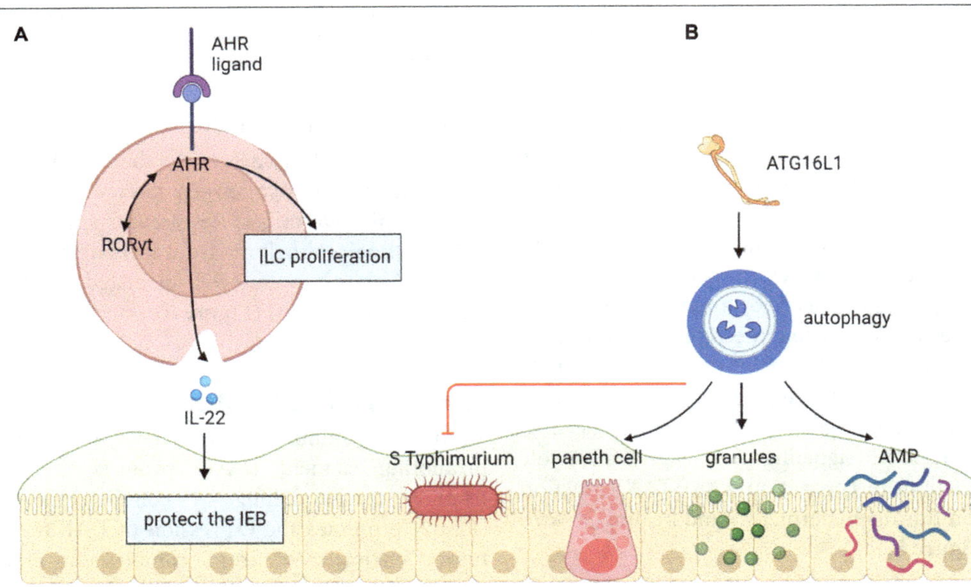

FIGURE 3 | (A) Aryl-hydrocarbon receptor (AHR) protect the intestinal epithelial barrier. Directly binding to the Il22 promoter, AHR induces Il22 expression in ILC3 by acting together with RORγt. **(B)** Autophagy (ATG16L1) protects the intestinal epithelial barrier. The decrease of Atg16l1 led to fewer Paneth cells and abnormal granule morphology, resulting in decreased expression of AMP, which growing inflammation and comprehensive translocation of bacteria.

show anti-apoptotic features containing EGFR phosphorylation and, in the case of P40, preventing the intestinal epithelium from triggered inflammation in a vitro model (Bauerl et al., 2020) (**Figure 4**).

STAT3

Recently, researchers identified a subset of genes specially mediated by STAT3 in answer to leptin, particularly the TRIB1 and inhibitor of cytokine signaling 3 (SOCS3) genes, which have

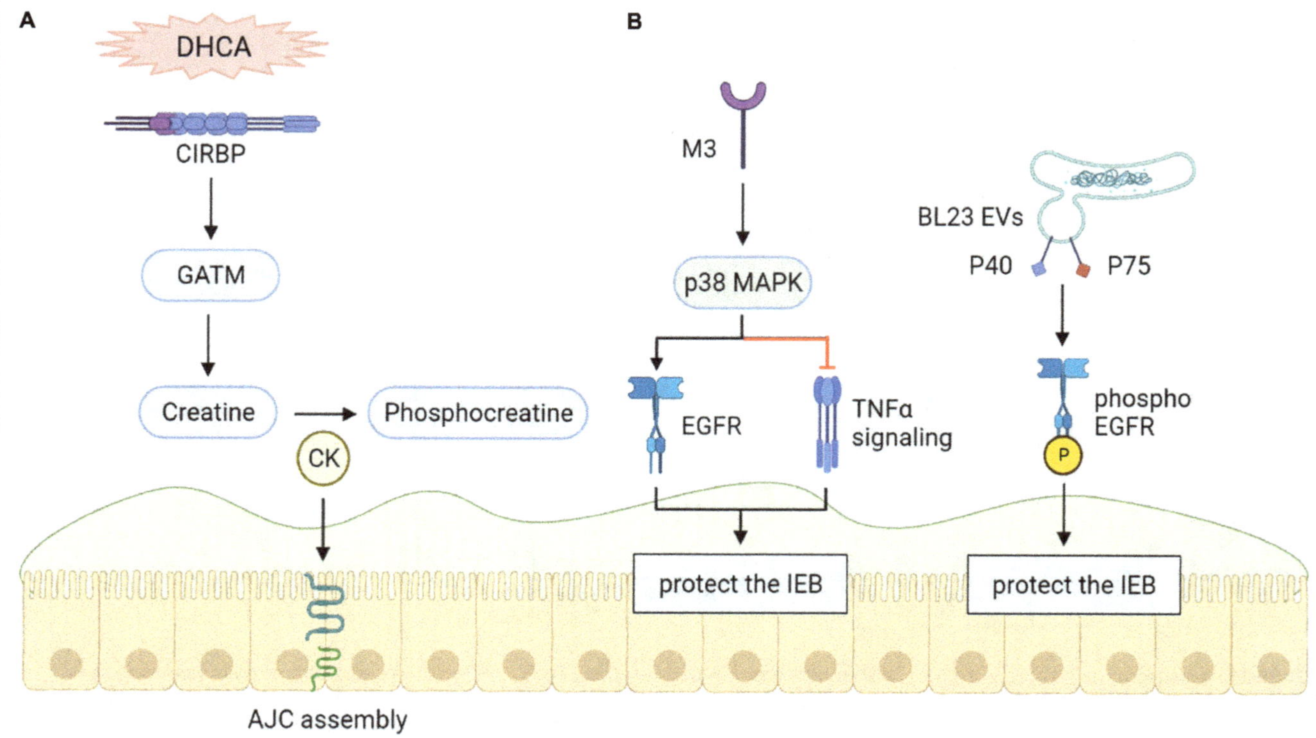

FIGURE 4 | (A) Cold-inducible RNA-binding protein (CIRBP) protect the intestinal epithelial barrier. CIRBP is related to intestinal epithelial barrier maintenance during DHCA. **(B)** Epidermal growth factor receptors (EGFR) protect the intestinal epithelial barrier. The maintenance of epithelial barrier function may be helped by M3 receptor-induced activation of p38 MAPK through activation of EGFR. Phosphorylation of the epidermal element factor receptor (EGFR) can be induced by lactobacillus casei BL23 extracellular vesicles (BL23 EVs) because of proteins P40 and P75 that are bound to their surface.

opposed effects in apoptosis control. Whole apoptotic genes were greatly accumulated in this gene set ($P < 1E\mu05$), boosting the assumption that protection results from leptin control of host apoptotic genes by STAT3.

Interestingly, *in vivo* researches on amebiasis show that both SHP-2 and STAT3 are necessary for leptin-mediated protection (Guo et al., 2011), suggesting the need for a more sophisticated protective signaling mechanism during infection of the intestinal epithelium (Marie et al., 2012). Additionally, the C-EPIYA-CagA-mediated JAK/STAT pathway promoting cell migration IL6 has a significant effect on the initial stage of intestinal wound healing, and gp130-mediated STAT1/3 signaling plays a protective role in the intestinal epithelium in correlation with STAT3 function in epithelial migration during epidermal wound healing *in vitro* (Sano et al., 1999) (**Figure 5**).

Lipopolysaccharide /CD14

Recent vitro model demonstrates that excess apoptosis and obstacle deficiencies triggered by Lipopolysaccharide (LPS) exposure are present *via* improved glucose uptake enterocytes. The reported research shows that a novel cell signaling path created by activating CD14 in intestinal epithelial cells, independently of TLR-4, may induce a SGLT-1-mediated glucose uptake to prevent the epithelium from LPS-caused apoptosis. This new pathway may help identify therapeutic targets in a variety of intestinal disorders (Yu et al., 2006) (**Figure 6**).

NADPH Oxidase 4

Vitro and animals models reveal that NADPH oxidase 4 (NOX4) as an effective reactive oxygen species generator was considered a direct miR-99b target. Researchers speculated that protecting the intestinal epithelium from oxidative stress-induced injury may be related to miR-99b-mediated NOX4 downregulation (Chandra et al., 2015) (**Figure 6**).

Krüppel-Like Factors

Vitro and animal models have documented the decay-promoting element (DAF) protects the intestinal mucosa from bystander killing by complement. Prostaglandin E2 (PGE2) promotes the expression of DAF that may prevent the tumor environment from complement attack. Krüppel-like factors (KLFs) are evolutionarily kept zinc finger-including transcription elements with various administrative functions in cell propagation and differentiation. DAF may have protective functions on both normal intestinal epithelium and intestinal neoplasia, so we can speculate that both KLF4 and KLF5 exert similar effects on regulating DAF expression. These novel findings provide insight into the functional role of the COX/PGE2system and KLF transcription factors in the gut and may contribute to new therapeutic strategies for a variety of intestinal disorders (Shao et al., 2008) (**Figure 6**).

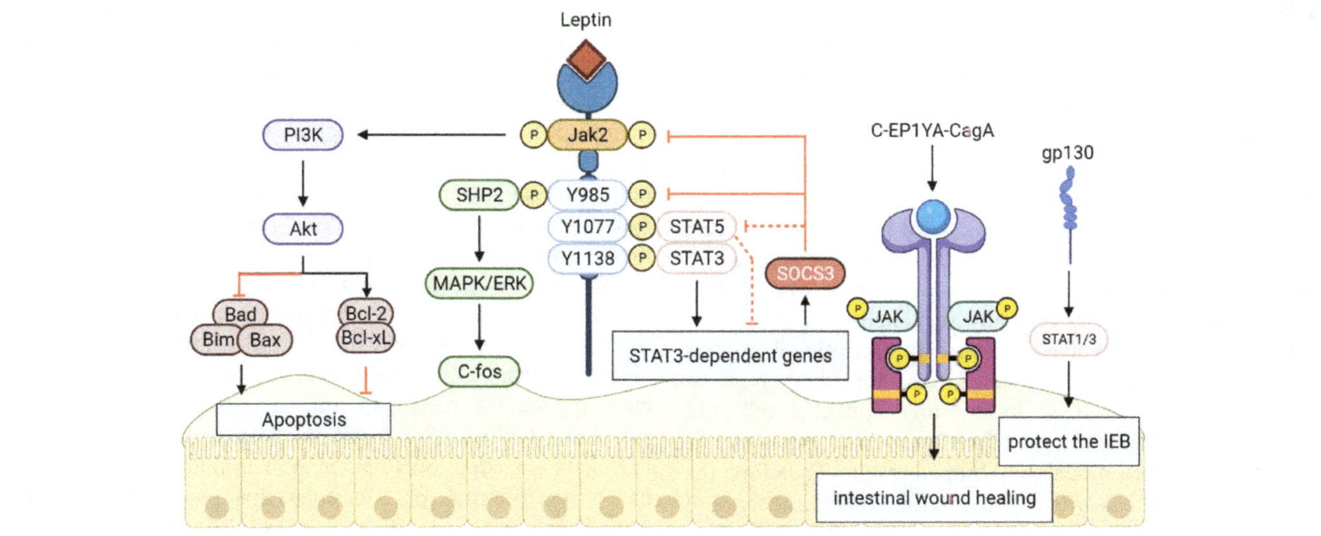

FIGURE 5 | Signal transducers and activators of transcription 3 (STAT3) protect the intestinal epithelial barrier. The activation and tyrosine phosphorylation of Janus kinase 2 (JAK2) are resulted in by Leptin binding to the long form LRb, which causes follow-up phosphorylation of downstream tyrosine residues (Tyr985, Tyr1077, and Tyr1138) in the intracellular tail of LRb. The phosphorylation-dependent activation of signal transducers and activators of transcription 3 (STAT3) are bound and mediated by Phosphorylated Tyr1138, activating transcription of suppressor of cytokine signaling 3 (SOCS3) and other positive effectors of leptin action. The SH2-containing tyrosine phosphatase SHP-2 is recruited by Phosphorylated Tyr985, activating the signaling path culminating in extracellular signal-regulated kinase (ERK) activation. The C-EPIYA-CagA-mediated JAK/STAT Pathway and gp130-mediated STAT1/3 signaling plays a protective role in intestinal epithelium.

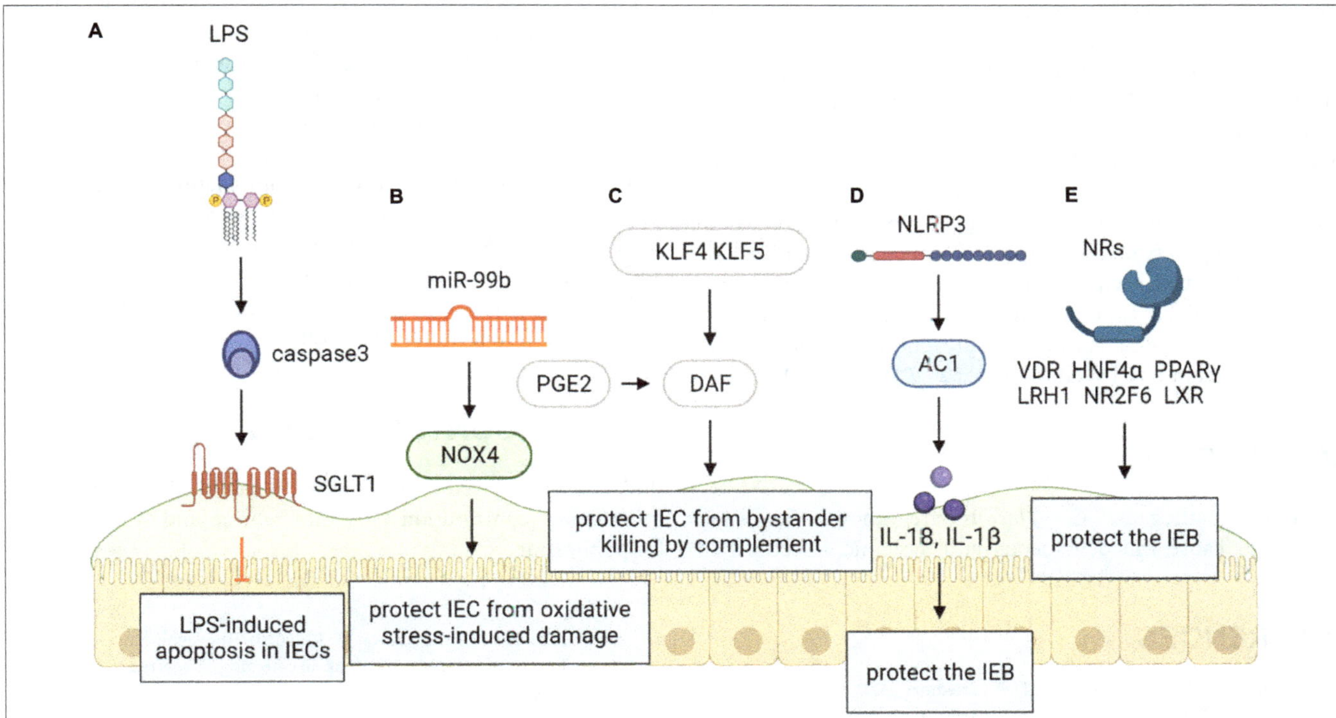

FIGURE 6 | (A) Lipopolysaccharide (LPS/CD1) protects the intestinal epithelial barrier. SGLT-1-mediated glucose uptake may be activated by CD14 activation in intestinal epithelial cells, independently of TLR-4, for the protection of the epithelium against LPS-induced apoptosis. **(B)** NADPH oxidase 4 (NOX4) protects the intestinal epithelial barrier. The intestinal epithelium from oxidative stress-caused damage may be protected by MiR-99b-mediated NOX4 downregulation. **(C)** Krüppel-like factors (KLF) protect the intestinal epithelial barrier. Rapid protection against complementary attacks may be provided by synergistic induction of DAF by COX/PGE2and KLF4/5 after intestinal mucosal injury. **(D)** Recombinant NLR Family, Pyrin Domain Containing Protein 3 (NLRP3) protects the intestinal epithelial barrier. NLRP3 induced generation of IL-18 in intestinal epithelial cells can be protective. **(E)** Nuclear receptors (NRs) protect the intestinal epithelial barrier. Within the intestinal epithelium, NRs including VDR, HNF4α, LXR, PPARγ, LRH1, and NR2F6 exert protective effect on intestinal epithelial integrity.

NLRP3

As an intracellular multiprotein signaling complex, Inflammasome is related to pathogen sensing and inflammatory response initiating in physiological and pathological states. The NLRP3 inflammasome, the most typical inflammasome, which has been identified as a sensor of cell stress that is strictly controlled in resting cells.

Nevertheless, changed management of the NLRP3 inflammasome is exhibited in many pathological states such as cancer and autoimmune diseases. It was proven that NLRP3 expression is post-transcriptionally regulated, and diversified miRNA have been involved in post transcriptional control of the inflammasome. Zaki et al. disclosed that NLRP3-induced generation of IL-18 in intestinal epithelial cells can be protective and leads to epithelium integrity in experimental colitis *in vitro*, animal models and human results. Therefore, the Nlrp3 inflammasome is critically involved in the maintenance of intestinal homeostasis and protection against colitis (Zaki et al., 2010; Tezcan et al., 2019) (**Figure 6**).

Nuclear Receptors

Gastrointestinal (GI) homeostasis is strongly dependent on nuclear receptor (NR) functions. They play a variety of roles ranging from nutrient uptake, sensing of microbial metabolites, regulation of epithelial intestinal cell integrity to shaping of the intestinal immune cell repertoire. Several NRs are associated with GI pathologies; therefore, systematic analysis of NR biology, the underlying molecular mechanisms, and regulation of target genes can be expected to help greatly in uncovering the course of GI diseases. Within the intestinal epithelium, nuclear receptors (NRs) including VDR, HNF4α, LXR, PPARγ, LRH1, and NR2F6 have a protective effect on intestinal epithelial integrity (Klepsch et al., 2019); declined mRNAs have also been recognized in intestinal samples from IBD patients (Ahn et al., 2008; Li et al., 2015). PRAP1 has been identified an intrinsically disordered protein that is highly expressed by the gastrointestinal epithelium and roles on exposed surfaces to prevent the obstacle from oxidative insult *in vivo* (Wolfarth et al., 2020) (**Figure 6**).

CONCLUSION

As fungal pathogens, *C. albicans* are known for their capacity to cause mucocutaneous and systemic infections in human hosts. Biofilm formation causes increasing *Candida* resistance to antifungal agents, which results in the failure of traditional antifungal agents' therapeutic measures. Current advances on different natural product, antifungal agents, microorganisms, protein and peptides inhibitors, lipid inhibitors and polysaccharide inhibitors have made a growth of the knowledge of the intricate mechanism underlying the biofilm resistance. Recently, phenotypic Screening and different nanoparticles have also appeared as antibiofilm agents against *C. albicans* and gaining momentum. Yet, a limitation of this method is that most of the results from *in vitro* and animal models. If we want to apply the method of inhibiting *C. albicans* biofilms to clinical treatment, one challenge is that few studies have applied the results to human trials so that we cannot rule out whether other intestinal flora and their metabolites may have an effect on these biofilm inhibitors. On the other hand, new progress on molecules that protect the intestinal epithelial barrier like MUC, TLR, AHR, ATG16L1, CIRBP, EGFR, LPS/CD14, NOX4, STAT3, KLF, NLRP3 and Nuclear Receptors have improved the understanding of protective mechanism of intestinal epithelium barrier. However, these molecular studies did not specifically target intestinal infections of *C. albicans.* Therefore, further studies need to be carried out in order to validate the mechanism of how the intestinal barrier prevents *C. albicans* invasion. A key strength of the review is that the combination of preventing *C. albicans* from forming biofilms and using molecules to protect the intestinal epithelial barrier has not previously been reported for the treatment of *C. albicans* intestinal infection. Thus, the combination of these two methods is a novel idea to improve the therapeutic effect of *C. albicans* intestinal invasion. In terms of directions for future research, further work could focus on which combinations of antibiofilm and mucosal enhancement treatments were more effective in inhibiting *C. albicans* biofilm formation and intestinal infection and what are the mechanisms. We hope to develop effective strategies to treat *C. albicans* drug resistance and invasive intestinal infections in the near future.

AUTHOR CONTRIBUTIONS

Both authors listed have made a substantial, direct, and intellectual contribution to the work, and approved it for publication.

REFERENCES

1. Ahmadi, M. S., Lee, H. H., Sanchez, D. A., Friedman, A. J., Tar, M. T., Davies, K. P., et al. (2016). Sustained nitric oxide-releasing nanoparticles induce cell death in *Candida albicans* yeast and hyphal cells, preventing biofilm formation in vitro and in a rodent central venous catheter model. *Antimicrob. Agents Chemother.* 60, 2185–2194. doi: 10.1128/AAC. 02659-15
2. Ahn, S. H., Shah, Y. M., Inoue, J., Morimura, K., Kim, I., Yim, S., et al. (2008). Hepatocyte nuclear factor 4 alpha in the intestinal epithelial cells protects against inflammatory bowel disease. *Inflamm. Bowel Dis.* 14, 908–920. doi: 10.1002/ibd.20413
3. Alem, M. A., and Douglas, L. J. (2004). Effects of aspirin and other nonsteroidal anti-inflammatory drugs on biofilms and planktonic cells of *Candida albicans*. *Antimicrob. Agents Chemother.* 48, 41–47. doi: 10.1128/AAC.48.1.41-47.2004
4. Alim, D., Sircaik, S., and Panwar, S. L. (2018). The significance of lipids to biofilm formation in *Candida albicans*: an emerging perspective. *J. Fungi* 4:140. doi: 10.3390/jof4040140
5. Andes, D. R., Safdar, N., Baddley, J. W., Playford, G., Reboli, A. C., Rex, J. H., et al. (2012). Impact of treatment strategy on outcomes in patients with candidemia and other forms of invasive candidiasis: a patient-level quantitative review of randomized trials. *Clin. Infect. Dis.* 54, 1110–1122. doi: 10.1093/cid/cis021
6. Arai, R., Sugita, T., and Nishikawa, A. (2005). Reassessment of the in vitro synergistic effect of fluconazole with the non-steroidal anti-inflammatory agent ibuprofen against *Candida albicans*. *Mycoses* 48, 38–41. doi: 10.1111/j.1439- 0507.2004.01052.x
7. Bachtiar, B. M., Srisawat, C., and Bachtiar, E. W. (2019). RNA aptamers selected against yeast cells inhibit *Candida albicans* biofilm formation in vitro. *Microbiologyopen* 8:e00812. doi: 10.1002/mbo 3.812

8. Barbosa, J. O., Rossoni, R. D., Vilela, S. F., de Alvarenga, J. A., Velloso Mdos, S., Prata, M. C., et al. (2016). *Streptococcus mutans* can modulate biofilm formation and attenuate the virulence of *Candida albicans*. *PLoS One* 11:e0150457. doi: 10.1371/journal.pone.0150457
9. Bauerl, C., Coll-Marques, J. M., Tarazona-Gonzalez, C., and Perez-Martinez, G. (2020). *Lactobacillus casei* extracellular vesicles stimulate EGFR pathway likely due to the presence of proteins P40 and P75 bound to their surface. *Sci. Rep.* 10:19237. doi: 10.1038/s41598-020-75930-9
10. Brown, G. D., Denning, D. W., Gow, N. A., Levitz, S. M., Netea, M. G., and White, T. C. (2012). Hidden killers: human fungal infections. *Sci. Transl. Med.* 4:165rv113. doi: 10.1126/scitranslmed.3004404
11. Cario, E. (2008). Barrier-protective function of intestinal epithelial Toll-like receptor 2. *Muc. Immunol.* 1, S62–S66. doi: 10.1038/mi.2008.47
12. Cauda, R. (2009). Candidaemia in patients with an inserted medical device. *Drugs* 69(Suppl. 1), 33–38. doi: 10.2165/11315520-000000000-00000
13. Chandra, J., and Ghannoum, M. A. (2018). CD101, a novel echinocandin, possesses potent antibiofilm activity against early and mature *Candida albicans* biofilms. *Antimicrob. Agents Chemother.* 62:AAC.01750-17. doi: 10.1128/AAC.01750-17
14. Chandra, J., Kuhn, D. M., Mukherjee, P. K., Hoyer, L. L., McCormick, T., and Ghannoum, M. A. (2001). Biofilm formation by the fungal pathogen *Candida albicans*: development, architecture, and drug resistance. *J. Bacteriol.* 183, 5385–5394. doi: 10.1128/JB.183.18.5385-5394.2001
15. Chandra, L. C., Kumar, V., Torben, W., Stouwe, C. V., Winsauer, P., Amedee, A., et al. (2015). Chronic administration of Delta(9)-tetrahydrocannabinol induces intestinal anti-inflammatory MicroRNA expression during acute simian immunodeficiency virus infection of rhesus macaques. *J. Virol.* 89, 1168–1181. doi: 10.1128/JVI.01754-14
16. Chelakkot, C., Choi, Y. W., Kim, D. K., Park, H. T., Ghim, J., Kwon, Y., et al. (2018). Akkermansia muciniphila-derived extracellular vesicles influence gut permeability through the regulation of tight junctions. *Exp. Mol. Med.* 50, 1–11. doi: 10.1038/emm.2017.282
17. Chen, L., Qu, S., Yang, K., Liu, M., Li, Y. X., Keller, N. P., et al. (2020). Perillaldehyde: a promising antifungal agent to treat oropharyngeal candidiasis. *Biochem. Pharmacol.* 180:114201. doi: 10.1016/j.bcp.2020. 114201
18. Conway, K. L., Kuballa, P., Song, J. H., Patel, K. K., Castoreno, A. B., Yilmaz, O. H., et al. (2013). Atg16l1 is required for autophagy in intestinal epithelial cells and protection of mice from *Salmonella* infection. *Gastroenterology* 145, 1347–1357. doi: 10.1053/j.gastro.2013.08.035
19. Cornely, O. A., Bassetti, M., Calandra, T., Garbino, J., Kullberg, B. J., Lortholary, O., et al. (2012). ESCMID* guideline for the diagnosis and management of *Candida* diseases 2012: non-neutropenic adult patients. *Clin. Microbiol. Infect.* 18(Suppl. 7), 19–37. doi: 10.1111/1469-0691.12039
20. Costa, E., Silva, S., Tavaria, F., and Pintado, M. (2014). Antimicrobial and antibiofilm activity of chitosan on the oral pathogen *Candida albicans*. *Pathogens* 3, 908–919. doi: 10.3390/pathogens3040908
21. Costerton, J. W., Stewart, P. S., and Greenberg, E. P. (1999). Bacterial biofilms: a common cause of persistent infections. *Science* 284, 1318–1322. doi: 10.1126/science.284.5418.1318
22. De Cremer, K., Delattin, N., De Brucker, K., Peeters, A., Kucharíková, S., Gerits, E., et al. (2014). Oral administration of the broad-spectrum antibiofilm compound toremifene inhibits *Candida albicans* and *Staphylococcus aureus* biofilm formation in vivo. *Antimicrob. Agents Chemother.* 58, 7606–7610. doi: 10.1128/AAC.03869-14
23. de Medina, F. S., Romero-Calvo, I., Mascaraque, C., and Martinez-Augustin, O. (2014). Intestinal inflammation and mucosal barrier function. *Inflamm. Bowel Dis.* 20, 2394–2404. doi: 10.1097/mib.000000000000 0204
24. de Quadros, A. U., Bini, D., Pereira, P. A., Moroni, E. G., and Monteiro, M. C. (2011). Antifungal activity of some cyclooxygenase inhibitors on *Candida albicans*: PGE2-dependent mechanism. *Folia Microbiol.* 56, 349–352. doi: 10. 1007/s12223-011-0049-6
25. Delattin, N., De Brucker, K., Craik, D. J., Cheneval, O., De Coninck, B., Cammue, B. P. A., et al. (2014). Structure-activity relationship study of the plant- derived decapeptide osip108 inhibiting *Candida albicans* biofilm formation. *Antimicrob. Agents Chemother.* 58, 4974–4977. doi: 10.1128/AAC.033 36-14
26. Ding, X. R., Liu, Z. Z., Su, J. R., and Yan, D. H. (2014). Human serum inhibits adhesion and biofilm formation in *Candida albicans*. *BMC Microbiol.* 14:80. doi: 10.1186/1471-2180-14-80
27. Donlan, R. M., and Costerton, J. W. (2002). Biofilms: survival mechanisms of clinically relevant microorganisms. *Clin. Microbiol. Rev.* 15, 167–193. doi: 10. 1128/CMR.15.2.167-193.2002
28. dos Santos, C. I., Franca, Y. R., Campos, C. D. L., Bomfim, M. R. Q., Melo, B. O., Holanda, R. A., et al. (2019). Antifungal and antivirulence activity of vaginal *Lactobacillus* Spp. products against candida vaginal Isolates. *Pathogens* 8:150. doi: 10.3390/pathogens8030150
29. Elce, A., Amato, F., Zarrilli, F., Calignano, A., Troncone, R., Castaldo, G., et al. (2017). Butyrate modulating effects on pro-inflammatory pathways in human intestinal epithelial cells. *Beneficial Microbes* 8, 841–847. doi: 10.3920/BM2016. 0197
30. Ells, R., Kock, J. L., Albertyn, J., Kemp, G., and Pohl, C. H. (2011). Effect of inhibitors of arachidonic acid metabolism on prostaglandin E2 production by *Candida albicans* and *Candida dubliniensis* biofilms. *Med. Microbiol. Immunol.* 200, 23–28. doi: 10.1007/s00430-010-0169-7
31. Fang, Y., Wu, C., Wang, Q., and Tang, J. (2019). Farnesol contributes to intestinal epithelial barrier function by enhancing tight junctions via the JAK/STAT3 signaling pathway in differentiated Caco-2 cells. *J. Bioenerg. Biomembr.* 51, 403–412. doi: 10.1007/s10863-019-09817-4
32. Feldman, M., Al-Quntar, A., Polacheck, I., Friedman, M., and Steinberg, D. (2014). Therapeutic potential of thiazolidinedione-8 as an antibiofilm agent against *Candida albicans*. *PLoS One* 9:e93225. doi: 10.1371/journal. pone.0093225
33. Fox, E. P., and Nobile, C. J. (2012). A sticky situation: untangling the transcriptional network controlling biofilm development in *Candida albicans*. *Transcription* 3, 315–322. doi: 10.4161/trns.22281
34. Frosali, S., Pagliari, D., Gambassi, G., Landolfi, R., Pandolfi, F., and Cianci, R. (2015). How the intricate interaction among toll-like receptors, microbiota, and intestinal immunity can influence gastrointestinal pathology. *J. Immunol. Res.* 2015:489821. doi: 10.1155/2015/489821
35. Gao, Y., Zhang, C., Lu, C., Liu, P., Li, Y., Li, H., et al. (2013). Synergistic effect of doxycycline and fluconazole against *Candida albicans* biofilms and the impact of calcium channel blockers. *FEMS Yeast Res.* 13, 453–462. doi: 10.1111/1567-1364.12048
36. Gaudier, E., Jarry, A., Blottiere, H. M., de Coppet, P., Buisine, M. P., Aubert, J. P., et al. (2004). Butyrate specifically modulates MUC gene expression in intestinal epithelial goblet cells deprived of glucose. *Am. J. Physiol. Gastrointest. Liver Physiol.* 287, G1168–G1174. doi: 10.1152/ajpgi.00219.2004
37. Goto, Y., and Kiyono, H. (2012). Epithelial barrier: an interface for the cross-communication between gut flora and immune system. *Immunol. Rev.* 245, 147–163. doi: 10.1111/j.1600-065X.2011.01078.x
38. Graham, C. E., Cruz, M. R., Garsin, D. A., and Lorenz, M. C. (2017). Enterococcus faecalis bacteriocin EntV inhibits hyphal morphogenesis, biofilm formation, and virulence of *Candida albicans*. *Proc. Natl. Acad. Sci. U.S.A.* 114, 4507–4512. doi: 10.1073/pnas.1620432114
39. Gulati, M., and Nobile, C. J. (2016). *Candida albicans* biofilms: development, regulation, and molecular mechanisms. *Microbes Infect.* 18, 310–321. doi: 10. 1016/j.micinf.2016.01.002
40. Guo, X., Roberts, M. R., Becker, S. M., Podd, B., Zhang, Y., Chua, S. C., et al. (2011). Leptin signaling in intestinal epithelium mediates resistance to enteric infection by *Entamoeba histolytica*. *Mucosal Immunol.* 4, 294–303. doi: 10.1038/mi.20 10.76
41. Haque, F., Alfatah, M., Ganesan, K., and Bhattacharyya, M. S. (2016). Inhibitory effect of sophorolipid on *Candida albicans* biofilm formation and hyphal growth. *Sci. Rep.* 6:23575. doi: 10.1038/srep23575
42. Hernandez-Delgadillo, R., Velasco-Arias, D., Martinez-Sanmiguel, J. J., Diaz, D., Zumeta-Dube, I., Arevalo-Nino, K., et al. (2013). Bismuth oxide aqueous colloidal nanoparticles inhibit *Candida albicans* growth and biofilm formation. *Int. J. Nanomed.* 8, 1645–1652. doi: 10.2147/IJN.S38708

43. Jacquelot, N., Luong, K., and Seillet, C. (2019). Physiological regulation of innate lymphoid cells. *Front. Immunol.* 10:405. doi: 10.3389/fimmu.2019.00405
44. Katragkou, A., Kruhlak, M. J., Simitsopoulou, M., Chatzimoschou, A., Taparkou, A., Cotten, C. J., et al. (2010). Interactions between human phagocytes and *Candida albicans* biofilms alone and in combination with antifungal agents. *J. Infect. Dis.* 201, 1941–1949. doi: 10.1086/652783
45. Ke, C. L., Liao, Y. T., and Lin, C. H. (2021). MSS2 maintains mitochondrial function and is required for chitosan resistance, invasive growth, biofilm formation and virulence in *Candida albicans*. *Virulence* 12, 281–297. doi: 10.1080/21505594. 2020.1870082
46. Khan, S. N., Khan, S., Iqbal, J., Khan, R., and Khan, A. U. (2017). Enhanced killing and antibiofilm activity of encapsulated cinnamaldehyde against *Candida albicans*. *Front. Microbiol.* 8:1641. doi: 10.3389/fmicb.2017.01641
47. Kiss, E. A., Vonarbourg, C., Kopfmann, S., Finke, D., Esser, C., and Diefenbach, A. (2012). Natural aryl hydrocarbon receptor ligands control organogenesis of intestinal lymphoid follicles. *Immunology* 137, 223–223. doi: 10.1126/science. 1214914
48. Klepsch, V., Moschen, A. R., Tilg, H., Baier, G., and Hermann-Kleiter, N. (2019). Nuclear receptors regulate intestinal inflammation in the context of IBD. *Front. Immunol.* 10:1070. doi: 10.3389/fimmu.2019.01070
49. Kojic, E. M., and Darouiche, R. O. (2004). *Candida* infections of medical devices. *Clin. Microbiol. Rev.* 17, 255–267. doi: 10.1128/cmr.17.2.255-267.2004
50. Lang, Y., Fu, F., Sun, D., Xi, C., and Chen, F. (2015). Labetalol prevents intestinal dysfunction induced by traumatic brain injury. *PLoS One* 10:e0133215. doi: 10.1371/journal.pone.0133215
51. Lee, J. H., Kim, Y. G., Gupta, V. K., Manoharan, R. K., and Lee, J. (2018). Suppression of fluconazole resistant *Candida albicans* biofilm formation and filamentation by methylindole derivatives. *Front. Microbiol.* 9:2641. doi: 10. 3389/fmicb.2018.02641
52. Lee, J. S., Cella, M., McDonald, K. G., Garlanda, C., Kennedy, G. D., Nukaya, M., et al. (2012). AHR drives the development of gut ILC22 cells and postnatal lymphoid tissues via pathways dependent on and independent of Notch. *Nat. Immunol.* 13, 144–U158. doi: 10.1038/ni.2187
53. Lepak, A., and Andes, D. (2011). Fungal sepsis: optimizing antifungal therapy in the critical care setting. *Crit. Care Clin.* 27, 123–147. doi: 10.1016/j. ccc.2010.11.001
54. Li, Y., Du, M., Chen, L. A., Liu, Y., and Liang, Z. (2016). Nosocomial bloodstream infection due to *Candida* spp. in China: species distribution, clinical features, and outcomes. *Mycopathologia* 181, 485–495. doi: 10.1007/ s11046-016-9997-3
55. Li, Y., Ma, Y., Zhang, L., Guo, F., Ren, L., Yang, R., et al. (2012). In vivo inhibitory effect on the biofilm formation of *Candida albicans* by liverwort derived riccardin D. *PLoS One* 7:e35543. doi: 10.1371/journal.pone.0035543
56. Li, Y. C., Chen, Y. Z., and Du, J. (2015). Critical roles of intestinal epithelial vitamin D receptor signaling in controlling gut mucosal inflammation. *J. Steroid Biochem. Mol. Biol.* 148, 179–183. doi: 10.1016/j.jsbmb.2015.01.011
57. Li, Y. N., Liu, M. Y., Gao, S. Z., Cai, L. M., Zhang, Q. N., Yan, S. J., et al. (2019). Cold-inducible RNA-binding protein maintains intestinal barrier during deep hypothermic circulatory arrest. *Int. Cardiovasc. Thorac. Surg.* 29, 583–591. doi: 10.1093/icvts/ivz147
58. Liu, R. H., Shang, Z. C., Li, T. X., Yang, M. H., and Kong, L. Y. (2017). In vitro antibiofilm activity of eucarobustol E against *Candida albicans*. *Antimicrob. Agents Chemother.* 61:e02707-16. doi: 10.1128/AAC.02707-16
59. Lohse, M. B., Gulati, M., Craik, C. S., Johnson, A. D., and Nobile, C. J. (2020). Combination of antifungal drugs and protease inhibitors prevent *Candida albicans* biofilm formation and disrupt mature biofilms. *Front. Microbiol.* 11:1027. doi: 10.3389/fmicb.2020.01027
60. Lohse, M. B., Gulati, M., Johnson, A. D., and Nobile, C. J. (2018). Development and regulation of single- and multi-species *Candida albicans* biofilms. *Nat. Revi. Microbiol.* 16, 19–31. doi: 10.1038/nrmicro.2017.107
61. Lortholary, O., Petrikkos, G., Akova, M., Arendrup, M. C., Arikan-Akdagli, S., Bassetti, M., et al. (2012). ESCMID* guideline for the diagnosis and management of *Candida* diseases 2012: patients with HIV infection or AIDS. *Clin. Microbiol. Infect.* 18(Suppl. 7), 68–77. doi: 10.1111/1469-0691. 12042
62. Luo, Y., McLean, D. T. F., Linden, G. J., McAuley, D. F., McMullan, R., and Lundy, F. T. (2017). The naturally occurring host defense peptide, LL-37, and its truncated mimetics KE-18 and KR-12 have selected biocidal and antibiofilm activities against *Candida albicans*, *Staphylococcus aureus*, and *Escherichia coli* In vitro. *Front. Microbiol.* 8:544. doi: 10.3389/fmicb.2017.00544
63. MacAlpine, J., Daniel-Ivad, M., Liu, Z., Yano, J., Revie, N. M., Todd, R. T., et al. (2021). A small molecule produced by *Lactobacillus* species blocks *Candida albicans* filamentation by inhibiting a DYRK1-family kinase. *Nat. Commun.* 12:6151. doi: 10.1038/s41467-021-26390-w
64. Madariaga-Venegas, F., Fernández-Soto, R., Duarte, L. F., Suarez, N., Delgadillo, D., Jara, J. A., et al. (2017). Characterization of a novel antibiofilm effect of nitric oxide-releasing aspirin (NCX-4040) on *Candida albicans* isolates from denture stomatitis patients. *PLoS One* 12:e0176755. doi: 10.1371/ journal.pone.017 6755
65. Manoharan, R. K., Lee, J. H., and Lee, J. (2017). Antibiofilm and antihyphal activities of cedar leaf essential oil, camphor, and fenchone derivatives against *Candida albicans*. *Front. Microbiol.* 8:1476. doi: 10.3389/fmicb.2017.01476
66. Manoharan, R. K., Lee, J. H., and Lee, J. (2018). Efficacy of 7-benzyloxyindole and other halogenated indoles to inhibit *Candida albicans* biofilm and hyphal formation. *Microb. Biotechnol.* 11, 1060–1069. doi: 10.1111/1751-7915.1 3268
67. Maraki, S., Hamilos, G., Dimopoulou, D., Andrianaki, A. M., Karageorgiadis, A. S., Kyvernitakis, A., et al. (2015). Study on the comparative activity of echinocandins on murine gut colonization by *Candida albicans*. *Med. Mycol.* 53, 597–602. doi: 10.1093/mmy/myv028
68. Marie, C. S., Verkerke, H. P., Paul, S. N., Mackey, A. J., and Petri, W. A. (2012). Leptin protects host cells from *Entamoeba histolytica* cytotoxicity by a STAT3-dependent mechanism. *Infect. Immun.* 80, 1934–1943. doi: 10.1128/IAI.061 40-11
69. Mermel, L. A., Allon, M., Bouza, E., Craven, D. E., Flynn, P., O'Grady, N. P., et al. (2009). Clinical practice guidelines for the diagnosis and management of intravascular catheter-related infection: 2009 update by the infectious diseases society of America. *Clini.Infect. Dis.* 49, 1–45. doi: 10.1086/599376
70. Miceli, M. H., Bernardo, S. M., and Lee, S. A. (2009). In vitro analyses of the combination of high-dose doxycycline and antifungal agents against *Candida albicans* biofilms. *Int. J. Antimicrob. Agents* 34, 326–332. doi: 10.1016/j. ijantimicag.2009.04.011
71. Miranda, L. N., van der Heijden, I. M., Costa, S. F., Sousa, A. P., Sienra, R. A., Gobara, S., et al. (2009). *Candida* colonisation as a source for candidaemia. *J. Hosp. Infect.* 72, 9–16. doi: 10.1016/j.jhin.2009.02.009
72. Mohammad, H., Eldesouky, H. E., Hazbun, T., Mayhoub, A. S., and Seleem, M. N. (2019). Identification of a phenylthiazole small molecule with dual antifungal and antibiofilm activity against *Candida albicans* and *Candida auris*. *Sci. Rep.* 9:18941. doi: 10.1038/s41598-019-55379-1
73. Morici, P., Fais, R., Rizzato, C., Tavanti, A., and Lupetti, A. (2016). Inhibition of *Candida albicans* biofilm formation by the synthetic lactoferricin derived peptide hLF1-11. *PLoS One* 11:e0167470. doi: 10.1371/journal.pone.0167470
74. Natividad, J. M., and Verdu, E. F. (2013). Modulation of intestinal barrier by intestinal microbiota: pathological and therapeutic implications. *Pharmacol. Res.* 69, 42–51. doi: 10.1016/j.phrs.2012.10.007
75. Nguyen, L. N., Lopes, L. C., Cordero, R. J., and Nosanchuk, J. D. (2011). Sodium butyrate inhibits pathogenic yeast growth and enhances the functions of macrophages. *J. Antimicrob. Chemother.* 66, 2573–2580. doi: 10.1093/jac/ dkr358
76. Nieminen, M. T., Novak-Frazer, L., Rautemaa, V., Rajendran, R., Sorsa, T., Ramage, G., et al. (2014). A novel antifungal is active against *Candida albicans* biofilms and inhibits mutagenic acetaldehyde production in vitro. *PLoS One* 9:e97864. doi: 10.1371/journal.pone.0097864
77. Noverr, M. C., and Huffnagle, G. B. (2004). Regulation of *Candida albicans* morphogenesis by fatty acid metabolites. *Infect. Immun.* 72, 6206–6210. doi: 10.1128/IAI.72.11.6206-6210.2004
78. Parizkova, R., Dostal, P., and Cerny, V. (1999). Fluconazole prophylaxis of systemic candida infection in non-neutropenic critically ill patients: a prospective randomized study. *Crit. Care* 3(Suppl. 1):P061. doi: 10.1186/ cc436

79. Pfaller, M. A., and Diekema, D. J. (2007). Epidemiology of invasive candidiasis: a persistent public health problem. *Clin. Microbiol. Rev.* 20, 133–163. doi: 10.1128/CMR.00029-06

80. Pierce, C. G., Chaturvedi, A. K., Lazzell, A. L., Powell, A. T., Saville, S. P., McHardy, S. F., et al. (2015). A novel small molecule inhibitor of *Candida albicans* biofilm formation, filamentation and virulence with low potential for the development of resistance. *Npj Biofilms Microbiomes* 1:15012. doi: 10.1038/npjbiofilms.20 15.12

81. Pierce, C. G., Saville, S. P., and Lopez-Ribot, J. L. (2014). High-content phenotypic screenings to identify inhibitors of *Candida albicans* biofilm formation and filamentation. *Pathog Dis.* 70, 423–431. doi: 10.1111/2049-632X.12161

82. Pongrácz, J., Juhász, E., Iván, M., and Kristóf, K. (2015). Significance of yeasts in bloodstream infection: Epidemiology and predisposing factors of Candidaemia in adult patients at a university hospital (2010-2014). *Acta Microbiol. Immunol. Hung.* 62, 317–329. doi: 10.1556/030.62.2015.3.9

83. Qiu, J., Heller, J. J., Guo, X. H., Chen, Z. M. E., Fish, K., Fu, Y. X., et al. (2012). The aryl hydrocarbon receptor regulates gut immunity through modulation of innate lymphoid cells. *Immunity* 36, 92–104. doi: 10.1016/j.immuni.2011.11.011

84. Ramage, G., Mowat, E., Jones, B., Williams, C., and Lopez-Ribot, J. (2009). Our current understanding of fungal biofilms. *Crit. Rev. Microbiol.* 35, 340–355. doi: 10.3109/10408410903241436

85. Ramage, G., Saville, S. P., Thomas, D. P., and López-Ribot, J. L. (2005). *Candida* biofilms: an update. *Eukaryotic Cell* 4, 633–638.

86. Ramamourthy, G., Park, J., Seo, C., Vogel, H. J., and Park, Y. (2020). Antifungal and antibiofilm activities and the mechanism of action of repeating lysine-tryptophan peptides against *Candida albicans*. *Microorganisms* 8:758. doi: 10.3390/microorganisms8050758

87. Reddy, G. K. K., and Nancharaiah, Y. V. (2020). Alkylimidazolium ionic liquids as antifungal alternatives: antibiofilm activity against *Candida albicans* and underlying mechanism of action. *Front. Microbiol.* 11:730. doi: 10.3389/fmicb.2020.00730

88. Reen, F. J., Phelan, J. P., Gallagher, L., Woods, D. F., Shanahan, R. M., Cano, R., et al. (2016). Exploiting interkingdom interactions for development of small-molecule inhibitors of *Candida albicans* biofilm formation. *Antimicrob. Agents Chemother.* 60, 5894–5905. doi: 10.1128/AAC.00190-16

89. Rene, H. D., José, M. S., Isela, S. N., and Claudio, C. R. (2014). Effects of ambroxol on *Candida albicans* growth and biofilm formation. *Mycoses* 57, 228–232.

90. Ricardo, E., Costa-de-Oliveira, S., Dias, A. S., Guerra, J., Rodrigues, A. G., and Pina- Vaz, C. (2009). Ibuprofen reverts antifungal resistance on *Candida albicans* showing overexpression of CDR genes. *FEMS Yeast Res.* 9, 618–625. doi: 10. 1111/j.1567-1364.2009.00504.x

91. Rocha, F. A. C., Alves, A., Rocha, M. F. G., Cordeiro, R. D., Brilhante, R. S. N., Pinto, A., et al. (2017). Tumor necrosis factor prevents *Candida albicans* biofilm formation. *Sci. Rep.* 7:1206. doi: 10.1038/s41598-017-01400-4

92. Round, J. L., Lee, S. M., Li, J., Tran, G., Jabri, B., Chatila, T. A., et al. (2011). The toll- like receptor 2 Pathway establishes colonization by a commensal of the human microbiota. *Science* 332, 974–977. doi: 10.1126/science.1206095

93. Rudramurthy, S. M., Colley, T., Abdolrasouli, A., Ashman, J., Dhaliwal, M., Kaur, H., et al. (2019). In vitro antifungal activity of a novel topical triazole PC945 against emerging yeast *Candida auris*. *J. Antimicrob. Chemother.* 74, 2943–2949. doi: 10.1093/jac/dkz280

94. Salvatori, O., Kumar, R., Metcalfe, S., Vickerman, M., Kay, J. G., and Edgerton, M. (2020). Bacteria modify *Candida albicans* hypha formation, microcolony properties, and survival within macrophages. *mSphere* 5, e00689-20. doi: 10.1128/mSphere.00689-20

95. Sánchez de Medina, F., Ortega-González, M., González-Pérez, R., Capitán-Cañadas, F., and Martínez-Augustin, O. (2013). Host-microbe interactions: the difficult yet peaceful coexistence of the microbiota and the intestinal mucosa. *Br. J. Nutr.* 109(Suppl. 2), S12–S20. doi: 10.1017/S0007114512004035

96. Sano, S., Itami, S., Takeda, K., Tarutani, M., Yamaguchi, Y., Miura, H., et al. (1999). Keratinocyte-specific ablation of Stat3 exhibits impaired skin remodeling, but does not affect skin morphogenesis. *EMBO J.* 18, 4657–4668. doi: 10.1093/ emboj/18.17.4657

97. Schiering, C., Wincent, E., Metidji, A., Iseppon, A., Li, Y., Potocnik, A. J., et al. (2017). Feedback control of AHR signalling regulates intestinal immunity. *Nature* 542, 242–245. doi: 10.1038/nature21080

98. Seddiki, S. M., Boucherit-Otmani, Z., Boucherit, K., Badsi-Amir, S., Taleb, M., and Kunkel, D. (2013). Assessment of the types of catheter infectivity caused by *Candida* species and their biofilm formation. First study in an intensive care unit in Algeria. *Int. J. Gen. Med.* 6, 1–7. doi: 10.2147/IJGM.S38065

99. Shang, L. M., Fukata, M., Thirunarayanan, N., Martin, A. P., Arnaboldi, P., Maussang, D., et al. (2008). Toll-like receptor signaling in small intestinal epithelium promotes B-cell recruitment and IgA production in lamina propria. *Gastroenterology* 135, 529–538. doi: 10.1053/j.gastro.2008.04.020

100. Shankar, J., Solis, N. V., Mounaud, S., Szpakowski, S., Liu, H., Losada, L., et al. (2015). Using Bayesian modelling to investigate factors governing antibiotic-induced *Candida albicans* colonization of the GI tract. *Sci. Rep.* 5:8131. doi: 10.1038/srep08131

101. Shao, J. Y., Yang, V. W., and Sheng, H. M. (2008). Prostaglandin E(2) and Kruppel-like transcription factors synergistically induce the expression of decay-accelerating factor in intestinal epithelial cells. *Immunology* 125, 397–407. doi: 10.1111/j.1365-2567.2008.02847.x

102. Sonnenberg, G. F., Fouser, L. A., and Artis, D. (2011). Border patrol: regulation of immunity, inflammation and tissue homeostasis at barrier surfaces by IL-22. *Nat. Immunol.* 12, 383–390. doi: 10.1038/ni.2025

103. Sun, H., Ni, X. Q., Song, X., Wen, B., Zhou, Y., Zou, F. Q., et al. (2016). Fermented Yupingfeng polysaccharides enhance immunity by improving the foregut microflora and intestinal barrier in weaning rex rabbits. *Appl. Microbiol. Biotechnol.* 100, 8105–8120. doi: 10.1007/s00253-016-7 619-0

104. Tezcan, G., Martynova, E. V., Gilazieva, Z. E., McIntyre, A., Rizvanov, A. A., and Khaiboullina, S. F. (2019). MicroRNA Post-transcriptional regulation of the NLRP3 Inflammasome in Immunopathologies. *Front. Pharmacol.* 10:451. doi: 10.3389/fphar.2019.00451

105. Tsang, P. W., Bandara, H. M., and Fong, W. P. (2012). Purpurin suppresses *Candida albicans* biofilm formation and hyphal development. *PLoS One* 7:e50866. doi: 10.1371/journal.pone.0050866

106. Uppuluri, P., Chaturvedi, A. K., Srinivasan, A., Banerjee, M., Ramasubramaniam, A.K., Köhler, J. R., et al. (2010). Dispersion as an important step in the *Candida albicans* biofilm developmental cycle. *PLoS Pathog.* 6:e1000828. doi: 10.1371/journal.ppat.1000828

107. Uppuluri, P., Srinivasan, A., Ramasubramanian, A., and Lopez-Ribot, J. L. (2011). Effects of fluconazole, amphotericin B, and caspofungin on *Candida albicans* biofilms under conditions of flow and on biofilm dispersion. *Antimicrob. Agents Chemother.* 55, 3591–3593. doi: 10.1128/AAC.01701-10

108. Uwada, J., Yazawa, T., Islam, M. T., Khan, M. R. I., Krug, S. M., Fromm, M., et al. (2017). Activation of muscarinic receptors prevents TNF-alpha-mediated intestinal epithelial barrier disruption through p38 MAPK. *Cell. Signal.* 35, 188–196. doi: 10.1016/j.cellsig.2017.04.007

109. Vila, T., Ishida, K., Seabra, S. H., and Rozental, S. (2016). Miltefosine inhibits *Candida albicans* and non-albicans *Candida* spp. biofilms and impairs the dispersion of infectious cells. *Int. J. Antimicrob. Agents* 48, 512–520. doi: 10.1016/j.ijantimicag.2016.07.022

110. Weiland-Brauer, N., Malek, I., and Schmitz, R. A. (2019). Metagenomic quorum quenching enzymes affect biofilm formation of *Candida albicans* and *Staphylococcus epidermidis*. *PLoS One* 14:e211366. doi: 10.1371/journal.pone. 0211366

111. Wenzel, R. P. (1995). Nosocomial candidemia: risk factors and attributable mortality. *Clin. Infect. Dis.* 20, 1531–1534. doi: 10.1093/clinids/20.6.1531

112. Willems, H. M. E., Stultz, J. S., Coltrane, M. E., Fortwendel, J. P., and Peters, A.M. (2019). Disparate *Candida albicans* biofilm formation in clinical lipid emulsions due to capric acid-mediated inhibition. *Antimicrob. Agents Chemother.* 63:e01394-19. doi: 10.1128/AAC.01394-19

113. Wisplinghoff, H., Bischoff, T., Tallent, S. M., Seifert, H., Wenzel, R. P., and Edmond, M. B. (2004). Nosocomial bloodstream infections in US hospitals: analysis of 24,179 cases from a prospective nationwide surveillance study. *Clin. Infect. Dis.* 39, 309–317. doi: 10.1086/421946

114. Wolfarth, A. A., Liu, X., Darby, T. M., Boyer, D. J., Spizman, J. B., Owens, J. A., et al. (2020). Proline-rich acidic protein 1 (PRAP1) protects the gastrointestinal epithelium from irradiation-induced apoptosis. *Cell. Mol. Gastroenterol. Hepatol.* 10, 713–727. doi: 10.1016/j.jcmgh.2020. 06.011
115. Xia, J., Qian, F., Xu, W., Zhang, Z., and Wei, X. (2017). In vitro inhibitory effects of farnesol and interactions between farnesol and antifungals against biofilms of *Candida albicans* resistant strains. *Biofouling* 33, 283–293. doi: 10.1080/08927014.2017.1295304
116. Xu, H., Sobue, T., Thompson, A., Xie, Z., Poon, K., Ricker, A., et al. (2014). Streptococcal co-infection augments *Candida* pathogenicity by amplifying the mucosal inflammatory response. *Cell Microbiol.* 16, 214–231. doi: 10.1111/cmi. 12216
117. Yan, L., Yang, C., and Tang, J. (2013). Disruption of the intestinal mucosal barrier in *Candida albicans* infections. *Microbiol. Res.* 168, 389–395. doi: 10.1016/j.micres. 2013.02.008
118. Yan, Y., Tan, F., Miao, H., Wang, H., and Cao, Y. (2019). Effect of shikonin against *Candida albicans* Biofilms. *Front. Microbiol.* 10:1085. doi: 10.3389/fmicb.2019. 01085
119. You, Z., Zhang, C., and Ran, Y. (2020). The effects of clioquinol in morphogenesis, cell membrane and ion homeostasis in *Candida albicans. BMC Microbiol.* 20:165. doi: 10.1186/s12866-020-01850-3
120. Yu, H., Liu, X., Wang, C., Qiao, X., Wu, S., Wang, H., et al. (2016). Assessing the potential of four cathelicidins for the management of mouse candidiasis and *Candida albicans* biofilms. *Biochimie* 121, 268–277. doi: 10.1016/j.biochi.2015. 11.028
121. Yu, L. C. H., Turner, J. R., and Buret, A. G. (2006). LPS/CD14 activation triggers SGLT-1-mediated glucose uptake and cell rescue in intestinal epithelial cells via early apoptotic signals upstream of caspase-3. *Exp. Cell Res.* 312, 3276–3286. doi: 10.1016/j.yexcr.2006.06.023
122. Zaki, M., Oukhrib, A., El Hakmaoui, A., Hiebel, M. A., Berteina-Raboin, S., and Akssira, M. (2019). Synthesis of novel 1,2,3-triazole-substituted tomentosins. *ZNaturforsch* 74, 273–281. doi: 10.1515/znb-2018-0225
123. Zaki, M. H., Boyd, K. L., Vogel, P., Kastan, M. B., Lamkanfi, M., and Kanneganti, T. D. (2010). The NLRP3 inflammasome protects against loss of epithelial integrity and mortality during experimental colitis. *Immunity* 32, 379–391. doi: 10.1016/j.immuni.2010.03.003
124. Zarnowski, R., Westler, W. M., Lacmbouh, G. A., Marita, J. M., Bothe, J. R., Bernhardt, J., et al. (2014). Novel entries in a fungal biofilm matrix encyclopedia. *mBio* 5:e01333-14. doi: 10.1128/mBio.01333-14
125. Zeise, K. D., Woods, R. J., and Huffnagle, G. B. (2021). Interplay between *Candida albicans* and lactic acid bacteria in the gastrointestinal tract: impact on colonization resistance, microbial carriage, opportunistic infection, and host immunity. *Clin. Microbiol. Rev.* 34:e0032320. doi: 10.1128/CMR.003 23-20
126. Zhang, L., Chang, W., Sun, B., Groh, M., Speicher, A., and Lou, H. (2011). Bisbibenzyls, a new type of antifungal agent, inhibit morphogenesis switch and biofilm formation through upregulation of DPP3 in *Candida albicans. PLoS One* 6:e28953. doi: 10.1371/journal.pone.0028953

19

Pathogenic Fungal Infection in the Lung

Zhi Li[1,2]*, Gen Lu*[1*] *and Guangxun Meng*[2*]

[1] *The Joint Center for Infection and Immunity, Guangzhou Women and Children's Medical Center, Guangzhou Institute of Pediatrics, Guangzhou, China,* [2] *The Joint Center for Infection and Immunity, Institute Pasteur of Shanghai, Chinese Academy of Science, Shanghai, China*

***Correspondence:**
Gen Lu
lugen5663330@sina.com
Guangxun Meng
gxmeng@ips.ac.cn

Respiratory fungal infection is a severe clinical problem, especially in patients with compromised immune functions. *Aspergillus, Cryptococcus, Pneumocystis,* and endemic fungi are major pulmonary fungal pathogens that are able to result in life-threatening invasive diseases. Growing data being reported have indicated that multiple cells and molecules orchestrate the host's response to a fungal infection in the lung. Upon fungal challenge, innate myeloid cells including macrophages, dendritic cells (DC), and recruited neutrophils establish the first line of defense through the phagocytosis and secretion of cytokines. Natural killer cells control the fungal expansion in the lung via the direct and indirect killing of invading organisms. Adaptive immune cells including Th1 and Th17 cells confer anti-fungal activity by producing their signature cytokines, interferon-γ, and IL-17. In addition, lung epithelial cells (LEC) also participate in the resistance against fungal infection by internalization, inflammatory cytokine production, or antimicrobial peptide secretion. In the host cells mentioned above, various molecules with distinct functions modulate the immune defense signaling: Pattern recognition receptors (PRRs) such as dectin-1 expressed on the cell surface are involved in fungal recognition; adaptor proteins such as MyD88 and TRAF6 are required for transduction of signals to the nucleus for transcriptional regulation; inflammasomes also play crucial roles in the host's defense against a fungal infection in the lung. Furthermore, transcriptional factors modulate the transcriptions of a series of genes, especially those encoding cytokines and chemokines, which are predominant regulators in the infectious microenvironment, mediating the cellular and molecular immune responses against a fungal infection in the lung.

Keywords: pulmonary fungal infection, pattern recognition receptor, inflammasome, cytokine, chemokine

INTRODUCTION

With the increasing number of immunocompromised patients, diseases caused by fungal infections remain a great threat in public health. Opportunistic fungi, including *Aspergillus* with invasive aspergillosis (1–3), *Cryptococcus* with cryptococcosis (4–6), *Pneumocystis* with pneumonia (7), and endemic fungi (8, 9) are the main sources of fungal infections in the lungs of humans. Although these infections are rarely found in the target organs in healthy people, they may result in life-threatening invasive diseases in patients with an impaired immune system. These individuals include patients suffering immunodeficiency disorders such as HIV/AIDS and cancer patients who undergo chemotherapy, as well as those patients who receive immunosuppressive therapy such as in bone marrow/stem cell transplantation. Pathogenic fungal infections in the lung has resulted in the incidence and infectious death of invasive mycoses especially in patients with severe defects of host immune responses (10, 11). Some fungal pathogens initiate the infection through surface proteins from pathogen-host interaction, finally leading to mycosis with multiple tissue lesions especially in immunocompromised patients as mentioned above. For example, *Cryptococcus* mainly infects the lung and invades the brain via circulation, leading to lethal cryptococcal meningitis. Cryptococcosis, including life-threatening Cryptococcal meningoencephalitis (12), afflicts about 1 million AIDS patients and causes more than 600,000 deaths worldwide annually.

To minimize the damage of fungal infections, the human body has shaped a set of unique and sophisticated defense mechanisms, in which host innate immunity plays a crucial role. Aiming to eliminate the fungal dissemination, two typical innate immune cells, namely macrophage and dendritic cells (DC) as defensive troops constitute the first line in multiple organs. Furthermore, these innate cells can connect innate and adaptive immunity by serving as specific antigen-presenting cells (APCs), by which fungal antigens could be presented, to prime naïve T cells. Upon recognition of fungal pathogens, innate immune cells are activated by pathogen associated molecular patterns (PAMPs) via specific pattern recognition receptors (PRRs) on the surface, for further intracellular signaling transduction. The PRRs involved in fungal detection identified to date include Toll like receptors (TLRs), C-type lectin receptors (CLRs) as well as NOD like receptors (NLRs) (13, 14). Although breakthroughs have been made in the field of host response to mycobiota, novel cellular and molecular mechanisms on antifungal immunity remain to be established for the control of fungal infection and associated organ damage (**Figure 1**).

Despite antifungal drugs conferring protection against the pulmonary fungal infection (15, 16), drug resistance is still a severe problem. To reduce the incidence and death rate of pulmonary mycoses, efforts for the investigation on the mechanisms of pulmonary fungal infection, remain a requirement from both basic and clinical aspects. In the current review, we will discuss concrete research progress made on mycobiota regarding respiratory tract infections including *Aspergillus, Cryptococcus, Pneumocystis,* and endemic mycoses.

PATHOGENS: THE CHARACTERIZATION OF FUNGAL INFECTIONS IN THE LUNG

Aspergillus and *Cryptococcus* Are the Major Fungal Pathogens in the Lung Infection

Aspergillus mold is one of the most common fungal species which can sufficiently sporulate with released airborne conidia. The produced conidia in the air are small enough (2 to 3 μm) to arrive at human airways and pulmonary alveoli, causing a spectrum of diseases including lethal infections in immunocompromised individuals and in atopic patients with asthma allergies (1, 17). In healthy individuals, inhaled conidia are engulfed by alveolar macrophages and killed in a phagocyte oxidase-dependent fashion (18–20). In immunocompromised individuals, incomplete killing of inhaled fungal conidia results in germination and tissue invasion by fungal hyphae (21).

Cryptococcosis is caused by *Cryptococcus* exposure to the lung after the airborne organisms' inhalation. As a subtype of *Cryptococcus, Cryptococcus neoformans* distribute widely, particularly in soil and avian habitats. The most severe outcome of *Cryptococcus* infection is cryptococcal meningitis. Because *C. neoformans* and *Cryptococcus gattii* can disseminate from the lung to invade the brain by crossing the blood brain barrier (BBB), the fungal cells directly penetrate the BBB via endothelial cells on the blood vessels of the brain, using a "Trojan horse" strategy that is involved in the transport of phagocytes (5). In animal experiments, by counting yeast cell numbers in the brain, *C. neoformans* still remained in the CNS where large scale colonization and tissue injury could occur in spite of defense mechanisms employed by the host (22).

Pneumocystis and Endemic Mycoses Cause Infections in the Lung

Pneumocystis pneumonia (PCP) induced by fungal pathogen species *Pneumocystis* such as *Pneumocystis jirovecii* is the most common AIDS-defining disease, and is also found in non-HIV immunocompromised patients with deficiency in adaptive immunity, or individuals taking prolonged high-dose systemic glucocorticoids (11). The *Pneumocystis* antigens are primary surface glycoprotein (Msg or glycoprotein A) (23) and *Pneumocystis* protease, kexin (Kex1, Prt1). As a potential therapeutic target, Kex is supposed to participate in the proteolytic processing of *Pneumocystis* surface antigens (24, 25).

Endemic mycoses usually occur in restricted geographic areas, which can result in severe and fatal cases in hospitalizations (26, 27). In immunocompromised patients, endemic mycoses can also cause more severe and disseminated disease resulting in higher mortality (28). In many regions, increased incidence of endemic mycoses is correlated with a growing population of immunocompromised subjects (29). Surprisingly, mortality is also high in non-immunocompromised hosts (27). In North America, three major endemic mycoses including coccidioidomycosis, histoplasmosis, and blastomycosis could present as community acquired pneumonias (CAP) (27). For coccidioidomycosis in the southwest endemic areas, half of the infected people are asymptomatic, as an acute infection resembles respiratory symptoms such as pneumonia and bronchitis and a fever, which then experiences a self-limited process, and only very few cases can develop into disseminated infection; histoplasmosis, caused by infectious agent *Histoplasma. capsulatum*, are considered a community acquired infection with an exposure history to soil containing bat or bird droppings, patients also present pneumonia symptoms described as acute/chronic pulmonary histoplasmosis, severe cases may culminate in respiratory failure and death. However, blastomycosis is less common than histoplasmosis and coccidioidomycosis. As to the management of the endemic

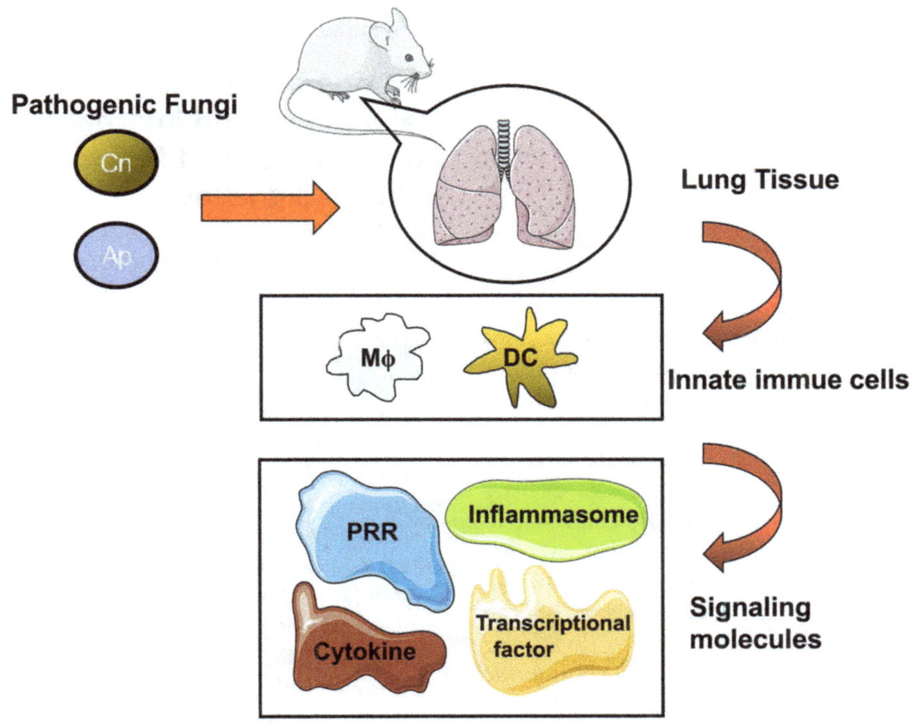

FIGURE 1 | General cellular and molecular components against fungal infection in the lung. The pathogenic fungi including *Cryptococcus neoformans* (Cn), *Aspergillus* (Ap), *Pneumocystis* (Pc), and Endemic fungi (Ef) can cause pulmonary infection in human and in murine models. Fungal pathogens could trigger host immune response upon inhalation, and lung tissue is the major infectious target of these pathogens. Once the infection is initiated, multiple cells either resident in the lung tissue or recruited from blood and lymphoid organs are motivated to clear the invasive fungi. After entering host cells, molecules including receptors, adaptors, or effectors come into play with divergent roles to respond to the fungal challenge.

mycoses-caused community acquired pneumonia, anti-fungal treatment is deemed important, and azole therapy such as oral fluconazole is preferred (30). In addition, paracoccidioidomycosis has its geographic distribution mainly in Latin America where Brazil largely accounts for the reported cases (31).

The Characteristics of Human Responses to Fungal Infection: A Distinct Susceptibility Between Immunocompromised and Immunocompetent Hosts

The clinical case of human cryptococcosis was first described by German scientists in a young woman with inflammatory symptoms on the tibia (32). The lungs and brain are the major target organs for the cryptococcus organism causing human cryptococcosis: pulmonary cryptococcosis is caused by the entry of pathogenic airborne spores (conidia) or dried yeast cells into the airway and lungs; Cryptococcal meningitis (CM) is the infection from occasional sits (preferentially from lung or elsewhere such as skin, liver etc.) disseminating during the early phase via circulation into other parts of the body especially to the central nervous system (CNS) (33–35). Cryptococci can utilize parasitized phagocytes (monocytes/macrophages) to cross the BBB into the brain via a Trojan horse smuggling the pathogens in Charlier et al. (36) and Kim (37). The outcome of fungal colonization in the lungs include either clearance of fungi by the host immune system or by establishing an infection from the asymptomatic infection at the latency period, into a virtual local inflammation characterized by pulmonary nodules and pneumonia, or the development of a subsequent dissemination into the systemic organs, preferentially into the CNS through the BBB. Therefore, cryptococci could be detected mainly from lung tissue/bronchoalveolar lavage/sputum or cerebrospinal fluid or blood, and CM has been considered to be severe and life-threatening mycosis with a high mortality derived from cryptococcosis. According to the immune reaction to a diverse polysaccharide capsule, the encapsulated cryptococcus could be classified into four serotypes (from A to D) and two major species are responsible for opportunistic cryptococcosis: the *C. neoformans* (serotype A and D) and *C. gattii* (serotypes B and C), which are infectious in immunocompromised hosts and in immunocompetent or immunologically normal individuals, respectively (32, 38–40). In the AIDS epidemic, $CD4^{+}$ T lymphocytes deficiency renders HIV/AIDS patients less resistant to cryptococci, thus the rise in cases of human cryptococcosis was followed by the increasing diagnosis and morbidity of HIV/AIDS despite the development of antiretroviral therapy (6, 41, 42). The lung serves as a target organ in patients with HIV/AIDS, but once infected, the likelihood of meningitis is high, possibly because the dissemination of cryptococcosis into the brain depends on the status of the pulmonary immune responses (43). On the other hand, highly encapsulated cryptococci are more frequently found in the respiratory tracts rather than the CNS, which resist phagocytes and reduce the rate of systemic dissemination (44). In a basic study, severe combined immunodeficiency (SCID) mice inoculated intravenously (i.v.) with *C. neoformans* viable cells could mimic systemic cryptococcosis in immunodeficient hosts (45). Of note, cryptococcosis in HIV-negative patients is not rare as previously thought (46), these individuals show no apparent immune deficiency but under immunosuppressive conditions including: immunosuppressive drug treatment (glucocorticoids or other immunosuppressants) and malignancies or hematologic disorders (chronic leukemia, lymphoma), the presence of systemic lupus erythematosus patients receiving immunosuppressive agents, and other diseases e.g., diabetes mellitus, cirrhosis, and chronic renal failure, also contribute to cryptococcosis in HIV-negative patients. Interestingly, although the development of the immune system in children is insufficient, pulmonary cryptococcosis in children is rarely observed but can often be a fatal disease (47) (**Figure 2A**).

Invasive pulmonary aspergillosis (IPA), an opportunistic mycosis, results from a pathogenic Aspergillus infection in the lung. Susceptible individuals on Ap are very similar to those on *C. neoformans*, *Aspergillus* poses a serious threat to immunocompromised individuals including hosts with cancer or hematological disorders (acute leukemia, neutropenia), patients with chronic granulomatous diseases, and immunosuppressed bone marrow/organ transplant recipients (48–51). In contrast, inhaled infectious propagules in immunocompetent hosts show no further significance because of the strong resistance to the fungus, with killing and clearance by the cells of the pulmonary immune system (**Figure 2B**).

Regarding the process of human infection establishment, we elucidate that as an "unbalanced fight between host and pathogen," on one hand, the immunocompromised host or patient with underlying immunosuppressive conditions are characterized by the cellular immune deficiency: impairing the killing and clearance ability of phagocytes e.g., resident alveolar macrophages and weakened cellular immunity mediated by CD4 T lymphocytes, both of which contribute to host resistance to the fungi; in addition, the encapsulated cryptococcus resist the phagocytosis and rapidly germinate into mature yeast cells to establish an infection in the host. For the therapy of mycoses, Amphotericin B (AMB), fluconazole and echinocandins are the optimal antifungal agents (52–54), but occasionally current antifungal drugs may not cure these diseases due to a growing antifungal drug resistance, toxicity and side effects which still pose a potential risks for patients (55). Therefore, novel strategies rather focus on host defense, which is a crucial determinant of fungal pathogenesis. At present, the biggest problem in the treatment of invasive mycoses, is the ineffectiveness of antifungal agents, and new therapeutic immunomodulators are urgently required. Additionally, poor prognosis is attributed to the severe CM in the brain. Therefore, therapeutic strategies for both enhancing the clearance of pathogens and the prevention of dissemination into the CNS, are still required (**Figure 2C**).

Pathogen-Host Communication: The Recognition and Internalization of the Fungi by the Host Innate Immune System in Initial Signaling

To understand the communication between host and fungal pathogens would be useful for the diagnosis and therapy of the trajectory of infection. However, the signaling events for fungal pathogens and mammalian host cell interactions remain poorly understood, although some research progress has been made in this direction (56).

The fungal surface is mainly comprised of chitin, modified glycoproteins and glucans. In most yeasts, mannose-containing polysaccharides are often covalently attached to proteins. *C. neoformans* is a unique pathogenic fungus which employs a specific polysaccharide component (glucuronoxylomannan, GXM) as a capsule outside the cell wall, with anti-phagocytosis activity (57). GXM in the *C. neoformans* capsule is essential for virulence of this pathogen. In fact, patients with cryptococcosis accumulate GXM in the cerebrospinal fluid and serum, where it is associated with several immunomodulatory properties.

The fungal surface polysaccharides or glycans are the main PAMPs recognized by host cells, while the surface proteins on different fungi are essential for their virulence and intracellular survival. It is known that phagocytosis by host cells is an early event in the host-microbe interaction (58). The precise mechanisms that mammalian host cells coordinately manipulate fungal phagocytosis and replication with, remain complex and elusive. Professional phagocytes, for instance macrophages, can internalize and kill large particulate material through the phagocytosis process involving reactive oxygen species (ROS) (59). Binding of pathogenic particles to cell surface receptors, with a reorganization of the membrane and intracellular elements, would result in the phagocytosis of particle and phagosome formation. Notably, engulfment of the fungal pathogens triggers the host autophagy initiation complex (AIC) and the upstream kinases, which can mediate autophagy in the *C. neoformans* infection (60).

HOST: CELL POPULATIONS IN HOST DEFENSE AGAINST PULMONARY FUNGAL INFECTIONS

Although mounting signal mechanisms driving the initiation and development of fungal infection investigated by independent groups have been clearly revealed, we have a relatively poor view about the integrated cellular and molecular systems involved in the process of distinct fungal infection in the lung. Thus, effective therapeutic approaches that tackle harmful pulmonary fungal infection and inflammation, remains limited. Host mechanisms of fungal infection in the lung derive from multiple cell types and numerous molecules including receptors, adaptors, kinases, and transcriptional factors in the fungal pathogen challenge. Here we highlight the representative components in the host response,

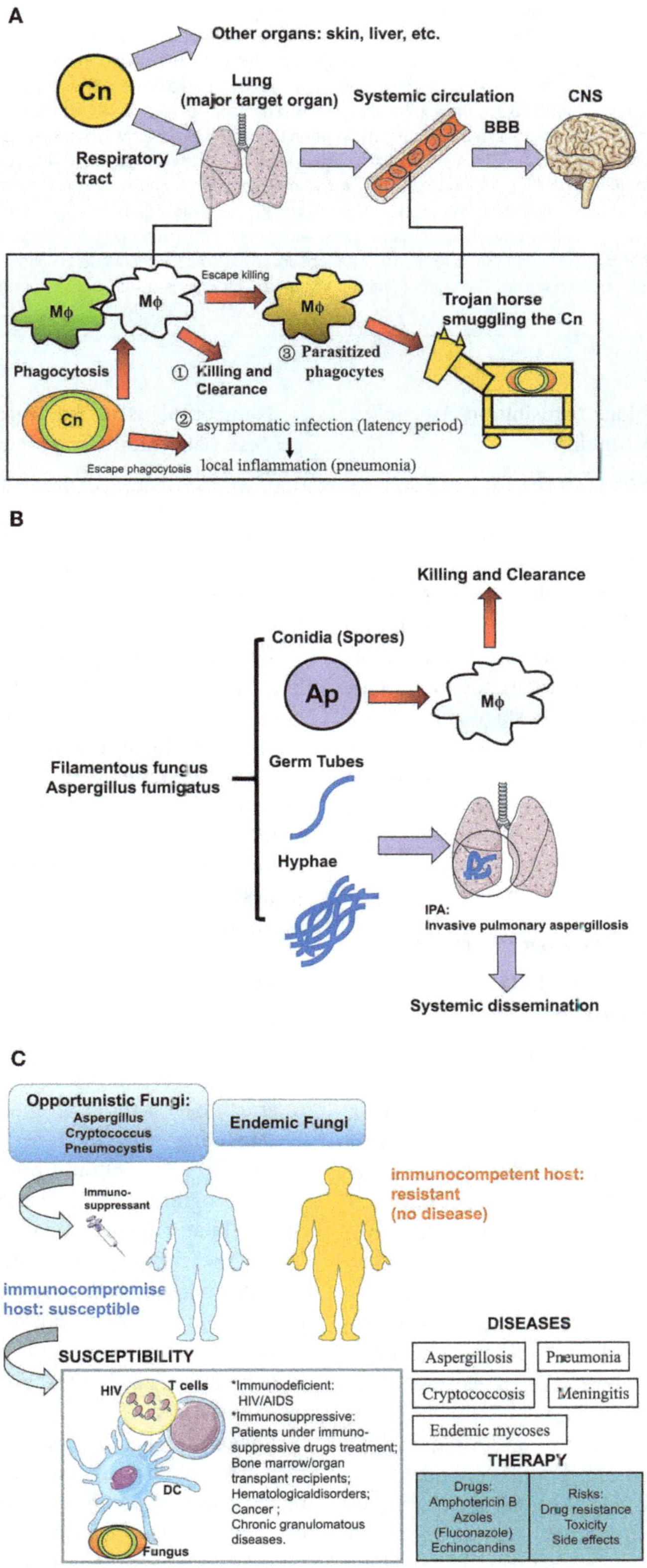

FIGURE 2 | The human fungal infections in the lung. **(A)** *Cryptococcus neoformans* infection in human: how the pathogens invade into the brains? *Cryptococcus neoformans* (Cn) is more aggressive to the lung via respiratory the tract compared with other organs (skin, liver, etc.), where outcomes of fungi-host communication

(Continued)

FIGURE 2 | are: (1) Killing and clearance by macrophages after phagocytosis, (2) The encapsulated fungi are able to escape from phagocytosis and further establish virtual infection after pulmonary colonization, from asymptomatic latency period to local inflammation such as pneumonia, (3) Part of the phagocytosed fungi could escape from killing and further parasitize the host cells like a Trojan horse which transports the pathogenic fungi into the CNS via circulation. After crossing the BBB, Cn that accumulates in the CNS would result in lethal cryptococcal meningitis especially in susceptible hosts. **(B)** *Aspergillus fumigatus* in humans: Hyphae are the pathogenicity for IPA. *Aspergillus fumigatus* are a Filamentous fungus in the lung. Inhaled conidia (spores) could be ingested and killed by the alveolar macrophage constructed first line of host defense, while the escaped conidia grow into mature germ tubes and hyphae, which not only cause Invasive pulmonary aspergillosis (IPA) but also lead to systemic dissemination by invading vessels. This pathological process is more common in patients who lack an effective host defense or hyphal defense. **(C)** Summary of fungal infection to human hosts. Opportunistic fungi or endemic fungi infection usually occurs in hosts with a profound immune deficiency involving HIV/AIDS, or immunosuppressed patients involving hematopoietic stem cells (HSCs) or solid organ transplantation, immunosuppressant, or glucocorticosteroid therapy, hematological disorders (leukemia, lymphoma), cancer, and chronic granulomatous diseases. In susceptible hosts, multiple factors participated in this process including HIV, Fungus, immune cells, or even the immunosuppressant and transplanted organs. Infection and subsequent inflammation result in diseases e.g., Mycoses, pneumonia, and meningitis. The current anti-fungal treatments include azoles and Amphotericin B, but potential risks also lead to the ineffectiveness of the drugs.

from cell biology and molecule biology focusing on *Aspergillus*, *Cryptococcus*, and other pulmonary fungi.

Innate immunity plays a central role in host protection against fungal infection. Invasive pulmonary aspergillosis (IPA), candidiasis, and other mycosis can develop in the hosts with specific immune deficiency defects, such as chemotherapy-induced prolonged neutropenia or functional defects in NADPH oxidase. After the initial sensing of fungi by pattern recognition receptors (PRRs), neutrophils, and macrophages are recruited as first line immune cells to the infected and inflammatory sites for their clearance of fungal pathogens at an early stage of infection. Fungal antigens can be presented by DCs which also bridge innate and adaptive immunity. In fact, the innate immune system orchestrates the first step to protect hosts against fungal infection in the lungs (**Figures 3A,B**).

Phagocytic and Myeloid Derived Innate Cells: Alveolar Macrophages and Neutrophils, and Dendritic Cells

Multiple innate immune cells as well as adaptive $CD4^+$ Th1 type cells constitute a precise network for the host defense against pulmonary fungal pathogens (61, 62), In particular, lung phagocytic leukocytes including the resident alveolar macrophages (AM), DC and neutrophils often functionally modulate the early immune response to pulmonary infection in the first line with an armored defense against pulmonary fungal infections (63). A study reported that $CD11c^+$ AM population is dominant in uninfected lungs as well as in the early days following infection with *C. neoformans*. Thereafter, the rapid expansion of $CD11c^+$ cells is attributed to accumulated pulmonary DCs. Here CD11b expression is used to distinguish between AM ($CD11c^+CD11b^{min}$) and conventional DC ($CD11c^+CD11b^+$). It was indicated that *C. neoformans* infected mice with AM/DC population depletion, displayed very rapid deterioration (63). Alveolar macrophages mount inflammatory responses to fungal pathogens with specific productions of tumor necrosis factor-α, interleukin-1α/β, interleukin-6, macrophage inflammatory proteins, and granulocyte-colony stimulating factor, which mainly depends on signaling from recognized fungal glucan via host PRR dectin-1 (64).

Neutrophil is a key orchestrator during the infectious process, particularly for polymorphonuclear neutrophils (PMN). These cells can initiate quick lung infiltration and arrive at inflammatory sites rapidly where it plays a vital role in eliminating pathogens and promoting tissue repair. At the early phase of *Aspergillus fumigatus* infection, AM and neutrophil coordinately shape the host defense and survival following fungal inhalation, AM is believed to kill conidia while neutrophil is postulated to restrict the tissue invasion of hyphae (61). A well-established method of transient neutrophil depletion by antibody RB6-8C5 (anti-Ly6G/Ly6C) can be used to evaluate the temporal demand for neutrophils in the host defense (65). RB6-8C5 induced neutropenia both in circulation and in single cell lung suspensions without affecting the CD11b+Ly6C+Ly6G− monocytes, leading to invasive aspergillosis and tissue damage (66). Mice who received RB6-8C5 prior to or within 3 h after infection were exacerbated as shown by higher susceptibility and death rates, and this is different from the depletion at later time points when the survival rates are nearly normal. Neutrophil is also a type of IL-17, producing cells in several mouse infectious models. Eric Pearlman demonstrated that in *A. fumigatus* infection Ly6G+Ly6C+ neutrophil is activated through IL-17A-IL-17RC interactions via multiple molecules including dectin-2 and IL-17 signaling related IL-6, IL-23, RORgt (67). In addition, neutrophil activity could be governed by TLRs in Aspergillosis (10). Recognition of *A. fumigatus* conidia trigger non-oxidative intracellular killing involving integrin CD11b/CD18 and PI3K. When the conidia germinate, by escaping from early killing, the extracellular destruction of the *Aspergillus* hyphae requires antibody mediated opsonization with the involvement of Fcγ receptors recognition and kinase signaling for downstream ROS related metabolites production via MPO and NADPH oxidase (68).

Phagocytic leukocytes in the lung such as AM, DC, and recruited neutrophils constitute the first immune defense by phagocytosing pathogens with yeast lysis upon *Aspergillus* or *Cryptococcus* exposure after inhalation. Therefore, the resident AM and DC contribute to the early innate immune response and later adaptive immunity regulation. Dendritic cells (DCs) are the classic innate immune cell type which potentially trigger and control adaptive immunity mainly mediated by T and B lymphocytes (69, 70). Pulmonary

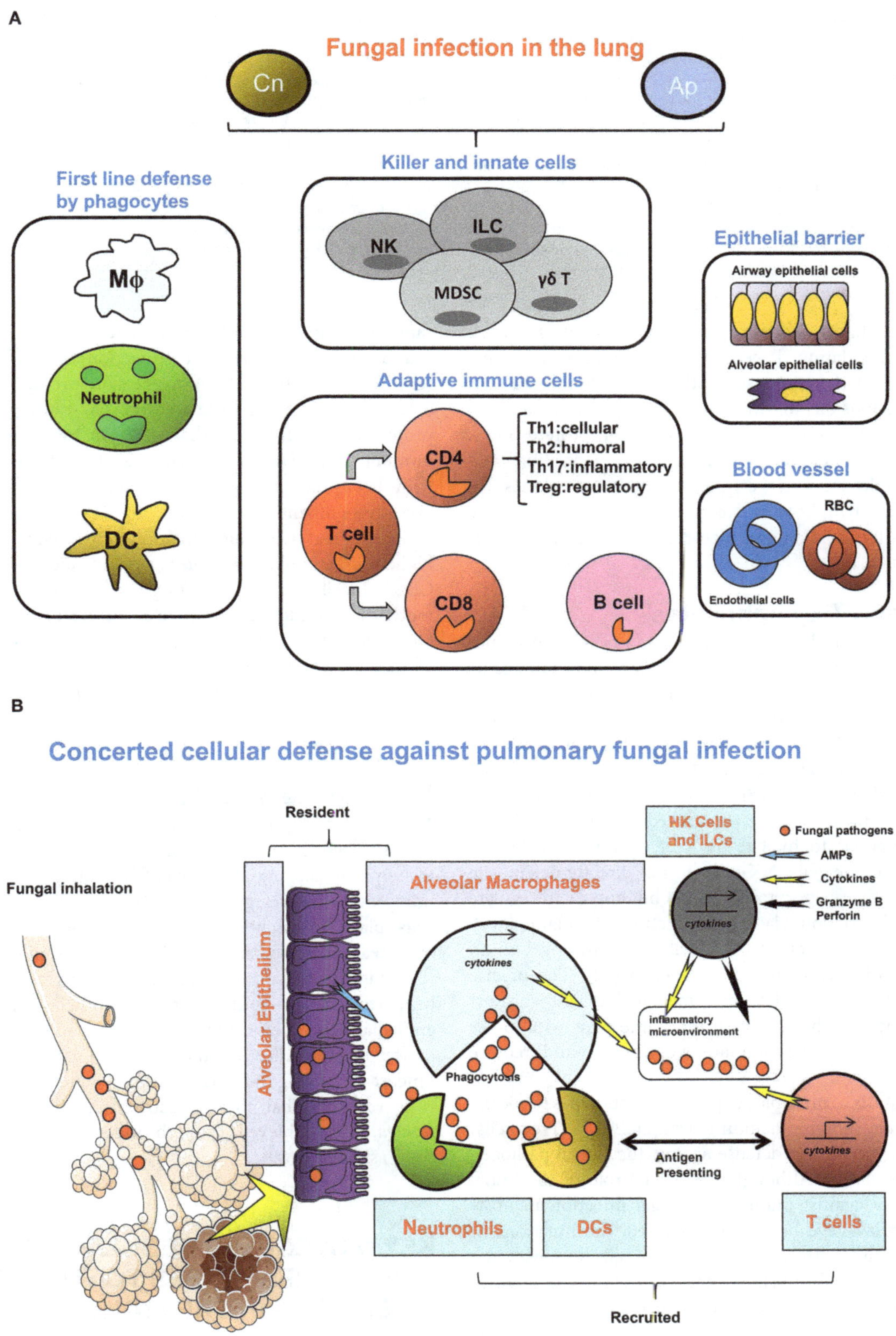

FIGURE 3 | The Network of cellular defense against fungal infections in the lung. **(A)** The cell category in host response to infection. **(B)** Divergent cell types orchestrate the host defense against fungal infection. To defend against pulmonary fungal infection as represented by Cn and Ap, divergent types of cells are involved *(Continued)*

FIGURE 3 | in the microenvironment of infectious sites in the lung. Among them, granulocytes and monocytes/macrophages as well as DCs are typical myeloid lineages which are the major cellular agents for innate immunity. In the respiratory tract, these cells constitute the first line of defense. Alveolar macrophages are monocytes/macrophages specifically resident in the lung; neutrophils are quickly infiltrated to the sites of inflammation to eradicate the pathogen and to promote tissue repair; pathogens, or antigens bind to DC-SIGN expressing immature DC to gain access to these cells for uptake, meanwhile activating DCs trigger adaptive immunity mediated by the T cells. The direct anti-pathogen effect of NK cells, as a central player of innate immunity, is ascribed to their direct cytotoxicity and cytotoxic cytokine and granule-secreting capacity. In addition, other innate like cells such as ILC and special T cell types such as γδT cells also contribute to the host defense by cytokine secretion or other approaches, with concomitant inducement of MDSC. For adaptive immunity, CD4+ T helper cell immunity is predominant in the pathogen-induced inflammation, IFN-γ produced by active Th1 cells is beneficial for Cryptococcus clearance, and the Th17 cells with IL-17/IL-23 secretion also plays important roles in a pulmonary fungal infection. In contrast, Th2 cells with anti-inflammatory cytokine secretion and Treg cells with immunosuppression activity participate in the host response upon fungal infection usually with negative outcomes. In the microenvironment, secreted cytokines further regulate innate and adaptive cells function through cytokine receptors. Besides multiple immune cells, the epithelial cells at the mucosal barrier also confer a protective activity against the invasive fungi in the respiratory tract. In the lung blood vessel, the roles of endothelial cells and RBC have rarely been reported.

DCs also coordinate adaptive immune responses to *A. fumigatus*, by regulating T cell proliferation and enhancing the protective Th1 response within the lung (71–73). Dectin-1 is responsible for immature DCs pro-inflammatory responses after exposure to *A. fumigatus* (74). Interestingly, the neutrophils could mediate lung DC maturation and efflux (75), while DCs trigger the secreted chemokines for neutrophils/Th1 lymphocytes recruitment (76), suggesting the cellular interplays among neutrophils, DCs as well as T lymphocytes.

NK Cells and T Lymphocytes

Natural killer (NK) cells in the innate immune system confer primary immune protection against tumor and pathogenic microbes through granule-mediated killing and effector IFN-γ release (77, 78). It was found that the adoptive transfer of $CD4^+$ T cells are responsible for the increase of NK cells with the activation marker NKG2D and the release of gamma interferon, granzyme B and perforin during *Pneumocystis* infection (79). Moreover, *A. fumigatus* can be recognized by human NK cells with direct response to hyphae but not conidia (80). Unlike other innate immune cells, NK cells do not directly defend the fungus through the phagocytosis process but appear to regulate their antifungal function via the production of inflammatory IFN-γ (81). Similarly, murine NK cells are antifungal to *C. neoformans* by directly killing the organism (82). The precise mechanisms of natural killer cells' mediated defense against respiratory fungal pathogens and their interplay with other immune cells during infection are still limited and require further study.

The homeostasis of multiple adaptive T cell subtypes (typically the CD4 positive subtype including Th1/Th2/Th17/Treg cells) are involved in the further defense against the fungal pathogen after innate immunity. Although the crucial role of the initial innate immunity against pulmonary fungal infection in host-pathogen interplay have been well-described, the functions of adaptive immunity with representative T cell responses are equally of importance for the host's defense. A Th1 type adaptive immune response with Th1 cytokines such as IFN-γ is often required for the clearance of *C. neoformans* yeast. For instance, previous studies have revealed that in mice infected with an IFN-gamma-producing *C. neoformans* strain, H99-gamma were resistant to a second infection in the lung by a lethal strain. In the respiratory tract, increased granulomatous formation, rapid inflammatory infiltration, and Th1 mediated adaptive immunity were involved in protection against cryptococcosis (83). Whereas, the mRNA expression profiles of pulmonary cytokines in infected mice were of Th2-type, IL-12 treatment protected mice from a lethal infection by regulating the host immune balance toward the Th1-state (84). Tregs and Th17 cells with IL-17 production exhibit opposite responses to aspergillosis, excessive inflammation driven by IL-17 augmentation and deduced Tregs with anti-inflammatory activity which led to a higher susceptibility to *A. fumigatus* in mice (85). In addition, $CD8^+$ T cell regulation in fungal infection and the molecular mechanisms underlying fungal clearance, remain rarely reported. One study displayed that the protective and antifungal memory of $CD8^+$ T cell responses are promoted by TLR3, which could serve as a new therapeutic target for aspergillosis in high-risk patients (86).

Lung Epithelial Cells

Multiple hematopoietic cells such as macrophages and DCs play crucial roles in anti-fungal immune resistance and tolerance to fungi, by keeping a balance of immunopathology/protective immunity. However, growing evidence implies that the epithelial cells play an essential role in infection and inflammation by releasing surfactant proteins and antimicrobial peptides (87). First, internalization of *A. fumigatus* conidia could also be performed by epithelial and endothelial cells (88). Lung epithelial cells (LEC) often play crucial roles in the crosstalk between fungal infection and mucosal immunity at pulmonary mucosa by linking the innate immunity and adaptive immunity. It is reported that the protective tolerance on LECs to *A. fumigatus* is achieved through indoleamine 2,3-dioxygenase (IDO) signaling (89).

KEY MOLECULES REGULATING HOST IMMUNE RESPONSES TO FUNGAL INFECTIONS IN THE LUNG

Various molecules play distinct but crucial roles in innate immunity, the present review also sheds light on molecular regulation, during immune tolerance, to pulmonary fungus (**Figure 4A**).

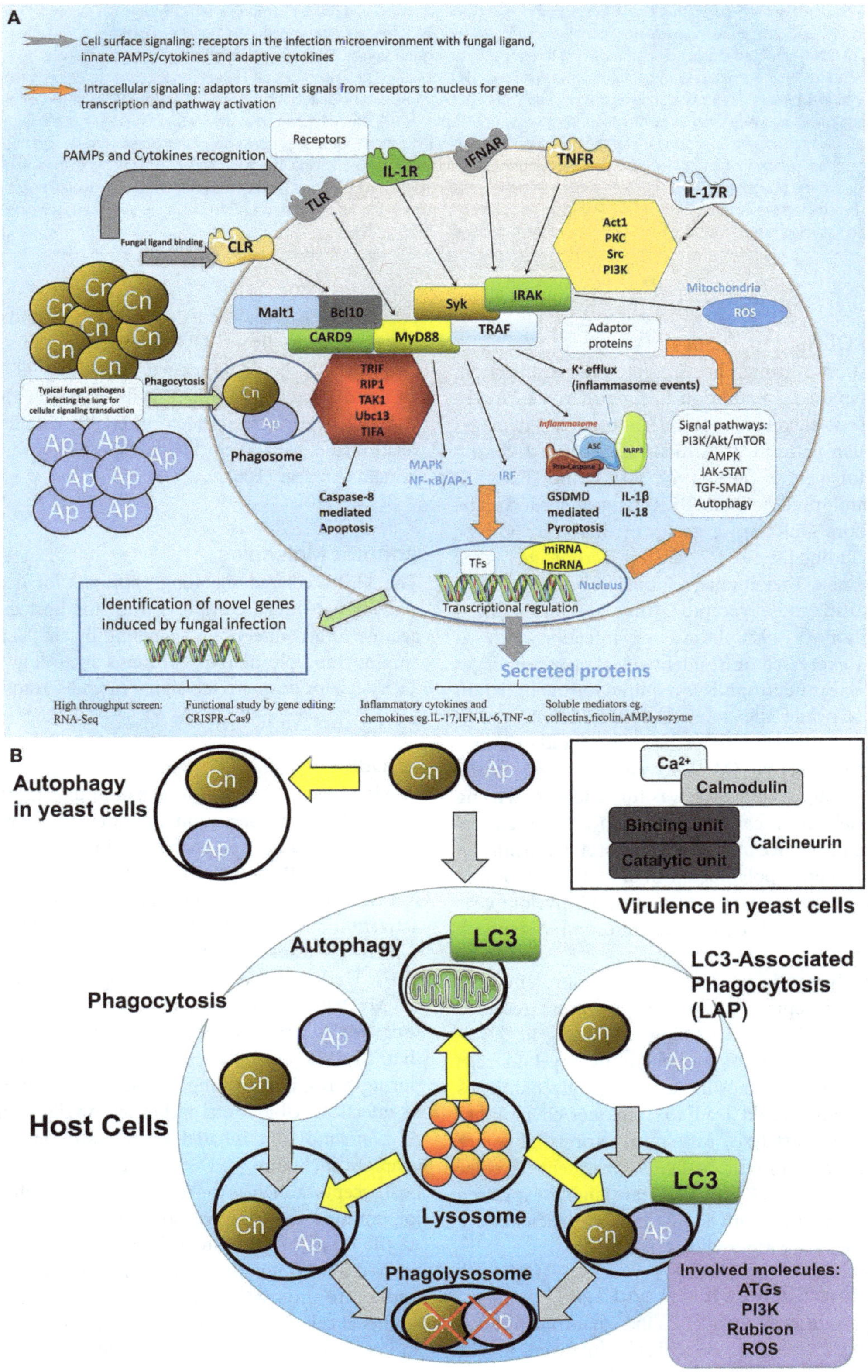

FIGURE 4 | The network of molecular defense against fungal infections in the lung. **(A)** General involved molecules and signal events in the aspects of cellular fungal infection. During fungal infection, various intracellular and extracellular molecules are involved in the host defense process. On the cell surface, receptors especially the

(Continued)

FIGURE 4 | PRRs are predominant regulators that sense fungal pathogens or fungal associated PAMPs, among them C-type lectin receptors (CLRs) including dectin-1, dectin-2, and MR play crucial roles in fungi-host communication. Toll-like receptors are also engaged in fungal recognition. In addition, cytokine receptors such as IL-1R, TNFR, or IFNAR are also activated upon receiving the respective cytokine signals. For intracellular signaling, adaptor proteins such as MyD88, CARD9, Syk, and TRAF are able to transmit signals via divergent pathways, leading to MAPK/NF-κB activation or type I interferon production, or inflammasome assembly. These pathways either trigger protein modulation such as translocation, phosphorylation, and ubiquitination, or at a transcriptional level, regulates gene expression in the nucleus, which leads to cell activities such as proliferation, apoptosis, or pyroptosis. **(B)** The autophagy and calcineurin signaling in the host-fungi during infection. The autophagy and calcineurin signaling are important for the host-fungi bioactivity regulation. Autophagy could occur in both yeast cells and host cells. LAP is a special intracellular autophagy which forms to defend extracellular fungi internalized and which combine the characters of both phagocytosis and autophagy: pathogens are entrapped in the phagosome with LC3 recruitment, and the lysosome is involved in the fusion of the phagolysosome where engulfed pathogens are killed by degradation. Key molecules including ATG, PI3K, Rubicon, and ROS play roles in the fungi induced LAP signaling. In yeast cells, calcineurin transmitting calcium and calmodulin signaling is responsible for the virulence of the pathogen.

Receptors

PRRs Including CLRs and TLRs

Signaling transductions, through host receptors in an infection, are relatively well-studied. In addition to fungi-derived PAMPs (13, 14), some host-encoded molecules, and the damage-associated molecular patterns (DAMPs), are secreted during fungal infection for host defense (90). CLRs and TLRs/IL-1R are classical and predominant PRRs engaged for fungal pathogen recognition. CLRs are a group of important C-type lectin receptors including Dectin-1/2, MCL (Macrophage C-type lectin, Clec4d), Mincle (macrophage inducible C-type lectin, Clec4e), and MR (mannose receptor) that are all involved in fungal recognition (91–94). In a lung infection with *A. fumigatus*, dectin-1 expressed on resident alveolar macrophages and polymorphonuclear neutrophils is required for the initiation of a host response while the role of dectin-2 is not well-investigated (95). Recently, a C-type lectin receptor, and melanin-sensing C-type lectin receptor (MelLec) was identified by the Gordon D. Brown group (96), which was fungicidal by sensing the unit of DHN-melanin in conidia of *A. fumigatus*. In animal models, MelLec was protective in the host against *A. fumigatus*. In humans, a single nucleotide polymorphism of MelLec was found to negatively regulate inflammatory responses by myeloid cells and increased the incidence of *A. fumigatus* infection in stem cell transplant recipients.

C. neoformans glycoantigens such as mannoproteins can bind different lectin receptors such as the mannose receptor and DC-specific non-ICAM3 grabbing nonintergrin (DC-SIGN) (97, 98). Phagocytosis of pathogens by myeloid cells upon PRRs recognition or exposure to soluble glycoantigens and cryptococcal DNA, would lead to cytokines/chemokines production as well as yeast lysis. Moreover, various TLRs are distributed on different immune cells, for resident alveolar macrophages, the TLR2, TLR4, and TLR7 are highly expressed and immunosuppressed mice with TLR2 or TLR4 deficiency are more susceptible to invasive aspergillosis (99–102).

Cytokine Receptors Such as IL-1R and TNFR

IL-1R shares a high homology with TLRs in signal transduction (103) and TLR/IL-1R signaling is critically involved in the response to fungal infections (104). The complex group of TNF ligands and the receptor superfamily serve as a rich source of drug targets on both innate and adaptive immunity. As a member of the tumor necrosis factor receptor superfamily, the T cell expressed glucocorticoid-induced TNFR-related protein (GITR, TNFRSF18), mediates both natural and acquired immune reactions by its ligand GITRL expressed on APC. Intriguingly, GITR-GITRL could influence TLR2 and TLR4 expression on DC and correlated with T cells response during candidiasis (105). However, the effect of GITRL-GITR on DC and its relationship with Treg during aspergillusis and cryptococcosis remains unclear (106).

Adaptor Molecules

The TLR-mediated signaling pathways for downstream NF-κB activation events are important for the host immune regulation against fungal infections. Signaling by the TLR or IL-1R could employ multiple adaptor proteins including MyD88, TRAF6, TRIF, etc. for downstream signal cascades transduction.

MyD88

Previous studies on the innate immune mechanisms revealed a remarkable conservation of molecular components of host defense signaling pathways, among them the Toll-IL-1R homology (TIR) domain, presented by adaptor protein MyD88, is essential for signal transduction. The downstream NF-κB activation could be induced by adaptor MyD88 via the human TLR/IL-1R transduction pathway, through IRAK and TRAF6 (107). Under the condition of fungal infection, overexpression of MyD88 was functionally enough for the induction of fungicidal peptide Drosomycin *in vitro*. A recent study found that MyD88 also plays an important role in host defense during a respiratory fungal attack especially at the first phase of infection. Ubiquitous airborne conidia could be formed by *A. fumigatus* and inhaled by humans daily, in this process, adaptors including MyD88 and CARD9 could be activated via host receptors such as the C-type lectin receptor, TLR, and IL-1R for immune signaling transduction as well as fungal clearance. At the onset of respiratory fungal infection, MyD88 depletion delayed neutrophil lung trafficking and chemokines production, leading to injury in the lung. MyD88 expressed on lung epithelial cells, was responsible for rapid neutrophil recruitment and chemokines production through IL-1R. Exogenous CXCL1 treatment also rescued mortality in MyD88-deficient mice. Altogether, host MyD88 associated pathways play essential roles in early responses to *A. fumigatus* in the lung, and fungal PAMPs to host PRRs triggered MyD88-NF-κB signaling, serves as a pivotal pathway for the pathogen-host interaction (100, 108).

TRAF Family Proteins

Besides MyD88, TRAF family proteins especially TRAF3 and TRAF6 play a crucial role in regulating innate and adaptive immune responses. In TLR signaling, TRAF6 is dependent on MyD88 instead of TRIF, and TRAF6 differently modulates MyD88- and IRAK-1-induced activation of NF-κB (109, 110). Although TRAF6 and TAK1 are implicated in candida infection (111), the roles of these molecules in CLRs initiated signaling have not been studied. In canonical CLR signaling, both TRAF6-TAK1 interaction and spleen tyrosine kinase (Syk) phosphorylation are initial events involved for CARD9-Bcl10-MALT1 complex and downstream MAPK and NF-κB activation, adaptor Syk-coupled CLRs including Dectin1/2 and Mincle mediate innate immunity against fungal infection (92, 93, 112). Another study proved that Dectin-1-Syk and autophagy contributed to maturation of the *A. fumigatus* phagosome (113).

The Inflammasome Complex

Inflammasomes

The inflammasomes comprised of several intracellular proteins are functionally essential for the regulation of innate immune responses. These complexes could be activated through PAMPs-PRRs or host derived DAMPs signaling and are intensely associated with various infectious pathogens including fungi, viruses, and bacteria especially in a respiratory tract infection (114, 115).

NLRP3 and AIM2

NLRP3 and AIM2 are typical inflammasomes which play critical roles, with distinct mechanisms, in innate immunity: The NLRP3 inflammasome activation is regulated by two major steps of signals. First is the priming step in which NLRP3 and pro-IL-1β expressions are induced in a NF-κB dependent manner by PAMPs e.g., TLR agonists (LPS as representative one) or key cytokines e.g., TNF-α and IL-1β. The second step is NLRP3 assembly activation by multiple stimuli e.g., ATP/nigericin. Further studies showed that phagocytosis of particulate matter triggers K^+ Efflux and activates NLRP3 (116, 117), but recently a K^+ independent activation mode was described via targeting mitochondria using small molecules (118). As to the biological function of the NLRP3 inflammasome in the field of infection, it plays a crucial protective role against several pulmonary fungi which could cause aspergillosis and cryptococcosis (90). Notably, NLRP3 inflammasome mediated signaling in the host immune defense is recognized by fungal polysaccharides such as curdlan (119, 120), and specifically *A. fumigatus* triggers the activation of the NLRP3 inflammasome via syk kinase activity and reactive oxygen species (121). In addition, fungal zymosan and mannan stimulated the NLRP3 inflammasome to induce macrophage and DC caspase-1 activity as well as IL-1β secretion, suggesting that conserved cell wall components are responsible for the ASC and NLRP3 inflammasome activation during fungal infection (122). In cellular and animal models, acapsular mutant strain of *C. neoformans* (Cap59) but not the encapsulated wild type strain (H99), activates the inflammasome (123). Moreover, the germline NLRP3 knock-out mice infected with biofilm of clinical *C. neoformans* strain HS1101 displayed a more severe infection and inflammation in the lung, which is also true for Casp1 KO and ASC KO mice. These data illustrate the importance of the NLRP3 inflammasome components in the host responses to the fungal challenge in the lung (124). Absence in melanoma 2 (AIM2) is a unique DNA-sensing receptor. Mice lacking AIM2 alone displayed similar susceptibility compared with wild type mice upon *Aspergillus* infection, meanwhile mice with double knockout of AIM2 and NLRP3 failed to control *Aspergillus* hyphae dissemination, thus they succumbed to the attack of fungus more rapidly than wild type mice or mice deficient in either AIM2 or NLRP3 (125). Until now, the functions and mechanisms of other novel host inflammasomes such as NLRP1, NLRC6, NLRP7 in controlling fungal infection, remain elusive.

NF-κB and MAPK

NF-κB plays a critical role in the control of infection as well as inflammation, it was shown that the macrophage dectin-1, TLR2 and TLR4 could specifically recognize the mature hyphal forms of *Aspergillus* but not spores, resulting in NF-κB dependent inflammatory cytokine secretion and antimicrobial ROS production (126). As to dectin-2, it was also reported that IκBα (inhibitor of the kappa B protein) phosphorylation and NF-κB activation following *A. fumigatus* stimulation occurs through dectin-2-Syk signaling (127). Interestingly, gliotoxin as a toxic metabolite synthesized by *A. fumigatus* specifically inhibits NF-κB activation and induces apoptosis (128). However, another paper found that ERK but not p38 was essential in the MAPK pathway responsible for defense against *A. fumigatus* on Alveolar Macrophages, whereas NF-κB activation appeared to play a secondary role. After phosphorylation by upstream molecules, the MAPKs translocated to the nucleus to phosphorylate downstream target molecules which transcriptionally regulated cytokine genes (129). Different from macrophage detecting *Aspergillus* via TLR2/4 to trigger TLR-MyD88-NF-κB-dependent synthesis of inflammation related molecules, pulmonary epithelial cells could sense sprouting but not resting spores of *A. fumigatus* to induce the synthesis of interleukin (IL)-8 through p38 MAPK and ERK1/2 and PI3 kinase, suggesting that MAPK is of importance both on phagocytes and epithelial cells (130). In addition, TLR4 in conjunction with CD14 could be involved in the host defense against *C. neoformans* capsular polysaccharide glucuronoxylomannan (GXM), stimulating NF-κB nuclear translocation in macrophages without MAPK activation and TNF-α release (131).

Cytokines and Chemokines

Cytokines/chemokines are predominant modulators secreted from the host in response to various fungal pathogen infections. An imbalance between pro- and anti-inflammatory cytokines might negatively result in infectious diseases due to an impaired host defense.

IL-1 Family Cytokines: IL-1β/IL-18/IL-33/IL-36

Until now, 11 members were identified in the IL-1 family of cytokines with seven proinflammatory cytokines including IL-1α, IL-1β, IL-18, IL-33, IL-36α, IL-36β, and IL-36γ and four with anti-inflammatory cytokines including IL-1Ra, IL-36Ra,

IL-37, and IL-38 (132). IL-1α and IL-1β are thought to be the primary drivers of inflammation in chronic granulomatous disease via decreased autophagy and increased inflammasome activation (133). With respect to immunopathogenic effects of IL-1 family members during invasive pulmonary aspergillusis (IPA), both lung homogenates and alveolar macrophages derived from Dectin-1 KO mice demonstrated reduced IL-1α/IL-1β and TNF, MIP, and KC *in vitro* (134), suggesting that IL-1α and IL-1β production is derived from Dectin-1 signaling. A recent study has indicated that IL-1R was critical for protection, as IL-1α was required for lymphocytes recruitment and IL-1β conferred resistance to fungal dissemination (135). Members of the IL-1 family often play protective roles in immune defense against the opportunistic mold *A. fumigatus*. Besides IL-1β, IL-18 is also protective against *C. neoformans* by inducing IFN-γ. IL-1β/IL-18 is primarily secreted in a NLRP3 inflammasome-dependent manner as mentioned above. In a recent study, the IL-1 family member IL-33 played regulatory roles in lung infection defense against *A. fumigatus*. IL-33 expression was detected in the lung and increased after exposure to the fungus, independent of Dectin-1. Mice lacking the receptor for IL-33 ($Il1rl1^{-/-}$) unexpectedly demonstrated enhanced lung clearance of the fungal pathogen while IL-33 administration to normal mice inhibited fungal-induced IL-17A and IL-22 through PGE2. Because normal mice produced less PGE2 after fungal exposure when administered IL-33, and PGE2 was significantly increased in fungal-exposed $Il1rl1^{-/-}$ mice, suggesting that IL-33-mediated regulation of IL-17A and IL-22 occurred at the level of PGE2 This was confirmed by cyclooxygenase 2 or PGE2 inhibition, which attenuated fungal-induced protective IL-17A and IL-22, as well as IL-1α, IL-1β, and IL-6 productions in $Il1rl1^{-/-}$ mice, resulting in impaired fungal clearance (132). Moreover, interleukin-36γ (IL-36γ) is a newly identified IL-1 family inflammatory cytokine which is highly expressed on epithelium and some myeloid cells. Studies in humans have shown that *A. fumigatus* strongly induces IL-36g and IL-36Ra, but not IL-36α, which were dependent on dectin-1 and TLR4 while Inhibiting IL-36 signaling was found to abrogate the induction of protective Th17 and Th1 responses (136).

IL-17 and IL-23

IL-17 (IL-17A) is a pleiotropic cytokine which has been implicated in the pathogenic development of autoimmune inflammations such as rheumatoid arthritis (RA) and multiple sclerosis (MS), but this cytokine has also been associated with protection against bacterial infections (137). IL-17 could be synthesized more rapidly by natural killer T cells (NKT cells), γδ T cells and innate lymphoid cells than Th17 cells due to RORγt expressed on these cell types (138). The roles of IL-17 and IL-23 in the host response against the pulmonary fungal infection have been of focus recently. In the airway lesions, IL-17 potently regulates neutrophil recruitment and homeostasis together with IL-23 (139, 140). Intriguingly, in the *A. fumigatus* model of pulmonary fungal infection via the intranasal route, IL-17 and IL-23 are highlighted with a regulatory role, by restricting the IL-12-IFN-γ mediated Th1 protective response or even by impairing the antifungal immune resistance (141). Thus, the multi-faceted roles of host inflammation are that it is not only critically responsible for the antifungal immunity, but that is also negatively controls proper immune reactions or even worsens fungal diseases when the balance between protection and pathogenesis is dysregulated in certain milieu.

Type I and Type II Interferons

IFN-α/β, as typical type I interferons, are produced by various cell types after many stimuli, especially viruses. Through binding to a specific IFN-α/β receptor termed IFNAR (IFN-α/βR), downstream IFN-stimulated genes (ISGs) can be induced. IFNAR is composed of two subtypes IFNAR1 and IFNAR2. Canonical type I IFN signaling triggers the Janus kinase (JAK)-signal transducer and activator of the transcription (STAT) pathway, resulting in ISG gene expression (142). Up to now, limited information is known about their roles in non-viral infections such as a pulmonary fungal infection (143). Nonetheless, signaling of type I interferons in the host defense against *C. neoformans* was addressed in a recent study. Using IFNAR1 defected mice (IFNAR1 KO), the authors found an increased fungal clearance by, with enhanced Th2 and Th17 responses after infection compared with control animals. In addition, IFNAR1 KO mice exhibited significantly higher MUC5AC expression in bronchoepithelial cells. Therefore, type I interferons might negatively control the early host defense to a fungal infection (144). In contrast, another group reported that mice lacking either the IFNAR or IFN-β succumbed to unrestrained pneumonia and encephalitis after an intratracheal or i.v. challenge with *C. neoformans* with increased Th2 cytokines IL-4/IL-10/IL-13, but reduced TNF-α, IFN-γ, iNOS, and CXCL10, suggesting that the protection by the type I IFN signaling is associated with type I cytokine polarization (145). In the case of *Aspergillus* infection, its conidia mainly induce the IFN-β signaling in respiratory epithelial cells. Resting conidia can be sensed by differentiated human bronchial epithelial cells (HBECs), leading to the production of IFN-β-inducible genes such as IP-10 (CXCL10). Activated T cells could be further regulated by IP-10 (146). Inhibition on the IFN-β signaling mediators such as RIP-1 (Receptor-interacting protein 1) and TBK-1 (TANK-binding kinase-1) by resveratrol could reduce the IFN-β/IP-10 expression (147). On the other hand, Th1 type cytokine IFN-γ has been proven effective in clearing *C. neoformans* infection. Recombinant IFN-γ administration reduces the fungal burden and increases survival rates, reinforcing the efficiency of fungicidal amphotericin B (148, 149). Pulmonary cytokine analysis demonstrated that there was a Th1-type pro-inflammatory cytokines bias, rather than Th2-type cytokines expression, in mice challenged with the IFN-γ releasing *C. neoformans* strain, compared to wild-type strain-treated mice (150).

Classical Cytokines and Chemokines: IL-6, IL-8 (KC), TNF-α, GM-CSF

Pro-inflammatory cytokines such as IL-6, IL-8 (KC), TNF-α, and GM-CSF are essential factors in innate immunity. In hematological patients combined with IPA, levels of interleukin (IL)-6 and IL-8 are elevated in both serum and bronchoalveolar lavage fluid (BALF) (151). Circulating IL-6 is able to induce acute phase responses to control local or systemic acute inflammation with its anti-inflammatory activity (152). IL-6 increases *Aspergillus*-induced IL-17 production from healthy and hematological controls, but not in IPA patients with impaired T cell responsiveness to IL-6 (153). The interleukin (IL)-8 as CXC chemokine (CXCL8) is produced from respiratory epithelial cells via the PI3K and MAPK pathways, instead of the TLR-MyD88 pathway, upon *Aspergillus* infection (130), similar to other CXC chemokines such as KC (keratinocyte-derived chemokine, CXCL1, specific for mice, it is the counter chemokine for IL-8 in humans) and MIP-2 (macrophage inflammatory protein 2, CXCL2/3). Markedly higher levels of KC and MIP-2 were observed in the lungs of animals transiently depleted of neutrophils upon challenge with AF; and the lung-specific overexpression of KC improved the outcome of mice in IPA by augmenting the host defense against *Aspergillus* (154, 155). Transcriptionally, proteases derived from *A. fumigatus* elicit cytokine production in human alveolar type II epithelium-like cells by enhancing IL-6 and IL-8 mRNA levels (156). However, some other study showed that limited levels of these cytokines (IL-6 and IL-8) are synthesized by A549 cells after contacting *A. fumigatus,* resulting in impaired recruitment of leukocytes to the lesion sites and pathogen escape from immune defense (157). There is also a study showing that *A. fumigatus* induced an acute inflammation regulated by neutrophils with pro-inflammatory cytokines (TNF-α, GM-CSF, and IL-1β) and chemokines (MIP-1a, MCP-1 and MIP-2) induction during the peak of infection in the lung, neutralizing TNF-α or GM-CSF decreased neutrophil influx and delayed fungal clearance (158). As for TNF-α, it enhances host responses to *A. fumigatus*, and inhibition on its function might increase susceptibility to aspergillosis (159). In a cytokine network assay in the lungs of immunocompetent mice, endogenous TNF-α, IFN-γ, IL-18, IL-12 are immunoreactive with antifungal activity in response to inhaled *A. fumigatus* (160). In addition, Granulocyte-macrophage colony-stimulating factor (GM-CSF) is a pleiotropic hematopoietic cytokine regulating the myeloid cell host response. The receptor of GM-CSF (GM-CSFR/Csf2r) is composed of GM-CSFRα/Csf2ra and GM-CSFRβ/Csf2rb (161). Mice lacking the GM-CSF receptor β chain (GM-CSFRβ) were susceptible with a higher mortality to *A. fumigatus* conidia infection in the lung, illustrating the functional role of GM-CSFRβ signaling against inhaled *A. fumigatus* (162). In response to *C. neoformans* yeast cells or the major capsular polysaccharide-glucuronoxylomannan (GXM), human polymorphonuclear leukocytes from normal subjects can release inflammatory cytokines (163). Of note, macrophage polarization status is critical for the control of *C. neoformans* infection, and IFN-γ and IL-4 are responsible for the M1/2 polarization associated fungicidal activity. This implies the communication of cell mediated immunity and cytokine mediated immunity (164). Considering clinical applications and translational medicine, targeting the cytokines might be a useful intervention approach for antifungal therapy. To this end, a fluoroquinolone drug moxifloxacin conferred protective effect on human monocytes infected with *A. fumigatus* by inhibiting inflammatory cytokine productions via NF-κB and MAPK inactivation (165).

Secreted Soluble Proteins in the Lung

Besides the cytokine/chemokines produced by infected immune cells, the innate immune system can form another barrier on the airway surface upon inhalation of fungal pathogens. Numerous soluble peptides and proteins in this barrier confer antimicrobial activity. This is fulfilled by lysozyme, lactoferrin, secretory leukocyte proteases, secretory phospholipase A2, defensins, and cathelicidins, which are largely secreted by cells in the airway submucosal glands or epithelial cells. Just like antimicrobial peptides (AMPs) in the gut, these soluble effector molecules have been shown to eliminate a spectrum of microorganisms with neutralizing, opsonization, antibiotic or direct killing activities (166, 167).

Collectin

The human collectin family (collagen-like or C-type lectin) include MBL (Mannan binding lectin) and pulmonary surfactant proteins A to D, which are important innate immune mediators for antifungal defense (168). The functions of collectins encompass opsonization, inflammation regulation, and the direct clearance of pathogens (169). A previous study has demonstrated the protective effect of MBL in a mouse model of aspergillosis (170). In the airways, surfactant proteins render immune protection against fungal accumulation and then clearance by phagocytosis (171). Among them, SP-A in respiratory tract binds to *C. neoformans* without enhancing phagocytosis, while SP-D may play an essential role at the early stage of infection with an increased rate of uptake and phagocytosis. Thus, a greater number of phagocytosed *C. neoformans* cells in wild-type mice than in SP-D KO mice were observed. However, SP-D enhances fungal survival in macrophages *in vitro*, and mice lacking SP-D are protected *in vivo* (171, 172). The reason for these opposing findings awaits further investigation. Collectins, ficolins, and pentraxins are circulating proteins which could serve as opsonins (173). Type II alveolar epithelial cells could secrete H-Ficolin as innate immune opsonin to involve in pulmonary defenses against fungal infection. Silke Schelenzls' group found that H-ficolin was involved in *A. fumigatus* defense through the activation of the lectin complement pathway, highlighting the interaction between host and fungus and the modulation of the immune response by ficolin (174). In addition, another human serum opsonin, L-ficolin was detected in the bronchoalveolar lavage (BAL) fluid from patients with a fungal infection. L-ficolin opsonization increased IL-8 production from A549 cells and enhanced conidial uptake and the killing of *A. fumigatus* by macrophages and neutrophils, with a reduced release of inflammatory cytokines (175).

Defensin and Lysozyme

The release of inflammatory mediators such as AMPs from epithelial cells is a critical step for the generation of protective activities, including inflammatory cell recruitment and generation of direct antimicrobial factors. Defensins, especially the b-defensin, are one type of AMP that are characterized by the presence of b-sheets stabilized by two disulfide bonds. In a study on HBECs, exposure to *A. fumigatus* led to the expression of b-defensin2 and b-defensin9 (hBD2 and hBD9) genes, suggesting that AMPs from respiratory epithelium are involved in host response during *Aspergillus* infection (176). In addition, as one of the most abundant antimicrobial proteins in the airway, lysozyme is a small enzyme synthesized by epithelium as well as resident macrophages in human lung tissues, which contributes to hyphal disruption of *A. fumigatus* (177, 178). In mice, two isoforms of lysozyme are identified: lysozyme P is expressed on Paneth cells in the small intestine and lysozyme M is predominantly expressed on alveolar macrophages, alveolar type II epithelial cells, and bronchoalveolar lavage fluid (BALF) in the lung (179, 180). However, whether lysozyme M renders protection against pathogenic fungi in the lung remains uncertain. The results of previous studies indicate that defensins and lysozymes might be involved in host defense against respiratory fungal infections.

Calcineurin Signaling

Calcineurin and related pathways are implicated in the controls of hyphal synthesis, morphological character, and virulence in *A. fumigatus*. Calcineurin is a heterodimer consisting of catalytic subunit A and Ca^{2+}/calmodulin binding unit, a mutant of *A. fumigatus* with depleted calcineurin A catalytic subunit displayed deficiency in morphology and decreased filamentation (181–183). Meanwhile, calcineurin mutant with decreased beta-glucan amounts promoted the fungicidal activity of cell wall inhibitors, implying that targeting calcineurin can be a potential synergistic therapy with other fungicides against *A. fumigatus* (184). The growth of *C. neoformans* was sensitive to CsA and FK506 which mediates the signal transduction on Ca^{2+}-regulated protein phosphatase calcineurin and the Calcineurin mutant strains failed to be infectious in an animal model of cryptococcosis in the brain, suggesting that calcineurin is necessary for pathogenicity of *C. neoformans* (185) (**Figure 4B**). In addition, host calcineurin signaling is also important for host immune responses to fungal infection, for instance, the intervention of neutrophils ability on *Aspergillus* species germination was inhibited in hematopoietic stem cell transplant (HSCT) individuals and this impairment was partly attributed to the administration of calcineurin inhibitors (186), suggesting that calcineurin signaling might affect neutrophils activity against *Aspergillus*, especially in immunocompromised hosts. Besides the neutrophils, calcineurin signaling has also been associated with macrophages. It was reported that calcineurin inhibitor tacrolimus impaired macrophage immune responses and the clearance of the major mold pathogen *A. fumigatus* from the airway, possibly due to the inhibition on *A. fumigatus*-induced phagosomal TLR9-BTK-Calcineurin-NFAT cascades independent of MyD88, and NFAT collaborating with NF-κB contributed to TNF-α production in primary alveolar macrophages (187). By targeting host and pathogen calcineurin signaling, cyclosporine (CsA) can be used as both an immunosuppressive drug and antimicrobial agent, which inhibits the protein phosphatase calcineurin. However, inhibition on host calcineurin or its downstream calcineurin-NFAT pathway by CsA, can protect transplant recipients from severe transplant rejection, and it might at the same time increase host immune defects leading to organ transplant-related invasive aspergillosis. Interestingly, the diversity of immunosuppressants determined different outcomes in transplant patients but the death rate appeared to be lower in individuals who received the calcineurin-inhibitors (188), the precise mechanisms on the host calcineurin signaling in cryptococcosis remains to be investigated.

Molecules Mediating Autophagy During Fungal Infection

Fungi often infect mammalian hosts where they interplay with each other: extracellular pathogens could be cleared by immune cells via a phagocytosis process, however, pathogens could also induce autophagy which is a lysosomal degradation process essential for host cell survival and homeostasis (189–192). The autophagy machinery may functionally protect against the threats of infection by eliminating invading pathogens (193, 194), specifically, LC3-associated phagocytosis (LAP) in phagocytes is a non-canonical autophagy pattern, and phagosomes with a single-membrane formation followed by a clearance of engulfed pathogens is required in this process. This LAP is started by the sensing of pathogens via PRRs and the LC3 recruitment to the phagosomal membrane. This precise xenophagy or LAP is utilized for the host defense against invading fungi (195–197).

In both host cells and in yeast cells, the extent of autophagy is critical for the control of pathogens as well as the survival of immune cells, mainly because autophagy is associated with cell programmed necrosis and apoptosis (191). In a reported study, the autophagy in macrophages with response to *C. neoformans* was assessed and the non-activated bone marrow derived macrophages (BMDMs) from Atg5 (Autophagy-related gene 5) knockout mice restricted *C. neoformans* growth with fungistatic activity; and *in vivo*, mice with myeloid depletion of Atg5 consistently displayed reduced susceptibility to *C. neoformans* (198). LAP is essential for host anti-fungal responses. Recent evidence indicated that LAP could be blocked by *A. fumigatus* cell wall melanin for its pathogenicity enhancement (199, 200). Canonical autophagy molecules are equally of importance for LAP, for instance, beclin-1 in the PI3K pathway is recruited to autophagosome after internalization of zymosan by phagocytes. The autophagy protein ATG7 is also important for LAP, because the loss of this molecule would lead to the abolishment of LC3 recruitment and reduced clearance of internalized pathogens and apoptotic cells (201, 202). The recently identified autophagy protein Rubicon (RUN domain protein as Beclin-1 interacting and cysteine rich containing) was essential for LAP but not autophagy induction, mediating clearance of A fumit (203). In addition, NADPH regulated ROS is required for LC3 recruitment (200).

Interestingly, the Atg genes in *C. neoformans* are also involved in the pathogenic mechanisms of infection, for instance, *C. neoformans* with an Atg7 mutant conferred lower survival but higher susceptibility to the killing machinery of different host phagocytes in mice with a pulmonary infection (204). It was proven that PI3K signaling with a defective formation of Atg8 labeled vesicles within the *C. neoformans* exhibited a dramatic virulence attenuation in a murine infection model (205) (**Figure 4B**).

PERSPECTIVE

In this review, we summarized the cellular and molecular mechanisms of the host defense against pathogenic fungal infections in the lung. Regarding cells, innate immune cells including macrophages, neutrophils as well as DCs are the first line cells in the host response to a fungal invasion; adaptive T lymphocytes are of importance to restrain fungal expansion; and the NK cells are responsible for the direct killing of fungal pathogens by its cytotoxicity. In regards to signaling molecules, receptors-adaptors-transcriptional factors form the cascade to maintain homeostasis upon fungal infection. Cytokine production is essential in the microenvironment as they can link intercellular communication via cytokine receptors. However, more mechanisms remain to be studied. Phagocytosis regulation, autophagy, miRNA/lncRNA or ROS signaling in pathogenic fungal infections are all currently understudied. Moreover, the identification of novel genes and pathways via high throughput screening or sequencing and further functional studies are still required to elucidate the host defense mechanisms against infectious fungal pathogens in the lung.

AUTHOR CONTRIBUTIONS

ZL and GM conceived, wrote, and edited the manuscript. GL and GM checked and approved the final version of the manuscript.

FUNDING

This work was supported by funding to the Postdoc program by the Joint Center for Infection and Immunity between Guangzhou Women and Children's Medical Center and the Institute Pasteur of Shanghai, Chinese Academy of Science. The lab of GM is supported by funding from the National Key Basic Research Programs of China (2015CB554302, 2016YFE0133500, 2018YFA0507300), Natural Science Foundation of China (81830049, 81761128012, 31570895) and Strategic Priority Research Program Grant (XDB29030303) of Chinese Academy of Sciences.

REFERENCES

1. Latge JP. *Aspergillus fumigatus* and aspergillosis. *Clin Microbiol Rev.* (1999) 12:310–50. doi: 10.1128/CMR.12.2.310
2. Yao Z, Liao W. Fungal respiratory disease. *Curr Opin Pulm Med.* (2006) 12:222–7. doi: 10.1097/01.mcp.0000219272.57933.01
3. Carvalho A, Cunha C, Iannitti RG, De Luca A, Giovannini G, Bistoni F, et al. Inflammation in aspergillosis: the good, the bad, and the therapeutic. *Ann N Y Acad Sci.* (2012) 1273:52–9. doi: 10.1111/j.1749-6632.2012.06754.x
4. Chayakulkeeree M, Perfect JR. Cryptococcosis. *Infect Dis Clin North Am.* (2006) 20:507–44, v-vi. doi: 10.1016/j.idc.2006.07.001
5. Kronstad JW, Attarian R, Cadieux B, Choi J, D'Souza CA, Griffiths EJ, et al. Expanding fungal pathogenesis: *Cryptococcus* breaks out of the opportunistic box. *Nat Rev Microbiol.* (2011) 9:193–203. doi: 10.1038/nrmicro2522
6. Mitchell TG, Perfect JR. Cryptococcosis in the era of AIDS—100 years after the discovery of *Cryptococcus neoformans*. *Clin Microbiol Rev.* (1995) 8:515–48. doi: 10.1128/CMR.8.4.515
7. Udwadia ZF, Doshi AV, Bhaduri AS. Pneumocystis pneumonia. *N Engl J Med.* (2004) 351:1262–3. doi: 10.1056/NEJM200409163511222
8. Lortholary O, Denning DW, Dupont B. Endemic mycoses: a treatment update. *J Antimicrob Chemother.* (1999) 43:321–31. doi: 10.1093/jac/43.3.321
9. Wheat J. Endemic mycoses in AIDS: a clinical review. *Clin Microbiol Rev.* (1995) 8:146–59. doi: 10.1128/CMR.8.1.146
10. Bellocchio S, Moretti S, Perruccio K, Fallarino F, Bozza S, Montagnoli C, et al. TLRs govern neutrophil activity in aspergillosis. *J Immunol.* (2004) 173:7406–15. doi: 10.4049/jimmunol.173.12.7406
11. José RJ, Brown JS. Opportunistic and fungal infections of the lung. *Medicine.* (2012) 40:335–9. doi: 10.1016/j.mpmed.2012.03.013
12. McMath WF, Hussain KK. Cryptococcal meningo-encephalitis. *Br Med J.* (1961) 2:91–3. doi: 10.1136/bmj.2.5244.91
13. Kawai T, Akira S. The role of pattern-recognition receptors in innate immunity: update on Toll-like receptors. *Nat Immunol.* (2010) 11:373–84. doi: 10.1038/ni.1863
14. Takeuchi O, Akira S. Pattern recognition receptors and inflammation. *Cell.* (2010) 140:805–20. doi: 10.1016/j.cell.2010.01.022
15. Howard SJ, Cerar D, Anderson MJ, Albarrag A, Fisher MC, Pasqualotto AC, et al. Frequency and evolution of Azole resistance in *Aspergillus fumigatus* associated with treatment failure. *Emerg Infect Dis.* (2009) 15:1068–76. doi: 10.3201/eid1507.090043
16. Lamoth F, Juvvadi PR, Soderblom EJ, Moseley MA, Asfaw YG, Steinbach WJ. Identification of a key lysine residue in heat shock protein 90 required for azole and echinocandin resistance in *Aspergillus fumigatus*. *Antimicrob Agents Chemother.* (2014) 58:1889–96. doi: 10.1128/AAC.02286-13
17. Yaguchi T. The genus *Aspergillus*. *Med Mycol J.* (2011) 52:193–7. doi: 10.3314/mmj.52.193
18. Ibrahim-Granet O, Philippe B, Boleti H, Boisvieux-Ulrich E, Grenet D, Stern M, et al. Phagocytosis and intracellular fate of *Aspergillus fumigatus* conidia in alveolar macrophages. *Infect Immun.* (2003) 71:891–903. doi: 10.1128/IAI.71.2.891-903.2003
19. Mouyna I, Sarfati J, Recco P, Fontaine T, Henrissatz B, Latge JP. Molecular characterization of a cell wall-associated beta(1-3)endoglucanase of *Aspergillus fumigatus*. *Med Mycol.* (2002) 40:455–64. doi: 10.1080/mmy.40.5.455.464
20. Bernard M, Mouyna I, Dubreucq G, Debeaupuis JP, Fontaine T, Vorgias C, et al. Characterization of a cell-wall acid phosphatase (PhoAp) in *Aspergillus fumigatus*. *Microbiology.* (2002) 148(Pt 9):2819–29. doi: 10.1099/00221287-148-9-2819
21. Latge JP. The pathobiology of *Aspergillus fumigatus*. *Trends Microbiol.* (2001) 9:382–9. doi: 10.1016/S0966-842X(01)02104-7
22. Blasi E, Barluzzi R, Mazzolla R, Bistoni F. Differential host susceptibility to intracerebral infections with *Candida albicans* and *Cryptococcus neoformans*. *Infect Immun.* (1993) 61:3476–81.
23. Lundgren B, Lipschik GY, Kovacs JA. Purification and characterization of a major human *Pneumocystis carinii* surface antigen. *J Clin Invest.* (1991) 87:163–70. doi: 10.1172/JCI114966
24. Kling HM, Shipley TW, Patil SP, Kristoff J, Bryan M, Montelaro RC, et al. Relationship of *Pneumocystis jiroveci* humoral immunity to prevention of colonization and chronic obstructive pulmonary disease in a primate model of HIV infection. *Infect Immun.* (2010) 78:4320–30. doi: 10.1128/IAI.00507-10
25. Morris A, Netravali M, Kling HM, Shipley T, Ross T, Sciurba FC, et al. Relationship of pneumocystis antibody response to severity of chronic obstructive pulmonary disease. *Clin Infect Dis.* (2008) 47:e64–8. doi: 10.1086/591701
26. Queiroz-Telles F, Fahal AH, Falci DR, Caceres DH, Chiller T, Pasqualotto AC. Neglected endemic mycoses. *Lancet Infect Dis.* (2017) 17:e367–77. doi: 10.1016/S1473-3099(17)30306-7
27. Hage CA, Knox KS, Wheat LJ. Endemic mycoses: overlooked causes of community acquired pneumonia. *Respir Med.* (2012) 106:769–76. doi: 10.1016/j.rmed.2012.02.004
28. Malcolm TR, Chin-Hong PV. Endemic mycoses in immunocompromised hosts. *Curr Infect Dis Rep.* (2013) 15:536–43. doi: 10.1007/s11908-013-0387-4

29. Le T, Wolbers M, Chi NH, Quang VM, Chinh NT, Lan NP, et al. Epidemiology, seasonality, and predictors of outcome of AIDS-associated *Penicillium marneffei* infection in Ho Chi Minh City, Viet Nam. *Clin Infect Dis.* (2011) 52:945–52. doi: 10.1093/cid/cir028
30. Galgiani JN, Catanzaro A, Cloud GA, Johnson RH, Williams PL, Mirels LF, et al. Comparison of oral fluconazole and itraconazole for progressive, nonmeningeal coccidioidomycosis. A randomized, double-blind trial. Mycoses Study Group. *Ann Intern Med.* (2000) 133:676–86. doi: 10.7326/0003-4819-133-9-200011070-00009
31. Brummer E, Castaneda E, Restrepo A. Paracoccidioidomycosis: an update. *Clin Microbiol Rev.* (1993) 6:89–117. doi: 10.1128/CMR.6.2.89
32. Boulware DR. Cryptococcus: from human pathogen to model yeast. *Lancet Infect Dis.* (2011) 11:434. doi: 10.1016/S1473-3099(11)70140-2
33. Pappalardo MC, Melhem MS. Cryptococcosis: a review of the Brazilian experience for the disease. *Rev Inst Med Trop.* (2003) 45:299–305. doi: 10.1590/S0036-46652003000600001
34. Sugar AM, Stern JJ, Dupont B. Overview: treatment of cryptococcal meningitis. *Rev Infect Dis.* (1990) 12(Suppl. 3):S338–48. doi: 10.1093/clinids/12.Supplement_3.S338
35. Park BJ, Wannemuehler KA, Marston BJ, Govender N, Pappas PG, Chiller TM. Estimation of the current global burden of cryptococcal meningitis among persons living with HIV/AIDS. *AIDS.* (2009) 23:525–30. doi: 10.1097/QAD.0b013e328322ffac
36. Charlier C, Nielsen K, Daou S, Brigitte M, Chretien F, Dromer F. Evidence of a role for monocytes in dissemination and brain invasion by *Cryptococcus neoformans. Infect Immun.* (2009) 77:120–7. doi: 10.1128/IAI.01065-08
37. Kim KS. Mechanisms of microbial traversal of the blood-brain barrier. *Nat Rev Microbiol.* (2008) 6:625–34. doi: 10.1038/nrmicro1952
38. Speed B, Dunt D. Clinical and host differences between infections with the two varieties of *Cryptococcus neoformans. Clin Infect Dis.* (1995) 21:28–34. doi: 10.1093/clinids/21.1.28
39. Sorrell TC, Ellis DH. Ecology of *Cryptococcus neoformans. Rev Iberoam Micol.* (1997) 14:42–3.
40. Kidd SE, Hagen F, Tscharke RL, Huynh M, Bartlett KH, Fyfe M, et al. A rare genotype of *Cryptococcus gattii* caused the cryptococcosis outbreak on Vancouver Island (British Columbia, Canada). *Proc Natl Acad Sci USA.* (2004) 101:17258–63. doi: 10.1073/pnas.0402981101
41. Jarvis JN, Boulle A, Loyse A, Bicanic T, Rebe K, Williams A, et al. High ongoing burden of cryptococcal disease in Africa despite antiretroviral roll out. *AIDS.* (2009) 23:1182–3. doi: 10.1097/QAD.0b013e32832be0fc
42. Chuck SL, Sande MA. Infections with *Cryptococcus neoformans* in the acquired immunodeficiency syndrome. *N Engl J Med.* (1989) 321:794–9. doi: 10.1056/NEJM198909213211205
43. Meyohas MC, Roux P, Bollens D, Chouaid C, Rozenbaum W, Meynard JL, et al. Pulmonary cryptococcosis: localized and disseminated infections in 27 patients with AIDS. *Clin Infect Dis.* (1995) 21:628–33. doi: 10.1093/clinids/21.3.628
44. Xie S, Sao R, Braun A, Bottone EJ. Difference in *Cryptococcus neoformans* cellular and capsule size in sequential pulmonary and meningeal infection: a postmortem study. *Diagn Microbiol Infect Dis.* (2012) 73:49–52. doi: 10.1016/j.diagmicrobio.2012.01.008
45. Silva EG, Paula CR, de Assis Baroni F, Gambale W. Voriconazole, combined with amphotericin B, in the treatment for pulmonary cryptococcosis caused by C. neoformans (serotype A) in mice with severe combined immunodeficiency (SCID). *Mycopathologia.* (2012) 173:445–9. doi: 10.1007/s11046-011-9499-2
46. Kiertiburanakul S, Wirojtananugoon S, Pracharktam R, Sungkanuparph S. Cryptococcosis in human immunodeficiency virus-negative patients. *Int J Infect Dis.* (2006) 10:72–8. doi: 10.1016/j.ijid.2004.12.004
47. Severo CB, Xavier MO, Gazzoni AF, Severo LC. Cryptococcosis in children. *Paediatr Respir Rev.* (2009) 10:166–71. doi: 10.1016/j.prrv.2009.06.009
48. Denning DW. Epidemiology and pathogenesis of systemic fungal infections in the immunocompromised host. *J Antimicrob Chemother.* (1991) 28(Suppl. B):1–16. doi: 10.1093/jac/28.suppl_B.1
49. Marr KA, Carter RA, Crippa F, Wald A, Corey L. Epidemiology and outcome of mould infections in hematopoietic stem cell transplant recipients. *Clin Infect Dis.* (2002) 34:909–17. doi: 10.1086/339202
50. Saral R. Candida and *Aspergillus* infections in immunocompromised patients: an overview. *Rev Infect Dis.* (1991) 13:487–92. doi: 10.1093/clinids/13.3.487
51. Warris A, Bjorneklett A, Gaustad P. Invasive pulmonary aspergillosis associated with infliximab therapy. *N Engl J Med.* (2001) 344:1099–100. doi: 10.1056/NEJM200104053441415
52. Saag MS, Graybill RJ, Larsen RA, Pappas PG, Perfect JR, Powderly WG, et al. Practice guidelines for the management of cryptococcal disease. Infectious Diseases Society of America. *Clin Infect Dis.* (2000) 30:710–8. doi: 10.1086/313757
53. Maesaki S, Kawamura S, Miyazaki Y, Tomono K, Tashiro T, Kohno S. Effect of sequential combination of amphotericin B and azole antifungal agents against *Aspergillus fumigatus. J Infect Chemother.* (1999) 5:125–9. doi: 10.1007/s101560050021
54. Manavathu EK, Ramesh MS, Baskaran I, Ganesan LT, Chandrasekar PH. A comparative study of the post-antifungal effect (PAFE) of amphotericin B, triazoles and echinocandins on *Aspergillus fumigatus* and *Candida albicans. J Antimicrob Chemother.* (2004) 53:386–9. doi: 10.1093/jac/dkh066
55. Pfaller MA. Antifungal drug resistance: mechanisms, epidemiology, and consequences for treatment. *Am J Med.* (2012) 125(Suppl. 1):S3–13. doi: 10.1016/j.amjmed.2011.11.001
56. Medzhitov R. Recognition of microorganisms and activation of the immune response. *Nature.* (2007) 449:819–26. doi: 10.1038/nature06246
57. Doering TL. How sweet it is! Cell wall biogenesis and polysaccharide capsule formation in *Cryptococcus neoformans. Annu Rev Microbiol.* (2009) 63:223–47. doi: 10.1146/annurev.micro.62.081307.162753
58. Underhill DM, Ozinsky A. Phagocytosis of microbes: complexity in action. *Annu Rev Immunol.* (2002) 20:825–52. doi: 10.1146/annurev.immunol.20.103001.114744
59. Philippe B, Ibrahim-Granet O, Prevost MC, Gougerot-Pocidalo MA, Sanchez Perez M, Van der Meeren A, et al. Killing of *Aspergillus fumigatus* by alveolar macrophages is mediated by reactive oxidant intermediates. *Infect Immun.* (2003) 71:3034–42. doi: 10.1128/IAI.71.6.3034-3042.2003
60. Pandey A, Ding SL, Qin QM, Gupta R, Gomez G, Lin F, et al. Global reprogramming of host kinase signaling in response to fungal infection. *Cell Host Microbe.* (2017) 21:637–49 e6. doi: 10.1016/j.chom.2017.04.008
61. Mircescu MM, Lipuma L, van Rooijen N, Pamer EG, Hohl TM. Essential role for neutrophils but not alveolar macrophages at early time points following *Aspergillus fumigatus* infection. *J Infect Dis.* (2009) 200:647–56. doi: 10.1086/600380
62. Cenci E, Mencacci A, Fe d'Ostiani C, Del Sero G, Mosci P, Montagnoli C, et al. Cytokine- and T helper-dependent lung mucosal immunity in mice with invasive pulmonary aspergillosis. *J Infect Dis.* (1998) 178:1750–60. doi: 10.1086/314493
63. Osterholzer JJ, Milam JE, Chen GH, Toews GB, Huffnagle GB, Olszewski MA. Role of dendritic cells and alveolar macrophages in regulating early host defense against pulmonary infection with *Cryptococcus neoformans. Infect Immun.* (2009) 77:3749–58. doi: 10.1128/IAI.00454-09
64. Steele C, Rapaka RR, Metz A, Pop SM, Williams DL, Gordon S, et al. The beta-glucan receptor dectin-1 recognizes specific morphologies of *Aspergillus fumigatus. PLoS Pathog.* (2005) 1:e42. doi: 10.1371/journal.ppat.0010042
65. Hestdal K, Ruscetti FW, Ihle JN, Jacobsen SE, Dubois CM, Kopp WC, et al. Characterization and regulation of RB6-8C5 antigen expression on murine bone marrow cells. *J Immunol.* (1991) 147:22–8.
66. Stephens-Romero SD, Mednick AJ, Feldmesser M. The pathogenesis of fatal outcome in murine pulmonary aspergillosis depends on the neutrophil depletion strategy. *Infect Immun.* (2005) 73:114–25. doi: 10.1128/IAI.73.1.114-125.2005
67. Taylor PR, Roy S, Leal SM Jr, Sun Y, Howell SJ, Cobb BA, et al. Activation of neutrophils by autocrine IL-17A-IL-17RC interactions during fungal infection is regulated by IL-6, IL-23, RORgammat and dectin-2. *Nat Immunol.* (2014) 15:143–51. doi: 10.1038/ni.2797
68. Gazendam RP, van Hamme JL, Tool AT, Hoogenboezem M, van den Berg JM, Prins JM, et al. Human neutrophils use different mechanisms to kill *Aspergillus fumigatus* conidia and hyphae: evidence from phagocyte defects. *J Immunol.* (2016) 196:1272–83. doi: 10.4049/jimmunol.1501811
69. Banchereau J, Steinman RM. Dendritic cells and the control of immunity. *Nature.* (1998) 392:245–52. doi: 10.1038/32588

70. Zitvogel L. Dendritic and natural killer cells cooperate in the control/switch of innate immunity. *J Exp Med.* (2002) 195:F9–14. doi: 10.1084/jem.20012040
71. Ramadan G. The proliferative response of T cells to *Aspergillus* antigen requires prior presentation on dendritic cells. *Egypt J Immunol.* (2004) 11:47–58.
72. Gafa V, Remoli ME, Giacomini E, Severa M, Grillot R, Coccia EM. Enhancement of anti-*Aspergillus* T helper type 1 response by interferon-beta-conditioned dendritic cells. *Immunology.* (2010) 131:282–8. doi: 10.1111/j.1365-2567.2010.03302.x
73. Bozza S, Perruccio K, Montagnoli C, Gaziano R, Bellocchio S, Burchielli E, et al. A dendritic cell vaccine against invasive aspergillosis in allogeneic hematopoietic transplantation. *Blood.* (2003) 102:3807–14. doi: 10.1182/blood-2003-03-0748
74. Mezger M, Kneitz S, Wozniok I, Kurzai O, Einsele H, Loeffler J. Proinflammatory response of immature human dendritic cells is mediated by dectin-1 after exposure to *Aspergillus fumigatus* germ tubes. *J Infect Dis.* (2008) 197:924–31. doi: 10.1086/528694
75. Park SJ, Burdick MD, Mehrad B. Neutrophils mediate maturation and efflux of lung dendritic cells in response to *Aspergillus fumigatus* germ tubes. *Infect Immun.* (2012) 80:1759–65. doi: 10.1128/IAI.00097-12
76. Gafa V, Remoli ME, Giacomini E, Gagliardi MC, Lande R, Severa M, et al. *In vitro* infection of human dendritic cells by *Aspergillus fumigatus* conidia triggers the secretion of chemokines for neutrophil and Th1 lymphocyte recruitment. *Microbes Infect.* (2007) 9:971–80. doi: 10.1016/j.micinf.2007.03.015
77. Vivier E, Tomasello E, Baratin M, Walzer T, Ugolini S. Functions of natural killer cells. *Nat Immunol.* (2008) 9:503–10. doi: 10.1038/ni1582
78. Mortha A, Diefenbach A. Natural killer cell receptor-expressing innate lymphocytes: more than just NK cells. *Cell Mol Life Sci.* (2011) 68:3541–55. doi: 10.1007/s00018-011-0803-6
79. Kelly MN, Zheng M, Ruan S, Kolls J, D'Souza A, Shellito JE. Memory CD4+ T cells are required for optimal NK cell effector functions against the opportunistic fungal pathogen *Pneumocystis murina. J Immunol.* (2013) 190:285–95. doi: 10.4049/jimmunol.1200861
80. Schmidt S, Tramsen L, Hanisch M, Latge JP, Huenecke S, Koehl U, et al. Human natural killer cells exhibit direct activity against *Aspergillus fumigatus* hyphae, but not against resting conidia. *J Infect Dis.* (2011) 203:430–5. doi: 10.1093/infdis/jiq062
81. Morton CO, Bouzani M, Loeffler J, Rogers TR. Direct interaction studies between *Aspergillus fumigatus* and human immune cells; what have we learned about pathogenicity and host immunity? *Front Microbiol.* (2012) 3:413. doi: 10.3389/fmicb.2012.00413
82. Hidore MR, Nabavi N, Sonleitner F, Murphy JW. Murine natural killer cells are fungicidal to *Cryptococcus neoformans. Infect Immun.* (1991) 59:1747–54.
83. Wozniak KL, Ravi S, Macias S, Young ML, Olszewski MA, Steele C, et al. Insights into the mechanisms of protective immunity against *Cryptococcus neoformans* infection using a mouse model of pulmonary cryptococcosis. *PLoS ONE.* (2009) 4:e6854. doi: 10.1371/journal.pone.0006854
84. Koguchi Y, Kawakami K. Cryptococcal infection and Th1-Th2 cytokine balance. *Int Rev Immunol.* (2002) 21:423–38. doi: 10.1080/0883018 0213274
85. Romani L, Fallarino F, De Luca A, Montagnoli C, D'Angelo C, Zelante T, et al. Defective tryptophan catabolism underlies inflammation in mouse chronic granulomatous disease. *Nature.* (2008) 451:211–5. doi: 10.1038/nature06471
86. Carvalho A, De Luca A, Bozza S, Cunha C, D'Angelo C, Moretti S, et al. TLR3 essentially promotes protective class I-restricted memory CD8(+) T-cell responses to *Aspergillus fumigatus* in hematopoietic transplanted patients. *Blood.* (2012) 119:967–77. doi: 10.1182/blood-2011-06-362582
87. Bals R. Epithelial antimicrobial peptides in host defense against infection. *Respir Res.* (2000) 1:141–50. doi: 10.1186/rr25
88. Paris S, Boisvieux-Ulrich E, Crestani B, Houcine O, Taramelli D, Lombardi L, et al. Internalization of *Aspergillus fumigatus* conidia by epithelial and endothelial cells. *Infect Immun.* (1997) 65:1510–4.
89. de Luca A, Bozza S, Zelante T, Zagarella S, D'Angelo C, Perruccio K, et al. Non-hematopoietic cells contribute to protective tolerance to *Aspergillus fumigatus* via a TRIF pathway converging on IDO. *Cell Mol Immunol.* (2010) 7:459–70. doi: 10.1038/cmi.2010.43
90. Cunha C, Carvalho A, Esposito A, Bistoni F, Romani L. DAMP signaling in fungal infections and diseases. *Front Immunol.* (2012) 3:286. doi: 10.3389/fimmu.2012.00286
91. Barreto-Bergter E, Figueiredo RT. Fungal glycans and the innate immune recognition. *Front Cell Infect Microbiol.* (2014) 4:145. doi: 10.3389/fcimb.2014.00145
92. Taylor PR, Tsoni SV, Willment JA, Dennehy KM, Rosas M, Findon H, et al. Dectin-1 is required for beta-glucan recognition and control of fungal infection. *Nat Immunol.* (2007) 8:31–8. doi: 10.1038/ni1408
93. Sato K, Yang XL, Yudate T, Chung JS, Wu J, Luby-Phelps K, et al. Dectin-2 is a pattern recognition receptor for fungi that couples with the Fc receptor gamma chain to induce innate immune responses. *J Biol Chem.* (2006) 281:38854–66. doi: 10.1074/jbc.M606542200
94. Richardson MB, Williams SJ. MCL and Mincle: C-Type lectin receptors that sense damaged self and pathogen-associated molecular patterns. *Front Immunol.* (2014) 5:288. doi: 10.3389/fimmu.2014.00288
95. Svirshchevskaya E, Zubkov D, Mouyna I, Berkova N. Innate immunity and the role of epithelial barrier during *Aspergillus fumigatus* infection. *Curr Immunol Rev.* (2012) 8:254–61. doi: 10.2174/1573395128006 71985
96. Stappers MHT, Clark AE, Aimanianda V, Bidula S, Reid DM, Asamaphan P, et al. Recognition of DHN-melanin by a C-type lectin receptor is required for immunity to *Aspergillus. Nature.* (2018) 555:382–6. doi: 10.1038/nature 25974
97. Mansour MK, Schlesinger LS, Levitz SM. Optimal T cell responses to *Cryptococcus neoformans* mannoprotein are dependent on recognition of conjugated carbohydrates by mannose receptors. *J Immunol.* (2002) 168:2872–9. doi: 10.4049/jimmunol.168.6.2872
98. Mansour MK, Latz E, Levitz SM. *Cryptococcus neoformans* glycoantigens are captured by multiple lectin receptors and presented by dendritic cells. *J Immunol.* (2006) 176:3053–61. doi: 10.4049/jimmunol.176.5.3053
99. Balloy V, Si-Tahar M, Takeuchi O, Philippe B, Nahori MA, Tanguy M, et al. Involvement of toll-like receptor 2 in experimental invasive pulmonary aspergillosis. *Infect Immun.* (2005) 73:5420–5. doi: 10.1128/IAI.73.9.5420-5425.2005
100. Bellocchio S, Montagnoli C, Bozza S, Gaziano R, Rossi G, Mambula SS, et al. The contribution of the Toll-like/IL-1 receptor superfamily to innate and adaptive immunity to fungal pathogens *in vivo. J Immunol.* (2004) 172:3059–69. doi: 10.4049/jimmunol.172.5.3059
101. Chignard M, Balloy V, Sallenave JM, Si-Tahar M. Role of toll-like receptors in lung innate defense against invasive aspergillosis. Distinct impact in immunocompetent and immunocompromized hosts. *Clin Immunol.* (2007) 124:238–43. doi: 10.1016/j.clim.2007.05.004
102. Netea MG, Ferwerda G, van der Graaf CA, Van der Meer JW, Kullberg BJ. Recognition of fungal pathogens by toll-like receptors. *Curr Pharm Des.* (2006) 12:4195–201. doi: 10.2174/138161206778743538
103. Suzuki N, Chen NJ, Millar DG, Suzuki S, Horacek T, Hara H, et al. IL-1 receptor-associated kinase 4 is essential for IL-18-mediated NK and Th1 cell responses. *J Immunol.* (2003) 170:4031–5. doi: 10.4049/jimmunol.170.8.4031
104. Bozza S, Zelante T, Moretti S, Bonifazi P, DeLuca A, D'Angelo C, et al. Lack of Toll IL-1R8 exacerbates Th17 cell responses in fungal infection. *J Immunol.* (2008) 180:4022–31. doi: 10.4049/jimmunol.180.6.4022
105. Vecchiarelli A, Pericolini E, Gabrielli E, Agostini M, Bistoni F, Nocentini G, et al. The GITRL-GITR system alters TLR-4 expression on DC during fungal infection. *Cell Immunol.* (2009) 257:13–22. doi: 10.1016/j.cellimm.2009.02.001
106. Nocentini G, Riccardi C. GITR: a multifaceted regulator of immunity belonging to the tumor necrosis factor receptor superfamily. *Eur J Immunol.* (2005) 35:1016–22. doi: 10.1002/eji.200425818
107. Medzhitov R, Preston-Hurlburt P, Kopp E, Stadlen A, Chen C, Ghosh S, et al. MyD88 is an adaptor protein in the hToll/IL-1 receptor family signaling pathways. *Mol Cell.* (1998) 2:253–8. doi: 10.1016/S1097-2765(00)80136-7
108. Bretz C, Gersuk G, Knoblaugh S, Chaudhary N, Randolph-Habecker J, Hackman RC, et al. MyD88 signaling contributes to early pulmonary responses to *Aspergillus fumigatus. Infect Immun.* (2008) 76:952–8. doi: 10.1128/IAI.00927-07
109. Gohda J, Matsumura T, Inoue J. Cutting edge: TNFR-associated factor (TRAF) 6 is essential for MyD88-dependent pathway but not toll/IL-1

receptor domain-containing adaptor-inducing IFN-beta (TRIF)-dependent pathway in TLR signaling. *J Immunol.* (2004) 173:2913–7. doi: 10.4049/jimmunol.173.5.2913
110. Muroi M, Tanamoto KI. TRAF6 distinctively mediates MyD88- and IRAK-1-induced activation of NF-kappaB. *J Leukoc Biol.* (2008) 83:702–7. doi: 10.1189/jlb.0907629
111. Gorjestani S, Darnay BG, Lin X. Tumor necrosis factor receptor-associated factor 6 (TRAF6) and TGFbeta-activated kinase 1 (TAK1) play essential roles in the C-type lectin receptor signaling in response to *Candida albicans* infection. *J Biol Chem.* (2012) 287:44143–50. doi: 10.1074/jbc.M112.414276
112. Hara H, Saito T. CARD9 versus CARMA1 in innate and adaptive immunity. *Trends Immunol.* (2009) 30:234–42. doi: 10.1016/j.it.2009.03.002
113. Kyrmizi I, Gresnigt MS, Akoumianaki T, Samonis G, Sidiropoulos P, Boumpas D, et al. Corticosteroids block autophagy protein recruitment in *Aspergillus fumigatus* phagosomes via targeting dectin-1/Syk kinase signaling. *J Immunol.* (2013) 191:1287–99. doi: 10.4049/jimmunol.1300132
114. Fang R, Tsuchiya K, Kawamura I, Shen Y, Hara H, Sakai S, et al. Critical roles of ASC inflammasomes in caspase-1 activation and host innate resistance to *Streptococcus pneumoniae* infection. *J Immunol.* (2011) 187:4890–9. doi: 10.4049/jimmunol.1100381
115. Skeldon A, Saleh M. The inflammasomes: molecular effectors of host resistance against bacterial, viral, parasitic, and fungal infections. *Front Microbiol.* (2011) 2:15. doi: 10.3389/fmicb.2011.00015
116. Munoz-Planillo R, Kuffa P, Martinez-Colon G, Smith BL, Rajendiran TM, Nunez G. K(+) efflux is the common trigger of NLRP3 inflammasome activation by bacterial toxins and particulate matter. *Immunity.* (2013) 38:1142–53. doi: 10.1016/j.immuni.2013.05.016
117. Katsnelson MA, Rucker LG, Russo HM, Dubyak GR. K+ efflux agonists induce NLRP3 inflammasome activation independently of Ca2+ signaling. *J Immunol.* (2015) 194:3937–52. doi: 10.4049/jimmunol.1402658
118. Gross CJ, Mishra R, Schneider KS, Medard G, Wettmarshausen J, Dittlein DC, et al. K(+) Efflux-independent NLRP3 inflammasome activation by small molecules targeting mitochondria. *Immunity.* (2016) 45:761–73. doi: 10.1016/j.immuni.2016.08.010
119. Huang Y, Hua M, Cui X. Fungal beta-Glucan Activates the NLRP3 inflammasome in human bronchial epithelial cells through ROS production. *Inflammation.* (2018) 41:164–73. doi: 10.1007/s10753-017-0674-6
120. Kumar H, Kumagai Y, Tsuchida T, Koenig PA, Satoh T, Guo Z, et al. Involvement of the NLRP3 inflammasome in innate and humoral adaptive immune responses to fungal beta-glucan. *J Immunol.* (2009) 183:8061–7. doi: 10.4049/jimmunol.0902477
121. Said-Sadier N, Padilla E, Langsley G, Ojcius DM. *Aspergillus fumigatus* stimulates the NLRP3 inflammasome through a pathway requiring ROS production and the Syk tyrosine kinase. *PLoS ONE.* (2010) 5:e10008. doi: 10.1371/journal.pone.0010008
122. Lamkanfi M, Malireddi RK, Kanneganti TD. Fungal zymosan and mannan activate the cryopyrin inflammasome. *J Biol Chem.* (2009) 284:20574–81. doi: 10.1074/jbc.M109.023689
123. Guo C, Chen M, Fa Z, Lu A, Fang W, Sun B, et al. Acapsular *Cryptococcus neoformans* activates the NLRP3 inflammasome. *Microbes Infect.* (2014) 16:845–54. doi: 10.1016/j.micinf.2014.08.013
124. Lei G, Chen M, Li H, Niu JL, Wu S, Mao L, et al. Biofilm from a clinical strain of *Cryptococcus neoformans* activates the NLRP3 inflammasome. *Cell Res.* (2013) 23:965–8. doi: 10.1038/cr.2013.49
125. Karki R, Man SM, Malireddi RKS, Gurung P, Vogel P, Lamkanfi M, et al. Concerted activation of the AIM2 and NLRP3 inflammasomes orchestrates host protection against *Aspergillus* infection. *Cell Host Microbe.* (2015) 17:357–68. doi: 10.1016/j.chom.2015.01.006
126. Gersuk GM, Underhill DM, Zhu L, Marr KA. Dectin-1 and TLRs permit macrophages to distinguish between different *Aspergillus fumigatus* cellular states. *J Immunol.* (2006) 176:3717–24. doi: 10.4049/jimmunol.176.6.3717
127. Sun H, Xu XY, Tian XL, Shao HT, Wu XD, Wang Q, et al. Activation of NF-κB and respiratory burst following *Aspergillus fumigatus* stimulation of macrophages. *Immunobiology.* (2014) 219:25–36. doi: 10.1016/j.imbio.2013.06.013
128. Pahl HL, Krauss B, Schulze-Osthoff K, Decker T, Traenckner EB, Vogt M, et al. The immunosuppressive fungal metabolite gliotoxin specifically inhibits transcription factor NF-kappaB. *J Exp Med.* (1996) 183:1829–40. doi: 10.1084/jem.183.4.1829
129. Dubourdeau M, Athman R, Balloy V, Huerre M, Chignard M, Philpott DJ, et al. *Aspergillus fumigatus* induces innate immune responses in alveolar macrophages through the MAPK pathway independently of TLR2 and TLR4. *J Immunol.* (2006) 177:3994–4001. doi: 10.4049/jimmunol.177.6.3994
130. Balloy V, Sallenave JM, Wu Y, Touqui L, Latgé JP, Sitahar M, et al. *Aspergillus fumigatus*-induced IL-8 synthesis by respiratory epithelial cells is controlled by the PI3 kinase, p38 MAPK and ERK1/2 pathways and not by the TLR-MYD88 pathway. *J Biol Chem.* (2008) 283:30513–21. doi: 10.1074/jbc.M803149200
131. Shoham S, Huang C, Chen JM, Golenbock DT, Levitz SM. Toll-like receptor 4 mediates intracellular signaling without TNF-alpha release in response to *Cryptococcus neoformans* polysaccharide capsule. *J Immunol.* (2001) 166:4620–6. doi: 10.4049/jimmunol.166.7.4620
132. Garth JM, Reeder KM, Godwin MS, Mackel JJ, Dunaway CW, Blackburn JP, et al. IL-33 signaling regulates innate IL-17A and IL-22 production via suppression of prostaglandin E2 during lung fungal infection. *J Immunol.* (2017) 199:2140–8. doi: 10.4049/jimmunol.1602186
133. de Luca A, Smeekens SP, Casagrande A, Iannitti R, Conway KL, Gresnigt MS, et al. IL-1 receptor blockade restores autophagy and reduces inflammation in chronic granulomatous disease in mice and in humans. *Proc Natl Acad Sci USA.* (2014) 111:3526–31. doi: 10.1073/pnas.1322831111
134. Werner JL, Metz AE, Horn D, Schoeb TR, Hewitt MM, Schwiebert LM, et al. Requisite role for the dectin-1 beta-glucan receptor in pulmonary defense against *Aspergillus fumigatus.* *J Immunol.* (2009) 182:4938–46. doi: 10.4049/jimmunol.0804250
135. Caffrey AK, Lehmann MM, Zickovich JM, Espinosa V, Shepardson KM, Watschke CP, et al. IL-1alpha signaling is critical for leukocyte recruitment after pulmonary *Aspergillus fumigatus* challenge. *PLoS Pathog.* (2015) 11:e1004625. doi: 10.1371/journal.ppat.1004625
136. Gresnigt MS, Rosler B, Jacobs CW, Becker KL, Joosten LA, van der Meer JW, et al. The IL-36 receptor pathway regulates *Aspergillus fumigatus*-induced Th1 and Th17 responses. *Eur J Immunol.* (2013) 43:416–26. doi: 10.1002/eji.201242711
137. Stockinger B, Veldhoen M. Differentiation and function of Th17 T cells. *Curr Opin Immunol.* (2007) 19:281–6. doi: 10.1016/j.coi.2007.04.005
138. Cua DJ, Tato CM. Innate IL-17-producing cells: the sentinels of the immune system. *Nat Rev Immunol.* (2010) 10:479–89. doi: 10.1038/nri2800
139. Miyamoto M, Prause O, Sjostrand M, Laan M, Lotvall J, Linden A. Endogenous IL-17 as a mediator of neutrophil recruitment caused by endotoxin exposure in mouse airways. *J Immunol.* (2003) 170:4665–72. doi: 10.4049/jimmunol.170.9.4665
140. Stark MA, Huo Y, Burcin TL, Morris MA, Olson TS, Ley K. Phagocytosis of apoptotic neutrophils regulates granulopoiesis via IL-23 and IL-17. *Immunity.* (2005) 22:285–94. doi: 10.1016/j.immuni.2005.01.011
141. Zelante T, De Luca A, Bonifazi P, Montagnoli C, Bozza S, Moretti S, et al. IL-23 and the Th17 pathway promote inflammation and impair antifungal immune resistance. *Eur J Immunol.* (2007) 37:2695–706. doi: 10.1002/eji.200737409
142. Ivashkiv LB, Donlin LT. Regulation of type I interferon responses. *Nat Rev Immunol.* (2014) 14:36–49. doi: 10.1038/nri3581
143. Bogdan C, Mattner J, Schleicher U. The role of type I interferons in non-viral infections. *Immunol Rev.* (2004) 202:33–48. doi: 10.1111/j.0105-2896.2004.00207.x
144. Sato K, Yamamoto H, Nomura T, Matsumoto I, Miyasaka T, Zong T, et al. *Cryptococcus neoformans* infection in mice lacking type I interferon signaling leads to increased fungal clearance and IL-4-dependent mucin production in the lungs. *PLoS ONE.* (2015) 10:e0138291. doi: 10.1371/journal.pone.0138291
145. Biondo C, Midiri A, Gambuzza M, Gerace E, Falduto M, Galbo R, et al. IFN-alpha/beta signaling is required for polarization of cytokine responses toward a protective type 1 pattern during experimental cryptococcosis. *J Immunol.* (2008) 181:566–73. doi: 10.4049/jimmunol.181.1.566
146. Dufour JH, Dziejman M, Liu MT, Leung JH, Lane TE, Luster AD. IFN-gamma-inducible protein 10 (IP-10; CXCL10)-deficient mice reveal a role for IP-10 in effector T cell generation and trafficking. *J Immunol.* (2002) 168:3195–204. doi: 10.4049/jimmunol.168.7.3195

147. Beisswenger C, Hess C, Bals R. *Aspergillus fumigatus* conidia induce interferon-beta signalling in respiratory epithelial cells. *Eur Respir J.* (2012) 39:411–8. doi: 10.1183/09031936.00096110
148. Joly V, Saint-Julien L, Carbon C, Yeni P. *In vivo* activity of interferon-gamma in combination with amphotericin B in the treatment of experimental cryptococcosis. *J Infect Dis.* (1994) 170:1331–4. doi: 10.1093/infdis/170.5.1331
149. Lutz JE, Clemons KV, Stevens DA. Enhancement of antifungal chemotherapy by interferon-gamma in experimental systemic cryptococcosis. *J Antimicrob Chemother.* (2000) 46:437–42. doi: 10.1093/jac/46.3.437
150. Wormley FL Jr, Perfect JR, Steele C, Cox GM. Protection against cryptococcosis by using a murine gamma interferon-producing *Cryptococcus neoformans* strain. *Infect Immun.* (2007) 75:1453–62. doi: 10.1128/IAI.00274-06
151. Heldt S, Eigl S, Prattes J, Flick H, Rabensteiner J, Pruller F, et al. Levels of interleukin (IL)-6 and IL-8 are elevated in serum and bronchoalveolar lavage fluid of haematological patients with invasive pulmonary aspergillosis. *Mycoses.* (2017) 60:818–25. doi: 10.1111/myc.12679
152. Xing Z, Gauldie J, Cox G, Baumann H, Jordana M, Lei XF, et al. IL-6 is an antiinflammatory cytokine required for controlling local or systemic acute inflammatory responses. *J Clin Invest.* (1998) 101:311–20. doi: 10.1172/JCI1368
153. Camargo JF, Bhimji A, Kumar D, Kaul R, Pavan R, Schuh A, et al. Impaired T cell responsiveness to interleukin-6 in hematological patients with invasive aspergillosis. *PLoS ONE.* (2015) 10:e0123171. doi: 10.1371/journal.pone.0123171
154. Mehrad B, Strieter RM, Moore TA, Tsai WC, Lira SA, Standiford TJ. CXC chemokine receptor-2 ligands are necessary components of neutrophil-mediated host defense in invasive pulmonary aspergillosis. *J Immunol.* (1999) 163:6086–94.
155. Mehrad B, Wiekowski M, Morrison BE, Chen SC, Coronel EC, Manfra DJ, et al. Transient lung-specific expression of the chemokine KC improves outcome in invasive aspergillosis. *Am J Respir Crit Care Med.* (2002) 166:1263–8. doi: 10.1164/rccm.200204-367OC
156. Borger P, Koeter GH, Timmerman JA, Vellenga E, Tomee JF, Kauffman HF. Proteases from *Aspergillus fumigatus* induce interleukin (IL)-6 and IL-8 production in airway epithelial cell lines by transcriptional mechanisms. *J Infect Dis.* (1999) 180:1267–74. doi: 10.1086/315027
157. Zhang Z, Liu R, Noordhoek JA, Kauffman HF. Interaction of airway epithelial cells (A549) with spores and mycelium of *Aspergillus fumigatus*. *J Infect.* (2005) 51:375–82. doi: 10.1016/j.jinf.2004.12.012
158. Schelenz S, Smith DA, Bancroft GJ. Cytokine and chemokine responses following pulmonary challenge with *Aspergillus fumigatus*: obligatory role of TNF-alpha and GM-CSF in neutrophil recruitment. *Med Mycol.* (1999) 37:183–94. doi: 10.1046/j.1365-280X.1999.00219.x
159. Roilides E, Dimitriadou-Georgiadou A, Sein T, Kadiltsoglou I, Walsh TJ. Tumor necrosis factor alpha enhances antifungal activities of polymorphonuclear and mononuclear phagocytes against *Aspergillus fumigatus*. *Infect Immun.* (1998) 66:5999–6003.
160. Brieland JK, Jackson C, Menzel F, Loebenberg D, Cacciapuoti A, Halpern J, et al. Cytokine networking in lungs of immunocompetent mice in response to inhaled *Aspergillus fumigatus*. *Infect Immun.* (2001) 69:1554–60. doi: 10.1128/IAI.69.3.1554-1560.2001
161. Fleetwood AJ, Cook AD, Hamilton JA. Functions of granulocyte-macrophage colony-stimulating factor. *Crit Rev Immunol.* (2005) 25:405–28. doi: 10.1615/CritRevImmunol.v25.i5.50
162. Kasahara S, Jhingran A, Dhingra S, Salem A, Cramer RA, Hohl TM. Role of granulocyte-macrophage colony-stimulating factor signaling in regulating neutrophil antifungal activity and the oxidative burst during respiratory fungal challenge. *J Infect Dis.* (2016) 213:1289–98. doi: 10.1093/infdis/jiw054
163. Retini C, Vecchiarelli A, Monari C, Tascini C, Bistoni F, Kozel TR. Capsular polysaccharide of *Cryptococcus neoformans* induces proinflammatory cytokine release by human neutrophils. *Infect Immun.* (1996) 64:2897–903.
164. Davis MJ, Tsang TM, Qiu Y, Dayrit JK, Freij JB, Huffnagle GB, et al. Macrophage M1/M2 polarization dynamically adapts to changes in cytokine microenvironments in *Cryptococcus neoformans* infection. *MBio.* (2013) 4:e00264–13. doi: 10.1128/mBio.00264-13
165. Shalit I, Halperin D, Haite D, Levitov A, Romano J, Osherov N, et al. Anti-inflammatory effects of moxifloxacin on IL-8, IL-1beta and TNF-alpha secretion and NFkappaB and MAP-kinase activation in human monocytes stimulated with *Aspergillus fumigatus*. *J Antimicrob Chemother.* (2006) 57:230–5. doi: 10.1093/jac/dki441
166. Moser C, Weiner DJ, Lysenko E, Bals R, Weiser JN, Wilson JM. beta-Defensin 1 contributes to pulmonary innate immunity in mice. *Infect Immun.* (2002) 70:3068–72. doi: 10.1128/IAI.70.6.3068-3072.2002
167. Ordonez SR, Veldhuizen EJA, van Eijk M, Haagsman HP. Role of soluble innate effector molecules in pulmonary defense against fungal pathogens. *Front Microbiol.* (2017) 8:2098. doi: 10.3389/fmicb.2017.02098
168. Colley KJ, Beranek MC, Baenziger JU. Purification and characterization of the core-specific lectin from human serum and liver. *Biochem J.* (1988) 256:61–8. doi: 10.1042/bj2560061
169. Panepinto JC, Komperda KW, Hacham M, Shin S, Liu X, Williamson PR. Binding of serum mannan binding lectin to a cell integrity-defective *Cryptococcus neoformans* ccr4Delta mutant. *Infect Immun.* (2007) 75:4769–79. doi: 10.1128/IAI.00536-07
170. Kaur S, Gupta VK, Thiel S, Sarma PU, Madan T. Protective role of mannan-binding lectin in a murine model of invasive pulmonary aspergillosis. *Clin Exp Immunol.* (2007) 148:382–9. doi: 10.1111/j.1365-2249.2007.03351.x
171. Geunes-Boyer S, Oliver TN, Janbon G, Lodge JK, Heitman J, Perfect JR, et al. Surfactant protein D increases phagocytosis of hypocapsular *Cryptococcus neoformans* by murine macrophages and enhances fungal survival. *Infect Immun.* (2009) 77:2783–94. doi: 10.1128/IAI.00088-09
172. Geunes-Boyer S, Beers MF, Perfect JR, Heitman J, Wright JR. Surfactant protein D facilitates *Cryptococcus neoformans* infection. *Infect Immun.* (2012) 80:2444–53. doi: 10.1128/IAI.05613-11
173. Litvack ML, Palaniyar N. Review: soluble innate immune pattern-recognition proteins for clearing dying cells and cellular components: implications on exacerbating or resolving inflammation. *Innate Immun.* (2010) 16:191–200. doi: 10.1177/1753425910369271
174. Bidula S, Sexton DW, Yates M, Abdolrasouli A, Shah A, Wallis R, et al. H-ficolin binds *Aspergillus fumigatus* leading to activation of the lectin complement pathway and modulation of lung epithelial immune responses. *Immunology.* (2015) 146:281–91. doi: 10.1111/imm.12501
175. Bidula S, Sexton DW, Abdolrasouli A, Shah A, Reed A, Armstrong-James D, et al. The serum opsonin L-ficolin is detected in lungs of human transplant recipients following fungal infections and modulates inflammation and killing of *Aspergillus fumigatus*. *J Infect Dis.* (2015) 212:234–46. doi: 10.1093/infdis/jiv027
176. Alekseeva L, Huet D, Femenia F, Mouyna I, Abdelouahab M, Cagna A, et al. Inducible expression of beta defensins by human respiratory epithelial cells exposed to *Aspergillus fumigatus* organisms. *BMC Microbiol.* (2009) 9:33. doi: 10.1186/1471-2180-9-33
177. Dubin RF, Robinson SK, Widdicombe JH. Secretion of lactoferrin and lysozyme by cultures of human airway epithelium. *Am J Physiol Lung Cell Mol Physiol.* (2004) 286:L750–5. doi: 10.1152/ajplung.00326.2003
178. Diamond RD, Krzesicki R, Epstein B, Jao W. Damage to hyphal forms of fungi by human leukocytes *in vitro*. A possible host defense mechanism in aspergillosis and mucormycosis. *Am J Pathol.* (1978) 91:313–28.
179. Cortopassi GA, Wilson AC. Recent origin of the P lysozyme gene in mice. *Nucleic Acids Res.* (1990) 18:1911. doi: 10.1093/nar/18.7.1911
180. Cross M, Mangelsdorf I, Wedel A, Renkawitz R. Mouse lysozyme M gene: isolation, characterization, and expression studies. *Proc Natl Acad Sci USA.* (1988) 85:6232–6. doi: 10.1073/pnas.85.17.6232
181. Steinbach WJ, Cramer RA Jr, Perfect BZ, Asfaw YG, Sauer TC, Najvar LK, et al. Calcineurin controls growth, morphology, and pathogenicity in *Aspergillus fumigatus*. *Eukaryot Cell.* (2006) 5:1091–103. doi: 10.1128/EC.00139-06
182. Cramer RA Jr, Perfect BZ, Pinchai N, Park S, Perlin DS, Asfaw YG, et al. Calcineurin target CrzA regulates conidial germination, hyphal growth, and pathogenesis of *Aspergillus fumigatus*. *Eukaryot Cell.* (2008) 7:1085–97. doi: 10.1128/EC.00086-08
183. da Silva Ferreira ME, Heinekamp T, Hartl A, Brakhage AA, Semighini CP, Harris SD, et al. Functional characterization of the

Aspergillus fumigatus calcineurin. *Fungal Genet Biol.* (2007) 44:219–30. doi: 10.1016/j.fgb.2006.08.004
184. Steinbach WJ, Cramer RA Jr, Perfect BZ, Henn C, Nielsen K, Heitman J, et al. Calcineurin inhibition or mutation enhances cell wall inhibitors against *Aspergillus fumigatus*. *Antimicrob Agents Chemother.* (2007) 51:2979–81. doi: 10.1128/AAC.01394-06
185. Odom A, Muir S, Lim E, Toffaletti DL, Perfect J, Heitman J. Calcineurin is required for virulence of *Cryptococcus neoformans*. *EMBO J.* (1997) 16:2576–89. doi: 10.1093/emboj/16.10.2576
186. Imbert S, Bresler P, Boissonnas A, Gauthier L, Souchet L, Uzunov M, et al. Calcineurin inhibitors impair neutrophil activity against *Aspergillus fumigatus* in allogeneic hematopoietic stem cell transplant recipients. *J Allergy Clin Immunol.* (2016) 138:860–8. doi: 10.1016/j.jaci.2016.02.026
187. Herbst S, Shah A, Mazon Moya M, Marzola V, Jensen B, Reed A, et al. Phagocytosis-dependent activation of a TLR9-BTK-calcineurin-NFAT pathway co-ordinates innate immunity to *Aspergillus fumigatus*. *EMBO Mol Med.* (2015) 7:240–58. doi: 10.15252/emmm.201404556
188. Singh N, Alexander BD, Lortholary O, Dromer F, Gupta KL, John GT, et al. *Cryptococcus neoformans* in organ transplant recipients: impact of calcineurin-inhibitor agents on mortality. *J Infect Dis.* (2007) 195:756–64. doi: 10.1086/511438
189. Levine B, Kroemer G. Autophagy in the pathogenesis of disease. *Cell.* (2008) 132:27–42. doi: 10.1016/j.cell.2007.12.018
190. Araya J, Hara H, Kuwano K. Autophagy in the pathogenesis of pulmonary disease. *Intern Med.* (2013) 52:2295–303. doi: 10.2169/internalmedicine.52.1118
191. Joven J, Guirro M, Marine-Casado R, Rodriguez-Gallego E, Menendez JA. Autophagy is an inflammation-related defensive mechanism against disease. *Adv Exp Med Biol.* (2014) 824:43–59. doi: 10.1007/978-3-319-07320-0_6
192. Racanelli AC, Kikkers SA, Choi AMK, Cloonan SM. Autophagy and inflammation in chronic respiratory disease. *Autophagy.* (2018) 14:221–32. doi: 10.1080/15548627.2017.1389823
193. Sharma V, Verma S, Seranova E, Sarkar S, Kumar D. Selective autophagy and xenophagy in infection and disease. *Front Cell Dev Biol.* (2018) 6:147. doi: 10.3389/fcell.2018.00147
194. Mao K, Klionsky DJ. Xenophagy: a battlefield between host and microbe, and a possible avenue for cancer treatment. *Autophagy.* (2017) 13:223–4. doi: 10.1080/15548627.2016.1267075
195. Romao S, Munz C. LC3-associated phagocytosis. *Autophagy.* (2014) 10:526–8. doi: 10.4161/auto.27606
196. Lai SC, Devenish RJ. LC3-Associated Phagocytosis (LAP): connections with host autophagy. *Cells.* (2012) 1:396–408. doi: 10.3390/cells1030396
197. Cunha LD, Yang M, Carter R, Guy C, Harris L, Crawford JC, et al. LC3-Associated phagocytosis in myeloid cells promotes tumor immune tolerance. *Cell.* (2018) 175:429–41 e16. doi: 10.1016/j.cell.2018.08.061
198. Nicola AM, Albuquerque P, Martinez LR, Dal-Rosso RA, Saylor C, De Jesus M, et al. Macrophage autophagy in immunity to *Cryptococcus neoformans* and *Candida albicans*. *Infect Immun.* (2012) 80:3065–76. doi: 10.1128/IAI.00358-12
199. Chamilos G, Akoumianaki T, Kyrmizi I, Brakhage A, Beauvais A, Latge JP. Melanin targets LC3-associated phagocytosis (LAP): a novel pathogenetic mechanism in fungal disease. *Autophagy.* (2016) 12:888–9. doi: 10.1080/15548627.2016.1157242
200. Akoumianaki T, Kyrmizi I, Valsecchi I, Gresnigt MS, Samonis G, Drakos E, et al. *Aspergillus* cell wall melanin blocks LC3-associated phagocytosis to promote pathogenicity. *Cell Host Microbe.* (2016) 19:79–90. doi: 10.1016/j.chom.2015.12.002
201. Sanjuan MA, Dillon CP, Tait SW, Moshiach S, Dorsey F, Connell S, et al. Toll-like receptor signalling in macrophages links the autophagy pathway to phagocytosis. *Nature.* (2007) 450:1253–7. doi: 10.1038/nature06421
202. Martinez J, Almendinger J, Oberst A, Ness R, Dillon CP, Fitzgerald P, et al. Microtubule-associated protein 1 light chain 3 alpha (LC3)-associated phagocytosis is required for the efficient clearance of dead cells. *Proc Natl Acad Sci USA.* (2011) 108:17396–401. doi: 10.1073/pnas.1113421108
203. Martinez J, Malireddi RK, Lu Q, Cunha LD, Pelletier S, Gingras S, et al. Molecular characterization of LC3-associated phagocytosis reveals distinct roles for Rubicon, NOX2 and autophagy proteins. *Nat Cell Biol.* (2015) 17:893–906. doi: 10.1038/ncb3192
204. Oliveira DL, Fonseca FL, Zamith-Miranda D, Nimrichter L, Rodrigues J, Pereira MD, et al. The putative autophagy regulator Atg7 affects the physiology and pathogenic mechanisms of *Cryptococcus neoformans*. *Fut Microbiol.* (2016) 11:1405–19. doi: 10.2217/fmb-2016-0090
205. Hu G, Hacham M, Waterman SR, Panepinto J, Shin S, Liu X, et al. PI3K signaling of autophagy is required for starvation tolerance and virulence of *Cryptococcus neoformans*. *J Clin Invest.* (2008) 118:1186–97. doi: 10.1172/JCI32053

Permissions

All chapters in this book were first published by Frontiers; hereby published with permission under the Creative Commons Attribution License or equivalent. Every chapter published in this book has been scrutinized by our experts. Their significance has been extensively debated. The topics covered herein carry significant findings which will fuel the growth of the discipline. They may even be implemented as practical applications or may be referred to as a beginning point for another development.

The contributors of this book come from diverse backgrounds, making this book a truly international effort. This book will bring forth new frontiers with its revolutionizing research information and detailed analysis of the nascent developments around the world.

We would like to thank all the contributing authors for lending their expertise to make the book truly unique. They have played a crucial role in the development of this book. Without their invaluable contributions this book wouldn't have been possible. They have made vital efforts to compile up to date information on the varied aspects of this subject to make this book a valuable addition to the collection of many professionals and students.

This book was conceptualized with the vision of imparting up-to-date information and advanced data in this field. To ensure the same, a matchless editorial board was set up. Every individual on the board went through rigorous rounds of assessment to prove their worth. After which they invested a large part of their time researching and compiling the most relevant data for our readers.

The editorial board has been involved in producing this book since its inception. They have spent rigorous hours researching and exploring the diverse topics which have resulted in the successful publishing of this book. They have passed on their knowledge of decades through this book. To expedite this challenging task, the publisher supported the team at every step. A small team of assistant editors was also appointed to further simplify the editing procedure and attain best results for the readers.

Apart from the editorial board, the designing team has also invested a significant amount of their time in understanding the subject and creating the most relevant covers. They scrutinized every image to scout for the most suitable representation of the subject and create an appropriate cover for the book.

The publishing team has been an ardent support to the editorial, designing and production team. Their endless efforts to recruit the best for this project, has resulted in the accomplishment of this book. They are a veteran in the field of academics and their pool of knowledge is as vast as their experience in printing. Their expertise and guidance has proved useful at every step. Their uncompromising quality standards have made this book an exceptional effort. Their encouragement from time to time has been an inspiration for everyone.

The publisher and the editorial board hope that this book will prove to be a valuable piece of knowledge for researchers, students, practitioners and scholars across the globe.

List of Contributors

Mitra Shourian
Division of Experimental Medicine, McGill University, Montreal, QC, Canada Meakins-Christie Laboratories, McGill University, Montreal, QC, Canada

Ben Ralph
Program in Infectious Diseases and Immunology in Global Health, Centre for Translational Biology, The Research Institute of the McGill University Health Center (RI-MUHC), Montreal, QC, Canada Department of Microbiology and Immunology, McGill University, Montreal, QC, Canada

Isabelle Angers
Meakins-Christie Laboratories, McGill University, Montreal, QC, Canada Program in Translational Research in Respiratory Diseases, Department of Critical Care, The Research Institute of the McGill University Health Center (RIMUHC), Montreal, QC, Canada

Salman T. Qureshi
Montreal, QC, Canada Meakins-Christie Laboratories, McGill University, Montreal, QC, Canada Program in Translational Research in Respiratory Diseases, Department of Critical Care, The Research Institute of the McGill University Health Center (RIMUHC), Montreal, QC, Canada Department of Medicine, McGill University, Montreal, QC, Canada

Donald C. Sheppard
Program in Infectious Diseases and Immunology in Global Health, Centre for Translational Biology, The Research Institute of the McGill University Health Center (RI-MUHC), Montreal, QC, Canada Department of Microbiology and Immunology, McGill University, Montreal, QC, Canada Department of Medicine, McGill University, Montreal, QC, Canada

Ofer Zimmerman, Lindsey B. Rosen, Muthulekha Swamydas, Elise M. N. Ferre, Mukil Natarajan, Steven M. Holland and Michail S. Lionakis
Laboratory of Clinical Infectious Diseases, National Institute of Allergy and Immunology, National Institutes of Health, Bethesda, MD, United States

Frank van de Veerdonk
Department of Internal Medicine, Radboud University Medical Center, Radboud Institute for Molecular Life Sciences (RILMS), Nijmegen, Netherlands

Marley C. Caballero Van Dyke, Ashok K. Chaturvedi, Sarah E. Hardison, Chrissy M. Leopold Wager, Natalia Castro-Lopez, Camaron R. Hole, Karen L. Wozniak and Floyd L. Wormley Jr.
Department of Biology, The University of Texas at San Antonio, San Antonio, TX, United States The South Texas Center for Emerging Infectious Diseases, The University of Texas at San Antonio, San Antonio, TX, United States

Fabrine Sales Massafera Tristão, Fernanda Agostini Rocha, Daniela Carlos, Natália Ketelut-Carneiro, Camila Oliveira Silva Souza, Cristiane Maria Milanezi and João Santana Silva
Department of Biochemistry and Immunology, School of Medicine of Ribeirão Preto, University of São Paulo, Ribeirão Preto, Brazil

John F. Kernien
Department of Medicine, University of Wisconsin, Madison, WI, United States

Brendan D. Snarr
Department of Microbiology and Immunology, McGill University, Montreal, QC, Canada

Jeniel E. Nett
Department of Medicine, University of Wisconsin, Madison, WI, United States Department of Medical Microbiology and Immunology, University of Wisconsin, Madison, WI, United States

Gloria Hoi Wan Tso, Jose Antonio Reales-Calderon and Norman Pavelka
Singapore Immunology Network (SIgN), Agency of Science, Technology and Research (A*STAR), Singapore

Tsokyi Choera
Department of Medical Microbiology and Immunology, University of Wisconsin-Madison, Madison, WI, United States

Teresa Zelante and Luigina Romani
Department of Experimental Medicine, University of Perugia, Perugia, Italy

Nancy P. Keller
Department of Medical Microbiology and Immunology, University of Wisconsin-Madison, Madison, WI, United States Department of Bacteriology, University of Wisconsin-Madison, Madison, WI, United States

Irene Zhang, Bridget M. Barker and Emily K. Cope
Pathogen and Microbiome Institute, Northern Arizona University, Flagstaff, AZ, United States

Steven D. Pletcher and Andrew N. Goldberg
Department of Otolaryngology Head and Neck Surgery, University of California, San Francisco, San Francisco, CA, United States

Pappanaicken R. Kumaresan and Thiago Aparecido da Silva
Department of Pediatrics, The University of Texas MD Anderson Cancer Center, Houston, TX, United States

Dimitrios P. Kontoyiannis
Department of Infectious Diseases, The University of Texas MD Anderson Cancer Center, Houston, TX, United States

Tara L. Croston, Angela R. Lemons and Brett J. Green
Allergy and Clinical Immunology Branch, Health Effects Laboratory Division, National Institute for Occupational Safety and Health, Centers for Disease Control and Prevention, Morgantown, WV, United States

Donald H. Beezhold
Health Effects Laboratory Division, National Institute for Occupational Safety and Health, Centers for Disease Control and Prevention, Morgantown, WV, United States

Arturo Casadevall and Carolina Coelho
Department of Molecular Microbiology and Immunology, Johns Hopkins School of Public Health, Baltimore, MD, United States

Alexandre Alanio
Department of Molecular Microbiology andImmunology, Johns Hopkins School of Public Health, Baltimore, MD,United States Institut Pasteur, Molecular Mycology Unit, CNRS UMR2000, Paris, France Laboratoire de Parasitologie- Mycologie, Hôpital Saint-Louis, Groupe Hospitalier Lariboisière, Saint-Louis, Fernand Widal, Assistance Publique-Hôpitaux de Paris (AP-HP), Université Paris Diderot, Sorbonne Paris Cité, Paris, France

Tania Costa
Department of Immunology, Institute of Biomedical Sciences, University of São Paulo, São Paulo, Brazil

Carmine Vacca, Paolo Puccetti and Luigina Romani
Department of Experimental Medicine, University of Perugia, Perugia, Italy

Zhenzong Fa
PLA Key Laboratory of Mycosis, Department of Dermatology and Venereology, Changzheng Hospital, Shanghai, China, Shanghai Key Laboratory of Molecular Medical Mycology, Shanghai Institute of Medical Mycology, Second Military Medical University, Shanghai, China Division of Pulmonary and Critical Care Medicine, Department of Internal Medicine, University of Michigan Health System, Ann Arbor, MI, United States

Qun Xie
Department of Anesthesiology and Intensive Care, Changhai Hospital, Second Military Medical University, Shanghai, China Key Laboratory of Nutrition and Metabolism, Institute for Nutritional Sciences, Shanghai Institutes for Biological Sciences, Chinese Academy of Sciences, Shanghai, China

Wei Fang
Shanghai Key Laboratory of Molecular Medical Mycology, Shanghai Institute of Medical Mycology, Second Military Medical University, Shanghai, China Department of Dermatology, Huashan Hospital, Fudan University, Shanghai, China

Haibing Zhang and Haiwei Zhang
Key Laboratory of Nutrition and Metabolism, Institute for Nutritional Sciences, Shanghai Institutes for Biological Sciences, Chinese Academy of Sciences, Shanghai, China

Jintao Xu
Division of Pulmonary and Critical Care Medicine, Department of Internal Medicine, University of Michigan Health System, Ann Arbor, MI, United States

Weihua Pan and Wanqing Liao
PLA Key Laboratory of Mycosis, Department of Dermatology and Venereology, Changzheng Hospital, Shanghai, China Shanghai Key Laboratory of Molecular Medical Mycology, Shanghai Institute of Medical Mycology, Second Military Medical University, Shanghai, China

Jinhua Xu
Department of Dermatology, Huashan Hospital, Fudan University, Shanghai, China

Michal A. Olszewski
Shanghai Key Laboratory of Molecular Medical Mycology, Shanghai Institute of Medical Mycology, Second Military Medical University, Shanghai, China Division of Pulmonary and Critical Care Medicine, Department of Internal Medicine, University of Michigan Health System, Ann Arbor, MI, United States

Xiaoming Deng
Department of Anesthesiology and Intensive Care, Changhai Hospital, Second Military Medical University, Shanghai, China

Florian Sparber and Salomé Leibund Gut-Landmann
Section of Immunology, Vetsuisse Faculty, University of Zürich, Zürich, Switzerland

Eliseu Frank de Araújo, Claudia Feriotti, Nayane Alves de Lima Galdino, Nycolas Willian Preite, Vera Lúcia Garcia Calich and Flávio Vieira Loures
Departamento de Imunologia, Instituto de Ciências Biomédicas, Universidade de São Paulo, São Paulo, Brazil

Waleed Elsegeiny and Peter R. Williamson
Laboratory of Clinical Immunology and Microbiology (LCIM), National Institute of Allergy and Infectious Diseases (NIAID), National Institutes of Health (NIH), Bethesda, MD, United States

Kieren A. Marr
Johns Hopkins University, Baltimore, MD, United States

Ritesh Agarwal, Puneet Saxena, Valliappan Muthu, Inderpaul Singh Sehgal, Sahajal Dhooria, Kuruswamy Thurai Prasad and Ashutosh Nath Aggarwal
Department of Pulmonary Medicine, Postgraduate Institute of Medical Education and Research (PGIMER), Chandigarh, India

Arunaloke Chakrabarti
Department of Medical Microbiology, Postgraduate Institute of Medical Education and Research (PGIMER), Chandigarh, India

Ziyao Peng and Jianguo Tang
Department of Trauma-Emergency and Critical Care Medicine, Shanghai Fifth People's Hospital, Fudan University, Shanghai, China

Zhi Li
The Joint Center for Infection and Immunity, Guangzhou Women and Children's Medical Center, Guangzhou Institute of Pediatrics, Guangzhou, China
The Joint Center for Infection and Immunity, Institute Pasteur of Shanghai, Chinese Academy of Science, Shanghai, China

Gen Lu
The Joint Center for Infection and Immunity, Guangzhou Women and Children's Medical Center, Guangzhou Institute of Pediatrics, Guangzhou, China

Guangxun Meng
The Joint Center for Infection and Immunity, Institute Pasteur of Shanghai, Chinese Academy of Science, Shanghai, China

Index

www.ingramcontent.com/pod-product-compliance
Lightning Source LLC
LaVergne TN
LVHW082150200926
848484LV00008B/2132